FUNDAMENTALS OF PROTEIN NMR SPECTROSCOPY

FOCUS ON STRUCTURAL BIOLOGY

Volume 5

Series Editor
ROB KAPTEIN
Bijvoet Center for Biomolecular Research,
Utrecht University, The Netherlands

Fundamentals of Protein NMR Spectroscopy

by

GORDON S. RULE
Department of Biological Sciences,
Carnegie Mellon University, Pittsburgh, PA, U.S.A.

and

T. KEVIN HITCHENS
Pittsburgh NMR Center for Biomedical Research,
Carnegie Mellon University, Pittsburgh, PA, U.S.A.

Springer

A C.I.P. Catalogue record for this book is available from the Library of Congress.

ISBN 978-90-481-6886-6
ISBN 1-4020-3500-4 (eBook)
ISBN 978-1-4020-3500-5 (eBook)

Published by Springer,
P.O. Box 17, 3300 AA Dordrecht, The Netherlands.

www.springeronline.com

Printed on acid-free paper

Contents

List of Figures

List of Tables

Chapter 1

NMR SPECTROSCOPY

Nuclear magnetic resonance (NMR) spectroscopy is a powerful technique that can be used to investigate the structure, dynamics, and chemical kinetics of a wide range of biochemical systems. Due to the complexity of this technique it can be challenging for the novice to become a practicing NMR spectroscopist. This text is primarily designed to provide a practical, as well as theoretical, guide to modern NMR spectroscopy.

A brief survey of the important features of NMR spectroscopy is provided in Chapter 1. The acquisition and processing of one-dimensional NMR spectra are described in the following two chapters. Sufficient theoretical background to understand modern multi-dimensional NMR spectroscopic methods is developed in Chapters 4 through 12. Three appendices provide information on Fourier transforms and other mathematical tools. The last appendix describes building blocks of NMR experiments, facilitating the analysis of complex NMR experiments. The resonance assignment process, an important step in the analysis of protein NMR spectra, is described in Chapters 13 and 14. Chapter 15 guides the reader through the process of acquiring and processing multi-dimensional NMR spectra.

Figure 1.1. Protein structure determined by NMR spectroscopy. Four structures of a 130 residue protein, derived from NMR constraints, are overlayed to highlight the accuracy of structure determination by NMR spectroscopy.

The remaining chapters of the text discuss several important applications of NMR spectroscopy: structure determination, molecular dynamics, and chemical kinetics. Protein structure determination by NMR spectroscopy is discussed in Chapter 16. This method can be be used to determine the atomic level structure of proteins *in solution*, often providing structures as accurate as those determined by X-ray diffraction (see Fig. 1.1). Although structure determination by NMR is limited to proteins with molecular weights less than 40-60 kDa, NMR techniques provide an alternative method for structure determination if a protein cannot be crystallized, or if there is concern that crystal packing has distorted the true structure in solution.

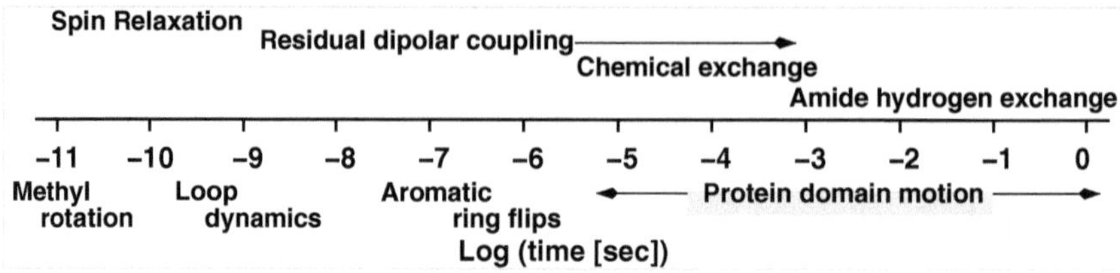

Figure 1.2. Characterization of molecular dynamics with NMR. The time scale of motions in proteins are indicated below the axis and NMR techniques that can be used to characterize motion on the various time scales are shown above the axis.

One of the more powerful attributes of NMR spectroscopy is its ability to detect molecular motion in proteins and other polymers. Other methods of detecting molecular motion, such as fluorescence spectroscopy, are limited by the small number of sites that can be probed and the narrow time scale over which the motion can be characterized. The ability to observe and characterize resolved NMR resonance lines from individual atoms provides information on dynamics from a large number of sites within the protein. Furthermore, it is possible to characterize the dynamic properties over a wide range of time scales. The time scales for a number of biological processes, as well as the NMR techniques that are applicable to these time scales, are shown in Fig. 1.2. The applications of chemical exchange and spin relaxation to characterize dynamics are discussed extensively in Chapters 18 and 19. Residual dipolar coupling is discussed in Chapter 16 and the reader is referred to [159] for the application of this technique to protein dynamics. Amide hydrogen exchange is discussed in Chapter 15, and the application of these measurements to the characterization of protein dynamics has been recently reviewed by Dempsey [50].

Lastly, NMR is useful in probing molecular interactions, such as protein-drug or protein-protein interactions. The binding site can often be detected from changes in NMR resonance frequencies that occur when the complex is formed. The kinetic rate constants for binding can be measured over a broad time scale using the same techniques that are used to measure molecular motion.

1.1 Introduction to NMR Spectroscopy

In this chapter we will employ a semi-classical model of the nuclear spins to obtain an intuitive understanding of many of the fundamental aspects of modern NMR spectroscopy. This treatment will highlight a number of important features of NMR spectroscopy, including:

1. How energy states are created by the magnetic field,
2. How resonance signals are detected,
3. How the detected signals are transformed into spectra,
4. How the relaxation properties of the excited state are affected by the environment,
5. How the relaxation properties affect the NMR lineshape,
6. How the absorption frequency of a nuclear spin is affected by its environment,
7. How the characteristic NMR absorption frequencies of amino acids are interpreted.

1.2 One Dimensional NMR Spectroscopy

1.2.1 Classical Description of NMR Spectroscopy

The basic phenomenon of NMR spectroscopy is similar to other forms of spectroscopy, such as visible spectroscopy. A photon of light causes a transition from the ground state to the excited state. In the case of visible spectroscopy an electron absorbs the energy, while in the case of NMR spectroscopy the absorbed photon promotes a nuclear spin from its ground state to its excited state.

NMR spectroscopy differs in a number of important aspects from other forms of spectroscopy. First, the generation of the ground and excited NMR states requires the existence of an external magnetic field. This requirement is a very important distinction of NMR spectroscopy because it allows one to change the characteristic frequencies of the transitions by simply changing the applied magnetic field strength. Second, the NMR excited state has a lifetime that is on the order of 10^9 times longer than the lifetime of excited electronic states. This difference in lifetimes follows directly from Einstein's law for spontaneous emission that relates the lifetime of the excited state, τ, to the frequency of the transition, ω:

$$\tau \propto \frac{1}{\omega^3} \tag{1.1}$$

The long lifetime of the excited state implies extremely narrow spectral lines since the ability to define the energy of a transition is proportional to the lifetime of the excited state[1]. In the case of small organic molecules, linewidths less than 1 Hz are easily attainable. Thus it is possible to detect small changes in absorption energies that arise from subtle differences in the environment of a nuclear spin. The persistence of the excited state also facilitates multi-dimensional spectroscopy, the cornerstone of modern multi-nuclear NMR studies on biopolymers, by allowing the resonance frequency information associated with one spin to be passed to another. Finally, the long lifetime of the excited state permits the measurement of molecular dynamics over a wide range of time scales.

1.2.2 Nuclear Spin Transitions

For all forms of spectroscopy it is necessary to have two or more different states of the system that differ in energy. In a system with two energy levels, the one with the lower energy level is often referred to as the ground state and the higher energy state as the excited state. In the case of nuclear magnetic resonance spectroscopy, the energies of these states arise from the interaction of a *nuclear magnetic dipole moment* with an intense external magnetic field. Excitation of transitions between these states is stimulated using radio-frequency (RF) electromagnetic radiation.

1.2.2.1 Magnetic Dipole

The nuclear magnetic dipole moment arises from the *spin angular momentum* of the nucleus. All nuclei with an odd mass number (e.g. ^{1}H, ^{13}C, ^{15}N) have spin

[1]This is one form of Heisenberg's uncertainty principle: $\Delta E \Delta t \geq \hbar/2$.

angular momentum because they have an unpaired proton. All nuclei with an even mass number and an odd charge (e.g. ^{2}H, ^{14}N) also have spin angular momentum.

The spin angular momentum, $\vec{S}$, is quantized (as is all angular momentum) and the different quantum states are indexed with the spin quantum number I. The total angular momentum of a nuclear spin is: $\vec{S} = \hbar\sqrt{I(I+1)}$. We will generally be interested in the z-component of the angular moment, S_z, which is restricted to integral steps of $\hbar$ ranging from $-I$ to $+I$. For example, a spin one-half nuclei would have two possible values of S_z: $+\frac{1}{2}\hbar$, and $-\frac{1}{2}\hbar$, corresponding to spin quantum numbers $m_z = +\frac{1}{2}$ and $m_z = -\frac{1}{2}$, respectively. The magnetic moment of a nuclear spin, $\vec{\mu}$, is proportional to its spin angular momentum, $\hbar\vec{I}$, by a factor, γ, which has units of radians $\sec^{-1}$ gauss^{-1}.

$$\vec{\mu}_n = \gamma_n \hbar \vec{I} \tag{1.2}$$

The magnitude of γ depends on the type of nuclei. Useful NMR properties of various nuclear spins, including values of γ, are shown in Table 1.1. NMR active isotopes of hydrogen, carbon, nitrogen, and phosphorus exist, thus it is possible to observe NMR signals from virtually every atom in biopolymers. Protons (^{1}H) and phosphorus are highly abundant in natural biopolymers, while in the case of carbon and nitrogen it is usually necessary to introduce the appropriate isotope into the sample by biosynthetic labeling. Also note, that with the exception of deuterium (^{2}H), all of these nuclei have a z-component of the spin angular momentum of $\hbar/2$. Consequently, the material presented in this text applies to all of the above atomic nuclei, except for deuterium. Deuterium with a spin quantum number $I = 1$ is a quadrapolar nuclei and in certain instances needs to be treated differently than spin-1/2 nuclei.

Table 1.1. Properties of NMR active nuclei.

Nuclei[1]	γ (rad × sec^{-1}× gauss^{-1})[†]	*I*	*Natural Abundance (%)*
^{1}H	26,753	1/2	99.980
^{2}H	4,106	1	0.016
^{19}F	25,179	1/2	100.000[2]
^{13}C	6,728	1/2	1.108[3]
^{15}N	-2,712	1/2	0.37[3]
^{31}P	10,841	1/2	100.00

[1]The term "Protons" is used interchangeably with ^{1}H in the text.
[2]Fluorine is not normally found in biopolymers, therefore it has to be introduced by chemical or biosynthetic labeling.
[3]These isotopes of carbon and nitrogen are normally found in low levels in biopolymers, therefore the levels of these two spins are generally enriched, often to 100%, by biosynthetic labeling.
[†]CGS units are used throughout the text.

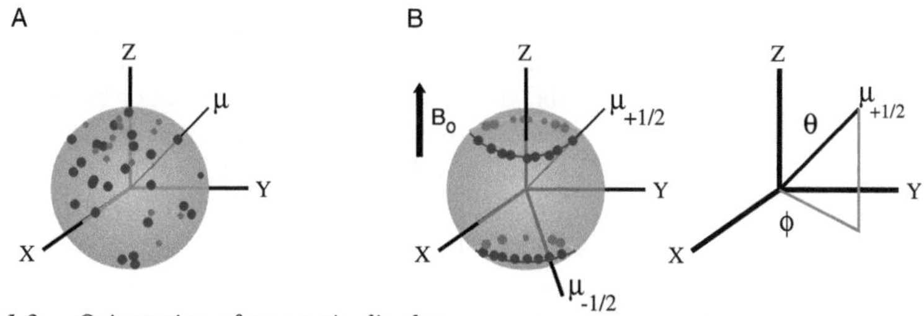

Figure 1.3. Orientation of magnetic dipoles.
A. Orientation of nuclear magnetic dipoles in the absence of a magnetic field. A unit sphere is shown and the dots on the surface illustrate the various orientations of the dipoles in space. The orientation of one dipole μ is indicated by a line drawn from the center of the sphere.
B. The orientation of the nuclear spin dipoles in a static magnetic field along the z-axis. Note that approximately one-half of the spins are pointed up and the other half are pointed down. Also note that they can assume any value of ϕ, but only two values of θ. ϕ and θ represent the orientation of the magnetic dipole in spherical coordinates, as shown on the right part of this figure.

1.2.2.2 Nuclear Dipole-Magnetic Field Interaction

When a collection of nuclear spins is observed in the absence of a magnetic field, all possible orientations of the magnetic dipole are possible, as shown in Fig. 1.3. However, once the spins are placed in a magnetic field, the direction of z-axis becomes defined by the direction of the field, and the magnetic moments of spin-1/2 nuclei assume two orientations, either in the same direction as, or opposed to, the magnetic field (see Fig. 1.3). Note that the magnetic moments cannot orient parallel to the magnetic field because of the restrictions placed on the value of μ_z by the quantum mechanical properties of the system.

The energy of a state depends on the interaction of the aligned magnetic dipole with an externally applied magnetic field. The size of this interaction can be found from classical electromagnetic theory. For example, consider a magnetic dipole in a static magnetic field, $\vec{B}$, with the magnetic field along the z-axis, as shown in panel B of Fig. 1.3. The energy required to change the angle, θ, is

$$\begin{aligned} E &= \int \Gamma d\theta = \int (\vec{\mu} \times \vec{B}) d\theta = |\mu||B| \int \sin(\theta) d\theta = -|\mu||B| \cos(\theta) \\ &= -\vec{\mu} \cdot \vec{B} \end{aligned} \tag{1.3}$$

where the torque, Γ, arises as the static field attempts to align the magnetic dipole of the nucleus. The energy of interaction, or Hamiltonian[2], between the field and the dipole is therefore given by the dot product of the two vectors: $\mathcal{H} = -\vec{\mu} \cdot \vec{B}$.

The actual magnetic field that is present at the nucleus is usually attenuated, or shielded, by the presence of electrons that surround the nucleus, giving a modified field at the nucleus, B:

$$B = (1 - \sigma) B_o \tag{1.4}$$

[2]The Hamiltonian is a quantum mechanical function, or operator, that can be used to determine the energy of the system, provided the quantum state of the system is known. See Chapter 4 for more details.

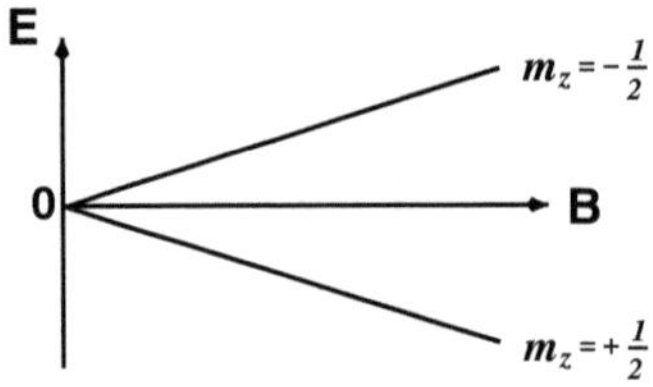

Figure 1.4 Effect of magnetic field on the energy levels of a spin-1/2 particle. The energy separation between the ground and excited state increases as the magnetic field strength is increased. The diagram shown applies to a spin with $\gamma > 0$. The diagram would be identical for $\gamma < 0$, except that the quantum numbers would be interchanged.

where σ represents the degree of shielding and B_o is the strength of the applied magnetic field. An extensive discussion of shielding effects is found in Section 1.5.

Assuming that the magnetic field is along the z-axis, the energy of each state is:

$$\mathcal{H} = -u_z B_z \tag{1.5}$$

At this point the result is entirely classical and it is applicable to *any* magnetic dipole that is placed into a magnetic field. To relate the energy of a nuclear spin to its quantum state, we make use of the relationship between the magnetic dipole and the z-component of the spin angular momentum, $\mu_z = \gamma\hbar m_z$, giving the energy for a spin in a quantum state, m_z:

$$\mathcal{H} = -\gamma\hbar B m_z \tag{1.6}$$

This is often referred to as the Zeeman Hamiltonian. Using Eq. 1.6 it is possible to draw an energy diagram for a spin-1/2 particle ($m_z = \pm 1/2$) as a function of magnetic field strength, as shown in Fig. 1.4.

The ground, or lower energy state, referred to as α, corresponds to $m_z = +1/2$, and the excited, or higher energy state, referred to as β, corresponds to $m_z = -1/2$. The energies of the two states can be calculated using Eq. 1.6,

$$E_\alpha = -\frac{\gamma\hbar B}{2} \qquad E_\beta = +\frac{\gamma\hbar B}{2} \tag{1.7}$$

The energy difference between the two states is easily computed,

$$\Delta E = E_\beta - E_\alpha = \gamma\hbar B \tag{1.8}$$

and using the relationship, $E = \hbar\omega$, gives the well known Larmor equation:

$$\omega_s = \gamma B \tag{1.9}$$

In the above equation, ω_s refers to the absorption, or resonance frequency, of the shielded nucleus, i.e. its *observed* resonance frequency. The Larmor equation is one of the *key* equations in NMR spectroscopy, it states that the absorption frequency of a transition is equal to γ multiplied by the strength of the magnetic field *at the nucleus*.

The energy of an NMR transition is quite low, requiring radiowaves to excite the spins. The small value of ΔE has two important consequences:

1. The population difference between the two energy levels is very small, on the order of 1 part in 10^6. The actual population difference can be easily calculated from Boltzmann's relationship:

$$\frac{N_\beta}{N_\alpha} = e^{\frac{-\gamma\hbar B}{kT}} \approx 1 - \frac{\gamma\hbar B}{kT} \tag{1.10}$$

The consequence of having a small population differences is that NMR spectroscopy is a relatively insensitive experimental technique because of the small excess of spins in the ground state. In any form of spectroscopy the presence of electro-magnetic radiation induces transitions from the ground to the excited state and vice versa. Consequently, the *net* absorption depends on the population difference between the two states. Due to the low sensitivity, it is common to increase the signal-to-noise of the spectrum by signal averaging. In addition, typical NMR experiments require protein concentrations on the order of 1 mM. However, in some cases concentrations in the range of 50 μM have be used.

2. The lifetime of the excited state can be quite long, on the order of msec to sec. As discussed above, a long lifetime provides three benefits: narrow resonance lines, experimental manipulation of the excited state in multi-dimensional experiments, and sensitivity to molecular motion over a wide time scale.

1.3 Detection of Nuclear Spin Transitions

We have seen how placing a nuclear spin in a static magnetic field generates a ground and an excited state. Irradiation of a sample with radiofrequency (RF) waves of the appropriate frequency, $\omega_s = \gamma B$, will excite transitions from the ground to the excited state due to the interaction of the magnetic dipole with the oscillating magnetic field component of the electromagnetic radiation. This excitation field, called $\vec{B}_1$, must be orthogonal to the direction of the magnetic dipoles, such that $\vec{B}_1 \times \vec{\mu} \neq 0$, to generate transitions of the nuclear spin state. The relative orientation of the static B_0 field and the oscillatory B_1 field are shown in Fig. 1.5.

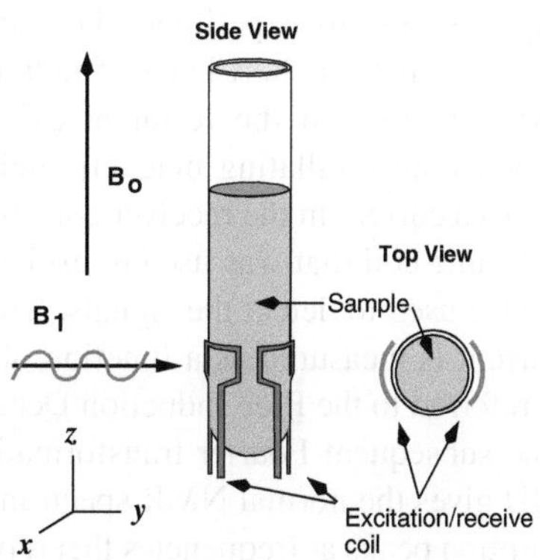

Figure 1.5. Geometry of magnetic fields around the sample. The sample is contained in a glass tube. The diameter of the sample tube is typically 5 mm and the sample volume is approximately 0.5 ml. The directions of the static B_0 field and the oscillating B_1 field are shown. The B_1 field is usually applied using two paddle shaped Helmholtz coils, shaded dark gray. These coils are also used to detect the magnetic field produced by the excited spins. Magnetic field strengths (B_o) that are commonly used range from 7 Tesla to 21 Tesla, corresponding to proton resonance frequencies (ν_H) of 300 MHz to 900 MHz, respectively.

The B_1 field can be applied to the sample in one of two ways, either by scanning through multiple wavelengths (continuous wave NMR), or as a short burst of high power RF that excites a broad range of transitions (pulsed NMR). Each of these methods are discussed below.

1.3.1 Continuous Wave NMR

In continuous wave (CW) spectroscopy the NMR spectrum is obtained by using a technique that is similar to traditional UV-visible spectroscopy, namely scanning the wavelength of the incident light and detecting the absorbance as a function of frequency. Prior to the introduction of pulsed methods in the early 1970's, all NMR spectra were acquired in this way. Low sensitivity and the general restriction to one-dimensional NMR experiments are the principal reasons why continuous wave spectroscopy is no longer used. Due to the low inherent sensitivity of NMR it is necessary to average signals. Consequently, the overall sensitivity depends on how fast each individual spectrum can be acquired. A continuous wave scan takes much longer than pulsed excitation because it is necessary to wait for the excited spins to return to the ground state while the spectrum is being scanned. Otherwise the signals from previously excited spins will interfere with the newly excited spins. In contrast, since all of the spins are excited at the same time with pulsed NMR, it is only necessary to allow the spins to relax for a single time period, between each excitation pulse.

1.3.2 Pulsed NMR

The simplest pulsed NMR experiment consists of a short RF-pulse followed by detection of the signal. This pulse sequence is shown in Fig. 1.6. In this experiment the nuclear spins are excited by a short burst of radiofrequency (RF) energy and the resultant excited states produce an oscillating magnetic field that induces a current in the receiver coil. In practice, the same coil that was used to excite the spins is also used to detect the signals. The induced current is measured as a function of time and is referred to the Free Induction Decay or FID. The subsequent Fourier transformation of the FID gives the normal NMR spectrum with absorption peaks at frequencies that represent the energy difference between the ground and excited states.

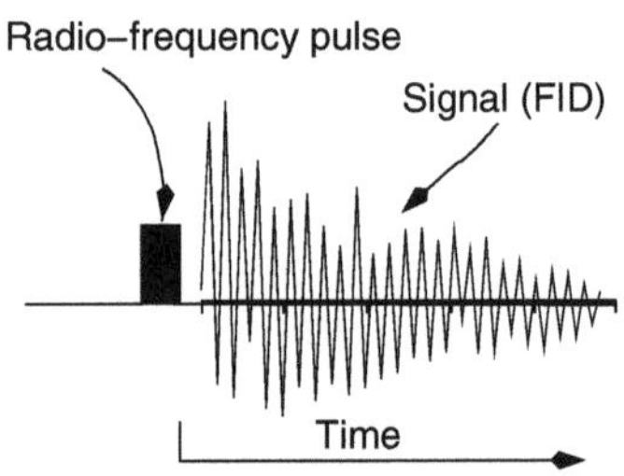

Figure 1.6. A simple one-pulse NMR experiment. The experiment begins with a short ($\approx$ 10 μsec) radiofrequency pulse. The induced signal (FID) is sampled as it evolves over time.

To understand how this procedure can generate an NMR spectrum, the motion of the spins during each segment of the experiment will be analyzed using classical mechanics. The one-pulse experiment will be divided into the following three discrete time intervals, and the evolution of the spins within each of these periods will be discussed in detail.

1. Preparation: Prior to the excitation pulse the spins are at thermal equilibrium and are subject to only the static B_o field.
2. Excitation: During the excitation pulse the spins are subject to the static B_o field plus the oscillatory excitation field, B_1.
3. Detection: The excited spins precess under the static B_o field, generating the free induction decay or FID. The spectrum is obtained by Fourier transformation of the FID.

1.3.2.1 Before the Pulse: Magnetization at Equilibrium

Since the nuclear spins possess angular momentum, the effect of applying *any* external field (B_o and/or B_1) to the spins is to generate a torque, Γ, on the spin. This torque will cause a change in angular momentum as described by the following classical equation:

$$\Gamma = \frac{dS}{dt} = \vec{\mu} \times \vec{B} \qquad (1.11)$$

Using $\vec{\mu} = \gamma \vec{S}$ we can write

$$\frac{d\mu}{dt} = \gamma \vec{\mu} \times \vec{B} \qquad (1.12)$$

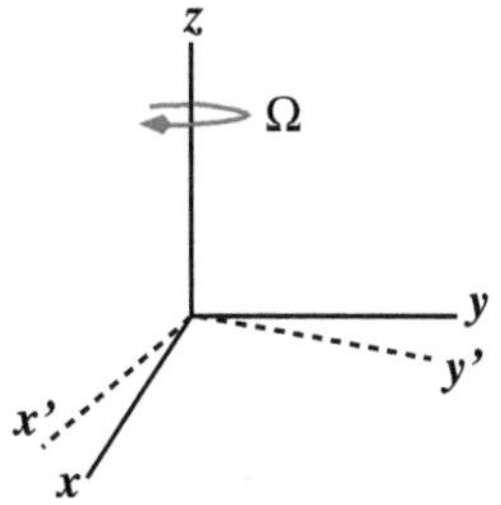

Figure 1.7. Rotating frame of reference. The coordinate system is rotating at a frequency $|\Omega|$ about the z-axis.

This equation could be solved by standard methods, but it is much more instructive to introduce a rotating frame of reference before attempting its solution. In the rotating reference frame, the basis vectors change their direction according to the following:

$$\frac{dx_i}{dt} = \vec{\Omega} \times x_i \qquad (1.13)$$

where x_i are the normal Cartesian basis vectors $(\hat{i}, \hat{j}, \hat{k})$ and $\vec{\Omega}$ is a vector that characterizes the rotation. The axis of rotation is along $\vec{\Omega}$ and the rate of rotation is given by $|\Omega|$. Although this may sound complicated, it is just the Cartesian frame rotating with a speed $|\Omega|$ around some axis in the direction of Ω, as shown in Fig. 1.7.

Equation 1.12 gives one expression for the change in μ with respect to time. It is also possible to write $d\mu/dt$ as a standard differential of two variables. In this case, the variables are the components of $\vec{\mu}$ *as well as* the coordinate axis since both are time dependent:

$$\begin{aligned} \frac{d\vec{\mu}}{dt} &= \sum x_i \frac{d\mu_i}{dt} + \sum \mu_i \frac{dx_i}{dt} \\ &= \frac{\delta u}{\delta t} + \sum \mu_i \vec{\Omega} \times x_i \\ &= \frac{\delta \mu}{\delta t} + \vec{\Omega} \times \vec{\mu} \end{aligned} \qquad (1.14)$$

Here, $\delta\mu/\delta t$ is the item of interest, the change of μ with respect to time *in the rotating frame*. Combining Eq. 1.12 and Eq. 1.14 gives the following ($\vec{A} \times \vec{B} = -\vec{B} \times \vec{A}$):

$$\begin{aligned} \frac{\delta \vec{\mu}}{\delta t} &= \frac{d\mu}{dt} - \vec{\Omega} \times \mu \\ &= \gamma \vec{\mu} \times \vec{B} - \vec{\Omega} \times \vec{\mu} \\ &= \gamma \vec{\mu} \times \vec{B} + \vec{\mu} \times \vec{\Omega} \\ &= \gamma \vec{\mu} \times (\vec{B} + \frac{\Omega}{\gamma}) \end{aligned} \qquad (1.15)$$

The above equation (Eq. 1.15) has the same form as $d\mu/dt$, but with the addition of a fictitious field Ω/γ that is generated solely by the change in the coordinate system. What happens when $\vec{B}+\vec{\Omega}/\gamma = 0$? In this case neither the direction nor the amplitude of μ changes, since $\delta\mu/\delta t = 0$; μ becomes a stationary vector in the rotating frame. If $B + \Omega/\gamma = 0$ then the rate of rotation of the coordinate frame must be:

$$\Omega = -\gamma B \tag{1.16}$$

Consequently, in the laboratory, or stationary coordinate frame, the magnetic dipole that is associated with a single spin must be rotating, or precessing, around the z-axis at a frequency of $\omega_s = \gamma B$; ω_s is again used to indicate that the field at the nucleus defines the precessional frequency. Note that this frequency is identical to the frequency of the spin transition that was obtained above using simple quantum mechanics (Eq. 1.9). The key conclusion is that the magnetic dipole precesses around $\vec{B}_o$ at a frequency ω_s and this frequency is *identical* to the resonance frequency.

In addition to following the evolution of a single spin, it is also useful to consider the evolution of the bulk, or average, magnetization during the experiment. The bulk magnetization of the sample is just the sum of the individual magnetic dipoles. The vector components of the bulk magnetization, $\vec{M}$, are defined as:

$$M_i = \sum_{i=1}^{All\ spins} \mu_i \tag{1.17}$$

In the presence of the static field, the sum of the z-components of each magnetic dipole will produce detectable bulk magnetization because there is a slight difference in the population of spins that are aligned in one direction versus spins aligned in the other direction (see Fig. 1.3). The net magnetization along z, referred to as the longitudinal magnetization, is therefore defined as:

$$M_z = M_o \tag{1.18}$$

In contrast, before the pulse, the distribution of the magnetic dipoles in the x-y plane is random. In other words, there is no relationship between the transverse (x-y) magnetization of one spin to another. The transverse magnetization is termed to be *incoherent*. Since the sum of a large collection of vectors aligned in random directions is zero, there is no bulk transverse magnetization at thermal equilibrium, i.e.:

$$M_x = M_y = 0 \tag{1.19}$$

1.3.2.2 Effect of the B_1 Pulse: Excitation of Nuclear Spins

The next step of the experiment, the application of the B_1 pulse, has to be considered. Assuming that the B_1 magnetic field oscillates in y-direction, it can be described as:

$$\vec{B_1} = |b_1| cos(\omega t)\hat{\jmath} \tag{1.20}$$

where b_1 is the amplitude of the applied field, ω is its frequency, and $\hat{\jmath}$ describes its direction. All three of these parameters; intensity, frequency, and direction, are under computer control in modern NMR instruments.

The total magnetic field in the rotating frame is the sum of both the static field and the oscillating B_1 field. Since the rotation rate of the rotating frame is always set to the frequency of the B_1 pulse, the B_1 field is stationary in the rotating frame. Therefore, the fields present in the rotating frame are,

$$\vec{B}_{rot} = \left[(B + \frac{\Omega}{\gamma})\hat{k} + B_1\hat{\jmath}\right] \tag{1.21}$$

and the change in the magnetic dipole in the rotating frame is then,

$$\frac{\delta\mu}{\delta t} = \gamma\mu \times \left[(B + \frac{\Omega}{\gamma})\hat{k} + B_1\hat{\jmath}\right] \tag{1.22}$$

If we make the following substitutions: $\Omega = -\omega$; $\gamma B = \omega_s$; $\omega_1 = \gamma B_1$, then the above equation is converted to:

$$\frac{\delta\mu}{\delta t} = \gamma\mu \times \left[(\frac{\omega_s - \omega}{\gamma})\hat{k} + \frac{\omega_1}{\gamma}\hat{\jmath}\right] \tag{1.23}$$

where ω_s is the resonance frequency of the *shielded* spin, while ω_1 is proportional to the *strength* of the B_1 field, not its frequency, which is ω.

Thus, there are two stationary fields in the rotating frame, one aligned along the z-axis and one along the y-axis. The vertical (z) field, consists of the applied B field and an opposing field, ω/γ, that arises solely from the change in the coordinate system. If the resonance frequency of the spin, ω_s, is equal to the rotational rate of the coordinate frame, then $\omega_s - \omega = 0$ and there is no magnetic field in the z-direction; the applied B_o field has been canceled by the fictitious field. Under these conditions the only field remaining in the rotating frame is the field provided by the pulse, B_1 (see Fig. 1.8).

The B_1 field will generate a torque on an individual spin:

$$\frac{\delta\vec{\mu}}{\delta t} = \gamma\mu \times \frac{\omega_1}{\gamma}\hat{\jmath} \tag{1.24}$$

Replacing a single spin with the bulk magnetization, represented as a unit vector along the z-axis, gives:

$$\hat{M} \times \vec{\omega_1} = \begin{vmatrix} i & j & k \\ 0 & 0 & 1 \\ 0 & \omega_1 & 0 \end{vmatrix} = -\omega_1\hat{\imath} \tag{1.25}$$

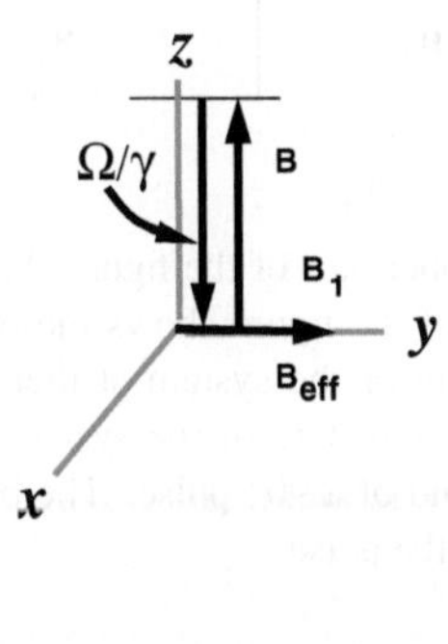

Figure 1.8 Magnetic fields present in the rotating frame. Shown are the magnetic fields present in the rotating frame of reference when $\Omega = \omega_s$. B is the magnetic field at the nucleus and is fixed along the z-axis. $\frac{\Omega}{\gamma}$ is the fictitious field that arises due to the change in reference frame. B_z is the resultant field in the rotating frame: $B_z = B + \frac{\Omega}{\gamma}$. In this case, B_z is zero and not shown. B_1 is the oscillating field that is only present during a pulse. B_{eff} is the vector sum of B_1 and B_Z and in this case it is equal to B_1.

therefore:

$$\frac{\delta \vec{M}}{\delta t} = -\omega_1 \hat{i} \tag{1.26}$$

Thus, $\vec{M}$ is tipped from the z-axis at a rate of ω_1 rad/sec. The direction of the change in $\vec{M}$, is given in Eq. 1.26, and is in the direction of the minus x-axis ($\hat{i}$) for this particular direction of the B_1 field (see Fig. 1.9). Although the direction of rotation (clockwise or counterclockwise) depends on γ, it is customary to use the 'right-hand-rule'; using the right hand, the thumb is placed along the B_1 axis and the direction of rotation of the magnetization is in the direction of one's fingers.

The extent of precession about B_1 depends on both the strength of the B_1 field (ω_1) and the length of time, τ, the pulse was applied. A pulse of length τ, applied at a field strength of ω_1 rad/sec will rotate the magnetization through a "tip", or "flip", angle of $\beta = \omega_1 \tau$. To specify both the direction of the B_1 field as well as the rotation angle, the following notation is used: $P_\beta^{\vec{u}}$, where $\vec{u}$ indicates the direction of the applied B_1 field and β refers to the rotation angle. For example, a 45° pulse applied along the y-axis would be represented as P_{45}^{y}.

After the application of a pulse with a tip angle of β, a component of the magnetization, $\cos\beta$, will remain along the z-axis and a component in the x-y plane equal to $\sin\beta$ will have been created by the pulse. For example, if a B_1 pulse is applied along the y-axis, the individual components of the magnetization after the pulse are:

$$M_z = M_o \cos(\beta) \qquad M_x = M_o \sin(\beta) \qquad M_y = 0 \tag{1.27}$$

This transformation is often abbreviated as:

$$M_z \xrightarrow{P_\beta^y} M_z cos(\beta) + M_x sin(\beta) \tag{1.28}$$

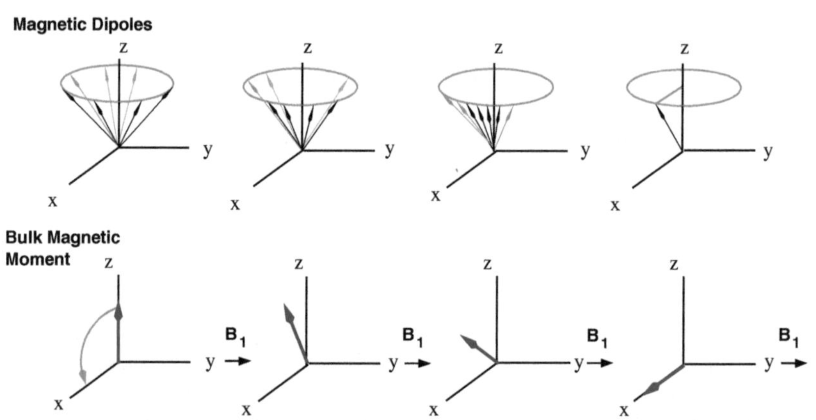

Figure 1.9. Effect of a B_1 y-pulse on the nuclear spins. The upper part of the figure shows a collection of individual magnetic dipoles while the lower part of the figure shows the bulk magnetization. The leftmost part of the panel illustrates the state of the system at thermal equilibrium, *prior* to the pulse. The subsequent 3 sections show the state of the system, in the rotating frame, near the beginning, at the middle, and at the end of a 90° pulse. The bulk magnetization remains in the z-x plane during the entire period of the pulse.

When the tip angle is $\pi/2$, the bulk magnetization is found only along the x-axis. This is referred to as a *90° pulse*. This tip angle generates the maximum amount of bulk magnetization in the x-y plane, and therefore generates the largest signal from a single scan.

Similar relationships exist for pulses along any other axis. For example, a pulse along the minus y-axis, P_β^{-y}, will produce the following magnetization, starting from $M_z = M_o$:

$$M_z = M_o \cos(\beta) \qquad M_x = -M_o \sin(\beta) \tag{1.29}$$

The effect of a y-pulse on the individual magnetic dipoles is shown in Fig. 1.9. Note that the dipoles are transformed from a random distribution about the z-axis to a distribution in which all of the dipoles have the same phase, aligned along the x-axis ($\phi = 0$). The distribution of the magnetic dipoles after the 90° pulse is referred to as a *coherent* state. In addition, the magnetization that is in the x-y plane is called transverse magnetization because it is orthogonal to the direction of the main field.

The net result of a 90° pulse is to turn the equilibrium bulk magnetization from the z-axis and place it in the x-y plane.

1.3.2.3 Detection of Resonance

After the B_1 pulse is turned off, the transverse magnetization precesses in the x-y plane around the B_o field, just as it did before the pulse. The key difference is that the transverse magnetization is now coherent and gives rise to a non-zero magnetic moment in the x-y plane.

The precession of the coherent magnetization in the x-y plane induces a time dependent current in the receiver coil. This signal is called the *free induction decay* (FID) and represents bulk magnetization that exists in the x-y plane. The frequency of the induced signal is *exactly* equal to the resonance frequency of the nuclear spin transition since the magnetization precesses around B_o at $\omega_s = \gamma B$.

Detection of the precessing magnetization is accomplished by analog circuits that actually measure the magnetization in the rotating frame, i.e. the observed frequency, ω', is $\omega_s - \omega$, where ω_s is the precessional frequency of the spin and ω is the rate of rotation of the coordinate frame, or equivalently, the frequency of the applied B_1 pulse.

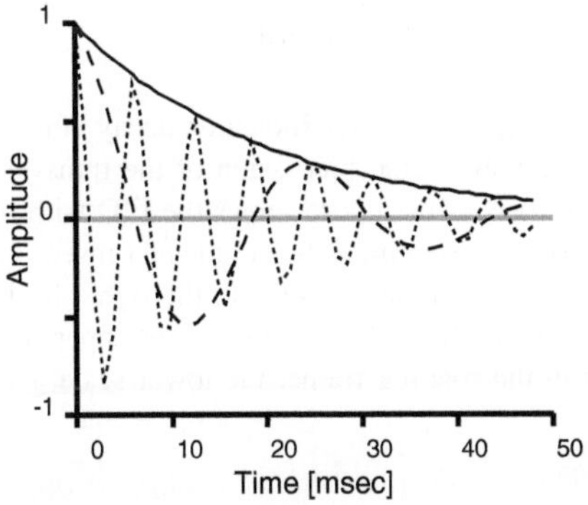

Figure 1.10 On- and off-resonance signals. The detected signals represent M_x in the rotating frame after an P_{90}^y pulse. The solid line is on-resonance spectral line ($\omega_s = \Omega$), the dashed line represents spins that are 150 Hz from Ω, and the dotted line represents spins whose resonance frequency is 650 Hz from Ω. All three resonances have the same T_2 of 20 msec.

A spin is considered to be on-resonance if its resonance frequency is the same as the frequency of the B_1 field ($\omega' = 0$). In this case, the signal does not oscillate, but decays exponentially with a time constant of T_2, the spin-spin relaxation time. T_2 is the characteristic time for the decay of transverse magnetization, and is discussed in more detail later in this chapter, as well as in Chapter 19. An off-resonance spin is one whose resonance frequency is not equal to ω. The x- and y-components of its signal will oscillate at the frequency difference $\omega_s - \omega$. Examples of on- and off-resonance signals are shown in Fig. 1.10.

It is possible to measure the bulk magnetization along both the x- and y-axis, giving independent measurements of M_x and M_y. This is accomplished using a technique called quadrature detection, which is discussed in more detail in Chapter 12. In the case of a 90° B_1 pulse along the y-axis, the initial bulk magnetization immediately after the pulse is $M_x = M_0$. As the spins precess around the z-axis, the individual components of the bulk magnetization will evolve as follows, and illustrated in Fig. 1.11:

$$M_x(t) = M_o cos(\omega' t)e^{-t/T_2} \qquad M_y(t) = M_o sin(\omega' t)e^{-t/T_2} \tag{1.30}$$

where ω' is the resonance frequency in the rotating frame, and e^{-t/T_2} represents the decay of the excited state due to relaxation, with a time constant of T_2.

These two signals are usually combined into a single complex number:

$$S(t) = M_x(t) + iM_y(t) = M_o e^{i\omega' t} e^{-t/T_2} \tag{1.31}$$

where the magnetization along the x-axis is arbitrarily chosen to be the real component and the magnetization along the y-axis is arbitrarily chosen to be the imaginary component.

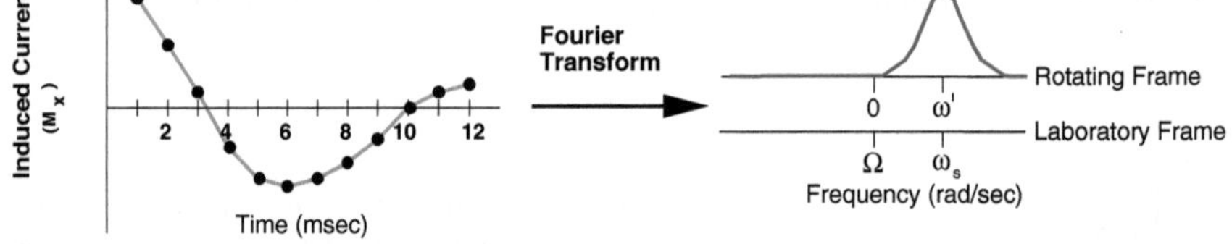

Figure 1.11. Fourier transform of the time domain signal. The free induction decay after the 90° pulse is shown. The upper section of the figure shows the precession of the transverse (i.e. x-y) magnetization after the pulse. The lower part of the figure shows the FID with the points indicating the data sampled during digitization, representing a dwell time of 1 msec. The subsequent resonance line obtained after Fourier transformation is shown to the right. In this case the pulse is slightly off-resonance and precesses in the rotating frame. The upper scale for the abscissa of the spectra gives frequencies in the rotating frame, the lower scale gives frequencies in the laboratory frame.

The frequencies that are present in the FID can be obtained by Fourier transformation of the time domain signal, as illustrated in Fig. 1.11. Since the Fourier transformation is performed with digital computers, it is necessary to sample the FID at fixed time intervals. The delay between each sampling is referred to as the dwell time (τ_{dw}).

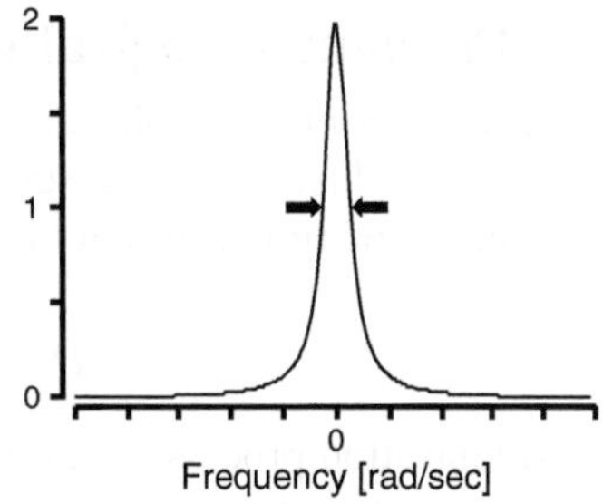

Figure 1.12. Lineshape of an NMR resonance. The Lorentzian lineshape, which is the real part of the Fourier transform of e^{-t/T_2}, is shown. The full width at half-height, $\Delta\nu$, is indicated.

The position of the resonance line in the spectrum depends on its precessional frequency. In the case of $e^{i\omega t}$, the Fourier transform[3] gives a delta function located at ω. The lineshape of a resonance depends on how the signal decays with time. The Fourier transform of the second function, e^{-t/T_2}, gives a complex function. The real part of this function is the Lorentzian lineshape:

$$F(\omega) = \frac{T_2}{1 + T_2^2\omega^2} \tag{1.32}$$

This lineshape is shown in Fig. 1.12. The full width of the line at half-height, $\Delta\nu$, is inversely proportional to the T_2: $\Delta\nu = 1/(\pi T_2)$.

Since the time domain signal is a product of two functions, $e^{i\omega' t}$ and e^{-t/T_2}, its Fourier transform will be the convolution of the Fourier transforms of each function. The final spectrum consists of a Lorentzian line located at ω', giving the final NMR spectrum shown in Fig. 1.11. Note that since detection of these frequencies occurs in the rotating frame, the origin of the frequency axis is zero in that frame, but ω, or the frequency of the applied B_1 field, in the laboratory frame.

1.3.3 Summary of the Process of Acquiring a One Dimensional Spectrum

The overall process of obtaining an NMR spectrum consists of four steps:

1. Prior to the pulse, the magnetic dipoles precess about B_o at a frequency that is equal to their resonance frequency, however there is no *net* magnetization in the x-y plane.
2. During the pulse, the bulk magnetic moment of the sample is tipped from the z-axis to the x-y plane.
3. After the pulse, the magnetization precesses about B_o at a frequency ω_s, inducing a current in the receiver coil.
4. The induced current is sampled at discrete times and digitized. The frequency domain spectrum is obtained from the time domain data by Fourier transformation.

[3]Properties of Fourier transforms are discussed in detail in Appendix A.

1.4 Phenomenological Description of Relaxation

The excited nuclear dipoles are subject to two different relaxation processes. In general, these relaxation processes follows a first-order rate equation, characterized by a characteristic time constant (T) or a rate constant (R = 1/T):

$$I(t) = I_o e^{-t/T} = I_o e^{-Rt} \tag{1.33}$$

The first relaxation process arises from a re-alignment of the bulk magnetic moment along the static field to regenerate the original Boltzmann population difference associated with M_z. This is referred to as spin-lattice relaxation because it involves the transfer of energy from the excited state to the surroundings or lattice. This process restores M_z to its equilibrium value, as indicated in Fig. 1.13. The rate of this process is characterized by a relaxation time, T_1, or by a relaxation rate, R_1 (=1/T_1). Typical T_1 values range from 1 sec for proteins to tens of seconds for small molecules.

The second relaxation process causes dephasing of the coherent transverse magnetization as it precesses around B_o. This process is also illustrated in Fig. 1.13. Since the dephasing does not affect the net population of excited spins, simply their phase coherence, the system is *not* returned to thermal equilibrium by this relaxation process. The time constant for this process is T_2, or in terms of rate constants, R_2. Dephasing, or loss of coherence between the spins, occurs by three main mechanisms, each of which are briefly discussed below.

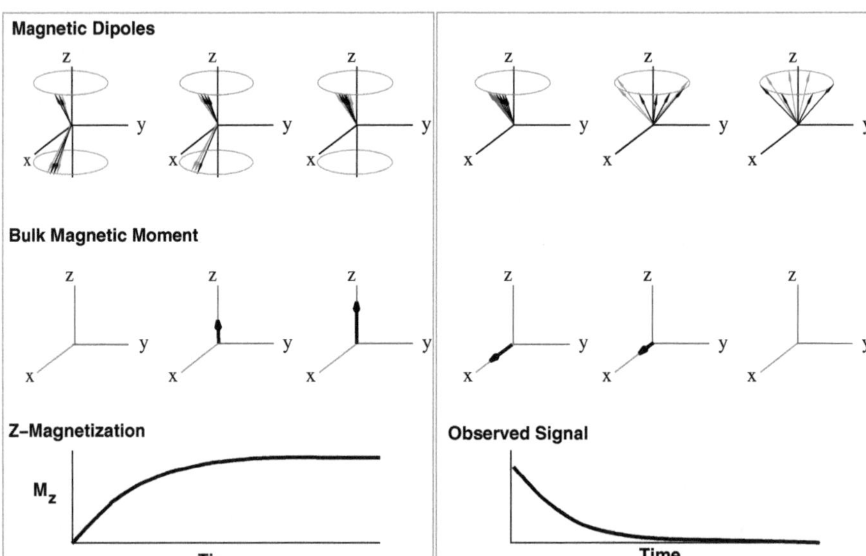

Figure 1.13. Relaxation of nuclear spins. Each panel shows, from top to bottom, the individual nuclear moments as a function of time, the bulk magnetization as a function of time, and the z-magnetization (M_z) or the bulk magnetization in the x-y plane ($\sqrt{M_y^2 + M_x^2}$), as a function of time. The left panel illustrates spin-lattice relaxation in the absence of spin-spin relaxation. The right panel shows spin-spin relaxation in the absence of spin-lattice relaxation.

1. The first process that contributes to dephasing is an inhomogeneous magnetic field. This inhomogeneity causes the precessional frequency of the spins to differ from location to location within the sample. The different precessional frequencies lead to dephasing of the magnetization while the data is being collected, as illustrated in Fig. 1.14. This relaxation mechanism clearly has nothing to do with the intrinsic properties of the molecules in the sample and is thus an annoyance. Fortunately, the contribution of magnetic field inhomogeneities to the relaxation rate is quite small in modern superconducting magnets, less than 1 sec^{-1}, provided the magnet is shimmed correctly. Consequently this mechanism can be ignored in studies on high molecular weight molecules.
2. The second process that contributes to dephasing of the transverse magnetization arises from dipolar coupling between magnetic dipoles. Consequently, this process is often referred to as spin-spin relaxation. The rate constant for this relaxation process is R_2^{DD}. Dipolar interactions between spins will be discussed in more detail in chapters 16 and 19. This mechanism of relaxation is sensitive to molecular motion and plays a dominate role in the relaxation of most spin-1/2 nuclei.
3. The third process that contributes to relaxation is an anisotropic electron density surrounding the nucleus. This is referred to as Chemical Shift Anisotropy (CSA) and will be discussed in more detail in Chapter 19. The rate constant for this process is R_2^{CSA}. This relaxation mechanism is important for nuclei in anisotropic bonding environments, such as carbon and nitrogen, and is also sensitive to molecular motion.

The observed relaxation rate of the transverse magnetization is given by the sum of all of the above mechanisms:

$$R_2^* = \frac{1}{T_2^*} = R_2^{\Delta B} + R_2^{DD} + R_2^{CSA} \tag{1.34}$$

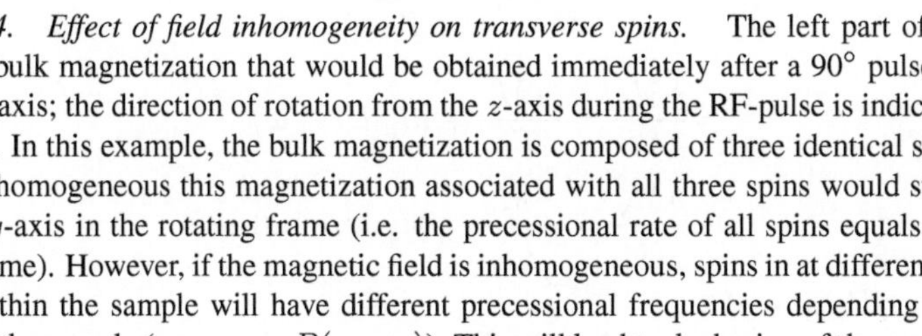

Figure 1.14. Effect of field inhomogeneity on transverse spins. The left part of the figure shows the bulk magnetization that would be obtained immediately after a 90° pulse along the *negative* x-axis; the direction of rotation from the z-axis during the RF-pulse is indicated by the gray arrow. In this example, the bulk magnetization is composed of three identical spins. If the field were homogeneous this magnetization associated with all three spins would stay aligned along the y-axis in the rotating frame (i.e. the precessional rate of all spins equals that of the rotating frame). However, if the magnetic field is inhomogeneous, spins in at different locations, (x,y,z), within the sample will have different precessional frequencies depending upon their location in the sample (e.g. $\omega = \gamma B(x, y, z)$). This will lead to dephasing of the magnetization over time (left to right), eventually producing a random distribution of the magnetic dipoles in the x-y plane. At this point $M_x = 0$ and $M_y = 0$.

This relaxation time constant is generally referred to as T_2 'star' , indicating a contribution from magnetic field inhomogeneity. The time constant for spin-spin relaxation due to the intrinsic characteristics of the molecule is referred to as T_2, and it is the sum of the relaxation due to spin-spin dipolar coupling and CSA effects:

$$\frac{1}{T_2} = R_2 = R_2^{DD} + R_2^{CSA} \tag{1.35}$$

Often, the contribution of CSA to T_2 is ignored, and the relaxation process is simply referred to as spin-spin relaxation.

1.4.1 Relaxation and the Evolution of Magnetization

For the moment we will treat T_1 and T_2 as measurable quantities, but we will avoid any discussion about their relationship to structure and dynamics. To develop analytical equations which describe the motion of the magnetic moment *in the presence of the static B_o field* we will use the following approach that was originally proposed by F. Bloch in 1946 [18]. This approach begins with the classical description of the evolution of the magnetization:

$$\frac{d\vec{M}}{dt} = \gamma \vec{M} \times B \tag{1.36}$$

The decay of magnetization due to relaxation is added to each component of magnetization:

$$\begin{aligned} \frac{dM_z}{dt} &= \frac{M_o - M_z}{T_1} + \gamma(M \times B)_z \\ \frac{dM_x}{dt} &= \frac{-M_x}{T_2} + \gamma(M \times B)_x \\ \frac{dM_y}{dt} &= \frac{-M_y}{T_2} + \gamma(M \times B)_y \end{aligned} \tag{1.37}$$

The Bloch equations describe a return of the *z*-magnetization to the equilibrium value, M_o, with a time constant of T_1, and a decay of the transverse magnetization with a time constant of T_2. In the rotating frame these equations become:

$$\frac{\delta M_z}{\delta t} = \frac{M_o - M_z}{T_1} \tag{1.38}$$

$$\frac{\delta M_x}{\delta t} = \frac{-M_x}{T_2} + M_y(\omega_s - \omega) \tag{1.39}$$

$$\frac{\delta M_y}{\delta t} = \frac{-M_y}{T_2} - M_x(\omega_s - \omega) \tag{1.40}$$

where $\omega_s = \gamma B, \omega = -\Omega$.

These equations are most easily solved by defining a function:

$$M^+ = M_x + iM_y \tag{1.41}$$

Adding eqs. 1.39 and 1.40 gives the following:

$$\frac{\delta M^+}{\delta t} = -M^+ \left[\frac{1}{T_2} + i\omega'\right] \tag{1.42}$$

$\omega' = \omega_s - \omega$, is the precessional rate of the spin in the rotating frame. In practice, this is actually the frequency of the signal obtained from the spectrometer.
The solution to Eq. 1.42 is:

$$M^+ = e^{-i\omega' t}\, e^{-t/T_2} \tag{1.43}$$

Taking into account the initial conditions immediately after a 90° x-pulse (M_x & M_z=0) gives the following final solutions:

$$\begin{aligned} M_x(t) &= \sin(\omega' t)\, e^{-t/T_2} \\ M_y(t) &= \cos(\omega' t)\, e^{-t/T_2} \\ M_z(t) &= M_o[1 - e^{-t/T_1}] \end{aligned} \tag{1.44}$$

The above shows that the magnetization along the *z*-axis will grow with an exponential time constant of T_1 while the transverse magnetization will decay with a time constant of T_2; exactly the behavior that is depicted in Fig. 1.13.

1.5 Chemical Shielding

A very important feature of NMR spectroscopy is that the absorption frequency of a nuclear spin depends on the magnetic field strength at the nucleus. This magnetic field usually differs slightly from the applied field, B_o, because the magnetic field at the nucleus is shielded by the electron density surrounding the nucleus. This shielding is due to precession of electrons under the influence of the applied magnetic field. This precession generates an additional magnetic field that usually opposes the externally applied magnetic field. The local magnetic field strength at the nucleus is given by:

$$B = (1 - \sigma) \cdot B_o \tag{1.45}$$

where σ represents the shielding of the nuclear spin. For an isotropic electron distribution the shielding is given by the Lamb formula [93]:

$$\sigma = \frac{e^2}{3mc^2} \int \frac{\rho(r)}{r} dr \tag{1.46}$$

As the electron density around the nucleus increases, the effective field decreases, leading to lower resonance frequencies. Since the resonance frequency is due to the chemical environment of the nuclear spin, the observed frequency is referred to as a *chemical shift*. Due to differences in shielding, different spins will experience different local magnetic fields, giving rise to shifts in their frequencies.

For anisotropic electron distributions the shielding is described by a tensor. A tensor is a concise mathematical expression of the anisotropic properties of a physical system in a three-dimensional space and has the following form:[4]

$$\sigma = \begin{bmatrix} \sigma_{xx} & 0 & 0 \\ 0 & \sigma_{yy} & 0 \\ 0 & 0 & \sigma_{zz} \end{bmatrix} \tag{1.47}$$

Equation 1.47 describes how the chemical shielding would look if the magnetic field was along the x-axis (σ_{xx}), y-axis (σ_{yy}), or the z-axis (σ_{zz}). Under conditions of rapid tumbling, which is generally the case in solution, an averaged shielding is observed:

$$\bar{\sigma} = \frac{1}{3}\left[\sigma_{xx} + \sigma_{yy} + \sigma_{zz}\right] \tag{1.48}$$

The dependence of the resonance frequency on the applied field, B_o, is removed by converting all frequencies to a dimensionless scale, known as the *chemical shift scale*. This scale is defined as:

$$\delta = \frac{\nu - \nu_o}{\nu_o} \times 10^6 \tag{1.49}$$

with units of ppm, or parts-per-million.

The conversion from frequency to chemical shift makes the position of the spectral line independent of the magnetic field strength (by dividing by ν_o). Thus, making it possible to directly compare the position of resonance lines in spectra that are obtained at different field strengths.

For example, a resonance line at 2 ppm would be 600 Hz higher in frequency than the reference line on a 300 MHz (ν_H) magnet. On a 900 MHz magnet, the same line would have a resonance frequency that is 1800 Hz higher than the reference compound.

The constant, ν_o, is a reference frequency, in units of Hertz (Hz). It is often the frequency of the line from a reference compound whose resonance is at one end of the spectrum. For example, tri-methyl silane is used to reference organic samples and its proton and carbon frequencies are set to zero ppm. In the case of protein solutions, the water line can be used as an approximate proton chemical shift reference point, with a chemical shift of about 4.70 ppm (see Section 3.7).

Conversion of a chemical shift back to an absorption frequency requires knowledge of the magnetic field strength, or spectrometer frequency, from which the spectra was acquired. In addition, to determine the absolute frequency of the spectral line it is also necessary to know ν_o, which can usually be approximated as the spectrometer frequency.

[4]This simple form, with all off-diagonal elements having the value of zero, is only found for one particular orientation of the molecule with respect to the magnetic field, called the principle axis system (PAS). Tensors are discussed in more detail in Appendix B.

1.6 Characteristic ^{1}H, ^{13}C and ^{15}N Chemical Shifts

The proton, carbon, and nitrogen chemical shifts found for amino-acids in proteins are presented in tables 1.2, 1.4, 1.3.

1.6.1 Effect of Electronic Structure on Chemical Shifts

The chemical shifts presented in tables 1.2 and 1.4 are clearly different from atom to atom. For example, amide protons resonate at $\approx$ 8 ppm, H_α protons at $\approx$ 4 ppm and methyl protons at $\approx$ 1 ppm. A similar trend in carbon shifts is observed for α- and β-carbons. These trends in chemical shifts can be largely explained by the electronegativity of the atoms that are chemically bonded to the atom of interest. The amide proton has a high chemical shift because the nitrogen atom is more electron withdrawing than carbon. The reduced electron density at the amide proton decreases the shielding and therefore increases the effective field and resonance frequency. Similarly the H_α shifts are higher than the methyl-H shifts because of the proximity of the α-protons to the electronegative nitrogen.

Note that within a residue, the relationship between atom type and chemical shift is similar for both carbon and proton shifts. For example, in the case of arginine the following ordering is found for both carbon and proton shifts: $\alpha > \delta > \beta > \gamma$ (see

Table 1.2. Proton chemical shifts. The average proton chemical shifts in proteins are shown. These data were obtained from BioMagResBank [52].

Residue	*NH*	H_α	H_β	*Others*
Gly	8.34	3.94		
Ala	8.20	4.26	1.38	
Val	8.29	4.16	1.99	0.84, 0.83(CH3)
Ile	8.26	4.20	1.80	1.30, 1.24 (CH2), 0.80 (γCH3), 0.70 (δCH3)
Leu	8.22	4.32	1.63,1.57	1.54 (γCH), 0.77, 0.76(δCH3)
Pro	-	4.41	2.05,2.05	1.93 (γCH2), 3.64, 3.63 (δCH2)
Ser	8.29	4.51	3.88	5.33 Hγ (OH)
Thr	8.27	4.48	4.17	1.16 (γCH3), 4.40 Hγ1 (OH)
Asp	8.33	4.61	2.74,2.70	
Glu	8.34	4.26	2.04	2.31 (γCH2)
Lys	8.22	4.28	1.79,1.78	1.38 (γCH2), 1.61 (δCH2), 2.93 (ϵCH2), 7.52 (ζNH3)
Arg	8.24	4.27	1.79	1.58 (γCH2), 3.13 (δCH2), 7.32, 6.74, 6.72 (NH)
Asn	8.37	4.70	2.80,2.78	7.27, 7.20 (δNH2)
Gln	8.22	4.28	2.05,2.04	2.32 (γCH2), 7.17, 7.07 (γNH2)
Met	8.26	4.39	2.03,2.01	2.44 (γCH2), 1.86 (ϵCH3)
Cys	8.42	4.73	2.95,2.98	1.66 -SH
Trp	8.35	4.74	3.32,3.18	6.68-7.17 (aromatic), 10.13 (NH)
Phe	8.42	4.62	2.97,2.99	6.89-6.91 (aromatic)
Tyr	8.37	4.63	1.91	6.86 (Hδ), 6.64 (Hϵ), 9.25 (-OH)
His	8.25	4.62	3.11,3.12	Hδ1 10.14(NH), Hδ2 7.08, Hϵ1 8.08, Hϵ2 10.43(NH)

Table 1.3. Nitrogen chemical shifts. The nitrogen chemical shifts for side-chain atoms are shown. The amide nitrogen chemical shifts are $\approx$ 120 ppm, with the exception of glycine, which is found at 109.9 ppm. Data from BioMagResBank [52].

Residue	*Shifts*	*Residue*	*Shifts*
Arg	89.8 (ϵ), 74.8 NH1, 75.8 NH2	His	190.7 ($\delta1$), 179.8 ($\epsilon2$)
Asn	112.8 (δ)	Lys	71.86 (ζ)
Gln	111.8 (ϵ)	Trp	129.5 (ϵ)

Table 1.4. Carbon chemical shifts. The average carbon chemical shifts were obtained from the BioMagResBank [52]. Carbonyl shifts have been omitted from this table since they are quite uniform at approximately 175 ppm.

Residue	C_α	C_β	*Others*
Gly	45.3		
Ala	53.1	18.9	
Val	62.5	32.6	21.3 (CH3)
Ile	61.6	38.6	27.6 ($\gamma1$), 17.3 (γCH3), 13.4 (δCH3)
Leu	55.7	42.3	26.8 (γ), 24.5 (δCH3)
Pro	63.3	31.8	27.1 (γ), 50.3 (δ)
Ser	58.6	63.8	
Thr	62.1	69.6	21.4 (γCH3)
Asp	54.5	40.7	178.41 (γ) sidechain
Glu	57.4	30.0	36.0 (γ), 181.9 (δ) sidechain
Lys	56.8	32.8	24.9 (γ), 28.8 (δ), 40 (ϵ)
Arg	56.9	30.7	27.3 (γ), 43.1 (δ), 159.0 (ζ)
Asn	54.5	40.7	178.41 (γ) sidechain
Gln	56.6	29.1	33.7 (γ), 179.7 (δ) sidechain
Met	56.1	32.9	32.1 (γ), 17.2 (ϵCH3)
Cys	57.4	34.1	
Trp	57.7	30.1	110-137 (aromatic)
Phe	58.2	40.0	129-138 (aromatic)
Tyr	58.0	39.1	117 (ϵC), 132 (δC), 156 (ζ)
His	56.4	30.0	119.8 ($\delta2$), 136 ($\epsilon1$)

Fig. 1.16). The identical ordering reflects a similar electronic environment for both the carbon and its attached proton.

The conformation, or secondary structure, of the polypeptide backbone alters the chemical shift of chemically equivalent atoms in a predictable fashion. For example, the C_α shift of alanine decreases by 1.3 ppm when this residue is found in a β-strand, and increases by 2.3 ppm when found in an α-helical configuration. Although the secondary structure of the backbone is only one of the factors that alter chemical shifts, it is possible to predict secondary structure by correlating the deviation of the chemical shift from random-coil values for a number of different atoms (e.g. C_α, H_α, etc). This topic is discussed in more detail in Chapter 17, Section 17.1.1.5.

In addition to electronegativity effects, a formal positive change (e.g. Lys ϵN) also withdraws electrons from adjacent atoms, decreasing the shielding and therefore increasing the chemical shift. A formal negative change will have the opposite effect.

1.6.2 Ring Current Effects

Chemical shifts are also perturbed by additional magnetic fields that arise from the precession of delocalized electrons in conjugated systems, such as aromatic rings or carboxylic acid groups. These *ring current effects* can be substantial. Several methods have been devised to calculate the effect of ring-currents on chemical shifts. The simplest method, introduced by Pople [134], is to consider the induced magnetic field as a point dipole at the center of the aromatic ring. The magnetic field that is generated by this dipole is given by the standard dipole equation:

$$\sigma = iB\frac{1 - 3cos^2\theta}{r^3} \tag{1.50}$$

where i is a ring-current factor that depends on the geometry of the ring. For the phenyl group $i = 1$. B is a constant of proportionality, on the order of 25 ppm. θ is the angle between the normal to the aromatic ring and the vector that joins the atom to the center of the aromatic ring.

This function is plotted in Fig. 1.15. The ring-current effect depends on both distance and orientation. The weak distance dependence of $1/r^3$ implies that ring current effects can significantly perturb the chemical shifts of near-by residues. Chemical shift perturbations also depend on the orientation of the atom with respect to the plane of the aromatic ring. For example, placement of a proton 4 Å directly above the center of the ring will decrease the chemical shift by approximately 0.75 ppm. In contrast, if the atom is placed in the plane of the ring at the same distance from the center, its chemical shift will be increased by approximately 0.33 ppm. Aromatic protons, because of their close proximity to the center of the ring, experience an increase of approximately 2 ppm due to ring current effects.

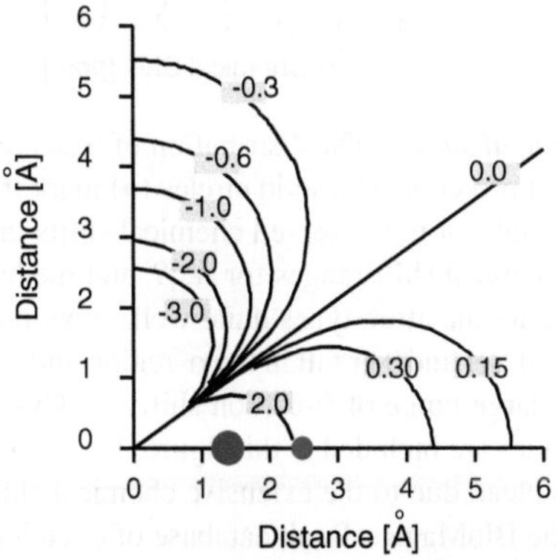

Figure 1.15 Ring current shifts. Calculated ring current shifts for a phenylalanine ring. The x-axis lies in the plane of the ring and the y-axis is perpendicular to the plane of the ring. The location of the carbon and its attached hydrogen are indicated by the large and small spheres, respectively. The large gray area represents the approximate Van der Waals radius of the phenyl group. The lines represent contours of iso-chemical shift changes.

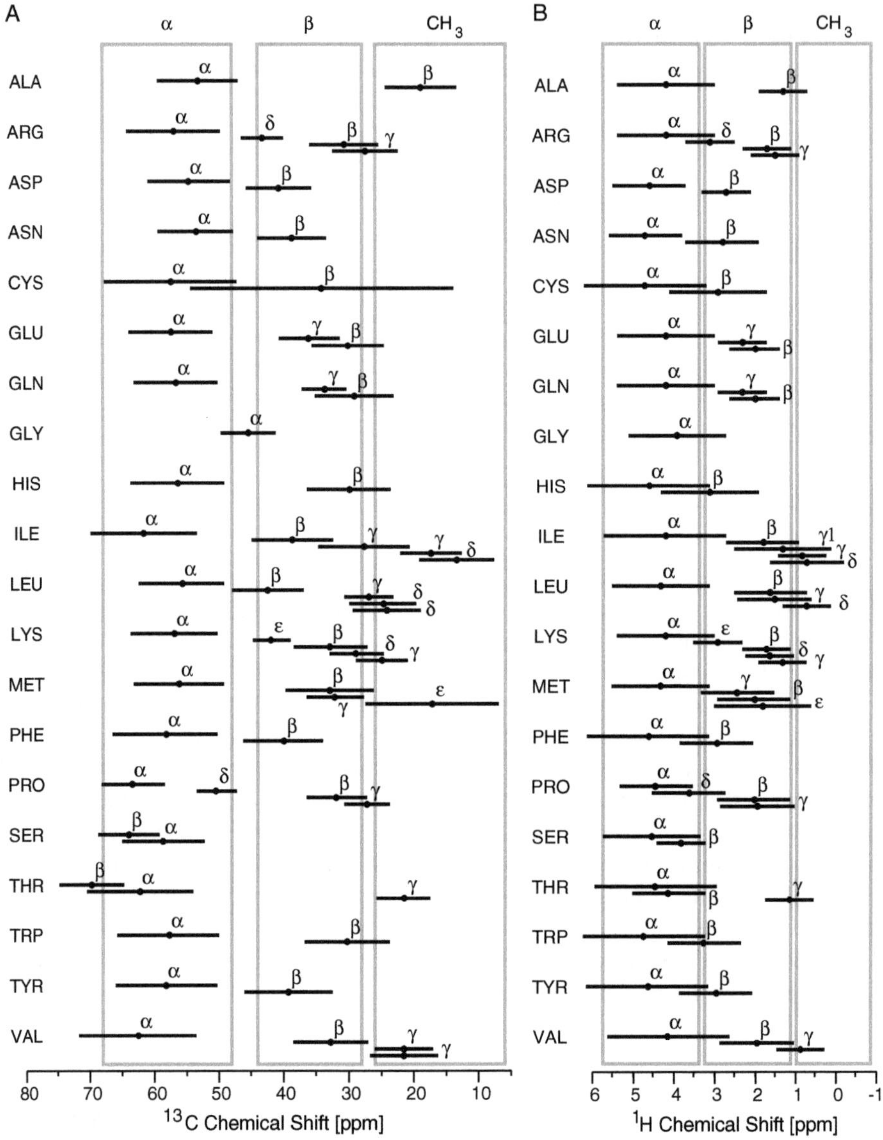

Figure 1.16. Distribution of carbon and proton chemical shifts. The distribution of observed carbon (A, left) and proton (B, right) chemical shifts in proteins. The solid circles (•) mark the average chemical shift. The solid lines indicate $\pm 3\sigma$; 95% of the observed chemical shifts fall within this range. The gray boxes indicate nominal chemical shift ranges for α, β, and methyl atoms. In the case of carbon shifts, these ranges separate the atom types quite well. Note that there are a few exceptions, for example, the β-carbons of Ser and Thr fall in the α-region and the α-carbon of Gly can fall in the β-carbon region. The large range of β-carbon shifts for Cys is due to the fact that both free and disulfide bonded residues are included in this figure. In the case of proton shifts, the separation by atom type is not as clean due to the extensive chemical shift overlap between the various atom types. Data from the BioMagResBank database of chemical shifts [52].

1.6.3 Effects of Local Environment on Chemical Shifts

The above discussion suggests that chemically equivalent atoms will have identical chemical shifts. For example, one might expect that all of the alanine methyl protons in a protein to resonance at 1.39 ppm. Fortunately, the tertiary structure in folded proteins generates sufficient diversity in the local environment such that different resonance frequencies are observed for otherwise identical groups. The environmental differences that cause diversity in chemical shifts include electro-negativity effects (e.g. in hydrogen bonded amide protons), as well as the electrostatic and magnetic (ring current) effects that were discussed previously. Typical ranges of carbon and proton chemical shifts for aliphatic atoms in a large number of proteins are shown in Fig. 1.16. The carbon chemical shifts are more disperse, covering a range of approximately 70 ppm, while the proton chemical shift range is about 6 ppm.

1.6.3.1 Degeneracy and Equivalent Chemical shifts

In a number of cases, two or more spins (e.g. protons) will have identical chemical shifts. These spins are said to have degenerate chemical shifts. This chemical shift degeneracy can occur for a number of reasons. First, the two spins may exist in an environment that causes the electronic shielding for both to be identical. Note that the two protons need not be chemically equivalent for this to occur[5].

Chemical shift degeneracy can also occur due to averaging of the environments of the spins, provided the rate of exchange between multiple environments is much faster than the difference in the resonance frequency of each environment (see Chapter 18). For example, the rapid rotation of the methyl group(s) in alanine, valine, isoleucine, and leucine will cause the methyl protons to experience an average environment. Similarly, the flipping of aromatic rings in tyrosine and phenylalanine will often lead to averaging of the chemical shifts of the δ and ϵ protons and carbons.

1.6.4 Use of Chemical Shifts in Resonance Assignments

A critical problem in applying NMR techniques to proteins is that of resonance assignment, or determining an atom's chemical shift (see Chapter 13). Once this has been accomplished the information obtained from the NMR resonance signal, such as relaxation rates, can be associated with a particular atom in the protein. During the assignment process, resonance signals that belong to the same amino acid are identified by a number of different NMR experiments that are discussed in subsequent chapters. This collection of resonance signals is called a *spin-system*. Identification of the residue type of a spin-system is helpful in the assignment process. In particular, if a certain amino acid only appears once in the primary sequence of a protein, then identification of the amino acid type of a spin-system leads directly to assignment. The residue type of a spin-system can often be discerned from the number of resonances and their chemical shifts. Inspection of Fig. 1.16 shows that there is a cleaner distinction between the type of atom and its carbon chemical shift versus the corresponding

[5]Chemical equivalency occurs when two protons exist in identical chemical environments because of the absence of a chiral center, e.g. the two α-protons in glycine.

proton shift. For example, the chemical shift ranges for the α-carbons of Trp, Tyr, and Val do not overlap the range of their β-carbon shifts. In contrast, these ranges overlap in the case of proton shifts. Therefore, carbon chemical shifts are generally more reliable for predicting the atom type in a spin-system.

1.6.5 Chemical Shift Dispersion & Multi-dimensional NMR

Another problem that we will address is how to generate resolved NMR spectra from complex biopolymers, such as proteins. The chemical shift ranges shown in Fig. 1.16 indicate that the NMR spectra of a polypeptide of modest size, say 50 residues, will have a complex NMR spectrum that will contain many overlapping peaks. One solution to this problem has been to increase the dimensionality of the NMR experiment, such that the positions of peaks are defined by two or more resonance frequencies. Two-, three- and four-dimensional experiments are routinely performed. Two dimensional NMR techniques are introduced in Chapter 9, and higher dimensional experiments are presented in subsequent chapters. With these techniques it is possible to generate resolved spectra of proteins in the 30-50 kDa range. Additional techniques can be applied to extend the use of NMR spectroscopy to larger proteins (see Sec. 15.5.4).

1.7 Exercises

1. Calculate the frequency separation between carbonyl carbons and alpha carbons assuming a magnetic field strength that generates a proton absorption frequency of 600 MHz. (see Table 1.4)
2. B_1 field strengths are usually reported in units of Hz, i.e. $\nu = \gamma B_1/2\pi$. For a B_1 field strength of 10 kHz, how long is a 90° pulse?
3. The following two pulses are applied to magnetization that is initially at thermal equilibrium: P^x_{90} followed by P^y_{90}. What is the direction of the bulk magnetization after these two pulses?
4. Suggest an explanation as to why the H_β shift for Cys is 3.00 ppm while that for Ser is 3.88 ppm (see Table 1.2).
5. A conformational change in a protein moves the methyl group of an alanine residue from a position that is in-plane with a phenyl ring to a position that is directly above the ring. Assuming that the distance from the alanine methyl to the center of the ring is 5.5 Å in both environments, calculate the chemical shift difference of the methyl group between these two conformations.
6. A ^{13}C NMR spectrum of an amino acid shows three resonance lines, one at 174 ppm, one at 52.4 ppm, and one at 23 ppm. What is the amino acid?

1.8 Solutions

1. Carbonyls resonate at 175 ppm and the average α-carbon chemical shift is approximately, 55 ppm, giving a difference of 120 ppm. Note that this difference is invariant with respect to the magnetic field strength. In order to calculate the actual frequency difference it is necessary to obtain ν_o for the carbon reference

line. Assuming a proton frequency of 600 MHz, the carbon resonance frequency is $(\gamma_C/\gamma_H) \times 600$ MHz, or 150.89 MHz. Therefore, the frequency difference is 18,106 Hz.

2. The required equation is: $\theta = \tau 2\pi(\gamma B_1)$. The 2π converts the units of B_1 from Hz to rad/sec. Simple algebra gives: $\tau = \theta/2\pi(\gamma B_1) = (\pi/2)/(2\pi(\gamma B_1)) = 1/(4\gamma B_1)$. $\tau = 1/(4 \times 10,000) = 25$ μsec.
3. The two pulses are equivalent to a single P^x_{90} pulse. The 90° pulse along the x-axis will tip the magnetization in the y-z plane such that it is aligned along the minus y-axis at the end of the pulse. The second pulse represents a B_1 field along the y-axis. Since this B_1 field is parallel to the magnetization it has no effect.
4. In Cys, the β-carbon is bonded to a sulfur atom while in Ser it is bonded to an oxygen. Oxygen is more electronegative than sulfur, hence it will withdraw more electron density from the β-protons on Ser. Since the electron density will be smaller, the magnetic field at the nucleus is higher, leading to a larger chemical shift.
5. The ring-current effect in the plane of the ring leads to an increase in chemical shift by 0.15 ppm at 5.5 Å. A proton placed 5.5 Å above the plane of the ring will show a change in chemical shift of -0.3 ppm, therefore the total change is 0.45 ppm.
6. The amino acid is Ala. There are three carbon signals, one from the carbonyl, one from the α-carbon, and one from the sidechain group. Of the amino acids, only Ala, Ser, and Cys have a single carbon in their sidechain. The chemical shift for the β-carbon of Ala, Ser, and Cys are 18.9, 63.8, and 34.1, respectively. Therefore Ala is the best match.

Chapter 2

PRACTICAL ASPECTS OF ACQUIRING NMR SPECTRA

This chapter provides a practical description of how to acquire a one-dimensional NMR spectrum. It begins with a description of the instrument, followed by a discussion of how to set the parameters for data acquisition, and ends with a description of a simple pulse program and a discussion of artifact suppression by phase cycling.

2.1 Components of an NMR Spectrometer

The basic design of an NMR spectrometer is shown in Fig. 2.1 and some of the more important details of the instrument are discussed below.

2.1.1 Magnet

The magnet is responsible for generating the intense B_o field. Modern spectrometers usually employ superconducting solenoids that require cooling with liquid helium and liquid nitrogen. There are three safety issues of importance for the general user:

1. Medical devices, such as neurostimulators and cardiac pacemakers can be affected by the stray (fringe) magnetic fields that surround the magnet. Although the stray field in the newer actively shielded magnets[1] is lower than in older magnets, interference with medical devices is still an issue.
2. Ferromagnetic objects, such as scissors, tools, gas cylinders, etc., are attracted to the magnet. Note that some grades of stainless steel are also ferromagnetic. The force of attraction is very non-linear and usually once a metal object is moving towards the magnet it is difficult to halt its movement. Users should remove themselves from the path between the flying object and the magnet.
3. The unexpected loss of superconductivity results in a quench of the magnet. This releases large quantities of helium gas that can displace the oxygen from the magnet room. The beginning of a quench is usually signaled by rapid release of helium gas from the magnet. Since asphyxiation can occur, viewing the quench from *outside* the magnet room is strongly encouraged.

[1] An actively shielded magnet has a second set of coils surrounding the main coil. The second set of coils are wound in the opposite direction, such that their field partially cancels the stray field from the main coil.

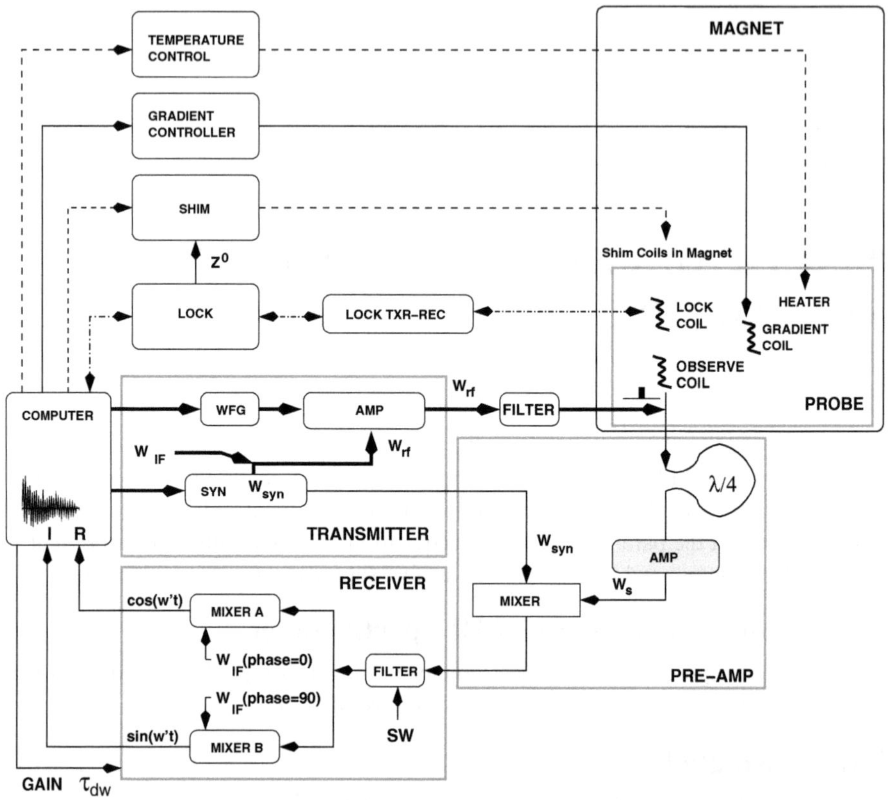

Figure 2.1. *Schematic diagram of a NMR spectrometer.* Circuit paths associated with the acquisition of the FID are shown in solid lines. The thicker solid lines show the path of the excitation pulse to the sample. Paths associated with control of the sample environment are shown as dotted lines. The blocks labeled 'probe' and 'pre-amp' represent discrete physical components of the spectrometer. The probe is inserted into the magnet such that the sample is placed at the center of the magnetic field (see Fig. 2.2). The pre-amp is found at the base of the magnet. The elements in the center of the diagram exist within a console adjacent to the computer. The transmitter generates the RF-pulse. W_{syn} is the frequency generated by the synthesizer. This is mixed with the intermediate frequency, W_{IF}, to produce the frequency of the B_1 field, W_{rf}, that is sent to the amplifier. The output of the amplifier can be modulated in amplitude and phase by the waveform generator (WFG). The amplified signal is subject to broadband filtering prior to entering the probe, e.g. the ^{1}H channel would be filtered to remove frequencies that may interfere with the deuterium lock. The resonance signals from the excited spins are indicated by W_s. After amplification, this signal is reduced in frequency by W_{syn}, and signals outside of the spectral width (SW) are removed by filtering. The real (R) and imaginary (I) components are generated by mixing the signal with the intermediate frequency. The R and I components are then sampled at time points spaced by τ_{dw}, digitized, and stored separately in the computer. The gain of the receiver and the dwell time (τ_{dw}) are controlled by the computer. Magnetic field stability is controlled by the lock circuitry. The lock consists of an independent deuterium transmitter and receiver that excites and receives signals from the deuterons in the lock solvent within the sample. The current through the Z^o shim is adjusted by the lock to maintain a defined field strength.

2.1.2 Computer

The computer is responsible for controlling the environment of the sample, the generation of the radio-frequency pulses and other experimental parameters. The computer is also responsible for storing the real and imaginary components of the free induction decay. Although data can be processed on the instrument computer, processing is often performed off-line using third-party software.

2.1.3 Probe

The probe is inserted into the bottom of the superconducting magnet and holds the sample in the center of the magnet, as illustrated in Fig. 2.2. Most probes provide a mechanism to allow spinning of the sample at 15-25 Hz about the z-axis. Spinning averages any magnetic field inhomogeneities in the x-y plane, which facilitates shimming of the magnet (see Section 2.2.4). Since spinning can introduce noise into multi-dimensional spectra, it is usually turned off after shimming is completed.

The probe also provides a heater for temperature control of the sample. Accurate control of the temperature during the experiment is essential; temperature fluctuations of the sample can cause changes in the field homogeneity as well as changes in the chemical shifts of the resonance lines in the spectrum. The probe also contains a deuterium locking coil to compensate for a small drift in the main magnetic field during the measurement.

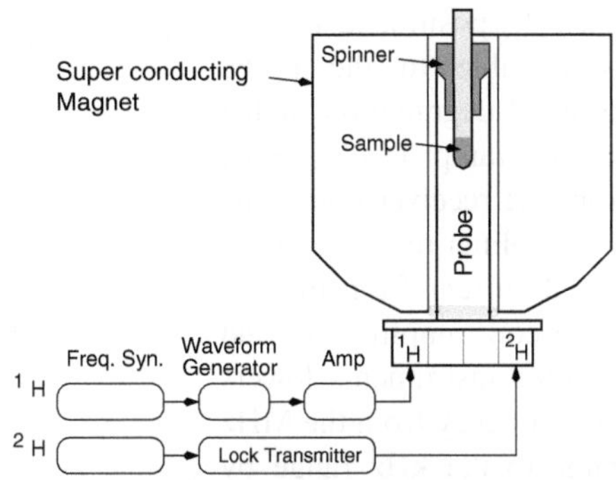

Figure 2.2. Experimental set-up for acquiring a proton spectrum. The location of the sample holder (probe) within the superconducting magnet is shown. The NMR sample tube is held in a spinner and is usually inserted from the top of the magnet using a pneumatic lift system. The cabling required for a single radio-frequency channel as well as the radio-frequency deuterium lock is also illustrated.

The radio-frequency (RF)-pulse that is used to excite the nuclear transitions is created by passing a current through the observe/receive coil. The same RF-coil is used to receive current induced by the precessing magnetic moment after excitation. Additional RF-coils can also be present around the sample for excitation of heteronuclear spins, such as ^{13}C, ^{15}N, or ^{31}P. The electrical properties, specifically capacitance and impedance, of all of these RF-coils are sensitive to the sample. Therefore it is necessary to tune the coils in the probe after the sample has been inserted into the probe. Tuning optimizes the transfer of power from the transmitter to the excitation coils in the probe.

There are two different arrangements of the proton and heteronuclear coils. In an inverse probe, which is designed for direct detection of protons, the proton RF-coil is closest to the sample and the heteronuclear RF-coil(s) surround the proton coil. In a heteronuclear probe, that is designed for the direct detection of heteronuclear spins, such as ^{13}C and ^{15}N, the proton coil is outside of the heteronuclear coil. Positioning the RF-coil of the nuclear spin of primary interest closest to the sample provides maximum sensitivity for both excitation of the spins and detection of the signal.

Finally, most modern probes include gradient coils that are used to induce spatial magnetic field gradients across the sample. The constructive use of pulsed magnetic field gradients will be discussed in Chapter 11.

2.1.4 Pre-amplifier Module

The signal pre-amplifier is usually positioned as close as possible to the probe to avoid degradation of the small, μamp level, signal from the receiver coil. The pre-amplifier receives the induced current from the observe coil, amplifies it, and in some instruments lowers the frequency from the MHz range to the kHz range by subtracting W_{syn}. The frequency of the signal is lowered because it is much easier, and therefore cheaper, to build electronic devices that operate at lower frequencies.

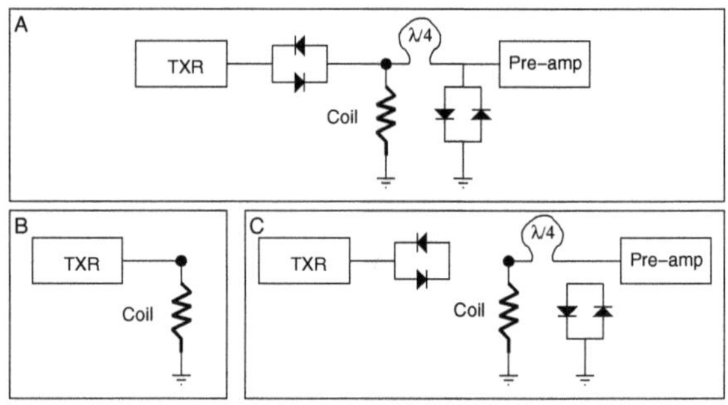

Figure 2.3. Signal routing to the probe and pre-amp. Panel A shows a simplified diagram of the connection from the transmitter (TXR) to the probe and preamp. The two pairs of diodes, one pair located between the transmitter and the coil, the second after the $\lambda/4$ cable, will pass high power pulses, but will appear as open circuits to the weak signal from the coil during detection. Panel B shows the effective circuit during the pulse. Because of the high power, the signal can sense the ground on the other side of the $\lambda/4$ cable. Due to the $\lambda/4$ cable, this connection appears as an infinite resistance to the pulse, hence all of the power goes to the coil for excitation of the spins. The effective circuit during acquisition is shown in Panel C. The small voltage from the coil during detection is blocked from returning to the transmitter by the left-most pair of diodes. Similarly, the weak signal cannot sense the ground connection through the right-most pair of diodes. Consequently, the $\lambda/4$ cable appears as if it was any other length of wire, and the signal goes to the pre-amp.

Since the same coil is used to both send the radio-frequency pulse and to detect the FID, it is necessary to route the RF-pulse to the coil during the pulse, while not allowing it to reach the pre-amplifier. Otherwise the rather intense RF power, usually on the order of 50 watts, will destroy sensitive components in the pre-amplifier. The circuit that accomplishes this routing is shown in Fig. 2.3. The key feature of this circuit is the grounding, via cross-diodes, of the circuit at 1/4 λ from the junction point (In many modern spectrometers the 1/4 λ cable has been replaced by equivalent circuitry). Under these conditions the pre-amplifier side of the circuit appears as an infinite resistance during the pulse and most of the RF-pulse goes to the sample coil. When the pulse is turned off, the weak signal from the probe can pass through to the pre-amplifier.

2.1.5 The Field-frequency Lock

All magnets drift in field strength, resulting in a small change in resonance frequencies over time (1-10 Hz/hour for ν_H). Since many NMR experiments require extended acquisition times, it is critical to compensate for the field drift and maintain the static magnetic field at a constant value. To adjust the B_o field, a small current is passed through a coil (the Z^0 shim, see below). The field-frequency locking circuitry controls the amount of current passed through this coil, essentially 'locking' the field to a constant value, B_o+Z^0, throughout the NMR experiment.

Changes in the magnetic field strength are detected by measuring the position of a deuterium resonance from the sample. Hence, it is usually necessary to include a deuterated 'lock-solvent' in the sample. A buffered solution of 95% H_2O/5% D_2O is commonly used for NMR studies on biomolecules. Both the absorptive and dispersive component of the deuterium resonance line are used in adjusting and maintaining field homogeneity and stability, as illustrated in Fig. 2.4. The absorptive component is utilized when optimizing the shim coils to obtain a homogeneous field while the dispersive component is used to correct the field strength during data acquisition. If the field is at the desired strength the deuterium signal will be on resonance and there is no intensity in the dispersive component at the frequency of the deuterium transmitter. However, if the field changes, then the position of the deuterium resonance will change and either a positive or negative value will be detected from the dispersive line. Therefore, the dispersive component gives both the direction and the magnitude of the required change in the current in the Z^0 shim coil to return the magnetic field to its initial value. In order for this locking mechanism to function properly the signal in the *real* channel of the deuterium receiver must have a pure absorption lineshape for shimming while the signal in the *imaginary* channel must have a pure dispersion

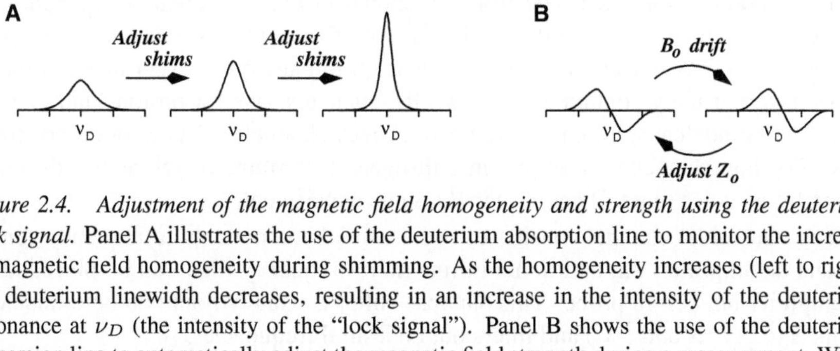

Figure 2.4. Adjustment of the magnetic field homogeneity and strength using the deuterium lock signal. Panel A illustrates the use of the deuterium absorption line to monitor the increase in magnetic field homogeneity during shimming. As the homogeneity increases (left to right) the deuterium linewidth decreases, resulting in an increase in the intensity of the deuterium resonance at ν_D (the intensity of the "lock signal"). Panel B shows the use of the deuterium dispersion line to automatically adjust the magnetic field strength during a measurement. When the field is locked (left spectrum), the magnetic field is adjusted such that the observed deuterium resonance position is equal to the deuterium transmitter frequency, ν_D. Consequently, the zero-crossing of the dispersion signal occurs at ν_D. If the field drifts, the zero-crossing point will shift and the dispersion signal will be non-zero at ν_D. In the example shown in the right part of Panel B, the magnetic field has increased, consequently the deuterium dispersion signal is greater than zero at ν_D. A positive value for the deuterium dispersive signal at ν_D will signal to the field-frequency lock circuitry to decrease the Z^0 shim, until the overall magnetic field is restored to its original value.

lineshape for locking. Hence it is important to adjust the phase of the lock receiver prior to beginning an experiment.

2.1.6 Shim System

The shim system consists of a number of coils which surround the area of the sample. Passage of current through these coils creates magnetic fields that are used to adjust the homogeneity of the static B_o field. There are two types of shim coils: superconducting shims that are adjusted during installation of the magnet, and room-temperature shims that are adjusted by the user prior to each experiment. The room-temperature shim coils are contained within the room-temperature bore of the magnet, between the main magnet and the probe.

There can be a large number of room-temperature shim coils, each of which generates a small magnetic field that is shaped like a spherical harmonic. Since these functions are orthogonal the field generated by each shim coil is, in theory, independent of the fields generated by the other shim coils. However, in practice there can be considerable interaction and it is usually necessary to adjust several coils interactively.

Axial shims, Z^0, Z^1, Z^2, etc. only change the field along the z-axis and are thus adjusted with the sample spinning at 12-25 Hz. The spinning averages out any field inhomogeneity in the x-y direction.

Radial shims (e.g. ZXY) depend on x- and y-coordinates and must be shimmed with a non-spinning sample. There are a larger number of radial shims (e.g.XY), which affect the field in the x-y plane, as well as mixed axial and radial shims, such as Z^2X, which modify the field in all three directions.

2.1.7 Transmitter & Pulse Generation

The diagram of the NMR spectrometer shown in Fig. 2.1 contains a single RF-channel, consisting of a transmitter and a RF-coil. Consequently, this particular spectrometer could only detect one type of nuclear spin in any given experiment. Modern spectrometers usually contain at least two RF-channels, one for protons and one for another heteronuclear spin. Up to 5 radio-frequency channels exist on some spectrometers, allowing the excitation of protons, nitrogen, deuterium, as well as two different carbon frequencies (e.g CO and C_α) in the same experiment.

Each transmitter channel consists of two essential components, a frequency generator and an amplifier. An additional in-line radio-frequency filter can be placed after the amplifier but before probe. This bandpass filter is usually found on heteronuclear channels (e.g. ^{15}N and ^{13}C) and filters out undesired frequencies.

The transmitter frequency, or *carrier frequency* of the RF-pulse, ω_{RF}, is generated by mixing together two frequencies. The first frequency that contributes to the carrier frequency is ω_{syn}. This frequency is adjustable and is generated from the frequency synthesizer (SYN) under computer control. The second is an internal constant frequency called the intermediate frequency, or ω_{IF}. Electronic mixing of these two frequencies generates signals at two frequencies, $\omega_{syn} + \omega_{IF}$ and $\omega_{syn} - \omega_{IF}$. The higher frequency is retained and sent to the amplifier. The purpose of this mode of frequency generation is to provide a means to reduce electronic noise in the spectrum by adding to the detected signal a known offset frequency, ω_{IF}, that can be used to separate the real experimental signal from the noise (see below).

The amplification and phase of the signal from the frequency synthesizer is under control of the computer. The amplification levels are specified in decibels of *attenuation*[2]. Decibels are defined as follows:

$$dB = 10\, log\, \frac{P_1}{P_2} = 20\, log \frac{V_1}{V_2} \tag{2.1}$$

Since the intensity of the B_1 field during the pulse depends on the *voltage* of the signal, not its power, the right-most definition (Eq. 2.1) is more useful. A 6 dB decrease in power results in a 2-fold decrease in the voltage applied to the sample, and therefore a 2-fold *increase* in the pulse length required to generate the same flip angle ($6 = 20\, log(2/1)$).

Although the transmitter power is specified in dB, its value in published NMR experiments (pulse sequences) is generally given as the strength of the B_1 field, in Hz. Consequently, it is necessary to inter-convert between the two units. The B_1 field can be calculated from the length of the 90° pulse length, as follows:

The flip angle, β, of a pulse of length τ, applied at a field strength of B_1, is:

$$\beta = \gamma B_1 \tau \tag{2.2}$$

Selecting a flip angle of 90° and converting γB_1 to frequency units of Hz gives:

$$\frac{\pi}{2} = 2\pi[\gamma B_1]\tau_{90} \tag{2.3}$$

Therefore, the field strength in Hz (γB_1) is:

$$\gamma B_1 = \frac{1}{4\tau_{90}} \tag{2.4}$$

For example, a 90°pulse of 10 μsec corresponds to a B_1 field strength of 25 kHz.

2.1.7.1 Shaped Pulses

Many RF-channels will contain an optional waveform generator, indicated as WFG in Fig. 2.1, that allows the production of pulses with arbitrary shapes. The pulse is divided into a number of small segments, and each segment possesses its own amplitude and phase. For example a Gaussian shaped pulse can be generated by simply varying the amplitude of each segment by the value of the Gaussian function. When the shaped pulse is applied, the phase and amplitude of each segment is used to control the output of the amplifier.

[2]NMR instrument manufactures use different definitions for power levels. For example, Bruker defines power changes as decibels of attenuation, with the highest voltage produced at a setting of -6 dB and the lowest voltage produced at 120 dB. Varian defines voltage levels in the opposite sense, i.e. 63 dB is the highest voltage, and 0 dB ifs the lowest voltage. Care must be taken to insure that the correct power levels are used, especially if experimental parameters are transferred between different types of NMR spectrometers.

2.1.8 Receiver

The induced transverse magnetization is detected using the same coil that carried the excitation pulse to the sample. The frequency of the induced RF is quite high (e.g. 500 MHz). This is an extremely high frequency and there are practical problems associated with operating various circuits, such as analog to digital converters, at this frequency. Instead of trying to sample the magnetization at 500 MHz the frequency of the signal is reduced by first mixing the signal with ω_{syn}; therefore, the frequency of the signal that exits the preamplifier module is $\omega_s - \omega_{syn}$, or the sum of ω_s and ω_{IF}.

The amplified signal is then sent to the dual channel analog-to-digital converters. Each channel contains a mixer that will lower the frequency of the incoming signal by the intermediate frequency, giving a final frequency of $\omega_s - \omega_{syn} - \omega_{IF}$, this is equivalent to $\omega_s - \omega_{RF}$, or the frequency in the *rotating frame*. These frequencies are quite low, e.g. in the Hz to kHz range, and therefore subject to pollution with electronic noise in the environment. Consequently, to reduce the noise in the data, the signal is kept at the intermediate frequency through most of the circuitry, and only reduced to lower frequencies at the final stage, prior to digitization by the computer.

The individual mixers in each channel differ in the phase of the intermediate frequency that is mixed with the signal from the pre-amp, one is phase shifted by 90° relative to the other. The signals that are output from each mixer are either $\cos(\omega t)$ or $\sin(\omega t)$ and can be considered to be equivalent to the projection of the transverse magnetization on to the x- and y-axis, respectively, namely M_x and M_y.

After mixing, the individual analog signals are digitized and added into two separate memory locations within the computer. The first memory location, which contains the data representing M_x, can be considered to be the real signal, while the second memory location, which contains the data representing M_y, can be considered the imaginary signal. Note that this data storage pattern is not fixed, rather it depends on the phase of the receiver, as discussed in Section 2.3.1.2. The combination of the real and imaginary signals produces a complex signal. The complex signal allows the discrimination of positive and negative frequencies, otherwise known as *quadrature detection.*

2.1.8.1 Quadrature Detection

Since the signal is detected in the rotating frame, the precession of the bulk magnetization about the z-axis can occur in either direction, depending on whether the resonance frequency, ω_s is lower or higher than the rotational rate of the rotating frame (Ω). If only one component of the transverse magnetization, such as M_x, is detected, it is not possible to distinguish between the two frequencies since both directions of rotation about the z-axis give exactly the same value for M_x (See Fig. 2.5). However, if both M_x and M_y are detected then it is possible to distinguish between the two directions of rotation. In the case of the spin that precesses clockwise ($\omega_s > \Omega$), M_y initially decreases with time. In contrast, a spin that precesses counterclockwise ($\omega_s < \Omega$), M_y increases with time (see Fig. 2.5).

For data acquired with quadrature detection, the FID is represented by combining both M_x and M_y signals into a single complex number. M_x is considered to be the real component of the complex number and M_y represents the imaginary component. If the magnetization begins aligned along the x-axis, then the FID is described as follows:

$$\begin{aligned} M(t) &= M_x + iM_y \\ &= \cos(\omega' t)e^{-t/T_2} + i\sin(\omega' t)e^{-t/T_2} \\ &= e^{i\omega' t}e^{-t/T_2} \end{aligned} \tag{2.5}$$

Fourier transform of this signal will give a complex function (see Appendix A). The real part is a Lorentzian line shape, located at $+\omega'$.

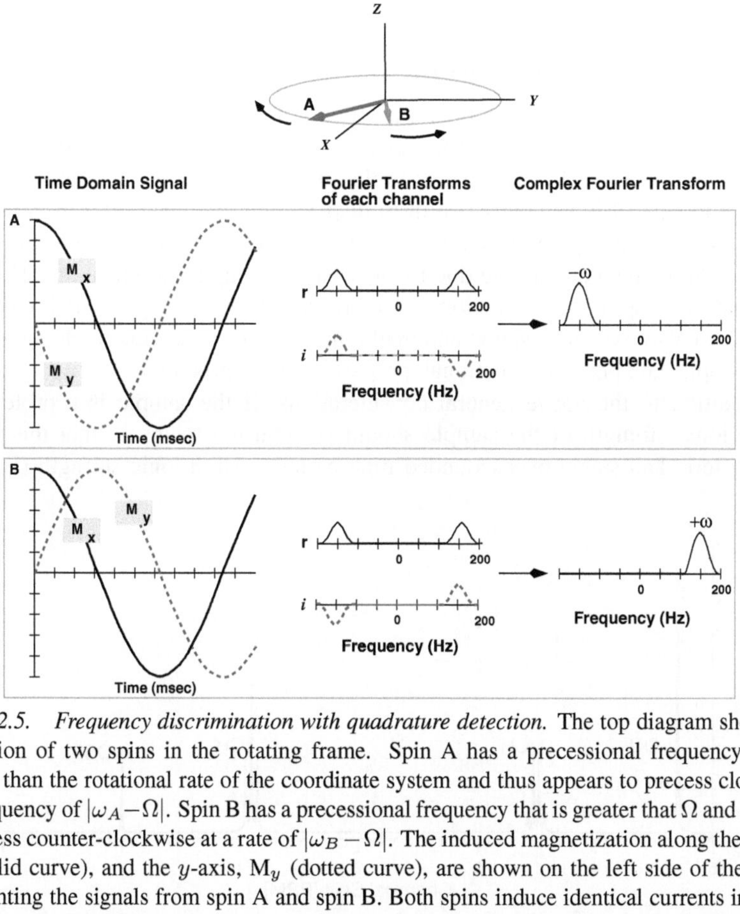

Figure 2.5. *Frequency discrimination with quadrature detection.* The top diagram shows the precession of two spins in the rotating frame. Spin A has a precessional frequency that is smaller than the rotational rate of the coordinate system and thus appears to precess clockwise at a frequency of $|\omega_A - \Omega|$. Spin B has a precessional frequency that is greater that Ω and appears to process counter-clockwise at a rate of $|\omega_B - \Omega|$. The induced magnetization along the x-axis, M_x (solid curve), and the y-axis, M_y (dotted curve), are shown on the left side of the panels representing the signals from spin A and spin B. Both spins induce identical currents in the x-direction, but of opposite sign in the y-direction. The magnetization along x is considered to be the real channel and the magnetization along y represents the imaginary channel. The Fourier transform of each of these signals are shown in the central part of the diagram. Each spectrum represents the separate transform of the signal for the real (r) and imaginary (i) channels. In practice, a complex Fourier transform is performed. This is equivalent to the sum of the two individual transforms, giving a single peak at the correct frequency, as shown on the right of the diagram.

2.2 Acquiring a Spectrum

To illustrate a number of aspects of instrument configuration, we will use the compound ethanol, whose structure and NMR spectrum are shown in Fig. 2.6. This spectrum contains two peaks, one from the two CH_2 protons from ethanol, located at 3.5 ppm, and a second peak, located at 1.1 ppm, that represents the three methyl protons. The remaining peaks are from residual HDO and DSS, a reference standard. For historical reasons, the chemical shift scale is reversed; high chemical shifts are found to the left of the spectrum. The relative intensities of the peaks are related to the number of protons that give rise to the signal. In the case of ethanol, both peaks show fine structure, or splitting. This fine structure is due to scalar coupling between the two types of protons and will be discussed in Chapter 7.

2.2.1 Sample Preparation

Careful preparation of the sample for NMR spectroscopy can dramatically improve the quality of the data collected. The sample should not contain any particulate matter, as this will cause inhomogeneities in the magnet field, causing linebroading via T_2^*. Additionally, it is generally useful to displace the dioxygen (O_2) in the sample with N_2. Dioxygen is paramagnetic due to its unpaired electron. The unpaired electron has a large magnetic dipole due to electron spin angular momentum. The large dipole enhances spin-spin relaxation of the nuclear spins, leading to line broadening. Oxygen can also oxidize cysteine and methionine residues, leading to changes in the spectroscopic, and possibly functional, properties of the protein.

In addition to the above general considerations, if the sample is a protein, the pH and ionic strength of the sample should be adjusted to insure that the protein is monomeric and stable over extended time periods. High ionic strength (> 0.2 M

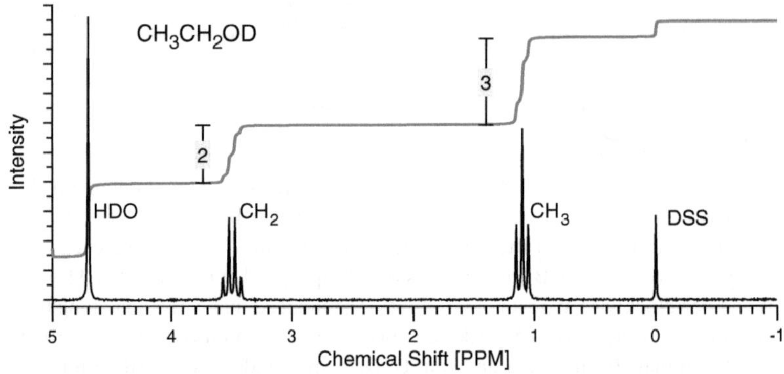

Figure 2.6. NMR spectrum of ethanol. The proton NMR spectrum of ethanol, in D_2O, is shown. The chemical shifts are approximate and are for illustrative purposes. The methyl protons give rise to the peak at 1.1 ppm and the methylene protons give rise to the peak at 3.5 ppm. The hydroxyl proton is absent in this example due to exchange with the solvent. The line at 4.7 ppm arises from residual protons in the solvent. The line at zero ppm is from the methyl groups of the reference compound (DSS, see Chapter 3). The gray line is the integral of the spectrum; the CH_2 and CH_3 peaks have a relative area of 2:3.

NaCl) should be avoided because of the generation of ion currents during RF-pulses. These currents decrease the efficiency of the pulses, especially towards the center of the sample, and lead to sample heating.

The sample should be contained in a high-quality NMR tube to avoid damage to the probe because of the close fit of the sample tube in the probe. There should also be sufficient sample volume in the NMR tube to completely fill the excitation/receiver coil, as well as to place the solvent-air interface at least a centimeter from this coil. The solvent-air interface will cause magnetic field inhomogeneities due to the different magnetic susceptibility of the two phases. Typical volumes for normal NMR tubes are on the order of 0.5 ml. If sample material is limited, as often occurs in biomolecular NMR, it is possible to use special NMR tubes that have a column of glass or polymer in the lower part of the tube as well as a glass/polymer plunger that rests above the sample. The magnetic susceptibility of the glass or polymer closely matches that of common biochemical buffers, thus the glass/polymer replaces the liquid outside of the coil. This allows a much smaller sample volume, of approximately 0.35 ml (see Chapter 15).

2.2.2 Beginning the Experiment

After the sample is inserted into the magnet the following steps should be performed before data acquisition occurs.

1. *Temperature equilibration:* Ideally, the sample should be in the probe for approximately 10 – 15 min to allow for thermal equilibrium to occur.
2. *Locking the spectrum on the deuterium line:* Locking may require adjustment of the lock frequency, power of the lock transmitter, and gain of the lock receiver.
3. *Adjustment of the lock phase:* This insures that a pure dispersion signal is available to the lock circuit for field stability.
4. *Shimming of the magnetic field:* This can be done either manually or with a computer automated algorithm.
5. *Tuning and matching of the probe:* This insures optimal transfer of the power from the transmitter to the sample.
6. *Calibration of pulses:* Transmitter power levels are measured to insure that the pulses are of the desired flip angle.

2.2.3 Temperature Measurement

The sample temperature that is measured by the instrument is only approximate because the temperature is obtained from a thermocouple that is outside of the sample. Additionally, heating of the sample by RF-pulses can cause the temperature to increase significantly above the set temperature. The temperature within the sample can be determined by measuring the separation between the hydroxyl and methyl resonances in methanol [135]. The following formula is valid over the range of 250 – 320 K:

$$T = 403.0 - 29.53\Delta\delta - 23.87[\Delta\delta]^2 \tag{2.6}$$

where T is the temperature in units of Kelvin and $\Delta\delta$ is the chemical shift difference between the two resonances.

2.2.4 Shimming

Adjustment of shim coils is termed shimming. The overall goal is to increase the homogeneity of the magnetic field. Most modern spectrometers have routines that can automatically shim a sample. However, these methods only work well on samples that are in H_2O (i.e. 95%H_2O/5%D_2O). If the sample is in 100% D_2O the automated methods may not converge and the magnet will have to be shimmed manually. The three steps in manual shimming are described below (see [43] for more details). The key principle to keep in mind is that the shim coils interact with each other, particularly the higher order shims. For example, altering the Z^4 shim coil will change the optimal setting of the Z^2 coil. Consequently, the shims have to be adjusted in an interactive fashion.

Step A. Select a Method to Assess Homogeneity: An increase in homogeneity can be determined by one of the following three methods:

1. *An increase in the height of the lock signal:* As the homogeneity increases, the deuterium linewidth decreases, hence the height of the deuterium absorption line increases. This method is the least sensitive to inhomogeneity in the field.
2. *The shape of the FID:* If the transmitter frequency is set to the frequency of the solvent, then the FID of the solvent should decay with a single exponential. Oscillations in this signal indicate that the solvent molecules are experiencing slightly different B_o fields at different locations in the sample, giving rise to slightly different frequencies.
3. *The line shape:* A resonance peak from a homogeneous magnet will be Lorentzian in shape. Any asymmetries in the peak likely arise from poor shimming. For example, misadjustment of Z^2 and Z^4 generally cause the formation of an asymmetric tail on one side of the peak or the other. Misadjusted Z^3 shims tend to produce a broad base in the lineshape. This method is the most sensitive to inhomogeneities in the magnetic field.

Usually the lock signal is used at the beginning of the process and the final shim settings are assessed by inspection of the lineshape.

Step B. Adjust Axial Shims:

1. Spin the sample at a rate between 12 and 20 Hz.
2. Do a grid search to find the best value of Z^2 by selecting values of Z^2 and adjusting Z^1 to maximize the homogeneity. It is helpful to find three values of Z^2 (A, B, C) such that the homogeneity increases from A to B and then decreases from B to C. The best value of Z^2 can be found by assuming a parabolic fit through these three points, as illustrated in Fig. 2.7.
3. Do another grid search by varying Z^3 systematically and adjusting Z^1 at each Z^3 value.
4. Do another grid search by varying Z^4 systematically and adjusting *both* Z^1 and Z^2 at each Z^4 value.
5. Do another grid search by varying Z^5 systematically and adjusting *both* Z^3 and Z^1 at each Z^5 value.

If Z^1 or Z^2 changes appreciably during steps 3-5, return to step 2.

Step C. Adjust Radial Shims:

1. Make sure the spinning is off.
2. Do a grid search to find the best value of ZX. Vary ZX systematically, and then adjust X and Z^1 to attain the best homogeneity. If large changes in Z^1 occur, re-adjust Z^1 and Z^2 as described above.
3. Do a grid search to find the best value of ZY. In this case vary ZY systematically and adjust with Y and Z^1 until no further changes occur.
4. Adjust XY and X^2-Y^2 interactively to maximize the homogeneity, adjust X and ZX interactively, followed by Y and ZY. Continue this cycle until no further change is obtained.
5. Do a grid search to find the best value of Z^2X. Vary ZX, Z^1, and X (in this order) to optimize the level for each value of Z^2X.
6. Do a grid search to find the best value of Z^2Y. Optimize, ZY, Z^1, and Y (in this order) for each value of Z^2Y.
7. Do a grid search on ZXY, for each value of ZXY optimize XY and X^2-Y^2 as described in step 4.
8. Do a grid search on $Z(X^2$-$Y^2)$. Optimize at each step as described in step 4.
9. Do a grid search on X^3. Optimize X and Y for each value of X^3.
10. Do a grid search on Y^3. Optimize X and Y for each value of Y^3.

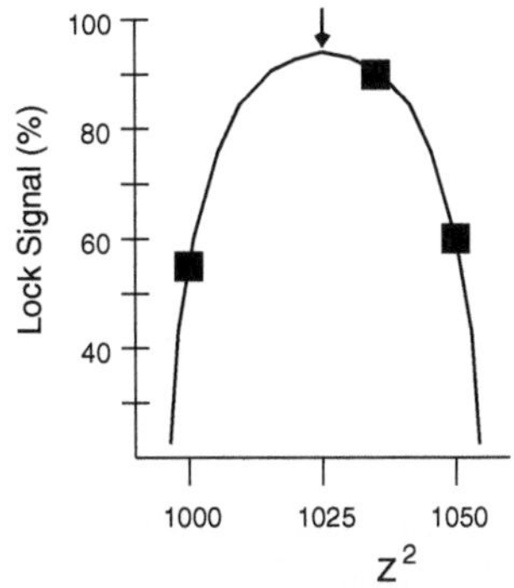

Figure 2.7. Grid search optimization of Z^2 shim. Three values of Z^2 were selected, and the homogeneity was optimized at each of these values by adjusting Z^1 to maximize the lock signal. The best value of Z^2 is approximately 1025 units, as indicated by the arrow.

Steps B and C can be quite tedious to complete as described. Often, the shims are close enough that it is not necessary to perform the suggested grid search. Instead, the interacting shims (e.g. Z^1 and Z^2) can be adjusted by simply varying the two in an interactive fashion to achieve the best homogeneity.

2.2.5 Tuning and Matching the Probe

Optimal transfer of power from the amplifiers to the sample occur when the probe has a complex impedance, at the frequency of the transmitter pulse, that matches the 50 Ω impedance of the amplifier circuits. Otherwise, a fraction of the RF-power will be reflected from the probe and not absorbed by the sample. Each RF-channel on the probe will have an adjustment for tuning and matching of the probe circuit. The impedance of the probe is assessed by measuring the amount of power reflected from the probe at different frequencies, as illustrated in Fig. 2.8. During this process it is essential that the sample be present in the magnet, as the solvent properties of the sample can have a large effect on the tuning of the probe.

Ideally, altering the tuning changes the position of the minimum in reflected power while changing the matching alters the amount of reflected power at the minimum. In practice, these two adjustments are inter-dependent and it is necessary to adjust them in an iterative fashion.

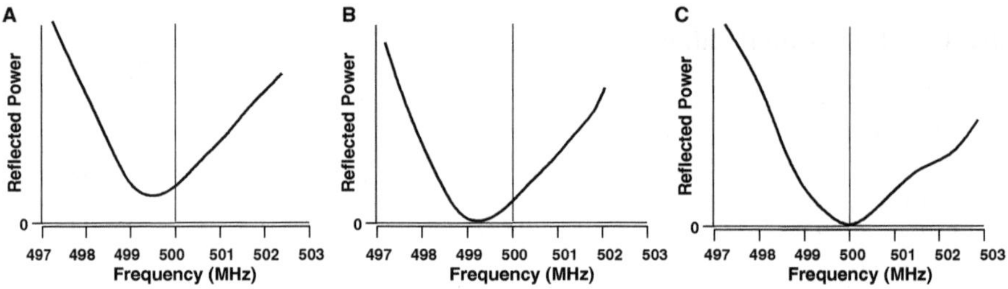

Figure 2.8. Probe tuning. The reflected power as a function of frequency is shown for the proton channel on a 500 MHz instrument. The vertical line at 500 MHz in each panel marks the proton resonance frequency. Panel A shows the tuning immediately after changing the sample or the temperature. Panel B illustrates the affect of adjusting the matching to 50 Ω impedance, while panel C shows the desired final result, after adjusting the tuning.

2.2.6 Adjusting the Transmitter

The next step in setting up the instrument is to define the frequency, power, and duration of the radio-frequency (RF) excitation pulse. The frequency of the RF-pulse is usually set in the center of the spectrum, which would be at 2 ppm for the example of the ethanol sample (Fig. 2.6). However, if the sample contains a strong solvent line (e.g. H_2O) it is better to place the transmitter on the solvent to avoid artifacts from digital processing. The highest available power for the B_1 field strength is generally used to give a uniform excitation as possible across the entire spectrum. If a single scan is to be acquired, the pulse length is adjusted to provide a 90° pulse because this gives the largest signal. In contrast, if multiple scans are recorded with a repetition rate less than approximately $3 \times T_1$, then a pulse angle of less than 90° will give a larger signal (see Section 2.2.10.1).

2.2.6.1 Frequency and Power of the Transmitter

The frequency of the transmitter, W_{RF}, defines Ω, the rotation rate of the rotating coordinate frame. In our discussions of pulses in Chapter 1, it was assumed that the frequency of the excitation pulse was the same as the frequency of the nuclear spin transition and the 90° pulse cleanly tipped the magnetization from the z-axis to the x-y plane. However, as the resonance line becomes further removed from the frequency of the RF-pulse, the efficiency of excitation decreases. This is due to the fact that the effective field, B_{eff}, becomes closer to the z-axis as resonance frequency becomes more distant from the transmitter frequency, as illustrated in Fig. 2.9.

The precession of the magnetization about the effective field is shown in Fig. 2.10. As before, for those spins that are on-resonance, the 90° x-pulse brings them to the y-axis. In contrast, those spins that are off-resonance ($\omega_s \neq \Omega$) will precess about B_{eff} for the same period of time, but the total angle that they precess will be always be greater than 90° because B_{eff} is always greater than B_1 ($B_{eff} = \sqrt{B_1^2 + B_z^2}$). After the pulse, the magnetic moment of off-resonance spins will be found either below or above the x-y plane, depending on the magnitude and orientation of B_{eff}. Therefore the amount of detectable magnetization in the x-y plane will decrease relative to an

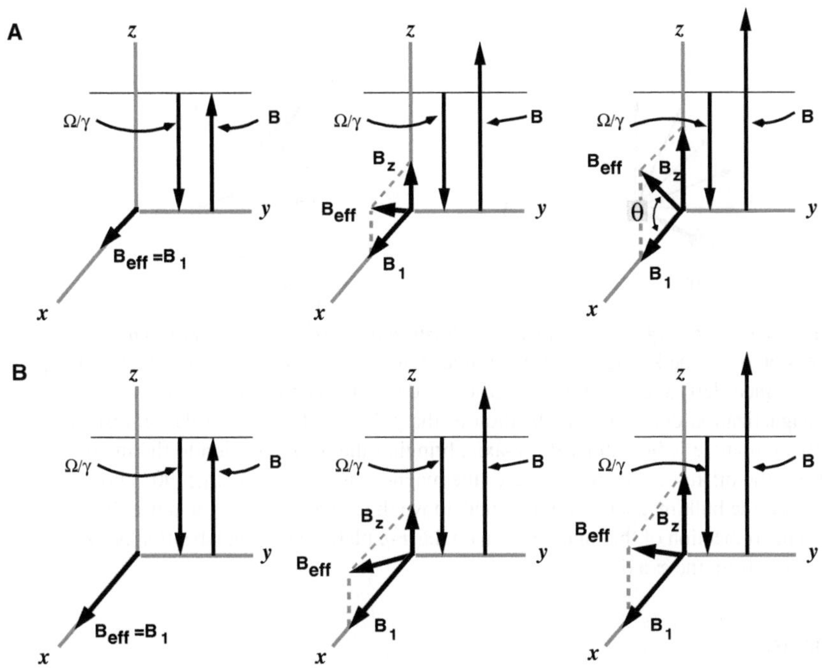

Figure 2.9. Effect of frequency offset and B_1 *field strength on* B_{eff} *in the rotating frame.* Panel A illustrates the effect of a weak B_1 field while panel B illustrates the effect of a strong B_1 field on B_{eff}. The leftmost section in each figure shows an on-resonance spin, the middle and right sections illustrate the effects of increasing the resonance offset on B_{eff}. The three fields present during the pulse are B_1, along the x-axis in this example, Ω/γ, the fictitious field generated by the rotation of the coordinate frame, and B, the field felt by the spin at its nucleus ($B = (1 - \sigma)B_o$). The effective field of the pulse, indicated by B_{eff}, is the vector sum of all three fields. Note that Ω/γ is constant in all cases and is defined by the transmitter frequency. In contrast, the magnetic field at the nucleus, B, varies according to the resonance position of the line. When the rate of rotation of the coordinate frame equals the resonance frequency, $\Omega = \omega_s$, the two fields along the z-axis cancel and the B_{eff} is equal to B_1 (left diagrams in both A and B). Off-resonance, B_{eff} is tilted towards the z-axis by an angle θ. Note that for any given frequency offset, the tilt of B_{eff} towards the z-axis becomes more pronounced as the strength of the B_1 field decreases. At large resonance offsets (B >> Ω/γ), B_{eff} will be aligned along the z-axis and is completely ineffective at rotating the magnetization to the x-y plane.

on-resonance spin. In addition to a loss of intensity, the bulk magnetization that is present in the x-y plane will be shifted away from the x-axis by an angle ϕ. Each of these effects will be discussed in more detail below.

Phase Effects: The shift of the transverse (x-y) magnetization from the x-axis by the angle ϕ, will produce a mixed lineshape that is a combination of absorption and dispersion lineshapes. This can be seen by considering the time domain signal associated with an off-resonance signal and its subsequent Fourier transform. The time domain

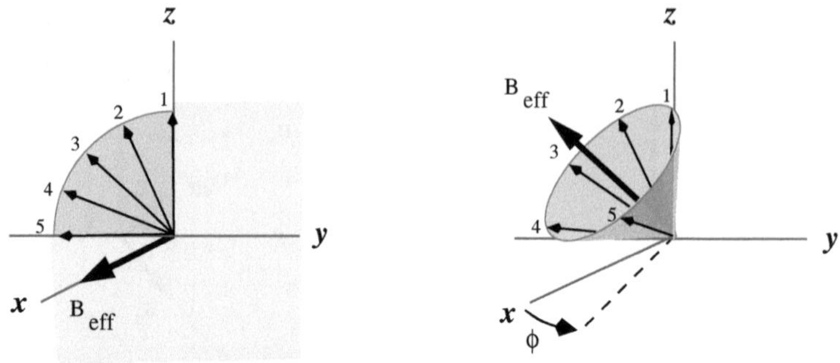

Figure 2.10. Effect of on-resonance and off-resonance 90° x-pulses on the bulk magnetization. Precession of the bulk magnetization around B_{eff} is shown when the frequency of the pulse in on-resonance (left) and when it is off resonance (right). The labeled arrows shown the path of the magnetization during the application of the pulse. In the case of the on-resonance pulse, the rotation angle is 90° about the x-axis, bringing the magnetization to the minus y-axis. In contrast, the off-resonance pulse rotates the magnetization by more than 90° about B_{eff}. Consequently, the bulk magnetization is found above the x-y plane after the pulse. The dotted line shows the projection of the final position on the x-y plane. The magnetization is also shifted by an angle ϕ from the x-axis.

signal is:

$$S(t) = e^{i\phi}e^{i\omega_s t}e^{-t/T_2} \tag{2.7}$$

$$= [cos(\phi) + isin(\phi)]e^{i\omega_s t}e^{-t/T_2} \tag{2.8}$$

where the term, $e^{i\phi}$, represents the shift in the phase of magnetization, ϕ.

The Fourier transform of the above FID will give the following:

$$F(\omega) = [cos(\phi) + isin(\phi)][\frac{T_2}{1 + T_2^2(\omega - \omega_s)^2} + i\frac{T_2^2(\omega - \omega_s)}{1 + T_2^2(\omega - \omega_s)^2}] \tag{2.9}$$

The real, or observable, part of this signal is:

$$cos(\phi)\frac{T_2}{1 + T_2^2(\omega - \omega_s)^2} - sin(\phi)\frac{T_2^2(\omega - \omega_s)}{1 + T_2^2(\omega - \omega_s)^2} \tag{2.10}$$

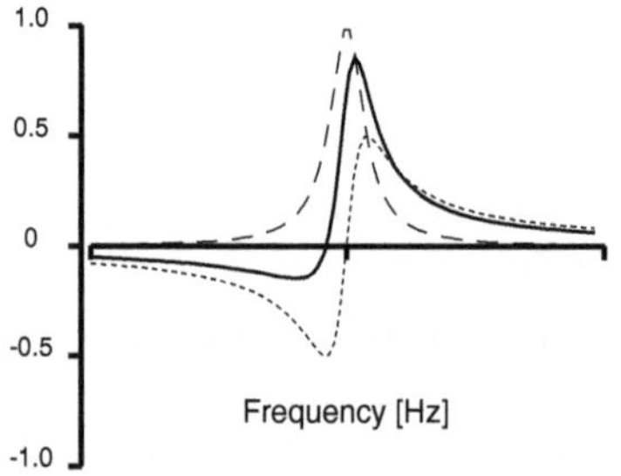

Figure 2.11 Effect of phase angle on the lineshape. Lineshapes are pure absorption (ϕ=0, dashed line), mixed (ϕ=45°, solid line), and pure dispersion (ϕ=90°, dotted line).

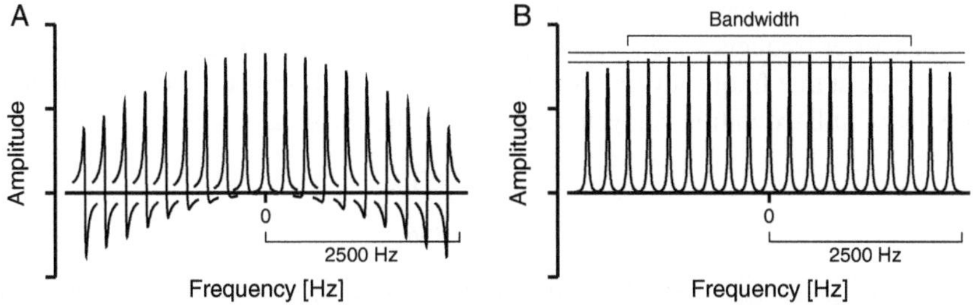

Figure 2.12. Effect of frequency offset on the phase and amplitude of resonances. Phase (Panel A) and amplitude changes (Panel B) as a function of the frequency difference between the resonance line and the carrier frequency. Panel (A) shows the NMR resonance line for different resonance offsets. When the offset is 2.5 kHz, the lineshape is pure dispersion, indicating that the magnetization is along the imaginary axis after the pulse. The right panel (B) shows the intensity of the resonance ($\sqrt{S_r^2 + S_i^2}$) as a function of the offset. The upper horizontal line indicates the maximum amplitude, which occurs when $\omega_s = \Omega = \omega_{RF}$. The lower horizontal line indicates the 95% level, which defines the bandwidth of the pulse. The 90° pulse length was 50 μsec.

Note that this contains both the absorption lineshape: $\frac{T_2}{1+T_2^2(\omega-\omega_s)^2}$, as well as the dispersion lineshape: $\frac{T_2^2(\omega-\omega_s)}{1+T_2^2(\omega-\omega_s)^2}$, as illustrated in Fig. 2.11. A pure absorption mode lineshape can be recovered by application of the appropriate phase correction, as discussed in Chapter 3.

Amplitude Effects: The resonance offset also affects the intensity of the line; resonances that are further away from the carrier will be excited to a lesser extent. As the resonance offset increases the effective field B_{eff} becomes increasingly tilted toward the z-axis as B_1 decreases. The closer B_{eff} is to the z-axis, the less efficient it is at reorienting the magnetization.

The frequency range over which acceptable excitation occurs is referred to as the *bandwidth* of the pulse. Although the cut-off level is arbitrary, when the excitation level for a line drops below 95%, that frequency is usually considered to be outside the bandwidth. An example of the excitation profile of 50 μsec pulse is shown in Fig. 2.12. The bandwidth of this pulse is approximately 4 kHz, not quite sufficient to cover the 5 kHz wide proton spectra on a 500 MHz instrument (assuming a 10 ppm spectral width).

The bandwidth of a pulse is inversely proportional to the pulse length; shorter pulses cause excitation over a larger bandwidth. Therefore, the shortest RF-pulses that can be produced by the instrument are generally used for excitation, with the exception of selective pulses (see below).

2.2.6.2 Selective Excitation

The drop in signal amplitude (see Fig. 2.12) as the position of the resonance line becomes further removed from the carrier frequency can be used for selective excitation of particular frequencies. This approach is often used to excite one region of the carbon spectrum, such as the C_α carbons (40-70 ppm), while leaving the magnetization of the carbonyl carbons (175 ppm) along the z-axis. At some frequency distant from the carrier frequency the effective pulse angle will be zero degrees and the spins that resonate at that frequency will not be excited by the RF-pulse. The location of the first point of null excitation is given by the following formula:

$$\nu_{null} = \pm\frac{\sqrt{15}}{4\tau_{90}} \tag{2.11}$$

where τ_{90} is the length of the 90° pulse in units of seconds. The derivation of this formula, as well as for a selective 180° pulse, is presented in Section 14.3.

If a very weak RF-field is used, e.g. 90° pulse lengths on the order of 1 msec, then only spins that have a resonance frequency within a few hundred Hz of the carrier frequency will be excited. This allows one to selectively excite a single spectral line, such as the solvent.

It is tempting to calculate excitation bandwidths of a pulse from the Fourier transform of the pulse. Since the pulse is a product of a harmonic function, $cos(\omega t)$, and a square pulse of length τ, its Fourier transform is a sinc function centered at ω. The overall shape of the sinc function is a reasonable representation of the excitation profile, namely an intense peak at ω and a series of less intense side lobes[3]. However, the sinc function only gives *approximate* null points because the analysis does not take into account evolution of the system during the pulse. Consequently, Eq. 2.11 should be used to determine the position of null excitation.

2.2.7 Calibration of the 90° Pulse Length

Errors in the pulse angles can cause the appearance of artifacts in the spectrum or the loss of signal intensity. Pulse calibration is especially important in multi-dimensional experiments because of the large number of RF-pulses that are required for these experiments.

The 90° pulse length for each RF-channel (^{1}H, ^{13}C, ^{15}N) should be calibrated with each new sample. It is not uncommon to have a 10-15% change in pulse length due to different sample conditions. Changes in RF-pulse lengths that are larger than this range suggest hardware problems or poor tuning of the sample.

The proton pulse length is calibrated by observing the effect of the pulse length on the solvent line, or any other reasonably intense line from the sample. Calibration of the other RF-channels utilizes indirect methods that will be discussed in more detail in Chapter 15.

Usually, the 90° pulse length is obtained by calibrating the length of the 360° pulse. If a 360° pulse is used for calibration then the magnetization is returned to a near equilibrium state by the pulse, avoiding the need to wait for spin-lattice relaxation before acquiring the FID with the next value of transmitter power or pulse length.

[3] See Appendix A for a more detailed description of the sinc function.

Calibration of a 360° Pulse

1. Place the transmitter on the strongest line in the spectrum.
2. Set the transmitter power to the desired level, usually full power.
3. Acquire a spectrum with a very short pulse length, $\approx 1\mu$sec.
4. Phase the spectrum to give a pure absorption line.
5. Increase the pulse length in small (1-2 μsec) steps until the maximum signal is observed, this is *approximately* a 90° pulse.
6. Apply a pulse that is four times the 90° pulse length. If the observed resonance is positive, the pulse was greater than 360°. If the resonance is negative, the pulse was shorter than 360°. Alter the pulse length accordingly until a null (zero) signal is obtained, indicating a 360° pulse.
7. Divide the 360° pulse length by four to obtain the 90° pulse length. A final accuracy of 0.1 μsec is usually sufficient.

Once the transmitter has been calibrated by determining the length of the 90° pulse length it is possible to calculate the power levels required for 90° pulses at different power levels, such as for selective pulses as defined by Eq. 2.11. Given a known 90° pulse length, τ_1, the attenuation required to produce a 90° pulse of another length, τ_2, is given by:

$$dB = 20\ log\ \frac{\tau_1}{\tau_2} \tag{2.12}$$

For example, if a 15 μsec 90° pulse was obtained using a transmitter setting of 0 dB, then the transmitter power should be reduced by 6 dB to produce a 90° pulse length of 30 μsec. Note that the use of Eq. 2.12 implies that the amplifier is linear over the power range. This is a reasonable approximation starting at about 5 or 6 dB below the highest setting. In contrast, for power levels close to the maximum level, it is generally necessary to calibrate the power of the transmitter directly.

2.2.7.1 Placement of the Carrier

Fig. 2.12 clearly shows that the frequency difference between the pulse and the position of the resonance line affects both the phase of the acquired signal as well as the amplitude. The effects on the phase are readily corrected. In contrast, the amplitude decrease will cause a loss in sensitivity. To insure equal excitation at both ends of the spectrum, the frequency of the transmitter is generally placed in the *center* of the spectrum. In the example of the ethanol spectrum shown in Fig. 2.6, the transmitter should be placed at approximately 2 ppm. However, in the case of biological NMR spectroscopy, it is necessary to modify the above approach because of the intense solvent (H_2O) line in the spectra. Digitization errors cause the appearance of artifacts at frequencies that are multiples of the distance of the line from the carrier frequency. The size of these artifacts are proportional to the original intensity of the line. Therefore, placing the transmitter at the same frequency as water will reduce the number of artifacts in the spectrum. Another advantage of placing the carrier frequency on the solvent is that the signal from any residual solvent signal will be on-resonance and will not oscillate. Non-oscillatory signals can be eliminated by the application of a high-pass filter during data processing.

2.2.8 Setting the Sweepwidth: Dwell Times and Filters

Once the pulse width is defined, the sweepwidth (SW), or equivalently the spectral width, of the spectrum has to be defined. The sweepwidth is the range of frequencies that will appear in the spectrum; the lowest frequency will be $\nu - SW/2$ and the highest frequency will be $\nu + SW/2$, where ν is the frequency of the excitation pulse in Hz. In the case of the ethanol sample shown in Fig. 2.6 the sweepwidth was set at 6 ppm, or 3000 Hz if the spectra was acquired on a 500 MHz instrument. Both of the CH_2 and CH_3 resonance signals from ethanol, as well as the residual water resonance and the chemical shift reference (DSS), fall within this sweepwidth.

2.2.8.1 Dwell Time

Setting the sweepwidth defines how frequently the free induction decay is sampled during conversion of the analog signal to a digital signal. The spacing between time points is referred to as the dwell time. Given a dwell time of τ_{dw}, the resultant spectral width, in hertz, is:

$$SW = 1/\tau_{dw} \tag{2.13}$$

This relationship arises entirely from the properties of Fourier transforms (see Appendix A) and is quite independent of the spectrometer frequency.

2.2.8.2 Folding and Aliasing

Digital sampling of the FID makes it impossible to determine all possible frequencies that may be contained in the spectrum. The Nyquist frequency, $f_N = 1/(2\tau_{dw})$, defines the highest frequency that can be uniquely sampled when the dwell time set to τ_{dw}. The sweepwidth ranges from $-f_N$ to $+f_N$.

A signal with a frequency that is higher than f_N will give the same intensities in the digital FID as a signal that is found between $-f_N$ and f_N. This effect is illustrated in Figure 2.13, which shows the FID from two spins that have different frequencies, but produce exactly the same *digital* FID, and therefore appear at the same position in

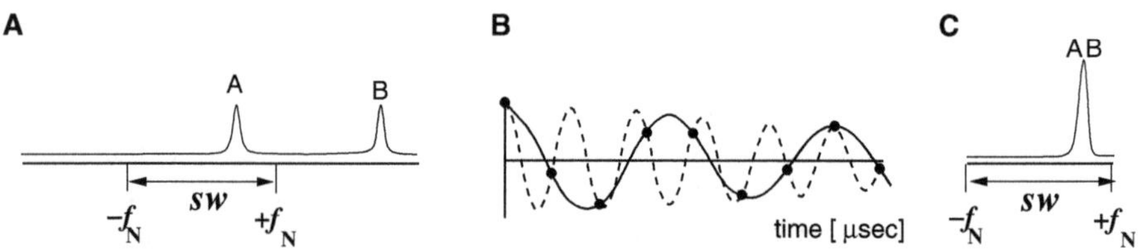

Figure 2.13. Nyquist frequency and folding of spectral lines. The true NMR spectrum of a sample is shown in Panel A (left). Two resonance lines are present, "A" and "B". The experimental spectral width, defined by the dwell time, is indicated below the spectrum. The lowest and highest frequencies correspond to the Nyquist frequency, f_N. The analog FID that would be detected from these two resonances is shown in Panel B (center). The solid line is the FID from resonance "A" and the dotted line is the FID from resonance "B". The dots indicate the points that are sampled during digitization of the FID. Since both resonances give exactly the same digital FID, they will have the same frequency in the transformed spectrum.

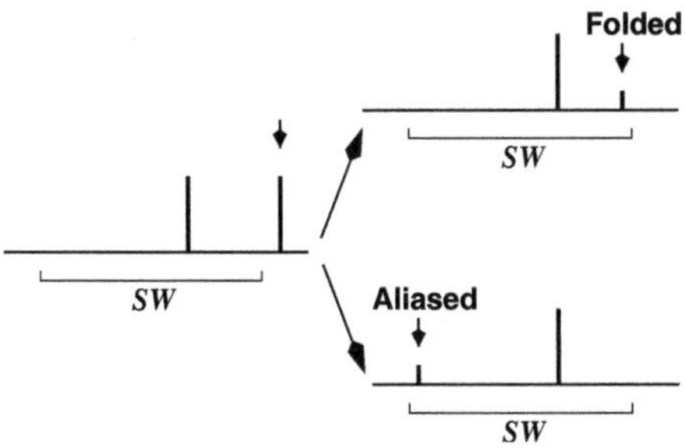

Figure 2.14 Effect of folding or aliasing on the position and intensity of resonance lines. The left section of the diagram shows an NMR spectrum with two peaks, both of equal intensity. One peak lies outside the sweepwidth and will either be folded (upper right) or aliased (lower right), depending on the type of data collection. Note that in the case of direct detection, the intensity of the folded/aliased peak will be reduced by the filters in the receiver.

the final spectrum. Note that $\nu_B > f_N$, but the observed frequency for spin B falls within the sweepwidth.

The observed position in the spectrum of a resonance that lies outside of the sweepwidth depends on whether the data were collected as a single channel, such as $cos(\omega t)$, or in quadrature ($S(t) = cos(\omega t) + isin(\omega t)$). Single channel detection will produce a folded peak, while quadrature detection will produce an aliased peak (see Fig. 2.14).

Folding and aliasing are defined as:

Folding: A peak with a frequency $f_N + \delta$ will appear at a frequency of $f_N - \delta$. This effect is called folding because the peak outside of the sweepwidth has been folded, or reflected, about the edge of the spectrum.

Aliasing: A peak with a frequency of $f_N + \delta$ will appear at a frequency of $-f_N + \delta$. This is referred to as aliasing. Note that in this case the relative frequencies of the peaks have been maintained, but they have been shifted by a multiple of the sweepwidth.

Whether a peak is folded or aliased is easily verified; its position in the spectrum will depend on the sweepwidth. Note that the intensity of aliased (or folded) peaks will be attenuated by filters in the receiver because the bandwidth of these filters is generally set to approximately $\pm SW/2$.
The position of folded or aliased peaks in the spectrum can be explained by three equivalent methods.

1. Equating the intensities of the sampled point in the time domain,
2. Consideration of the relationship between the phase evolution of the transverse magnetization and the dwell time, as shown in Fig. 2.15,
3. From the convolution properties of Fourier transforms, as shown in Fig. 2.16.

The first method will be used to explain the position of folded peaks, while the positions of aliased peaks will be explained using all three methods.

1. Equating Intensities: Assume that the true resonance frequency of a peak is $f_N + \delta$. In the case of folded peaks we need to show that the real ($cos(\omega t)$) digital FID has exactly the same intensity for each point as the FID produced by a resonance with a frequency of $f_N - \delta$. For this to be true the following identity must be true for all points in the FID:

$$cos[2\pi(f_N - \delta)n\tau_{dw}] = cos[2\pi(f_N + \delta)n\tau_{dw}] \tag{2.14}$$

where n is the nth data point and τ_{dw} is the dwell time. Substituting $f_N = 1/(2\tau_{dw})$ gives:

$$cos[n\pi - 2\pi\delta n\tau_{dw}] = cos[n\pi + 2\pi\delta n\tau_{dw}] \tag{2.15}$$

expanding both sides using double angle formula gives:

$$\begin{aligned} cos[n\pi]cos[2\pi\delta n\tau_{dw}] &+ sin[n\pi]sin[2\pi\delta n\tau_{dw}] \\ &= cos[n\pi]cos[2\pi\delta n\tau_{dw}] - sin[n\pi]sin[2\pi\delta n\tau_{dw}] \end{aligned} \tag{2.16}$$

since $sin(n\pi) = 0$, the above identity holds and therefore the frequency of the folded peak is $f_N - \delta$.

In the case of aliased peaks we need to show that the *sampled* complex FID from the peak with a frequency $f_N + \delta$ has exactly the same intensity as the FID produced by a resonance with a frequency of $-f_N + \delta$. For this to be true, the following two identities must hold for every point in the FID:

$$\begin{aligned} cos[2\pi(-f_N + \delta)n\tau_{dw}] &= cos[2\pi(f_N + \delta)n\tau_{dw}] \\ sin[2\pi(-f_N + \delta)n\tau_{dw}] &= sin[2\pi(f_N + \delta)n\tau_{dw}] \end{aligned} \tag{2.17}$$

since both the real and imaginary components of the FID must be equal. Both of these identities can be shown to be true, using the approach shown above, verifying that the frequency of the aliased peak is $-f_N + \delta$.

2. *Phase Evolution:* Sampling of the free induction decay implies taking snapshots of the magnetization at different periods of time. The position of the magnetization at the time of sampling depends on the precessional rate of the magnetization in the rotating frame. The angle, θ, that the spin precesses during the dwell time is just the product of its precessional rate and the dwell time: $\theta = (\omega_s - \Omega)\tau_{dw}$. For example, a peak on the very edge of the spectral width, such as spin B in Fig. 2.15, will precesses exactly -180° during a dwell time since:

$$\begin{aligned} \theta &= (\omega_b - \Omega)\tau_{dw} \\ &= 2\pi(\nu_b - \nu)\tau_{dw} \\ &= 2\pi(-SW/2)\tau_{dw} \\ &= 2\pi(-SW/2)(1/SW) \\ &= -\pi \end{aligned} \tag{2.18}$$

Therefore, if the frequency of a resonance line is within the spectral width, such as resonances B or C in Fig. 2.15, its total rotation angle after one dwell time (τ_{dw}) must lie between -π and π.

In contrast, if a resonance line is outside the spectral width by an amount δ, such as spin A in Fig. 2.15, then it will precess a total angle of $\pi + \delta\tau_{dw}$ (clockwise) after the first dwell time. This position is equivalent to precession in the *opposite* direction by an angle $\pi - \delta\tau_{dw}$. Therefore, the apparent frequency of A in the spectrum is $SW/2 - \delta$, which is equal to the true frequency of A plus one sweepwidth.

3. Convolution of Fourier Transforms: The position of aliased peaks can also be explained from the properties of Fourier transforms (see Fig. 2.16). The free induction decay before digitization is a continuous function; its Fourier transform is the *unaliased* NMR spectrum. However, if the FID is sampled at a series of time points, the resultant digital signal is the product of a comb function, with spacing between the teeth of τ_{dw}, and the free induction decay. The resultant digital spectrum will be the convolution between the transform of the comb function and the transform of the free induction decay. The transform of a comb function is simply another comb function, with the teeth spaced $1/\tau_{dw}$, or the sweepwidth, apart. The convolution of this comb function with the unaliased spectrum will generate a series of spectra, each spaced a sweepwidth apart. Peaks that are outside the sweepwidth in one spectrum will appear in the adjacent spectrum.

As the dwell time is shortened, the spacing between the teeth in the comb function will become larger since $SW = 1/\tau_{dw}$. As the spacing increases the image of each spectrum is moved further apart, eventually reaching the point where there is no over-

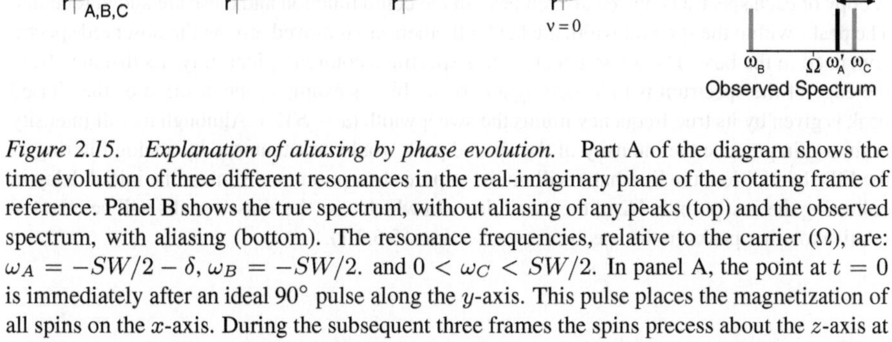

Figure 2.15. ***Explanation of aliasing by phase evolution.*** Part A of the diagram shows the time evolution of three different resonances in the real-imaginary plane of the rotating frame of reference. Panel B shows the true spectrum, without aliasing of any peaks (top) and the observed spectrum, with aliasing (bottom). The resonance frequencies, relative to the carrier (Ω), are: $\omega_A = -SW/2 - \delta$, $\omega_B = -SW/2$. and $0 < \omega_C < SW/2$. In panel A, the point at $t = 0$ is immediately after an ideal 90° pulse along the y-axis. This pulse places the magnetization of all spins on the x-axis. During the subsequent three frames the spins precess about the z-axis at a frequency that is the difference between ω_s and the rotational rate of the coordinate frame, Ω. Sampling of the magnetization at the first time point occurs in the right-most frame, at the dwell time, $\tau_{dw} = 1/SW$. At this time the position of the magnetization determines the position of its line in the spectrum; the corresponding position in the spectrum is indicated on the outside of the right-most circle. At the first dwell time resonance B has precessed 180° as it is at the edge of the spectrum, C has precessed less than 180° counter-clockwise, and resonance A has precessed greater than 180° clockwise. Consequently, the resonance line for A is aliased, and found a distance δ from the right edge of the spectrum, with an apparent frequency of ω'_A= SW/2-$\delta = \omega_A + SW$

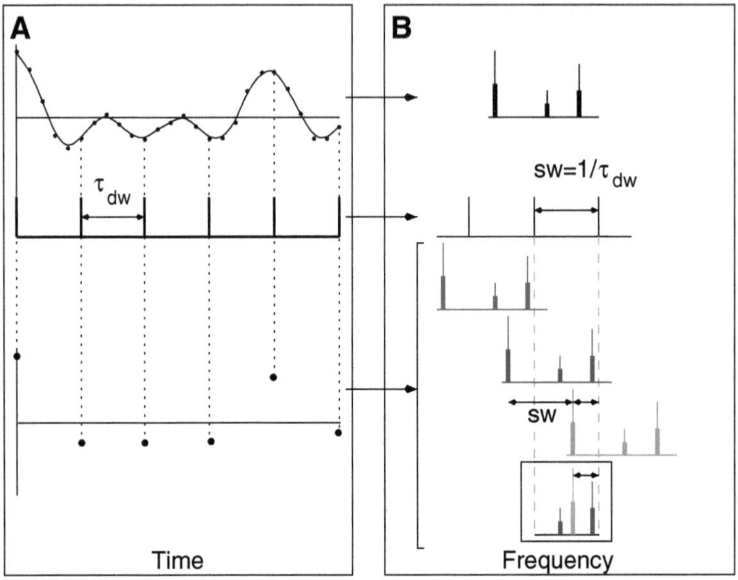

Figure 2.16. Explanation of aliasing using Fourier transforms. Panel A shows time domain signals. The top graph is the continuous FID, the middle is the comb function that represents sampling of the FID every τ_{dw} seconds. The lower graph shows the digital points that are used for the transform. Panel B shows the resultant Fourier transforms of each of the functions in Panel A. The top is the unaliased spectra that was obtained from the continuous FID. The middle panel is the Fourier transform of the comb function. It is another comb function with teeth spaced $1/\tau_{dw}$, or the sweepwidth, apart. The bottom section shows how the observed spectrum is constructed from the convolution of the unaliased spectrum and the comb function. A copy of each spectra is placed at each peak in the comb function and these are added together. The peaks within the spectral width are kept, all others are removed, giving the observed spectra contained in the box. The aliased peak in this spectra is colored lighter gray. Its distance from the edge of the spectrum is indicated by an arrow. In this example, the position of the aliased peak is given by its true frequency minus the sweepwidth ($\omega - SW$). Although its full intensity is shown, in practice the intensity of the aliased peak would be attenuated by analog filtering of the FID. In addition, the lineshape will not be in pure absorption mode unless the initial delay in collecting the first point is 1/2 of the dwell time. In which case the aliased peaks will be inverted relative to other peaks in the spectrum (see Section 15.6.3.1).

lap between adjacent images of the spectra, producing an observed spectrum without aliased peaks.

2.2.8.3 Receiver Filter

Defining the sweepwidth sets the bandwidth of analog filters in the receiver (see Fig. 2.1). The purpose of these filters is to reduce the noise in the spectrum by rejecting any signals (and noise) from frequencies that fall outside the spectral window. Any frequencies that are within the spectral window are passed without distortion.

Aliased peaks will be very weak in the spectrum that is obtained from Fourier transformation of the directly detected FID, because this signal is attenuated by the analog filters in the receiver channel. In contrast, in multi-dimensional NMR the additional frequency dimensions are obtained by indirect detection. Since the filters do not operate on these signals, the aliased peaks will appear at full intensity at their aliased positions (see Chapter 15).

2.2.9 Setting the Receiver Gain

The analog signal from the dual mixers is converted to a digital signal before it is stored in the computer. The size of this signal depends on the amplification, or gain, that was applied to the signal in the receiver. It is important to accurately represent the analog signal in digital form so that an undistorted spectrum is obtained after Fourier transformation. If the gain is too high, there will not be enough bits in the digitizer to represent the signal and the signal will be clipped at the largest value of the digitizer. The effect of clipping on the Fourier transform will be discussed in more detail in Chapter 3.

Alternatively, if the gain is too small, the signal will only utilize the lowest order bits of the digitizer. The result is an inaccurate conversion of the signal to its digital form because of round-off errors. The round-off error introduces additional noise in the spectrum, as shown in Fig. 2.17. In addition to introducing noise, the error in digitizing a low gain signal also introduces harmonics of resonance lines that are present in the spectrum (see [96]). For example, if a resonance line is 100 Hz from the carrier frequency, digitization errors will produce artifacts at 200 Hz, 300 Hz, etc. The intensity of these artifacts will be directly proportional to the intensity of the original signal. Placing the strongest signal, e.g. the solvent, on resonance will prevent the occurrence of these spurious signals.

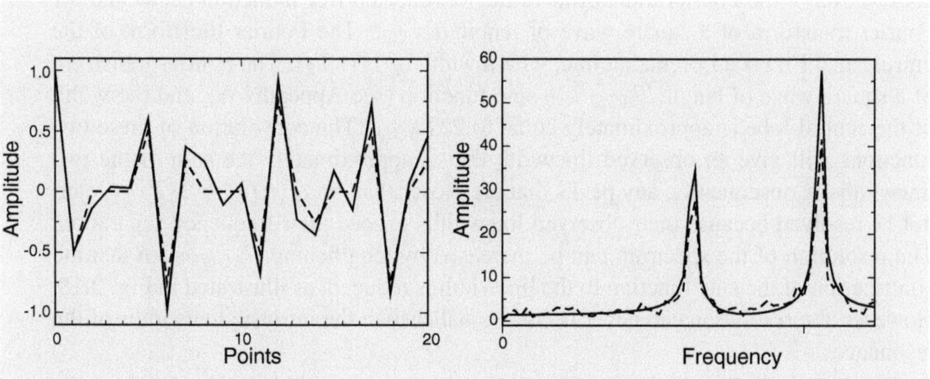

Figure 2.17. Effect of digitizer resolution on the spectrum. The left-hand panel shows the first 20 points of the free induction decay. The solid line is the undistorted signal while the dotted line shows the effect of using only the lowest four bits of the the digitizer. Notice that each point is rounded to some multiple of 0.25. The right panel shows the resultant Fourier transform of the two signals. The noise introduced by poor digitization is readily apparent.

To insure adequate representation of the signal, the receiver gain should be set such that the intensity of the signal is from 1/2 to 2/3 of the range of the digitizer. For example, the largest signal that a 16-bit digitizer can process is approximately 32,000. This number is 2^{15}, 15 bits are used for recording the number and one bit is required for the sign. To avoid clipping of the signal, the gain should be set such that the amplitude of the FID is less than $\approx$ 20,000.

Digitization errors can still occur if there is a large difference in the intensity of resonance lines in the spectrum. For example, in a typical protein sample in H_2O the concentration of the protons in the solvent is $\approx$ 110 M while the concentration of the protons on the protein is $\approx$ 1 mM. The ratio of these two concentrations is 10^5, which is greater than the largest number that can be produced from a 16-bit digitizer. Therefore, if the gain is properly set to avoid clipping of the water signal, then only one or two bits will be available to represent the protein signal. As discussed above, this causes the introduction of significant noise in the spectrum, making it impossible to observe the resonances from the protein. Consequently, it is necessary to have effective methods to suppress the solvent signal in biomolecular NMR experiments. These methods are discussed in detail in Chapter 15.

2.2.10 Spectral Resolution and Acquisition Time of the FID

The total data acquisition time, T_{ACQ}, defines the highest possible resolution that can be obtained in the spectrum. The spectral resolution is the inverse of the total acquisition time. For example, if a FID is acquired for 1 sec, the spectral resolution is 1 Hz. In this case, it would be possible to resolve lines that are separated by 1 Hz, provided that the intrinsic linewidth of the resonances is much less than one Hz.

The true, or actual, resolution of the spectrum is defined by the *observed* linewidths of the resonances, which depends on both the intrinsic linewidths of the resonances *and* the spectral resolution defined by T_{ACQ}. The observed lineshape is obtained by convolution of the Fourier transforms of the untruncated free induction decay and the Fourier transform of a square wave of length T_{ACQ}. The Fourier transform of the untruncated FID is a Lorentzian line, with a width of $1/(\pi T_2)$. The Fourier transform of a square wave of length T_{ACQ} is a sinc function (see Appendix A), and the width of the central lobe is approximately equal to $2/T_{ACQ}$. The convolution of these two functions will give an observed linewidth that is approximately the sum of the two linewidths. Consequently, any peaks that are closer than $\approx 1/(\pi T_2) + 2/T_{ACQ}$ cannot be resolved because their observed linewidths exceed their frequency separation. The resolution of the spectrum can be increased by lengthening T_{ACQ}, such that the contribution of the sinc function to the linewidth is reduced, as illustrated in Fig. 2.18. However, the resolution can never become smaller than the intrinsic linewidths of the resonances.

The *digital resolution* of a spectrum is simply the sweepwidth divided by the number of data points. If n points are sampled in the time domain, then a total of n points will be used to represent the spectrum in the frequency domain. For a given acquisition time, the digital resolution is constant, and equal to the highest possible resolution ($1/T_{ACQ}$), regardless of the number of points (or the spectral width) that are used to acquire the FID (see question 2 in exercises).

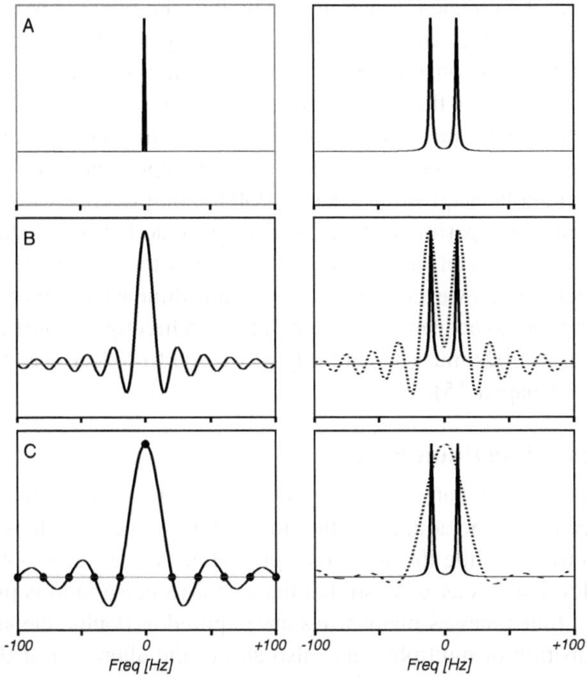

Figure 2.18. *Acquisition time and spectral resolution.* The effect of the acquisition time on the spectral resolution is illustrated. A two line spectrum is convoluted with various sinc functions, corresponding to increasingly shorter acquisition times. The peak separation is 20 Hz and the intrinsic linewidths of each peak is 2 Hz. Rows A, B, and C represent acquisition times of 2 sec, 100 msec, and 50 msec, respectively. The left panels show the Fourier transform of the square waves that corresponds to each acquisition time. The right panels show the true spectra as solid lines and the convolution of the spectra with the sinc function as dotted lines. When the acquisition time is 2 sec (A), there is essentially no additional broadening of the resonances by the convolution with the sinc function. When an acquisition time of 100 msec is used (B), the lines are broadened by approximately 20 Hz, but they are still resolved. When the signal is acquired for only 50 msec, the additional broadening is approximately 40 Hz and the individual lines are no longer resolved. The 'wiggle' patterns that are introduced to the spectra by the transform of the square wave may not be observed in the final spectrum because the sampling frequency of the spectrum (SW/n) is such that only the zero-crossing points of the sinc function are sampled (see Panel C). Increasing the digital resolution, by adding more points, will cause these wiggles to appear in the spectrum.

The digital resolution of the spectrum can be increased as much as desired by simply adding more zero points to the end of the FID post-acquisition. This is equivalent to increasing T_{ACQ}. In most cases, this process has very little effect on the true resolution of the spectrum, but simply provides more digital points to represent the lineshape (see Sec. 3.2).

Data Collection Times: Data should be collected for as long as necessary to reduce the contribution of the sinc function to the linewidth. In practice, an acquisition time is chosen such that the broadening introduced by the sinc function is small compared to the *intrinsic* line width. This approach also greatly attenuates the ripples in the spectrum that come from convolution with the sinc function (see Fig. 2.18). The intrinsic linewidth is related to the spin-spin relaxation time of the spin: $\Delta\nu = 1/(\pi T_2)$. Therefore, an acquisition time that is 3 to 4 times T_2, will produced a negligible distortion of the spectrum. For a typical 10 kDa protein, typical proton T_2 values are 50 to 100 msec, giving total acquisition times of 250 to 500 msec.

In one-dimensional experiments there is absolutely no harm in acquiring data for too long a time period, it is an easy matter to throw away the noise in the tail end of the FID during processing. However, in the case of multi-dimensional experiments only a small number of time points can be collected in the indirectly sampled time domains. Consequently, these FIDs must be carefully processed to reduce the sinc wiggles in the spectrum (see Chapter 15).

2.2.10.1 Scan Repetition Rate

In all likelihood, more than one scan will be acquired and stored in memory. Multiple scans are acquired and added together in memory to increase the signal-to-noise ratio of the spectrum. While the summed signal after N scans is proportional to N, the total noise level scales as $\sqrt{N}$, so that the signal-to-noise ratio is proportional to $\sqrt{N}$. Therefore, four times as many scans are required to double the signal-to-noise ratio. The acquisition of multiple scans also allows the alteration of the pulses and receiver phase from scan to scan which can eliminate artifacts from the spectrum, as discussed in Section 2.19 and in more detail in Chapter 11.

It is important to repeat each scan at a slow enough rate such that the magnetization returns to, or near to, thermal equilibrium before the next scan is started. If the pulse rate is too fast, spin-lattice relaxation cannot restore the population difference between the ground and excited states. Consequently, the *net* magnetization along the z-axis prior to each scan drops with each successive scan and the population of the ground and excited states may become equal. This is referred to as saturation of the transition.

Saturation can be avoided by waiting 4 to 5 times T_1 between scans to allow sufficient time for spin-lattice relaxation to restore thermal equilibrium. Although this approach will give the largest signal *per* RF-pulse, it is inefficient due to the long delay between pulses. A more efficient strategy is to scan more frequently, but to use less than a 90° pulse to reduce the effect of saturation. The optimal pulse angle depends on the ratio of the scan repetition rate to spin-lattice relaxation time according to the following equation, derived by Ernst et al [53]:

$$\cos\beta_{opt} = e^{-T/T_1} \tag{2.19}$$

where β is often referred to as the Ernst angle.

In most 2D experiments the pulses must be 90°. Consequently, the above analysis does not apply and it is necessary to balance a loss in sensitivity due to partial saturation with the gain in sensitivity due to the larger number of scans acquired in a given time. If a 90° pulse is employed at the beginning of the scan, and the interscan delay time is T, then the steady state signal that is obtained from each scan is :

$$A \approx A_o(1 - e^{-T/T_1}) \tag{2.20}$$

where A_o represents the intensity obtained from the system at thermal equilibrium. Typically the inter-scan delay, T, is set to be approximately equal to the spin-lattice relaxation time, T_1.

Dummy Scans: Since it is generally not possible to have the magnetization return to its equilibrium value at the beginning of each scan, it is often useful to collect a number of "dummy scans" prior to the actual data accumulation. These scans will place the magnetization at the steady-state value that will be present during the experiment, reducing potential artifacts from changes in the signal intensity during the first few scans. If decoupling is used in the experiment, which causes sample heating, then the dummy-scans will also bring the sample to thermal equilibrium. Typically 8 dummy-scans are sufficient to bring the magnetization to its steady state values while 64-128 dummy-scans are required for temperature equilibration if decoupling has been applied during the experiment.

2.3 Experimental 1D-pulse Sequence: Pulse and Receiver Phase

The pulse sequence code that would be used to obtain a one-dimensional spectrum is shown in Fig. 2.19. The sequence is quite simple and the comments next to each step are self-explanatory. The program loops executes the following commands for as many scans as requested by the user:

```
d1
(p1 ph1):H
acq phR
```

Each scan begins with a delay (d1) to allow spin-lattice relaxation to occur. It is followed by a single pulse, of length p1 and phase $ph1$, on the proton channel (H), and then acquisition of the FID using a receiver phase of phR. The index of pulse and receiver phases is incremented after each scan. For example, the first scan would use a phase of 0, the second scan a phase of 1, etc. The index is reset to 1 at the end of the list, therefore the fifth scan would use a phase of 0. In the phase list a '0' means a B_1 field along the x-axis, a '1' along the y-axis, a '2' along the minus x-axis, and a '3' along the minus y-axis. Thus the four element sequence: 0 1 2 3, represents a pulse

```
1 zero          ;Zero the memory.
2 d1            ;Inter-scan relaxation delay (T)(e.g. 1 sec)
  (p1 ph1):H    ;Pulse of length p1, phase=ph1 on proton (H) channel.
  acq phR       ;Acquire(acq) the FID, receiver phase = phR.
  go to 2       ;loop to 2 until n scans are done
  wr            ;Write the FID to disk.
exit            ;End of pulse sequence.
ph1=0 1 2 3     ;Phase of excitation pulse (0=x, 1=y, 2=-x, 3=-y)
phR=0 1 2 3     ;Phase of receiver.
```

Figure 2.19. NMR pulse sequence to acquire a one-dimensional spectrum. The instrument commands are listed in the left column and the text to the right of the semi-colon briefly describes each step of the pulse sequence.

along x, y, $-x$, and $-y$. This is equivalent to relative phase shifts of the B_1 field by 0, $\pi/2$, π, and $3\pi/2$.

In addition to providing the pulse sequence, the user would also have to specify the following parameters.

1. Inter-scan delay time (d1).
2. Power level of transmitter.
3. Pulse length for 90° pulse (p1).
4. Number of data points to acquire. The number of points must be a power of 2 in order to process the data using the fast Fourier transform (FFT).
5. Dwell time between data points (or alternatively, the sweepwidth).
6. The number of dummy-scans.
7. Number of scans, which must be a multiple of the phase cycle.

2.3.1 Phase Cycle

The process of acquiring data with different pulse and receiver phases is called phase cycling. The phase of the excitation pulse and the phase of the receiver are changed in a defined way over a number of scans. The phase cycle for this experiment is indicated in Fig. 2.19 at the bottom of the pulse sequence by the lines that begin with 'ph1' and 'phR', which are the pulse phase and receiver phase, respectively. Summation of the data from all of the scans in the cycle will result in the suppression of spectral artifacts. In this simple case, the phase cycle suppresses instrumentation errors associated with quadrature detection. Much more involved phase cycles are used to suppress unwanted signals in multi-dimensional NMR experiments, as discussed in Chapter 11.

2.3.1.1 Phase of the RF-pulse

In the NMR pulse sequence shown above, the direction of the RF-pulse is along the x-axis for the first scan. The B_1 magnetic field oscillating along the x-axis can be represented as the sum of two circularly polarized waves, one going clockwise and the other going counterclockwise:

$$P_x = cos(\omega t) = \frac{1}{2}\left[e^{i\omega t} + e^{-i\omega t}\right] \tag{2.21}$$

If the phase of the RF-pulse is shifted by 90° ($\pi/2$), then the pulse becomes:

$$\begin{aligned} P_{x+\pi/2} &= \frac{1}{2}\left[e^{i(\omega t+\pi/2)} + e^{-i(\omega t+\pi/2)}\right] \\ &= \frac{1}{2}\left[i\, e^{i\omega t} - i\, e^{-i\omega t}\right] \\ &= sin(\omega t) \end{aligned} \tag{2.22}$$

The above shows that a 90° phase change in the pulse can be represented by the *difference* in the two circularly polarized waves, instead of the sum. The resultant B_1 field is shown in the lower part of Fig. 2.20, this field is clearly along the y-axis, therefore a 90° phase shift of an x-pulse generates a y-pulse.

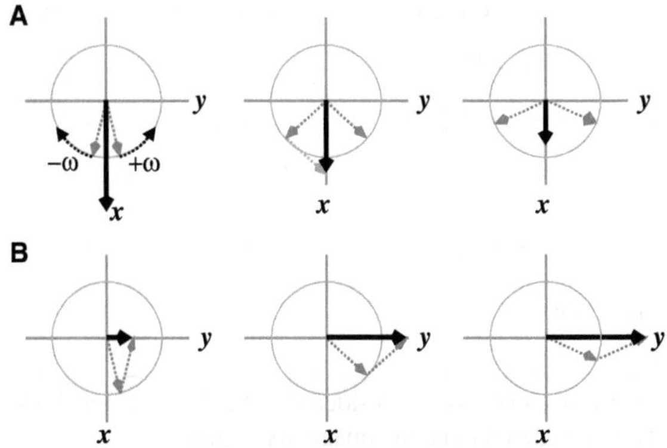

Figure 2.20. Phase shift of radio-frequency pulses. The phase shift of pulses can be explained by considering the pulse as a superposition of circularly polarized waves. Panel A shows the addition of two polarized waves to give a pulse along the x-axis. Time increases from left to right. Panel B shows the result of the subtraction of the same two waves to give a pulse along the y-axis.

Although it is convenient to associate the direction of the pulse with a particular axis in space, there is no absolute coordinate system for the direction of the x- and y-pulses and it is not possible to specify in absolute terms that a particular direction in space is the x-axis. Rather, it is the *relative* phase relationship between the pulses that defines their relative directions in space. For example, if an RF-pulse of $cos(\omega t)$ is applied to the sample, followed by an RF-pulse of $cos(\omega t + \frac{\pi}{2})$, then the orientation of the B_1 field of the second pulse will be shifted by 90°, relative to the B_1 field of the first pulse. The effect of 90° pulses of various phases on the initial magnetization are given in Table 2.1.

Table 2.1. Effect of the pulse phase on the initial magnetization. The first column indicates the phase of the 90° pulse, the second column the resultant effect of this pulse on the magnetization, M_o initially aligned along the z-axis. For example, a pulse along the y-axis tips the magnetization from the z-axis to the x-axis. This convention follows the 'right-hand rule', the direction of the pulse is along the thumb and the spins are tipped in the same direction as the curl of ones fingers. The last two columns indicate the resultant signals that would be observed along the x-axis and the y-axis while the magnetization precesses about B_o. This assumes a counter-clockwise precession of the bulk magnetization. Whether these signals would be interpreted as the real or imaginary component of the signal would depend on the receiver phase.

Phase of Pulse	*Effect of Pulse*	M_x	M_y
0 [= x]	z → -y	$M_o sin\omega t$	$-M_o cos\omega t$
1 [= y]	z → x	$M_o cos\omega t$	$M_o sin\omega t$
2 [= -x]	z → y	$-M_o sin\omega t$	$M_o cos\omega t$
3 [= -y]	z → -x	$-M_o cos\omega t$	$-M_o sin\omega t$

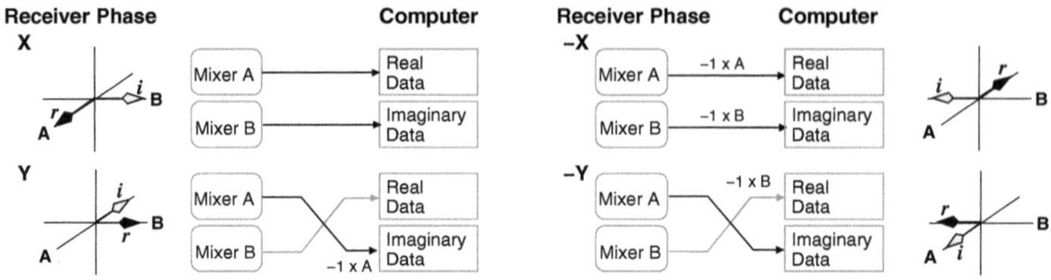

Receiver Phase	*Implementation*
0 [= x]	The signal from mixer A is considered to be the real signal and the signal from mixer B is considered to be the imaginary signal.
1 [= y]	The signal from mixer A is multiplied by minus one and considered to be the imaginary signal while the signal from mixer B becomes the real signal (i.e. the output from the two mixers are switched and one is multiplied by minus one.)
2 [= -x]	The same signal pathway is used as above for a phase of x, except that both the real and imaginary signals are multiplied by minus one
3 [= -y]	Same as with a phase of y, except that the signal from mixer B is multiplied by minus one.

Figure 2.21. Implementation of receiver phase. The signal routing paths for receiver phase values of x, y, $-x$, and $-y$ are shown in the center section of the diagram. The outer sections of the diagram indicate the magnetization measured by the two quadrature mixers, A and B. Mixer A is arbitrarily defined to always measure M_x and mixer B always measures M_y. These signals are routed to the computer memory differently, depending on the setting of the receiver phase. For each phase setting of the receiver the closed arrowhead indicates the direction of the real axis (r) while the open arrowhead shows the direction of the imaginary axis.

2.3.1.2 Receiver Phase

The receiver phase defines the relationship between the real and imaginary data channels of the digitized FID and a particular coordinate axis in the rotation frame. As with the phase of RF-pulses, the direction defined by a receiver phase of zero is arbitrary. It is the change in the receiver phase from scan to scan that is of importance. For illustrative purposes a receiver phase of zero will imply that the real component of the magnetization will represent the magnetization along the x-axis (M_x) and the imaginary component will represent the magnetization along the y-axis (M_y). If the phase of the receiver is shifted by 90°, the magnetization along the y-axis will be sent to the real data channel and magnetization along the negative x-axis will be sent to the imaginary data channel (see Fig. 2.21).

Changes in the receiver phase can be implemented in two ways, by direct alteration of the phase of the received signal by varying the phase of W_{IF} that enters the two mixers in the receiver (see Fig. 2.1), or by changing the routing of the signal from the two mixers to the computer (see Fig. 2.21). The first method is used for receiver phase shifts that are not multiples of 90°. The second method is generally used to produce phase shifts in the receiver that are a simple multiple of 90°.

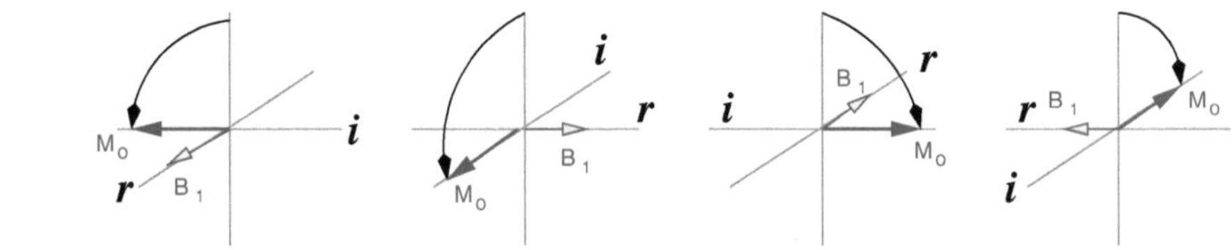

Figure 2.22. Four-step phase cycle. The pulse and receiver phases that are used in the four-step phase cycle are shown. The r and i indicate the real and imaginary data channels of the computer, respectively. The open-headed arrow represents the applied B_1 field. Note that the pulse is defined to be along the x-axis for the first scan, the y-axis for the next, and so on. The solid-head arrow indicates the bulk magnetization immediately after the pulse. The curved arrow indicates the trajectory of the magnetization during the pulse. After the pulse, the magnetization precesses counter-clockwise in the x-y plane.

2.3.2 Phase Cycle and Artifact Suppression

Once the phase of the pulse and the receiver have been defined it is possible to discuss the overall effect of the phase cycle on the recorded signal. First, we will consider how the phase cycle defines how the signals are added to the real and imaginary data signals that are stored in the computer. Then we will show that the particular phase cycle in the one-pulse experiment removes artifacts due to both a DC offset in the FID as well as an imbalance of the quadrature detection channels.

In the case of the one-pulse experiment shown in Fig. 2.19, the receiver phase is the same as the pulse phase in any given scan. Therefore any phase shifts in the signal that are caused by the change in the phase of the pulse will be compensated by a change in the receiver phase, consequently we expect that the real and imaginary signals that are added after each scan will be identical. The relationship between the transverse magnetization and the receiver phase is shown in Fig. 2.22. Immediately after the pulse, the magnetization is always aligned along the negative imaginary axis, therefore the imaginary signal after four scans will be $-4M_o cos(\omega t)$. If we assume that the spins will precess in a counter-clockwise direction in the rotating frame then the real signal will be $4M_o sin(\omega t)$ after four scans. Therefore, the total signal will be:

$$S(t) = 4M_o[-sin(\omega t) + icos(\omega t)] \tag{2.23}$$

Multiplying this signal by $-i$, which is equivalent to simple 180° phase shift of the signal, gives: $S(t) = 4M_o[cos(\omega t) + isin(\omega t)]$, the Fourier transform of which is a resonance line at ω with an intensity of $4M_o$.

2.3.2.1 Suppression of DC Offsets

When the entire FID is displaced from the origin by a constant amount, it is said to have a DC offset. When a signal with a DC offset is subject to Fourier transformation the resultant spectrum will have a signal at zero frequency due to the fact that a DC offset is a signal with zero frequency. Since the DC offset will have the same value at all times, irrespective of the phase of the pulse or the receiver, cycling of the pulse and

Table 2.2. Suppression of DC offsets with phase cycling.

Data	*Real Data*	*Imaginary Data*
Scan 1 (Pulse & Rec. phase= x)	-$M_o sin\omega t$ +DC	$M_o cos\omega t$ +DC
Scan 2 (Pulse & Rec. phase= $-x$)	-$M_o sin\omega t$ -DC	$M_o cos\omega t$ -DC
Sum	-2$M_o sin\omega t$	2$M_o cos\omega t$

receiver phase by x (=0) and $-x$ (=2) will cancel the DC offset. During the first scan (phase = x) the DC offset will be added with the true signal. However the scan with a phase of $-x$ will invert the signal from the precessing spins, but not affect the DC offset. Since the receiver phase is also set to $-x$, the real and imaginary channels will be inverted before addition to the FID from the first scan. The inversion of the signals will also change the sign of the DC offset. Therefore, when the second scan is added to the first the desired signals add, and the DC offset is canceled.

2.3.2.2 Suppression of Imbalance in Quadrature Detection

Quadrature detection requires splitting of the signal from the probe and then passing each individual signal through two independent mixing circuits to generate real and imaginary signals. Assume that one circuit measures $cos(\omega t)$ with a gain of $(1 + \eta)$ while the other measures $sin(\omega t)$ with a gain of 1. When these two signals are combined for complex Fourier transformation the residual cosine term will give rise to peaks at both $+\omega$ and $-\omega$, as indicated in Eq. 2.24, where η is the amplitude of the excess cosine signal. The artifact that arises from the imbalance of the channels are called quadrature images.

$$(1 + \eta)cos\omega t + isin\omega t = \eta cos\omega t + e^{i\omega t} \tag{2.24}$$

Quadrature images are effectively removed by cycling the phase of the pulse and receiver by $x(= 0)$ and $y(= 1)$. The shift in the receiver phase cause the signals from the probe to be routed through opposite mixers, i.e. the magnetization that is destined for the real data channel is actually passed through the mixer that was used for the imaginary data when the receiver phase is zero. Similarly, the signal that is destined for the imaginary data is sent through the mixer that previously processed the real data. This averages any imbalance in the channels.

As an example, consider that mixing channel B has a higher gain, by an amount η, than mixing channel A. The resulting real and imaginary data from the first two scans of the experiment is described in Table 2.3, giving an overall signal of:

$$S(t) = (2 + \eta)M_o[-sin\omega t + icos\omega t] \tag{2.25}$$

Table 2.3. Suppression of quadrature imbalance with phase cycling.

Data	*Real Data*	*Imaginary Data*
Scan 1 (Pulse & Rec. phase = x)	-$M_o sin\omega t$	$(1+\eta)M_o cos\omega t$
Scan 2 (Pulse & Rec. phase = y)	-$(1+\eta)M_o sin\omega t$	$M_o cos\omega t$
Sum	-$(2+\eta)M_o sin\omega t$	$(2+\eta)M_o cos\omega t$

2.3.2.3 Cyclops: Suppression of DC offsets and Quad Imbalance

Shifting the pulse and receiver phases by π will remove DC offsets from the overall FID and a phase shift of $\pi/2$ will remove the effects of Quadrature imbalance. These two phase shifts can be combined into a complete phase cycle involving the following four-steps of the receiver phase and pulse phases: $0(=x)$, $\pi/2(=y)$, $\pi(=-x)$, $3\pi/2(=-y)$. Note that this set of phases combines the two phase cycles such that for each phase of one artifact suppression phase cycle, all of the phases of the other artifact suppression phase cycle are applied.That is, aphase shift of 0 and $\pi/2$ for suppression of quadrature imbalance has been applied to *each* of the phase shifts that were usedto remove DC offsets. The name of this simple phase cycleis *cyclops*, because the'single eye'of the receiver follows the change in the direction of the magnetization that results from an equivalent phase shift of the pulse.

2.4 Exercises

1 (a) The spectrum of ethanol shown in Fig. 2.6 was acquired with a sweep width of 6 ppm. Assuming that this spectrum was acquired on a spectrometer with a proton frequency of 500 MHz, what was the dwell time between points?

(b) A spectrum of ethanol was acquired on the same instrument with a dwell time of 200 μsec. Calculate the spectral width associated with this dwell time and describe the appearance of the spectrum.

(c) Same as part B, but for a dwell time of 400 μsec.

2 Show that the digital resolution, for a fixed acquisition time T_{ACQ}, is independent for the number of points, n, that are used to sample the FID (Hint: Keep T_{ACQ} constant, vary n, and calculate the sweepwidth and the Hz/point for different n values.)

3 A transmitter power setting of 0 dB generates a 10 μsec 90° pulse.

(a) What power level will be required to produce a 15 μsec 90° pulse?

(b) It is often useful to insert a radio-frequency filter between the output of the transmitter and the coil in the probe. The insertion of such a filter causes a 1 dB decrease in power to the probe. How would you change the length of the 90° pulse to accommodate this filter?

2.5 Solutions

1 (a) The sweepwidth is 500 MHz × 6 ppm = 3000 Hz. Therefore the dwell time is 1/3000 = 333 μsec.

(b) A dwell time of 200 μsec corresponds to sweepwidth of 5000 Hz (1/200 μsec), or 10 ppm on a 500 MHz spectrometer. Therefore, the spectrum will look the same and none of the resonance lines will be aliased.

(c) A dwell time of 400 μsec corresponds to a sweep width of 2500 Hz, or 5 ppm. Since the transmitter was set at 2 ppm, the range of the spectrum is from 4.5 ppm to -0.5 ppm. Since the water resonance, at 4.8 ppm, is outside this window, it will be aliased and appear at -0.2 ppm. This frequency is the resonance frequency, minus the sweep width.

2 Setting T_{ACQ}=1 sec. If 128 points are collected, the dwell time is 7.81 msec, giving a spectral width of 128 Hz, or a digital resolution of 1 Hz/point. If 1024 points are collected, the dwell time is 0.976 msec, giving a spectral width of 1024 Hz, or a digital resolution of 1 Hz/point.

3 (a) $20\, log(10/15) = 20 \times -0.176 = -3.52$ dB. Therefore the power will have to be *decreased* by 3.52 dB.

(b)

$$\begin{aligned} 1 &= 20\, log \frac{\tau}{10\ \mu\text{sec}} \\ 0.05 &= log \frac{\tau}{10\ \mu\text{sec}} \\ 1.122 &= \frac{\tau}{10\ \mu\text{sec}} \\ \tau &= 11.2\ \mu\text{sec} \end{aligned}$$

An 11.2 μsec pulse will generate a 90° pulse when the filter is present.

Chapter 3

INTRODUCTION TO SIGNAL PROCESSING

The sequence of steps that are required to generate a spectrum from the time domain data (FID) are discussed in this chapter and summarized in Fig. 3.1. Although the focus of this chapter will be primarily on processing one-dimensional spectra,

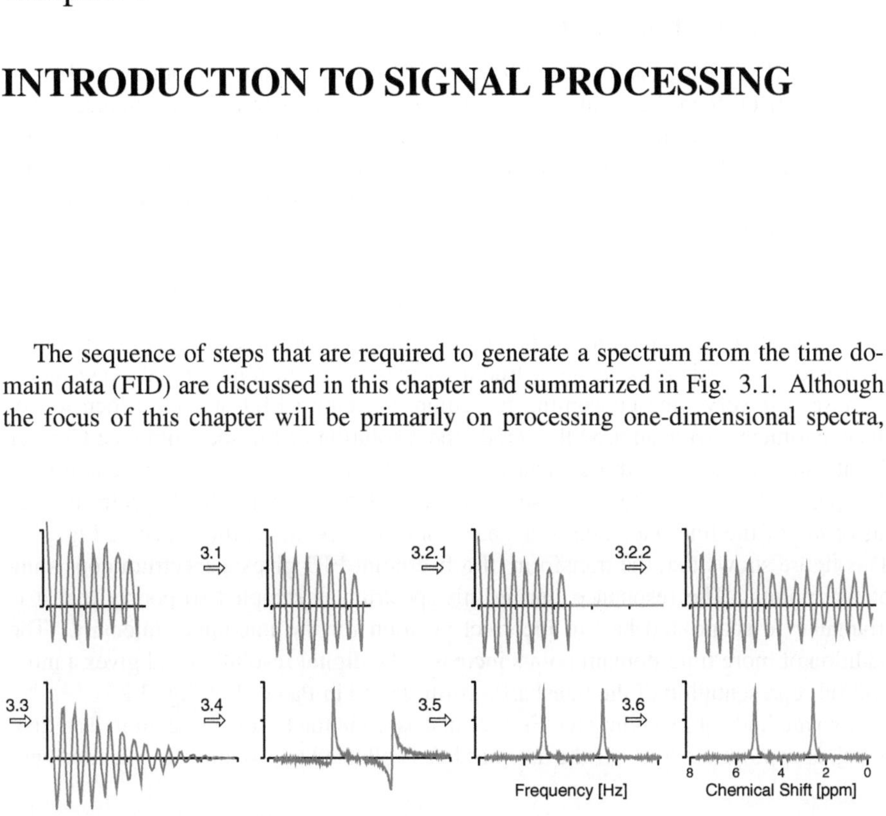

Figure 3.1. Overview of data processing. The steps involved in data processing are illustrated, beginning with the free induction decay (FID, upper left), of N data points, and ending with the final phased and referenced spectra, consisting of $2N$ data points (lower right). Each step is briefly described below and additional details are given in the indicated section of the text:

Sec. 3.1.	Remove the DC offset.
Sec. 3.2.1	Increase the resolution of the spectrum by zero-filling. N additional points are added to the end of the acquired data, giving $2N$ points.
Sec. 3.2.2	Linear prediction (LP) replaces the zero-filled points with extrapolated data.
Sec. 3.3	Apply an apodization function to remove the discontinuity from the end of the FID.
Sec. 3.4	Fourier Transformation. The solvent line is removed by Linear prediction and errors in the initial data points are corrected prior to transformation.
Sec. 3.5	Phase correct the spectrum to generate pure absorption peaks.
Sec. 3.6	Reference the spectra using a chemical shift standard.

many of the aspects of processing are also directly applicable to higher-dimensional experiments. In addition to data processing, the effect of artifacts in the free induction decay (FID) on the appearance of the final spectra will also be discussed. Many of these artifacts arise in the acquisition of data, consequently it is essential to be able to identify the source of the spectral distortion and to correct the FID post-acquisition.

3.1 Removal of DC Offset

If the FID is displaced from the baseline by a constant amount, otherwise known as a DC offset, then the resultant spectrum will contain a spurious resonance at the center of the spectrum (see Chapter 2). The DC offset is usually removed by phase cycling during data acquisition. It can also be removed during processing by taking the average value of the last 10-20% points of the FID and then subtracting this value from each point in the FID.

3.2 Increasing Resolution by Extending the FID

In many cases the sampling of the FID is terminated, or truncated, before the signal decays to zero. This is particularly true in the case of multi-dimensional experiments because it is very time consuming to acquire the entire FID in each dimension of the experiment. Truncation will degrade the resolution of the spectrum (see Chapter 2). It can introduce spectral artifacts, and can also result in a poor representation of the spectral lineshape due to the small number of points in the final spectrum. The distortion of the lineshape due to a small number of points is illustrated in Fig. 3.2. This figure shows that the transform of a five-point FID gives a spectrum consisting of five points. The resonance line in this spectrum is sampled so poorly that it is triangular in shape and has an incorrect position for the maximum intensity. The addition of more time-domain points increases the digital resolution and gives a more faithful representation of the lineshape, as illustrated in Panel B of Fig. 3.2.

Two methods of extending the FID are discussed in the following section. The first, which is appropriate to use if the data has been collected to 3-4 times T_2 is called zero

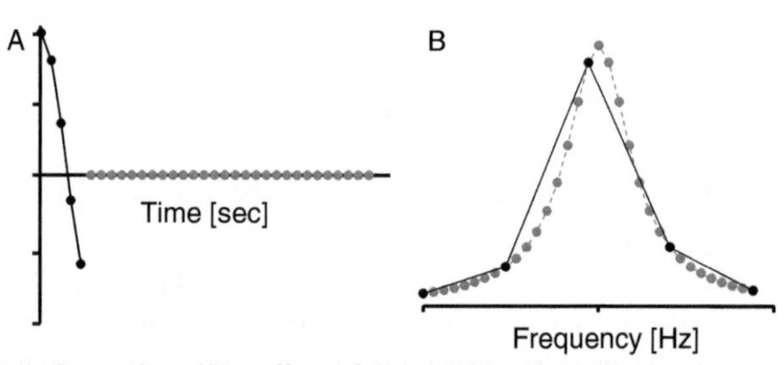

Figure 3.2. Digital sampling. The effect of the number of acquired points on the digital sampling of the spectrum. Panel A shows a FID consisting of five data points (black dots). Panel B shows the resultant spectrum from the Fourier transform of the five data point FID, drawn as a solid black line connecting the five points in the spectrum. The gray dots in panel A are additional zero-filled points that have been added to the end of the FID. Panel B shows the resultant spectrum (dotted line) after transforming the zero-filled FID. The additional interpolated points in the spectrum are colored gray.

filling because the FID is simply extended with zeros. The second method, which is appropriate to use if the data acquisition time is approximately equal to T_2 is called linear prediction. This method uses the acquired data to predict the additional data points.

Although extending the FID is usually employed to increase the resolution of the spectrum, this technique is also useful if the number of acquired points, N, is not a power of 2. If $2^{m-1} < N < 2^m$ then the Fast Fourier Transform (FFT) cannot be used for data processing. Extending the FID to 2^m allows the use of the FFT in processing.

3.2.1 Increasing Resolution by Zero-filling

Zero-filling always increases the *digital* resolution of the spectrum because the sweepwidth is represented by more points. Zero-filling can also increase the true resolution of the spectrum under certain conditions, as illustrated in Fig. 3.3. If the FID has not decayed significantly at the end of data collection, then additional resolution can be gained by doubling the size of the FID by zero-filling. It may seem paradoxical that the addition of zeros can provide additional information. The additional information comes from the fact that the FID is collected in quadrature, thus a total of $2N$ data points are collected for a complex FID of N points. Consequently when a FID of N *complex* data points is transformed to a spectrum of N data points, half of the data has not really been used. Doubling the size of the FID generates a spectrum of $2N$ data points, which utilizes all of the data that was acquired. Additional rounds of

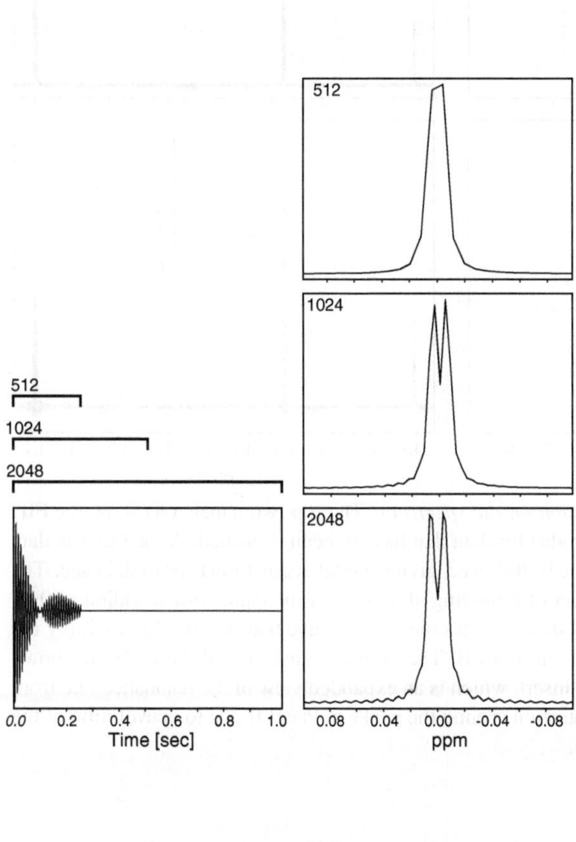

Figure 3.3 Increasing resolution by zero-filling. Zero-filling can increase the observed resolution. The left section shows a FID that was truncated to 512 points. The corresponding spectrum is shown on the top right. Zero-filling to 1024 points generates the middle spectrum, and zero-filling to 2048 points gives the bottom spectrum. Note that the unresolved doublet in the 512 point spectrum is resolved after adding 512 zeros to the FID. The additional doubling of the number of points to 2048 only increases the digital resolution, producing a better representation of the lineshape, but no further increase in actual resolution.

zero-filling, such as increasing the FID size to $4N$ complex points, will only increase the digital resolution.

3.2.1.1 Truncation Artifacts Due to Zero-Filling

Artifacts in the spectrum can occur if the FID does not decay to zero by the end of data collection and the digital resolution has been increased by zero-filling. These artifacts arise from the discontinuity in the signal introduced by zero-filling and are often referred to as truncation artifacts.

The origin of the truncation artifacts can be understood by considering the zero-filled FID to be the product of the original FID multiplied by a square wave that is the length of the acquisition time, as illustrated in Fig. 3.4. Consequently, the observed spectrum will be the convolution of the two respective transforms. The Fourier transform of the square wave is a sinc function (see Appendix A). The ripples, or wiggles, that are associated with the sinc function are convoluted with the undistorted spectral

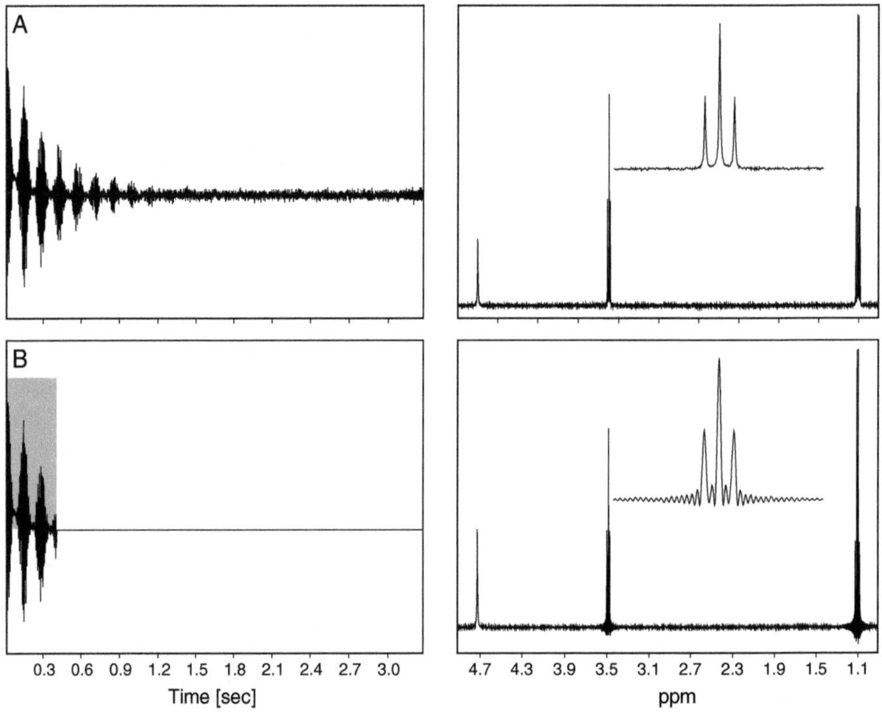

Figure 3.4. Effect of FID truncation on the spectrum. The top two panels (A) show the FID (left) and the resultant spectrum (right) for data that has not been truncated. A total of 16k data points were acquired, the dwell time is 200 μsec, giving a total acquisition time of 3.28 sec. The lower panels (B) illustrate the effect of removing all but 2k of the data, corresponding to 0.41 sec of acquisition time. The gray box illustrates the square wave that was used to multiply the FID in panel A to generate the FID in panel B. The Fourier transforms of the FIDs are shown on the right. Both spectra show an insert, which is an expanded view of the resonance line from the triplet at 3.5 ppm. Note the distortion around the peaks in Panel B due to convolution of the true lineshape with a sinc function.

lines, generating ripples that emanate from each peak. The distortion of the spectrum is readily apparent in Fig. 3.4. The intensity of the ripples will be proportional to the size of the original signal. Consequently, artifacts from intense resonances in the spectrum, such as the solvent, can easily interfere with the detection of weaker signals. This truncation artifact can be removed by apodization, which is discussed below, in Sec. 3.3.

3.2.2 Increasing Resolution by Linear Prediction (LP)

Extrapolation of the time-domain points can also be accomplished by a technique called linear prediction. A comprehensive discussion of this technique is provided by Barhuijsen *et al.* [7]. This technique is so named because an extrapolated data point is obtained from a linear combination of existing data points that preceded it:

$$x_n = a_1 x_{n-1} + a_2 x_{n-2} + ... a_M x_{n-M} \tag{3.1}$$

once the nth point is predicted, it can then be used to predict the next point, and so on.

This approach assumes that the FID is represented as a sum of damped cosine functions:

$$x_n = \sum_{k=1}^{K} c_k e^{-n\Delta t/T_{2k}} cos(\omega_k n\Delta t + \phi_k) \tag{3.2}$$

where $n\Delta t$ represents the time that the x_n point is sampled.

Two LP coefficients are required to represent a single resonance line. If a spectrum contains N lines, then a minimum of $2N$ coefficients are required to represent the spectrum. For example, if the spectrum contained a single spectral line, it is possible to show the following relationship between the spectral parameters and the LP coefficients.

$$\begin{aligned} a_1 &= 2e^{-\Delta t/T_2} cos(\omega\Delta t) \\ a_2 &= -e^{-2\Delta t/T_2} \end{aligned} \tag{3.3}$$

Equation 3.3 is valid, regardless of the value of n. Once these coefficients have been determined *correctly*, they will be valid for the prediction of any number of points. You can convince yourself that this is true by substituting a_1 and a_2 for any two adjacent points of the FID and show that you obtain the value for the next point, i.e.[1]

$$\begin{aligned} e^{-n\Delta t/T_2} cos(\omega\, n\Delta t) &= a_1 e^{-(n-1)\Delta t/T_2} cos(\omega\, (n-1)\Delta t) \\ &+ a_2 e^{-(n-2)\Delta t/T_2} cos(\omega\, (n-2)\Delta t) \end{aligned}$$

3.2.2.1 Determination of LP Coefficients

Before the data can be extended by linear prediction it is necessary to determine the coefficients, $a_1, a_2, etc.$ If N points are used to find M coefficients, then the following represents the set of simultaneous equations that must be solved for a:

$$Xa = x \tag{3.4}$$

[1]The solution requires application of the following identity: $cos\alpha \times cos\beta = \frac{1}{2}[cos(\alpha-\beta)+cos(\alpha+\beta)]$.

X is a matrix of known data, with a dimension $(N - M) \times M$, a is a vector of length M, and x is a vector of known data with a length $N - M$. Both X, and x, are known, while a is not. For example, consider determining a_1 and a_2 ($M = 2$) from five known data points ($N = 5$), x_5, x_4, x_3, x_2, x_1. The following equations can be used to solve for a_1 and a_2:

$$\begin{aligned} a_1x_4 + a_2x_3 &= x_5 \\ a_1x_3 + a_2x_2 &= x_4 \\ a_1x_2 + a_2x_1 &= x_3 \end{aligned}$$

or, in matrix form:

$$\begin{bmatrix} x_4 & x_3 \\ x_3 & x_2 \\ x_2 & x_1 \end{bmatrix} \times \begin{bmatrix} a_1 \\ a_2 \end{bmatrix} = \begin{bmatrix} x_5 \\ x_4 \\ x_3 \end{bmatrix}$$

The values of a_n can be found using singular value decomposition. This is a least squares technique that finds the best values for a_n using all of the data. The procedure begins by decomposing the X matrix into a product of three matrices:

$$X = U\Lambda\tilde{V} \tag{3.5}$$

where Λ is diagonal with zero or positive elements. U and V are orthogonal in the sense that their columns are orthonormal, and V is a square matrix so that $V\tilde{V} = 1$.

This form of X allows direct determination of a from the existing data:

$$\begin{aligned} Xa &= x \\ U\Lambda\tilde{V}a &= x \\ \Lambda\tilde{V}a &= U^{-1}x \\ \tilde{V}a &= \Lambda^{-1}U^{-1}x \\ a &= V\Lambda^{-1}U^{-1}x \end{aligned} \tag{3.6}$$

Information on the number of significant coefficients can be obtained from the Λ matrix. The product of Λ with its transpose, gives a square diagonal matrix, $\lambda = \Lambda^{-1}\Lambda$, with diagonal elements $\lambda_1, \lambda_2 \ldots \lambda_M$. The diagonal elements are referred to as singular values and they contain information on the number and intensity of the spectral components that were present in the initial data. If there are K lines in the spectrum then $2K$ of the diagonal values of Λ will be non-zero. The remaining values will be zero, at least in the absence of noise.

As an example, suppose a spectrum contains two resonance lines, one that is twice as intense as the other. Since the number of resonance lines is not known beforehand, one might try to represent the signal with a total of 6 coefficients, representing three spectral lines. The original equations that define a are:

$$x_{12} = a_1x_{11} + a_2x_{10} + a_3x_9 + a_4x_8 + a_5x_7 + a_6x_6$$

$$\vdots$$

$$x_7 = a_1x_6 + a_2x_5 + a_3x_4 + a_4x_3 + a_5x_2 + a_6x_1$$

The λ matrices that would be obtained in the absence and presence of noise are shown below:

$$\text{Without Noise} \qquad \lambda = \begin{bmatrix} 2 & 0 & 0 & 0 & 0 & 0 \\ 0 & 2 & 0 & 0 & 0 & 0 \\ 0 & 0 & 1 & 0 & 0 & 0 \\ 0 & 0 & 0 & 1 & 0 & 0 \\ 0 & 0 & 0 & 0 & 0 & 0 \\ 0 & 0 & 0 & 0 & 0 & 0 \end{bmatrix} \qquad \text{With Noise} \qquad \lambda = \begin{bmatrix} 2.05 & 0 & 0 & 0 & 0 & 0 \\ 0 & 1.98 & 0 & 0 & 0 & 0 \\ 0 & 0 & .97 & 0 & 0 & 0 \\ 0 & 0 & 0 & .99 & 0 & 0 \\ 0 & 0 & 0 & 0 & .01 & 0 \\ 0 & 0 & 0 & 0 & 0 & .02 \end{bmatrix}$$

In the absence of noise, only the first four elements of λ are non-zero, indicating that the data can be represented by four coefficients, corresponding to the two spectral lines. For illustrative purposes, the two fold difference in the peak intensity has been reflected in the coefficients; the value of the first two coefficients are twice that of the second two. When noise is present, the 5th and 6th elements are non-zero. In this example, these values arise from the noise in the data, however, they could equally represent a third line in the spectrum that is much weaker than the other two lines.

The above discussion assumed that there was sufficient data to obtain all 6 coefficients. However, the number of coefficients that can be obtained is limited by the number of existing data points. For a spectrum that has fewer points than twice the number of signals there is insufficient information to determine all of the coefficients. In this case, only the strongest peaks will be predicted while the weaker peaks will lost. Linear prediction is still useful in this case because the intense lines in the spectrum generate the most intense artifacts. Suppression of truncation artifacts from these signals can be quite beneficial.

If more than $2N$ data points are available, it is generally beneficial to use more than the minimum number of data points to reduce the influence of noise on the determination of the coefficients.

3.2.2.2 Suppression of Artifacts in Linear Prediction

If the signal-to-noise ratio of the spectrum is high, and the peaks are roughly of the same intensity, then the prediction of subsequent points in the FID using linear prediction is very robust. Unfortunately, the noise present in typical data generates errors in the linear prediction coefficients. These errors in coefficients will result in errors in the predicted points. Two methods of error suppression will be discussed: root-reflection and the use of forward-backward prediction.

Root Reflection: One of the most offensive errors in linear prediction is a negative value for T_2. This results in the prediction of an exponentially *growing* signal that is clearly incorrect. To avoid this effect, it is possible to constrain the values of the coefficients such that only decaying signals are predicted.

Constraining the coefficients is accomplished by a technique called *root-reflection*. Assuming that there is a single resonance line in the spectrum we can use the analytical expressions for a_1 and a_2 (Eq. 3.3) to solve for T_2 and ω, giving [7]:

$$e^{-\Delta t/T_2} e^{\pm i\omega\Delta t} = \frac{1}{2}\left[a_1 \pm \sqrt{a_1^2 + 4a_2}\right] \qquad (3.7)$$

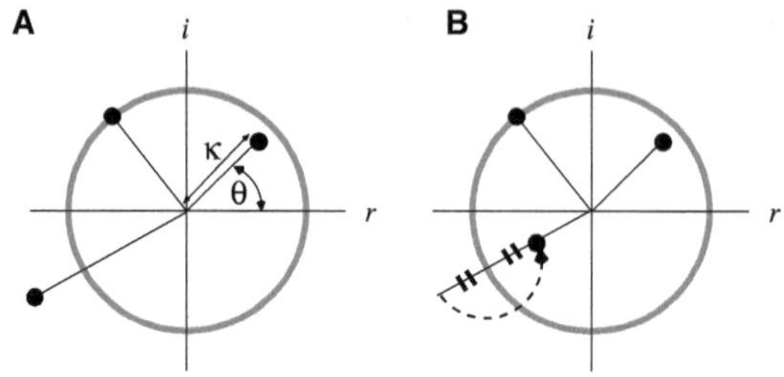

Figure 3.5. *Root reflection in Linear Prediction.* Three complex roots are shown as solid dots on an Argand diagram (See Appendix B). The amplitude (κ) and phase (Θ) of one of these roots are indicated on the left (A) diagram. The unit circle is drawn in gray. Panel A shows the placement of the roots prior to root reflection. Those roots that lie within the unit circle, corresponding to $\kappa < 1$, are acceptable because they represent a T_2 relaxation time that is greater than zero. In contrast, the roots lying outside the unit circle are not physically possible, since they represent a negative T_2. Panel B shows the effect of reflection of one unacceptable root into the unit circle, reversing the sign of the T_2.

The left side of this equation is just a complex number $\kappa e^{i\theta}, (\kappa = e^{-\Delta t/T_2},\ \theta = \omega\Delta t)$ while the right side is clearly the formula for the roots of a quadratic equation. Consequently, the roots of the following complex equation are equal to $\kappa e^{i\theta}$.

$$z^2 - a_1 z - a_2 = 0 \qquad (3.8)$$

Once the roots are determined, then the predicted T_2 can be obtained from κ. If κ is less than one the T_2 will be positive [2]. A signal that does not decay ($T_2 = \infty$) will have a value of κ equal to one. A growing signal, with a *negative* T_2, will have a κ value of greater than one. The sign of T_2 can be reversed by a process called root reflection, which is illustrated in Fig. 3.5. After the correction of the roots, the new coefficients will represent a decaying signal.

Use of Forward and Backward Prediction: The above discussion has focused on the forward prediction of data points. It is also possible to use linear prediction to predict earlier time points. Returning to the previous example of five known data points and two coefficients, it is also possible to write the following set of linear equations:

$$\begin{aligned} x_1 &= a_1 x_2 + a_2 x_3 \\ x_2 &= a_1 x_3 + a_2 x_4 \\ x_3 &= a_1 x_4 + a_2 x_5 \end{aligned}$$

In the absence of noise, the above coefficients will be identical to those that were obtained from forward prediction. However, noise will introduce small differences between the otherwise equal coefficients. Simply averaging of the two coefficients increases the stability of the prediction. The effect of root-reflection and forward-backward prediction on the accuracy of linear prediction is illustrated in Fig. 3.6.

[2] If $T_2 > 0$ then $e^{-x} < 1$. If $T_2 < 0$, then $e^{+x} > 1$, where $x = -\Delta t/T_2$. If $T_2 = \infty$, then x=0 and $e^0 = 1$.

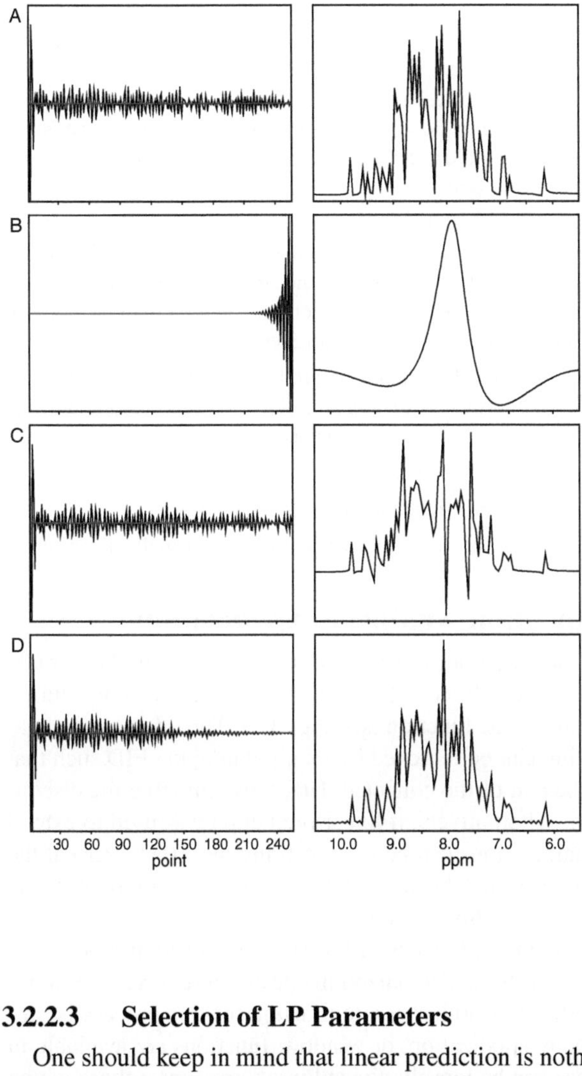

Figure 3.6 Improvement of linear prediction with root-reflection and forward-backward prediction. The left panels show the free induction decay (FID) while the right panels show the resultant spectrum. The spectrum is the amide region of a 130 residue protein. Row A shows the FID acquired with 256 points and the resultant transform, using all 256 points. Rows B-D show the result of linear prediction of the 2nd half of the FID, using the first half of the acquired data. A total of 60 coefficients were predicted from the data. Row B shows the result that is obtained without root reflection or the use of forward-backward averaging. Note the large increase in the predicted points towards the end of the FID, indicative of a negative T_2. Row C employed root-reflection while Row D employed both root reflection and forward-backward prediction, giving a spectrum similar to the original spectrum.

3.2.2.3 Selection of LP Parameters

One should keep in mind that linear prediction is nothing more than a least-squares fitting procedure that operates in the time domain. Consequently, the use of noisy data will result in incorrect coefficients that can lead to changes in the signal intensity, linewidth, and frequency! With particularly noisy data it is possible to generate peaks that are solely artifacts of the linear prediction. Therefore, linear prediction should be used with caution. The most sensible use of linear prediction is to extend the FID by 20-40% and then use an apodization function (see below) to reduce the influence of the predicted points on the final spectrum, as illustrated in Fig. 3.8. This approach will enhance the contribution of the acquired data points that are towards the end of the FID, thereby increasing the resolution in the spectrum, without undue influence from the predicted points.

Most NMR processing software packages have linear prediction routines. A brief description of typical parameters is presented in Table 3.1.

Table 3.1. Linear prediction parameters.

Parameter	*Description*
N	Number of data points. Points 1 through N will be used to determine M coefficients. You should use all of the data, unless the FID is particularly noisy at later time points.
M	Number of coefficients, this cannot exceed N/2. The number of coefficients should be equal to twice the number of signals in the FID. If too few coefficients are chosen then only the strongest signals will be used in the prediction. If too many coefficients are chosen then the noise in the spectrum will be represented by artifactual spectral lines.
First point	This is the 1st point that should be predicted, normally it is the next point after the end of the existing FID; however, data acquired at the end of the FID can certainly be replaced.
Last point	This is the last point to be predicted.
Root reflection	Implementation of root-reflection is usually an option.
Forward-backward	Implementation of forward-backward prediction is usually an option.

3.3 Removal of Truncation Artifacts: Apodization

A discontinuity in the free induction decay will lead to artifacts in the final transformed spectrum. In the case of a zero-filled FID, the final spectrum is a convolution of the normal NMR spectrum and a sinc function (see Fig. 3.4). The effects of truncation of the FID due to zero-filling can be removed by manipulating the FID such that it smoothly decays to zero at the end of the collected data, thus removing the discontinuity in the time-domain data. Alternatively, if linear prediction was used to extend the time-domain, the same manipulations can be used to bring the FID to zero at the last point generated by linear prediction. This procedure is useful in that it reduces the influence of the predicted points on the final spectrum.

The FID is brought to zero by multiplying the FID by an apodization function to give a modified FID. This process is termed apodization because it removes the 'feet', or ripples, from the resonance lines. Another common name for this process is 'applying a window-function'. Many apodization, or window, functions are available in data processing packages. These can be loosely divided into three types: those which maintain the lineshape (exponential multiplication), those that change the lineshape in a well defined matter (Lorentzian to Gaussian) and those which simply bring the FID to zero at the end of the acquired data. Of the latter class, there are many different types [53]. However, the differences between them are subtle and often not discernible with typical data. Consequently, we will focus on one of the more commonly applied versions of this class of window functions, the trigonometric window functions.

3.3.1 Effect of Apodization on Resolution and Noise

In addition to removing truncation artifacts, the apodization functions will also affect the resolution and the signal-to-noise ratio of the spectra. For example, exponential multiplication leads to line broadening and a reduction in noise, while certain trigonometric functions produce narrowing of the spectral line with an increase in the noise. In general, these two effects are correlated. Efforts to increase the resolution of the spectra will also increase the noise.

The relationship between changes in resolution and changes in noise levels can be understood by considering the FID to be the sum of the true signal and a noise function. As discussed in Chapter 2 the resolution of the final spectrum depends on how long the FID is collected. Therefore, data points acquired at later times can be considered to provide high-resolution information while earlier data points provide low-resolution information. The true signal decays with time while the noise is uniform during the entire data collection period. Apodization functions, such as exponential multiplication, that suppress signals late in the FID, will decrease the contribution of the noise to the Fourier transform. Unfortunately, they will also decrease the resolution of the spectrum. Conversely, functions that amplify signals late in the FID will increase the final resolution at a cost of increasing the noise.

3.3.1.1 Exponential Multiplication

The most common modification of the FID is exponential multiplication (EM), otherwise known as line broadening (LB). The EM apodization function simply multiplies the original FID by $e^{-t/a}$ to give the following apodized FID (see Fig.3.7):

$$g(t) = e^{-t/T_2} e^{-t/a} = e^{-t[\frac{1}{T_2}+\frac{1}{a}]} \tag{3.9}$$

The effect of this apodization function on the spectral lines can be seen by calculating the Fourier transform of the modified FID. The Fourier transform of this function will produce a Lorentzian line, but with a modified T_2 that is shorter than the original T_2,

$$\frac{1}{T_2'} = \frac{1}{T_2} + \frac{1}{a} \tag{3.10}$$

Note that the EM apodization function always increases the resonance linewidth. In the example shown in Fig. 3.7, the linewidth is doubled, from 10 Hz to 20 Hz. Exponential multiplication reduces the noise in the spectrum because it suppresses the contribution from the right-hand side of the FID, which contains proportionally more noise than the beginning part of the FID.

Usually the parameter $1/a$ (in Hz) is specified in the LB routines provided with the processing software. Depending on the choice of a, it is possible to bring the FID to zero at the end of the acquisition. The optimal value of LB is equal to $\Delta\nu_{1/2}$, which is just the average width of the lines in the spectra.

3.3.1.2 Lorentz to Gaussian transform

In addition to suppressing truncation artifacts this function converts a Lorentzian lineshape to a Gaussian lineshape. Gaussian lineshapes approach the baseline faster than Lorentzian lineshapes, providing some enhancement in the apparent resolution of the spectrum, making it easier to fit overlapping peaks. The apodization function is:

$$g_{GL}(t) = e^{+t/a} e^{-\frac{l^2}{2}t^2} \tag{3.11}$$

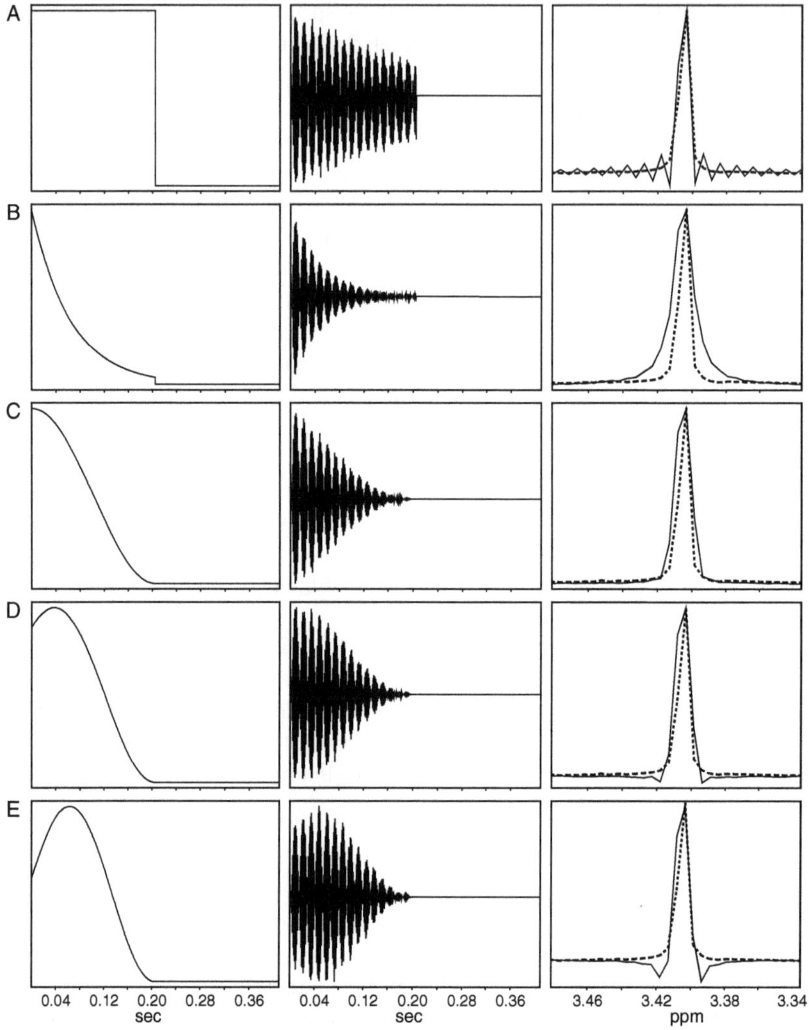

Figure 3.7. *Effect of apodization functions on spectral lineshapes.* The left column shows the apodization function, the center column shows the FID after multiplication by the apodization function, and the right column shows the transformed spectrum, drawn as a solid line. The dotted line in the right-hand panels show the spectrum that would be obtained if the data were collected until it decayed to zero and no apodization function was applied. The natural linewidth of the resonance line, $\Delta\nu$, is 10 Hz. Row A shows the effect of zero-filling. The spectrum is distorted by the convolution with the sinc function. Row B shows the effect of applying a exponential multiplication (EM) with a time constant of 100 msec, equivalent to a broadening of 10 Hz. Note that the sinc ripples are removed, at the expense of an increased linewidth. Rows C, D, and E illustrate the effect of a shifted sine squared window, with phase shifts of 90°, 70°, and 50°, respectively. The 90° shifted $sine^2$ curve is equivalent to a $cosine^2$ function. Note the decrease in the linewidth of the spectrum as the phase shift of the $sine^2$ function decreases. A 50° shifted $sine^2$ function (row E) produces a resonance line with nearly the natural line width. The two negative features on either side of the line arise from the convolution of the Fourier transform of the $sine^2$ function with the undistorted line. These lobes can be reduced by using linear prediction to increase the apparent acquisition time (see Fig. 3.8).

If a is chosen to be equal to T_2, then the modified FID has the following form:

$$g(t) = e^{-\frac{l^2}{2}t^2} \tag{3.12}$$

The Fourier transform of a Gaussian time domain is a Gaussian line:

$$G(\omega) = e^{-\frac{l^2}{2}\omega^2} \tag{3.13}$$

Usually, l is specified by a ratio of the linewidth, $l = r \times \Delta\nu_{1/2}$ typical values of r are 0.2 to 0.7. Note that if the widths of the lines in the spectrum vary, then most of the lines will have a lineshape that is a mixture of Gaussian and Lorentzian character.

3.3.1.3 Trigonometric Windows

This group of window functions are based on trigonometric functions. They remove truncation artifacts by bringing the FID to zero at the end of the acquisition time. These functions can also enhance resolution by increasing the contribution of later time-points to the transform. Both cosine and sine curves, as well as their squares can be used. The sine functions are more appropriate for signals that begin at zero, such as in a DQF-COSY experiment, while cosine functions are more useful for signals that have their maximum at t = 0 and decay as t increases, i.e. a normal free induction decay.

The effect of a cosine squared window (Eq. 3.14) on the FID and subsequent spectrum is shown in Fig. 3.7, Panel C. Note that the ripples due to the sinc function have been removed with only a modest increase in linewidth.

$$g(t) = cos^2(\frac{\pi}{2T_{ACQ}}t) \tag{3.14}$$

where τ_{ACQ} is the total acquisition time.

3.3.2 Using LP & Apodization to Increase Resolution

When apodization functions are applied, they generally degrade the resolution of the spectrum by introducing additional line-broadening of the resonances. This additional broadening occurs because the data points towards the end of the FID are suppressed by the apodization function. The resolution can be restored by adjusting the apodization function such that the contribution of the latter points of the FID are emphasized. The most common way of achieving this is to simply shift the $cos^2\tau$ apodization function[3] such that the maximum occurs at some time point other than the origin of the FID. Examples of these functions are shown in Fig. 3.7, rows D and E. As the function's maximum is shifted to the right, the linewidth narrows, almost to the point of returning the resonance line to the natural linewidth (Fig. 3.7 row E). The enhancement of resolution is not without cost, as the resolution is increased the degree of baseline distortions also increase (compare row D to row E, in Fig. 3.7). This

[3]These shifted window functions are generally referred to as shifted $sin^2\tau$ functions. A 90° shifted $sin^2\tau$ function is the same as a $cos^2\tau$ function.

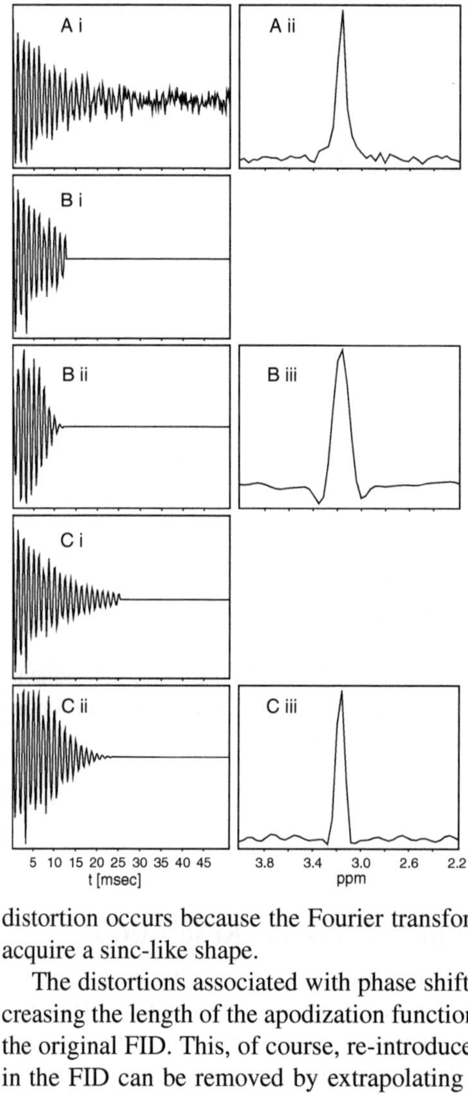

Figure 3.8 Using linear prediction to enhance resolution. Panel A*i* shows the FID from a single resonance line, acquired with 256 points. The spectrum is shown on the right (A*ii*). Panel B*i*, shows the same FID, except that only 64 data points were collected. A 45° shifted $\sin^2\tau$ function that reached zero at 64 points has been applied to remove truncation artifacts (B*ii*), giving the final spectrum shown in B*iii*. Section C shows the effect of the prediction of an additional 64 points using the first 64 points (C*i*), followed by apodization with a 45° shifted $\sin^2\tau$ function that is zero at 128 points, giving the FID shown in Panel C*ii*. The resultant transform (C*iii*) is almost the same as the original signal that was acquired for 4 times longer (256 versus 64 points).

distortion occurs because the Fourier transform of the apodization function begins to acquire a sinc-like shape.

The distortions associated with phase shifted sin^2 functions can be reduced by increasing the length of the apodization function such at it becomes zero *past* the end of the original FID. This, of course, re-introduces truncation artifacts. The discontinuity in the FID can be removed by extrapolating the FID, using linear prediction, to the point that the apodization function is zero, as illustrated in Fig. 3.8.

3.4 Solvent Suppression

Linear prediction can be used to remove strong solvent lines by predicting the signal due to the solvent and then subtracting this signal from the original FID. Transformation of the modified FID will produce a spectrum that lacks the solvent line. The linear prediction coefficients that represent the solvent line can be identified by virtue of the fact that they will be associated with the two largest elements of the λ matrix because the solvent resonance is generally the largest in the spectrum. These two coefficients are then used to generate the FID that represents the solvent line. This method can be quite effective at removing the solvent, as illustrated in Fig. 3.9.

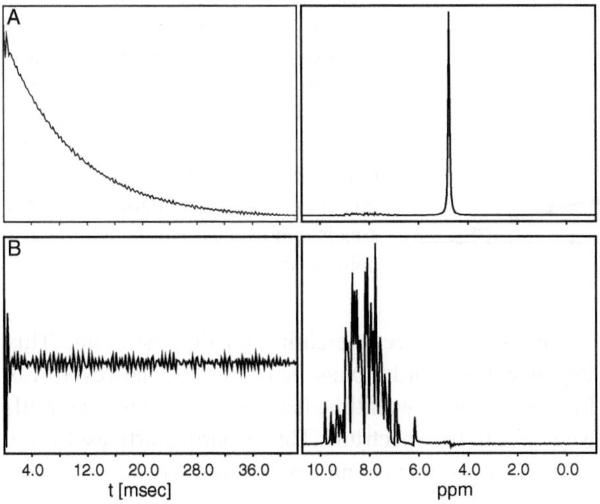

Figure 3.9 Solvent suppression using linear prediction. The top row shows the FID (left) and spectrum (right) when the water signal is $\approx$ 1000 times that of the protein in the sample. The water is on resonance, hence the non-oscillatory appearance of the FID. The lower panels show the FID and resultant spectrum after water suppression. The resultant spectrum shows almost no residual signal from the water.

3.5 Spectral Artifacts Due to Intensity Errors

In this section we will consider the effect of intensity errors at the beginning of the FID on the final spectrum. The least severe and most easily correctable of these errors is caused by the digital Fourier transform itself. Minor, and usually correctable, distortions of the signal can arise due to hardware limitations. In some cases the first few points of the FID are not collected at all, leading to severe distortions of the spectrum if the missing points are not replaced by predicted ones.

3.5.1 Errors from the Digital Fourier Transform

Consider the following real Fourier transform of a free induction decay, in integral form:

$$S(\omega) = \int F(t)cos(\omega t)dt \tag{3.15}$$

Since the FID is sampled at discrete time points, the above integral is evaluated as the following sum:

$$S(k\Delta\omega) = \sum_{i=0}^{n-1} F_i \cos([k\Delta\omega][i\Delta t])\Delta t \tag{3.16}$$

Here, F_i represents each time point in the free induction decay, indexed with the variable i, and $S(k\Delta\omega)$ represents the spectrum, indexed by the variable k.

The sum in Eq. 3.16 is really the total area of a series of rectangles, as indicated in Fig. 3.10. The contribution of the first rectangle to the overall sum is twice as large as it should be. Consequently the value of the first point should be reduced by 50% to correct its contribution to the Fourier transform. Many of the commercial software packages apply this correction as a default.

Alternatively, the correct summation can be obtained by delaying the acquisition of the first point by 1/2 of a dwell time. This delay shifts all of the rectangles shown in

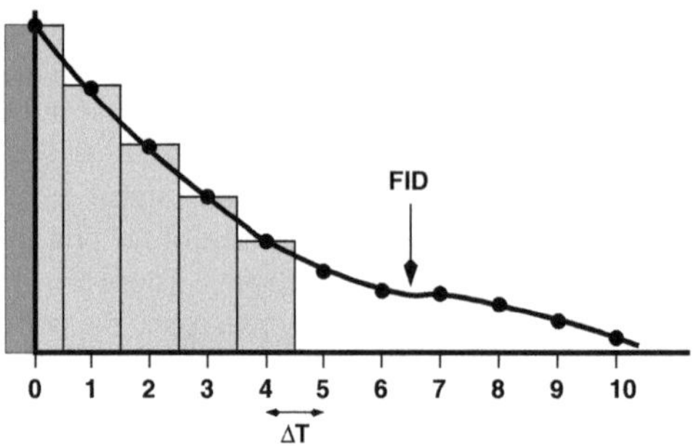

Figure 3.10 Digital Fourier transformation. The summation in Eq. 3.16 is represented by the area of the shaded triangles, shown here for the first five points of the FID. The digitized points are shown as filled circles. The additional area that is incorrectly added to this sum is shaded in dark gray. This area is removed by multiplying the first point by 0.5.

Fig. 3.10 to the right by $\Delta t/2$, giving the correct weighting for the first point. This shift also causes a frequency dependent first-order phase shift of 180°; however, it is easily corrected. Shifting the FID by 1/2 of a dwell time has other advantages as well; aliased peaks appear as negative peaks in the spectrum. This is particularly useful for the identification of aliased peaks in multi-dimensional spectra and will be discussed in more detail in Chapter 15.

3.5.2 Effect of Distorted and Missing Points

The first few points of the FID can be distorted due to residual effects from the excitation pulse at the beginning of the digitization. The contribution of the initial points to the final spectrum can be ascertained from the digital form of the Fourier transform:

$$\begin{aligned} S(k\Delta\omega) &= \sum_{i=0}^{n-1} F_i \cos([k\Delta\omega][i\Delta t])\Delta t \\ &= F_0 cos([k\Delta\omega][0 \times \Delta t])\Delta t + F_1 cos([k\Delta\omega][1 \times \Delta t])\Delta t + \ldots \\ &= F_0 + F_1 cos([k\Delta\omega][1 \times \Delta t])\Delta t + \ldots \end{aligned} \tag{3.17}$$

In this summation, F_0 is the first point of the FID. As shown by the last line of the above equation, this point contributes to the baseline offset of the resultant spectrum. Hence, distortion of the first point will only shift the spectrum up or down, an easily correctable artifact.

Equation 3.17 indicates that the subsequent points of the FID are related to increasingly higher frequencies in the spectrum, i.e. F_1 is associated with $cos(\omega t)$ and F_2 is associated with $cos(2\omega t)$, etc. Consequently, distortion of these points lead to low frequency oscillations in the baseline of the spectrum. Some of the effect of distortion of the first points of the FID on the spectrum are illustrated in Fig. 3.11.

As shown in part C of Fig. 3.11, distortion of the first few points leads to severe distortions in the baseline. With crowded spectra, it would be difficult to correct this distortion in the spectrum using baseline correction routines. However, it may be possible to correct the original error in the FID by using linear prediction to generate the first points of the FID, as illustrated in part D of Fig. 3.11. Since it is necessary to predict only the low frequency components of the spectrum, it is possible to use a small number of coefficients, in this example only 16 coefficients were used, even though the spectrum contains over 100 lines.

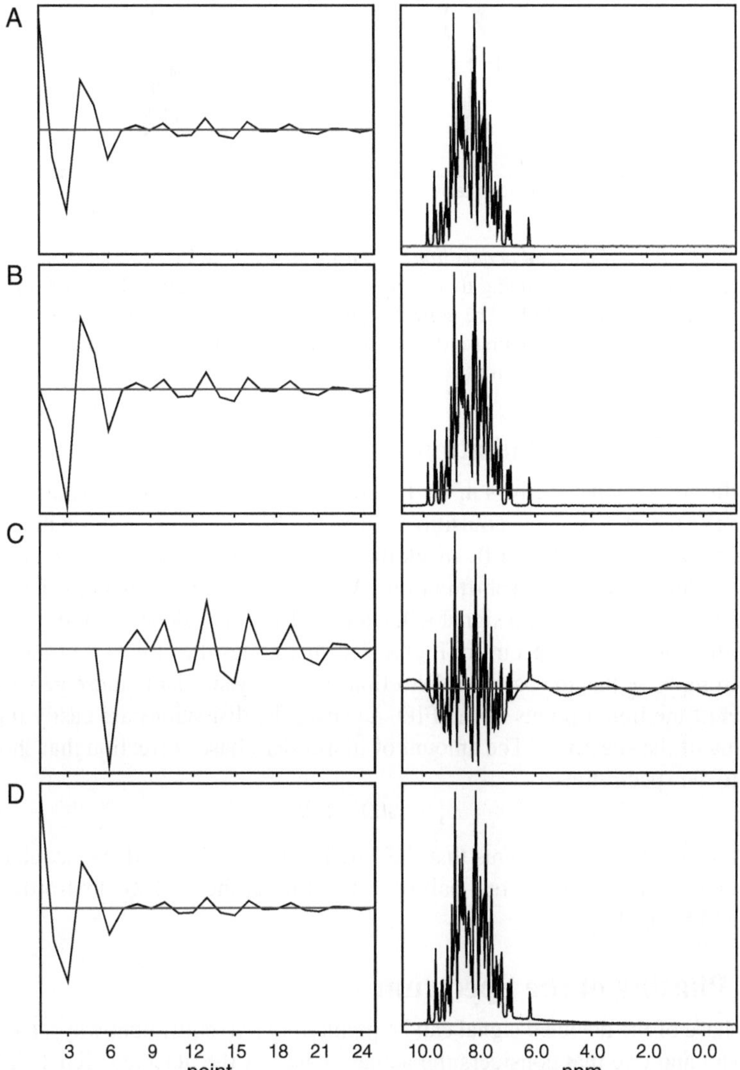

Figure 3.11. Spectral distortion due to errors in the initial points. Row A shows the first 25 points of an undistorted FID (left) and the resultant spectrum from the FID (right). Row B shows the effect of setting the first point to zero. The spectrum is undistorted, but shifted below the origin. Row C shows the effect of setting the first five points to zero. The resultant spectrum is distorted by low frequency oscillations, referred to as a "rolling baseline". Row D illustrates the effect of using linear prediction to predict the previously nulled points. Sixty-four points (i.e. points 6 through 70) were used to generate 16 coefficients. Root reflection and forward-backward linear prediction was not used in this example. The suppression of the artifacts in the final spectrum is remarkable.

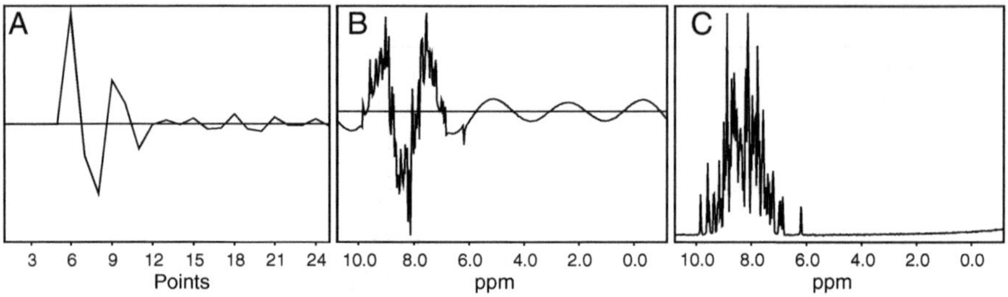

Figure 3.12. Effect of delayed acquisition on the spectrum. The FID (A) has been shifted to the right by 5 points, corresponding to a delay of $\approx$ 660 μsec (sweepwidth = 6000 Hz). The resultant spectrum is shown in B. The correctly phased spectrum is shown in Panel C. A zero-order phase shift of -180° and a first-order phase shift of +1800° has been applied to give the phase-corrected spectrum shown in C.

3.5.3 Delayed Acquisition

In addition to intensity errors in the first points, it is also possible to have errors in the timing of the start of data acquisition. Errors of this type will occur on instruments that use digital filtering of the FID in addition to analog filters. In this case there is no distortion of the signal, only a shift in time. As discussed in Section 3.6, a delay in the acquisition of the time domain signal will lead to a frequency dependent, or first-order, phase shift applied to the spectrum. Figure 3.12 illustrates the effect of a time shift on the spectrum. Note that the baseline distortions look very similar to those generated by distortion of the initial points of the FID, however the distortions are easily removed by phasing of the spectrum. The amount of first-order phase correction that should be applied to the spectrum is:

$$\phi_1 = 360^\circ \times N \tag{3.18}$$

where N is the number of points that the FID has been shifted. If the acquisition of the FID has been delayed by one-half of a dwell time, then a 180° first-order phase shift should be applied.

3.6 Phasing of the Spectrum

The phase of the detected signal can be pictured in two equivalent ways, a geometrical model and one that considers the actual phases of the detected signal. The two different descriptions of the receiver phase are illustrated in Fig. 3.13. The geometrical model will be used first to discuss the effect of the signal phase on the observed spectrum and how these effects are removed by phasing.

The geometrical model defines the phase of the detected signal by the position of the bulk magnetization in the transverse plane at the initiation of signal detection. In this case, the coordinate frame is *defined* by the real and imaginary axis of the receiver. If the signal is aligned along the real axis of the receiver then it has a phase shift of zero and a pure absorption spectrum will be observed. Conversely, if the signal is aligned along the imaginary axis of the receiver, then it has a phase shift of 90° and a dispersion curve will be observed. Signals that lie elsewhere will have a mixed lineshape. The effect of the signal phase on its appearance is illustrated in Fig. 3.14.

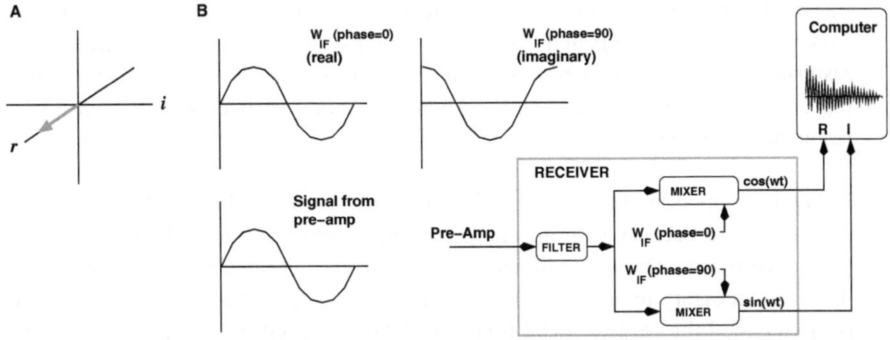

Figure 3.13. Geometric and electronic description of the signal phase. Panel A illustrates the geometric description of signal phases. Here, the real axis of the receiver has been arbitrarily defined to be along the x-axis. The bulk magnetization, illustrated with the gray arrow, is aligned along the x-axis. This signal will have a phase shift of 0° and will give a pure absorption mode lineshape. In B, the actual signals that enter the receiver are shown. The two top signals represent the two separate intermediate frequencies (W_{IF}) that are mixed with the signal (FID) from the pre-amp, as indicated in the spectrometer diagram on the right. If the phase of the pre-amp signal is the same as the phase of W_{IF} (phase=0), then the phase shift of the signal is defined to be zero.

The geometrical model is very useful for visualizing the detection process when the magnetization is represented by a vector precessing in the transverse plane. However, the notion that the receiver phase actually defines the orientation of a coordinate frame, although convenient and of practical use, is incorrect.

The phase of the signal is correctly described as the phase difference between the signal that enters the receiver and the intermediate frequency (W_{IF}). When the signal from the pre-amp enters the receiver it is split into two signals. One of these is mixed with W_{IF} that has not been phase shifted, while the other portion is mixed with W_{IF} that has been shifted by 90° (see Fig. 3.13). If the detected signal is in-phase with the non-phase shifted W_{IF} then its phase shift is zero. Alternatively, if the detected signal is 90° out of phase with the non-shifted W_{IF} and in-phase with the shifted W_{IF}, then its phase is 90°.

3.6.1 Origin of Phase Shifts

There are several factors that produce a phase shift of the signal:

1. Electronic effects: As the signal progresses from the probe to the digitizer it is amplified and mixed with other signals. These steps introduce a shift in the phase of the signal that will depend on, among other things, the receiver gain and the frequency of the signal. To first-order, these shifts are linear with frequency.

2. Off-resonance effects: Consider the effect of a 90° pulse along the minus x-axis for on- and off-resonance spins. The pulse will rotate the on-resonance spins to the y-axis. If we assume the y-axis of the receiver is the real axis, the phase of on-resonance spins will be zero. Spins that resonate at a lower frequency than the

transmitter will precess about a larger effective field, and will end up behind the y-axis after the pulse, equivalent to a negative phase shift (see Fig. 2.10). Spins that resonate at a higher frequency will end up in front of the y-axis after the pulse, representing a positive phase shift. Again, the phase shift is proportional to the resonance frequency.

3. Delay in acquisition of the signal: It is impossible to collect the signal immediately after the pulse because of transient signals that remain in the probe after a high-power RF-pulse. These transients are several times as intense as the real signal, thus it is necessary to wait 10-20 μsec before acquisition of the signal. This delay is often called the receiver dead time. A delay of τ, will lead to a phase shift of $\omega_s\tau$ in the detected signal, as illustrated in Fig. 3.15.

The above three factors combine together to produce a phase shift in the detected magnetization that can be characterized as having a frequency independent term (zero-order) and a frequency dependent term (first-order). It is assumed that the frequency dependence is linear, thus:

$$\phi(\omega) = \phi_o + k\omega \tag{3.19}$$

Since the phase delay of the time domain signal of a resonance line depends on the location of the line in the spectrum, the actual lineshape, absorbance or dispersive, of the resonance will depend on its position in the spectrum, as illustrated in Fig. 3.15.

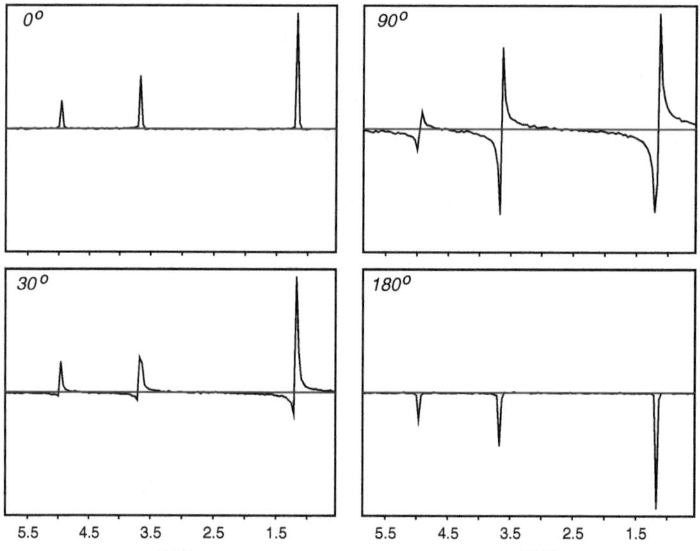

Figure 3.14. Effect of signal phase on the final spectrum. The four panels show the effect of a 0° (pure absorption), a 30°, a 90° (pure dispersion lineshape), and a 180° phase shift on the observed spectrum. Note that the spectrum obtained with a phase shift of 180° is also pure absorption, but inverted.

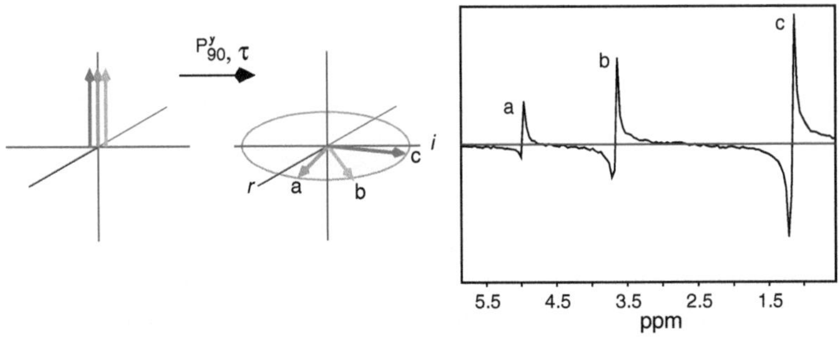

Figure 3.15. *Origin of first-order phase shifts.* This figure illustrates the origin of frequency dependent phase shift that arise from the receiver dead time. Spins "a", "b", and "c", represent three different resonance frequencies with $\omega_c > \omega_b > \omega_a$. Immediately after the pulse, all three spins are placed on the x, or real, axis. Sampling of the signal occurs after a delay τ. At this time resonance "a" shows the smallest phase shift, while "b" and "c" show increasing larger phase shifts since their precessional frequencies are correspondingly larger. The effect of a phase shift on the lineshape is to produce a mixed lineshape that is the combination of absorption and dispersion lineshapes. The right side of the figure shows the resultant spectrum. Note that the lineshape of "a" is almost pure absorption while that of "c" is almost a pure dispersion curve.

3.6.2 Applying Phase Corrections

The goal is to apply a frequency dependent phase shift to the spectrum such that the phases of all of the peaks are returned to 0°, yielding a pure absorption lineshape for all resonances in the spectrum. Phase corrections are performed with the software by forming the appropriate linear combination of the real and imaginary signals.

The signal in the time domain with the phase shift incorporated as $e^{i\phi(\omega)}$ is:

$$S(t) = e^{i\omega t} e^{-t/T_2} e^{i\phi(\omega)} \tag{3.20}$$

The observed real, h_r, and imaginary, h_i, components of the spectrum are:

$$\begin{aligned} h_r(\omega) &= \cos(\phi) g_r - \sin(\phi) g_i \\ h_i(\omega) &= \sin(\phi) g_r + \cos(\phi) g_i \end{aligned} \tag{3.21}$$

where g_r and g_i represent absorption and dispersion lineshapes, respectively, and ϕ represents the phase shift.

To obtain pure absorptive line shapes for all the peaks in the spectrum it is necessary to apply a phase correction that removes both ϕ_o (zero-order phase correction), as well as the frequency dependent term, $k\omega$ (first-order phase correction). These corrections are specified in degrees to the processing software. The software, in turn, computes the phase corrected spectrum by adding $\cos(\phi) h_r(\omega)$ to $\sin(\phi) h_i(\omega)$ to give :

$$\begin{aligned} &cos^2(\phi) g_r - \cos(\phi)\sin(\phi) g_i + sin^2(\phi) g_r + \cos(\phi)\sin(\phi) g_i \\ &= g_r \end{aligned} \tag{3.22}$$

Note that it is necessary to have both the real and imaginary components of the spectrum in order to phase the spectrum.

3.7 Chemical Shift Referencing

Once the spectrum has been phased it is necessary to define the chemical shift scale or "reference the spectrum". The chemical shift is given by the following formula:

$$\delta = \frac{\nu - \nu_o}{\nu_o} \times 10^6 \tag{3.23}$$

where ν_o is the resonance frequency of the reference compound.

Compounds that are used to reference proton, carbon and nitrogen spectra are listed in Table 3.2. The chemical structures of some of these compounds are shown in Fig. 3.16. The most commonly used referencing compounds, TMS, DSS, and DSA, have methyl groups attached to a silicon atom. Since silicon has a smaller electronegativity than carbon (1.9 versus 2.5), the electron density on the methyl groups is increased in these compounds. Consequently the proton and carbon spins are highly shielded and they resonate at a lower frequency than most other resonances in the spectrum.

TMS is insoluble in aqueous solution while the other compounds listed in Table 3.2 have limited solubility in organic solvents. Consequently TMS is generally used to reference spectra taken in organic solvents while any of the other compounds can be used for aqueous solutions. Of the aqueous reference compounds, DSS is the compound of choice because of its negligible sensitivity to pH and temperature changes (see [166] for more details). DSA was developed as a replacement for DSS in samples containing components that interact with DSS, such as cationic peptides [119]. Either DSS or DSA are excellent compounds to use for both proton and carbon chemical shift referencing in biomolecular NMR. In contrast, the chemical shift of one of the most frequently used proton reference compounds, H_2O or HDO, has a high sensitivity to both pH and temperature. This sensitivity must be taken into account when using HDO as a reference compound.

Compounds used for referencing can either be employed as internal or external standards. In the case of internal standards a small amount of the compound is dissolved directly in the NMR sample. In cases where the reference compound is insoluble in the solvent, or reacts with components in the sample, the reference compound can be used externally by inserting a sealed capillary containing the reference compound coaxially into the sample. Internal standards are generally preferred because the capillary required for external referencing can alter the static magnetic field, leading to small changes in the observed shifts.

Referencing ^{15}N spectra with liquid ammonia presents some challenges because it is difficult to prepare sealed capillaries with liquid ammonia for external referenc-

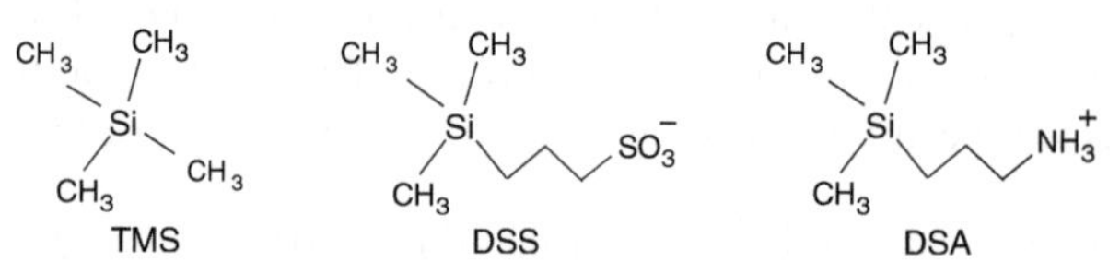

Figure 3.16. Chemical shift reference compounds. Compounds for the referencing of chemical shifts in organic (tetramethylsilane,TMS) and aqueous (4,4-dimethyl-4-silapentane-1-sulfonate, DSS; 4,4-dimethyl-4-silapentane-1-ammonium, DSA) solvents are shown.

Table 3.2. Compounds for chemical shift referencing. The chemical shift of each compound relative to DSS is indicated, as is the effect of temperature and pH on the resonance line position. This information was obtained from [166].

Compound	Chemical Shift	Effect of pH	Effect of Temperature
^{1}H Reference Compounds			
TMS	0.000	none	none
DSS/DSA	0.000	none	none
Acetone	2.218	none	none
HDO (25 °C)	4.766	-2 ppb/pH	-11.9 ppb/°C
^{13}C Reference Compounds			
TMS	1.70	none	-4 ppb/°C
DSS/DSA	0.00	none	none
NaAcetate	26.10	none	none
Acetone	33.00	none	none
^{15}N Reference Compound			
NH_3 (liq., 25 °C)	0.0	n.a.	40 ppb/°C

ing. Consequently, ^{15}N spectra are often referenced indirectly, based on the measured proton shift of DSS. This approach is based on the fact that the *ratio* of the proton frequency at zero ppm to that of the nitrogen frequency at zero ppm should be independent of the spectrometer field strength. This ratio has been measured at several field strengths and is approximately 0.101329118 ([166]).

3.8 Exercises

1. Prove that sinc ripples from a resonance line in the center of the spectrum ($\nu = 0$) do not appear in the spectrum if the FID is not extended by zero-filling.

2. Assume that the phase of the intermediate frequency (W_{IF}) is shifted by 30°. How would this change affect the phase correction that would be applied to the spectrum?

3. A proton spectrum of a sample of DSS was acquired with a transmitter frequency of 500 MHz. After setting the methyl proton line from DSS to 0 ppm the chemical shift at the center of the spectrum was 5 ppm. What is the absolute frequency of the DSS methyl line and what is the absolute nitrogen frequency that corresponds to zero ppm on the nitrogen frequency scale.

3.9 Solutions

1. The observed spectrum is the convolution of the Fourier transform of the complete FID and the Fourier transform of a square wave of length, τ_{acq}, the acquisition time. The Fourier transform of the square wave is a sinc function (see Appendix A). We will show that the null, or zero, values of the sinc function, with the exception of $\nu = 0$, fall exactly on the sample data points in the spectrum, therefore the sinc function becomes equivalent to a delta function, $\delta(x)$, at $\nu = 0$. The convolu-

tion of a delta function with the spectra line does not distort the appearance of the line.

The spacing between the points in the spectra is equal to $1/\tau_{dw}$, where τ_{dw}, is the dwell time, or the spacing between the points in the FID.

The transform of a square wave of length $2a$ is a sinc function:

$$F(\omega) = \frac{\sqrt{2}sin(\omega a)}{\omega\sqrt{\pi}} \quad (3.24)$$

This function has null points when $\omega a = k\pi; k \in I, k \neq 0$. Replacing ω with the frequency in hertz, gives the following for the frequencies of the null points:

$$\nu = k\frac{1}{2a} \quad (3.25)$$

The length of the original square wave, $2a$, is just the total acquisition time, which is equal to $N \times \tau_{dw}$. Therefore nulls occur at:

$$\nu = k\frac{1}{N\tau_{dw}} \quad (3.26)$$

since k can be any integer, except zero, it can be set to a multiple of N. The null points will be:

$$\nu = \frac{1}{\tau_{dw}}, \frac{2}{\tau_{dw}}, \frac{3}{\tau_{dw}} \quad (3.27)$$

These frequencies are identical to the spacing of the points in the final spectrum.

2. This would require the application of a 30° shift to all of the lines in the spectrum, corresponding to a zero-order phase shift of 30°. The first order phase correction would be unchanged.
3. The frequency of the DSS line is 5 ppm lower than the transmitter frequency. At this field strength, this ppm change corresponds to 2500 Hz. Therefore the frequency of the DSS line is 500,000,000 - 2500 = 499,997,500 Hz. The zero point of the nitrogen spectrum is just this number multiplied by 0.101329118, or 50,664,306 Hz.

Chapter 4

QUANTUM MECHANICAL DESCRIPTION OF NMR

We have seen in the first few chapters that a classical description of NMR provides useful insight into the behavior of the spins. However, a classical description is completely inadequate in describing multidimensional NMR. Quantum mechanical approaches are required to describe these more complex experiments. This chapter provides an introduction to quantum mechanics and describes its application to isolated nuclear spins. Subsequent chapters explore the interaction between nuclear spins. The overall goal of these chapters is to provided the reader with a simple set of rules that describe the evolution and coupling of spins. These rules can be used to readily interpret complex multi-dimensional NMR experiments.

4.1 Schrödinger Equation

The basic tenet of quantum mechanics is that the properties, or states, of a system can be described in terms of a wave function. The wave function is an expression of the probability of finding a single system in a particular state. The wave functions, Ψ, are defined by the total energy of the system from the well known Schrödinger equation:

$$i\hbar\frac{d\Psi}{dt} = -\frac{\hbar^2}{2m}\frac{d^2\Psi}{dx^2} + V_o\Psi, \tag{4.1}$$

$$i\hbar\frac{d\Psi}{dt} = \mathcal{H}\Psi, \tag{4.2}$$

where $\mathcal{H}$ represents the total energy of the system. For a classical system, $\mathcal{H}$ corresponds to the sum of the kinetic energy ($p^2/2m$) and the potential energy ($V(x)$). The above equations suggest that the wavefunctions are functions of Cartesian space, e.g. $\Psi(x, y, z)$. Although this is appropriate for many different systems, such as those which describe the electron orbitals around atoms or a particle in a potential well, this is an unnecessary restriction. In the case of NMR, the wavefunctions that describe the system are not functions of Cartesian space. Consequently, it will be necessary to develop an alternative formalism to describe them.

In the above equation, $\mathcal{H}$, or the Hamiltonian, can also be described as an *operator*. Operators are mathematical representations of observables, such as energy, that are applied to wavefunctions. The Hamiltonian is a very important operator. Not only can it provide the energy associated with a state of the system, but it also determines how the system evolves in time (see Eq. 4.2).

The application of an operator onto a wavefunction returns the values of the observable and another wavefunction. If the *same* wavefunction is returned, then these wavefunctions are given a special name, eigenfunctions[1], and the observable is called an eigenvalue. For example, the application of the Hamiltonian to one of its eigenfunctions will return the same eigenfunction multiplied by the energy of that state:

$$\mathcal{H}\Psi = E_\Psi \Psi \tag{4.3}$$

Here, Ψ is an eigenfunction of $\mathcal{H}$, and is associated with an eigenvalue of E_Ψ. Wavefunctions that are eigenfunctions of the total energy are referred to as *stationary* states because only their phase changes during time evolution:

$$\begin{aligned} i\hbar\frac{d\Psi}{dt} &= E_\Psi \Psi \\ d\Psi &= \frac{1}{i\hbar} E_\Psi \Psi dt \\ \Psi(t) &= \Psi(0)e^{-i\frac{E_\Psi}{\hbar}t} \end{aligned} \tag{4.4}$$

The quantum mechanical description of NMR will utilize wavefunctions to represent the current state of the magnetization at any point in an experiment. The Hamiltonian that is present at that time will describe how the magnetization (wavefunction) will change with time. Therefore our goal is to describe the wavefunction associated with the initial state of the NMR experiment, and then use the Hamiltonian to determine the evolution of the magnetization as the experiment progresses, up to and including, the detection of the final signal.

4.1.1 Vector Spaces and Properties of Wavefunctions

The wavefunction, Ψ, that describes the properties of the system can be described as a sum of orthonormal basis functions, u_i, that form a vector space. These basis vectors are usually the eigenvectors of some operator and are capable of describing any arbitrary wavefunction. The dimensionality of the system, n, is defined by the number of basis vectors that are required to span the vector space, or equivalently, to define any arbitrary state of the system.

Basis vectors represent wavefunctions in much the same way that the Cartesian basis vectors , $\hat{\imath}, \hat{\jmath}, \hat{k}$ can be used to describe any vector in normal three-dimensional physical space, for example the vector $\vec{V}$:

$$\vec{V} = a\hat{\imath} + b\hat{\jmath} + c\hat{k}$$

[1] *Eigen* is from German. Loosely translated it means 'of one's own'.

Correspondingly, any arbitrary state of the system can be described as:

$$\Psi = \sum_{i=1}^{n} c_i u_i \tag{4.5}$$

where u_i are the basis vectors.

It is important to remember that there is no unique set of basis functions, in just the same way that there is no unique Cartesian coordinate system. However, the basis vectors that are most commonly used are the eigenvectors of the energy operator, i.e. $\mathcal{H}u_k = E_k u_k$. These basis vectors are stationary states and thus a convenient set of basis states to use.

The values of c_i can be found using the fact that the basis vectors are orthonormal, in much the same way $\hat{\imath}$ is orthonormal to $\hat{\jmath}$, e.g. $\hat{\imath}\cdot\hat{\imath} = 1,\ \ \hat{\imath}\cdot\hat{\jmath} = 0$. In the case of basis functions, the dot product is replaced by integration over all space. Since these basis functions can be complex, the complex conjugate of one of the integrands is taken, such that the final answer is real. The orthonormality is expressed as :[2]

$$\delta_{mn} = \int_{-\infty}^{+\infty} u_m^* u_n d\chi \tag{4.6}$$

This expression is often referred to as the *scalar product* because of the analogy to the dot product between two vectors, which generates a scalar. The term $d\chi$ represents the fact that the integral extends over the space that is associated with the wavefunction. For example, if Ψ is a function in Cartesian space, $\Psi(x, y, z)$, then $d\chi = dxdydz$.

The c_m^{th} coefficient is found as follows.

$$\begin{aligned} c_m &= \int u_m^* \Psi d\chi \\ &= \int u_m^* \sum c_n u_n d\chi \\ &= \sum c_n \int u_m^* u_n d\chi \\ &= \sum c_n \delta_{mn} \\ &= c_m \end{aligned} \tag{4.7}$$

Thus, c_m is the projection of the state of the system onto the mth basis vector. This is analogous to finding the value of one of the components of a vector by forming the dot product between a vector and the basis vectors in normal 3D space (e.g. $\vec{V}\cdot\hat{\imath} = a$).

The probability, P_m, that any given system can be found in any particular basis state m is given by:

$$P_m = c_m^* c_m \tag{4.8}$$

If a number of individual systems are sampled, each state will be found in one, and only one, of the basis states.

[2] δ_{mn} is the Dirac delta function, it is zero if $m \neq n$ and 1 if $m = n$.

Equation 4.8 implies that the scalar product of a wavefunction with itself must equal unity, since the probabilities of all possible states must sum to 1:

$$\begin{aligned}\int_{-\infty}^{+\infty} \Psi^*\Psi dx &= \int_{-\infty}^{+\infty} \sum_m c_m^* u_m^* \sum_n c_n u_n d\chi \\ &= \sum_{mn} c_m^* c_n \int_{-\infty}^{+\infty} u_m^* u_n d\chi \\ &= \sum_{mn} c_m^* c_n \delta_{mn} \\ &= \sum_m c_m^* c_m \\ &= 1 \end{aligned} \tag{4.9}$$

Therefore, $\Psi^*\Psi$ is equivalent to the probability density of the particle. Integration of this function over a range gives the probability of finding a particle within that range. Integration over all space gives a value of 1, since the particle must be found somewhere.

4.1.2 Particle in a Box

To illustrate some of the above points, consider the simple system of a particle in a one dimensional box, of width $2a$. The wavefunctions that are eigenfunctions of the Hamiltonian satisfy the boundary conditions $u_n(a) = u_n(-a) = 0$, and are:

$$u_n(x) = \frac{1}{\sqrt{a}} sin \frac{n\pi x}{2a};\ n = 2, 4, .. \qquad u_n(x) = \frac{1}{\sqrt{a}} cos \frac{n\pi x}{2a};\ n = 1, 3, .. \tag{4.10}$$

The energy associated with each wavefunction is:

$$E_n = \frac{\hbar^2 \pi^2}{8ma^2} n^2 \tag{4.11}$$

These wavefunctions form a orthonormal basis set since $\int u_n^* u_m dx = \delta_{nm}$ and any arbitrary state of the system can be represented by a linear combination of these basis

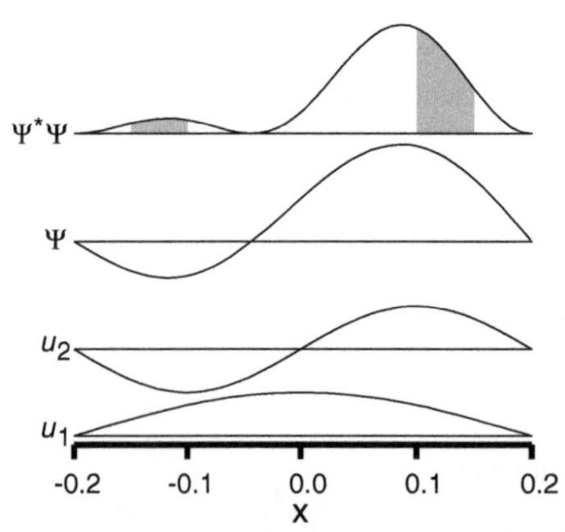

Figure 4.1 Wavefunctions associated with a particle in a box. The two lowest energy eigenfunctions, u_1 and u_2, of a particle in a box are shown. In addition, a mixed state, Ψ, that is a linear combination of these two eigenfunctions is also shown. The scalar product, $\Psi^*\Psi$ is shown in the top graph. The integrals discussed in the text are shaded.

vectors. Fig. 4.1 shows the first two basis functions, as well as a wavefunction that is a linear combination of these two:

$$\Psi = \frac{1}{\sqrt{5}}\frac{1}{\sqrt{a}}cos\frac{n\pi x}{2a} + \frac{2}{\sqrt{5}}\frac{1}{\sqrt{a}}sin\frac{n\pi x}{2a} \tag{4.12}$$

(The factors $1/\sqrt{5}$ and $2/\sqrt{5}$ insure that $\int \Psi^*\Psi dx = 1$).

The probability of finding the particle in interval between x and $x + \Delta x$ is:

$$P_{x \rightarrow x+\Delta x} = \int_x^{x+\Delta a} \Psi^*\Psi dx \tag{4.13}$$

The shaded areas in Fig. 4.1 show this integral for the ranges of -0.15 to -0.10 and from 0.10 to 0.15. The probability of finding the particle in the interval on the right side is approximately 5 times the probability of finding the particle in the left interval.

4.2 Expectation Values

In addition to the Hamiltonian, operators exist that correspond to other observables, such as the position of a particle (x), its linear momentum ($p = -i\hbar d/dx$), its total spin angular momentum (I), and the component of its angular momentum along an axis (e.g. I_x). In using quantum mechanics to describe NMR experiments we will be interested in two types of operators. The first is the Hamiltonian, because it drives the evolution of the system. The second type are the operators that describe spin-angular momentum because they will correspond to radio-frequency pulses. The spin angular momentum operators will also be used to determine the various states of the magnetization in the system during an NMR experiment.

The average, or measured, value of any observable can be extracted from the wavefunction of the system by calculating the *expectation value*. For example, the expectation, or observed, value of the energy of a system is given by:

$$< E >= \int \Psi^*\mathcal{H}\Psi d\chi \tag{4.14}$$

and that for the x-component of the spin angular momentum:

$$< I_x >= \int \Psi^* I_x \Psi d\chi \tag{4.15}$$

If the system is in a pure state, $\Psi = u_k$, such that its wavefunction is an eigenfunction of the operator whose expectation value is being computed, then the expectation value will be the same as the eigenvalue. For example:

$$\begin{aligned} < E > &= \int \Psi^* \mathcal{H} \Psi d\chi \\ &= \int u_k^* \mathcal{H} u_k d\chi \\ &= \int u_k^* E_k u_k d\chi \\ &= E_k \int u_k^* u_k d\chi \\ &= E_k \end{aligned} \qquad (4.16)$$

If the system is described as a mixture of the basis states, $\Psi = \sum c_k u_k$, then the expectation value is the weighted average, using the energy of the system as an example:

$$\begin{aligned} < E > &= \int \Psi^* \mathcal{H} \Psi d\chi \\ &= \int \sum_k c_k^* u_k^* \mathcal{H} \sum_j c_j u_j d\chi \\ &= \int \sum_k c_k^* u_k^* \sum_j E_j c_j u_j d\chi \\ &= \sum_k c_k^* \sum_j E_j c_j \int u_k^* u_j d\chi \\ &= \sum_k c_k^* \sum_j E_j c_j \delta_{jk} \\ &= \sum_k c_k^* E_k c_k \\ &= \sum_k P_k E_k \end{aligned} \qquad (4.17)$$

4.3 Dirac Notation

The above example of a particle in a box used wavefunctions that were functions of x, the coordinate of the particle. In the case of NMR, the wavefunctions will represent different states of the nuclear spin. Since there are no spatial coordinates associated with these wavefunctions it is necessary to develop a different representation of wavefunctions, their basis vectors, operators, and associated expectation values. This representation was developed by Dirac and is often referred to as the 'Bra-Ket'notation.

4.3.1 Wavefunctions in Dirac Notation

The individual basis functions, u_m, are represented as $|u_m >$. The complex conjugate of the basis functions are written as $< u_m|$. The orthonormality relationship is written as:

$$\delta_{nm} = < u_n | u_m > = \int u_n^* u_m d\chi \tag{4.18}$$

Any arbitrary wave function is represented as:

$$\Psi = \sum_{i=1}^{n} c_i | u_i > \tag{4.19}$$

The individual coefficients, c_m are obtained in the usual way:

$$c_m = \int u_m^* \Psi d\chi = < u_m | \Psi > \tag{4.20}$$

In systems with a finite number of eigenstates it is convenient to represent wavefunctions as vectors. In this representation, the basis vectors will have a single non-zero component, for example, u_2 would be written as:

$$u_2 = \begin{bmatrix} 0 \\ 1 \\ 0 \\ \vdots \\ 0 \end{bmatrix} \tag{4.21}$$

An arbitrary wavefunction is written in the same way, with the coefficients, c_m, forming the elements of the vector. For example, an arbitrary wavefunction in a $n-$dimensional space would be an $n \times 1$ matrix:

$$\Psi = \begin{bmatrix} < u_1 | \Psi > \\ < u_2 | \Psi > \\ < u_3 | \Psi > \\ < u_4 | \Psi > \\ \vdots \\ < u_n | \Psi > \end{bmatrix} = \begin{bmatrix} c_1 \\ c_2 \\ c_3 \\ c_4 \\ \vdots \\ c_n \end{bmatrix} \tag{4.22}$$

The adjoint[3] of a wavefunction is just a row of numbers, or an $1 \times n$ matrix, with the complex conjugate of each element taken:

$$\Psi^\dagger = [< u_1 | \Psi >< u_2 | \Psi >< u_3 | \Psi > \ldots\ldots < u_n | \Psi >] \tag{4.23}$$

$$= [c_1^* \ c_2^* \ c_3^* \ c_4^* \ldots c_n^*] \tag{4.24}$$

[3]The adjoint of a matrix is the complex conjugate of the transpose of the matrix. The transpose of a matrix is obtained by interchanging the row and column of each element, e.g. if the elements of the matrix $\tilde{A} = a_{ij}$ then the transpose of $\tilde{A}$ is $A^T = a_{ji}$. This is described in more detail in Section 4.4.

4.3.2 Scalar Product in Dirac Notation

The formation of the scalar product between two states, $< \Psi|\Psi >$, is just the product of a row matrix and a column matrix. The result of this operation is a single number, which is a scalar. For example, consider the possible scalar products between u_1 and u_1, or u_2 in a two-dimensional vector space:

$$< u_1|u_1 >= \begin{bmatrix} 1 & 0 \end{bmatrix} \begin{bmatrix} 1 \\ 0 \end{bmatrix} = 1 \qquad < u_1|u_2 >= \begin{bmatrix} 1 & 0 \end{bmatrix} \begin{bmatrix} 0 \\ 1 \end{bmatrix} = 0$$

For an arbitrary wavefunction in an n-dimensional vector space:

$$\int \Psi^*\Psi d\chi = [c_1^* \;\; c_2^* \;\; c_3^* \;\; c_4^* \ldots c_n^*] \begin{bmatrix} c_1 \\ c_2 \\ c_3 \\ c_4 \\ \vdots \\ c_n \end{bmatrix} = \sum_{m=1}^{n} c_m^* c_m \tag{4.25}$$

4.3.3 Operators in Dirac Notation

The Bra-Ket representation can be extended to operators. In this case, operators are represented by a matrix, whose elements are defined as:

$$A_{ij} =< u_i|A|u_j > \tag{4.26}$$

For example, the Hamiltonian operator for a particle in a box is:

$$\mathcal{H} = \begin{bmatrix} E_1 & 0 & 0 & . \\ 0 & E_2 & 0 & . \\ 0 & 0 & E_3 & . \\ . & . & . & . \end{bmatrix} \tag{4.27}$$

The individual elements of this matrix were calculated as follows:

$$\begin{aligned} \mathcal{H}_{ij} &= < u_i|\mathcal{H}|u_j > \\ &= < u_i|E_j|u_j > \\ &= E_j < u_i|u_j > \\ &= E_j\delta_{ij} \end{aligned} \tag{4.28}$$

4.3.4 Expectation Values in Dirac Notation

Using the example of the energy of the system, the expectation value of an operator is given as:

$$< E > = \int \Psi^*\mathcal{H}\Psi d\chi \tag{4.29}$$

$$= [c_1^* \; c_2^* \; c_3^* \ldots] \begin{bmatrix} A_{11} & A_{12} & A_{13} & \ldots \\ A_{21} & A_{22} & A_{23} & \ldots \\ A_{31} & A_{32} & A_{33} & \ldots \\ \vdots & \vdots & \vdots & \end{bmatrix} \begin{bmatrix} c_1 \\ c_2 \\ c_3 \\ \vdots \end{bmatrix} \tag{4.30}$$

As an example, the expectation value of the energy of the particle in the box that is represented by $\Psi = \frac{1}{\sqrt{5}}u_1 + \frac{2}{\sqrt{5}}u_2$ is as follows:

$$
\begin{aligned}
<E> &= \begin{bmatrix} \frac{1}{\sqrt{5}} & \frac{2}{\sqrt{5}} & 0\ldots \end{bmatrix} \begin{bmatrix} E_1 & 0 & 0 & \ldots \\ 0 & E_2 & 0 & \ldots \\ 0 & 0 & E_3 & \ldots \\ \vdots & \vdots & \vdots & \end{bmatrix} \begin{bmatrix} \frac{1}{\sqrt{5}} \\ \frac{2}{\sqrt{5}} \\ 0 \\ \vdots \end{bmatrix} \\
&= \begin{bmatrix} \frac{1}{\sqrt{5}} & \frac{2}{\sqrt{5}} & 0\ldots \end{bmatrix} \begin{bmatrix} E_1\frac{1}{\sqrt{5}} \\ E_2\frac{2}{\sqrt{5}} \\ 0 \\ \vdots \end{bmatrix} \\
&= \frac{1}{5}E_1 + \frac{4}{5}E_2 \qquad (4.31)
\end{aligned}
$$

This value is the average energy that would be obtained by measuring the energy of a large number of particles in a box. If the energy of individual particles were measured, only two values of energy would be obtained, E_1, or E_2. On average, one out of five particles would have an energy of E_1 and four out of five would have an energy of E_2.

4.4 Hermitian Operators

Hermitian operators are those operators whose eigenvalues are real. Since the eigenvalues are real, the expectation values are also real. Therefore Hermitian operators correspond to physically observable properties of the system.

An operator is Hermitian if it is equal to its *adjoint*:

$$A = A^\dagger \qquad (4.32)$$

The adjoint of a matrix is obtained by transposing columns and rows and taking the complex conjugate of each element:

$$A^\dagger_{ij} = A^*_{ji} \qquad (4.33)$$

Clearly, the Hamiltonian operator shown in Eq. 4.27 is Hermitian.

4.4.1 Determining Eigenvalues

Until this point we have assumed that the basis vectors are eigenvectors of the Hamiltonian. As such, $\mathcal{H}u_k = E_k u_k$. This assumption implies that the matrix form of $\mathcal{H}$ is diagonal, as presented in Eq. 4.27. This form of the Hamiltonian is convenient in that the energies of the different basis states can be read from the diagonal elements.

This simple diagonal form of the Hamiltonian exists only in one particular coordinate frame. In this frame the eigenfunctions of the Hamiltonian also have a simple form, e.g. $u_3^\dagger = [0\ 0\ 1\ 0\ldots]$ and the eigenvalues form the diagonal elements of the operator. For example, consider the following example of a Hamiltonian and associated wave functions for a spin 1/2 system:

$$\mathcal{H} = \begin{bmatrix} 1 & 0 \\ 0 & -1 \end{bmatrix}, u_1 = \begin{bmatrix} 1 \\ 0 \end{bmatrix}, u_2 = \begin{bmatrix} 0 \\ 1 \end{bmatrix} \tag{4.34}$$

The eigenvalues (λ_n) are immediately seen to be: $\lambda_1 = 1, \lambda_2 = -1$.

Often it is necessary to work in a different reference frame to describe the state of the system. In this case, the basis vectors will no longer have a simple form and the matrix representation of the Hamiltonian will be non-diagonal. For example, if the coordinate frame is rotated by 45°, the above Hamiltonian appears as follows:

$$\mathcal{H}' = \begin{bmatrix} 0 & -1 \\ -1 & 0 \end{bmatrix} \tag{4.35}$$

where the prime (') indicates the form of the operator in the rotated coordinate system.

This particular reference frame is inconvenient in the sense that it is no longer possible to obtain the eigenvalues directly from the diagonal of the operator. In such cases the eigenvalues of any operator, A, can be obtained from its eigenvalue equation:

$$A|u> = \lambda|u> \tag{4.36}$$

This equation can be solved to give both eigenvectors and eigenvalues of the operator. We assume that the eigenvectors, u, are linear combinations of some other basis vectors:

$$u = \sum_j c_j v_j \tag{4.37}$$

In much the same way as a rotated coordinate system can be expressed as a linear combination of the original, unrotated basis vectors.

The eigenvalue equation becomes:

$$\begin{aligned} A|u> &= \lambda|u> \\ A|\sum_j c_j v_j> &= \lambda|\sum_j c_j v_j> \end{aligned} \tag{4.38}$$

multiplying through by $< v_i|$,

$$\begin{aligned} < v_i|A|\sum_j c_j v_j> &= < v_i\lambda|\sum_j c_j v_j> \\ \sum_j < v_i|A|v_j > c_j &= \sum_j c_j\lambda < v_i|v_j> \\ \sum_j < v_i|A|v_j > c_j &= c_i\lambda \\ \sum_j A_{ij}c_j &= c_i\lambda \\ \sum_j [A_{ij} - \lambda\delta_{ij}]c_j &= 0 \end{aligned} \tag{4.39}$$

This equation has a non-trivial solution if the determinant is zero:

$$Det\,[A - \lambda I] = 0 \tag{4.40}$$

The roots of Eq. 4.40 are the eigenvalues of the operator, A. The eigenvectors for each eigenvalue are found by substituting the eigenvalue back into the eigenvalue equation: $A|u>= \lambda|u>$.

Returning to our previous example:

$$\mathcal{H}' = \begin{bmatrix} 0 & -1 \\ -1 & 0 \end{bmatrix} \tag{4.41}$$

To find the eigenvalues of the Hamiltonian, it is necessary to form the $A - \lambda I$ matrix:

$$\begin{bmatrix} 0-\lambda & -1 \\ -1 & 0-\lambda \end{bmatrix} = \begin{bmatrix} -\lambda & 1 \\ 1 & -\lambda \end{bmatrix} \tag{4.42}$$

The determinant of this matrix is:

$$-\lambda(-\lambda) - (-1)(-1) \tag{4.43}$$

Setting this expression to zero gives:

$$\lambda^2 - 1 = 0 \tag{4.44}$$

The two eigenvalues are therefore +1 and -1. To find the eigenvectors of $\mathcal{H}'$ in this particular reference frame we assume an arbitrary vector to represent the eigenvectors, $u' = \begin{bmatrix} a \\ b \end{bmatrix}$:

$$\mathcal{H}'\Psi = \lambda u' \tag{4.45}$$

$$\begin{bmatrix} 0 & -1 \\ -1 & 0 \end{bmatrix} \begin{bmatrix} a \\ b \end{bmatrix} = \lambda \begin{bmatrix} a \\ b \end{bmatrix} \tag{4.46}$$

$$\begin{bmatrix} -b \\ -a \end{bmatrix} = \begin{bmatrix} \lambda a \\ \lambda b \end{bmatrix} \tag{4.47}$$

Taking the first eigenvalue, $\lambda = 1$, implies that if $a = +1$, then $b = -1$. Since the eigenvector must be normalized: $u_1' = \frac{1}{\sqrt{2}} \begin{bmatrix} 1 \\ -1 \end{bmatrix}$. Similarly, using $\lambda = -1$ we find that if $a = 1$, then $b = 1$, giving after normalization: $u_2' = \frac{1}{\sqrt{2}} \begin{bmatrix} 1 \\ 1 \end{bmatrix}$.

The complete representation of the operator and associated eigenvalues and vectors in the rotated coordinate system is therefore:

$$\mathcal{H}' = \begin{bmatrix} 0 & -1 \\ -1 & 0 \end{bmatrix} \tag{4.48}$$

$$\lambda_1 = 1: \quad u_1' = \frac{1}{\sqrt{2}} \begin{bmatrix} 1 \\ -1 \end{bmatrix} \tag{4.49}$$

$$\lambda_2 = -1: \quad u_2' = \frac{1}{\sqrt{2}} \begin{bmatrix} 1 \\ 1 \end{bmatrix} \tag{4.50}$$

4.5 Additional Properties of Operators

The following are a series of properties and concepts that will be important for applying quantum mechanics to NMR.

4.5.1 Commuting Observables

The commutator of two operators is defined as:

$$[A, B] = AB - BA \tag{4.51}$$

When two observables (A and B) commute, they they must share the same eigenvectors, but clearly not the same eigenvalues.

4.5.2 Time Evolution of Observables

The time evolution of an observable, $< A >=< \Psi|A|\Psi >$, is obtained in a straight-forward fashion:

$$\frac{d < A >}{dt} =< \frac{d\Psi}{dt}|A|\Psi > + < \Psi|\frac{dA}{dt}|\Psi > + < \Psi|A|\frac{d\Psi}{dt} > \tag{4.52}$$

using:

$$\frac{d\Psi}{dt} = \frac{1}{i\hbar}\mathcal{H}\Psi$$

$$\begin{aligned}
\frac{d < A >}{dt} &= < \frac{1}{i\hbar}\mathcal{H}\Psi|A|\Psi > + < \Psi|\frac{dA}{dt}|\Psi > + \frac{1}{i\hbar} < \Psi|A|\mathcal{H}\Psi > \\
&= \frac{-1}{i\hbar} < \Psi|\mathcal{H}A|\Psi > + < \Psi|\frac{dA}{dt}|\Psi > + \frac{1}{i\hbar} < \Psi|A\mathcal{H}|\Psi > \\
&= \frac{1}{i\hbar} < \Psi|A\mathcal{H} - \mathcal{H}A|\Psi > + < \Psi|\frac{dA}{dt}|\Psi > \\
&= \frac{1}{i\hbar} < [A, \mathcal{H}] > + \frac{dA}{dt}
\end{aligned} \tag{4.53}$$

Therefore, if A commutes with the Hamiltonian, and has no intrinsic time dependence ($dA/dt = 0$), its value will be constant with time. In contrast, if A does not commute with the Hamiltonian, its observable will evolve in time.

In the case of NMR, we shall see that the z-component of the spin-angular momentum commutes with the Hamiltonian that is defined by the external magnetic field, B_o. Therefore, the z-component of the magnetization will not change in time. In contrast, the operators that describe the transverse (x, y) components of the spin-angular momentum do not commute with this Hamiltonian and therefore the x- or y-component of the magnetization will evolve in time, as predicted from the Bloch equations.

4.5.3 Trace of an Operator

The trace of an operator is the sum of its diagonal elements and is defined as:

$$Trace[A] = \sum_{1}^{n} < u_i|A|u_i > \tag{4.54}$$

The trace of a product of operators is invariant when a cyclic permutation is performed:

$$Trace[AB] = Trace[BA] \tag{4.55}$$

$$Trace[ABC] = Trace[BCA] = Trace[CAB] \tag{4.56}$$

4.5.4 Exponential Operator

The operator, e^A, is defined as:

$$e^A = \sum_{n=0}^{\infty} \frac{A^n}{n!} \tag{4.57}$$

If A is Hermitian, then e^A is also Hermitian. Furthermore, if the eigenvalue of A is λ, then the eigenvalue of e^A is e^λ. If two operators, A and B, commute then the following is true:

$$e^A e^B = e^B e^A = e^{A+B} \tag{4.58}$$

4.5.5 Unitary Operators

An operator is unitary if its inverse and its adjoint are equal:

$$U^\dagger U = UU^\dagger = UU^{-1} = 1 \tag{4.59}$$

Unitary operators do not affect the lengths of vectors, just their direction. Therefore unitary operators perform rotations on wavefunctions and other operators. Consider the application of a Unitary operator on two wavefunctions:

$$|\Psi_1> = U|\Phi_1>$$

$$|\Psi_2> = U|\Phi_2>$$

The scalar product of these two functions is invariant to the unitary transformation:

$$<\Psi_1|\Psi_2> = <\Phi_1|U^\dagger U|\Phi_2> = <\Phi_1|\Phi_2> \tag{4.60}$$

4.5.6 Exponential Hermitian Operators

If an operator A is Hermitian, then the operator W, defined as,

$$W = e^{iA} \tag{4.61}$$

is unitary since:

$$\begin{aligned} W &= e^{iA} \\ W^\dagger &= e^{-iA^\dagger} \\ W^\dagger &= e^{-iA} \\ W^\dagger W &= e^{-iA} e^{iA} = 1 \end{aligned} \tag{4.62}$$

4.6 Hamiltonian and Angular Momentum Operators for a Spin-1/2 Particle

The quantization of energy and spin angular momentum was first discovered by Stern and Gerlach in 1922. The original experiment was only sensitive enough to detect the electron spin, which is $\approx$ 2000 fold stronger than nuclear spin angular momentum. Nevertheless, the conclusions derived from electron spins are directly applicable to nuclear spins.

In this experiment, a beam of silver atoms is passed through a inhomogeneous magnetic field. Since the field is inhomogeneous, the atoms will experience a force that depends on the z-component of the spin angular momentum, S_z:

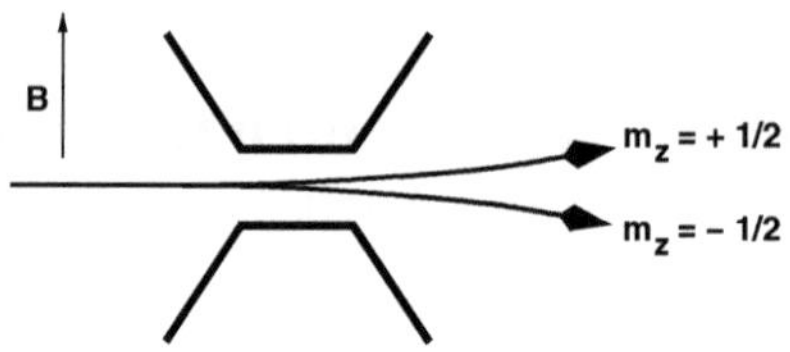

Figure 4.2. The Stern and Gerlach experiment. A beam of unpolarized silver atoms is passed through an inhomogeneous magnetic field. In the presence of the field, the atoms deflection is proportional to the z-component of the spin angular momentum. Only two discrete values are found, showing that spin is quantized.

$$\begin{aligned} F &= \nabla(M_z B_z) \\ &= \nabla(\gamma S_z B_z) \end{aligned}$$

The value of M_z is obtained from the trajectory of the silver atoms. If a beam of randomly oriented silver atoms is passed through this magnet then one might expect to observe a continuous distribution of atoms. However, the actual outcome of this experiment is the production of two populations of silver ions, those that were deflected up and those that were deflected down.

The interpretation of this experiment is that, in the presence of a magnetic field, the z-component of the spin angular momentum of the silver atoms assumes only two values, $+\hbar/2$ and $-\hbar/2$. Therefore, we anticipate that only two basis vectors will be required to describe any arbitrary wavefunction. We can write these wavefunctions in Dirac notation as:

$$u_{+1/2} = \begin{bmatrix} 1 \\ 0 \end{bmatrix} \qquad u_{-1/2} = \begin{bmatrix} 0 \\ 1 \end{bmatrix} \tag{4.63}$$

These are clearly orthonormal:

$$< u_{+1/2}|u_{+1/2} >= \begin{bmatrix} 1 & 0 \end{bmatrix} \begin{bmatrix} 1 \\ 0 \end{bmatrix} = 1 \qquad < u_{+1/2}|u_{-1/2} >= \begin{bmatrix} 1 & 0 \end{bmatrix} \begin{bmatrix} 0 \\ 1 \end{bmatrix} = 0 \tag{4.64}$$

Using these wavefunctions, the matrix form of the operator, S_z, is:

$$\begin{aligned} S_z &= \begin{bmatrix} <u_{+1/2}|S_z|u_{+1/2}> & <u_{+1/2}|S_z|u_{-1/2}> \\ <u_{-1/2}|S_z|u_{+1/2}> & <u_{-1/2}|S_z|u_{-1/2}> \end{bmatrix} \\ &= \begin{bmatrix} <u_{+1/2}|+\hbar/2|u_{+1/2}> & <u_{+1/2}|-\hbar/2|u_{-1/2}> \\ <u_{-1/2}|+\hbar/2|u_{+1/2}> & <u_{-1/2}|-\hbar/2|u_{-1/2}> \end{bmatrix} \\ &= \begin{bmatrix} +\hbar/2<u_{+1/2}|u_{+1/2}> & -\hbar/2<u_{+1/2}|u_{-1/2}> \\ +\hbar/2<u_{-1/2}|u_{+1/2}> & -\hbar/2<u_{-1/2}|u_{-1/2}> \end{bmatrix} \\ &= \begin{bmatrix} +\hbar/2\cdot 1 & -\hbar/2\cdot 0 \\ +\hbar/2\cdot 0 & -\hbar/2\cdot 1 \end{bmatrix} \\ &= +\frac{\hbar}{2}\begin{bmatrix} 1 & 0 \\ 0 & -1 \end{bmatrix} \end{aligned} \tag{4.65}$$

The other two operators for angular momentum are obtained by the use of the raising and lowering operators, which are defined as follows:

$$J_+ = J_x + iJ_y \qquad J_- = J_x - iJ_y \tag{4.66}$$

S_x and S_y can be obtained from a linear combination of the raising and lowering operators:

$$S_x = \frac{1}{2}[J_+ + J_-] \quad S_y = \frac{1}{2i}[J_+ \ - \ J_-] \tag{4.67}$$

The raising operator increases the z-component of the angular momentum by one unit and is defined by the following equation:

$$J_+|j,m>= \hbar\sqrt{j(j+1)-m(m+1)}|j,m+1> \tag{4.68}$$

while the lowering operator lowers the z-component by one unit:

$$J_-|j,m>= \hbar\sqrt{j(j+1)-m(m-1)}|j,m-1> \tag{4.69}$$

In these equations, $|j,m>$ represents a wavefunction with j as the quantum number for the total spin angular momentum and m is the quantum number of the z-component of the spin angular momentum.

For a spin-1/2 system ($j = m = 1/2$) these simplify to:

$$J_+|u_{-1/2}>= \hbar|u_{+1/2}> \tag{4.70}$$

and,

$$J_-|u_{+1/2}>= \hbar|u_{-1/2}> \tag{4.71}$$

The raising operator cannot increase m_z higher than +1/2 so, $J_+|u_{+1/2}>$ gives a null vector, represented by $\emptyset$. Similarly, since the lowest value of m_z is -1/2, $J_-|u_{-1/2}>$ also gives a null vector.

The matrix representation of the raising and lowering operators are obtained in the same fashion as S_z:

$$J_+ = \begin{bmatrix} < u_{+1/2}|J_+|u_{+1/2} > & < u_{+1/2}|J_+|u_{-1/2} > \\ < u_{-1/2}|J_+|u_{+1/2} > & < u_{-1/2}|J_+|u_{-1/2} > \end{bmatrix} \quad (4.72)$$

$$= \hbar \begin{bmatrix} < u_{+1/2}|\emptyset > & < u_{+1/2}|u_{+1/2} > \\ < u_{-1/2}|\emptyset > & < u_{-1/2}|u_{+1/2} > \end{bmatrix} \quad (4.73)$$

$$= \hbar \begin{bmatrix} 0 & 1 \\ 0 & 0 \end{bmatrix} \quad (4.74)$$

$$J_- = \begin{bmatrix} < u_{+1/2}|J_-|u_{+1/2} > & < u_{+1/2}|J_-|u_{-1/2} > \\ < u_{-1/2}|J_-|u_{+1/2} > & < u_{-1/2}|J_-|u_{-1/2} > \end{bmatrix} \quad (4.75)$$

$$= \hbar \begin{bmatrix} < u_{+1/2}|u_{-1/2} > & < u_{+1/2}|\emptyset > \\ < u_{-1/2}|u_{-1/2} > & < u_{-1/2}|\emptyset > \end{bmatrix} \quad (4.76)$$

$$= \hbar \begin{bmatrix} 0 & 0 \\ 1 & 0 \end{bmatrix} \quad (4.77)$$

For S_y, this gives:

$$S_y = \frac{1}{2i}[J_+ - J_-]$$

$$= \frac{\hbar}{2i}\left[\begin{bmatrix} 0 & 1 \\ 0 & 0 \end{bmatrix} - \begin{bmatrix} 0 & 0 \\ 1 & 0 \end{bmatrix}\right]$$

$$= \frac{\hbar}{2}\begin{bmatrix} 0 & -i \\ i & 0 \end{bmatrix} \quad (4.78)$$

S_x is obtained in a similar manner:

$$S_x = \frac{\hbar}{2}\begin{bmatrix} 0 & 1 \\ 1 & 0 \end{bmatrix} \quad (4.79)$$

In summary, the matrix representations for the three Cartesian components of angular momentum are:

$$S_x = \frac{\hbar}{2}\begin{bmatrix} 0 & 1 \\ 1 & 0 \end{bmatrix} \quad S_y = \frac{\hbar}{2}\begin{bmatrix} 0 & -i \\ i & 0 \end{bmatrix} \quad S_z = \frac{\hbar}{2}\begin{bmatrix} 1 & 0 \\ 0 & -1 \end{bmatrix} \quad (4.80)$$

The matrix representation of the Hamiltonian operator is proportional to S_z:

$$\mathcal{H} = -\frac{\hbar\omega_s}{2}\begin{bmatrix} 1 & 0 \\ 0 & -1 \end{bmatrix} = -\omega_s S_z \quad (4.81)$$

Note that S_z and $\mathcal{H}$ are diagonal using these basis vectors. Therefore the basis vectors $u_{+1/2}$ and $u_{-1/2}$ are eigenvectors of both of these operators. The eigenvalues are simply the diagonal elements of the matrix form of the operator, for example:

$$S_z|u_{+1/2} >= \frac{\hbar}{2}\begin{bmatrix} 1 & 0 \\ 0 & -1 \end{bmatrix}\begin{bmatrix} 1 \\ 0 \end{bmatrix} = +\frac{\hbar}{2}\begin{bmatrix} 1 \\ 0 \end{bmatrix} = \frac{\hbar}{2}|u_{+1/2} >$$

Since the operators S_z and $\mathcal{H}$ clearly share the same eigenvectors they must also commute with each other. Consequently the expectation value of S_z is time invariant under the influence of this Hamiltonian. In contrast, these basis vectors are not eigenvectors of the operators for transverse magnetization, S_x and S_y, as can be seen from the following:

$$S_x|u_{+1/2}> = \frac{\hbar}{2}\begin{bmatrix}0 & 1\\ 1 & 0\end{bmatrix}\begin{bmatrix}1\\ 0\end{bmatrix} = +\frac{\hbar}{2}\begin{bmatrix}0\\ 1\end{bmatrix} = \frac{\hbar}{2}|u_{-1/2}>$$

Consequently, the operators for transverse magnetization do not commute with the Hamiltonian operator, thus the expectation value for S_x and S_y will evolve with time.

4.7 Rotations

All NMR RF-pulse sequences (experiments) can be described as a series of rotations (i.e. pulses) applied to the system followed by time evolution of the system under the influence of various Hamiltonians. For example, the simple one pulse experiment involves a rotation of the magnetization by a 90° pulse followed by rotation of the transverse magnetization about the z-axis due to evolution of the system under the Hamiltonian, $\mathcal{H}$. Therefore, it is important to develop operators that describe rotations. In doing so, we are developing a method of representing pulses as well as the free precession of spins in an NMR experiment.

4.7.1 Rotation Groups

A rotation can be characterized by an axis of rotation and an amount of rotation. For example, the operator that describes a rotation of α degrees about the z-axis can be written as:

$$R_z(\alpha) = \alpha\hat{k} \tag{4.82}$$

The group of rotations is said to be closed, that is the application of two rotations produces a third rotation which is part of the same group. Most importantly, this group is not commutative.

$$R_a(\alpha)R_b(\beta) \neq R_b(\beta)R_a(\alpha) \tag{4.83}$$

For example, consider the rotation:

$$R_z(90)R_x(90) \text{ versus } R_x(\mathit{90})R_z(\mathit{90}) \tag{4.84}$$

These two rotations will place a three dimensional object in a different orientation, as shown in Fig. 4.3.

However, two rotations performed along the same axis always commute,

$$R_u(\alpha)R_u(\beta) = R_u(\beta)R_u(\alpha) \tag{4.85}$$

giving a net rotation angle of $\alpha + \beta$.

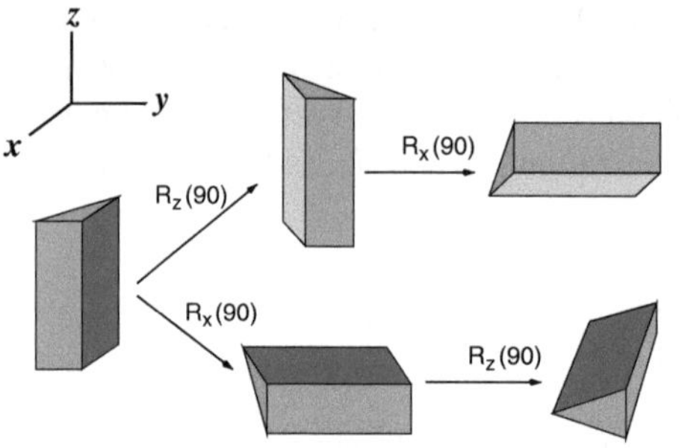

Figure 4.3 Rotations about different axis are not commutative. A wedge shaped object is rotated by 90°, first about the z-axis and then about the x-axis (top) or about the x-axis first and then about the z-axis (bottom). The final orientation after these transformations is different.

4.7.2 Rotation Operators

Rotation operators change the orientation of wavefunctions and operators with respect to a *fixed* coordinate system. This should not be confused with the rotating frame of reference. Here, the wavefunctions rotate, not the coordinate frame.

The effect of rotations on wavefunctions is fairly easy to calculate (see [142]). First, the rotation operator for an infinitesimal rotation will be determined. This operator will then be integrated to give the rotation operator for a finite rotation. In this derivation we will assume that a wavefunction is a function of Cartesian coordinates, $\Psi(x, y, z)$. This restriction is easily removed for non-Cartesian based wavefunctions.

The change in Ψ, due to a rotation of the wavefunction about the z-axis, can be written as a complete differential:

$$\Psi'(x,y) = \Psi(x+dx, y+dy, z) = \Psi(x,y,z) + \frac{\partial\psi}{\partial x}dx + \frac{\partial\Psi}{\partial y}dy \tag{4.86}$$

where Ψ' is the new value of the wavefunction at the coordinates (x, y, z).

The difficult part is determining the relationship between the change in Cartesian coordinates (i.e. dx and dy) to the change in the angular coordinate, $d\alpha$. In the case of a rotation of the wavefunction, Ψ, the changes in dx and dy are opposite in sign with respect to the changes that would occur if the *coordinates* were rotated. This is illustrated in Fig. 4.4. In this example, an arbitrary two dimensional wavefunction has been rotated +10°(counterclockwise). The value of the *rotated* wavefunction at x, y is equal to the value of the unrotated wavefunction at $x = x + yd\alpha$, $y = y - xd\alpha$, therefore:

$$dx = yd\alpha \qquad dy = -xd\alpha \tag{4.87}$$

giving,

$$\begin{aligned}\Psi_{rot} &= \Psi(x,y,z) + y\frac{d\Psi}{dx}d\alpha - x\frac{d\Psi}{dy}d\alpha \\ &= \Psi(x,y,z) - d\alpha[x\frac{d}{dy} - y\frac{d}{dx}]\Psi\end{aligned} \tag{4.88}$$

The expression:

$$\left[x\frac{d}{dy} - y\frac{d}{dx}\right] \tag{4.89}$$

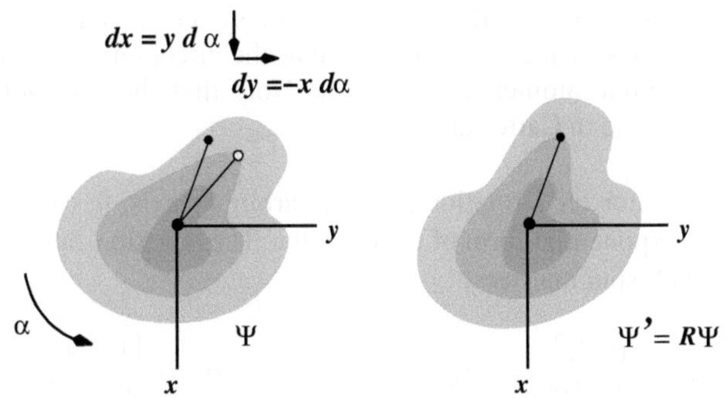

Figure 4.4. Effect of rotations on wavefunctions. An arbitrary two dimensional wavefunction, $\Psi(x, y)$ is shown on the left. The effect of the application of a rotation of $d\alpha = +10°$ on the wavefunction is shown on the right, generating Ψ'. In each diagram the filled circles mark the location of $\Psi(x, y)$ (left) and $\Psi'(x, y)$(right). The value of Ψ that is equal to $\Psi'(x, y)$ is marked by the open circle. The coordinates of this point are $x = x + yd\alpha,\ y = y - xd\alpha$, i.e., $\Psi'(x, y) = \Psi(x + yd\alpha, y - xd\alpha)$.

is proportional to the quantum mechanical description of the angular momentum:

$$L = r \times P_{op} \tag{4.90}$$

where P is the operator for linear momentum.

Specifically, the z-component of the angular momentum is equal to:

$$L_z = \frac{\hbar}{i}\left[x\frac{d}{dy} - y\frac{d}{dx}\right] \tag{4.91}$$

Therefore the infinitesimal rotation operator is given by:

$$R(d\alpha)\Psi = [1 - \frac{i}{\hbar}L_z d\alpha]\Psi \tag{4.92}$$

To find the finite rotational operator, R_z, we can integrate the above equation with respect to $d\alpha$. First noting that the rotation about an axis by the sum of two rotations is equivalent to applying two rotations sequentially:

$$R_z(\alpha + d\alpha) = R_z(\alpha)R_z(d\alpha) \tag{4.93}$$

Expanding the left hand side as the differential:

$$R_z(\alpha) + \frac{dR_z}{d\alpha}d\alpha = R_z(\alpha)\,[1 - \frac{i}{\hbar}L_z d\alpha] \tag{4.94}$$

Gives:

$$\frac{dR_z}{d\alpha} = -\frac{i}{\hbar}L_z R_z \tag{4.95}$$

Which can be integrated to give the rotation operator for finite rotation:

$$R_z(\alpha) = e^{\frac{-i}{\hbar}L_z\alpha} \tag{4.96}$$

The above equation shows that the rotation of the system by an angle α about any axis is given by an exponential operator and that the direction of the rotation axis is defined by the angular momentum operator. Note that the operator for angular momentum need not be in Cartesian form.

The rotation operator for a rotation about an arbitrary axis, $\hat{u}$, can be obtained by expanding out the exponential form of the operator. The equation can be simplified by introducing the Pauli spin matrices:

$$\sigma_x = \begin{bmatrix} 0 & 1 \\ 1 & 0 \end{bmatrix} \quad \sigma_y = \begin{bmatrix} 0 & -i \\ i & 0 \end{bmatrix} \quad \sigma_z = \begin{bmatrix} 1 & 0 \\ 0 & -1 \end{bmatrix} \tag{4.97}$$

These matrices are proportional to the matrix representation of angular momentum, e.g. $S_x = (\hbar/2)\sigma_x$. They are convenient to use because they are equal to one when squared, i.e.:

$$\sigma^2 = \begin{bmatrix} 1 & 0 \\ 0 & 1 \end{bmatrix} \tag{4.98}$$

The rotation operator for a rotation about $\vec{u}$ becomes:

$$R_u(\alpha) = e^{-i\sigma_u \frac{\alpha}{2}} \tag{4.99}$$

This equation can be expanded as a series, giving:

$$R_u(\alpha) = 1 + [-i\sigma_u \frac{\alpha}{2}] + \frac{[-i\sigma_u \frac{\alpha}{2}]^2}{2!} + \frac{[-i\sigma_u \frac{\alpha}{2}]^3}{3!} + \frac{[-i\sigma_u \frac{\alpha}{2}]^4}{4!} \tag{4.100}$$

Collecting terms with even and odd powers of σ_u:

$$\begin{aligned} R_u(\alpha) &= 1 + \frac{[-i\sigma_u \frac{\alpha}{2}]^2}{2!} + \frac{[-i\sigma_u \frac{\alpha}{2}]^4}{4!} + ... + [-i\sigma_u \frac{\alpha}{2}] + \frac{[-i\sigma_u \frac{\alpha}{2}]^3}{3!} + .. \\ &= 1 + \frac{(\frac{\alpha}{2})}{2!} + ... + [-i\sigma_u \frac{\alpha}{2}] + \frac{[-i\sigma_u (\frac{\alpha}{2})^3]}{3!} \\ &= cos\frac{\alpha}{2}\tilde{1} - i\sigma_u sin\frac{\alpha}{2} \end{aligned} \tag{4.101}$$

where $\tilde{1} = \begin{bmatrix} 1 & 0 \\ 0 & 1 \end{bmatrix}$, the unity matrix.
It
is convenient to define σ_u in polar coordinates:

$$\sigma_u = cos\theta\, \sigma_z + sin\theta\, [cos\phi\, \sigma_x + sin\phi\, \sigma_y] \tag{4.102}$$

$$= cos\theta \begin{bmatrix} 1 & 0 \\ 0 & -1 \end{bmatrix} + sin\theta\{cos\phi \begin{bmatrix} 0 & 1 \\ 1 & 0 \end{bmatrix} + sin\phi \begin{bmatrix} 0 & -i \\ i & 0 \end{bmatrix}\} \tag{4.103}$$

$$= \begin{bmatrix} cos\theta & 0 \\ 0 & -cos\theta \end{bmatrix} + sin\theta\{\begin{bmatrix} 0 & cos\phi - isin\phi \\ cos\phi + isin\phi \end{bmatrix}\} \tag{4.104}$$

$$= \begin{bmatrix} cos\theta & sin\theta e^{-i\phi} \\ sin\theta e^{i\phi} & -cos\theta \end{bmatrix} \tag{4.105}$$

Where θ and ϕ are the polarangles which describe the direction of the rotation axis. In Dirac notation, the rotation operator is:

$$R_u(\alpha) = \begin{bmatrix} cos\frac{\alpha}{2} - i\ cos\theta sin\frac{\alpha}{2} & -i\ sin\theta sin\frac{\alpha}{2}e^{-i\phi} \\ -i\ sin\theta sin\frac{\alpha}{2}e^{i\phi} & cos\frac{\alpha}{2} + i\ cos\theta sin\frac{\alpha}{2} \end{bmatrix} \tag{4.106}$$

4.7.2.1 Rotation Operator is Unitary

Rotations, by their very nature, do not affect any lengths or physical properties of the system, they simply rotate the system to a new orientation. Therefore, we expect the rotation operator to be unitary, $R^\dagger R = RR^\dagger = 1$. This can be easily seen by taking the product of R_u and $R_u^\dagger$.

$$R_u^\dagger(\alpha) = \begin{bmatrix} cos\frac{\alpha}{2} + i\ cos\theta sin\frac{\alpha}{2} & +i\ sin\theta sin\frac{\alpha}{2}e^{-i\phi} \\ +i\ sin\theta sin\frac{\alpha}{2}e^{i\phi} & cos\frac{\alpha}{2} - i\ cos\theta sin\frac{\alpha}{2} \end{bmatrix} \tag{4.107}$$

$$R_u R_u^\dagger = \begin{bmatrix} 1 & 0 \\ 0 & 1 \end{bmatrix} \tag{4.108}$$

This result should have been anticipated due to the fact that the operators, L, for angular momentum are Hermitian, therefore e^{iL} must be unitary.

4.7.3 Rotations of Wave Functions and Operators

Assume that our system can be described using a set of basis vectors, $|u_n>$, with eigenvalues for an operator , A, of a_n:

$$A|u_n>= a_n|u_n> \tag{4.109}$$

If we perform a rotation on the system, a new basis set is generated in the same coordinate system:

$$|u_n'>= R|u_n> \tag{4.110}$$

These new basis vectors will be eigenvalues of the operator in the new form of the wavefunction:

$$A'|u_n'>= a_n|u_n'> \tag{4.111}$$

The effect of rotations on operators can be obtained by considering the effect of rotations on observables. Rotation of the system should not change the outcome of any measurement we make, we expect the same expectation values regardless of the orientation of the system. Therefore:

$$\begin{aligned} A'|u_n'> &= a_n|u_n'> \\ A'R|u_n> &= a_nR|u_n> \\ <u_n|R^\dagger A'R|u_n> &= a_n<u_n|R^\dagger R|u_n> \\ <u_n|R^\dagger A'R|u_n> &= a_n<u_n|u_n> \end{aligned} \tag{4.112}$$

(R is a unitary operator $R^\dagger R = 1$)

$$a_n<u_n|u_n>=<u_n|A|u_n> \tag{4.113}$$

Therefore,

$$A = R^{\dagger} A' R \tag{4.114}$$

giving the form of the operator after the rotation:

$$A' = R A R^{\dagger} \tag{4.115}$$

4.7.3.1 Rotations and Commuting Observables

If an operator commutes with one of the components of angular momentum, then that operator will be invariant to rotations about that axis. Using rotations about the z-axis as an example:

$$R_z(\alpha) = e^{\frac{-i}{\hbar} L_z \alpha} \tag{4.116}$$

$$R_z(\alpha) \simeq 1 - \frac{i}{\hbar} d\alpha L_z \tag{4.117}$$

$$A' = (1 - \frac{i}{\hbar} d\alpha L_z) A (1 + \frac{i}{\hbar} d\alpha L_z) \tag{4.118}$$

To first order in $d\alpha$, this equation gives:

$$A' = A - \frac{i}{\hbar} d\alpha [L_i, A] \tag{4.119}$$

Therefore $A' = A$, if $[L_i, A] = 0$.

4.7.3.2 Example: Rotations about the *x*-axis

Rotations about the x-axis are equivalent to RF-pulses applied along the x-axis. In this case, the axis of rotation is specified as: $\theta = 90°$, $\phi = 0°$, giving the following rotation matrix:

$$R_x(\alpha) = \begin{bmatrix} cos(\frac{\alpha}{2}) & -isin(\frac{\alpha}{2}) \\ -isin(\frac{\alpha}{2}) & cos(\frac{\alpha}{2}) \end{bmatrix} \tag{4.120}$$

90° Rotation: R_u is:

$$R_x(\pi/2) = \frac{1}{\sqrt{2}} \begin{bmatrix} 1 & -i \\ -i & 1 \end{bmatrix} \tag{4.121}$$

If this operator was applied to $u_{+1/2}$:

$$\begin{aligned} u'_{+1/2} &= R_x(\pi/2) u_{+1/2} \\ &= \frac{1}{\sqrt{2}} \begin{bmatrix} 1 & -i \\ -i & 1 \end{bmatrix} \begin{bmatrix} 1 \\ 0 \end{bmatrix} \\ &= \frac{1}{\sqrt{2}} \begin{bmatrix} 1 \\ -i \end{bmatrix} \end{aligned} \tag{4.122}$$

This rotation has converted the original wavefunction, $u_{+1/2}$ to a linear combination of $u_{+1/2}$ and $u_{-1/2}$, indicating that a transition in the system has occurred due to the rotation, as would be expected for a 90° pulse.

The effect of the 90° rotation on the Hamiltonian operator is:

$$\begin{aligned}
\mathcal{H}' &= R\mathcal{H}R^{\dagger} \\
&= -\frac{\hbar\omega_o}{2}\frac{1}{\sqrt{2}}\begin{bmatrix}1 & -i\\ -i & 1\end{bmatrix}\begin{bmatrix}1 & 0\\ 0 & -1\end{bmatrix}\frac{1}{\sqrt{2}}\begin{bmatrix}1 & +i\\ +i & 1\end{bmatrix} \\
&= -\frac{\hbar\omega_o}{2}\frac{1}{2}\begin{bmatrix}1 & -i\\ -i & 1\end{bmatrix}\begin{bmatrix}1 & +i\\ -i & -1\end{bmatrix} \\
&= -\frac{\hbar\omega_o}{2}\begin{bmatrix}0 & i\\ -i & 0\end{bmatrix} \qquad (4.123)
\end{aligned}$$

Note that the rotated $\mathcal{H}$ is no longer diagonal. Regardless, the eigenfunctions of $\mathcal{H}'$ are the rotated wavefunctions, u', determined above.

$$\begin{aligned}
\mathcal{H}'|u'_{+1/2}> &= E_{+1/2}|u'_{+1/2}> \\
&= -\frac{\hbar\omega_o}{2}\begin{bmatrix}0 & i\\ -i & 0\end{bmatrix}\frac{1}{\sqrt{2}}\begin{bmatrix}1\\ -i\end{bmatrix} \\
&= -\frac{\hbar\omega_o}{2}\frac{1}{\sqrt{2}}\begin{bmatrix}1\\ -i\end{bmatrix} \\
&= -\frac{\hbar\omega_o}{2}|u'_{+1/2}> \qquad (4.124)
\end{aligned}$$

180° Rotation: $\alpha = \pi$

$$R_x(\pi) = \begin{bmatrix}0 & -i\\ -i & 0\end{bmatrix} \qquad (4.125)$$

Applying this rotation to $u_{+1/2}$:

$$\begin{aligned}
u'_{+1/2} &= R_x(\pi)u_{+1/2} \\
&= \begin{bmatrix}0 & -i\\ -i & 0\end{bmatrix}\begin{bmatrix}1\\ 0\end{bmatrix} = \begin{bmatrix}0\\ -i\end{bmatrix} \\
&= -iu_{-1/2} \qquad (4.126)
\end{aligned}$$

The 180° rotation completely converts one eigenstate, $u_{+1/2}$, to the other eigenstate, $u_{-1/2}$. This is the expected behavior of a 180° pulse, exchanging the populations of one state for another.

360° Rotation: $\alpha = 2\pi$
We expect this rotation to leave any operator or wavefunction unchanged, $R_u(2\pi) = \tilde{1}$, the unity matrix. This result should always be obtained, regardless of the direction of the rotation axis:

$$\begin{aligned}
R_u(2\pi) &= \tilde{1}cos\pi - i\sigma_u sin\pi \\
&= \tilde{1}(-1) \\
&= -1\begin{bmatrix}1 & 0\\ 0 & 1\end{bmatrix} \qquad (4.127)
\end{aligned}$$

The negative sign simply represents a phase shift of the wavefunction and does not change any expectation values.

4.8 Exercises

1. Determine $u'_{-1/2}$ for a rotation of 90° about the x-axis. Show that this wavefunction is orthonormal to $u'_{1/2}$.
2. Show, using rotation matrices, that the sequential application of the following two rotations leave the system unchanged: $R_y(\pi/2)$, followed by $R_y(-\pi/2)$.

4.9 Solutions

1. The rotation matrix for $R_x(\pi/2)$ was defined in the text:

$$R_x(\pi/2) = \frac{1}{\sqrt{2}}\begin{bmatrix} 1 & -i \\ -i & 1 \end{bmatrix} \quad (4.128)$$

If this operator was applied to $u_{-1/2}$:

$$\begin{aligned} u'_{-1/2} &= R_x(\pi/2)u_{-1/2} \\ &= \frac{1}{\sqrt{2}}\begin{bmatrix} 1 & -i \\ -i & 1 \end{bmatrix}\begin{bmatrix} 0 \\ 1 \end{bmatrix} \\ &= \frac{1}{\sqrt{2}}\begin{bmatrix} -i \\ 1 \end{bmatrix} \end{aligned} \quad (4.129)$$

To show that $u'_{-1/2}$ is orthogonal to $u'_{+1/2}$:

$$\begin{aligned} < u'_{-1/2}|u'_{+1/2} > &= [\frac{1}{\sqrt{2}}]^2 \begin{bmatrix} i & 1 \end{bmatrix}\begin{bmatrix} 1 \\ -i \end{bmatrix} \\ &= 0 \end{aligned} \quad (4.130)$$

2. A rotation about the y-axis is defined by the following operator:

$$\begin{aligned} R_y(\alpha) &= cos\frac{\alpha}{2}\tilde{1} - i\sigma_y sin\frac{\alpha}{2} \\ &= \begin{bmatrix} cos(\alpha/2) & 0 \\ 0 & cos(\alpha/2) \end{bmatrix} - isin(\alpha/2)\begin{bmatrix} 0 & -i \\ i & 0 \end{bmatrix} \\ &= \begin{bmatrix} cos(\alpha/2) & -sin(\alpha/2) \\ sin(\alpha/2) & cos(\alpha/2) \end{bmatrix} \end{aligned} \quad (4.131)$$

$$R_y(-\pi/2) = \begin{bmatrix} 1/\sqrt{2} & 1/\sqrt{2} \\ -1/\sqrt{2} & 1/\sqrt{2} \end{bmatrix} \quad R_y(\pi/2) = \begin{bmatrix} 1/\sqrt{2} & -1/\sqrt{2} \\ 1/\sqrt{2} & 1/\sqrt{2} \end{bmatrix} \quad (4.132)$$

$$R_y(-\pi/2)R_y(\pi/2) = \begin{bmatrix} 1 & 0 \\ 0 & 1 \end{bmatrix} \quad (4.133)$$

The final result is the identity matrix. This result is not surprising since the two rotations described a 90° pulse about the y-axis, followed by a -90° pulse along the same axis. The latter pulse should restore the system to its original state.

Chapter 5

QUANTUM MECHANICAL DESCRIPTION OF A ONE PULSE EXPERIMENT

In the first chapter a classical model of the nuclear magnetism was used to describe a one-dimensional NMR experiment. In this model, the magnetization was represented as a bulk magnetic moment and the pulses were used to rotate the bulk magnetism to the transverse plane. Precession of the bulk magnetization in the x-y plane induces a current in the coil, which is then digitized to produce the acquired free induction decay.

The same experiment can be described using quantum mechanics, as illustrated in Fig. 5.1. In this case the magnetization is represented by a wavefunction, pulses

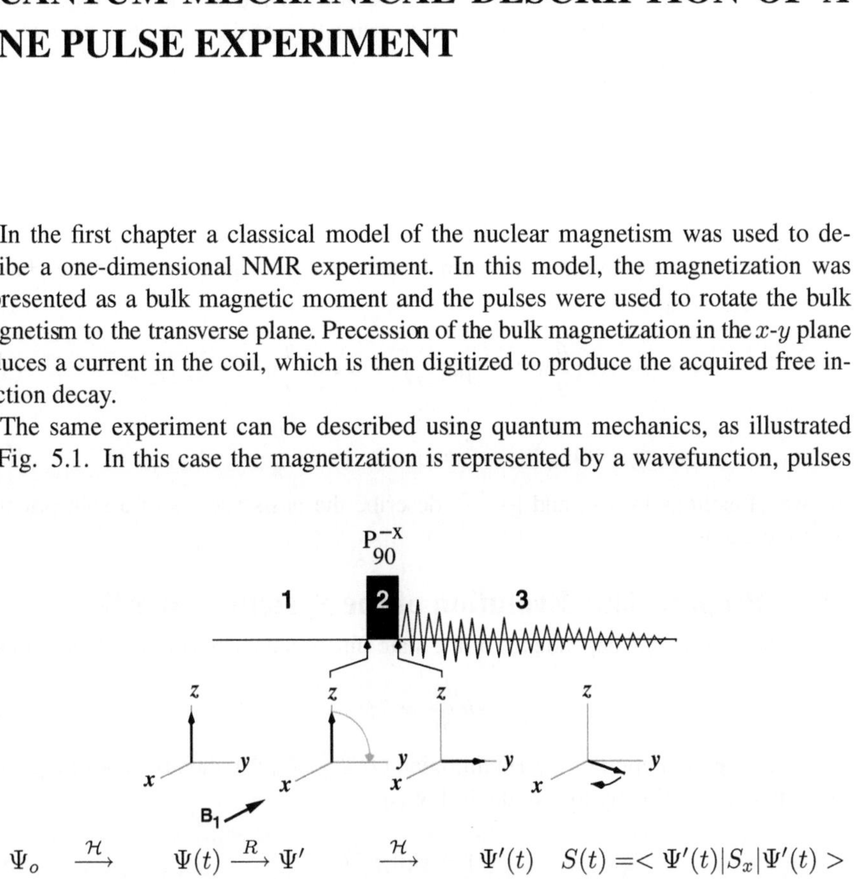

Figure 5.1. Quantum and classical description of a one-pulse experiment. The pulse sequence and resultant FID are shown in the top section of the figure. The experiment is divided into three sections: 1) preparation period, prior to the pulse, 2) excitation by a 90° RF-pulse along the minus x-axis, and 3) the detection period. The middle section of the diagram shows a classical vector model. The final detected signal (far right) is from the bulk magnetization precessing in the transverse plane. The lower part of the panel shows the quantum mechanical description. Prior to the pulse, the system is represented by Ψ_o. This wavefunction evolves with time, under the influence of the Hamiltonian, to give $\Psi(t)$. The pulse causes a rotation of $\Psi(t)$ to produce Ψ'. The rotated wavefunction evolves under the Hamiltonian to give $\Psi'(t)$. The final detected signal is the expectation value of the x-component of the spin angular momentum.

become rotation operators, and the detected signal is the expectation value of the x- and y-components of the spin angular momentum.

The NMR experiment will be subdivided into three distinct periods: 1) preparation, 2) excitation, and 3) detection. To maintain generality and correspondence with the classical solution, we will consider a wavefunction that represents a spin, $\vec{u}$, that is oriented in space with polar angles θ and ϕ. Using the classical vector model as a guide, the initial direction of $\vec{u}$, prior to the pulse, would be along the z-axis. After a 90° pulse, we would anticipate $\vec{u}$ to lie in the x-y plane.

In order to write the wavefunction for this spin we will determine the eigenvectors of the operator for angular momentum in the direction of the spin. The operator for the spin angular momentum along $\vec{u}$ is:

$$S_u = S_x \sin(\theta)\cos(\phi) + S_y \sin(\theta)\sin(\phi) + S_z \cos(\theta) \tag{5.1}$$

$$S_u = \frac{\hbar}{2}\begin{bmatrix} cos(\theta) & sin(\theta)e^{i\phi} \\ sin(\theta)e^{-i\phi} & -\cos(\theta) \end{bmatrix} \tag{5.2}$$

This operator has eigenvectors (written in terms of the eigenvectors of the Hamiltonian, $u_{+1/2}$, $u_{-1/2}$):

$$|+>_u = \cos(\frac{\theta}{2})e^{-i\phi/2}|u_{+1/2}> + \sin(\frac{\theta}{2})e^{i\phi/2}|u_{-1/2}> \tag{5.3}$$

$$|->_u = -\sin(\frac{\theta}{2})e^{-i\phi/2}|u_{+1/2}> + \cos(\frac{\theta}{2})e^{i\phi/2}|u_{-1/2}> \tag{5.4}$$

The wavefunctions $|+>_u$ and $|->_u$ describe the basis vectors of a spin oriented along the u axis.

5.1 Preparation: Evolution of the System Under B$_o$

In general, the time dependence of the state function depends on the Hamiltonian:

$$i\hbar\frac{d\psi}{dt} = \mathcal{H}\psi \tag{5.5}$$

If Ψ is an eigenfunction of the Hamiltonian ($\mathcal{H}\Psi = E_\Psi\Psi$), then the above equation can be integrated directly to give the following;

$$\psi(t) = \psi(0)e^{\frac{-iEt}{\hbar}} \tag{5.6}$$

For the eigenfunctions of a spin one-half particle, the time dependence of the eigenfunctions are explicitly:

$$u_{+1/2}(t) = u_{+1/2}(0)e^{\frac{+i\omega_S t}{2}} \qquad u_{-1/2}(t) = u_{-1/2}(0)e^{\frac{-i\omega_S t}{2}} \tag{5.7}$$

The time dependence of the $|+>$ state is:

$$\begin{aligned} |+(t)> &= \cos(\frac{\theta}{2})e^{-i\phi/2}e^{+i\frac{\omega_S t}{2}}|u_{+1/2}> + \sin(\frac{\theta}{2})e^{i\phi/2}e^{-i\frac{\omega_S t}{2}}|u_{-1/2}> \\ &= \cos(\frac{\theta}{2})e^{-i(\phi-\omega_S t)/2}|u_{+1/2}> + \sin(\frac{\theta}{2})e^{i(\phi-\omega_S t)/2}|u_{-1/2}> \end{aligned} \tag{5.8}$$

Comparing Eq. 5.3 to Eq. 5.8 shows that:

$$\theta(t) = \theta_0 \qquad \phi(t) = \phi_0 - \omega_s t \tag{5.9}$$

Thus, the angle of $\vec{u}$ with respect to the field, θ, does not change in time, but $\vec{u}$ precesses around $\hat{k}$ with an angular velocity ω_s, the absorption frequency of the transition. Note that this precession is identical to the classical precession of the magnetic moment about the B_o field.

The expectation values of the z- and x-components of the angular momentum of the system are easily calculated for the basis functions:

$$\begin{aligned}
|+(t)> &= \begin{bmatrix} \cos(\frac{\theta}{2})e^{-i(\phi_o-\omega_S t)/2} \\ \sin(\frac{\theta}{2})e^{i(\phi_o-\omega_S t)/2} \end{bmatrix} = \begin{bmatrix} \cos(\frac{\theta}{2})e^{-i\phi(t)/2} \\ \sin(\frac{\theta}{2})e^{i\phi(t)/2} \end{bmatrix} \\
< S_z > &= < +(t)|S_z|+(t) > \\
&= \begin{bmatrix} \cos(\frac{\theta}{2})e^{+i\phi(t)/2} & \sin(\frac{\theta}{2})e^{-i\phi(t)/2} \end{bmatrix} \frac{\hbar}{2} \begin{bmatrix} 1 & 0 \\ 0 & -1 \end{bmatrix} \begin{bmatrix} \cos(\frac{\theta}{2})e^{-i\phi(t)/2} \\ \sin(\frac{\theta}{2})e^{i\phi(t)/2} \end{bmatrix} \\
&= \begin{bmatrix} \cos(\frac{\theta}{2})e^{+i\phi(t)/2} & \sin(\frac{\theta}{2})e^{-i\phi(t)/2} \end{bmatrix} \frac{\hbar}{2} \begin{bmatrix} cos\frac{\theta}{2}e^{-i\phi(t)/2} \\ -sin\frac{\theta}{2}e^{i\phi(t)/2} \end{bmatrix} \\
&= \frac{\hbar}{2}\left[cos^2\frac{\theta}{2} - sin^2\frac{\theta}{2}\right] \\
&= \frac{\hbar}{2}cos\theta
\end{aligned} \tag{5.10}$$

Similarly, the x-component is:

$$\begin{aligned}
< S_x > &= \begin{bmatrix} \cos(\frac{\theta}{2})e^{+i\phi(t)/2} & \sin(\frac{\theta}{2})e^{-i\phi(t)/2} \end{bmatrix} \frac{\hbar}{2} \begin{bmatrix} 0 & 1 \\ 1 & 0 \end{bmatrix} \begin{bmatrix} \cos(\frac{\theta}{2})e^{-i\phi(t)/2} \\ \sin(\frac{\theta}{2})e^{i\phi(t)/2} \end{bmatrix} \\
&= \begin{bmatrix} \cos(\frac{\theta}{2})e^{+i\phi(t)/2} & \sin(\frac{\theta}{2})e^{-i\phi(t)/2} \end{bmatrix} \frac{\hbar}{2} \begin{bmatrix} sin\frac{\theta}{2}e^{i\phi(t)/2} \\ cos\frac{\theta}{2}e^{-i\phi(t)/2} \end{bmatrix} \\
&= \frac{\hbar}{2}\left[cos\frac{\theta}{2}sin\frac{\theta}{2}e^{i\phi(t)} + cos\frac{\theta}{2}sin\frac{\theta}{2}e^{-i\phi(t)}\right] \\
&= \frac{\hbar}{2}2cos\frac{\theta}{2}sin\frac{\theta}{2}\left[e^{+i\phi(t)-i\phi(t)}\right] \\
&= \frac{\hbar}{2}sin\theta cos\phi(t) \\
&= \frac{\hbar}{2}sin\theta\cos(\phi_o - \omega_s t)
\end{aligned} \tag{5.11}$$

Note that the z-component of the angular momentum is *not* time dependent. Since the energy of the system is directly proportional to S_z, the energy of the system is also time invariant, as one would expect for a spin in a static magnetic field. In contrast, the x-component is time dependent, oscillating with a frequency equal to ω_s. The spin is precessing about the static magnetic field at its Larmor frequency, as expected from the classical model. Since the spins are incoherent, the net x-component of the magnetization, averaged over a collection of spins, will be zero.

5.2 Excitation: Effect of Application of B_1

In the presence of the static field, a single spin will precess about the static field with no change in its energy. Transitions in energy are caused by the application of the oscillating B_1 field. To determine how the system evolves during the pulse it is necessary to the determine the form of the Hamiltonian while the pulse is being applied. The Hamiltonian operator in the presence of the B_1 field is given by [1]:

$$\begin{aligned}\mathcal{H} &= -\gamma\vec{S}\cdot[\vec{B_o}+\vec{B_1(t)}] \\ &= \omega_s S_z + \omega_1[\cos(\omega t)S_x + \sin(\omega t)S_y]\end{aligned} \tag{5.12}$$

where the minus sign has been absorbed into ω_s and ω_1. Note that ω_s and ω_1 are proportional to the magnetic field strengths, while ω is the frequency of the applied RF-pulse.

Using the definitions of the S_x, S_y, and S_z operators it is possible to express the Hamiltonian in matrix form as the following:

$$\mathcal{H} = \frac{\hbar}{2}\begin{bmatrix}\omega_s & \omega_1 e^{-i\omega t} \\ \omega_1 e^{+i\omega t} & -\omega_s\end{bmatrix} \tag{5.13}$$

As before, the time evolution of ψ is given by:

$$i\hbar\frac{d\psi}{dt} = H\psi \tag{5.14}$$

If we evaluate the time dependence of the following arbitrary wavefunction:

$$\psi = a_+u_{+1/2} + a_-u_{-1/2} = \begin{bmatrix}a_+ \\ a_-\end{bmatrix} \tag{5.15}$$

Equation 5.14 becomes:

$$i\frac{da_+}{dt} = \frac{\omega_s}{2}a_+ + \frac{\omega_1}{2}e^{-i\omega t}a_- \tag{5.16}$$

$$i\frac{da_-(t)}{dt} = \frac{\omega_1}{2}e^{i\omega t}a_+ - \frac{\omega_s}{2}a_- \tag{5.17}$$

Equations 5.16 and 5.17 indicate that the wave function now becomes time dependent, i.e. the probability that the system will be found in one of the two eigenfunctions of the Hamiltonian changes with time. This is due to the fact that the Hamiltonian during the pulse is not diagonal, consequently the original eigenfunctions are no longer eigenfunctions of the new Hamiltonian.

These time dependent equations are difficult to solve in their current form. However, we know from the classical analysis that a rotating frame of reference would be

[1] For convenience we assume that the B_1 field is described as rotating in the x-y plane, therefore it has an x-component of $\cos\omega t$ and a y-component of $\sin\omega t$.

useful. The operator for rotation about the z-axis at a rate of $\omega_r t$ is:

$$R_z(-\omega_r t) = cos\frac{-\omega_r t}{2}\tilde{1} - i\sigma_z sin\frac{-\omega_r t}{2} \tag{5.18}$$

$$= \begin{bmatrix} cos\frac{-\omega_r t}{2} - isin\frac{-\omega_r t}{2} & 0 \\ 0 & cos\frac{-\omega_r t}{2} + isin\frac{-\omega_r t}{2} \end{bmatrix} \tag{5.19}$$

$$= \begin{bmatrix} e^{+i\frac{\omega_r}{2}t} & 0 \\ 0 & e^{-i\frac{\omega_r}{2}t} \end{bmatrix} \tag{5.20}$$

(The rotation angle here is negative, since the coordinate frame is under rotation, not the wave function).

The effect of this rotation on the wave function is:

$$\psi' = R_z(-\omega_r t)\psi \tag{5.21}$$

$$= \begin{bmatrix} e^{+i\frac{\omega_r}{2}t} & 0 \\ 0 & e^{-i\frac{\omega_r}{2}t} \end{bmatrix} \begin{bmatrix} a_+ \\ a_- \end{bmatrix} \tag{5.22}$$

$$= \begin{bmatrix} a_+ e^{+i\frac{\omega_r}{2}t} \\ a_- e^{-i\frac{\omega_r}{2}t} \end{bmatrix} \tag{5.23}$$

and the coefficients in the rotating frame are:

$$a_+^r = e^{i\omega_r t/2} a_+ \qquad a_-^r = e^{-i\omega_r t/2} a_- \tag{5.24}$$

Substituting the expressions for $a_\pm^r$ into the time dependent differential equations (Eqs. 5.16, 5.17) gives the following:

$$\begin{aligned}
i\frac{da_+}{dt} &= \frac{\omega_S}{2}a_+ + \frac{\omega_1}{2}e^{-i\omega t}a_- \\
i\frac{de^{-i\frac{\omega_r}{2}t}a_+^r}{dt} &= \frac{\omega_S}{2}e^{-i\frac{\omega_r}{2}t}a_+^r + \frac{\omega_1}{2}e^{-i\omega t}e^{+i\frac{\omega_r}{2}t}a_-^r \\
i\left[-i\frac{\omega_r}{2}e^{-i\frac{\omega_r}{2}t}a_+^r + e^{-i\frac{\omega_r}{2}t}\frac{da_+^r}{dt}\right] &= \frac{\omega_S}{2}e^{-i\frac{\omega_r}{2}t}a_+^r + \frac{\omega_1}{2}e^{-i\omega t}e^{+i\frac{\omega_r}{2}t}a_-^r \\
\frac{\omega_r}{2}a_+^r + i\frac{da_+^r}{dt} &= \frac{\omega_S}{2}a_+^r + \frac{\omega_1}{2}e^{-i\omega t}e^{i\omega_r t}a_-^r \\
i\frac{da_+^r}{dt} &= \frac{(\omega_S - \omega_r)}{2}a_+^r + \frac{\omega_1}{2}e^{-i(\omega-\omega_r)t}a_-^r \\
similarly: \\
i\frac{da_-^r}{dt} &= \frac{\omega_1}{2}e^{+i(\omega-\omega_r)t}a_+^r(t) + -\frac{(\omega_S - \omega_r)}{2}a_-^r
\end{aligned}$$

The coefficients of the a_+^r and a_-^r terms can be used to define the effective Hamiltonian in the rotating frame. In the laboratory frame, the Hamiltonian is:

$$\mathcal{H} = \frac{\hbar}{2}\begin{bmatrix} \omega_s & \omega_1 e^{i\omega t} \\ \omega_1 e^{-i\omega t} & -\omega_s \end{bmatrix}$$

which gives the following time dependence of the coefficients:

$$i\frac{da_+}{dt} = \frac{\omega_S}{2}a_+ + \frac{\omega_1}{2}e^{-i\omega t}a_- \qquad i\frac{da_-(t)}{dt} = \frac{\omega_1}{2}e^{i\omega t}a_+ - \frac{\omega_S}{2}a_-$$

In the rotating frame, the equivalent expressions are:

$$\begin{aligned} i\frac{da_+^r}{dt} &= \frac{(\omega_S - \omega_r)}{2}a_+^r + \frac{\omega_1}{2}e^{-i(\omega-\omega_r)t}a_-^r \\ i\frac{da_-^r}{dt} &= \frac{\omega_1}{2}e^{+i(\omega-\omega_r)t}a_+^r(t) + -\frac{(\omega_S - \omega_r)}{2}a_-^r \end{aligned}$$

Comparing the expressions for the rotating frame to the equivalent expressions in the laboratory frame gives the form of the Hamiltonian in the rotating frame:

$$\mathcal{H}_{eff} = \frac{\hbar}{2}\begin{bmatrix} \Delta\omega & \omega_1 \\ \omega_1 & -\Delta\omega \end{bmatrix} \tag{5.25}$$

Here, we assume that the rotation rate of the coordinate system is equal to the frequency of the B_1 pulse, i.e. $\omega = \omega_r$, and we define $\Delta\omega = \omega_s - \omega_r$. If we further assume that the spin is on resonance, then the effective Hamiltonian simplifies further:

$$\mathcal{H}_{eff} = \frac{\hbar}{2}\begin{bmatrix} 0 & \omega_1 \\ \omega_1 & 0 \end{bmatrix} = \omega_1 S_x \tag{5.26}$$

The effective Hamiltonian can be used to calculate the time evolution of the system during the RF-pulse:

$$\begin{aligned} i\hbar\frac{d\Psi}{dt} &= \mathcal{H}\Psi \\ \frac{d\Psi}{\Psi} &= \frac{\omega_1 S_x}{i\hbar}dt \\ \Psi(t) &= e^{(-i/\hbar)S_x\omega_1 t}\Psi(0) = R_x(\omega_1\tau)\Psi(0) \end{aligned} \tag{5.27}$$

The above expressions shows that the time evolution of the system during a pulse is equivalent to a rotation. The direction of the rotation axis is defined by the direction of the pulse and the rotation angle is given by $\omega_1\tau$, where τ is the duration of the pulse.

5.2.1 The Resonance Condition

Application of the B_1 field to the system causes transitions from one state of the system to another, with the absorption of energy. The extent of the transition is described by the changes in the coefficient of each wavefunction, in this case a_- and a_+. Since the evolution of the system during the pulse is simply described by a rotation operator, it is easy to calculate the state of the system after the pulse, given the state of the system before the pulse. Taking the specific case of a system beginning in the $u_{+1/2}$ state ($a_+ = 1$), the effects of a 90°, 180°, or 360° pulse on this state are as follows (See Eqs. 4.121, 4.126, and 4.127, in Chapter 4):

$$\begin{bmatrix}1\\0\end{bmatrix} \xrightarrow{R_x(\pi/2)} \begin{bmatrix}\frac{1}{\sqrt{2}}\\ \frac{1}{\sqrt{2}}\end{bmatrix} \tag{5.28}$$

$$\begin{bmatrix}1\\0\end{bmatrix} \xrightarrow{R_x(\pi)} \begin{bmatrix}0\\ -i\end{bmatrix} \tag{5.29}$$

$$\begin{bmatrix}1\\0\end{bmatrix} \xrightarrow{R_x(2\pi)} \begin{bmatrix}-1\\ 0\end{bmatrix} \tag{5.30}$$

The probability of finding the system in the $u_{-1/2}$ after the pulse, $P_{-1/2}$, is just $a_-^* a_-$, giving:

$$\begin{aligned} P_{-1/2} : 0 &\xrightarrow{R_x(\pi/2)} 1/2 \\ 0 &\xrightarrow{R_x(\pi)} 1 \\ 0 &\xrightarrow{R_x(2\pi)} 0 \end{aligned} \tag{5.31}$$

The probability of finding the system in the $u_{-1/2}$ state, $P_- = a_-^* a_-$, can be calculated using the effective Hamiltonian in the rotating frame. The dependence of P_- on the frequency offset ($\Delta\omega = \omega_s - \omega$) and pulse length, τ, is given by Eq. 5.32 (see Cohen-Tannoudji *et al.* [42] for more details). This transition probability is shown in Fig. 5.2 for an on-resonance pulse and a pulse that is 6000 Hz off-resonance. The field strength of the pulse, ω_1 was 2500 Hz, giving a 90° pulse of 100 μsec. Note that the transition probability oscillates with time. In the case of the on-resonance pulse, there is complete conversion of the system from the $u_{+1/2}$ state to the $u_{-1/2}$ state in 200 μsec. This corresponds to a 180° pulse, or complete inversion of the magnetization. At four times the 90° pulse length, or 400 μsec, the system is returned to the original state, corresponding to a 360° pulse. The transition probabilities for the on-resonance pulses in Fig. 5.2 (solid curve) are identical to those presented in Eqs. 5.31, as expected.

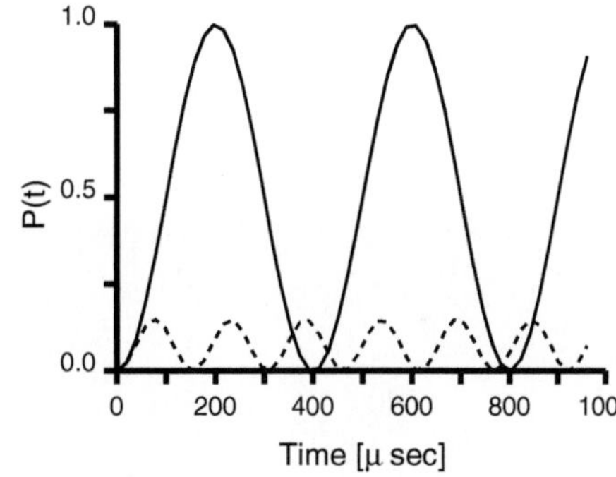

Figure 5.2. Effect of frequency offset on transition probabilities. The probability of finding the system in the $u_{-1/2}$ state, after starting entirely in the $u_{+1/2}$ state at t=0 is shown for an on-resonance pulse (solid line) and a pulse that is 6000 Hz off-resonance (dashed line).

$$P_-(t) = \frac{\omega_1^2}{\omega_1^2 + (\Delta\omega)^2} \sin^2\left[\sqrt{\omega_1^2 + (\Delta\omega)^2}\,\frac{\tau}{2}\right] \tag{5.32}$$

In the case of an off-resonance pulse, the extent of conversion between states is significantly smaller, showing that an off-resonance pulse is less efficient at causing transitions. Also note that the frequency of the oscillations of $P(t)$ has increased for the off-resonance pulse. This is due to the fact that the effective magnetic field is larger than for the on-resonance spins, causing a faster transition rate, or a faster precessional rate in the classical model.

5.3 Detection: Evolution of the System Under B_o

After the pulse, the magnetization will precess about B_o, as it did during the preparation period. The difference between the state of the system prior to the pulse and after the pulse is that the system will have been rotated to a new direction, $\vec{u'}$ by the application of the RF-pulse. As before, the z-component of the spin angular momentum of the system will be time invariant. In contrast, the expectation value of S_x will evolve as $cos(\omega_s t)$, while the expectation value of S_y will evolve as $sin(\omega_s t)$, generating the $cosine$ and $sine$ modulated signals in the detection coil.

Chapter 6

THE DENSITY MATRIX & PRODUCT OPERATORS

The previous chapter showed how it is possible to keep track of the evolution of a single spin during an NMR experiment by calculating the effect of pulses and free precession on its wavefunction. Using this approach, the system at the beginning of the experiment would be represented by linear combinations of the basis vectors:

$$\Psi(0) = \sum c_i(o)u_i \tag{6.1}$$

where $c_i(0)$ represent the coefficients that describe the system at t = 0.

At the end of the NMR experiment the system is also described as a linear combination of the basis vectors, but with different coefficients. The coefficients change from the initial values as a result of the various pulses and time periods that have been experienced by the spin during the experiment.

$$\Psi(t) = \sum c_i(t)u_i \tag{6.2}$$

Regardless of the change in the coefficients, it is possible to calculate the expectation value of any observable. For example, the signal detected in the real channel of the receiver, S_x, is given by:

$$\begin{aligned} < S_x > &= < \Psi|S_x|\Psi > \\ &= \sum_{ij} c_i^*(t)c_j(t) < u_i|S_x|u_j > \end{aligned} \tag{6.3}$$

However, in most NMR experiments, the detected signal arises from an ensemble of N spins, therefore, the average expectation value of S_x is the actual observable:

$$\overline{< S_x >} = \sum_N \sum_{ij} c_i^*(t)c_j(t) < u_i|S_x|u_j > \tag{6.4}$$

This ensemble average can be obtained in one of two ways. The first method involves calculating the evolution of all N wavefunctions through the experiment and then

$$\begin{array}{llllll} \Psi_o & \xrightarrow{\mathcal{H}} & \Psi(t) \xrightarrow{R} \Psi' & \xrightarrow{\mathcal{H}} & \Psi'(t) & S(t) =< \Psi'(t)^*|S_x|\Psi'(t) > \\ \rho_o & \xrightarrow{\mathcal{H}} & \rho(t) \xrightarrow{R} \rho' & \xrightarrow{\mathcal{H}} & \rho'(t) & S(t) = Trace[\rho S_x] \end{array}$$

Figure 6.1. Evolution of the wavefunction and density matrix during a one-pulse NMR experiment. The initial state of a single wavefunction is represented by Ψ_o, this evolves to $\Psi'(t)$ at the end of the experiment. The detected signal is the expectation value of S_x. In the case of the density matrix, the initial state of an ensemble of spins is represented by ρ_o. At the end of the experiment the final state is $\rho'(t)$ and the observed signal is calculated from the density matrix. $\mathcal{H}$ represents the Hamiltonian associated with B_o and R represents a radio-frequency pulse, in the form of a rotation operator.

averaging over all N spins. This procedure is very tedious. The second method, which is discussed in this chapter, is to use a description of the system that captures the nature of the ensemble in a more concise and manipulatable fashion. Such a representation is provided by the density matrix.

The density matrix is a matrix whose elements contain information on the average probability of all possible states of the ensemble. Instead of following the evolution of the different wavefunctions through the experiment, the evolution of the density matrix is followed. At the end of the experiment, the average expectation values are obtained from the final density matrix. The close correspondence between the wavefunction and the density matrix is illustrated in Fig. 6.1.

6.1 Introduction to the Density Matrix

The information on the ensemble average of a system is contained in the ensemble average of the pairwise products of the coefficient that are associated with the basis vectors, for example:

$$\begin{aligned} \overline{< S_x >} &= \sum_N \sum_{ij} c_i^*(t)c_j(t) < u_i|S_x|u_j > \\ &= \sum_{ij} \overline{c_i^*(t)c_j(t)} < u_i|S_x|u_j > \end{aligned} \tag{6.5}$$

Therefore, if we specified the value of:

$$\overline{c_i^*(t)c_j(t)}$$

for all possible values of i and j, then the properties of the system are completely determined. The density matrix, ρ, consists of all such products.

The mth, n^{th} element of the density matrix (ρ_{mn}) for an arbitrary, time-dependent, wavefunction is obtained as follows:

$$\begin{aligned} \rho_{mn}(t) &=< u_m|\psi(t) >< \psi(t)|u_n > \\ &=< u_m|\psi(t) > c_n^*(t) \\ &= c_m(t)c_n^*(t) \end{aligned} \tag{6.6}$$

For example, in a system that has two basis vectors, the density matrix is:

$$\rho = \begin{bmatrix} c_1 c_1^* & c_1 c_2^* \\ c_2 c_1^* & c_2 c_2^* \end{bmatrix} \tag{6.7}$$

The matrix elements of the density matrix, as defined by Eq. 6.6 are of the same form as the matrix elements of an operator. Hence the density matrix is an operator, and will be transformed by rotations with the same rules as any other operator:

$$\rho' = R\rho R^\dagger \tag{6.8}$$

6.1.1 Calculation of Expectation Values From ρ

The expectation value for an observable:

$$\begin{aligned} A &= <\psi|A|\psi> \\ &= \sum_{k,p} < c_k^* u_k|A|c_p u_p > \\ &= \sum_{k,p} c_p c_k^* < u_k|A|u_p > \\ &= \sum_{k,p} \rho_{pk} < u_k|A|u_p > \\ &= \sum_{k,p} < u_p|\rho|u_k >< u_k|A|u_p > \\ &= \sum_{p} < u_p|\rho A|u_p > \quad [note\ \ i] \\ &= \sum_{p} \rho_{pp} A_{pp} \\ &= \text{Trace}[\rho A] \quad [note\ \ ii] \end{aligned} \tag{6.9}$$

Notes: i) Closure, $1 = \sum_k |u_k >< u_k|$, is used to remove one of the sums.
ii) The trace of a matrix is simply the sum of its diagonal elements.

The above result implies that if we know the density matrix of the system at anytime, we can easily calculate the expectation value of an operator from that density matrix. For example, the signal detected for the real data channel would be:

$$M_x = \text{Trace}(\rho S_x) \tag{6.10}$$

6.1.2 Density Matrix for a Statistical Mixture

For the case of a single isolated spin, representation of the system by either its wavefunction or density matrix are essentially equivalent and equally tedious. However, if there is a statistical mixture of states, such as the large number of molecules in a typical NMR sample, the density matrix is much more convenient.

For an ensemble of spins, the total state of the system is given by:

$$\Psi = \sum_{k=0}^{N} p_k \psi_k \tag{6.11}$$

where p_k is the *statistical* probability of finding a spin in a particular state. p_k is the fraction of the kth mixed state or sub-system in the sample, and ψ_k corresponds to the wavefunction for that sub-system. Each ψ_k is a linear combination of the basis functions:

$$\psi_k = \sum_i c_i^k u_i \tag{6.12}$$

where c_i^k are the normal quantum mechanical probabilities associated with finding the wavefunction in the ith state.

The expectation value of any operator, A, for the kth sub-system is:

$$A_k = < \psi_k | A | \psi_k > \tag{6.13}$$

The average value of this observable for the entire system is the expectation value associated with a particular state, multiplied by the statistical probability of that particular state.

$$\overline{A} = \sum_k p_k A_k \tag{6.14}$$

The expectation value of the system in the kth sub-state is:

$$A_k = \text{Trace } (\rho^k A) = \sum_j \rho_{jj}^k A_{jj} \tag{6.15}$$

Therefore, the average expectation value over the entire ensemble of sub-states is:

$$\begin{aligned}
\overline{A} &= \sum_k p_k \text{Trace}(\rho^k A) \\
&= \sum_k p_k \sum_j \rho_{jj}^k A_{jj} \\
&= \sum_j \sum_k p_k \rho_{jj}^k A_{jj} \\
&= \sum_j \bar{\rho}_{jj} A_{jj} \\
&= \text{Trace}[\bar{\rho} A]
\end{aligned} \tag{6.16}$$

The important result is that the average expectation value can be obtained from the average density matrix in exactly the same fashion as the expectation value of a single spin was obtained from its density matrix (see Eq. 6.9).

As an example consider the wavefunction for a spin that is oriented at an angle θ and ϕ in polar coordinates:

$$\Psi = \cos(\frac{\theta}{2}) e^{-i\phi/2} | u_{+1/2} > + \sin(\frac{\theta}{2}) e^{i\phi/2} | u_{-1/2} > \tag{6.17}$$

The density matrix for this general state is:

$$\rho = \begin{bmatrix} cos^2(\theta/2) & cos(\theta/2)sin(\theta/2)e^{-i\phi} \\ cos(\theta/2)sin(\theta/2)e^{+i\phi} & sin^2(\theta/2) \end{bmatrix} \tag{6.18}$$

The expectation value for S_x is:

$$< S_x >= \text{Trace}(\rho S_x) \tag{6.19}$$

Calculating ρS_x first:

$$\begin{aligned} \rho S_x &= \begin{bmatrix} cos^2(\theta/2) & cos(\theta/2)sin(\theta/2)e^{-i\phi} \\ cos(\theta/2)sin(\theta/2)e^{+i\phi} & sin^2(\theta/2) \end{bmatrix} \frac{\hbar}{2}\begin{bmatrix} 0 & 1 \\ 1 & 0 \end{bmatrix} \\ &= \frac{\hbar}{2}\begin{bmatrix} cos(\theta/2)sin(\theta/2)e^{-i\phi} & cos^2(\theta/2) \\ sin^2(\theta/2) & cos(\theta/2)sin(\theta/2)e^{+i\phi} \end{bmatrix} \end{aligned} \tag{6.20}$$

$$\begin{aligned} < S_x > &= \text{Trace}(\rho S_x) = \frac{\hbar}{2}\left[cos\frac{\theta}{2}sin\frac{\theta}{2}e^{-i\phi} + cos\frac{\theta}{2}sin\frac{\theta}{2}e^{+i\phi}\right] \\ &= \frac{\hbar}{2}\,2\,cos\frac{\theta}{2}sin\frac{\theta}{2}\left[\frac{e^{i\phi}+e^{-i\phi}}{2}\right] \\ &= \frac{\hbar}{2}\,2\,cos\frac{\theta}{2}sin\frac{\theta}{2}cos\phi \end{aligned} \tag{6.21}$$

If the spin is oriented such that its magnetic moment is along the x-axis, then $\theta = 90$, and $\phi = 0$. The expectation value of S_x is then:

$$< S_x >= \frac{\hbar}{2}\,2\,\frac{1}{\sqrt{2}}\frac{1}{\sqrt{2}} = \frac{\hbar}{2} \tag{6.22}$$

as expected. Calculation of the expectation value of S_y for this example yields $< S_y >= 0$.

The conclusion from the above calculation is that a single spin can have a non-zero value of $< S_x >$. However, in a large population of *incoherent* spins at thermal equilibrium, there will be a random distribution of the angle ϕ. Under these conditions, the *average* density matrix is:

$$\overline{<\rho>} = \begin{bmatrix} cos^2(\theta/2) & cos(\theta/2)sin(\theta/2)\int_{\phi=0}^{\phi=2\pi} e^{-i\phi} \\ cos(\theta/2)sin(\theta/2)\int_{\phi=0}^{\phi=2\pi} e^{+i\phi} & sin^2(\theta/2) \end{bmatrix} \tag{6.23}$$

Since $\int_{\phi=0}^{\phi=2\pi} e^{i\phi} = 0$, the average density matrix is:

$$\overline{<\rho>} = \begin{bmatrix} cos^2(\theta/2) & 0 \\ 0 & sin^2(\theta/2) \end{bmatrix} \tag{6.24}$$

and the average expectation value of S_x is now zero:

$$\begin{aligned} \overline{< S_x >} &= \text{Trace}(\rho S_x) \\ &= \text{Trace}\left(\frac{\hbar}{2}\begin{bmatrix} cos^2(\theta/2) & 0 \\ 0 & sin^2(\theta/2) \end{bmatrix}\begin{bmatrix} 0 & 1 \\ 1 & 0 \end{bmatrix}\right) \\ &= \frac{\hbar}{2}\text{Trace}\left(\begin{bmatrix} 0 & cos^2(\theta/2) \\ sin^2(\theta/2) & 0 \end{bmatrix}\right) = 0 \end{aligned} \tag{6.25}$$

6.2 One-pulse Experiment: Density Matrix Description

A one-pulse experiment is simply a single 90° pulse followed by detection of the signal. The evolution of the system in this experiment can be followed by the density matrix, and the expectation value of any observable can be calculated from ρ (see Fig. 6.2).

To begin, the density matrix that describes an ensemble of spins at thermal equilibrium, or ρ_o, is required. This density matrix should reflect the fact that the average value of $< S_x >$ and $< S_y >$ are zero, and it should also reflect the population difference between the ground and the excited state at thermal equilibrium.

The population difference at thermal equilibrium is given by the Boltzmann distribution:

$$\frac{e^{-E/kT}}{Z} \tag{6.26}$$

Using the Boltzmann distribution as an operator gives the desired form of the density matrix. A single element of the density matrix is:

$$\rho_{pn} = < u_p|\frac{e^{-\mathcal{H}/kT}}{Z}|u_n >$$

and the entire density matrix is:

$$\begin{aligned}
\rho &= \frac{1}{Z}\begin{bmatrix} < u_1|e^{-\mathcal{H}/kT}|u_1 > & < u_1|e^{-\mathcal{H}/kT}|u_2 > \\ < u_2|e^{-\mathcal{H}/kT}|u_1 > & < u_2|e^{-\mathcal{H}/kT}|u_2 > \end{bmatrix} \\
&= \frac{1}{Z}\begin{bmatrix} e^{-E_1/kT} < u_1|u_1 > & e^{-E_2/kT} < u_1|u_2 > \\ e^{-E_1/kT} < u_2|u_1 > & e^{-E_2/kT} < u_2|u_2 > \end{bmatrix} \\
&= \frac{1}{Z}\begin{bmatrix} e^{-E_1/kT} & 0 \\ 0 & e^{-E_2/kT} \end{bmatrix} \\
&= \frac{1}{Z}\begin{bmatrix} e^{+\hbar\omega_o/2kT} & 0 \\ 0 & e^{-\hbar\omega_o/2kT} \end{bmatrix}
\end{aligned} \tag{6.27}$$

Expanding the exponential as a series and taking only the first term ($e^a \approx 1 + a$) we find:

$$\rho = \frac{1}{Z}\begin{bmatrix} 1+\gamma & 0 \\ 0 & 1-\gamma \end{bmatrix} = \frac{1}{Z}\begin{bmatrix} 1 & 0 \\ 0 & 1 \end{bmatrix} + \frac{1}{Z}\frac{\omega_o}{kT}\frac{\hbar}{2}\begin{bmatrix} 1 & 0 \\ 0 & -1 \end{bmatrix} \tag{6.28}$$

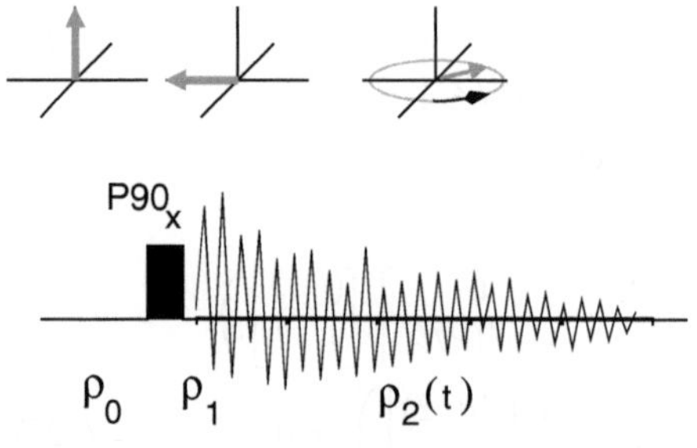

Figure 6.2 Density matrix description of the one-pulse experiment. The top part of the figure shows the classical description of a one-pulse NMR experiment. The density matrix ρ_o defines the system prior to the pulse, ρ_1 is the system immediately after the pulse, and $\rho_2(t)$ defines the time evolution of the system during detection of the free induction decay.

where $\gamma = \hbar\omega_o/2kT$.

One can see that this matrix is composed of the unit matrix plus S_z. Since the unit matrix has no effect on any of the common observables we can ignore it [1]. Furthermore, since we are only interested in the changes in the amplitudes and the time evolution of the individual elements of the density matrix, we can also ignore constants. Therefore, the density operator for a system under thermal equilibrium can be written:

$$\rho_0 = \frac{\hbar}{2}\begin{bmatrix} 1 & 0 \\ 0 & -1 \end{bmatrix} = S_z \tag{6.29}$$

To simplify the calculations even further, $\hbar$ can be removed from the expression using the following representation of angular momentum operators:

$$I_x = \frac{S_x}{\hbar} = \frac{1}{2}\begin{bmatrix} 0 & 1 \\ 1 & 0 \end{bmatrix} \quad I_y = \frac{S_y}{\hbar} = \frac{1}{2}\begin{bmatrix} 0 & -i \\ i & 0 \end{bmatrix} \quad I_z = \frac{S_z}{\hbar} = \frac{1}{2}\begin{bmatrix} 1 & 0 \\ 0 & -1 \end{bmatrix} \tag{6.30}$$

Thus, ρ_o is equal to the matrix I_z.

6.2.1 Effect of Pulses on the Density matrix

In the last chapter we showed that pulses, or B_1 fields, can be represented as rotation operators. Since the density matrix is also an operator, it will have the following transformation properties under the effect of a pulse:

$$\begin{aligned} \rho\prime &= R\rho R^\dagger \\ &= e^{-iP}\rho e^{iP} \end{aligned} \tag{6.31}$$

where the operator P describes a rotation of the system about any axis. For pulses in the x-y plane (θ=90°) the rotation operator for a rotation of α degrees is given by:

$$R_\phi(\alpha) = e^{-iP} = \begin{bmatrix} cos(\frac{\alpha}{2}) & -isin(\frac{\alpha}{2})e^{-i\phi} \\ -isin(\frac{\alpha}{2})e^{i\phi} & cos(\frac{\alpha}{2}) \end{bmatrix} \tag{6.32}$$

the direction of the pulse in the x-y plane is defined by the ϕ angle. For example, a pulse along the x-axis corresponds to $\phi = 0°$, giving the following rotation matrix for a 90° pulse:

$$R_x(\pi/2) = \begin{bmatrix} \frac{1}{\sqrt{2}} & \frac{-i}{\sqrt{2}} \\ \frac{-i}{\sqrt{2}} & \frac{1}{\sqrt{2}} \end{bmatrix} = \frac{1}{\sqrt{2}}\begin{bmatrix} 1 & -i \\ -i & 1 \end{bmatrix}$$

The density matrix after this pulse is given by:

$$\begin{aligned} \rho_1 &= R_x I_z R_x^\dagger \\ &= \frac{1}{\sqrt{2}}\begin{bmatrix} 1 & -i \\ -i & 1 \end{bmatrix}\frac{1}{2}\begin{bmatrix} 1 & 0 \\ 0 & -1 \end{bmatrix}\frac{1}{\sqrt{2}}\begin{bmatrix} 1 & i \\ i & 1 \end{bmatrix} \\ &= \frac{1}{2}\begin{bmatrix} 0 & i \\ -i & 0 \end{bmatrix} \\ &= -I_y \end{aligned} \tag{6.33}$$

[1] Specifically, the trace of the product of the unit matrix with S_x, S_y, and S_z is zero.

If we calculate the expectation value of S_x, S_y, and S_z for this density matrix, we find:

$$< S_x >= 0 \qquad < S_y >= -1 \qquad < S_z >= 0 \tag{6.34}$$

Note that the off-diagonal elements of the density matrix after the pulse are clearly non-zero. This implies that the ensemble average of this element of the density matrix is non-zero. Therefore, after the pulse, the distribution of the spins about the z-axis is no longer random - a preferred direction of the spins has been induced by the pulse. Specifically, the ensemble of spins has become *coherent*, with each spin having the same value of ϕ (compare to Eq. 6.24).

6.2.1.1 Free Precession: Time evolution of the Density Matrix

The time evolution of any wavefunction is defined by the evolution operator:

$$\Psi(t) = e^{-i\mathcal{H}t/\hbar}\Psi(0)$$

Under influence of the B_o field, or free precession, the Hamiltonian is simply: $-\omega_s I_z$, giving the following expression for the time evolution:

$$\Psi(t) = e^{i\omega_S I_z t}\Psi(0)$$

Comparing this equation to the rotation operator for z-rotations:

$$R_z(\alpha) = e^{-i\frac{L_z}{\hbar}\alpha} \tag{6.35}$$

shows that the evolution of the spins under free precession is equivalent to rotation of the system about the z-axis with a rotation angle equal to $\omega_s t$.

The time evolution of the density matrix formed after the 90°_x pulse is:

$$\rho_2(t) = e^{+i\omega_S t I_z}\rho_1 e^{-i\omega_S t I_z} \tag{6.36}$$

Conversion of $e^{i\omega_S t I_z}$ to its matrix form is simple since the basis vectors are eigenvalues of I_z. Therefore, we can directly write:

$$e^{+i\omega_S t I_z} = \begin{bmatrix} e^{+i\omega_S t/2} & 0 \\ 0 & e^{-i\omega_S t/2} \end{bmatrix} \tag{6.37}$$

$$\begin{aligned}
\rho(t) &= \frac{1}{2}\begin{bmatrix} e^{+i\omega_S t/2} & 0 \\ 0 & e^{-i\omega_S t/2} \end{bmatrix}\begin{bmatrix} 0 & i \\ -i & 0 \end{bmatrix}\begin{bmatrix} e^{-i\omega_S t/2} & 0 \\ 0 & e^{+i\omega_S t/2} \end{bmatrix} \\
&= \frac{1}{2}\begin{bmatrix} 0 & ie^{+i\omega_S t} \\ -ie^{-i\omega_S t} & 0 \end{bmatrix} \\
&= \begin{bmatrix} 0 & i[cos(\omega_S t) + isin(\omega_S t)] \\ -i[cos(\omega_S t) - isin(\omega_S t)] & 0 \end{bmatrix} \\
&= \begin{bmatrix} 0 & i\,cos(\omega_s t) \\ -i\,cos(\omega_s t) & 0 \end{bmatrix} + \begin{bmatrix} 0 & -sin(\omega_s t) \\ -sin(\omega_s t) & 0 \end{bmatrix} \\
&= \frac{1}{2}\left[-I_y cos(\omega_s t) - I_x sin(\omega_s t)\right]
\end{aligned} \tag{6.38}$$

This density matrix represents a spin precessing clockwise in the x-y plane at an angular velocity of ω_s, beginning from the minus y-axis.

6.2.1.2 Detection of the Signal

Equation 6.38 gives the density matrix that represents the FID during detection. Quadrature detection of the FID is defined as measuring $I^+ = I_x + iI_y$. In matrix notation I^+ is:

$$I^+ = I_x + iI_y = \frac{1}{2}\begin{bmatrix} 0 & 1 \\ 1 & 0 \end{bmatrix} + \frac{i}{2}\begin{bmatrix} 0 & -i \\ i & 0 \end{bmatrix} = \frac{1}{2}\begin{bmatrix} 0 & 1 \\ 0 & 0 \end{bmatrix} \quad (6.39)$$

The expectation value of I^+ is given by the trace of $I^+\rho$ (ignoring constants):

$$\begin{aligned} < I^+ > &= \text{Trace}\left(\begin{bmatrix} 0 & 1 \\ 0 & 0 \end{bmatrix}\begin{bmatrix} 0 & i\,e^{+i\omega_S t} \\ -i\,e^{-i\omega_S t} & 0 \end{bmatrix}\right) \\ &= \text{Trace}\left(\begin{bmatrix} -i\,e^{-i\omega_S t} & 0 \\ 0 & 0 \end{bmatrix}\right) \\ &= -ie^{-i\omega_S t} \end{aligned} \quad (6.40)$$

The Fourier transform of this function will give a dispersion lineshape found at ω_S. This result is completely consistent with a classical description of the effect of a 90° pulse applied along the x-axis, and detection such that the real channel is defined to be along the x-axis and the imaginary channel along the y-axis.

6.3 Product Operators

From the above analysis of a one-pulse experiment we see that it is possible to write ρ as a matrix that is *proportional* to one, or more, angular momentum operators:

- The initial density matrix, ρ_o, is proportional to S_z.
- The density matrix after the pulse, ρ_1, is proportional to S_y.
- The density matrix during detection is given by a combination of S_y and S_x.

This representation of the density matrix also provide a good deal of intuition to the quantum mechanical description of the system. For example, prior to the pulse, the bulk magnetization is aligned along the z-axis. After the 90° pulse on the x-axis, the magnetization is along the y-axis. During detection, the bulk magnetization will precess in the transverse plane.

The use of Cartesian angular momentum operators to represent the density matrix is referred to as the *product operator notation.* The origin of this name will become apparent when two coupled spins are analyzed, as products of angular momentum operators will be required to describe the density matrix. A detailed description of the product operator formalism can be found in [153].

For a single isolated spin, it is possible to represent any density matrix using a linear combination of the following four operators:

$$\tilde{1} \qquad I_x \qquad I_y \qquad I_z \quad (6.41)$$

where $\tilde{1}$ is the identity operator[2].

[2]The identity operator has no effect on a wavefunction. In the case of a spin-1/2 particle, $\tilde{1} = \begin{bmatrix} 1 & 0 \\ 0 & 1 \end{bmatrix}$.

Although the use of product operators provides a very convenient and concise algebraic representation of the density matrix, keep in mind that the various angular momentum operators simply *represent* the density matrix.

6.3.1 Transformation Properties of Product Operators

Since the density matrix can be represented by angular momentum operators, the effects of pulses and free precession on the density matrix can be determined by evaluating the effect of rotations on the angular momentum operators. In this context, pulses are represented by rotations about the x- or y-axis with a rotational angle equal to the flip angle of the pulse (β). Similarly, free precession is represented by a rotation about the z-axis with an angle ωt. The evolution of the density matrix due to rotations about the x-, y-, or z-axis are given in Table 6.1 and illustrated in Fig. 6.3.

These rotations follow the right-hand rule. In general, rotation of a product operator about an orthogonal axis gives the original product operator times $cos\beta$ plus the other orthogonal product operator, times $sin\beta$. For example, a z-rotation applied to $\rho = I_x$ generates the following:

$$I_x \xrightarrow{R_z(\beta)} I_x cos\beta + I_y sin\beta \tag{6.42}$$

These transformation laws are exactly as one would predict from the classical description of the system. For example, a 45° pulse applied along the y-axis would leave the bulk magnetic moment half-way between the z- and x-axis with a observed magnetic moment in the z-direction of $cos(45°)$ and the observed magnetic moment in the x-axis of $sin(45°)$.

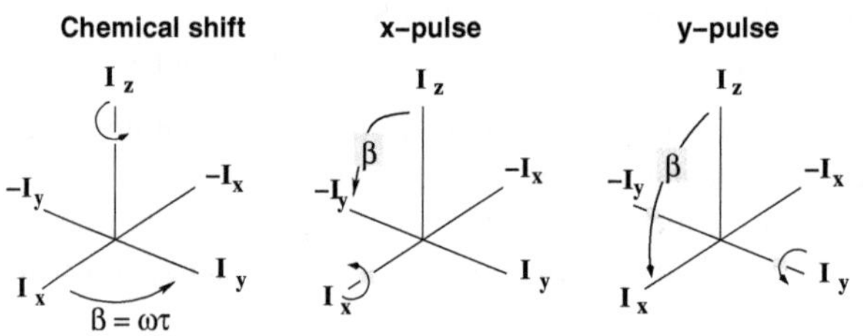

Figure 6.3 Evolution of single-spin product operators. A graphical representation of the effect of chemical shift evolution (rotation about the z-axis) and pulses on the single-spin density matrix is shown.

Table 6.1. Transformation of product operators for a single spin. Excitation pulses are represented as rotations about the x-axis (R_x) or the y-axis (R_y), with a flip angle of β. Evolution of the system under the static magnetic field, B_o, is represented by a rotation about the z-axis (R_z), with an overall rotation angle of ωt. In all cases, a right-handed rotation is used. The actual direction of evolution can be either clockwise or counter-clockwise, depending on the sign of γ.

ρ_o	R_x	R_y	R_z
I_x	I_x	$I_x cos(\beta) - I_z sin(\beta)$	$I_x cos(\omega t) + I_y sin(\omega t)$
I_y	$I_y cos(\beta) + I_z sin(\beta)$	I_y	$I_y cos(\omega t) - I_x sin(\omega t)$
I_z	$I_z cos(\beta) - I_y sin(\beta)$	$I_z cos(\beta) + I_x sin(\beta)$	I_z

6.3.1.1 Derivation of Transformation Laws

The effect of rotations on the density matrix can be derived in two ways. The first way utilizes the rotation operator derived earlier:

$$R_u(\beta) = cos\frac{\beta}{2}\tilde{1} - i\sigma_u sin\frac{\beta}{2} \tag{6.43}$$

This operator can be applied to the various starting representations of ρ (e.g. I_x, I_y, or I_z), to obtain the resultant density matrix.

Alternatively, the rotation of the density matrix can be written as a function (see Slichter [151] for more details):

$$f(\beta) = e^{-i\beta I_z} I_x e^{i\beta I_z} \tag{6.44}$$

and $f(\beta)$ is determined by trying to find a differential equation which provides a solution for $f(\beta)$. Taking the first derivative of $f(\beta)$ with respect to β gives:

$$\begin{aligned} \frac{df}{d\beta} &= -ie^{-i\beta I_z}\left[I_z I_x - I_x I_z\right] e^{i\beta I_z} \\ &= e^{-i\beta I_z}\left[I_y\right] e^{i\beta I_z} \end{aligned}$$

where the above step used the following commutator for angular momentum operators: $[I_x, I_z] = iI_y$. The second derivative is:

$$\begin{aligned} \frac{d^2 f}{d\beta^2} &= ie^{-i\beta I_z}\left[-I_z I_y + I_y I_z\right] e^{i\beta I_z} \\ &= -e^{-i\beta I_z}\left[I_x\right] e^{i\beta I_z} \\ &= -f(\beta) \end{aligned} \tag{6.45}$$

The solution to this equation is of the form:

$$f(\beta) = A\cos(\beta) + B\sin(\beta) \tag{6.46}$$

The constants, A and B, are obtained from $f(0)$ and $df/dt(0)$:

$$f(0) = I_x \qquad f'(0) = I_y$$

Thus:

$$f(\beta) = ie^{-i\beta I_z} I_x e^{i\beta I_z} = I_x\cos(\beta) + I_y\sin(\beta) \tag{6.47}$$

6.3.2 Description of the One-pulse Experiment

The product operator notation shown in Table 6.1 permits the rapid evaluation of the outcome of NMR experiments. The simple one-pulse experiment, consisting of a 90° degree pulse along the x-axis, is evaluated as follows (see Fig. 6.4):

$$I_z \xrightarrow{R_x(\pi/2)} -I_y \xrightarrow{R_z(\omega t)} -I_y cos(\omega t) + I_x sin(\omega t) \tag{6.48}$$

The observed signal is obtained in the usual fashion:

$$\begin{aligned} \text{Signal(t)} &= \text{Trace}[I^+ \rho_2(t)] \\ &= \text{Trace}[I^+(-I_y cos(\omega t) + I_x sin(\omega t))] \\ &= -ie^{i\omega t} \end{aligned} \tag{6.49}$$

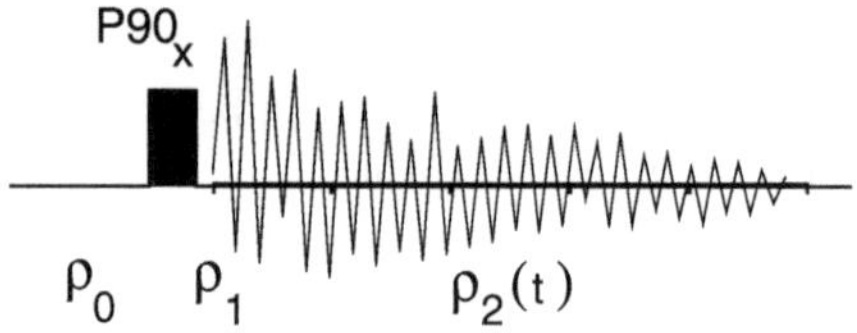

$$\rho_o = \xrightarrow{R_x(\pi/2)} \rho_1 = \xrightarrow{R_z(\omega_s t)} \rho_2(t) =$$

$$\frac{1}{2}\begin{bmatrix} 1 & 0 \\ 0 & -1 \end{bmatrix} \qquad \frac{1}{2}\begin{bmatrix} 0 & i \\ -i & 0 \end{bmatrix} \qquad \frac{1}{2}\begin{bmatrix} 0 & ie^{+i\omega_S t} \\ -ie^{-i\omega_S t} & 0 \end{bmatrix}$$

$$I_z \qquad -I_y \qquad -I_y cos(\omega_s t) + I_x sin(\omega_s t)$$

Figure 6.4. One-pulse experiment, representation by the density matrix and product operators. Equivalent description of a one pulse experiment using the density matrix (top), or product operator notation (bottom). Note that the sign of the frequency, ω_S, differs between the two methods of analysis because of the adoption of the right-hand-rule for the evolution of product operators.

6.3.3 Evaluation of Composite Pulses

The product operator notation is also useful for evaluating the effect of multiple pulses on the magnetization. A useful class of these experiments are composite pulses. These are a series of pulses whose overall effect is to produce a specific flip angle with reduced sensitivity to non-ideal pulse lengths or rotation angles. For example, a widely used composite 180° (π) pulse is:

$$[\frac{\pi}{2}]_x - [\pi]_y - [\frac{\pi}{2}]_x \tag{6.50}$$

The effect of this pulse on magnetization beginning along the z-axis (i.e. $\rho_o = I_z$) is rapidly obtained as follows:

$$I_z \xrightarrow{\pi/2_x} -I_y \xrightarrow{\pi_y} -I_y \xrightarrow{\pi/2_x} -I_z \tag{6.51}$$

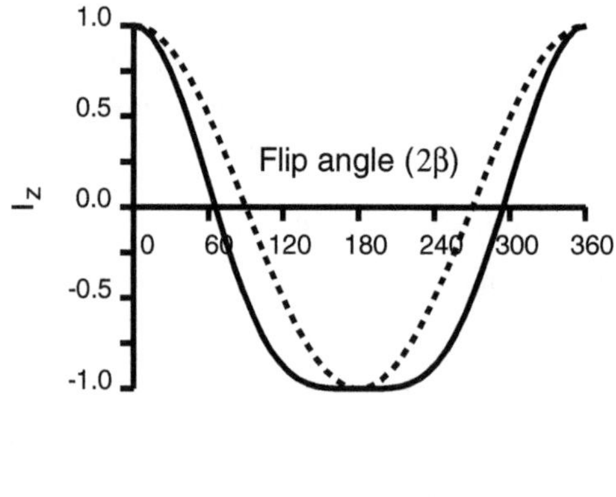

Figure 6.5 Excitation profile of 180° composite pulse. The effect of varying the flip angle of a single π pulse (dashed line) and a composite π pulse on the z-component of the magnetization are shown (solid line). The x-axis indicates the flip angle of the π pulse, one half of this value would be used for the $\pi/2$ pulses in the composite pulse. The y-axis shows the I_z component of the density matrix.

Now assume that the flip angles are set to $\beta < 90°$, then the final density matrix after this pulse is:

$$\rho = I_z[cos^2\beta\ cos2\beta - sin^2\beta] - I_y[cos\beta sin\beta cos2\beta + sin\beta cos\beta] + I_x[cos\beta sin2\beta] \tag{6.52}$$

The effect of the flip angle on the inversion profile of the simple π pulse and the composite pulse are shown in Fig. 6.5. It is clear that the composite pulse is capable of inverting the magnetization over a much wider range of flip angles and is thus less affected by having $\beta \neq 90°$.

The above analysis varied only the flip angle and assumed that the pulses were along the x- or y-axis. As the frequency of the pulses are moved away from the resonance frequency of the spin, the pulse length will become non-ideal *and* the rotation axis for the pulses will be displaced from the x-y plane. The greater the difference between the frequency of the pulse and the spin, the less effective the inversion will become. It can be shown that the composite pulse is also less sensitive to these resonance offset effects and is able to invert the magnetization over a wider frequency range than a simple π pulse.

6.4 Exercises

1. The NMR sample volume generally extends beyond the excitation/receive coil. This region of the magnetic field is less homogeneous, thus signals from this region will display a poor lineshape. This region of the sample experiences very weak B_1 fields during pulses, because it is outside of the coil. Show that the following composite pulse will suppress signals from this region, but produce the equivalent of a 90° pulse for flip angles close to 90°, i.e for spins within the sample coil.

$$\beta_x\ \beta_y\ \beta_{-x}\ \beta_{-y} \tag{6.53}$$

   where, β is the flip angle.

2. Using the product operator formalism determine the effect of the following composite pulse: $[\frac{\pi}{2}]_x - \pi_y - [\frac{\pi}{2}]_x$ with $\beta < 90°$, on $\rho = I_z$. That is, show that Eq. 6.52 holds true.

3. Show, using product operators, that there is no evolution of the chemical shift at the end of the following sequence: $\tau - P^x_{180} - \tau$. Assume that the density matrix is equal to I_x at the beginning of the first τ period.

6.5 Solutions

1. The effect of this composite pulse can be easily evaluated using product operators. The effect of each pulse on the density matrix is shown in the following table for two possible flip angles (β is assumed to be small, such that $cos(\beta) = 1$ and $sin(\beta) \approx \beta$.

Analysis of Composite Pulses Using Product Operators.

Pulse	$\beta = 90°$ Effect on ρ	$\beta \ll 90°$ Effect on ρ
x	$-I_y$	$I_z - \beta I_y$
y	$-I_y$	$I_z + \beta I_x - \beta I_y$
$-x$	$+I_z$	$I_z + \beta I_x$
$-y$	$-I_x$	I_z

After the completion of all four pulses, spins within the coil are represented by the density matrix $-I_x$, indicating that the magnetization is in the x-y plane. Thus the composite pulse behaved as if it were a single 90° pulse. In contrast, spins that were outside of the excitation coil are returned to the z-axis by the four-step sequence, thus they do not produce a detectable signal.

2.

$$
\begin{aligned}
& I_z \xrightarrow{\beta_x} I_z cos\beta - I_y sin\beta \\
\xrightarrow{2\beta_y}\; & cos\beta[I_z cos2\beta + I_x sin2\beta] - I_y sin\beta \\
\xrightarrow{\beta_x}\; & I_z[cos^2\beta cos2\beta - sin^2\beta] - I_y[cos\beta sin\beta cos2\beta + sin\beta cos\beta] \\
& \quad + I_x[cos\beta sin2\beta]
\end{aligned}
$$

3. During the first τ period, I_x evolves by precession about the z-axis, which is equivalent to a rotation about the z-axis by an angle $\omega\tau$:

$$I_x \rightarrow I_x cos(\omega\tau) + I_y sin(\omega\tau)$$

the 180° pulse along the x-axis causes the following transformation of the density matrix:

$$I_x cos(\omega\tau) + I_y sin(\omega\tau) \rightarrow I_x cos(\omega\tau) - I_y sin(\omega\tau)$$

Note that only the I_y term is affected by the pulse along the x-axis.

During the second τ period each term of the density matrix evolves as follows:

$$
\begin{aligned}
I_x cos(\omega\tau) - I_y sin(\omega\tau) \rightarrow \quad & cos(\omega\tau)[I_x cos(\omega\tau) + I_y sin(\omega\tau)] \\
& -sin(\omega\tau)[I_y cos(\omega\tau) - I_x sin(\omega\tau)] \\
= \quad & I_x[cos^2(\omega\tau) + sin^2(\omega\tau)] \\
& +I_y[cos(\omega\tau)sin(\omega\tau) - cos(\omega\tau)sin(\omega\tau)] \\
= \quad & I_x
\end{aligned}
$$

Since the density matrix has been returned, or refocused, to its original state, the sequence: $\tau - P_{180} - \tau$ is often referred to as a refocusing element. The evolution of the chemical shift during the first τ period is reversed during the second τ period.

Chapter 7

SCALAR COUPLING

Scalar couplings arise from spin-spin interactions that occur via bonding electrons. Consequently, they provide information on the chemical *connectivity* between atoms. Therefore, these couplings can be utilized to correlate NMR signals of atoms that are chemically bonded to one another, providing chemical shift assignments if the molecular structure is known. In particular, the scalar coupling across the peptide bond permits the linkage of spins within one amino acid to those of its neighbors, as discussed in Chapter 13.

In addition to providing information on chemical connectivities, the sizes of three bond scalar couplings are sensitive to the electron distribution of the intervening bonds, consequently these couplings can provide information on the conformation of rotatable bonds in proteins.

In this chapter we will first explore the origin of scalar couplings between nuclear spins, understanding the effect of this coupling on the resultant NMR spectrum from a classical perspective. The coupling will then be analyzed using quantum mechanics to fully evaluate the effect of the coupling on the frequency and intensity of resonance lines in the NMR spectrum of coupled spins. Finally, a density matrix treatment of coupled spins will be introduced in the subsequent Chapter as a prelude to analyzing the effect of scalar coupling in more complex multi-dimensional NMR spectra.

7.1 Introduction to Scalar Coupling

Scalar, or J-coupling, occurs between nuclei which are connected by chemical bonds. This coupling causes splitting of the spectral lines for both coupled spins by an amount J, or the coupling constant (See Fig. 7.1). The nomenclature that is used to describe the coupling is as follows:

$$^{n}J_{AB}$$

where n refers to the number of intervening bonds, and A and B identify the two coupled spins. For example, the coupling constant between the amide nitrogen and the C_β carbon would be written as: $^{2}J_{NC_\beta}$. The value of J is usually given in Hz and is the observed frequency separation between the split resonance lines of the coupled spins (see below). A resonance line that is split due to J-coupling is generally referred to as a multiplet. The spectrum shown in Fig. 7.1 is an example of a doublet. If the resonance line was split into three signals, it would be called a triplet. Finally, splitting into four lines generates a quartet.

The effect of J-coupling on the spectrum depends on the frequency separation of the coupled spins. If the two coupled spins differ greatly in their resonance frequencies ($\Delta\nu > J$), then the system is referred to as an AX system, where the X signifies the fact that the two chemical shifts are quite different. All coupling between different atom types, or heteronuclear spins, are AX couplings because of the large difference in the frequencies of coupled spins. Examples include, J_{NH}, J_{CH}, and J_{NC}. AX couplings can be analyzed using a classical analysis, similar to that depicted in Fig. 7.1. When two coupling spins have nearly equivalent resonance frequencies ($\Delta\nu \leq J$) then the system is referred to as an AB system. For example, the coupling between two H_β protons on an amino acid is an example of an AB system. Accurate analysis of AB systems require a detailed quantum mechanical treatment. Lastly, when the coupled spins have the identical resonance frequencies, the observed coupling disappears entirely. This is most often seen when multiple protons have equivalent environments, such as the three protons on a methyl group.

7.2 Basis of Scalar Coupling

Scalar coupling arises from the interaction of the nuclear magnetic moment with the electrons involved in the chemical bond. The nuclear spin polarization of one atom affects the polarization of the surrounding electrons. The electron polarization subsequently produces a change in the magnetic field that is sensed by the coupled spin. For example, consider a C-H group in a molecule, as illustrated in Fig. 7.1. The proton nuclear spin polarizes the electron in the σ bonding orbital. This polarization alters the magnetic field at the carbon nucleus. Since there are two possible spin states for the proton magnetic dipole, the effective field at the carbon nucleus is increased

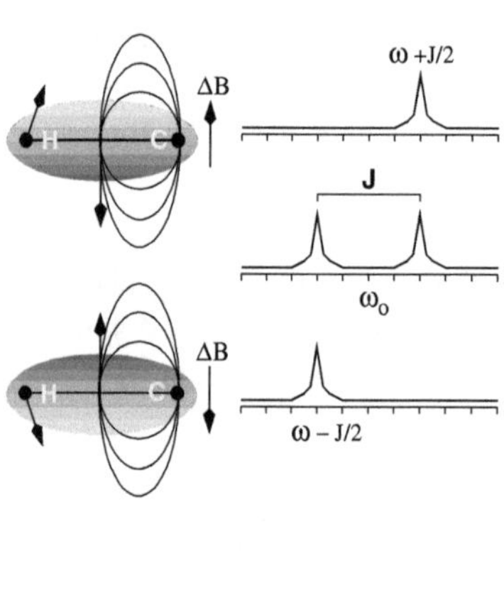

Figure 7.1 Nuclear spin coupling in a ^{13}C-H group. Two C-H groups in separate, but otherwise identical molecules, are shown. The two molecules differ only in the spin state of the proton. The bonding electrons in the σ orbital become polarized from the proton magnetic dipole, as indicated by the oval field lines. The polarization of the electrons can either increase or decrease the magnetic field at the ^{13}C nucleus, depending on the proton spin state. One orientation of the proton spin increases the magnetic field (upper molecule), while the other orientation causes a decrease in the apparent field (lower molecule), causing either an upfield or downfield shift in the resonance line, depending on the spin-state of the attached proton. The observed carbon spectrum is the sum of these two resonances, giving two peaks that are separated by the J-coupling constant, as shown in the middle spectrum.

or decreased, depending on the spin state of the attached proton. Since the population differences between the two orientations of the proton spin are approximately equal, one-half of the attached carbons will experience an increase in the local magnetic field while the other half will experience a decrease. This difference in the local magnetic field at the carbon nucleus will lead to a shift of the carbon resonance frequency. Since there are two possible proton spin states, the carbon spectrum is split into two lines, with the separation between the lines equal to the J-coupling.

The change in the magnetic field induced by the proton spins is proportional to γ_H. The corresponding change in the carbon resonance frequency is proportional to the product of this field change and the gyromagnetic ratio of the carbon spin, i.e.:

$$\Delta\omega \propto \pm\gamma_H\gamma_C \tag{7.1}$$

The effect of the carbon spin on the proton spin is calculated in the same way. The change in the local magnetic field at the proton nucleus is proportional to $\pm\gamma_C$, giving rise to a frequency shift of the protons of $\pm\gamma_C\gamma_H$. Consequently, the proton spin will experience exactly the same shift in frequency as its coupled partner, giving rise to exactly the same splitting of the proton resonance line as the resonance line from the attached carbon. Also note that the observed frequency shift only depends on the product of the gyromagnetic ratios of the coupled spins; the scalar coupling constant is independent of the applied magnetic field (B_o).

Values of J-coupling constants that are important in biomolecular NMR are shown in Table 7.1. The strength of the J-coupling depends on several factors, including the gyromagnetic ratio of the coupled spins, the number of bonds connecting the coupled spins, and the conformation of the intervening bonds in the case of multiple bond couplings. The series of single bond heteronuclear couplings (Table 7.1, left column) illustrates the effect of the gyromagnetic ratio on the coupling constant; the coupling constant increases with increasing γ. Scalar coupling through multiple bonds greatly attenuates the coupling. For example, the strong single bond H-C coupling of 130 Hz is reduced to 5 Hz when an additional carbon-carbon bond is inserted between the two coupled spins (Table 7.1).

In the case of multiple bond couplings, the conformation of the coupled atoms affects the coupling constant. For example, the three bond proton-proton coupling in

Table 7.1. Homonuclear and heteronuclear coupling constants. Homonuclear (proton-proton) and heteronuclear coupling constants that are commonly found in biopolymers are listed. The values in this table are approximate; the coupling constants will also be affected by the electronic environment of the associated spins.

Couplings Involving Heteronuclear (^{13}C or ^{15}N) Spins		*Proton-Proton Couplings*	
C-N	14 Hz	H-C-H	-12 to -15 Hz
C-C	35 Hz	H-C-C-H	2-14 Hz
H-N	92 Hz	H-C=C-H	10 (cis)/17 (trans)
H-C	130 Hz	H-N-C-H	1-10 Hz
H-C-C	5 Hz (two bond coupling)		(3 Hz α-helix)
			(10 Hz β-strand)

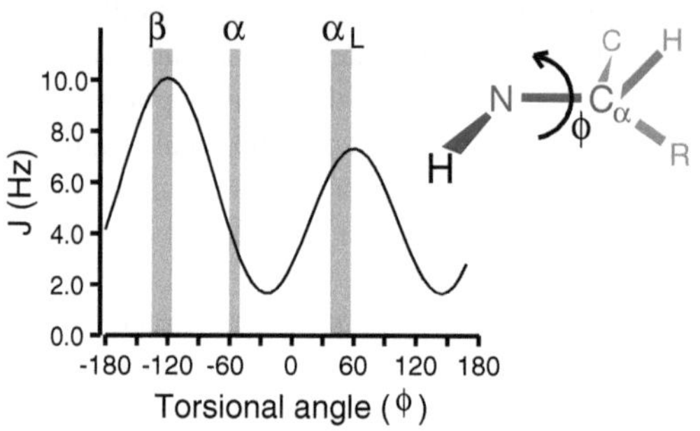

Figure 7.2 Karplus curve for a peptide group. The relationship between J and the ϕ torsional angle in polypeptides is shown. The ϕ angles for regular secondary structures are indicated by the vertical gray bars. The ϕ torsional angle is defined by the relative orientation of the H-N bond vector to the C_α-CO bond vector. The molecular fragment to the right of the plot has a ϕ angle of 180°. The actual curve plotted is: $J = 6.98\, cos^2(\phi - 60) - 1.38\, cos(\phi - 60) + 1.72$ (from Ref. [162]).

the H-C-C-H group ranges from 2 to 14 Hz. The relationship between the coupling constant and the torsional angle is represented by the Karplus relationship [80]:

$$J = Acos^2\theta + Bcos\theta + C \tag{7.2}$$

where A, B, and C are empirical constants. For example, the ϕ angle in the peptide bond affects the strength of the coupling between the amide proton and the alpha proton, as illustrated in Fig. 7.2.

7.2.1 Coupling to Multiple Spins

The coupling between a carbon and a hydrogen in a ^{13}C-H group results in the splitting of both the proton and carbon spectral line by an amount J_{CH} Hz. If the carbon atom is coupled to more than one *equivalent* proton[1], such as in a $^{13}CH_2$ or $^{13}CH_3$ group, then a more complex splitting pattern is observed.

In the case of a $^{13}CH_2$ group a triplet of lines is observed in the carbon NMR spectrum, as illustrated in Fig. 7.3. This pattern arises because there are four possible combinations of the spin-states of the two coupled protons and the local magnetic field at the carbon nucleus will be the sum of the field changes induced by each proton. When the magnetic moment of both protons point upwards, in the direction of B_o, the frequency shift of the resonance line will be $2 \times J/2$ Hz, or J Hz. When the proton spins are both pointing in the opposite direction the shift is $-J$ Hz. When the direction of the proton spins are opposite to each other, the local fields at the carbon cancel, resulting in a zero frequency shift of the carbon spin. The intensity of the lines in the carbon spectrum is proportional to the number of molecules in the sample having one of the four possible proton spin-states. Since the state in which the proton spins point in opposing directions occurs twice as frequently as the other two states, the central line of the triplet will have twice the intensity of the outer lines, giving an observed intensity ratio of 1:2:1, as illustrated in Fig. 7.3.

The coupling between a carbon atom and three equivalent protons, such as in a methyl group ($^{13}CH_3$), can be analyzed in exactly the same way. The change in the

[1]Equivalent protons are generally considered to be a collection of protons that are attached to a single carbon atom and have the same chemical shift. Equivalency is most often a result of free rotation of the group, which averages the local environments of all of the protons.

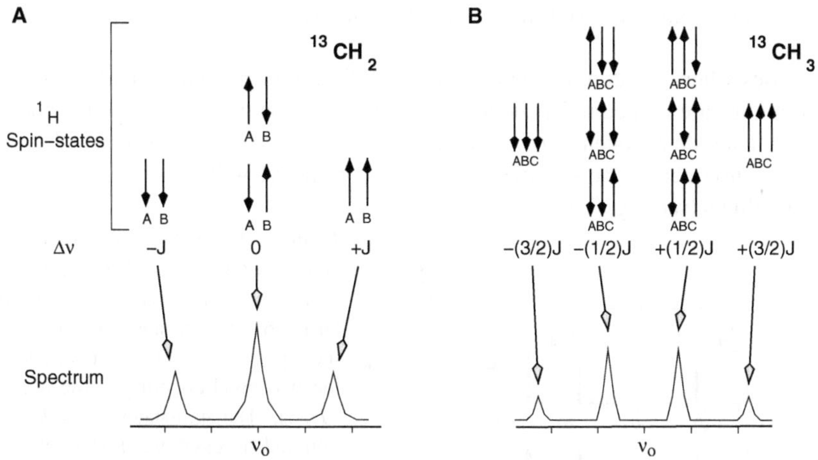

Figure 7.3. Scalar coupling to multiple equivalent protons. The effect of scalar coupling to two (A) or three (B) equivalent protons on the carbon spectrum is shown. The possible arrangements of two (labeled A B) or three protons (labeled A B C) are shown in the upper part of each panel (^{1}H spin-states). The resultant shift in the frequency of the attached carbon is indicated by $\Delta\nu$. The final carbon spectrum is shown in the lower part of each panel. In both cases the splitting, or separation between the lines, is equal to $^1J_{CH}$. The intensity of each line depends on the number of molecules in the sample with a particular spin state; a 1:2:1 ratio will be found for two coupled protons and a 1:3:3:1 ratio is found for three coupled protons.

local field that occurs when the magnetic dipoles from all of the protons are aligned in the same direction is $\pm\frac{3}{2}J$. When the magnetic dipoles of two protons are oriented in the same direction, while the third is pointing in the opposite direction, the frequency shift is $\pm\frac{1}{2}J$ since the opposing pairs cancel each other's effect on the local magnetic field at the carbon nucleus. Again, the relative intensity of each line is proportional to the number of atoms that give a particular frequency shift, in this case the four lines in the quartet will have a relative intensities of 1:3:3:1.

Homonuclear, proton-proton, couplings are analyzed in a similar way. For example, consider the NMR spectrum of ethanol, shown in Fig. 2.6. In this example, all of the carbons are ^{12}C and are therefore NMR inactive. The two CH_2 protons are split into a quartet by the three equivalent methyl protons. Likewise, the three methyl protons are split into a triplet due to coupling to the two equivalent CH_2 protons.

The effect of coupling to multiple spins on an NMR spectral line can be easily obtained from Pascal's triangle, as illustrated in Fig. 7.4. Each row of the triangle

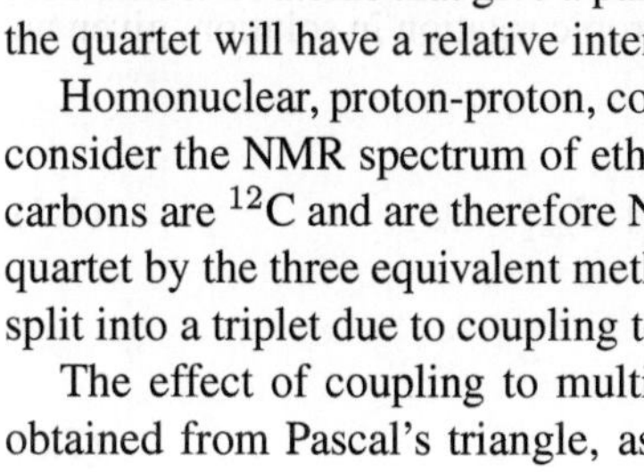

Figure 7.4 Analysis of J-coupling using Pascal's triangle. Pascal's triangle can be used to readily evaluate the effect of coupling to multiple equivalent spins on the appearance of a resonance line. The top of the triangle represents the resonance line from a spin with no coupling partner ($n = 0$) and each subsequent row represents the spectra that would be obtained as a result of coupling to one, two, or three additional spins.

indicates the location of each line in the multiplet as well as the relative intensity of the line.

In cases where an atom is coupled to two different, or non-equivalent, spins, then the couplings are treated independently. For example, the carbonyl carbon is coupled to both the amide nitrogen ($^1J_{NC'} \approx 12$ Hz) as well as the alpha carbon, ($^1J_{C'C_\alpha} \approx$ 55 Hz), consequently the spectra line from the carbonyl will be a quartet, showing both couplings (see Fig. 7.5).

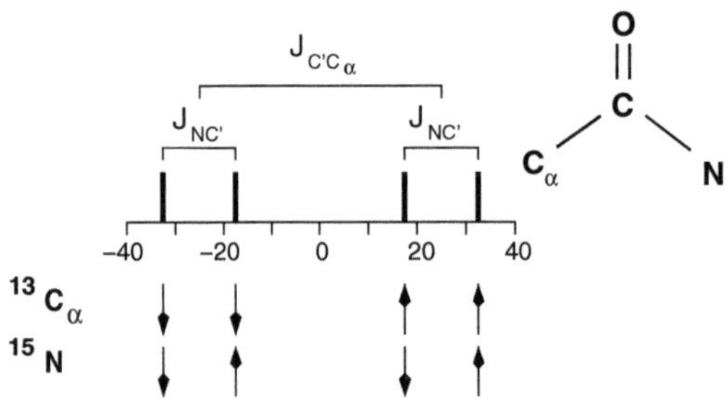

Figure 7.5 Scalar coupling to non-equivalent spins. The spectrum of a carbonyl carbon is shown. A quartet is observed because the coupling to the alpha carbon (J_{CC_α}) is larger than the two bond coupling to the nitrogen (J_{NC}). The spin-states of the C_α carbon and nitrogen are shown below the spectrum. Since these four states are equally likely, the ratio of the intensities of the lines in the quartet are 1:1:1:1.

7.3 Quantum Mechanical Description

The Hamiltonian that describes scalar coupling between spins is given by:

$$\mathcal{H} = \vec{I}\ \tilde{J}\ \vec{S} \tag{7.3}$$

The scalar coupling is represented by a tensor quantity[2] which will be diagonal in some coordinate frame:

$$\tilde{J} = \begin{bmatrix} J_{xx} & 0 & 0 \\ 0 & J_{yy} & 0 \\ 0 & 0 & J_{zz} \end{bmatrix} \tag{7.4}$$

Usually, the different elements are averaged by isotropic rotation in solution, given an observed average value which is a scalar:

$$< J >= \frac{1}{3}[J_{xx} + J_{yy} + J_{zz}] \tag{7.5}$$

This simplifies the Hamiltonian to:

$$\mathcal{H} = J\vec{I} \cdot \vec{J} \tag{7.6}$$

7.3.1 Analysis of an AX System

The dot product, $\vec{I} \cdot \vec{S}$, expands to:

$$I \cdot S = I_x S_x + I_y S_y + I_z S_z \tag{7.7}$$

[2]See Appendix B for a comparison of scalars, vectors, and tensors.

In general, it would be necessary to use the complete expression for the dot product when analyzing the effect of the coupling on the energy states. However, if the frequency difference between the coupled spins is larger than the J-coupling, as is the case in an AX spin-system, then terms involving transverse operators can be dropped. This leads to a simplified representation of the Hamiltonian for a pair of coupled spins:

$$\mathcal{H} = -\omega_I I_z - \omega_S S_z + 2\pi J I_z S_z \tag{7.8}$$

or in frequency units:

$$\mathcal{H} = -\nu_I I_z - \nu_S S_z + J I_z S_z \tag{7.9}$$

Before using this Hamiltonian to calculate the energy levels of the system it is necessary to write expressions for the basis states of the system. Four new states are generated by taking all possible combinations of the original basis vectors that were associated with each spin. These states are[3]:

$$\phi_1 = |\alpha\alpha> \quad \phi_2 = |\beta\alpha> \quad \phi_3 = |\alpha\beta> \quad \phi_4 = |\beta\beta> \tag{7.10}$$

These wavefunctions are eigenfunctions of the *uncoupled* Hamiltonian. The first character (α or β) refers to the I spin while the second character refers to the S spin. In both cases α is associated with an m_z of +1/2, and β is associated with an m_z of -1/2. For example, $|\alpha\beta>$ is a wavefunction in which the I spin has an m_z value of +1/2 and the S spin has an m_z value of -1/2. The four basis states form an ortho-normal basis set, e.g. $<\alpha\alpha|\alpha\alpha>= 1, \; <\alpha\alpha|\alpha\beta>= 0$.

The energy of each of these states is calculated directly from the Hamiltonian ($\mathcal{H}|\Psi>= E_\Psi|\Psi>$):

$$\begin{aligned}
\mathcal{H}|\phi_1> &= -\nu_I/2 - \nu_S/2 + J/4|\alpha\alpha> & (7.11)\\
\mathcal{H}|\phi_2> &= +\nu_I/2 - \nu_S/2 - J/4|\beta\alpha> & (7.12)\\
\mathcal{H}|\phi_3> &= -\nu_I/2 + \nu_S/2 - J/4|\alpha\beta> & (7.13)\\
\mathcal{H}|\phi_4> &= +\nu_I/2 + \nu_S/2 + J/4|\beta\beta> & (7.14)
\end{aligned}$$

The four energy levels are illustrated in Fig. 7.6. Six different transitions are possible with this system. Observable transitions are those that have a non-zero value for the scalar product of the ground and excited state with $I_x + S_x$. The actual probability that the transition will occur is given by the square of the scalar product:

$$P_{g\rightarrow e} = |<\psi_e|I_x + S_x|\psi_g>|^2 \tag{7.15}$$

Of the six possible transitions, the following four are observable with equal intensity, with the indicated energies:

[3]Normally, the convention would be to write these states as: $\phi_2 = |\alpha\beta>, \phi_3 = \beta\alpha>$, however, the order used here facilitates solving an AB coupled system, as described in Section 7.3.2.

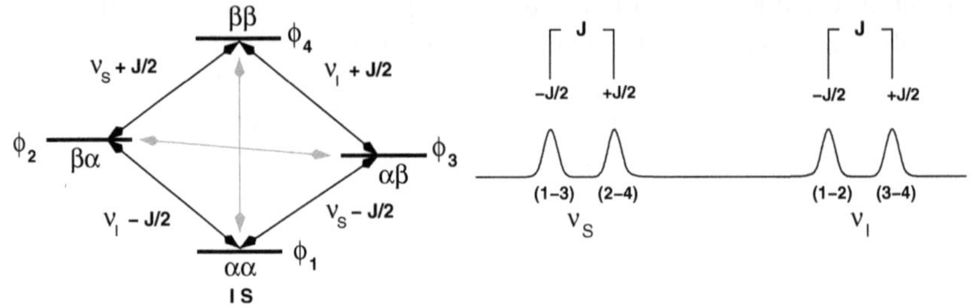

Figure 7.6. Energy levels and resultant spectrum for two coupled spins. The energy level diagram on the left shows the relative energy of all four states. The observable single quantum transitions are indicated by black arrows and the observed frequency of these transitions, in Hz, are also given. Unobservable zero- and double-quantum transitions are indicated with the gray arrows. The right side of the diagram gives the resultant spectrum. The two states that are involved in each transition are indicated underneath each peak. For example, the left-most peak arises from a transition between ϕ_1 and ϕ_3.

$$E_{1\to 2} = \nu_I - J/2 \quad (7.16)$$
$$E_{1\to 3} = \nu_S - J/2 \quad (7.17)$$
$$E_{2\to 4} = \nu_S + J/2 \quad (7.18)$$
$$E_{3\to 4} = \nu_I + J/2 \quad (7.19)$$

These transitions represent a normal 1D NMR spectrum between two coupled spins, as shown in Fig. 7.6.

In addition to the above single quantum transitions there is also a double quantum transition: $E_{\alpha\alpha\to\beta\beta} = \omega_I + \omega_S$, and a zero quantum transition: $E_{\alpha\beta\to\beta\alpha} = \omega_I - \omega_S$, neither of which are directly observable since they involve an overall change in m_z of ± 2 or zero.

7.3.2 Analysis of an AB system

When the chemical shift difference between the two coupled spins is of the same order as the coupling constant it is then necessary to perform a more complete quantum mechanical treatment. This analysis will show that the original eigenstates of the coupled spins are no longer eigenstates of the complete Hamiltonian. It will be necessary to find new eigenstates and eigenvalues (energies) of the system. A remarkable outcome of this analysis is that for the case of completely equivalent spins, such as the protons in a -CH_3 group, the proton-proton coupling vanishes and the quartet collapses into a single line.

We will begin by writing the complete Hamiltonian (in frequency units):

$$\mathcal{H} = -\nu_I I_z - \nu_s S_z + J(I_x S_x + I_y S_y + I_z S_z) \quad (7.20)$$

It is very useful to replace the transverse spin operators (e.g. I_x) by raising and lowering operators:

$$I_x = \frac{1}{2}(I^+ + I^-) \qquad I_y = \frac{1}{2i}(I^+ - I^-) \tag{7.21}$$

$$\begin{aligned} I_x S_x + I_y S_y &= \frac{1}{4}(I^+S^+ + I^+S^- + I^-S^+ + I^-S^-) \\ &\quad -\frac{1}{4}(I^+S^+ - I^+S^- - I^-S^+ + I^-S^-) \\ &= \frac{1}{2}[I^+S^- + I^-S^+] \end{aligned} \tag{7.22}$$

The Hamiltonian has the following form using the previous basis vectors:

$$\mathcal{H} = \begin{array}{ccccc} & |\alpha\alpha> & |\beta\alpha> & |\alpha\beta> & |\beta\beta> \\ <\alpha\alpha| & -\sum\nu/2 + J/4 & 0 & 0 & 0 \\ <\beta\alpha| & 0 & \Delta\nu/2 - J/4 & J/2 & 0 \\ <\alpha\beta| & 0 & J/2 & -\Delta\nu/2 - J/4 & 0 \\ <\beta\beta| & 0 & 0 & 0 & +\sum\nu/2 + J/4 \end{array} \tag{7.23}$$

where $\Delta\nu = \nu_I - \nu_S$ and $\sum\nu = \nu_I + \nu_S$.

The wavefunctions $|\alpha\alpha>$ and $|\beta\beta>$ are still eigenvectors of the complete Hamiltonian since the off-diagonal elements are all zero for these two wavefunctions. In contrast, the $|\alpha\beta>$ and $|\beta\alpha>$ wavefunctions have become mixed by the coupling. It is necessary to find a linear combination of these two functions that are eigenvectors of the complete Hamiltonian. These are obtained by finding the eigenvalues and eigenvectors of the central part of the above matrix (See Cohen-Tannoudji *et al.*[42] for an excellent description of the diagonization of 2x2 matrices. Their results are summarized in this discussion.):

$$\begin{array}{ccc} & |\beta\alpha> & |\alpha\beta> \\ <\beta\alpha| & \Delta\nu/2 - J/4 & J/2 \\ <\alpha\beta| & J/2 & -\Delta\nu/2 - J/4 \end{array} \tag{7.24}$$

The eigenvalues are found by solving the characteristic equation, and they are found to be:

$$E_2 = -\frac{J}{4} + \frac{1}{2}\sqrt{\Delta\nu^2 + J^2} \qquad E_3 = -\frac{J}{4} - \frac{1}{2}\sqrt{\Delta\nu^2 + J^2} \tag{7.25}$$

Finding the eigenvectors of the Hamiltonian is equivalent to finding the rotation that will diagonalize the Hamiltonian. It is convenient to define the rotation angle θ as:

$$tan\theta = \frac{J}{\Delta\nu} \tag{7.26}$$

where $\theta = 0$ when $\Delta\nu >> J$ and $\theta = \pi/2$ when $\Delta\nu << J$.

The eigenvectors, expressed as linear combinations of the original eigenvectors, are:

$$\phi_2 = cos\frac{\theta}{2}|\beta\alpha> + sin\frac{\theta}{2}|\alpha\beta> \qquad \phi_3 = -sin\frac{\theta}{2}|\beta\alpha> + cos\frac{\theta}{2}|\alpha\beta> \tag{7.27}$$

Table 7.2. Observed transitions for two AB coupled spins.

Transition	*Energy*	*Intensity*
$1 \rightarrow 2$	$\bar{\nu} + \frac{1}{2}\sqrt{\Delta\nu^2 + J^2} - \frac{1}{2}J$	$\frac{1}{4}(1 + sin\theta)$
$1 \rightarrow 3$	$\bar{\nu} - \frac{1}{2}\sqrt{\Delta\nu^2 + J^2} - \frac{1}{2}J$	$\frac{1}{4}(1 - sin\theta)$
$2 \rightarrow 4$	$\bar{\nu} - \frac{1}{2}\sqrt{\Delta\nu^2 + J^2} + \frac{1}{2}J$	$\frac{1}{4}(1 + sin\theta)$
$3 \rightarrow 4$	$\bar{\nu} + \frac{1}{2}\sqrt{\Delta\nu^2 + J^2} + \frac{1}{2}J$	$\frac{1}{4}(1 - sin\theta)$

$\bar{\nu} = (\nu_I + \nu_S)/2$ is the frequency at the center of the quartet.

The four observable transitions, their energies and intensities, are shown in Table 7.2. The transition probabilities are calculated as described in Eq. 7.15. For example, the probability of the $2 \rightarrow 1$ transition is obtained as follows:

$$
\begin{aligned}
P_{2-1} &= | < \phi_1 | I_x + S_x | \phi_2 > |^2 \\
&= | < \alpha\alpha | I_x + S_x | \left[cos\frac{\theta}{2} |\beta\alpha> + sin\frac{\theta}{2} |\alpha\beta> \right] |^2 \\
&= \frac{1}{4} | < \alpha\alpha | I^+ + I^- + S^+ + S^- | \left[cos\frac{\theta}{2} |\beta\alpha> + sin\frac{\theta}{2} |\alpha\beta> \right] |^2 \\
&= \frac{1}{4} | < \alpha\alpha | \left[sin\frac{\theta}{2} |\beta\beta> + cos\frac{\theta}{2} |\alpha\alpha> + cos\frac{\theta}{2} |\beta\beta> + sin\frac{\theta}{2} |\alpha\alpha> \right] |^2 \\
&= \frac{1}{4} | \left[cos\frac{\theta}{2} + sin\frac{\theta}{2} \right] |^2 \\
&= \frac{1}{4}(1 + sin\theta)
\end{aligned}
\tag{7.28}
$$

The effect of the frequency separation between the coupled spins on the appearance of the spectra is shown in Fig. 7.7. In this simulation the J-coupling constant was 10 Hz, a typical value for proton-proton couplings in amino acids. The lower curve shows the expected spectra for a frequency separation of 100 Hz, or 0.2 ppm on a 500 MHz spectrometer. Although $\Delta\nu = 10J$, some effects of the mixing of the states are

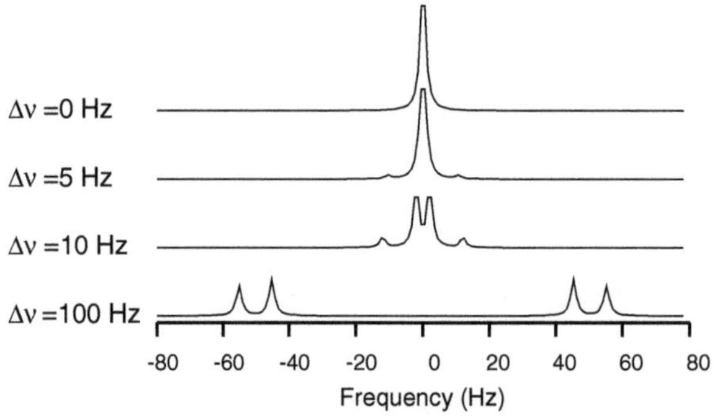

Figure 7.7 Effect of frequency separation on observed J-coupling. Simulated spectra are shown to illustrate the collapse of observed coupling as $\Delta\nu$ becomes smaller than J. The J-coupling constant is 10 Hz, and the separation between the lines is decreased from 100 Hz (bottom spectrum) to 0 Hz (top spectrum).

already apparent. The outer transitions show a lower intensity and the positions of the lines are shifted by a few 10^{ths} of a Hertz. As the frequency separation between the two coupled spins decreases, the intensity of the outer lines decreases and the lines move towards the average frequency. Finally, when the spins are equivalent ($\nu_I = \nu_S$; $\Delta\nu = 0$), there is no observable coupling.

7.4 Decoupling

Scalar coupling leads to splitting of spectra lines and therefore reduces the signal-to-noise in the spectrum by spreading the intensity over all of the peaks in the multiplet. This loss in signal-to-noise can be restored by collapsing the multiplet to a single line by *decoupling*. Since the splitting arises from the influence of the magnetic state of one spin on the other, decoupling can be accomplished by simply inverting the spin-state of the coupled partner during detection. For example, in the case of a ^{13}CH group, if the carbon magnetization is inverted rapidly, the protons in the sample no longer sense two distinct carbon spin-states, but a single averaged state. One way of achieving this inversion is to simply apply a train of 180° pulses to the carbon, as illustrated in Fig. 7.8. More effective decoupling is obtained as the rate of inversion is increased by reducing both the inter-pulse spacing and the 180° pulse length.

7.4.1 Experimental Implementation of Decoupling

Decoupling is incorporated into one-dimensional proton NMR experiments as illustrated in Fig. 7.9. In the case of proton decoupling during proton acquisition it is necessary to gate the decoupling off during digitization of the data, otherwise the decoupling pulses would enter the receiver. In the case of heteronuclear decoupling, it is possible to decouple the *non*-observed spin continuously while acquiring the proton signal.

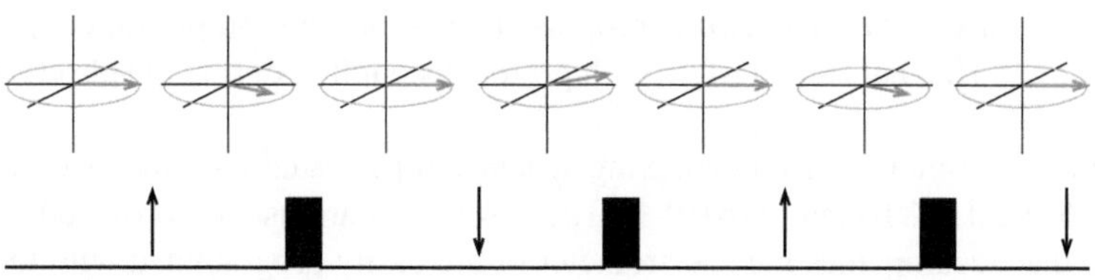

Figure 7.8. Averaging of spin states during decoupling. A series of π pulses can average spin states, removing the effects of spin-spin scalar coupling on the NMR spectrum (decoupling). The upper part of the figure shows the precession of a *single* proton in a CH or NH group. The lower segment shows a train of π pulses that are applied to the heteronuclear spin (^{13}C or ^{15}N). The black vertical arrows represent the spins state of the heteronuclear spin. The proton originally precesses in a clockwise direction. Inversion of the heteronuclear spin reverses the direction. If the inter-pulse delay is small compared to the J-coupling, the proton simply oscillates around the y-axis without undergoing any evolution due to J-coupling. Consequently its position in the spectrum is defined solely by its chemical shift and no splitting of the line occurs.

Figure 7.9. Decoupling in one-dimensional pulse sequences. The experimental set-up for homonuclear (A) and heteronuclear (B) decoupling are shown. In the case of heteronuclear decoupling a separate radio-frequency channel is required for the heteronuclear spin. In contrast, homonuclear ^{1}H-^{1}H decoupling would utilize the same circuit as the excitation pulse. The top part of each panel shows a standard one-dimensional pulse sequence and resultant FID, sampled at the times indicated by the filled circles. The lower part shows the decoupling pulses, as gray rectangles. Homonuclear decoupling is gated, or turned off, during digitization of each point, otherwise the intense decoupling power would corrupt the signal. In the case of heteronuclear decoupling, the pulses can be applied continuously because the decoupling enters the probe on a completely separate circuit, hence little power is transferred to the proton detection channel. In addition, the large frequency difference between the detected proton signal (e.g. 500 MHz) versus that of the decoupling pulse (e.g. 50 MHz) facilitates removal of the decoupling signal by filtering.

7.4.2 Decoupling Methods

Due to resonance off-set effects, the simple inversion of the coupled partner by the application of 180° pulses is very ineffective if the resonance frequency of the decoupled spin differs from the carrier frequency. Because of the wide frequency range of ^{13}C and ^{15}N spins it has been necessary to design more elaborate pulse sequences for decoupling of these nuclei.

Decoupling schemes can be divided into two broad categories, those which were designed for isolated X-H systems, where there are no other significant couplings, and those that were designed for systems where the coupled partner shows coupling to atoms of the same type. An example of the former is the NH group, where the ^{15}N has no other coupling partners except for the amide proton. An example of the latter is the $C_\alpha H_2$ group in glycine. The two α-protons show strong proton-proton coupling, on the order of 15 Hz, that can interfere with proton decoupling during the observation of carbon.

The three commonly used decoupling schemes for isolated H-C and H-N spins are MLEV-16, WALTZ-16, and GARP-1. These schemes are usually provided with the software that accompanies the spectrometer or are easily programmed with the pulse program software. A detailed description of the development and properties of these schemes can be found in Freeman [56]. MLEV-16 is an early decoupling scheme whose properties are inferior to WALTZ and GARP and is only included here for purposes of comparison.

The schemes that have been designed for the decoupling of scalar coupled spins consist of a collection of three related schemes: DIPSI-1, DIPSI-2, and DIPSI-3 (Decoupling In the Presense of Scalar Interations). The basic pulse elements that make up each of the DIPSI sequences, along with an indication of their performance levels are presented in Table 7.3. The three DIPSI sequences differ in the length of the fundamental rotation operator (see below) to accommodate timing limitations that may occur in pulse sequences. DIPSI-1 is the shortest sequence and comparable to WALTZ-16 in overall length.

Each of these decoupling schemes are composed of a fundamental rotation operator, R, which is applied with various phases during decoupling. Each rotation operator, R, can be considered to be equivalent to a 180° pulse, causing inversion of the decoupled spin. In these decoupling schemes the rotation operators have been designed to be insensitive to resonance offset effects. The MLEV rotation operator is one of the simplest; a composite 180° pulse (see Sec. 6.3.3, pg. 132). The different rotation operators associated with each decoupling scheme are shown in Table 7.3. Additional insensitivity to resonance offset effects is obtained by forming a four element cycle of the basic rotation operator:

$$RR\bar{R}\bar{R} \tag{7.29}$$

where $\bar{R}$ is the inverse of R. In the case of MLEV and WALTZ decoupling, cyclic permutations of this basic cycle are combined to give a 16 step super-cycle,

$$RR\bar{R}\bar{R}\ \bar{R}RR\bar{R}\ \bar{R}\bar{R}RR\ R\bar{R}\bar{R}R \tag{7.30}$$

which further compensates for resonance-offset effects. The number of R elements in the super-cycle are often indicated in name of the decoupling scheme, e.g. MLEV-4=$RR\bar{R}\bar{R}$ and MLEV-16=$RR\bar{R}\bar{R}\ \bar{R}RR\bar{R}\ \bar{R}\bar{R}RR\ R\bar{R}\bar{R}R$.

Decoupling Scheme	*R Element*	Ξ	*Residual Line Broadening*	*Ref.*
MLEV-16	$[\pi/2]_x[\pi]_y[\pi/2]_x$	1.5	Large	[94]
WALTZ-16	$[\pi/2]_x[\pi]_{-x}[3\pi/2]_x$	1.8	Small	[147]
DISPI-1	$365\ \overline{295}\ 65\ \overline{305}\ 350$	0.8	Small	[148]
DISPI-2	$320\ \overline{410}\ 290\ \overline{285}\ 30\ \overline{245}\ 375\ \overline{265}\ 370$	1.2	Small	[148]
DISPI-3	$\overline{245}\ 395\ \overline{250}\ 275\ \overline{30}\ 230\ \overline{360}\ 245\ \overline{370}\ 340$ $\overline{350}\ 260\ \overline{270}\ 30\ \overline{225}\ 365\ \overline{255}\ 395$	1.6	Very Small	[148]
GARP-1	$R = \bar{P}Q\overline{PQ}$, overall cycle is $RR\bar{R}\bar{R}$ $P = 27.1\ \overline{57.6}\ 122.0$ $Q = \overline{120.8}\ 262.8\ \overline{65.9}\ 64.6$ $\overline{87.0}\ 90.0\ \overline{137.2}\ 256.2\ \overline{71.6}\ 51.1$	4.8	Moderate	[146]

Table 7.3. Decoupling schemes. The properties of a number of common decoupling schemes are shown. The R element is the basic rotation operator that is used to form the $RR\overline{RR}$ decoupling element. Pulse angles and phases are indicated. In the case of DIPSI and GARP decoupling, the pulses are along the x-axis or along the minus x-axis if the angle is overlined. Angles are given in degrees. The figure of merit, Ξ, for each sequence is also provided. Finally, the quality of the decoupling is indicated by the residual line broadening; higher quality sequences have a smaller residual line broadening.

7.4.3 Performance of Decoupling Schemes

The frequency range over which the decoupling is effective is characterized by the *bandwidth.* Empirically, decoupling is considered to be effective if the intensity of the collapsed multiplet has at least 80% of the intensity of the fully decoupled signal. The bandwidth can be increased by using shorter pulses in the decoupling sequence. However, the additional power will cause sample heating, and in extreme cases can lead to equipment failure. Since the decoupling bandwidth is proportional to the field strength of the decoupling pulses, it is convenient to define a figure of merit, Ξ, as:

$$\Xi = \frac{2\pi\Delta F}{\gamma B} \tag{7.31}$$

where ΔF is the region over which the decoupling is effective (in Hz), and γB is the strength of the decoupling field (in units of rad/sec). Schemes that have a higher figure of merit can decouple a larger bandwidth for the same amount of RF power.

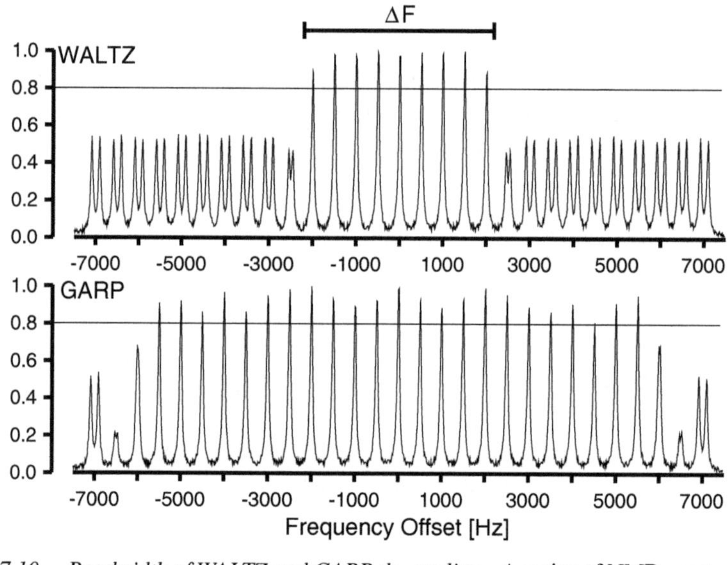

Figure 7.10. Bandwidth of WALTZ and GARP decoupling. A series of NMR spectra of a single amide proton, attached to an ^{15}N spin, are shown for different decoupler frequencies. The decoupler frequency is given relative to the frequency of the ^{15}N resonance. This frequency was varied in 500 Hz steps, ranging from 7000 Hz below (left) to 7000 Hz above the nitrogen frequency (right). A single line indicates collapse of the proton doublet to a singlet. For frequencies far outside the bandwidth, a doublet is observed, with a splitting of 92 Hz ($J_{NH} = 92$ Hz). The horizontal line marks a relative height of 0.8 for the proton line, defining the bandwidth. Outside the bandwidth, the residual linewidth broadens the line, decreasing the signal height. In this illustration the bandwidth of the WALTZ decoupling is 4 kHz, while that for GARP decoupling is 11 kHz. In the case of GARP decoupling, the height of the proton lines are not uniform within the bandwidth due to a variation in the residual decoupling that depends on the frequency offset. In contrast, the height of the lines for WALTZ decoupling are uniform, indicating a small dependence of the residual coupling on the frequency offset.

The Ξ values for MLEV-16, WALTZ-16, DIPSI-*n*, and GARP-1 decoupling are shown in Table 7.3. MLEV-16 has the smallest bandwidth while WALTZ-16 has a slightly larger bandwidth, but the quality of the decoupling is much higher (see below). GARP-1 has the largest bandwidth of the three schemes, but is of lower quality than WALTZ-16. The bandwidths of the DIPSI-*n* sequences depend on the length of the sequence with DIPSI-3 providing the largest bandwidth of the three. However, its bandwidth is still smaller than that of WALTZ-16.

In addition to differing in bandwidth, decoupling schemes also differ in the amount of residual coupling that remains in effect in the presence of the decoupling. The residual coupling will increase the apparent linewidth of the unresolved multiplet, causing a decrease the intensity of the observed peak. The amount of residual coupling depends on the decoupling scheme (see Table 7.3 and Fig. 7.10). In the case of H-X decoupling schemes, WALTZ-16 has a much smaller residual bandwidth than GARP or MLEV-16, and therefore gives high quality decoupling. Of the DIPSI decoupling schemes, DIPSI-3 produces excellent decoupling in the case of scalar coupled systems, out-performing WALTZ-16. A guide to the selection of decoupling schemes is given in Table 7.4

Table 7.4. Guide to decoupling schemes.

Situation	*Decoupling Scheme*	*Rational*
Decouple ^{15}N, observe protons.	WALTZ-16.	Narrow bandwidth of ^{15}N, typically 30 ppm (ΔF = 1.8 kHz) with no ^{15}N-^{15}N coupling, WALTZ-16 provides excellent line-narrowing.
Decouple ^{13}C, observe protons.	GARP-1 for natural abundance.	Require high bandwidth to cover carbon spectrum. Typical bandwidth is 80 ppm ($\Delta\ F = 12$ kHz).
	DIPSI-3 for uniformly labeled samples, provided sufficient bandwidth can be generated. This will depend on the hardware and the desired bandwidth.	^{13}C-^{13}C couplings can interfere with GARP-1 decoupling.
Decouple ^{1}H, observe carbon or nitrogen.	DISPI-2 or DIPSI-3. Timing constraints may force the use of DIPSI-2 in triple resonance experiments.	Moderate proton bandwidth needed, 4 ppm ($\Delta\ F$ = 2.4 kHz) for amides, 6 ppm for aliphatics ($\Delta\ F$ = 3.6 kHz). ^{1}H-^{1}H couplings will interfere with WALTZ-16 and GARP-1.

7.5 Exercises

1. What was the power of the decoupler field strength, in units of Hz, for the series of spectra that were shown in Fig. 7.10?

2. The ^{15}N chemical shift of amide groups range from 100 to 135 ppm. Assuming a magnetic field strength that gives a proton frequency 500 MHz, would WALTZ decoupling at the decoupler power level used in Fig. 7.10 provide satisfactory decoupling?

3. What *change* in power level would be required to adjust the bandwidth of the WALTZ decoupling in Fig. 7.10 to 1750 Hz.

4. Aliphatic carbon frequencies range from 0 to 70 ppm. Assuming a magnetic field strength that gives a proton frequency of 800 MHz, calculate the decoupler field strength required to decouple this bandwidth using WALTZ decoupling. What is the equivalent 90° pulse width for this field strength? Is is possible to use WALTZ decoupling to decouple carbons at this field strength?

5. Given that the 90° pulse length for ^{15}N spin is 40 μsec at a power level of 0 dB, calculate the power required to decouple over a bandwidth of 1800 Hz using WALTZ decoupling.

7.6 Solutions

1 The bandwidth for WALTZ decoupling was 4 kHz. The figure of merit for WALTZ is 1.8, $\Xi = 2\pi\Delta F/\gamma B$, therefore:

$$
\begin{aligned}
1.8 &= \frac{2\pi 4000}{\gamma B} \\
\gamma B &= \frac{2\pi 4000}{1.8} \\
&= 13,962 \text{ rad/sec} = 2,222 \text{ Hz}
\end{aligned}
$$

2 A 35 ppm ^{15}N chemical shift range corresponds to 1750 Hz at this magnetic field strength. The 4 kHz bandwidth will be more than adequate.

3 This requires a reduction in the decoupler field strength by a factor of 2.3 (4,000/1,750). Therefore the attenuation will be:

$$
\begin{aligned}
\text{dB} &= 20\, log \frac{V_1}{V_2} \\
&= 20\, log\, 2.3 \\
&= 7.18
\end{aligned}
$$

4 A 70 ppm is equivalent to 14 kHz (70 ppm x 200 MHz). Given a figure of merit of 1.8 for WALTZ decoupling, a decoupler field strength of approximately 7,800 Hz is required. At this field strength, a 90° pulse would require 32 μsec. This

pulse length is close to the length that would be considered at full power. Hence WALTZ decoupling of carbon at this magnetic field strength would certainly lead to sample heating and would likely result in equipment failure.

5 The figure of merit for WALTZ is 1.8, therefore a 1 kHz decoupler field strength would be sufficient. The field strength at 0 dB can be obtained from the 90° pulse length:

$$\begin{aligned} \gamma B_1[\mathrm{Hz}] &= \frac{1}{4\tau} \\ &= 6250\ \mathrm{Hz} \end{aligned}$$

The change in power is calculated as follows:

$$\mathrm{dB} = 20\ log \frac{1000}{6250} = 20 \times (-0.795) = -15.9\ \mathrm{dB}. \tag{7.32}$$

Therefore, the power level should be decreased by 15.9 dB.

Chapter 8

COUPLED SPINS: DENSITY MATRIX AND PRODUCT OPERATOR FORMALISM

In the previous chapter the coupling between two spins was analyzed using the wavefunctions associated with each of the quantum states. This type of analysis is quite satisfactory for determining the appearance of the one-dimensional NMR spectrum of the coupled spins. However, this approach is completely impractical for calculations on an ensemble of spins that are subject to multiple excitation pulses in typical NMR experiments. Consequently, we return to the density matrix to analyze the evolution of the coupled spins. In this analysis we will only assume weak coupling, of the AX kind. This assumption is not a severe restriction since heteronuclear couplings, such as between protons and carbon or carbon and nitrogen, are those that are used for magnetization transfer by various NMR experiments.

This chapter begins with the analysis of a one-pulse experiment using the density matrix. The analysis follows the same treatment as given in Chapter 6 for an uncoupled spin. The density matrix will then be represented by product operators and a series of rules that describe the transformation of the product operators by pulses and free evolution will be introduced. These rules will greatly simplify the analysis of complicated pulse sequences.

8.1 Density Matrix for Two Coupled Spins

For two coupled spins the wavefunctions are a linear combination of the four basis vectors:

$$\Psi = c_1|\alpha\alpha> + c_2|\alpha\beta> + c_3|\beta\alpha> + c_4|\beta\beta> \tag{8.1}$$

and the density matrix is now a 4x4 matrix, of the following form:

$$\rho = \begin{bmatrix} c_1c_1^* & c_1c_2^* & c_1c_3^* & c_1c_4^* \\ c_2c_1^* & c_2c_2^* & c_2c_3^* & c_2c_4^* \\ c_3c_1^* & c_3c_2^* & c_3c_3^* & c_3c_4^* \\ c_4c_1^* & c_4c_2^* & c_4c_3^* & c_4c_4^* \end{bmatrix} \tag{8.2}$$

where each element represents the average of the coefficients over the ensemble.

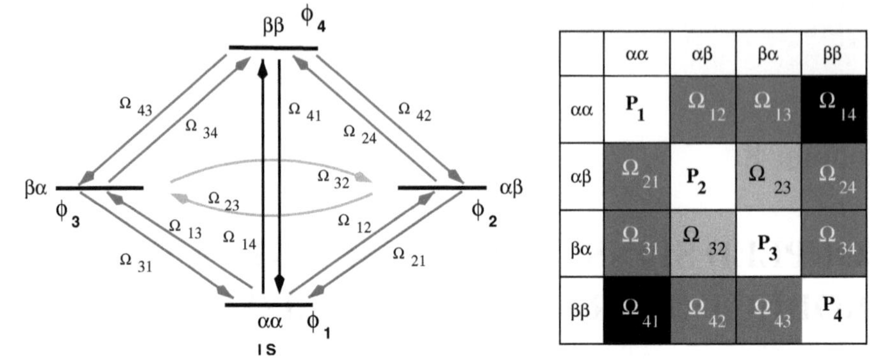

Figure 8.1. Transitions for a coupled two-spin system. Each individual transition is indicated by a uni-directional arrow. Each transition has a corresponding element in the density matrix, as indicated in the right side of the figure. The four populations are indicated with white squares. The two zero quantum transitions, with frequencies of Ω_{23} and Ω_{32} are indicated with light gray squares. The eight single quantum transitions are indicated in a darker gray. The two double quantum transitions, with frequencies Ω_{14} and Ω_{41} are indicated by the black squares.

Each of the 16 elements in this density matrix provides information on various states of the system. The nature of the information contained in each element can be identified from its time evolution under the Hamiltonian. This time dependence is given by:

$$i\hbar\frac{d\rho_{nm}}{dt} = (E_n - E_m)\rho_{nm} \tag{8.3}$$

or in frequency units:

$$\frac{d\rho_{nm}}{dt} = -i(\Omega_n - \Omega_m)\rho_{nm} \tag{8.4}$$

For two coupled spins, the individual elements of the density matrix will evolve with the frequencies given in Table 8.1. For example, given an initial value of $c_1c_3^*(t=0)$, the values of this element of the density matrix at some future time is:

$$c_1c_3^*(t) = e^{i(\omega_I - \pi J)t}c_1c_3^*(t=0) \tag{8.5}$$

Table 8.1. Time evolution of the elements of the density matrix. The frequency at which an element of the density matrix evolves is indicated. The row indicates the ground state and the column indicates the excited state. For example, the element of the density matrix that represents the double quantum transition from the $\alpha\alpha$ ground state to the $\beta\beta$ double quantum state will evolve at a frequency of $\omega = \omega_I + \omega_S$.

	$\alpha\alpha$	$\alpha\beta$	$\beta\alpha$	$\beta\beta$
$\alpha\alpha$	0	$\omega_S - \pi J$	$\omega_I - \pi J$	$\omega_I + \omega_S$
$\alpha\beta$	$-\omega_S + \pi J$	0	$\omega_I - \omega_S$	$\omega_I + \pi J$
$\beta\alpha$	$-\omega_I + \pi J$	$-\omega_I + \omega_S$	0	$\omega_S + \pi J$
$\beta\beta$	$-\omega_I - \omega_S$	$-\omega_I - \pi J$	$-\omega_S - \pi J$	0

The 16 elements of the density matrix can be divided into four distinct groups. The four diagonal elements, $c_i c_i^*$, do not evolve with time and therefore refer to the population of the ψ_i state. The off-diagonal elements indicate the presence of coherently excited states or transitions. These transitions can be divided into zero quantum, single quantum, and double quantum transitions:

- Zero quantum transitions, or coherences, are those whose frequency is given by the difference in the resonance frequency between the I and S spins. These transitions connect the $\alpha\beta$ and $\beta\alpha$ states. Zero quantum transitions correspond to spin-spin flips, or an interchange of m_z values. There is no net change in the overall quantum number. The zero-quantum transitions are indicated as Ω_{23} and Ω_{32} in Fig. 8.1.
- Single quantum transitions, or coherences, change m_z for one of the two spins. These transitions form the outer set of arrows in Fig. 8.1.
- Double quantum transitions, or coherences, involve a change in m_z for both spins. The two double quantum transitions connect the $\beta\beta$ state to the $\alpha\alpha$ state.

8.2 Product Operator Representation of the Density Matrix

In the case of a single spin it was possible to describe the density matrix in terms of a group of four product operators:

$$E \qquad I_x \qquad I_y \qquad I_z$$

where E is the identity matrix. This set of four density matrices forms a closed group, as such they are adequate to describe the evolution of *any* arbitrary density matrices under pulses or free precession.

For two uncoupled spins, I and S, we can also write two sets of product operators that can be used to represent the density matrices associated with the two spins:

$$\begin{array}{cccc} E & I_x & I_y & I_z \\ & S_x & S_y & S_z \end{array}$$

It is clear that the set of seven different density matrices will not be sufficient to describe a coupled system because the Hamiltonian for the scalar coupling itself is the product of two operators:

$$\mathcal{H} = 2\pi J I_z S_z$$

A closed group of product operators that can be used to describe any arbitrary density matrix for two coupled spins is generated by taking all possible products of the single spin operators. This forms a set of 16 product operators:

$$\begin{array}{cccc} 1/2E & I_x & I_y & I_z \\ S_x & 2I_xS_x & 2I_yS_x & 2I_zS_x \\ S_y & 2I_xS_y & 2I_yS_y & 2I_zS_y \\ S_z & 2I_xS_z & 2I_yS_z & 2I_zS_z \end{array} \tag{8.6}$$

The matrix form of these density matrices is found by forming the *tensor* or *direct product* between pairs of single-spin 2x2 matrices. This is equivalent to taking all possible combinations of the elements in each 2x2 matrix. The direct product of two single spin operators is calculated as follows.

Consider two operators, $\mathcal{O}_A$ and $\mathcal{O}_B$:

$$\mathcal{O}_A = \begin{bmatrix} a & b \\ c & d \end{bmatrix}, \mathcal{O}_B = \begin{bmatrix} e & f \\ g & h \end{bmatrix} \tag{8.7}$$

then the direct product of the two operators, $\mathcal{O}_A \otimes \mathcal{O}_B$, is obtained as follows:

$$\begin{aligned} \mathcal{O}_{AB} &= \mathcal{O}_A \otimes \mathcal{O}_B \\ &= \begin{bmatrix} a & b \\ c & d \end{bmatrix} \otimes \begin{bmatrix} e & f \\ g & h \end{bmatrix} = \begin{bmatrix} a\begin{bmatrix} e & f \\ g & h \end{bmatrix} & b\begin{bmatrix} e & f \\ g & h \end{bmatrix} \\ c\begin{bmatrix} e & f \\ g & h \end{bmatrix} & d\begin{bmatrix} e & f \\ g & h \end{bmatrix} \end{bmatrix} \\ &= \begin{bmatrix} ae & af & be & bf \\ ag & ah & bg & bh \\ ce & cf & de & df \\ cg & ch & dg & dh \end{bmatrix} \end{aligned} \tag{8.8}$$

Any pair of operators can be combined to give the resultant direct product. For example:

$$I_x S_y = I_x \otimes S_y = \frac{1}{2}\begin{bmatrix} 0 & 1 \\ 1 & 0 \end{bmatrix} \otimes \frac{1}{2}\begin{bmatrix} 0 & -i \\ i & 0 \end{bmatrix} = \frac{1}{4}\begin{bmatrix} 0 & 0 & 0 & -i \\ 0 & 0 & i & 0 \\ 0 & -i & 0 & 0 \\ i & 0 & 0 & 0 \end{bmatrix} \tag{8.9}$$

Single spin operators can also be written in the same representation as the two coupled spins by taking the direct product with the identity matrix:

$$I_z = I_z \otimes 1 = \frac{1}{2}\begin{bmatrix} 1 & 0 \\ 0 & -1 \end{bmatrix} \otimes \begin{bmatrix} 1 & 0 \\ 0 & 1 \end{bmatrix} = \begin{bmatrix} 1 & 0 & 0 & 0 \\ 0 & 1 & 0 & 0 \\ 0 & 0 & -1 & 0 \\ 0 & 0 & 0 & -1 \end{bmatrix} \tag{8.10}$$

The complete set of 16 product operators are listed in Table 8.2.

8.2.1 Detectable Elements of ρ

The density matrix contains information on populations, zero-quantum, single-quantum, and double-quantum transitions. Only a sub-set of these can be detected by the instrument. Quadrature detection measures $M_x + iM_y$, or equivalently, $[I_x + S_x] + i[I_y + S_y]$, which is equal to $I^+ + S^+$. Consequently, the detected signal is given by: Trace($\rho\,[I^+ + S^+]$).

The above trace is only non-zero for the density matrices I_x, I_y, S_x, and S_y. In addition, the density matrices represented by the lowering operators, I^- and S^- can

Table 8.2. Product operators for two coupled spins.

$$\frac{1}{2}E = \frac{1}{2}\begin{bmatrix} 1 & 0 & 0 & 0 \\ 0 & 1 & 0 & 0 \\ 0 & 0 & 1 & 0 \\ 0 & 0 & 0 & 1 \end{bmatrix} \qquad 2I_zS_z = \frac{1}{2}\begin{bmatrix} 1 & 0 & 0 & 0 \\ 0 & -1 & 0 & 0 \\ 0 & 0 & -1 & 0 \\ 0 & 0 & 0 & 1 \end{bmatrix}$$

$$I_x = \frac{1}{2}\begin{bmatrix} 0 & 0 & 1 & 0 \\ 0 & 0 & 0 & 1 \\ 1 & 0 & 0 & 0 \\ 0 & 1 & 0 & 0 \end{bmatrix} \qquad S_x = \frac{1}{2}\begin{bmatrix} 0 & 1 & 0 & 0 \\ 1 & 0 & 0 & 0 \\ 0 & 0 & 0 & 1 \\ 0 & 0 & 1 & 0 \end{bmatrix}$$

$$I_y = \frac{1}{2}\begin{bmatrix} 0 & 0 & -i & 0 \\ 0 & 0 & 0 & -i \\ i & 0 & 0 & 0 \\ 0 & i & 0 & 0 \end{bmatrix} \qquad S_y = \frac{1}{2}\begin{bmatrix} 0 & -i & 0 & 0 \\ i & 0 & 0 & 0 \\ 0 & 0 & 0 & -i \\ 0 & 0 & i & 0 \end{bmatrix}$$

$$I_z = \frac{1}{2}\begin{bmatrix} 1 & 0 & 0 & 0 \\ 0 & 1 & 0 & 0 \\ 0 & 0 & -1 & 0 \\ 0 & 0 & 0 & -1 \end{bmatrix} \qquad S_z = \frac{1}{2}\begin{bmatrix} 1 & 0 & 0 & 0 \\ 0 & -1 & 0 & 0 \\ 0 & 0 & 1 & 0 \\ 0 & 0 & 0 & -1 \end{bmatrix}$$

$$2I_xS_x = \frac{1}{2}\begin{bmatrix} 0 & 0 & 0 & 1 \\ 0 & 0 & 1 & 0 \\ 0 & 1 & 0 & 0 \\ 1 & 0 & 0 & 0 \end{bmatrix} \qquad 2I_yS_y = \frac{1}{2}\begin{bmatrix} 0 & 0 & 0 & -1 \\ 0 & 0 & 1 & 0 \\ 0 & 1 & 0 & 0 \\ -1 & 0 & 0 & 0 \end{bmatrix}$$

$$2I_xS_z = \frac{1}{2}\begin{bmatrix} 0 & 0 & 1 & 0 \\ 0 & 0 & 0 & -1 \\ 1 & 0 & 0 & 0 \\ 0 & -1 & 0 & 0 \end{bmatrix} \qquad 2I_yS_z = \frac{1}{2}\begin{bmatrix} 0 & 0 & -i & 0 \\ 0 & 0 & 0 & i \\ i & 0 & 0 & 0 \\ 0 & -i & 0 & 0 \end{bmatrix}$$

$$2I_zS_x = \frac{1}{2}\begin{bmatrix} 0 & 1 & 0 & 0 \\ 1 & 0 & 0 & 0 \\ 0 & 0 & 0 & -1 \\ 0 & 0 & -1 & 0 \end{bmatrix} \qquad 2I_xS_y = \frac{1}{2}\begin{bmatrix} 0 & 0 & 0 & -i \\ 0 & 0 & i & 0 \\ 0 & -i & 0 & 0 \\ i & 0 & 0 & 0 \end{bmatrix}$$

$$2I_zS_y = \frac{1}{2}\begin{bmatrix} 0 & -i & 0 & 0 \\ i & 0 & 0 & 0 \\ 0 & 0 & 0 & i \\ 0 & 0 & -i & 0 \end{bmatrix} \qquad 2I_yS_x = \frac{1}{2}\begin{bmatrix} 0 & 0 & 0 & -i \\ 0 & 0 & -i & 0 \\ 0 & i & 0 & 0 \\ i & 0 & 0 & 0 \end{bmatrix}$$

also be detected. All other density matrices will give no detectable signal. However, we shall see later that density matrices that contain one transverse operator and one longitudinal operator, for example, $2I_xS_z$ can evolve into a detectable signal due to J-coupling.

Whether a particular product operator representation of the density matrix can be detected is determined by simply calculating Trace($\rho\,[I^+ + S^+]$). The matrix form of I^+ and S^+ are obtained by adding $I_x + iI_y$ and $S_x + iS_y$, giving:

$$I^+ = \begin{bmatrix} 0 & 0 & 1 & 0 \\ 0 & 0 & 0 & 1 \\ 0 & 0 & 0 & 0 \\ 0 & 0 & 0 & 0 \end{bmatrix} \qquad S^+ = \begin{bmatrix} 0 & 1 & 0 & 0 \\ 0 & 0 & 0 & 0 \\ 0 & 0 & 0 & 1 \\ 0 & 0 & 0 & 0 \end{bmatrix} \tag{8.11}$$

As examples, we first show that the density matrices, represented by I_x or I^-, give a detectable signal.

$$\begin{aligned} \text{Signal}(I_x) &= \text{Trace}(\rho\,[I^+ + S^+]) = \text{Trace}(I_x\,[I^+ + S^+]) \\ &= \text{Trace}\left(\begin{bmatrix} 0 & 0 & 1 & 0 \\ 0 & 0 & 0 & 1 \\ 1 & 0 & 0 & 0 \\ 0 & 1 & 0 & 0 \end{bmatrix}\begin{bmatrix} 0 & 1 & 1 & 0 \\ 0 & 0 & 0 & 1 \\ 0 & 0 & 0 & 1 \\ 0 & 0 & 0 & 0 \end{bmatrix}\right) = \text{Trace}\left(\begin{bmatrix} 0 & 0 & 0 & 1 \\ 0 & 0 & 0 & 0 \\ 0 & 1 & 1 & 0 \\ 0 & 0 & 0 & 1 \end{bmatrix}\right) \\ &= 0 + 0 + 1 + 1 \\ &= 2 \end{aligned} \tag{8.12}$$

$$\begin{aligned} \text{Signal}(I^-) &= \text{Trace}(\rho\,[I^+ + S^+]) = \text{Trace}(I^-\,[I^+ + S^+]) \\ &= \text{Trace}\left(\begin{bmatrix} 0 & 0 & 0 & 0 \\ 0 & 0 & 0 & 0 \\ 1 & 0 & 0 & 0 \\ 0 & 1 & 0 & 0 \end{bmatrix}\begin{bmatrix} 0 & 1 & 1 & 0 \\ 0 & 0 & 0 & 1 \\ 0 & 0 & 0 & 1 \\ 0 & 0 & 0 & 0 \end{bmatrix}\right) = \text{Trace}\left(\begin{bmatrix} 0 & 0 & 0 & 0 \\ 0 & 0 & 0 & 0 \\ 0 & 0 & 1 & 0 \\ 0 & 0 & 0 & 1 \end{bmatrix}\right) \\ &= 0 + 0 + 1 + 1 \\ &= 2 \end{aligned} \tag{8.13}$$

In contrast, the density matrices represented by I^+, $2I_xS_z$, or $2I_xS_y$, do not give a detectable signal:

$$\begin{aligned} \text{Signal}(I^+) &= \text{Trace}(\rho\,[I^+ + S^+]) = \text{Trace}(I^+\,[I^+ + S^+]) \\ &= \text{Trace}\left(\begin{bmatrix} 0 & 0 & 1 & 0 \\ 0 & 0 & 0 & 1 \\ 0 & 0 & 0 & 0 \\ 0 & 0 & 0 & 0 \end{bmatrix}\begin{bmatrix} 0 & 1 & 1 & 0 \\ 0 & 0 & 0 & 1 \\ 0 & 0 & 0 & 1 \\ 0 & 0 & 0 & 0 \end{bmatrix}\right) = \text{Trace}\left(\begin{bmatrix} 0 & 0 & 0 & 1 \\ 0 & 0 & 0 & 0 \\ 0 & 0 & 0 & 0 \\ 0 & 0 & 0 & 0 \end{bmatrix}\right) \\ &= 0 + 0 + 0 + 0 \\ &= 0 \end{aligned} \tag{8.14}$$

$$
\begin{aligned}
\text{Signal}(2I_xS_z) &= \text{Trace}(\rho\,[I^+ + S^+]) = \text{Trace}(2I_xS_z\,[I^+ + S^+]) \\
&= \text{Trace}\left(\begin{bmatrix} 0 & 0 & 1 & 0 \\ 0 & 0 & 0 & -1 \\ 1 & 0 & 0 & 0 \\ 0 & -1 & 0 & 0 \end{bmatrix}\begin{bmatrix} 0 & 1 & 1 & 0 \\ 0 & 0 & 0 & 1 \\ 0 & 0 & 0 & 1 \\ 0 & 0 & 0 & 0 \end{bmatrix}\right) = \text{Trace}\left(\begin{bmatrix} 0 & 0 & 0 & 1 \\ 0 & 0 & 0 & 0 \\ 0 & 1 & 1 & 0 \\ 0 & 0 & 0 & -1 \end{bmatrix}\right) \\
&= 0 + 0 + 1 - 1 \\
&= 0
\end{aligned}
\tag{8.15}
$$

$$
\begin{aligned}
\text{Signal}(2I_xS_y) &= \text{Trace}(\rho\,[I^+ + S^+]) = \text{Trace}(2I_xS_y\,[I^+ + S^+]) \\
&= \text{Trace}\left(\begin{bmatrix} 0 & 0 & 0 & -i \\ 0 & 0 & i & 0 \\ 0 & -i & 0 & 0 \\ i & 0 & 0 & 0 \end{bmatrix}\begin{bmatrix} 0 & 1 & 1 & 0 \\ 0 & 0 & 0 & 1 \\ 0 & 0 & 0 & 1 \\ 0 & 0 & 0 & 0 \end{bmatrix}\right) = \text{Trace}\left(\begin{bmatrix} 0 & 0 & 0 & 0 \\ 0 & 0 & 0 & i \\ 0 & 0 & 0 & -i \\ 0 & i & i & 0 \end{bmatrix}\right) \\
&= 0 + 0 + 0 + 0 \\
&= 0
\end{aligned}
\tag{8.16}
$$

In summary, the relationships between the density matrix and the detected signals are:

1 Density matrices that are represented by product operators consisting of one transverse operator, e.g. I_x, represent single quantum transitions and yield a detectable signal. In the case of quadrature detection, I^- and S^-, are the detected signals.

2 Density matrices that are represented by product operators consisting of one transverse operator and one z-operator, e.g. $2I_xS_z$, are not directly detectable, but can evolve into detectable magnetization due to J-coupling. Product operators of this type represent undetectable single-quantum transitions.

3 Density matrices that are represented by product operators consisting entirely of z-operators, e.g. I_z, S_z, I_zS_z represent populations or zero quantum transitions and therefore cannot be detected.

4 Density matrices that represent product operators consisting of two transverse terms, e.g. $2I_xS_y$, represent double quantum transitions that do not give rise to detectable transitions.

8.3 Density Matrix Treatment of a One-pulse Experiment

As a simple example, the evolution of the density matrix for two coupled spins during a one pulse experiment will be described. In this example I and S represent two coupled protons. Both spins are excited by a 90° pulse along the x-axis and the resulting signal is detected after excitation. The evolution of the density matrix during this experiment can be represented as follows:

$$\rho_o \xrightarrow{P_X} \rho_1 \xrightarrow{\mathcal{H}} \rho_1(t) \tag{8.17}$$

The excitation pulse is represented as a rotation of the density matrix, using the rotation operator, generating the density matrix after the pulse, ρ_1:

$$\rho_1 = R_x(\pi/2)\rho_o R_x^\dagger(\pi/2) \tag{8.18}$$

During detection, the density matrix will evolve under the complete Hamiltonian, $\mathcal{H} = -\omega_I I_z - \omega_S S_z + 2\pi J I_z S_z$, as follows:

$$\rho_1(t) = e^{-i\mathcal{H}t}\rho_1 e^{+i\mathcal{H}t} \tag{8.19}$$

With the final detected signal given by:

$$S(t) = \text{Trace}(\rho_1(t)\ [I^+ + S^+]) \tag{8.20}$$

To calculate $\rho_1(t)$ it is necessary to define the initial density matrix, ρ_o, the rotation operator that represents the pulse, $R_x(\pi/2)$, as well as the evolution operator, $e^{-i\mathcal{H}t}$.

The simplest method of obtaining ρ_o and $R_x(\pi/2)$ is to take the direct product of the corresponding one-spin 2x2 matrices to generate the 4x4 matrix that describes the coupled spins.

$$\begin{aligned}
\rho_o &= I_z + S_z \\
&= I_z \otimes 1 + 1 \otimes S_z = \frac{1}{2}\begin{bmatrix} 1 & 0 & 0 & 0 \\ 0 & 1 & 0 & 0 \\ 0 & 0 & -1 & 0 \\ 0 & 0 & 0 & 1 \end{bmatrix} + \frac{1}{2}\begin{bmatrix} 1 & 0 & 0 & 0 \\ 0 & -1 & 0 & 0 \\ 0 & 0 & 1 & 0 \\ 0 & 0 & 0 & -1 \end{bmatrix} \\
&= \begin{bmatrix} 1 & 0 & 0 & 0 \\ 0 & 0 & 0 & 0 \\ 0 & 0 & 0 & 0 \\ 0 & 0 & 0 & -1 \end{bmatrix}
\end{aligned} \tag{8.21}$$

Note that this result could have also been obtained by direct addition of the 4x4 matrix forms of I_z and S_z.

The rotation matrix, $R_x(\pi/2)$ is obtained from the single-spin rotation matrix for an x-rotation (see Eq. 4.121) in a similar fashion:

$$\begin{aligned}
R_x^{IS}(\pi/2) &= R_x^I(\pi/2) \otimes R_x^S(\pi/2) \\
&= \frac{1}{\sqrt{2}}\begin{bmatrix} 1 & -i \\ -i & 1 \end{bmatrix} \otimes \frac{1}{\sqrt{2}}\begin{bmatrix} 1 & -i \\ -i & 1 \end{bmatrix} \\
&= \frac{1}{2}\begin{bmatrix} 1 & -i & -i & -1 \\ -i & 1 & -1 & -i \\ -i & -1 & 1 & -i \\ -1 & -i & -i & 1 \end{bmatrix}
\end{aligned} \tag{8.22}$$

Calculation of ρ_1: The density matrix immediately after the 90° pulse is obtained by applying the rotation operator to ρ_o, giving:

$$\begin{aligned}
\rho_1 &= R_x(\pi/2)\rho_o R_x^\dagger(\pi/2) \\
&= \left(\frac{1}{2}\right)^2 \begin{bmatrix} 1 & -i & -i & -1 \\ -i & 1 & -1 & -i \\ -i & -1 & 1 & -i \\ -1 & -i & -i & 1 \end{bmatrix} \begin{bmatrix} 1 & 0 & 0 & 0 \\ 0 & 0 & 0 & 0 \\ 0 & 0 & 0 & 0 \\ 0 & 0 & 0 & -1 \end{bmatrix} \begin{bmatrix} 1 & i & i & -1 \\ i & 1 & -1 & i \\ i & -1 & 1 & i \\ -1 & i & i & 1 \end{bmatrix} \\
&= \left(\frac{1}{2}\right)^2 \begin{bmatrix} 0 & 2i & 2i & 0 \\ -2i & 0 & 0 & 2i \\ -2i & 0 & 0 & 2i \\ 0 & -2i & -2i & 0 \end{bmatrix} = \frac{1}{2} \begin{bmatrix} 0 & i & i & 0 \\ -i & 0 & 0 & i \\ -i & 0 & 0 & i \\ 0 & -i & -i & 0 \end{bmatrix} \\
&= -I_y - S_y
\end{aligned} \tag{8.23}$$

Note that the effect of the pulse is to convert the initial density matrix, $\rho_o = I_z + S_z$ to $\rho_1 = -I_y - S_y$. This is exactly the same transformation that was observed with a single uncoupled spin. This result is not surprising since the coupling has not yet entered into the calculations.

Evolution of the Density Matrix during Detection: The time dependence of ρ is obtained by:

$$\rho_1(t) = e^{-i\mathcal{H}t}\rho_1 e^{+i\mathcal{H}t} \tag{8.24}$$

Evolution of individual single-quantum elements of the density matrix can be determined using the information provided in Table 8.1, giving the following:

$$\rho_1(t) = \begin{bmatrix} 0 & ie^{i(+\omega_S-\pi J)t} & ie^{i(+\omega_I-\pi J)t} & 0 \\ -ie^{i(-\omega_S+\pi J)t} & 0 & 0 & ie^{i(+\omega_I+\pi J)t} \\ -ie^{i(-\omega_I+\pi J)t} & 0 & 0 & ie^{i(+\omega_S+\pi J)t} \\ 0 & -ie^{i(-\omega_I-\pi J)t} & -ie^{i(-\omega_S-\pi J)t} & 0 \end{bmatrix} \tag{8.25}$$

The final detected signal is given by Trace(ρ $[I^+ + S^+]$). To simplify the calculation, the signal associated with only the I spin will be calculated. We first calculate ρI^+:

$$\begin{aligned}
\rho I^+ &= \\
&\begin{bmatrix} 0 & ie^{i(+\omega_S-\pi J)t} & ie^{i(+\omega_I-\pi J)t} & 0 \\ -ie^{i(-\omega_S+\pi J)t} & 0 & 0 & ie^{i(+\omega_I+\pi J)t} \\ -ie^{i(-\omega_I+\pi J)t} & 0 & 0 & ie^{i(+\omega_S+\pi J)t} \\ 0 & -ie^{i(-\omega_I-\pi J)t} & -ie^{i(-\omega_S-\pi J)t} & 0 \end{bmatrix} \begin{bmatrix} 0 & 0 & 1 & 0 \\ 0 & 0 & 0 & 1 \\ 0 & 0 & 0 & 0 \\ 0 & 0 & 0 & 0 \end{bmatrix} \\
&= \begin{bmatrix} 0 & 0 & 0 & ie^{i(+\omega_S-\pi J)t} \\ 0 & 0 & -ie^{i(-\omega_S+\pi J)t} & 0 \\ 0 & 0 & -ie^{i(-\omega_I+\pi J)t} & 0 \\ 0 & 0 & 0 & -ie^{i(-\omega_I-\pi J)t} \end{bmatrix}
\end{aligned} \tag{8.26}$$

The trace gives the detected signal (ignoring the sign of ω):

$$\text{Trace}(\rho\, I^+) = -i[e^{i(\omega_I+\pi J)t} + e^{i(\omega_I-\pi J)t}] \tag{8.27}$$

The amplitude "i" simply represents a phase shift of the signal and can be ignored. The remaining part of the expression represents two spectral lines, centered around ω_I and separated by a total of $2\pi J$ rad/sec. Repeating this calculation using S^+ would yield a similar result, with ω_I replaced by ω_S. The final detected signal contains four resonance lines, doublets at ω_I and ω_S, with each doublet split by J Hz:

$$S(t) = e^{i(\omega_I+\pi J)t} + e^{i(\omega_I-\pi J)t} + e^{i(\omega_S+\pi J)t} + e^{i(\omega_S-\pi J)t} \quad (8.28)$$

8.4 Manipulation of Two-spin Product Operators

The above manipulations of the 4x4 density matrix are even more tedious than the manipulation of the simpler 2x2 density matrix for uncoupled spins. The representation of the density matrix by a linear combination of product operators greatly simplifies the calculation of ρ at various positions in the NMR experiment. As with the analysis of the single isolated spin, we develop a set of rules to describe the effect of different propagators, such as pulses and free evolution, on the system. The transformation rules that were discussed in Chapter 6 for a single spin are repeated here in Table 8.3.

Since there are now two coupled spins, we also have to determine how to apply these rules to products of operators. In addition, it is also necessary to consider the effect of the J-coupling term in the Hamiltonian, $2\pi J I_z S_z$, on the evolution of the density matrix under free precession. Rules for determining how the density matrix is modified by pulses and free precession are as follows:

1. If operators P and Q commute (i.e. belong to different spins) , then a rotation operator associated with one spin has no effect on the density matrix that corresponds to the other spin:

$$\begin{aligned} e^{-i\theta P} Q e^{i\theta P} &= Q e^{-i\theta P} e^{i\theta P} \\ &= Q \end{aligned} \quad (8.29)$$

Table 8.3. Product operator transformations for a single spin. The left column lists the value of the density matrix prior to application of the rotation operator. The second two columns show the effect of pulses along the x- or y-axis, with a flip angle of β degrees, on the respective density matrix. The right column shows the effect of free precession on the density matrix. In this case the rotation angle is ωt.

	Rotation Operator		
ρ	*Pulses*		*Free Precession*
	$R_x(\beta)$	$R_y(\beta)$	$R_z(\omega t)$
I_x	I_x	$I_x \cos\beta - I_z \sin\beta$	$I_x \cos\omega t + I_y \sin\omega t$
I_y	$I_y \cos\beta + I_z \sin\beta$	I_y	$I_y \cos\omega t - I_x \sin\omega t$
I_z	$I_z \cos\beta - I_y \sin\beta$	$I_z \cos\beta + I_x \sin\beta$	I_z

For example, a proton pulse has no effect on the density matrix that represents magnetization associated with carbon or nitrogen spins.

2. The one spin component of an evolving product operator can be treated separately:

$$2I_yS_z \stackrel{\beta(I_x+S_x)}{\longrightarrow} 2\left[I_y cos(\beta) + I_z sin(\beta)\right]\left[S_z cos(\beta) - S_y sin(\beta)\right] \tag{8.30}$$

For example, a 90°_x pulse, applied to both spins, gives:

$$2I_yS_z \longrightarrow -2I_zS_y \tag{8.31}$$

3. One-spin operators (e.g. I_x), that are found as part of the rotation angle are taken as constants in the expression. The usual application of this rule is to evaluate the effect of J-coupling on evolution of the density matrix. For example, the effect of J-coupling on the evolution of protons (I spins) is equivalent to a rotation about the z-axis by an angle $2\pi JS_zt$,

$$e^{-i2\pi JI_zS_z} = e^{-i(2\pi JS_zt)I_z} \tag{8.32}$$

For example, the evolution of I_x due to J coupling is:

$$I_x \stackrel{2\pi JI_zS_zt}{\longrightarrow} I_x cos(\pi Jt2S_z) + I_y sin(\pi Jt2S_z) \tag{8.33}$$

The difficulty with this expression is that there is now an operator, S_z, that is part of an argument for a trigonometric function. The operators are taken out of the argument using the series expansion for each term:

$$cos\theta = 1 - \frac{\theta^2}{2!} + \frac{\theta^4}{4!}.... \qquad sin\theta = \frac{\theta}{1!} - \frac{\theta^3}{3!} + \frac{\theta^5}{5!}... \tag{8.34}$$

$$\begin{aligned} cos(\pi Jt2S_z) &= 1 - \frac{(\pi J2t)^2}{2!}S_z^2 + \frac{(\pi J2t)^4}{4!}S_z^4... \\ &= 1 - \frac{(\pi J2t)^2}{2!}\left[\frac{1}{4}\right] + \frac{(\pi J2t)^4}{4!}\left[\frac{1}{4}\right]^2... \\ &= cos(\pi Jt) \end{aligned} \tag{8.35}$$

In a similar fashion:

$$\begin{aligned} sin(\pi Jt2S_z) &= \frac{\pi J2t}{1}S_z - \frac{(\pi J2t)^3}{3!}S_z^3 + \frac{(\pi J2t)^5}{5!}S_z^5 \\ &= \frac{\pi J2t}{1}S_z - \frac{(\pi J2t)^3}{3!}S_z^2S_z + \frac{(\pi J2t)^5}{5!}S_z^4S_z \\ &= \frac{\pi J2t}{1}S_z - \frac{(\pi J2t)^3}{3!}\left[\frac{1}{4}\right]S_z + \frac{(\pi J2t)^5}{5!}\left[\frac{1}{4}\right]^2S_z \\ &= 2S_z\left[\frac{\pi Jt}{1} - \frac{(\pi Jt)^3 2^2}{3!}\left[\frac{1}{4}\right] + \frac{(\pi Jt)^5 2^4}{5!}\left[\frac{1}{4}\right]^2\right] \end{aligned} \tag{8.36}$$

$$= 2S_z sin(\pi Jt) \tag{8.37}$$

Thus,

$$I_x \xrightarrow{2\pi JtI_zS_z} I_x cos(\pi Jt) + 2I_yS_z sin(\pi Jt) \tag{8.38}$$

4. The evolution of a product of two operators under J-coupling would be evaluated using a combination of rule 2 and rule 3. First consider the evolution of $2I_yS_z$:

$$\begin{aligned} 2I_yS_z \xrightarrow{J} & \; 2[I_y cos(\pi JtS_z) - I_x sin(\pi JtS_z)] \times S_z \\ & = 2[I_y cos(\pi Jt) - I_xS_z sin(\pi Jt)] \times S_z \\ & = 2I_yS_z cos(\pi Jt) - 2I_xS_z^2 sin(\pi Jt) \\ & = 2I_yS_z cos(\pi Jt) - I_x sin(\pi Jt) \end{aligned} \tag{8.39}$$

The product of two transverse operators is evaluated in exactly the same way:

$$\begin{aligned} 2I_xS_y \xrightarrow{J} & \; 2\,[I_x cos(\pi Jt) + 2I_yS_z sin(\pi Jt)] \times \\ & [S_y cos(\pi Jt) - 2I_zS_x sin(\pi Jt)] \end{aligned} \tag{8.40}$$

It can be shown that product operators containing two transverse terms, such as $2I_xS_y$, do not evolve under J-coupling (see Section 10.2.1.1).

8.5 Transformations of Two-spin Product Operators

The transformations of product operators that are associated with evolution, pulse, and scalar coupling can be obtained by application of the above four rules. These transformations can be summarized in a graphical form that readily permits the calculation of the evolution of the density matrix in product operator format (see Fig. 8.2). The following are a series of example transformations that are evaluated using

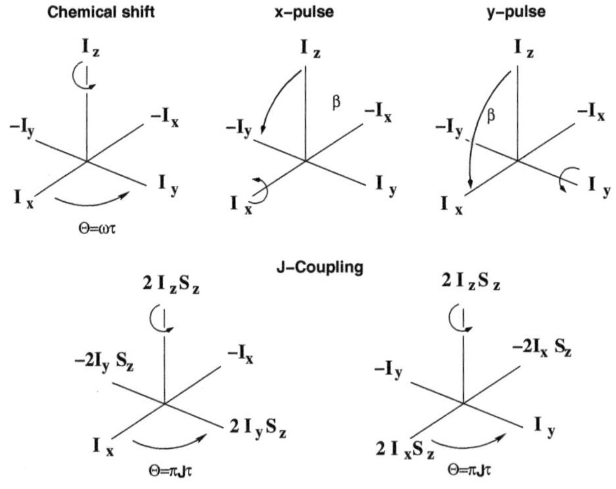

Figure 8.2 Manipulation of the density matrices using the product operator representation. The upper left section of the figure shows the effect of chemical shift evolution (e.g. $\mathcal{H} = \omega_I I_z$), with a rotation angle of ωt. The effects of pulses, with a flip angle of β degrees, are also shown on the top of the figure for pulses along the x-axis (middle), or y-axis (right). The effects of J-coupling on the density matrix are shown on the lower part of the figure, for the density matrix represented by I_x (left), or I_y (right). Here, the rotation angle is πJt.

this diagram. In all cases the product operator to the left of the arrow represents the density matrix before the transformation while the product operator to the right of the arrow represents the density matrix after the transformation, for example:

$$\begin{array}{lcl} I_y & \xrightarrow{\omega_I t} & I_y cos(\omega_I t) - I_x sin(\omega_I t) \\ I_y & \xrightarrow{\beta_x} & I_y cos(\beta) + I_z sin(\beta) \\ I_z & \xrightarrow{\beta_y} & I_z cos(\beta) + I_x sin(\beta) \\ I_y & \xrightarrow{2\pi J I_z S_z} & I_y cos(\pi J t) - 2I_x S_z sin(\pi J t) \end{array}$$

Note that in all cases the new density matrix, ρ, is a linear combination of the *cosine* weighted initial density matrix, ρ_i, plus the *sine* weighted density matrix that is advanced by 90°, ρ_{90}:

$$\rho_i \rightarrow \rho_i cos(\alpha t) + \rho_{90} sin(\alpha t) \quad (8.41)$$

As time passes, the system will pass through all four forms of the density matrices that are in the same plane within Fig. 8.2. For example, the evolution of I_x under J-coupling proceeds as follows:

$$I_x \rightarrow 2I_y S_z \rightarrow -I_x \rightarrow -2I_y S_z \rightarrow I_x \rightarrow 2I_y S_z ... \quad (8.42)$$

Transverse magnetization that is associated with a single spin operator, for example I_x, is often referred to as in-phase magnetization. In-phase magnetization evolves under the influence of J-coupling to anti-phase magnetization, $2I_yS_z$. A vector model that represents this evolution is illustrated in Fig. 8.3. This representation shows why anti-phase magnetization cannot be detected (see Section 8.2.1); the individual vector components of the anti-phase magnetization cancel each other.

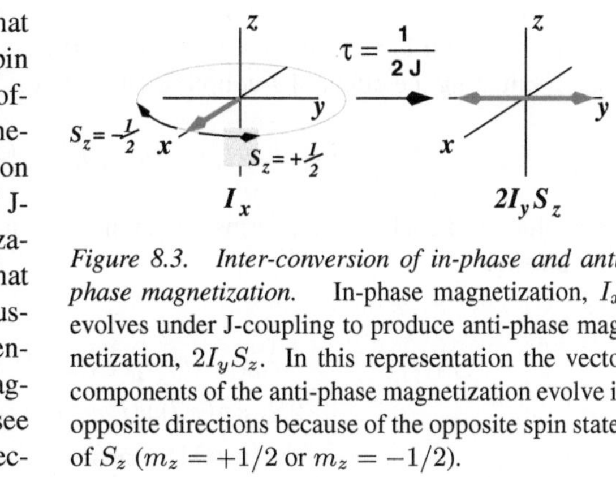

Figure 8.3. Inter-conversion of in-phase and anti-phase magnetization. In-phase magnetization, I_x, evolves under J-coupling to produce anti-phase magnetization, $2I_yS_z$. In this representation the vector components of the anti-phase magnetization evolve in opposite directions because of the opposite spin states of S_z ($m_z = +1/2$ or $m_z = -1/2$).

8.6 Product Operator Treatment of a One-pulse Experiment

The product operator treatment of the one-pulse experiment for two coupled spins is similar to that utilized for a single isolated spin. As before, the NMR experiment transforms the initial density matrix, ρ_o, to $\rho_1(t)$ and the final detected signal is extracted from the final density matrix. The only complication is that it is necessary to keep track of two spins, I and S. Analysis of each step of the one-pulse experiment is discussed below.

Initial Product Operator: As discussed above, $\rho_o = I_z + S_z$.

Effect of 90° Pulse: We assume that this is a homonuclear experiment, therefore the pulse is applied to both spins. Assuming a perfect 90° pulse:

$$\rho_o \quad \rightarrow \quad \rho_1 \tag{8.43}$$

$$I_z + S_z \quad \xrightarrow{P_x(\pi/2)} \quad -I_y - S_y \tag{8.44}$$

Free Precession: The free precession of the spins causes the density matrix to evolve according to:

$$\begin{aligned}
\rho_1(t) &= e^{-i\mathcal{H}t}\rho_1 e^{i\mathcal{H}t} \\
&= e^{-i(\omega_I I_z + \omega_S S_z + 2\pi J I_z S_z)t}\rho_1 e^{+i(\omega_I I_z + \omega_S S_z + 2\pi J I_z S_z)t} \\
&= e^{-i\omega_I I_z t}e^{-i\omega_S S_z t}e^{-i2\pi J I_z S_z t}\rho_1 e^{+i\omega_I I_z t}e^{+i\omega_S S_z t}e^{+i2\pi J I_z S_z t} \\
&= e^{-i\omega_I I_z t}e^{-i\omega_S S_z t}\left[e^{-i2\pi J I_z S_z t}\rho_1 e^{+i2\pi J I_z S_z t}\right]e^{+i\omega_I I_z t}e^{+i\omega_S S_z t}
\end{aligned} \tag{8.45}$$

In the above expression the order of the terms in the Hamiltonian has been changed such that evolution of the system due to J-coupling (i.e. $e^{-i2\pi J I_z S_z t}\rho_1 e^{i2\pi J I_z S_z}$) is evaluated first, and then the evolution of the system due to each chemical shift is subsequently evaluated. This reordering of the terms greatly facilitates the analysis by product operators. This rearrangement of the Hamiltonian is legitimate since all operators, I_z, S_z, and $I_z S_z$, commute with each other because I and S refer to different spins. Evaluating the effect of J-coupling on ρ_1 gives:

$$-I_y - S_y \xrightarrow{J} [-I_y cos(\pi J t) + 2I_x S_z sin(\pi J t)] + [-S_y cos(\pi J t) + 2S_x I_z(\pi J t)]$$

The evolution of each of these terms due to the chemical shift part of the Hamiltonian is as follows:

$$\begin{aligned}
\rho_1(t) \quad = & \quad -cos(\pi J t)\left[I_y cos(\omega_I t) - I_x sin(\omega_I t)\right] \\
+ & \quad 2S_z sin(\pi J t)\left[I_x cos(\omega_I t) + I_y sin(\omega_I t)\right] \\
- & \quad cos(\pi J t)\left[S_y cos(\omega_S t) - S_x sin(\omega_S t)\right] \\
+ & \quad 2I_z sin(\pi J t)\left[S_x cos(\omega_I t) + S_y sin(\omega_I t)\right]
\end{aligned} \tag{8.46}$$

Collecting terms:

$$\begin{aligned}
\rho_1(t) \quad = & \quad -I_y cos(\omega_I t)cos(\pi J t) + I_x sin(\omega_I t)cos(\pi J t) \\
- & \quad S_y cos(\omega_S t)cos(\pi J t) + S_x sin(\omega_S t)cos(\pi J t) \\
+ & \quad 2I_x S_z cos(\omega_I t)sin(\pi J t) + 2I_y S_z sin(\omega_I t)sin(\pi J t) \\
+ & \quad 2S_x I_z cos(\omega_S t)sin(\pi J t) + 2S_y I_z sin(\omega_S t)sin(\pi J t)
\end{aligned} \tag{8.47}$$

Since the only density matrices that give rise to detectable signal are represented by single transverse operators it is necessary to only focus on the first two lines of the above equation. Additional simplification is obtained if the transverse operators are written in the form of raising and lowering operators:

$$\begin{aligned}\rho_1(t) &= \left[-\frac{1}{2i}(I^+ - I^-)cos(\omega_I t) + \frac{1}{2}(I^+ + I^-)sin(\omega_I t)\right] cos(\pi Jt) \\ &+ \left[-\frac{1}{2i}(S^+ - S^-)cos(\omega_S t) + \frac{1}{2}(S^+ + S^-)sin(\omega_S t)\right] cos(\pi Jt)\end{aligned} \tag{8.48}$$

Since the only components of the density matrix that gives rise to detectable signal are I^- and S^-, it is only necessary to consider these two density matrices.

$$\begin{aligned}\rho_1(t) &= I^- \left[\frac{cos(\omega_I t)}{2i} + \frac{sin(\omega_I t)}{2}\right] cos(\pi Jt) \\ &+ S^- \left[\frac{cos(\omega_S t)}{2i} + \frac{sin(\omega_S t)}{2}\right] cos(\pi Jt) \\ &= I^- \frac{1}{2i} e^{i\omega_I t} cos(\pi Jt) + S^- \frac{1}{2i} e^{i\omega_S t} cos(\pi Jt)\end{aligned} \tag{8.49}$$

Therefore, the detected signal is:

$$S(t) = \frac{1}{2i} e^{i\omega_I t} cos(\pi Jt) + \frac{1}{2i} e^{i\omega_S t} cos(\pi Jt) \tag{8.50}$$

The Fourier transform of $e^{i\omega_I t}$ gives a single peak at ω_I and the $e^{i\omega_s t}$ term gives a single peak at ω_S. Each of these terms in the time domain is multiplied by $cos(\pi Jt)$, therefore the peaks at ω_I and ω_S will be convoluted with the Fourier transform of $cos(\pi Jt)$, resulting in the splitting of each peak by $2\pi J$. This gives the normal four line AX type spectrum, with a separation of $2\pi J$ between each set of doublets.

Chapter 9

TWO DIMENSIONAL HOMONUCLEAR J-CORRELATED SPECTROSCOPY

Multi-dimensional NMR experiments generate a spectrum in which the position of a spectral line, or peak, is defined by two or more frequencies. The existence of such a peak indicates that the participating spins are coupled to one other by scalar (J) coupling through chemical bonds or via dipolar coupling through space. The position of the peak is defined by the chemical shifts, or resonance frequencies, of the coupled spins.

In this chapter we will focus solely on the generation of multi-dimensional spectra via homonuclear J-coupling. In general, all of the protons within an amino acid residue belong to the same network of scalar coupled spins, or a *spin-system*. The exceptions to this rule are the aromatic residues, in which the aromatic protons form a separate spin-system because of the small coupling between the H_β proton and the protons on the aromatic ring. The spins that belong to a spin-system can be identified by multi-dimensional J-correlated spectroscopy. The identification of residue type on the basis of the properties of the spin system, such as the number and type of chemical shifts, is an important step in the assignment of resonance lines to individual atoms in the protein.

In addition to providing information for resonance assignments, the J-coupling constants can often be extracted from these spectra, providing structural information on the torsional angles. Finally, the increased dimensionality of the experiment also increases the resolution of the spectrum, permitting the observation of resolved lines in large systems.

This chapter begins with a general introduction to multi-dimensional NMR spectroscopy and then features a discussion of three important homonuclear two dimensional experiments: COSY, DQF-COSY, and TOCSY, each of which are used to elucidate scalar couplings between spins within a spin-system. Experiments that elucidate heteronuclear couplings will be discussed in Chapter 10.

9.1 Multi-dimensional Experiments

Multidimensional NMR experiments consist of an interleaved combination of chemical shift labeling periods and magnetization transfer, or mixing, periods. The mixing periods serve to transfer the chemical shift information from one spin to its coupled partner. For example, in a two-dimensional (2D) NMR experiment, represented by:

$$A \longrightarrow B,$$

the magnetization begins on spin "A", is frequency labeled with the chemical shift of "A", and then it is transferred during a mixing period to spin "B". The magnetization on spin "B" is detected in the usual way, as a signal in the receiver coil. The final detected signal is now dependent on *two* time domains, the first was used to record ω_A while the second time domain is used to measure ω_B. The detected signal at the end of the two dimensional experiment is given by:

$$S(t_1, t_2) = \eta e^{i(\omega_A t_1)} e^{i(\omega_B t_2)} \tag{9.1}$$

where η represents the efficiency of magnetization transfer between the two spins. The final detected signal that is shown in Eq. 9.1 indicates that the directly detected FID, $e^{i(\omega_B t_2)}$ is modulated by a term, $e^{i(\omega_A t_1)}$ that contains information of the frequency of the coupled spin. The Fourier transform of this signal will produce a two-dimensional spectrum that contains a single peak, located at (ω_a, ω_b). This type of peak is termed a *crosspeak* because the two frequencies that define its position are different. In some experiments, notably homonuclear proton multi-dimensional experiments, peaks exist that have the same frequency in all dimensions. These peaks are referred to as *self-peaks*, *autopeaks*, or *diagonal peaks* and represent magnetization that was not transferred to another spin during the experiment, hence the recorded frequencies are the same in all dimensions.

In a three-dimensional experiment two labeling and mixing segments are used, resulting in a path of magnetization flow between three spins as: $A \rightarrow B \rightarrow C$, giving

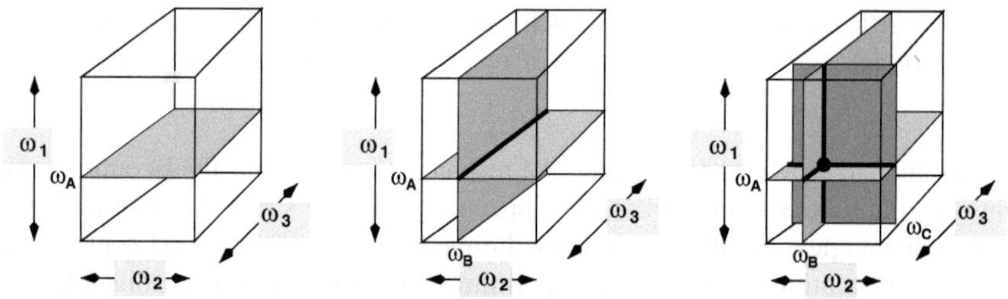

Figure 9.1. Peak location in a three-dimensional spectrum. The location of a crosspeak in a three dimensional spectrum is defined by the intersection of three orthogonal planes. The first plane is the locus of all points that have a frequency of ω_A in the first frequency dimension. The second plane is the locus of all points that have a frequency of ω_B in the second frequency dimension. The intersection of these two planes is a line, as indicated in the center diagram. The third plane is defined by all points that have a frequency of ω_C in the third frequency dimension. This plane intersects the line at a single point, which is location of the crosspeak.

the following signal.

$$S(t_1, t_2, t_3) = \eta e^{i(\omega_A t_1)} e^{i(\omega_B t_2)} e^{i(\omega_C t_3)} \quad (9.2)$$

In this case, the amplitude of the detected FID is modulated by terms that provide information on the frequencies of the other two coupled spins. Fourier transformation of this signal will generate a peak whose position within a three-dimensional cube is defined by ω_A, ω_B, and ω_C in each dimension, as illustrated in Fig. 9.1.

In multi-dimensional experiments, the intensity of the crosspeaks will be proportional to η. In the case of J-coupled spins η can approach unity. In the case of dipolar coupled spins η is related to the distance between the two spins providing a means to measure inter-atomic distances.

9.1.1 Elements of Multi-dimensional NMR Experiments

Generalized pulse sequences for two-dimensional and a three-dimensional experiments are shown in Fig. 9.2. Any two-dimensional NMR experiment can be divided into four basic elements: preparation, evolution, mixing, and detection. These elements are described in detail in Table 9.1. In a three dimensional experiment, the evolution period and mixing period would be repeated an additional time.

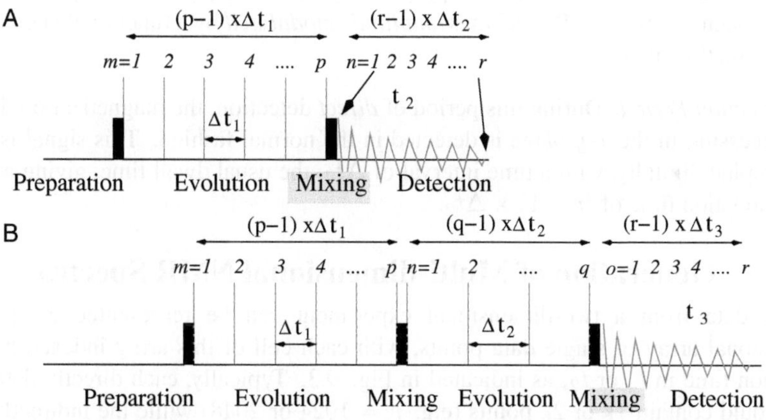

Figure 9.2. Generalized two-dimensional and three-dimensional pulse sequences. Panels A and B show a two-dimensional or a three-dimensional experiment, respectively. Both experiments begin with an excitation pulse that is followed by an evolution period, t_1, and then a mixing period. In a two-dimensional experiment the FID is collected after the mixing period. In the case of a three-dimensional experiment, another evolution and mixing period follow before acquisition of the FID. Initially, the length of the t_1 period is set to zero (or $\Delta t_1/2$) and the first ($m = 1$) FID containing r points is collected. Note that this FID usually consists of multiple scans, all of which are summed to the same memory location. Subsequently, t_1 is incremented by a fixed amount, Δt_1 (the dwell time in t_1), and a second ($m = 2$) FID is collected and stored in a different memory location. This process is repeated a total of p times until the desired evolution time is attained. In the case of the three-dimensional experiment (B), the t_1 and t_2 evolution periods are sampled *independently*. For every t_1 time, q t_2 times would be acquired, leading to a total of $p \times q$ separate FIDs. Note that the increment in t_1 (Δt_1) need not equal the increment in t_2 (Δt_2), nor does p necessarily equal q.

Table 9.1. Elements of a two dimensional NMR experiment.

1. *Preparation period:* The length of this period is fixed and is usually employed to allow the spins to return to, or near, thermodynamic equilibrium. This period typically ends with a single 90° pulse that excites the first spin ('A').

2. *Evolution period (t_1):* This time period is used to encode the chemical shift of 'A' in the density matrix due to evolution under the Hamiltonian: $\mathcal{H} = \omega_A I_{AZ}$. This period is referred to as the *indirectly* detected domain, or dimension, because the excited state of spin 'A' is not directly detected by the receiver coil. Rather, the evolution of the system is sampled digitally, i.e. t_1 begins at zero and then is incremented by a constant amount, Δt_1, with a separate FID acquired at each increment of t_1. A total of p FIDs are acquired, generating a total acquisition time in t_1 of $(p-1) \times \Delta t_1$.

3. *Mixing period:* This event causes the magnetization that is associated with spin 'A' to become associated with spin 'B'. This period leads to the transfer of the chemical shift information of spin 'A' to spin 'B'. The mixing can be evoked by either J-coupling or dipolar coupling. The *key* point is that the amount of magnetization transferred from A to B is proportional to $cos(\omega_A t_1)$ or $sin(\omega_A t_1)$. Hence the magnetization of 'B' becomes *amplitude modulated* by a function that contains information about ω_A.

4. *Detection Period:* During this period of *direct* detection, the magnetization that is precessing in the x-y plane is detected in the normal fashion. This signal is also sampled digitally, with a time interval of Δt_2, the usual dwell time, giving a total acquisition time of $(r-1) \times \Delta t_2$.

9.1.2 Generation of Multi-dimensional NMR Spectra

The data from a two-dimensional experiment can be represented as a two-dimensional array of single data points, with each cell of this array indexed by the evolution time in t_1 or t_2, as indicated in Fig. 9.3. Typically, each directly detected FID would contain 1k or 2k points (e.g. $r = 1024$ or 2048) while the indirectly detected dimension would contain between 128 and 1k points (e.g. $p = 128$ to 1024), depending on the nature of the experiment. Processing of this time domain data into a two-dimensional spectrum requires calculation of a two-dimensional Fourier transform:

$$\Omega(\omega_1, \omega_2) = \int\int S(t_1, t_2) e^{i\omega t_1} e^{i\omega t_2} dt_1 dt_2 \tag{9.3}$$

where Ω represents the final spectrum and $S(t_1, t_2)$ represents the initial matrix of data points.

In practice, this transform is computed one dimension at a time, usually beginning with the transform of the data as a function of t_2, followed by transformation as a

$t_1 \downarrow$ \ $t_2 \longrightarrow$	1	2	3	4	5	6	7	8	.	r
1	x	x	x	x	x	x	x	x	.	.
2	x	x	x	x	x	x	x	x	.	.
3	x	x	x	x	x	x	x	x	.	.
4	x	x	x	x	x	x	x	x	.	.
5	x	x	x	x	x	x	x	x	.	.
.	.	.	.	.	.	.	.	.	.	.
p	.	.	.	.	.	.	.	.	.	.

Figure 9.3 Data structure for two-dimensional data. The data structure for a two-dimensional data set is shown. Each row corresponds to a FID of r points that was acquired at the indicated t_1 value. There are a total of p t_1 values. Each FID may result from the sum of more than one scan, but all scans would be acquired with the same t_1 value.

function of t_1.

$$F(t_1,\omega_2) = \int S(t_1,t_2)e^{i\omega t_2}dt_2$$

$$\Omega(\omega_1,\omega_2) = \int F(t_1,\omega_2)e^{i\omega t_1}dt_1 \tag{9.4}$$

With reference to Fig. 9.3, the processing software would read and transform p rows, corresponding to the directly detected FIDs, to produce $F(t_1,\omega_2)$, a mixed data matrix. The software would then load r columns and perform the Fourier transform in the t_1 direction to generate the final data matrix or spectrum, $\Omega(\omega_1,\omega_1)$. These steps are illustrated in Fig. 9.4. In the case of a 3-dimensional experiment, these steps would proceed as t_3, followed by t_2, and then t_1.

9.2 Homonuclear J-correlated Spectra

In this section we will look with some detail at two common two-dimensional homonuclear J-correlated experiments, the COSY (**CO**rrelated **S**pectroscop**Y**) and the DQF-COSY (**D**ouble-**Q**uantum **F**iltered COSY). The COSY experiment was first presented by Jeener in 1971 [78] and was given its current name in 1980 [91]. The DQF-COSY experiment is a specific example of a multiple-quantum filtered COSY experiment [131].

The focus in this chapter will be on understanding how frequency labeling occurs and how this information is passed from one spin to its coupled partner. The application of these experiments to protein NMR spectroscopy will be discussed in subsequent chapters.

9.2.1 COSY Experiment

The pulse sequence, and the corresponding pulse program, for the COSY experiment are shown in Fig. 9.5. The COSY experiment is the simplest of all multi-dimensional NMR experiments, consisting of an excitation pulse, a frequency labeling period, and a second pulse that serves to transfer the magnetization from one coupled spin to another.

The pulse program will acquire n scans using the first t_1 time. Each of these scans are summed prior to storing the FID in memory. The t_1 delay is the incremented by Δt_1 (id0), and then n scans are acquired at the new t_1 time. This process is repeated a

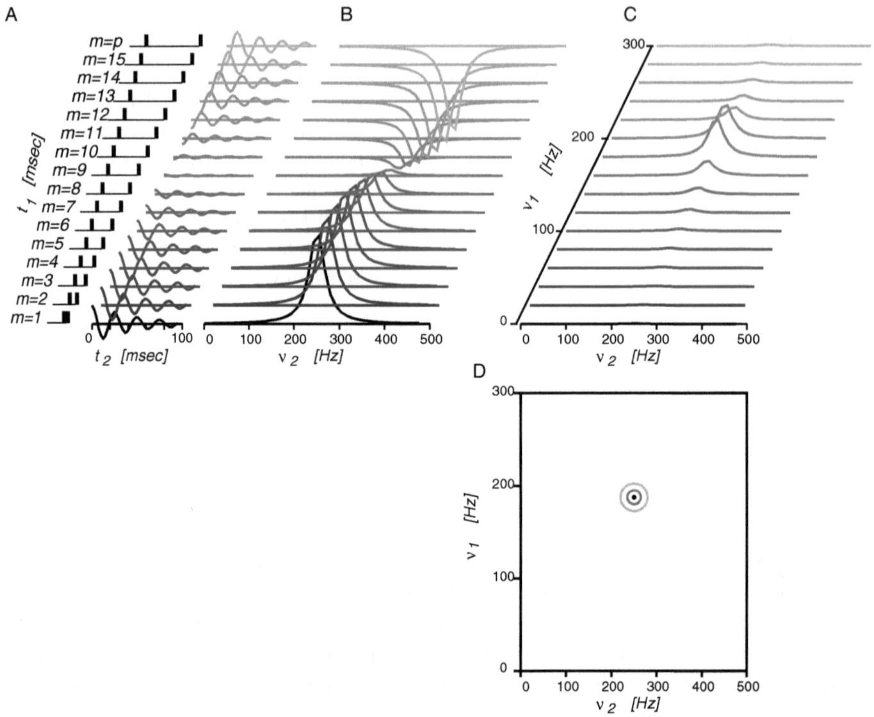

Figure 9.4. *Generation of a two-dimensional spectrum.* In this example the frequency of the two coupled spins are 190 and 250 Hz. Note that only one magnetization path is considered here (i.e. $A \rightarrow B$), therefore only one peak is present in the spectrum, located at $\nu_A = 190$ Hz and $\nu_B = 250$ Hz.

Panel A shows, at the extreme left, the series of pulse sequences that are used to obtain evolution of the first time domain. The sequence at the bottom has a t_1 time of zero; the mixing pulse immediately follows the excitation pulse. The t_1 evolution time is incremented by a fixed amount, Δt_1, producing a series of separate experiments, arranged from the bottom to the top of the diagram. The FIDs that are obtained for each value of these experiments are shown in the right part of panel A. Each FID corresponds to a single t_1 value. Note that at early t_1 times the first points of the FID are greater than zero, these points become negative at later t_1 times due to the fact the FID is equal to $cos(\omega_A t_1)e^{i\omega_B t_2}$.

Panel B shows the Fourier transform of each of the FIDs, hence the horizontal axis is converted from the time domain in panel A to the frequency domain in B. The vertical axis is still in units of time. Note that the intensity of the resonance line at 250 Hz, is positive for short t_1 values, but becomes negative at longer t_1 values, as anticipated from the FIDs shown in panel A.

Panel C shows the results from the second Fourier transform, along t_1. This transform was obtained by taking a column of data at each ν_2 frequency and computing the transform of the data in t_1. A single peak is found at $\nu_A = 190$ Hz and $\nu_B = 250$ Hz.

Panel D shows a contour, or topographical plot, of the same spectrum that is displayed in panel C. In this illustration, contour lines that join points of height intensity are more darkly shaded than lines that join regions of low intensity. The location of the two-dimensional peak is readily apparent in such a plot.

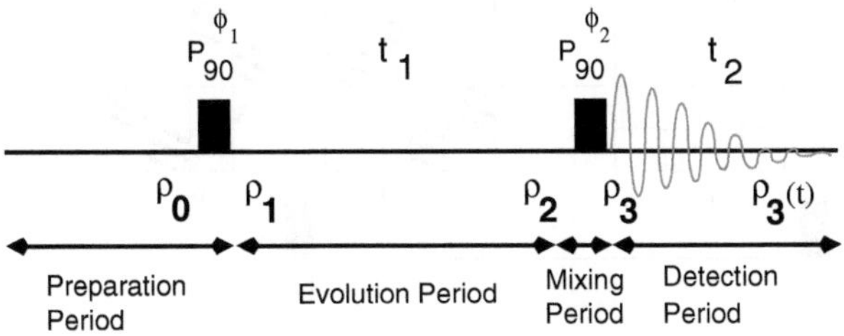


```
1 ze                  ;Zero the memory.
2 d1                  ;Inter-scan relaxation delay (1 sec)
3 (p1 ph1):f1         ;Pulse of length p1, phase=ph1 on channel 1.
  d0                  ;t1 evolution time
  (p1 ph2):f1         ;Second 90 pulse, with phase=ph2 (mixing pulse)
  go=2 ph31           ;Acquire FID (Receiver phase=31) go to 2, n-times
  write               ;Write the FID to disk.
  id0                 ;increment t1 time
  lo to 3 times td1 ;go to 3 p times (total number of t1 values)
exit                  ;End of pulse sequence.
ph1 =0 0 0 0 1 1 1 1 2 2 2 2 3 3 3 3 ;Phase of 1st excitation pulse.
ph2 =0 1 2 3 0 1 2 3 0 1 2 3 0 1 2 3 ;Phase of 2nd pulse.
ph31=0 2 0 2 3 1 3 1 2 0 2 0 1 3 1 3 ;Phase of receiver.
```

Figure 9.5. COSY pulse sequence. The upper part of this figure shows the COSY pulse sequence while the lower part shows the corresponding pulse program that is used to represent the sequence. The text to the right of the semi-colon in the pulse program briefly describes each step of program. The COSY experiment consists of two 90° pulses that bracket the t_1 evolution time. The phase of these pulses, as well as the phase of the receiver, are cycled as indicated in the last three lines of the pulse program. A total of p t_1 times are acquired, each of which consisting of n-scans. The second pulse serves to transfer the magnetization from one coupled spin to the other. The density matrices at various points in the pulse sequence are indicated by ρ_i. The density matrix immediately after the second pulse is ρ_3, which evolves during detection of the FID, giving $\rho_3(t)$.

total of p times. This phase cycle is more involved than the cyclops phase cycle used in the one-pulse experiment (Fig. 2.19), and requires a total of 16 scans to complete. Therefore, this experiment would have to be acquired with a multiple of 16 scans for each t_1 point. A more complicated phase cycle is required to eliminate unwanted signals from the spectrum. The generation of these cycles will be discussed in more detail in Chapter 11.

9.2.1.1 Overall Change of ρ During the COSY Experiment

The changes in the density matrix that occur during the COSY experiment are illustrated in Fig. 9.6. The density matrix begins with only the diagonal elements as non-zero. After the first pulse, single quantum transitions for both the I and S spins are created. These elements of the density matrix are labeled with their appropriate frequencies, ω_I and ω_S, respectively. The mixing pulse interchanges the single quantum elements that are associated with the I spins by the elements that were associated

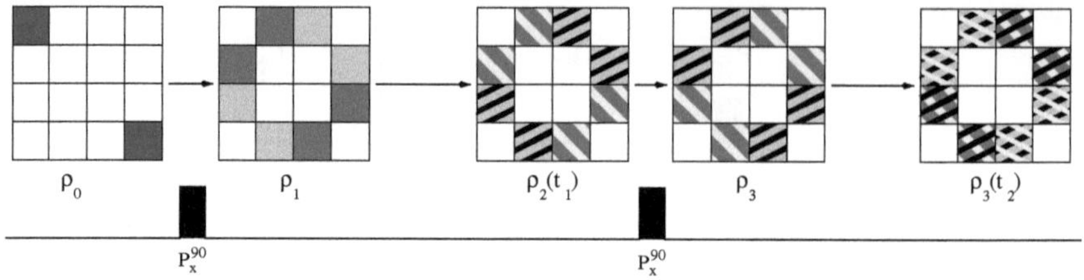

Figure 9.6. *Pictorial representation of density matrix changes during the COSY experiment.* The changes that occur in the density matrix during the COSY experiment are illustrated in this figure. The non-zero elements of the density matrix are shaded. A solid shading indicates that no time evolution has occurred. Squares shaded with a light gray background are initially associated with I spins and squares shaded in darker gray are initially associated with S spins. These associations are interchanged by the mixing pulse. Right-slanted black-lines () indicate evolution of the element of the density matrix at the chemical shift of spin I. Left-slanted white-lines () indicate evolution of the element of the density matrix at the chemical shift of spin S. Double cross-hatched squares indicate evolution at ω_I in t_1 and ω_S in t_2 () or at ω_S in t_1 and ω_I in t_2 (). The pulse sequence is shown below the 4×4 density matrices. For reference, each element of the density matrix evolves according to the following table shown to the right.

Evolution of the Density Matrix

	$\alpha\alpha$	$\alpha\beta$	$\beta\alpha$	$\beta\beta$
$\alpha\alpha$	P_1	Ω_S	Ω_I	Ω_{IS}
$\alpha\beta$	Ω_S	P_2	Ω_0	Ω_I
$\beta\alpha$	Ω_I	Ω_0	P_3	Ω_S
$\beta\beta$	Ω_{IS}	Ω_I	Ω_S	P_4

Entries containing the symbol Ω evolve with time according to the indicated frequency. For example, the element marked with Ω_S will evolve as $e^{-i\omega_S t}$ during free precession. Ω_{IS} evolves at the sum of the resonance frequencies and Ω_0 evolves at the difference between ω_I and ω_s. The diagonal elements correspond to populations and do not evolve. The contribution of the J-coupling to the evolution has been ignored here.

with the S spins, and vice versa. This mixing causes the chemical shift information from one spin to be transferred to the other spin.

9.2.1.2 Density Matrix/Product Operator Analysis of the COSY Experiment

The goal is to begin with the initial density matrix, ρ_0, and follow it through the entire sequence to find $\rho_3(t)$ from which the observed signal can be obtained.
The initial density matrix of the system is:

$$\rho_o = I_z + S_z \tag{9.5}$$

The first pulse is a 90° pulse along the x-axis. Since this is a homonuclear experiment this pulse is applied to *both* spins, bringing the magnetization from the z-axis to the minus y-axis. The transformation of the density matrix is:

$$\rho = e^{-i\beta I_x} e^{-i\beta S_x} \rho_o e^{i\beta I_x} e^{i\beta S_x} \tag{9.6}$$

where the term $e^{-i\beta I_x}e^{-i\beta S_x}$ is the rotation operator for a rotation about the x-axis.

$$\begin{aligned}\rho_1 &= e^{-i\beta I_x} I_z e^{i\beta I_x} \; + \; e^{-i\beta S_x} S_z e^{i\beta S_x} \\ &= I_z cos\beta - I_y sin\beta + S_z cos\beta - S_y sin\beta \\ &= -\left[I_y + S_y\right] \qquad (\beta = 90°)\end{aligned} \tag{9.7}$$

During t_1 the density matrix evolves under the complete Hamiltonian, by applying the following transformation to ρ_1:

$$R = e^{i(\omega_I t_1 I_z + \omega_S t_1 S_z + \pi J 2 I_z S_z t_1)} \tag{9.8}$$

Since all of these operators commute, they can be considered separately and rearranged for convenience. In this case we will evaluate evolution due to chemical shift of the S spin first, followed by evolution due to the chemical shift of the I spin, followed lastly by J-coupling. This gives the following transformation of the density matrix:

$$\rho_2 = e^{i\pi J 2 I_z S_z t_1}\left[e^{i\omega_I t_1 I_z}\left[e^{i\omega_S t_1 S_z}\rho_1 e^{i\omega_S t_1 S_z}\right]e^{i\omega_I t_1 I_z}\right]e^{i\pi J 2 I_z S_z t_1} \tag{9.9}$$

We can use the diagram presented in Chapter 8 to evaluate the evolution of the density matrix due to chemical shift (i.e. $e^{i\omega t}$). The relevant portion of the diagram is reproduced in Fig. 9.7.

$$\begin{aligned}-I_y &\rightarrow -\left[I_y cos(\omega_I t_1) - I_x sin(\omega_I t_1)\right] \\ -S_y &\rightarrow -\left[S_y cos(\omega_S t_1) - S_x sin(\omega_S t_1)\right]\end{aligned} \tag{9.10}$$

The effect of scalar coupling on each of the terms in the above equation is:

$$I_y \rightarrow I_y cos(\pi J t_1) - 2 I_x S_z sin(\pi J t_1) \tag{9.11}$$

$$S_y \rightarrow S_y cos(\pi J t_1) - 2 I_z S_x sin(\pi J t_1) \tag{9.12}$$

$$I_x \rightarrow I_x cos(\pi J t_1) + 2 I_y S_z sin(\pi J t_1) \tag{9.13}$$

$$S_x \rightarrow S_x cos(\pi J t_1) + 2 I_z S_y sin(\pi J t_1) \tag{9.14}$$

The combined effect of these two transformations is:

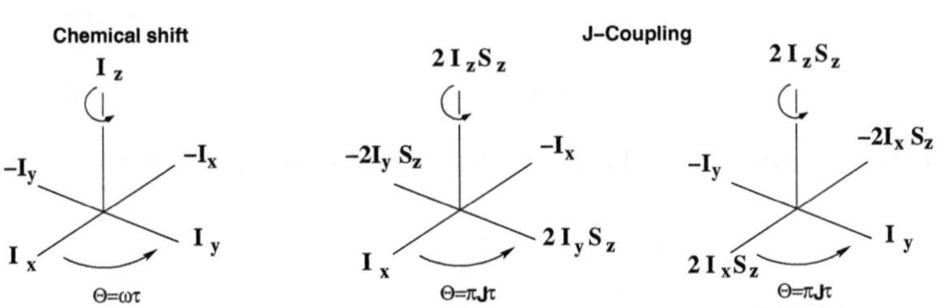

Figure 9.7. Transformation of ρ by chemical shift and J-coupling.

$$\begin{aligned}\rho_2 = \quad & -I_y cos(\omega_I t_1) cos(\pi J t_1) + 2I_x S_z cos(\omega_I t_1) sin(\pi J t_1) \\ & -S_y cos(\omega_S t_1) cos(\pi J t_1) + 2I_z S_x cos(\omega_S t_1) sin(\pi J t_1) \\ & +I_x sin(\omega_I t_1) cos(\pi J t_1) + \underline{2I_y S_z sin(\omega_I t_1) sin(\pi J t_1)} \\ & +S_x sin(\omega_S t_1) cos(\pi J t_1) + \underline{2I_z S_y sin(\omega_S t_1) sin(\pi J t_1)} \end{aligned} \quad (9.15)$$

The underlined terms will ultimately produce the crosspeaks in the spectrum, as discussed below.

The effect of the second 90° x-pulse on the various terms in the above equation is shown to the right of the arrows (the trigonometric terms have been ignored temporarily):

$$\begin{array}{ll} I_y \rightarrow I_z & S_y \rightarrow S_z \\ I_x \rightarrow I_x & S_x \rightarrow S_x \\ 2I_x S_z \rightarrow -2I_x S_y & 2I_z S_x \rightarrow -2I_y S_x \\ \underline{2I_y S_z \rightarrow -2I_z S_y} & \underline{2I_z S_y \rightarrow -2I_y S_z} \end{array}$$

The first line, containing I_z and S_z, corresponds to a density matrix with only diagonal matrix elements, representing undetectable magnetization. Terms in the second line (I_x and S_x) are detectable, but they will only produce diagonal peaks because the same spin is transverse before and after the mixing pulse. Therefore these elements of the density matrix will evolve with the same frequency during t_1 and t_2. The third line contains terms that represent the creation of double quantum coherence after the second pulse. These cannot be detected in the experiment. The last line contains the two terms of interest, those which will generate crosspeaks.

9.2.1.3 Origin of COSY Crosspeaks

The crosspeaks in the COSY spectrum arise from the following product operators:

$$2I_y S_z \stackrel{90_x}{\rightarrow} -2I_z S_y \qquad 2I_z S_y \stackrel{90_x}{\rightarrow} -2I_y S_z$$

The transformation by the 90° pulse causes the transverse magnetization that was associated with one spin to be transferred to the other spin. This can be seen by inspection of the density matrix which corresponds to these product operators, for example:

$$2I_y S_z sin(\omega_I t_1) sin(\pi J t_1) = \begin{bmatrix} 0 & 0 & -i\,\Omega_I sin(\pi J t_1) & 0 \\ 0 & 0 & 0 & i\,\Omega_I sin(\pi J t_1) \\ i\,\Omega_I sin(\pi J t_1) & 0 & 0 & 0 \\ 0 & -i\,\Omega_I sin(\pi J t_1) & 0 & 0 \end{bmatrix}$$

where $sin(\omega_I t_1)$ has been replaced by Ω_I. After the 90° pulse this density matrix becomes:

$$\begin{bmatrix} 0 & -i\,\Omega_I sin(\pi J t_1) & 0 & 0 \\ i\,\Omega_I sin(\pi J t_1) & 0 & 0 & 0 \\ 0 & 0 & 0 & -i\,\Omega_I sin(\pi J t_1) \\ 0 & 0 & i\,\Omega_I sin(\pi J t_1)) & 0 \end{bmatrix}$$

The matrix elements that represent single quantum transitions of one spin, such as $-i\Omega_I sin(\pi Jt_1)$, have been moved to elements of the density matrix that evolve with the frequency of the other spin (Ω_S) during t_2 (see Fig. 9.6). Therefore the amplitude of the density matrix element that evolves at a frequency of ω_S during t_2 is $sin(\omega_I t_1)sin(\pi Jt_1)$. The complete expression for the density matrix that describes the crosspeaks is:

$$\begin{aligned}\rho_3 &= -2I_zS_y sin(\omega_I t_1)sin(\pi Jt_1) \\ &\quad -2I_yS_z sin(\omega_S t_1)sin(\pi Jt_1)\end{aligned} \tag{9.16}$$

These product operators are not directly detectable, however they evolve into detectable magnetization due to the J-coupling term in the Hamiltonian. Temporarily neglecting amplitude factors (e.g. $sin(\omega_I t_1)sin(\pi Jt_1)$), the evolution of these product operators are:

$$\rho_3 \xrightarrow{J} \rho_3' \tag{9.17}$$
$$-2I_zS_y \rightarrow -2I_zS_y cos(\pi Jt_2) + S_x sin(\pi Jt_2) \tag{9.18}$$
$$-2I_yS_z \rightarrow -2I_yS_z cos(\pi Jt_2) + I_x sin(\pi Jt_2) \tag{9.19}$$

Of these, only I_x and S_x will be detectable, thus it is only necessary to evaluate how these terms will evolve due to chemical shift. Again, neglecting amplitude factors:

$$\begin{aligned}&\rho_3' \xrightarrow{\Omega} \rho_3(t) \\ &S_x \rightarrow S_x cos(\omega_S t_2) + S_y sin(\omega_S t_2) \\ &I_x \rightarrow I_x cos(\omega_I t_2) + I_y sin(\omega_I t_2)\end{aligned} \tag{9.20}$$

Forming I^- and S^-, and incorporating the amplitude factors from evolution in t_1 as well as evolution due to J-coupling in t_2, gives the following for the detectable portion of the density matrix:

$$\begin{aligned}\rho_3(t) = I^-&[sin(\omega_s t_1)sin(\pi Jt_1)sin(\pi Jt_2)e^{i\omega_I t_2}] \\ + S^-&[(sin(\omega_I t_1)sin(\pi Jt_1)sin(\pi Jt_2)e^{i\omega_S t_2}]\end{aligned} \tag{9.21}$$

The detected signal is obtained by evaluating Trace$[\rho(I^+ + S^+)]$:

$$\begin{aligned}S(t_1,t_2) = &\, sin(\omega_S t_1)sin(\pi Jt_1)sin(\pi Jt_2)e^{i\omega_I t_2} \\ &+ sin(\omega_I t_1)sin(\pi Jt_1)sin(\pi Jt_2)e^{i\omega_S t_2}\end{aligned} \tag{9.22}$$

9.2.1.4 Origin of COSY Diagonal Peaks

The diagonal peaks in the COSY spectrum arise from the I_x and S_x terms that are present after the second 90° pulse. Choosing to focus on the I spin only, and temporarily ignoring amplitude factors from evolution during t_1, the evolution under J-coupling is:

$$I_x \xrightarrow{J} I_x cos(\pi Jt_2) + I_yS_z sin(\pi Jt_2) \tag{9.23}$$

Only the I_x term will be detectable, thus its evolution under chemical shift is:

$$I_x \stackrel{\Omega}{\rightarrow} I_x cos(\omega_I t_2) + I_y sin(\omega_I t_2) \tag{9.24}$$

Incorporating the amplitude factors associated with I_x from evolution during t_1 give the following:

$$\begin{aligned} \rho_3(t) = & I_x sin(\omega_I t_1) cos(\pi J t_1) cos(\pi J t_2) cos(\omega_I t_2) \\ & + I_y sin(\omega_I t_1) cos(\pi J t_1) cos(\pi J t_2) sin(\omega_I t_2) \end{aligned} \tag{9.25}$$

Substituting $I_x = \frac{1}{2}[I^+ + I^-]$ and $I_y = \frac{1}{2i}[I^+ - I^-]$ gives the amplitude of the I^- part of the density matrix:

$$\rho_3(t) = I^- [sin(\omega_I t_1) cos(\pi J t_1) cos(\pi J t_1) e^{i\omega_I t_2}] \tag{9.26}$$

Therefore the signal that is associated with the diagonal peak is:

$$S(t_1, t_2) = sin(\omega_I t_1) cos(\pi J t_1) cos(\pi J t_1) e^{i\omega_I t_2} \tag{9.27}$$

9.2.1.5 Appearance of the COSY Spectrum

A schematic representation of the COSY spectrum is shown in Fig. 9.8 and a portion of a simulated COSY spectrum is shown in Fig. 9.9.

The signal that gives rise to the crosspeaks is:

$$\begin{aligned} S(t_1, t_2) = & sin(\omega_S t_1) sin(\pi J t_1) sin(\pi J t_2) e^{i\omega_I t_2} \\ & + sin(\omega_I t_1) sin(\pi J t_1) sin(\pi J t_2) e^{i\omega_S t_2} \end{aligned} \tag{9.28}$$

while the signal that is associated with the diagonal peaks is:

$$\begin{aligned} S(t_1, t_2) = & sin(\omega_I t_1) cos(\pi J t_1) cos(\pi J t_1) e^{i\omega_I t_2} \\ & + sin(\omega_S t_1) cos(\pi J t_1) cos(\pi J t_1) e^{i\omega_S t_2} \end{aligned} \tag{9.29}$$

The position of these peaks in the 2D-spectrum is determined by the terms that contain chemical shift information. For example, $sin(\omega_S t_1) e^{i\omega_I t_2}$ specifies a crosspeak at (ω_S, ω_I). Therefore, Eq. 9.28 represents the two crosspeaks and Eq. 9.29 represents the two diagonal peaks, at (ω_I, ω_I) and (ω_S, ω_S), as expected.

The additional terms in eqs. 9.28 and 9.29, such as $sin(\pi J t_1)$, are responsible for generating the splitting of each peak by the J-coupling. Since the time domain signal is a product of two functions in each dimension, its Fourier transform will be the convolution of the individual transforms with each other. For example,

$$sin(\pi J t) e^{i\omega t} \stackrel{FT}{\rightarrow} FT\left[(sin(\pi J t))\right] \otimes FT\left[(e^{i\omega t})\right]$$

In the case of the crosspeak, the resonance peak is convoluted with the Fourier transform of $sin(\pi J t)$. This produces an anti-phase doublet, with a negative peak at $-\pi J$ and a positive peak at $+\pi J$. Note that the overall splitting between these two peaks is 2π J rad/sec (or J Hz), as expected. Note that this splitting occurs in *both*

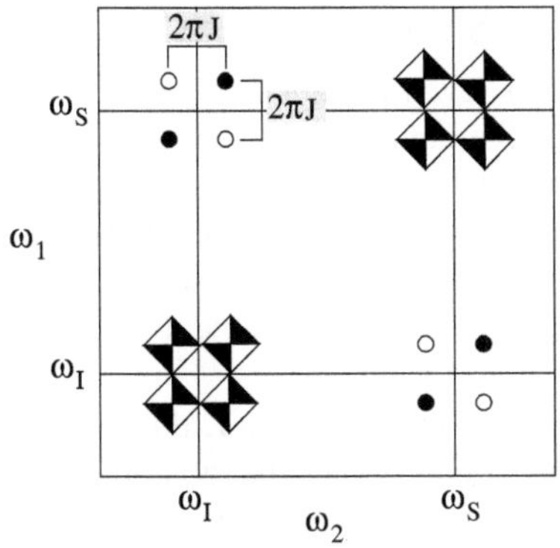

Figure 9.8 Sketch of an AX COSY spectrum. A schematic diagram of a COSY spectrum is shown for two coupled spins, I and S. The circles represent peaks with an absorption mode lineshape. Filled circles are positive and empty circles are negative. The spectrum has been phased such that the crosspeaks are absorption mode, thus the diagonal peaks have a dispersion lineshape in both dimensions, which is represented by the symbol ◆.

dimensions, thus forming a quartet of peaks. In addition to the introduction of the anti-phase splitting of the line, the convolution with the Fourier transform of $sin(\pi Jt)$ also causes the spectrum to be complex, since the transform of $sine$ is imaginary. Therefore, the crosspeaks will have a dispersion lineshape.

The diagonal peaks are modulated by $cos(\pi Jt)$, whose transform is doublet of real and positive peaks at $\pm\pi J$. Convolution of this function with $e^{i\omega t_2}$ will give an in-phase (i.e. both positive) doublet that will have an absorption mode lineshape. Again, this splitting occurs in both dimensions, leading to a quartet of peaks with an absorption lineshape.

Since the crosspeaks are usually of interest, they are phased to generate absorption lineshapes. Clearly, the same phase correction has to be applied to the diagonal peaks as well. Consequently, the diagonal peaks will be 90° out-of-phase and will have a dispersion lineshape, as indicated in Fig. 9.9.

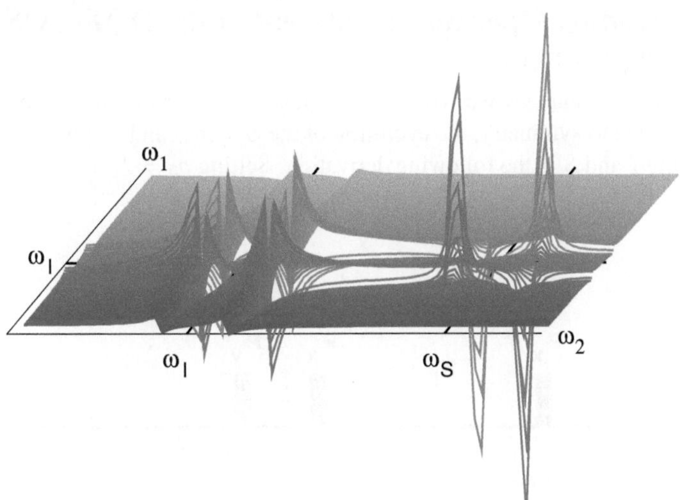

Figure 9.9. Lineshape in the COSY spectrum. A more detailed view of lower half of the COSY spectrum that was shown in Fig. 9.8 is presented here. Note that the diagonal peak (left) is a dispersion lineshape, while the crosspeaks have an absorption lineshape.

9.3 Double Quantum Filtered COSY (DQF-COSY)

The COSY experiment has several drawbacks, even though it is one of simplest two-dimensional experiments. First, the dispersive nature of the diagonal selfpeaks can cause considerable distortion of crosspeaks that are found near the diagonal of the spectrum. Second, the solvent peak (e.g. water), is not suppressed in the experiment. In the case of protein spectra acquired in H_20, the solvent peak can be several orders of magnitude larger than the protein resonances, causing a considerable dynamic range problem.

The double quantum filtered COSY experiment, or DQF-COSY, does not suffer from these deficiencies. This experiment filters out any signals that do not arise from coupled spins. Since the protons in water are equivalent, they behave as if they are not coupled and will be absent from the DQF-COSY spectrum. An additional benefit of the DQF-COSY experiment is that both the diagonal and the crosspeaks can be phased to be in pure absorption mode, producing a much cleaner spectrum in the diagonal region.

The DQF-COSY experiment is shown in Fig. 9.10. It is very similar to the COSY sequence, with the exception that the single mixing pulse in the COSY experiment has been replaced by two 90° pulses in the DQF-COSY. The first of these pulses converts the single quantum states to double quantum states. The last pulse returns this double quantum magnetization to detectable single quantum magnetization. The experiment filters out any elements of the density matrix that does not pass though a double quantum state. This filtering occurs as a result of the phase cycle. A more detailed analysis will be presented in Chapter 11. For the meantime we will assume that it occurs. The overall evolution of the elements of the density matrix are illustrated in Fig. 9.11.

9.3.1 Product Operator Treatment of the DQF-COSY Experiment

To simplify the analysis we will focus on just the I spins at the beginning of the experiment. Due to symmetry, the evolution of the S spins can be easily calculated by interchanging I and S in the following derivation. Setting $\rho_0 = I_z$, gives the following

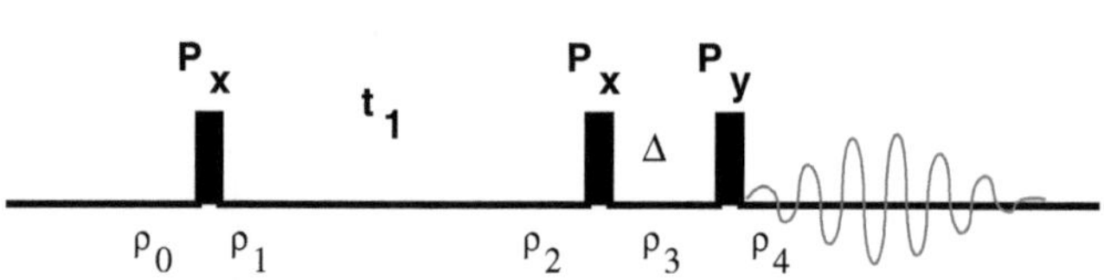

Figure 9.10. Double quantum filtered COSY (DQF-COSY) pulse sequence. All pulses are 90° pulses. The delay Δ is just long enough to change the phase of the transmitter, or about 10 μsec. Consequently evolution of the density matrix during this period can be ignored. The detected signal is given by Trace$[\rho_4(t)I^+]$.

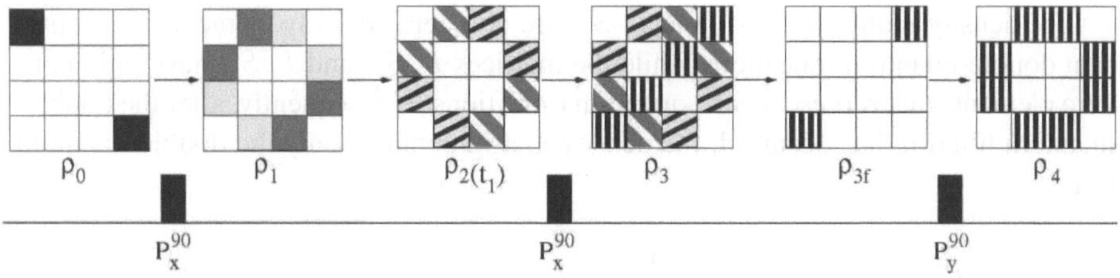

Figure 9.11. The density matrix during a DQF-COSY experiment. The evolution of the elements of the density matrix in the DQF-COSY experiment is shown. The density matrix just before the second pulse is identical to that in the COSY experiment. The second pulse generates double and zero quantum states, as indicated by the symbol (▥). These states also exist in the COSY experiment, but were not detectable. The next density matrix, ρ_{3f} is the filtered density matrix with non-zero values for *only* the double quantum elements of the density matrix. These elements are converted to detectable single quantum states by the last pulse. The filtering is accomplished by either phase cycling or pulsed field gradients.

value for the density matrix just before the second pulse:

$$\begin{aligned}\rho_2 = &-I_y cos(\omega_I t_1)cos(\pi J t_1) + 2I_x S_z cos(\omega_I t_1)sin(\pi J t_1)\\ &+ I_x sin(\omega_I t_1)cos(\pi J t_1) + 2I_y S_z sin(\omega_I t_1)sin(\pi J t_1)\end{aligned} \quad (9.30)$$

The second pulse produces the following transformations of the individual density matrices:

$$-I_y \rightarrow -I_z \quad (9.31)$$
$$I_x \rightarrow I_x \quad (9.32)$$
$$2I_x S_z \rightarrow -2I_x S_y \quad (9.33)$$
$$2I_y S_z \rightarrow -2I_z S_y \quad (9.34)$$

In the COSY experiment, it was the last of the above terms, $-2I_z S_y$, that gave rise to the crosspeak. In the DQF-COSY experiment, the only term that survives the double-quantum filter is the $2I_x S_y$. This particular density matrix actually contains both double-quantum and zero-quantum elements:

$$2I_x S_y = \frac{1}{2}\begin{bmatrix} 0 & 0 & 0 & -1 \\ 0 & 0 & 1 & 0 \\ 0 & 1 & 0 & 0 \\ -1 & 0 & 0 & 0 \end{bmatrix} \quad (9.35)$$

The removal of the zero-quantum terms can be accomplished by writing the above density matrix in terms of the raising and lowering operators:

$$\begin{aligned}2I_x S_y &= 2\frac{1}{2}(I^+ + I^-)\frac{1}{2i}(S^+ - S^-)\\ &= \frac{1}{2i}[I^+S^+ - I^+S^- + I^-S^+ - I^-S^-]\end{aligned} \quad (9.36)$$

The density matrices I^+S^+ and I^-S^- are non-zero for only elements that represent double-quantum transitions while the matrices I^+S^- and I^-S^+ have only non-zero elements that represent zero-quantum transitions. Consequently, after the double-quantum filtering has occurred, the density matrix contains only the double-quantum terms:

$$\begin{aligned} \rho_{3f} &= \frac{1}{2i}[I^+S^+ - I^-S^-] \\ &= \frac{1}{2i}[(I_x + iI_y)(S_x + iS_y) - (I_x - iI_y)(S_x - iS_y)] \\ &= \frac{1}{4}[2I_xS_y + 2I_yS_x] \end{aligned} \tag{9.37}$$

The last pulse, P_y^{90}, transforms ρ_{3f} to $\rho(4)$:

$$[2I_xS_y + 2I_yS_x] \xrightarrow{P_y^{90}} -[2I_zS_y + 2I_yS_z] \tag{9.38}$$

During detection, these terms evolve due to J-coupling to give detectable single-quantum states:

$$\begin{aligned} 2I_zS_y &\xrightarrow{J} 2I_zS_y cos(\pi J t_2) - S_x sin(\pi J t_2) \\ 2I_yS_z &\xrightarrow{J} 2I_yS_z cos(\pi J t_2) - I_x sin(\pi J t_2) \end{aligned} \tag{9.39}$$

The detectable single operator terms evolve with their respective chemical shifts:

$$\begin{aligned} -S_x &\xrightarrow{\omega_S} -S_x cos(\omega_S t_2) - S_y sin(\omega_S t_2) \\ -I_x &\xrightarrow{\omega_I} -I_x cos(\omega_I t_2) - I_y sin(\omega_I t_2) \end{aligned}$$

If we include the amplitude factor that was generated during the t_1 evolution time, as well as the trigonometric terms from above, then the final signal is:

$$\begin{aligned} S(t_1, t_2) &= cos(\omega_I t_1) sin(\pi J t_1) sin(\pi J t_2) e^{i\omega_S t_2} \\ &+ cos(\omega_I t_1) sin(\pi J t_1) sin(\pi J t_2) e^{i\omega_I t_2} \end{aligned} \tag{9.40}$$

The first of these two terms represents the crosspeak at (ω_I, ω_S) and the second represents the selfpeak at (ω_I, ω_I). The spectrum will also contain another crosspeak, at (ω_S, ω_I), and selfpeak, at (ω_S, ω_S), which would be generated if the analysis was started with $\rho_o = S_z$ instead of I_z.

Note that both the crosspeak and the selfpeak have exactly the same evolution due to J-coupling, specifically $sin(\pi J t_1) sin(\pi J t_2)$. Consequently, the selfpeak and the crosspeak will be found as anti-phase doublets that can both be phased to give pure absorption lineshapes. This feature leads to a remarkable improvement in the appearance of the DQF-COSY spectrum over that of the COSY spectrum, especially near the diagonal (see Fig. 9.12).

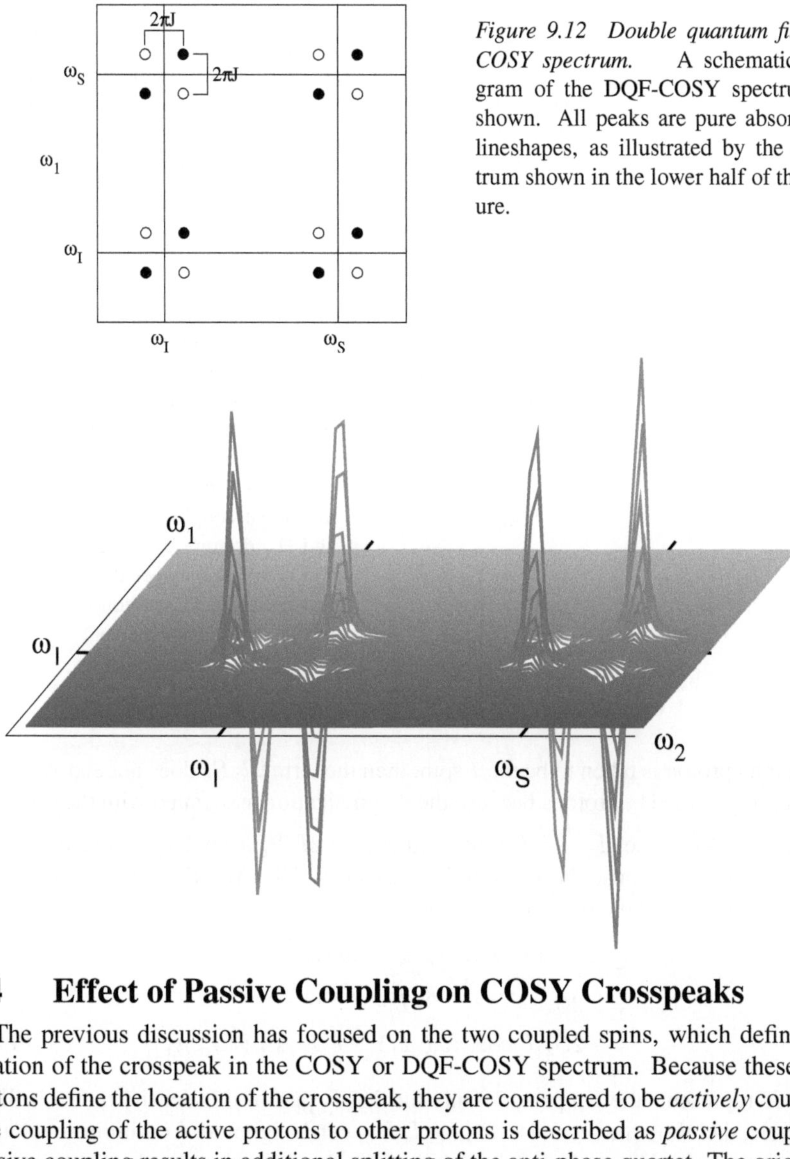

Figure 9.12 Double quantum filtered COSY spectrum. A schematic diagram of the DQF-COSY spectrum is shown. All peaks are pure absorption lineshapes, as illustrated by the spectrum shown in the lower half of the figure.

9.4 Effect of Passive Coupling on COSY Crosspeaks

The previous discussion has focused on the two coupled spins, which define the location of the crosspeak in the COSY or DQF-COSY spectrum. Because these two protons define the location of the crosspeak, they are considered to be *actively* coupled. The coupling of the active protons to other protons is described as *passive* coupling. Passive coupling results in additional splitting of the anti-phase quartet. The origin of this additional splitting can be easily seen by analyzing the influence of the passively coupled spin on the evolution of the density matrix during t_1 or t_2. To simplify the analysis, the density matrix associated with the COSY experiment will be used. The same result is obtained for the DQF-COSY experiment.

As an example, consider the effect of passive $J_{\alpha\beta}$ coupling on the crosspeak that is generated from active coupling between an amide proton (S) and an alpha proton (I). During t_1, the two components of the density matrix that ultimately give rise to the crosspeaks are:

$$2I_zS_y \qquad \text{and} \qquad 2I_yS_z \tag{9.41}$$

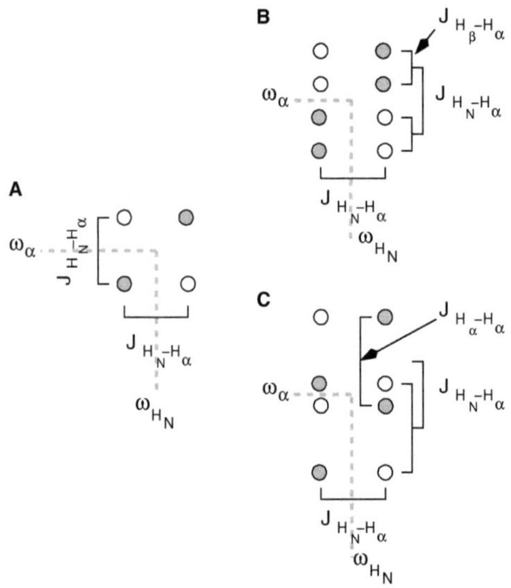

Figure 9.13 Effect of passive coupling on H_N-H_α COSY peaks. Section A shows the anti-phase quartet located at ω_α and ω_{HN}. The active coupling between these two protons is 9 Hz. Section B illustrates the effect of a 4 Hz passive coupling between the H_α proton and the H_β proton. Note the in-phase splitting of the peaks in the ω_α dimension. There is no splitting in the H_N dimension because the amide proton is not passively coupled to any protons. Section C shows the effect of a large passive coupling on the appearance of the cross peak. This often occurs in glycine residues where the passive coupling between the two geminal H_α proton ($\approx$ 15 Hz) exceeds the active coupling between the amide and H_α proton. The effect of the passive coupling between the amide proton and the other H_α proton on the cross-peak pattern is not shown.

If the alpha proton is taken to be the I spin, then the term $2I_zS_y$ does not evolve due to the coupling to the H_β protons because the magnetization associated with the α proton is along the z-axis (i.e. $I_z \overset{J\alpha\beta}{\rightarrow} I_z$). In contrast, the $2I_yS_z$ term does evolve due to the H_α-H_β coupling because the H_α spin is transverse. The evolution of this part of the density matrix under the passive α-β coupling is as follows:

$$
\begin{aligned}
\rho' &= e^{+i\pi J_{\alpha\beta}I_zK_z}2I_yS_ze^{-i\pi J_{\alpha\beta}I_zK_z} \\
&= 2S_ze^{+i\pi J_{\alpha\beta}I_zK_z}I_ye^{-i\pi J_{\alpha\beta}I_zK_z} \\
&= 2S_z[I_ycos(\pi J_{\alpha\beta}t_1) - I_xK_zsin(\pi J_{\alpha\beta}t_1)] \qquad (9.42)
\end{aligned}
$$

where K represents the H_β proton. Only the $2S_zI_y$ part of this density matrix will be detectable, therefore the second term $(2S_zI_xK_z)$ can be ignored. Combining evolution due to active coupling and chemical shift gives the following for the final detected signal of the crosspeak that originated with $\rho = 2I_yS_z$:

$$S(t_1,\, t_2) = cos(\pi J_{\alpha\beta}t_1)sin(\pi J_{\alpha H_N}t_1)sin(\omega_It_1)sin(\pi J_{\alpha H_N}t_2)e^{i\omega_St_2} \qquad (9.43)$$

and for the other crosspeak that originated from $\rho = 2I_zS_y$:

$$S(t_1,\, t_2) = sin(\pi J_{\alpha H_N}t_1)sin(\omega_St_1)cos(\pi J_{\alpha\beta}t_2)sin(\pi J_{\alpha H_N}t_2)e^{i\omega_It_2} \qquad (9.44)$$

Note the association of the $cos(\pi J_{\alpha\beta}t)$ term with the chemical shift evolution of the I proton in both time domain signals. The Fourier transform of this function generates

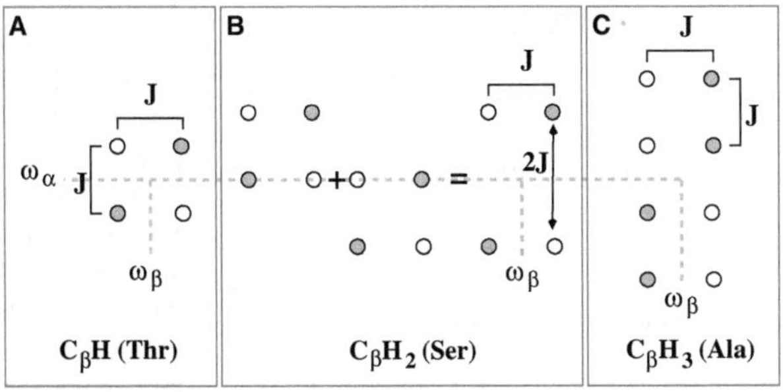

Figure 9.14. COSY crosspeaks of CH_2 and CH_3 groups. The COSY crosspeak for a proton coupled to one (A), two (B) or three (C) *equivalent* protons. If the two actively coupled protons are the α and β protons, then panel A corresponds to a threonine residue, panel B to a serine residue, and panel C to an alanine residue. In the case of coupling to two equivalent protons (panel B), the in-phase splitting generates two anti-phase doublets that overlap. Consequently, the observed spectrum is the sum of these two. In the left section of this panel the two separate anti-phase quartets have been displaced horizontally to show this cancellation. In the case of coupling to a CH_3 group (panel C), the third proton induces an additional in-phase splitting, generating an octet of peaks with equal spacing.

an *in-phase doublet*, separated by $J_{\alpha\beta}$ Hz, causing an additional splitting of the COSY crosspeak at the H_α frequency, as shown in Fig. 9.13. Normally the passive coupling is smaller than the active coupling. However, the passive coupling between the two alpha protons on glycine is often larger than the active coupling to the amide proton. Consequently, the COSY crosspeaks alternate in intensity, as shown in part C of Fig. 9.13. This feature provides a useful way of identifying glycine residues in COSY spectra.

Passive coupling is also observed when a proton is coupled to two or more equivalent protons. In this case the coupling to one of the equivalent protons is considered to be the active coupling, generating the anti-phase quartet crosspeak, and the coupling to the other proton(s) is considered to be passive, generating additional in-phase splittings. This situation leads to a distinctive pattern of uniform peak spacings for coupling to equivalent CH_2 and CH_3 groups, as shown in Fig. 9.14. The appearance of the crosspeak in the latter case provides a way of identifying resonance associated with methyl groups in COSY spectra.

9.5 Scalar Correlation by Isotropic Mixing: TOCSY

In the case of smaller proteins (<8 kDa), the COSY experiment can be used to identify the complete network of coupled protons within an amino acid residue by the detection of pair-wise interactions. However, as the protein size increases the region of the spectrum that contains correlations between side-chain protons is often quite crowded, making it difficult to identify all of the coupled protons. In addition, the anti-phase nature of the crosspeaks leads to a reduction in the signal-to-noise since the individual anti-phase peaks within the COSY crosspeak destructively interfere with each other. This problem becomes more severe as the size of the protein increases due

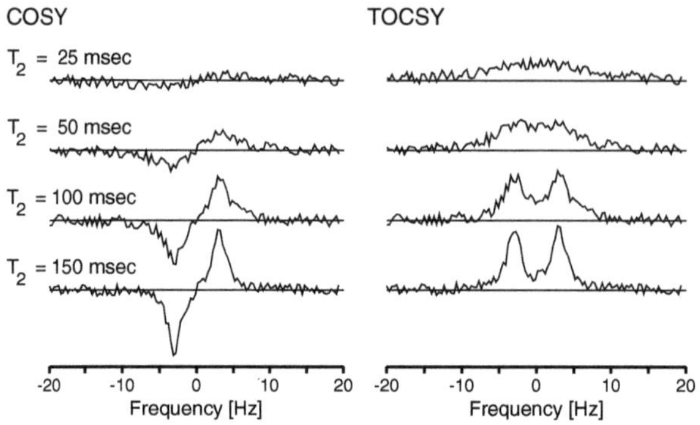

Figure 9.15 COSY versus TOCSY Lineshapes. A cross-section through an anti-phase (COSY) and an in-phase (TOCSY) doublet is shown for a J-coupling of 7 Hz. The lines broaden as the spin-spin relaxation time (T_2) decreases. The lowest curve corresponds to a ≈8 kDa protein while the highest curve corresponds to a ≈50 kDa protein.

to an increase in linewidth, as illustrated in Fig. 9.15. For small couplings, such as the H_N-H_α coupling in α-helices, it may be difficult to observe crosspeaks in COSY experiments when the molecular weight of the protein exceeds 10 kDa.

The TOCSY, or **TO**tal **C**orrelation **S**pectroscop**Y**,, introduced by Braunschweiler and Ernst [24], solves both of the deficiencies associated with COSY spectra. First, the crosspeaks are composed of lines that are all positive absorption mode, thus preventing the loss of signal via destructive interference (see Fig. 9.15). Second, the chemical shift information of one proton within a spin-system is relayed to all other protons within the spin system. This relay occurs by the *sequential* transfer of magnetization through the coupled network of spins. For example, a TOCSY peak between the H_N and H_β spin would occur via a two step process. The magnetization that is labeled with the chemical shift of the amide proton would first be passed to the H_α proton via $J_{H_N H\alpha}$ coupling, and then to the H_β proton via $J_{H\alpha H\beta}$ coupling. During t_2, this magnetization would precess at the chemical shift of the H_β proton, generating a crosspeak at (ω_{HN}, $\omega_{H\beta}$). Consequently, crosspeaks associated with the side-chain protons are moved into the relatively sparse amide region of the proton spectrum where individual resonances can be more readily identified.

The TOCSY experiment can also be applied to other spins besides protons. For example, it is possible to exchange magnetization between coupled carbon spins using this technique. Carbon TOCSY experiments play an important role in obtaining chemical shift assignments of sidechain carbons and protons. In the following sections we will investigate the transfer process with little emphasis on the implementation of actual pulse sequences until Chapter 13.

9.5.1 Analysis of TOCSY Pulse Sequence

A simple version of the TOCSY pulse sequence is shown in Fig. 9.16. This pulse sequence consist of an initial 90° pulse, a frequency labeling time (t_1), followed by a long series of 180° pulses that are applied to *all of the coupled spins*[1]. The 180° pulses prevent evolution of the magnetization by chemical shift during the mixing time. Con-

[1]Decoupling, in contrast, involves the application of pulses to only *one* of the coupled partners.

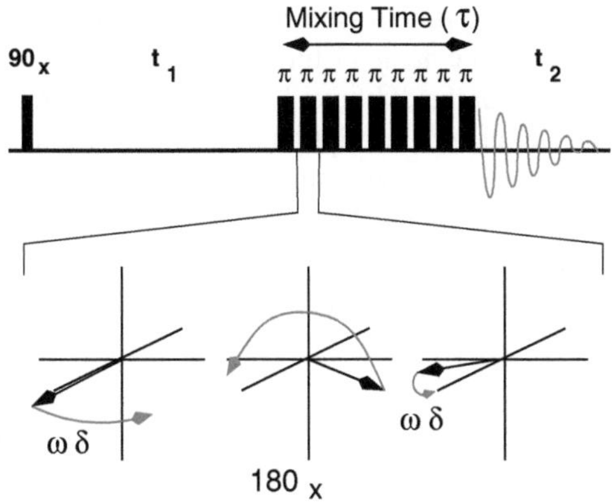

Figure 9.16 TOCSY pulse sequence. The TOCSY pulse sequence is shown in the top of the diagram. Like all other two-dimensional sequences it contains a preparation, t_1 frequency labeling, mixing, and t_2 detection periods. The length of the mixing period is τ and is illustrated here as a train of 180° pulses. In practice it is a series of phase alternated pulses whose net effect is a rotation of the magnetization by 180°. The lower half of the diagram illustrates the fact that each 180° pulse refocuses chemical shift evolution, removing this term from the Hamiltonian.

sequently, the system only evolves under J-coupling during this period, causing transfer of magnetization between coupled spins.

The suppression of chemical shift evolution during the mixing period is accomplished by a spin-echo sequence: $\delta - 180° - \delta$. During the first δ period the transverse spins will precess by an angle $\omega_o\delta$:

$$M_x = cos(\omega_o\delta), \qquad M_y = sin(\omega_o\delta), \text{ or in complex notation: } M = e^{i\omega_o\delta}$$

The 180° pulse rotates the magnetization about the x-axis, placing it in the fourth quadrant. In the complex representation, the 180° pulse negates the phase angle, giving the following for the magnetization after the 180° pulse:

$$M_x = cos(\omega_o\delta), \qquad M_y = -sin(\omega_o\delta), \qquad M = e^{-i\omega_o\delta}$$

During the second δ period, the magnetization precesses through an angle $\omega_o\delta$, bringing it back to its original starting point, along the x-axis. In complex notation the overall process can be written as:

$$e^{i0} \xrightarrow{\delta} e^{+i\omega_o\delta} \xrightarrow{180°} e^{-i\omega_o\delta} \xrightarrow{\delta} e^{-i\omega_o\delta}e^{+i\omega_o\delta} = 1 \tag{9.45}$$

These transformations are illustrated in the lower part of Fig. 9.16.

Since the time evolution of the system depends on the Hamiltonian:

$$\frac{d\Psi}{dt} \approx H\Psi \tag{9.46}$$

the lack of precession due to chemical shift implies that the term of the Hamiltonian that drives chemical shift evolution is *effectively* zero during the mixing time. Therefore, the only remaining term in the Hamiltonian is the scalar coupling between spins:

$$H = 2\pi J I \cdot S = 2\pi J[I_xS_x + I_yS_y + I_zS_z] \tag{9.47}$$

Previously, in the analysis of the COSY experiments, we had dropped the I_xS_x and I_yS_y terms because the J-coupling was a small contribution to the overall energy of the system. However, during the TOCSY mixing time it is necessary to keep all three (x, y, z) terms of the above dot product, hence the name "isotropic" mixing.

To describe the evolution of the system under the effective Hamiltonian it is necessary to develop new eigenvectors and the associated description of the density matrix. This derivation is beyond the scope of this text, but the results are summarized here for two coupled spins. A more complete description can be found in [53].

The density matrix can be described by new 'product' operators that are formed by taking combinations of the familiar Cartesian product operators:

$$\begin{aligned}\sum_{\alpha} &= \frac{1}{2}[I_\alpha + S_\alpha] \qquad \Delta_\alpha = \frac{1}{2}[I_\alpha - S_\alpha] \\ \sum_{\alpha\beta} &= [I_\alpha S_\beta + I_\beta S_\alpha] \qquad \Delta_{\alpha\beta} = [I_\alpha S_\beta - I_\beta S_\alpha]\end{aligned} \tag{9.48}$$

where $\alpha, \beta, \gamma = x,\ y,\ z$, e.g. $\Delta_{xy} = [I_x S_y - I_y S_x]$.

The evolution of Δ_α under J-coupling is as follows:

$$\Delta_\alpha \rightarrow \Delta_\alpha cos(2\pi J\tau) + \Delta_{\beta\gamma} sin(2\pi J\tau) \tag{9.49}$$

where τ is the entire mixing period. In contrast, neither $\sum_\alpha$ or $\sum_{\alpha\beta}$ evolve under J-coupling.

9.5.1.1 Evolution of Magnetization

At the end of the t_1 period the density matrix associated with spin I can be represented, using Cartesian product operators, as:

$$I_x sin(\omega_1 t_1) cos(\pi J t_1) \tag{9.50}$$

I_x can be converted to the new Δ, $\sum$ representation as follows:

$$I_x = \sum_x + \Delta_x \tag{9.51}$$

The evolution of I_x during the mixing time, τ, is given by:

$$\sum_x + \Delta_x \rightarrow \sum_x + \Delta_x cos(2\pi J\tau) + \Delta_{yz} sin(2\pi J\tau) \tag{9.52}$$

Converting this expression back to the Cartesian form at the end of the mixing period gives:

$$\begin{aligned} I_x &= \frac{1}{2} I_x [1 + cos(2\pi J\tau)] + \frac{1}{2} S_x [1 - cos(2\pi J\tau)] + (I_y S_z - I_z S_y) sin(2\pi J\tau) \\ &= I_x cos^2(\pi J\tau) + S_x sin^2(\pi J\tau) + (I_y S_z - I_z S_y) sin(2\pi J\tau) \end{aligned} \tag{9.53}$$

The above shows that during the mixing time, magnetization has been transferred from one spin (I_x) to the other coupled spin (S_x). The transfer is weighted by the original amplitude factor of I_x (Eq. 9.50):

$$[sin(\omega_1 t_1) cos(\pi J t_1)] \tag{9.54}$$

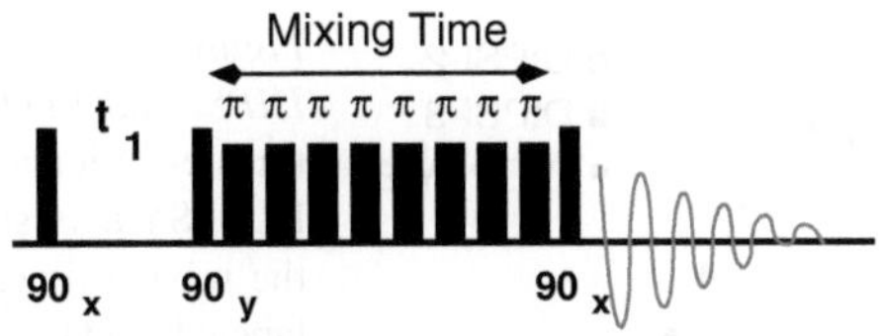

Figure 9.17. TOCSY - isotropic mixing of I_z. This pulse sequence will transfer magnetization from spin I to S by mixing of magnetization along the z-axis. This experiment is a simplified version of the experiment presented by Rance and Cavanagh [136]. The y-pulse at the end of t_1 converts $I_x cos\omega_I t_1$ to $I_z cos\omega_I t_1$. This magnetization is subsequently transferred to the S spin, giving $S_z cos(\omega_I t_1)$ at the end of the mixing time. The third 90° pulse converts S_z to $-S_y$ prior to detection. This sequence will also convert I_y, which is also present at the end of t_1, to S_y. The transverse component would have to be removed by phase cycling.

So the density matrix of the second spin, after the mixing time, is represented by:

$$S_x[sin(\omega_I t_1)\ cos(\pi J t_1)]sin^2(\pi J \tau) \tag{9.55}$$

where the set of terms in square brackets represent the original amplitude modulation of the I_x term and the $sin^2(\pi J \tau)$ represents the magnetization transfered during the mixing time.

S_x will evolve under chemical shift and J coupling to give as a final detected signal:

$$S(t_1, t_2) = sin^2(2\pi J \tau)[sin(\omega_I t_1)cos(\pi J t_1)]cos(\pi J t_2)e^{i\omega_S t_2} \tag{9.56}$$

Fourier transformation of this signal will give a crosspeak at $(\omega_I,\ \omega_S)$. Note that in contrast to the COSY experiment, the J-coupling term now appears as $cos(2\pi Jt)$ in both time dimensions. Since the Fourier transform of *cosine* gives a pair of positive peaks, the entire TOCSY crosspeak is positive, as indicated in Fig. 9.15.

The above analysis demonstrates transfer of magnetization from the x-component of spin I to the x-component of spin S. However, by simply changing the indices (e.g. replace x with y) it should be clear that exactly the same transfer would occur between the y- or z-components of the magnetization. This behavior is predicted from the effective Hamiltonian, which has no preferred direction. Consequently, the pulse train in the mixing time is usually referred to as an *isotropic mixing sequence* because it is capable of transferring magnetization along any axis. For example, the sequence shown in Fig. 9.17 would generate crosspeaks by causing transfer of magnetization from I_z to S_z during the mixing time.

9.5.2 Isotropic Mixing Schemes

Efficient isotropic mixing requires that the pulse sequence used to generate the effective Hamiltonian ($2\pi J I \cdot S$) is independent of the chemical shifts of the coupled spins. Since the series of π pulses used in Fig. 9.16 can also behave as decoupling sequences, it is not surprising then that the decoupling schemes discussed in Section 7.4 also function as isotropic mixing sequences with the same relative efficiency and bandwidth. The performance of various pulse sequences for isotropic mixing has been carefully analyzed by Glaser and Drobny [62]. For proton isotropic mixing, the

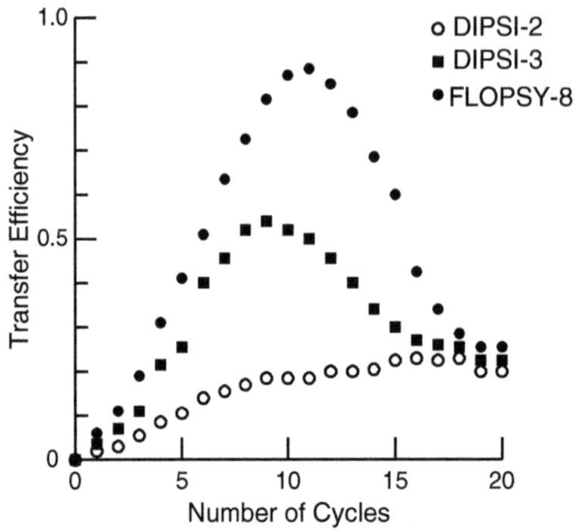

Figure 9.18 Transfer efficiency of DIPSI and FLOPSY sequences. The transfer efficiency of DIPSI-2, -3, and FLOPSY-8 is shown as a function of the mixing time. The sample was ^{13}C labeled acetate, a B_1 field strength of 7.93 kHz was used, and the transmitter was placed halfway between the C=O and methyl lines. The J-coupling constant for these two spins is 53 Hz, therefore 1/(2J) = 9.4 msec. Ten cycles corresponds to isotropic mixing times of 36.4 msec, 68.2 msec, and 29.7 msec, for DIPSI-2, DIPSI-3, and FLOPSY-8, respectively. Data was obtained from [74].

DIPSI-2 sequence has superior performance over WALTZ-16 and should be used in any proton-proton TOCSY experiments. In addition, DIPSI-2 can be used to transfer either transverse (I_x) or longitudinal magnetization (I_z) [139].

DIPSI-2 and DIPSI-3 sequences have been widely used for carbon TOCSY experiments, see for example Kay et al [85] or Bax et al [9]). Shaka [74] has developed an isotropic mixing sequence, FLOPSY-8, that gives optimal transfer of longitudinal magnetization. Although the figure of merit[2] of FLOPSY-8 is approximately 1.0, which is somewhat worse than DIPSI-2 or -3, the efficiency of transfer is considerably better, as illustrated in Fig. 9.18. In addition, the sequence is less sensitive to off-set effects within its bandwidth (see below).

9.5.3 Time Dependence of Magnetization Transfer by Isotropic Mixing

The optimal transfer time can, in principle, be obtained from the transfer function:

$$sin^2(\pi J\tau) \tag{9.57}$$

In practice, the *effective* J-coupling between coupled spins depends on the frequencies of the two coupled spins relative to the transmitter. In general, as the frequency difference between the coupled spins increases the effective J-coupling decreases, therefore longer mixing times are required for optimal transfer. In the case of isotropic mixing using DIPSI-3, the effective J-coupling, $J_{effective}$, is approximately [148, 9]:

$$J_{effective} \approx J\left[1-(1-cos(\theta))\sqrt{\frac{8}{3}}\right] \tag{9.58}$$

where θ is the angle between the spins in the rotating frame (see the discussion on the rotating frame, Chapter 1). This function is plotted in Fig. 9.19. The effective

[2]See Section 7.4.3.

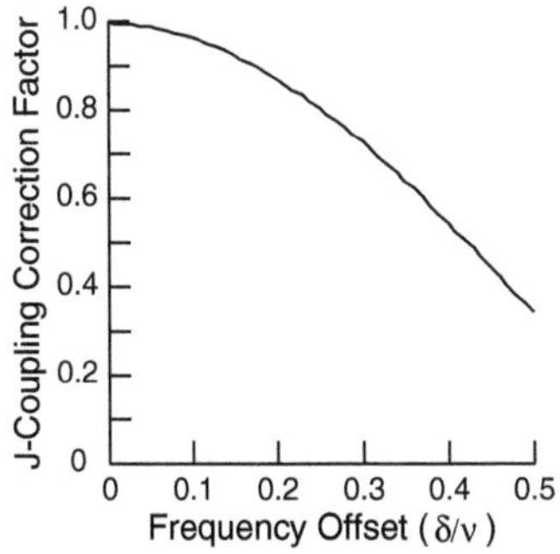

Figure 9.19 Effective J-coupling during isotropic mixing. The effective J-coupling is obtained by multiplying the true J-coupling by the indicated correction factor. δ is the frequency difference between the position of the resonance line and the transmitter. ν is the intensity of the B_1 field that is used for isotropic mixing. This plot assumes that the transmitter is placed half-way between the two coupled spins. For example, if a 10 kHz B_1 field was used and the resonance lines for the coupled spins were 4 kHz apart, δ would be 2 kHz, and δ/ν would be 0.2, giving $J_{effective} \approx 0.87$ J.

J-coupling values are higher for FLOPSY-8 than for DIPSI-3 [74]. Hence, the transfer times will be shorter if FLOPSY-8 is used for isotropic mixing (see Fig. 9.18).

9.5.3.1 TOCSY Transfer Times in Amino Acids

The time dependence of proton-proton transfer in alanine is shown in Fig. 9.20 and that for carbon-carbon transfer in threonine is shown in Fig. 9.21. The time dependence of transfer for additional amino acid residues can be found in [29] for proton-proton transfers, and in [9, 35] for carbon-carbon transfers.

Both figures 9.20 and 9.21 show that it is possible to efficiently transfer magnetization to distant spins within the same residue (spin system). As anticipated from the size of the coupling constants, transfer of magnetization between coupled protons requires a longer mixing time than the transfer between coupled carbons. Consequently, the transfer of magnetization via proton-proton coupling will be relatively less efficient in larger proteins due to the shorter T_2 relaxation time (Compare the right side of Fig. 9.20 to dashed lines on Fig. 9.21).

The secondary structure of a residue has a large effect on the ability to transfer magnetization from the amide proton to the side-chain. Residues in an α-helical conformation possess a small J-coupling between the amide and the H_α proton. This weak coupling greatly inhibits the transfer of magnetization from the amide proton to the remaining protons (compare the solid and dotted lines in Fig. 9.20). In contrast, since the carbon-carbon couplings are insensitive to secondary structure, the transfer of magnetization in carbon-carbon TOCSY is also insensitive to secondary structure.

In summary, a proton-proton TOCSY can be used to obtain a large number of correlations between the amide proton and the sidechain protons for smaller proteins, up to a size of $\approx$ 12-15 kDa. For proteins in the range of 20-25 kDa it would be possible to observe such cross peaks for residues in β-sheets. However, a poor signal-to-noise ratio may prevent the observation of transfers between distant spins, such as between the amide proton and the H_δ protons in isoleucine. In contrast, the transfer of carbon magnetization by isotropic mixing is much more robust than with protons due to the larger and more uniform carbon-carbon coupling constants.

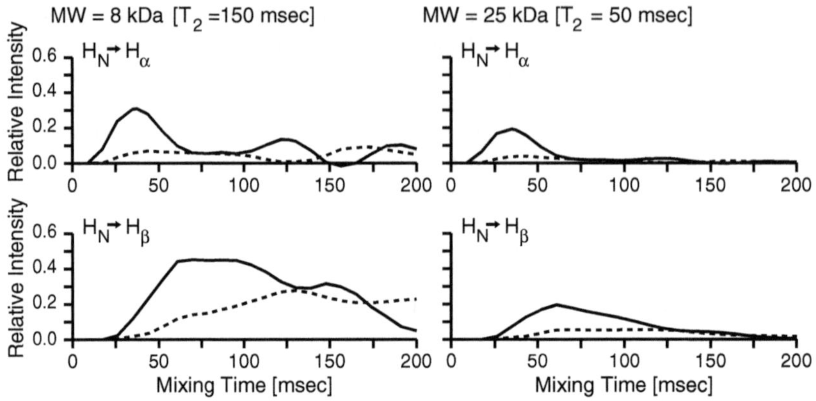

Figure 9.20. Effect of TOCSY mixing time on transfer efficiency. The effect of mixing time on the transfer of magnetization from the amide proton to the H_α proton (upper two panels) and H_β protons (lower two panels) of alanine are presented. The x-axis is the length of the isotropic mixing sequence, and the y-axis is the relative intensity of the crosspeak. The original intensity of the amide proton was 1.0. The solid line shows the transfer efficiency when $J_{HN-H\alpha}$ is 10 Hz (i.e. β-strand) and the dashed line shows the transfer efficiency when J $_{HN-H\alpha}$ is 4 Hz (i.e. α-helix). The left-hand panels show typical data for a $\approx$ 8 kDa protein and the right-hand panels represent transfer efficiencies for a $\approx$ 25 kDa protein. The transfer efficiencies, in the absence of relaxation, are from [29]. The effect of relaxation was simulated by multiplying the data by e^{-t/T_2}.

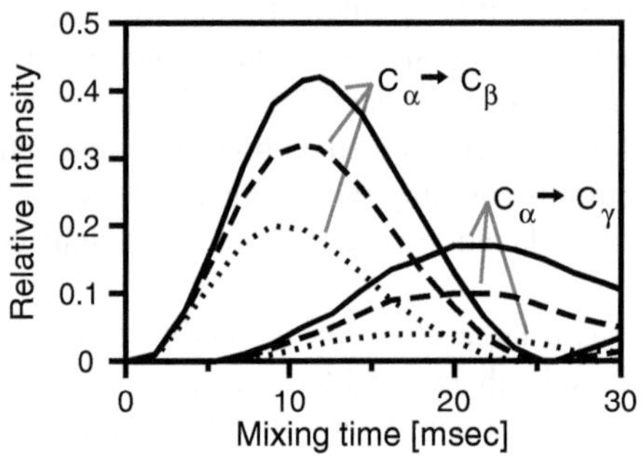

Figure 9.21 Carbon-carbon TOCSY transfer. The transfer rate of magnetization from the α-carbon of threonine to its β- or γ-carbon is shown for an 8 kDa (solid line), a 25 kDa (dashed line) or a 50 kDa protein (dotted line). The transfer efficiencies, in the absence of relaxation, were from Bax *et al* [9]: $J_{C\alpha C\beta}$ = 35 Hz, $J_{C\beta C\gamma}$ = 19 Hz. The effect of relaxation was simulated by multiplying the data by e^{-t/T_2}.

9.6 Exercises

1 Sketch the complete COSY spectra of the following tri-peptide: Gly-Ala-Val, assuming that it is contained within a larger polypeptide sequence. Assume that the protein is in H_2O and that all passive couplings are zero, with the exception of equivalent protons in methyl groups or the two H_α protons. You can also assume that all couplings are 10 Hz, except that for the geminal protons of glycine, which are 15 Hz. The following table gives the chemical shifts of the protons for these three residues:

Residue	H_N	H_α	H_β	H_γ
Gly	7.00	4.00, 4.50		
Ala	8.00	5.00	1.25	
Val	6.00	3.75	3.50	0.75, 1.00

2 Add to your sketch the crosspeaks that would appear in a TOCSY spectrum.

3 How would the COSY or TOCSY spectra change if the sample was placed in D_2O?

9.7 Solutions

1 The two-dimensional COSY spectrum is shown in Fig. 9.22.

2 The TOCSY spectrum is shown in Fig. 9.22.

3 The amide protons would replaced by deuterons. Therefore all crosspeaks that involved the amide proton would disappear from the spectrum.

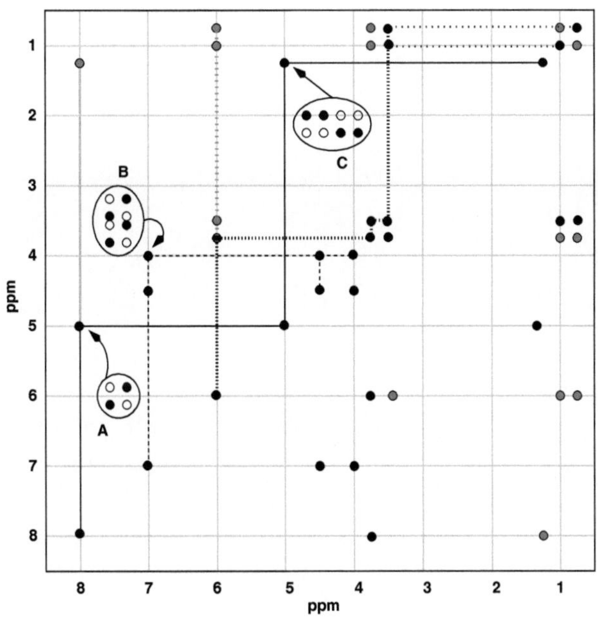

Figure 9.22 COSY Spectrum. The location of the self- and cross-peaks in the COSY spectrum are indicated by solid circles. TOCSY peaks are colored gray. The solid lines join peaks from the alanine residue, the dashed line joins peaks from the glycine residue, and the dashed line joins peaks associated with the valine residue. The fine structure of selected crosspeaks have been enlarged and labeled. Crosspeak A corresponds to a pair of spins with no passive coupling. Crosspeak B is typical of glycine H_N-H_α crosspeaks, and reflects the passive coupling between the two H_α protons. Crosspeak C corresponds to the peak between the H_α proton and the methyl protons of alanine. This peak is split due to passive coupling with the H_β protons.

Chapter 10

TWO DIMENSIONAL HETERONUCLEAR J-CORRELATED SPECTROSCOPY

10.1 Introduction

The experiments discussed in the previous chapter were directed at elucidating correlations between J-coupled homonuclear spins, such as protons or carbons. It is also possible to detect correlations between protons and their attached heteronuclear spins, such as nitrogen and carbon, using heteronuclear J-correlated spectroscopy.

Advantages of Heteronuclear NMR: There are a number of advantages associated with combining information from heteronuclear spins with that from protons in biomolecular NMR studies, including:

1. The chemical shifts of the heteronuclear spins are more disperse than proton chemical shifts, thus reducing the overlap of spectral peaks (see chapters 1 and 13).

2. The relaxation properties of the heteronuclear spin can be easier to interpret than the relaxation of the proton spins, facilitating measurements of protein dynamics (see Chapter 19).

3. The fixed distance between the proton and the heteronuclear spin facilitates the measurement of bond orientations using residual dipolar couplings, providing an important constraint in structure determination (see Chapter 16).

4. The one-bond coupling between the amide nitrogen and the carbonyl carbon provides a means to link amino acid spin-systems, facilitating resonance sequential assignments (see Chapter 14).

Sensitivity of Heteronuclear NMR: Unfortunately, the resonance signal from heteronuclear spins is significantly less intense than that from protons. The relative sensitivity of the signal from heteronuclear spins is given by [53]:

$$\gamma_X^{\frac{5}{2}}/\gamma_H^{\frac{5}{2}} \qquad (10.1)$$

The dependence of the sensitivity on γ is due to two factors. First, a lower γ reduces the population difference between the ground and excited quantum states by a factor that is proportional to γ, as indicated by the Boltzmann distribution. Second, the detection of the precessing magnetization is more sensitive for higher frequency spins, by a factor of $\gamma^{\frac{3}{2}}$.

In most applications of heteronuclear NMR, the protein or nucleic acid in the sample is enriched by biosynthetic incorporation of ^{13}C glucose and ^{15}N ammonium to an isotopic enrichment level of 100%. With this level of isotopic enrichment, the sensitivity for ^{13}C is 32 fold less than protons. The situation is worse for ^{15}N, with a 300 fold reduction in sensitivity. Consequently, most heteronuclear NMR experiments are designed to transfer the intense spin polarization of the proton to the heteronuclear spin. The sensitivity of the experiment is increased further by returning the magnetization back to the proton for detection at the end of the experiment.

The sensitivity of these experiments can be further increased by decoupling the proton-heteronuclear interaction during periods of chemical shift evolution. The decoupling collapses the quartet crosspeak into a singlet, in principal increasing the signal by a factor of four. Unfortunately, the expected intensity gains that arise from decoupling are not always realized. Each of the peaks in the quartet relaxes at a different rate. The decoupling process mixes, or averages, these relaxation rates. Consequently, decoupling decreases the average spin-spin (T_2) relaxation time of the observed peak. Shorter T_2 times result in increased linewidths of the peaks, degrading both resolution as well as the signal-to-noise. In the case of smaller proteins (<30 kDa), this effect is small and decoupling leads to an almost four fold increase in intensity. In larger proteins ($\geq$50 kDa), averaging of the relaxation rates leads to a significant increase in linewidth and loss of signal during magnetization transfer (mixing) periods. Recently, methods have been devised to select only the narrowest member of the quartet for detection, thus increasing the sensitivity and resolution of heteronuclear experiments in large proteins. The general name for this technique is **T**ransverse **R**elaxation **O**ptimized **S**pectroscop**Y**, or TROSY. An introductory analysis of this important experimental technique is presented in Chapter 15.

10.2 Two Dimensional Heteronuclear NMR Experiments

There are three two-dimensional heteronuclear correlation experiments in common use: the heteronuclear multiple quantum coherence (HMQC) experiment [10], the heteronuclear single quantum coherence (HSQC) experiment [21], and the refocused-HSQC experiment [12]. An example of a two-dimensional ^{1}H-^{15}N HSQC spectrum of a 130 residue protein is shown in Fig. 10.1.

Although these three experiments all generate crosspeaks that correlate the proton and heteronuclear chemical shifts, the experiments differ in the state of the density matrix that evolves during the t_1 (heteronuclear) labeling period. In the case of the HMQC experiment a double-quantum state evolves (I_xS_x), the HSQC experiment evolves as an anti-phase single-quantum state (I_zS_x), while in the refocused-HSQC experiment the heteronuclear magnetization evolves as pure in-phase magnetization (S_x). Since each of these terms relax at different rates, the linewidths of the resonance peaks in the heteronuclear dimension differ. In the case of proteins, the refocused-

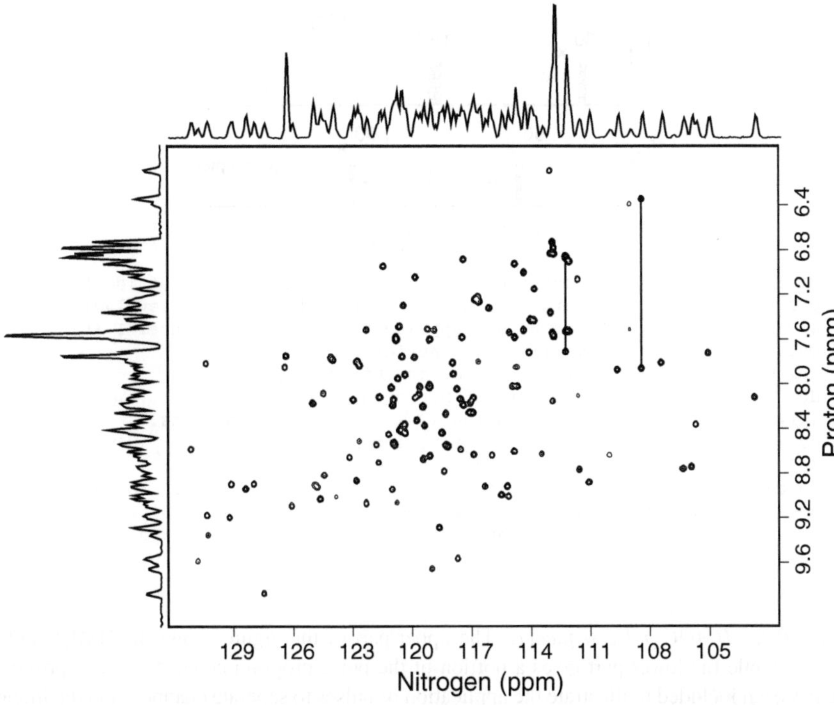

Figure 10.1. Two-dimensional ^1H-^{15}N-HSQC spectrum. A two-dimensional HSQC spectrum of a 130 residue protein is represented as a contour plot. The one-dimensional proton and nitrogen spectra are shown on the left side and top of the plot, respectively. Each crosspeak represents a signal from a single N-H pair. Each peak is a singlet because there is no evolution due to J-coupling during both t_1 and t_2 periods. The pairs of peaks connected by vertical lines indicate NH_2 groups from Gln and Asn groups. The two amide protons share the same nitrogen, consequently both peaks have the same nitrogen shift. The HMQC experiment would be almost identical in appearance.

HSQC experiment produces the narrowest lines, followed by the HSQC experiment, and lastly the HMQC experiment. Nevertheless, the HMQC experiment can be the most sensitive of the three because it has the fewest number of RF-pulses.

Each of the above experiments will be discussed in more detail using the product operator notation. In this analysis, the proton spin will be represented by I and the heteronuclear spin by S.

10.2.1 HMQC Experiment

The heteronuclear multiple-quantum correlation experiment (HMQC) is shown in Fig. 10.2. The most substantial technical difference between this experiment and the homonuclear experiments discussed in Chapter 9 is the use of RF-pulses that independently alter the proton or heteronuclear magnetization. Independent excitation of the different spins allows greater flexibility in manipulation of the spins during the experiment.

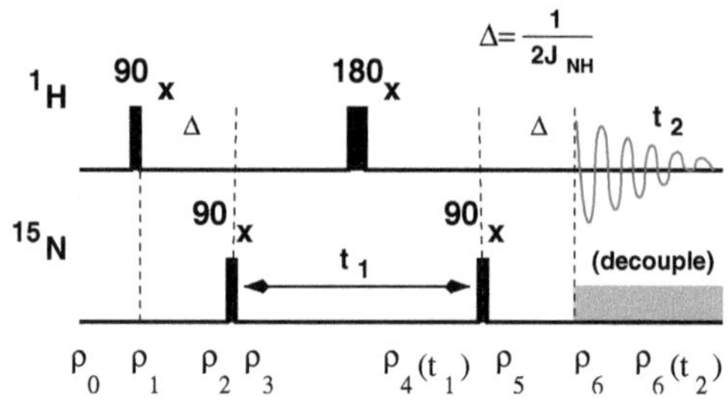


```
2  d1 do:f2           ;Relaxation Delay(d1), Nitrogen decoup. off(do:f2)
3  pl2:f2             ;Return N-15 transmitter to high power (level=pl2).
4  (p1 ph0):f1        ;First Pulse, applied to protons on channel 1 (f1)
5  Delta              ;Polarization Transfer Delay
6  (p2 ph0):f2        ;Nitrogen pulse, on channel 2 (f2).
7  d0                 ;First Half of Nitrogen evolution time.
8  (p1*2 ph0):f1      ;Refocusing 180 (2 x 90) pulse on protons.
9  d0                 ;Second half of Nitrogen evolution time.
10 (p2 ph0):f2        ;Second Nitrogen pulse
11 Delta pl12:f2      ;Second Delta, N-15 power lowered to Pl12 for decoupling.
12 go=2 ph31 cpd2:f2  ;Acquire FID with decoupling applied to Nitrogen
                      ;Loop to label 2 until n scans are acquired.
```

Figure 10.2. HMQC pulse sequence. The upper part of this figure shows the HMQC pulse sequence while the lower part gives a portion of the pulse program code. The pulse program code has been included to illustrate the application of pulses to separate channels and the implementation of decoupling. In the pulse sequence (upper diagram), the upper series of pulses are applied to the proton (I) spins while the lower series of pulses are applied to the heteronuclear (S) spins, which are ^{15}N in this example. The sequence would be identical for ^{13}C spins, with the appropriate change of the delay Δ. The narrow bars correspond to 90° pulses and the wide bar is a 180° pulse. The time period Δ is nominally set to $\frac{1}{2J}$. Decoupling is applied during acquisition of the proton signal in t_2. The decoupling scheme is usually WALTZ-16 in the case of ^{15}N or GARP-1 in the case of ^{13}C.
In the pulse program code, *f1* is the proton frequency channel and *f2* is the nitrogen frequency channel. In this example, all pulses are applied along the x-axis. In practice they would be phase cycled. The command *pl12:f2*, on line 11, changes the power level (*pl*) of the nitrogen channel from high power to a lower level that is appropriate for decoupling. The command *cpd2:f2* , on line 12, turns on the decoupler during acquisition of the proton FID. After the FID is acquired, the decoupler is turned off at the beginning of the relaxation delay of the next scan (line 2) and the power of the nitrogen channel is reset to high power level for the two nitrogen pulses (line3).

10.2.1.1 Analysis of the HMQC Experiment

The HMQC sequence can be divided into five distinct steps:

A	ρ_0	Initial density matrix.
B	$\rho_1 \rightarrow \rho_3$	Transfer of proton polarization to the nitrogen.
C	$\rho_3 \rightarrow \rho_4(t_1)$	Recording the nitrogen chemical shift during t_1.
D	$\rho_5 \rightarrow \rho_6$	Transfer of nitrogen magnetization back to the proton.
E	$\rho_6 \rightarrow \rho_6(t_2)$	Detection of proton magnetization, amplitude modulated by $e^{i\omega_N t_1}$.

A: Initial Density Matrix.

The initial density matrix is:

$$\rho_o = \gamma_H I_z + \gamma_N S_z \tag{10.2}$$

where I refers to the protons and S the heteronuclear spin, which is nitrogen in this example. The gyromagnetic ratios have be introduced here to account for the difference in the proton and heteronuclear spin populations that result from the different energy separation between the ground and the excited states. A careful analysis of the evolution of the $\gamma_N S_z$ throughout this pulse sequence will show that this term does not generate any detectable magnetization. The gyromagnetic ratio for nitrogen, γ_N, is 1/10 the size of γ_H, thus the contribution from the initial nitrogen magnetization to the final signal is generally minimal, and will be ignored here.

B: Transfer of Proton Polarization to the Nitrogen.

The first 90° pulse on the protons will generate transverse proton magnetization, this will evolve under heteronuclear J-coupling, as follows:

$$\gamma_H I_z \xrightarrow{P_{90}} -\gamma_H I_y \xrightarrow{J} -\gamma_H I_y cos(\pi J\Delta) + \gamma_H 2I_x S_z sin(\pi J\Delta) \tag{10.3}$$

If Δ is set to $\frac{1}{2J}$, then:

$$cos(\pi J\Delta) = 0 \qquad\qquad sin(\pi J\Delta) = 1 \tag{10.4}$$

consequently the cosine term disappears, leaving:

$$\rho_2 = \gamma_H 2I_x S_z \tag{10.5}$$

This density matrix also evolves due to the proton chemical shift because the proton magnetization is transverse, giving:

$$\rho_2 = \gamma_H 2S_z[I_x cos\omega_I\Delta + I_y sin\omega_I\Delta] \tag{10.6}$$

However, this evolution will be canceled by the symmetrically placed, equivalent delay, on the other side of the proton π pulse (see below). Consequently, chemical shift evolution of the protons will be ignored in this interval and only the $\gamma_H 2I_x S_z$ term will be followed.

The second 90° pulse, applied only to the nitrogen spins gives:

$$\rho_3 = \gamma_H 2I_x S_y \tag{10.7}$$

The presence of the S_y term indicates that the nitrogen has become excited. The intensity of the excited nitrogen state depends on the initial *proton* polarization, representing transfer of the proton polarization to the nitrogen spin. Note that this product operator represents a multiple (double) quantum transition, hence the name of this experiment.

C: Recording the Nitrogen Chemical Shift.

The density matrix ρ_3 has the potential to evolve during t_1 under the influence of all three terms of the Hamiltonian: J-coupling, proton chemical shift, and nitrogen chemical shift. It will be shown below that there is no evolution of double quantum terms via J-coupling. In addition, the proton chemical shift evolution will be refocused by the proton 180° pulse. Thus, the only *net* change of ρ_3 is due to evolution by the nitrogen chemical shift.

Effect of J-coupling on ρ_3: The general rule is that density matrices that are represented by the product of two transverse operators, such as, $2I_xS_x$, $2I_yS_y$, $2I_xS_x$, etc. do not evolve due to J coupling, specifically:

$$\begin{aligned}\rho' &= e^{iJ\pi I_z S_z}\rho\, e^{-iJ\pi I_z S_z}\\ &= \rho\end{aligned}$$

To show that the above holds, consider the example, $\rho = 2I_xS_y$. Evolution of this density matrix under J-coupling for a period arbitrary time τ gives:

$$\begin{aligned}2I_xS_y &\xrightarrow{J} 2(I_xcos\phi + 2I_yS_zsin\phi)(S_ycos\phi - 2I_zS_xsin\phi)\\ &= 2[I_xS_ycos^2\phi - 2I_xI_zS_xcos\phi sin\phi\\ &+ 2I_yS_zS_ysin\phi cos\phi - 4I_yI_zS_zS_xsin^2\phi]\end{aligned} \tag{10.8}$$

($\phi = \pi J\tau$). This expression can be simplified by reducing the three and four product operators (e.g. $I_yI_zS_zS_x$) to pairs of product operators. For example,

$$S_zS_x = \frac{1}{2}\begin{bmatrix}1 & 0\\ 0 & -1\end{bmatrix}\frac{1}{2}\begin{bmatrix}0 & 1\\ 1 & 0\end{bmatrix} = \frac{1}{4}\begin{bmatrix}0 & 1\\ -1 & 0\end{bmatrix} = -i\frac{1}{2}S_y \tag{10.9}$$

The general expression for the above is: $I_\alpha I_\beta = \frac{i}{2}I_\gamma$. Where $\alpha,\ \beta,\ \gamma$ are cyclic permutations of $x,\ y,\ z$. If the order is reversed, the sign changes, $I_\beta I_\alpha = -\frac{i}{2}I_\gamma$. Using these expressions:

$$\begin{aligned}4I_yI_zS_zS_x &= 4(\frac{i}{2}I_x)(\frac{i}{2}S_y) = -I_xS_y\\ 2I_xI_zS_x &= 2(-\frac{i}{2}I_y)S_x = -iI_yS_x\\ 2I_yS_zS_y &= 2I_y(-\frac{i}{2}S_x) = -iI_yS_x\end{aligned}$$

Substituting all of the above into the original equation gives,

$$\begin{aligned}2I_xS_y &\xrightarrow{J} 2[I_xS_ycos^2\phi - 2I_xI_zS_xcos\phi sin\phi\\ &+ 2I_yS_zS_ysin\phi cos\phi - 4I_yI_zS_zS_xsin^2\phi]\\ &= 2[I_xS_ycos^2\phi + iI_yS_xcos\phi sin\phi - iI_yS_xsin\phi cos\phi + I_xS_ysin^2\phi]\\ &= 2I_xS_y\end{aligned}$$

Therefore, the ρ_3 is unchanged by scalar coupling during t_1.

Proton Evolution During t_1 - The Spin Echo: The general rule is that a 180° pulse placed in the middle of an interval will result in no *net* evolution of the density matrix via chemical shift. In the case of the HMQC sequence, the 180° pulse in the t_1 evolution time is placed symmetrically; the evolution time before this pulse equals the evolution time after the pulse. Therefore the 180° pulse will refocus any proton chemical shift evolution. This pulse sequence element is called a spin-echo, because the diverging transverse proton magnetization de-phases during the period before the pulse and then re-phases by the end of the time period after the pulse, forming an echo of the original magnetization.

Consider an arbitrary time period, τ, followed by a 180° pulse, and then another τ period: $\tau - 180° - \tau$. Beginning with $\rho_3 = 2I_xS_y$, the evolution due to the chemical shift portion of the Hamiltonian prior to the 180° pulse is given by (setting $\phi = \omega_I\tau$):

$$\begin{aligned}\rho_3' &= e^{i\omega_I I_z\tau}\rho_3 e^{-i\omega_I I_z\tau} \\ &= 2S_y e^{i\omega_I I_z\tau} I_x e^{-i\omega_I I_z\tau} \\ &= 2S_y[I_x cos(\phi) + I_y sin(\phi)]\end{aligned} \quad (10.10)$$

The 180° pulse applied along the x-axis to the protons will give:

$$2S_y[I_x cos(\phi) + I_y sin(\phi)] \stackrel{P_{180}}{\rightarrow} 2S_y[I_x cos(\phi) - I_y sin(\phi)] \quad (10.11)$$

This product operator will evolve during the period after the pulse as follows:

$$\begin{aligned}2S_y[I_x cos(\phi) - I_y sin(\phi)] \stackrel{\omega_I}{\rightarrow} & \; 2S_y cos(\phi)\,[I_x cos(\phi) + I_y sin(\phi)] \\ & -2S_y sin(\phi)\,[I_y cos(\phi) - I_x sin(\phi)]\end{aligned} \quad (10.12)$$

This can be simplified as follows:

$$\begin{aligned}\rho_3' &= 2S_y[I_x(cos^2(\phi) + sin^2(\phi))] - 2S_y[I_y cos(\phi) sin(\phi) - I_y cos(\phi) sin(\phi)] \\ &= 2I_xS_y\end{aligned} \quad (10.13)$$

This result is equivalent to the original density matrix, therefore no net evolution due to the proton chemical shift has occurred during the $\tau - 180° - \tau$ period. Since τ can be any arbitrary length, this analysis applies to proton evolution during the Δ periods as well as during t_1.

Evolution due to Nitrogen Chemical Shift: The density matrix, $\rho_3 = 2I_xS_y$ evolves with the nitrogen chemical shift as follows:

$$2I_xS_y \stackrel{\omega_S t_1}{\rightarrow} 2I_x[S_y cos(\omega_S t_1) - S_x sin(\omega_S t_1)] \quad (10.14)$$

Therefore the final density matrix after t_1 is simply:

$$\rho_4(t_1) = 2I_xS_y cos(\omega_S t_1) - 2I_xS_x sin(\omega_S t_1) \quad (10.15)$$

Note that the density matrix is now amplitude modulated, carrying information on the nitrogen frequency.

D: Transfer of Magnetization back to the Proton.

The second 90° degree pulse that is applied to the nitrogen produces ρ_5,

$$\rho_5 = 2I_xS_z cos(\omega_S t_1) - 2I_xS_x sin(\omega_S t_1) \quad (10.16)$$

The term, $2I_xS_x$ represents double quantum coherence which cannot be detected and will therefore be ignored. The refocusing period, Δ, will transform the single-quantum elements of the density matrix as follows:

$$2I_xS_z cos(\omega_S t_1) \rightarrow [2I_xS_z cos(\pi J\Delta) + I_y sin(\pi J\Delta)] \, cos(\omega_S t_1) \quad (10.17)$$

Recall that since $\Delta = \frac{1}{2J}$, the only term that remains at the beginning of signal detection is:

$$\rho_6 = I_y cos(\omega_S t_1) \quad (10.18)$$

Again, the proton chemical shift evolution during this period can be ignored. Specifically, the proton chemical shift evolution that occurred during the first Δ period is refocused during the second Δ period because of the 180° proton pulse in the center of the t_1 period.

E: Detection.

Decoupling prevents the evolution of the system under J-coupling, thus during the detection period, the density matrix evolves as:

$$\rho_6(t) = cos(\omega_S t_1)[I_y cos(\omega_I t_2) - I_x sin(\omega_I t_2)] \quad (10.19)$$

Giving a detected signal of:

$$S(t_1, t_2) = \gamma_H cos(\omega_S t_1) e^{i\omega_I t_2} \quad (10.20)$$

10.2.2 HSQC Experiment

The principal difference between the heteronuclear single-quantum correlation experiment (HSQC) and the HMQC experiment discussed above is the fact that the density matrix during the t_1 evolution period is represented by $2I_zS_x$ instead of $2I_xS_x$. The $2I_zS_x$ term represents a singly excited state, hence the name of this sequence.

The HSQC pulse sequence is shown in Fig. 10.3. This sequence can be divided into the same five segments that were used to describe the HMQC experiment.

A: Initial Density Matrix.

The initial density matrix is: $\rho_o = \gamma_H I_z + \gamma_N S_z$. For reasons stated in the analysis of the HMQC experiment, the contribution from the heteronuclear spin, $\gamma_N S_z$, can be ignored. Furthermore, the γ_H constant will also be dropped.

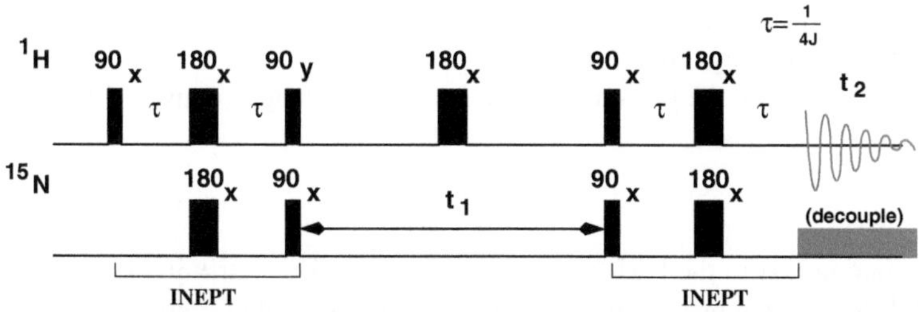

Figure 10.3. *HSQC pulse sequence.* The top set of pulses are applied to the protons and the lower set of the pulses are applied to the heteronuclear spins (^{15}N in this illustration) via a separate radio-frequency channel. Narrow bars correspond to 90° pulses and wider bars represent 180° pulses. The delay τ is nominally set to $\frac{1}{4J}$. Polarization transfer periods (INEPT) are labeled and include the simultaneous proton and nitrogen 90° pulse.

10.2.2.1 B: Polarization Transfer - First INEPT.

The first polarization transfer period is referred to as an INEPT[1] transfer and the second polarization period is often referred to as a reverse-INEPT transfer because the magnetization associated with the insensitive spins is transfered back to the attached proton by the same mechanism. For simplicity, the term INEPT will be used to describe this method of magnetization transfer, regardless of the direction of magnetization transfer. The first 90° proton pulse converts I_z (ρ_o) to $-I_y$. Evolution of the proton chemical shift during the remaining part of this period can be ignored because it is a spin-echo sequence ($\tau - 180° - \tau$). Therefore, it is only necessary to consider evolution by J-coupling. The 180° pulses in the middle of the INEPT period are usually applied to both the proton and the heteronuclear spin simultaneously. However, they are considered to occur sequentially in the analysis below. The evolution of the density matrix is as follows ($\phi = \pi J \tau$):

$$
\begin{aligned}
-I_y &\xrightarrow{J} -[I_y cos(\phi) - 2I_x S_z sin(\phi)] \\
&\xrightarrow{180^x_I} -[-I_y cos(\phi) - 2I_x S_z sin(\phi)] \xrightarrow{180^x_S} -[-I_y cos(\phi) + 2I_x S_z sin(\phi)] \\
&\xrightarrow{J} cos(\phi)[I_y cos(\phi) - 2I_x S_z sin(\phi)] - sin(\phi)[2I_x S_z cos(\phi) + I_y sin(\phi)] \\
&= I_y[cos^2\phi - sin^2\phi] - 2I_x S_z[2sin(\phi)cos(\phi)] \\
&= I_y cos(2\phi) - 2I_x S_z sin(2\phi) \qquad (10.21)
\end{aligned}
$$

where the trigonometric identities: $cos(2\phi) = cos^2\phi - sin^2\phi$ and $sin(2\phi) = 2sin\phi cos\phi$ were used in the last step.

The delay, τ, is set to be equal to $\frac{1}{4J}$, giving $sin(2\phi) = sin(2\pi J/[4J]) = sin(\pi/2) = 1$, therefore the density matrix just before the 90° pulses at the end of the INEPT period is:

$$\rho = -2I_x S_z \qquad (10.22)$$

[1]INEPT is an acronym for **I**nsensitive **N**uclei **E**nhanced by **P**olarization **T**ransfer.

Therefore the INEPT period cause the conversion of in-phase proton magnetization ($-I_y$) to anti-phase magnetization ($2I_xS_z$).

The 90° y-pulse on the proton and x-pulse on the nitrogen convert this to:

$$\rho = -2I_zS_y \tag{10.23}$$

Note that in contrast to the HMQC sequence, the density matrix in the HSQC experiment is $2I_zS_y$, i.e. the proton magnetization is longitudinal (I_z) and the nitrogen spin is transverse (S_y). Furthermore, note that the total time for polarization transfer is the same in both experiments, $\frac{1}{2J}$.

C: Evolution During t_1.

During this period the density matrix can potentially evolve under the influence of the proton chemical shift, the heteronuclear chemical shift, and J-coupling.

Evolution of Proton Chemical Shift: Evolution under the proton chemical shift does not occur since the proton state is I_z:

$$\begin{aligned}
\rho'(t_1) &= -e^{iI_z\omega_I\tau}2I_zS_ye^{-iI_z\omega_I\tau} \\
&= -2S_ye^{iI_z\omega_I\tau}I_ze^{-iI_z\omega_I\tau} \\
&= -2S_yI_ze^{iI_z\omega_I\tau}e^{-iI_z\omega_I\tau} \\
&= -2I_zS_y
\end{aligned}$$

J-Coupling Evolution: Evolution due to J-coupling is refocused by the proton 180° pulse during t_1. This can be seen with the following analysis ($\zeta = \pi J\frac{t_1}{2}$):

$$\begin{aligned}
-2I_zS_y &\overset{t_1/2}{\rightarrow} -2I_zS_ycos(\zeta) + S_xsin(\zeta) \\
&\overset{\pi_I}{\rightarrow} +2I_zS_ycos(\zeta) + S_xsin(\zeta) \\
&\overset{t_1/2}{\rightarrow} +cos(\zeta)[2I_zS_ycos(\zeta) - S_xsin(\zeta)] + sin(\zeta)[S_xcos(\zeta) + 2I_zS_ysin(\zeta)] \\
&= -S_x[cos(\zeta)sin(\zeta) - cos(\zeta)sin(\zeta)] + 2I_zS_y[cos^2(\zeta) + sin^2(\zeta)] \\
&= 2I_zS_y
\end{aligned}$$

Hence, the only effect of J-coupling on the density matrix during the t_1 period is a change in the sign. This is a general feature of applying a 180° pulse to one of the two coupled spins within a symmetrical interval, there is no net evolution of the density matrix due to J-coupling.

Evolution of Nitrogen Chemical Shift: The net evolution of the density matrix during t_1 is solely due the nitrogen chemical shift:

$$\begin{aligned}
2I_zS_y &\overset{\omega_St_1}{\rightarrow} 2I_z[S_ycos(\omega_St_1) - S_xsin(\omega_St_1)] \\
&= 2I_zS_ycos(\omega_St_1) - 2I_zS_xsin(\omega_St_1)
\end{aligned} \tag{10.24}$$

D: Polarization transfer back to Protons - The Reverse INEPT.

This segment begins with the pair of x-pulses that are applied to both the proton and heteronuclear spins, interchanging the state of the proton and heteronuclear spins:

$$\begin{aligned} 2I_zS_ycos(\omega_St_1) &\rightarrow \quad 2I_yS_zcos(\omega_St_1) \\ 2I_zS_xsin(\omega_St_1) &\rightarrow \quad -2I_yS_xsin(\omega_St_1) \end{aligned} \tag{10.25}$$

Note that the $-2I_yS_xsin(\omega_St_1)$ represents double quantum magnetization that cannot be detected during the t_2 period, hence it will be ignored. The subsequent part of the INEPT period ($\tau - 180^\circ_{(H,N)} - \tau$) will refocus the $2I_yS_z$ term to give the density matrix at the beginning of t_2, i.e.:

$$2I_yS_zcos(\omega_St_1) \rightarrow I_xcos(\omega_St_1) \tag{10.26}$$

E: Detection.

During the detection period, heteronuclear decoupling is applied so that only a single resonance line is detected for each I-S spin pair, giving the following signal, assuming quadrature detection in t_2.

$$S(t_1,\, t_2) = cos(\omega_St_1)e^{i\omega_It_2} \tag{10.27}$$

An HSQC spectrum of a ^{15}N labeled protein is shown in Fig. 10.1.

10.2.3 Refocused-HSQC Experiment

The refocused-HSQC experiment generates pure heteronuclear single quantum coherence, such as S_x or S_y during the t_1 evolution period. The refocused-HSQC experiment is shown in Fig. 10.4. This experiment consists of two back-to-back INEPT transfer periods, a period of nitrogen chemical shift evolution, followed by two additional INEPT periods that return the magnetization to the coupled proton.

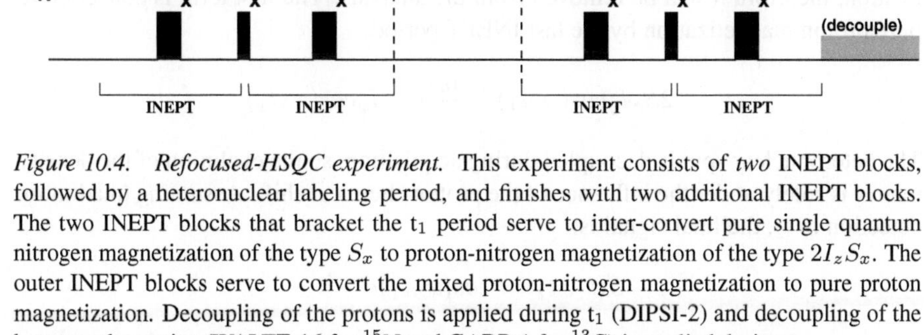

Figure 10.4. Refocused-HSQC experiment. This experiment consists of *two* INEPT blocks, followed by a heteronuclear labeling period, and finishes with two additional INEPT blocks. The two INEPT blocks that bracket the t_1 period serve to inter-convert pure single quantum nitrogen magnetization of the type S_x to proton-nitrogen magnetization of the type $2I_zS_x$. The outer INEPT blocks serve to convert the mixed proton-nitrogen magnetization to pure proton magnetization. Decoupling of the protons is applied during t_1 (DIPSI-2) and decoupling of the heteronuclear spins (WALTZ-16 for ^{15}N and GARP-1 for ^{13}C) is applied during t_2.

Although the refocused-HSQC experiment appears complex, it can be quickly analyzed by keeping in mind the following rules that describe how the density matrix evolves during an INEPT period:

1. The 180° pulses that are applied to both the proton and the heteronuclear spin will refocus any chemical shift evolution that occurred during the first τ delay.

2. Since 180° pulses are applied to both spins, evolution of the density matrix due to J-coupling will occur during the entire 2τ period. If $\tau = \frac{1}{4J}$ then an in-phase transverse term will evolve to an anti-phase term:

$$I_x \rightarrow 2I_yS_z \tag{10.28}$$

Likewise, an anti-phase term will evolve into in-phase magnetization:

$$2I_yS_z \rightarrow -I_x \tag{10.29}$$

Beginning from the initial density matrix ($\rho_o = I_z$), the evolution of the density matrix to the start of the t_1 period is as follows:

$$I_z \xrightarrow{90^I_x} -I_y \xrightarrow{\tau-180^\circ-\tau} 2I_xS_z \xrightarrow{90^I_y 90^S_x} 2I_zS_y \xrightarrow{\tau-180^\circ-\tau} S_x \tag{10.30}$$

During the t_1 labeling period the density matrix becomes:

$$S_x \xrightarrow{t_1} S_x cos(\omega_S t_1) + S_y sin(\omega_S t_1) \tag{10.31}$$

The density matrix does not evolve under J-coupling during this period since the protons have been decoupled. The first INEPT after the t_1 period changes the above density matrix to:

$$S_x cos(\omega_S t_1) \xrightarrow{\tau-180^\circ-\tau} 2S_yI_z cos(\omega_S t_1) \xrightarrow{90^I_x 90^S_x} -2S_zI_y cos(\omega_S t_1) \tag{10.32}$$

$$S_y sin(\omega_S t_1) \xrightarrow{\tau-180^\circ-\tau} -2S_xI_z sin(\omega_S t_1) \xrightarrow{90^I_x 90^S_x} 2S_xI_y sin(\omega_S t_1) \tag{10.33}$$

The second of these terms ($2S_xI_y sin(\omega_s t_1)$) will not develop into observable magnetization, therefore it will be removed from the analysis. The first term is converted to pure proton magnetization by the last INEPT period:

$$-2S_zI_y cos(\omega_S t_1) \xrightarrow{\tau-180^\circ-\tau} I_x cos(\omega_S t_1)$$

The heteronuclear spin is decoupled during acquisition, so the evolution of the density matrix is solely under the influence of the proton chemical shift. Assuming quadrature detection in t_2, the final signal is:

$$S(t_1, t_2) = cos(\omega_S t_1)e^{i\omega_I t_2}$$

10.2.4 Comparison of HMQC, HSQC, and Refocused-HSQC Experiments

The HMQC, HSQC, and refocused-HSQC experiments differ in several attributes, including: the linewidth of resonance lines in the heteronuclear dimension, inherent sensitivity, and the behavior of NH_2, CH_2, and CH_3 groups versus NH and CH groups.

10.2.4.1 Comparison of Linewidth

The observed linewidths in the HMQC, HSQC and refocused-HSQC experiments are different due to different relaxation rates and well as to the presence of additional unresolved couplings (see [12] for additional details). The different relaxation rates are due to the presence of different quantum states during t_1. The magnetization during t_1 is represented by $2I_xS_x$ in the HMQC experiment, as $2I_zS_x$ in the HSQC experiment, and as S_x in the refocused-HSQC experiment. In general, the order of transverse relaxation rates for these terms is: $2I_xS_x > 2I_zS_x > S_x$. In some systems, such as nucleic acids the $2I_zS_x$ term may relax more rapidly, reversing this order. Therefore, the HMQC experiment will have broadest line in the heteronuclear dimension, while the refocused-HSQC will have the narrowest line.

The linewidth of the peaks in the HMQC experiment are broadened further due to proton-proton couplings that are active during the t_1 evolution time. For example, the coupling between the H_N proton and the H_α proton will contribute to the linewidth in ω_1. Although the 180° pulse in the middle of the t_1 evolution period will refocus both proton evolution and heteronuclear coupling, it cannot refocus homonuclear J-couplings because the 180° pulse effects all protons. Therefore, during t_1 the system evolves as:

$$\rho(t_1) = 2S_y\left[I_x cos(\pi J_{HH}t_1) + I_yK_z sin(\pi J_{HH}t_1)\right] \tag{10.34}$$

where K_z represents a second proton that is coupled to the proton attached to the heteronuclear spin. This coupling will produce an in-phase splitting of the crosspeaks in the ω_1 frequency domain, thus reducing sensitivity.

In summary, for protein NMR, the HMQC spectrum will have the broadest lines while the refocused-HSQC will produce the narrowest lines.

10.2.5 Sensitivity in 2D-Heteronuclear Experiments

All of these experiments utilize polarization transfer periods to relay the magnetization between the proton and the heteronuclear spin. During these periods, the magnetization also decays due to relaxation. The total transfer time, τ_{total}, in the HMQC and the HSQC experiments are identical, at 1/J: $[2 \times 1/(2J)]$ for the HMQC, $[4 \times 1/(4J)]$ for the HSQC experiment. In contrast, the refocused-HSQC experiment requires twice as long, a total of 2/J. The amount of magnetization that is lost due to relaxation during an HMQC, or INEPT, transfer period is given by the following:

$$e^{-\tau_{total}/\bar{T_2}} \tag{10.35}$$

where, $\bar{T_2}$ represents the averaged relaxation rate of the magnetization during the transfer period τ_{total}. The loss of intensity due to relaxation is of little significance for a

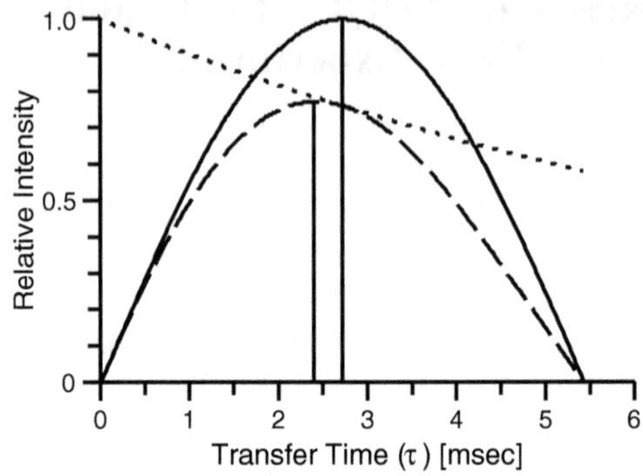

Figure 10.5. Optimization of HSQC delays. The transfer of magnetization during a single INEPT sequence of a ^{1}H-^{15}N HSQC experiment as a function of the delay τ is shown. The function plotted is $sin(\pi J2\tau)e^{-2\tau/T_2}$. The coupling constant was set to 90 Hz, i.e. $1/(4J) = 2.7$ msec. The overall loss of signal is the square of these curves since two INEPT periods are utilized in this experiment. The solid line shows the transfer function for a small protein while the dashed line shows the transfer function for a ≈50 kDa protein. The dotted line shows the decay of magnetization for the 50 kDa protein, assuming a T_2 of 20 msec. The optimal delay, τ, for the large protein is 2.4 msec.

small protein and quite severe for larger proteins. For example, the amide nitrogens in a 10 kDa protein will have an average T_2 of approximately 150 msec, resulting in a signal loss of 7% and 14% for a proton-nitrogen HSQC and a refocused HSQC, respectively. In contrast, the amide nitrogens in a 50 kDa protein have an average T_2 of approximately 20 msec, leading to signal losses of 36% and 77%, respectively, for both experiments, respectively.

Optimal sensitivity in these experiments is obtained when the transfer function in $sin(\pi Jt)e^{-t/T_2}$ is at a maximum. As shown in Fig. 10.5, the optimal value for the τ delay in the HSQC experiment is always shorter that $\frac{1}{4J}$.

10.2.6 Behavior of XH_2 Systems in HSQC-type Experiments

The above discussion referred to systems in which one proton is attached to one heteronuclear spin (i.e. CH or NH). In many cases two or more protons are attached to heteronuclear spins (e.g. CH_2, CH_3, NH_2). The presence of the additional heteronuclear-proton couplings have to be considered in the analysis of the above sequences.

In the case of both the HMQC and HSQC sequences, the additional proton has no effect on the experiment because the heteronuclear spin is transverse only during the t_1 evolution period. During this time the coupling between the heteronuclear spin and the protons is removed by the 180° refocusing pulse.

In contrast, in the refocused-HSQC experiment the heteronuclear spin is transverse during the *second* INEPT period. During this time it will evolve due to J-coupling

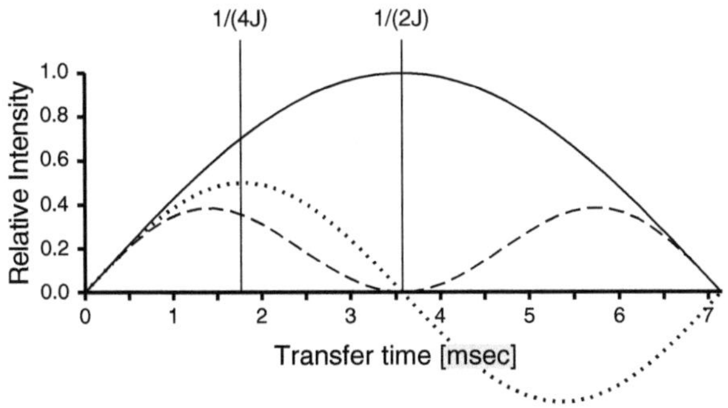

Figure 10.6 Transfer function for a refocused HSQC experiment. Transfer function for CH (solid), CH_2 (dotted) and CH_3 (dashed) groups in a refocused HSQC, assuming a coupling constant of J = 140 Hz. The intensity of the magnetization transferred during the second INEPT period of the pulse sequence is shown as a function of total length of the INEPT period (2τ).

with *both* protons. Labeling the protons as I_1 and I_2 and considering evolution due to coupling to the first proton gives:

$$2I_{1z}S_y \xrightarrow{J_1} 2I_{1z}S_y cos(\pi\tau J_1) + S_x sin(\pi\tau J_1) \qquad (10.36)$$

Each of these terms will also evolve due to heteronuclear coupling to the second proton to give the following terms:

$$\begin{aligned} 2I_{1z}S_y cos(\pi\tau J_1)cos(\pi\tau J_2) \quad &+ \quad 2I_{2z}S_x cos(\pi\tau J_1)sin(\pi\tau J_2) \\ +S_x sin(\pi\tau J_1)cos(\pi\tau J_2) \quad &+ \quad 2I_{2z}S_y sin(\pi\tau J_1)sin(\pi\tau J_2) \end{aligned} \qquad (10.37)$$

Since the only term which results in the formation of the crosspeaks in the spectra arises from the S_x term, the amplitude of the crosspeaks will be:

$$sin(\pi\tau J_1)cos(\pi\tau J_2) \qquad (10.38)$$

A similar analysis shows that for three coupled spins (e.g. CH_3) the intensity is:

$$sin(\pi\tau J_1)cos(\pi\tau J_2)cos(\pi\tau J_3) \qquad (10.39)$$

The amount of magnetization transfered by the second INEPT period for CH, CH_2, and CH_3 groups are shown in Fig. 10.6. The dependence of the intensity versus transfer time of NH and NH_2 groups would be the same as the solid and dotted lines, except that the optimal times would be calculated using J = 90 Hz.

This figure shows that a delay of $\tau = 1/(4\text{J})$ is optimal for a CH group while a delay of $\tau = 1/(8\text{J})$ is optimal for a CH_2 group. Furthermore, when the length of the INEPT period is set to 1/(2J), the resonance lines from CH_2 and CH_3 groups disappear from the spectra. If the INEPT period is set to longer than 1/(2J), the signals from the CH_2 groups will actually become inverted relative to the CH groups.

The dependence of the intensities of signals from the NH_2, CH_2, and CH_3, groups on the length of the INEPT period in refocused-HSQC experiments provides a means to identify such groups. In the case of NH_2 groups on asparagine and glutamine residues, the resonance lines will be absent in a refocused-HSQC experiment if the INEPT delay is set to 1/(2J).

Chapter 11

COHERENCE EDITING: PULSED-FIELD GRADIENTS AND PHASE CYCLING

Coherence editing is used to remove unwanted signals from NMR spectra. For example, in the double quantum filtered COSY experiment (Chapter 9), coherence editing is used to suppress signals from uncoupled protons, such as the intense solvent line, while retaining signals from coupled spins. This filtering is accomplished by distinguishing between the different quantum states of the magnetization during the experiment: uncoupled spins can only attain a single-quantum state while coupled spins can be in a double-quantum state during the experiment.

Coherence editing is accomplished by encoding different states with a unique phase during the pulse sequence. The selection of a desired state is attained by detecting only those signals with the appropriate phase. The phase encoding of the signal can be accomplished by either phase cycling of RF-pulses or by the application of short pulses of spatially varying magnetic fields, otherwise known as pulsed-field gradients (PFG), during evolution delays.

In the case of phase cycling, multiple scans are acquired with the phase of pulses and the receiver altered in a systematic fashion such that the desired signals (e.g. double-quantum) are always co-added while the undesired signals (e.g. single-quantum states) add to zero at the end of the phase cycle, as illustrated in Panel A of Fig. 11.1.

In the case of using pulsed-field gradients for coherence editing, a spatially varying magnetic field is used to impart a characteristic phase shift on both the desired and undesired signals. The amount of phase shift depends on the quantum level of the spins. For example, the phase shift encoded in a single-quantum term would be half of that encoded in a double-quantum term. The retention of the desired signal and discarding of the undesired signal is simply accomplished by refocusing (returning the phase shift to zero) the desired signal, as illustrated in Panel B of Fig. 11.1.

Both phase cycling and magnetic field gradients can effectively suppress undesirable signals. Magnetic field gradients tend to be more effective than phase cycling methods because they can be applied in a single scan, thus scan-to-scan reproducibility of the instrument is not a critical factor. In addition, suppression of strong solvent signals can also be accomplished within a single scan, greatly reducing the dynamic

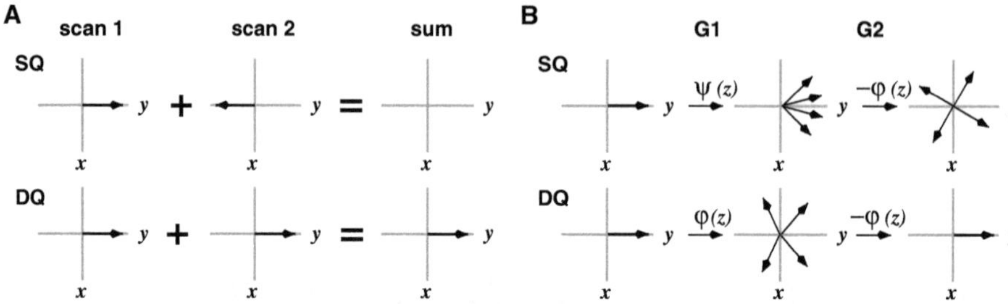

Figure 11.1. Coherence editing in the DQF-COSY experiment. Panel A illustrates coherence editing by phase cycling while Panel B shows editing using pulsed magnetic field gradients. In both cases, the signal associated with the double quantum (DQ) state is retained while that from the single quantum state (SQ) is rejected. In the case of phase cycling (A), two scans are acquired and summed. The phase of the signal associated with the DQ state differs from that of the SQ such that when the signals are added, the DQ signals add while the SQ signals cancel. In the case of coherence editing using pulsed-field gradients, the magnetization associated with the DQ state receives a phase shift of $\varphi(z)$ due to the application of the first field gradient (G1). The actual phase shift depends on the z-coordinate of the nuclear spin in the sample. In contrast, the SQ state acquires a different phase shift ($\psi(z)$). A second gradient (G2) is designed to reverse the phase shift associated with the DQ state, rendering it detectable. In contrast, the magnetization associated with the SQ state is not refocused, such that the signal averaged over the ensemble of spins is zero. Note that the coherence editing is accomplished within a single scan.

range of the acquired signals, resulting in more accurate conversion of the analog signal to digital form. In addition, the requirement of acquiring multiple scans with phase cycling can lead to an undesirable increase in the length of the experiment, especially in the case of three- or four-dimensional experiments. There are two significant disadvantages associated with pulsed-field gradients. First, additional delays have to be incorporated into the pulse sequence for the application of gradients. In some cases this is not possible due to timing constraints. Second, there can be an inherent reduction in sensitivity because the portion of the magnetization that is rejected by the field gradients may contain useful signal. Consequently, most experiments utilize a combination of pulsed-field gradients and phase cycling to remove undesirable signals from the spectrum.

11.1 Principals of Coherence Selection

11.1.1 Spherical Basis Set

The product operator representation of the density matrix using Cartesian angular momentum operators, such as I_x, has proven to be very convenient for the analysis of the effect of RF-pulses on the density matrix during an NMR experiment. However, the Cartesian representation is cumbersome for analyzing the effect of phase cycling or field gradients on the density matrix. Instead, we will adopt a different representation of the density matrix in which the basis set will represent *individual* transitions, or coherences, of the system. This basis set is often referred to as a spherical basis set.

Table 11.1. Spherical and Cartesian Basis Set. The relationship between the spherical and Cartesian basis set is presented. The property of the system that is associated with each of the spherical basis elements is also given.

Spherical	*Cartesian Representation*	*Description*
σ^0	I_z	Population difference P_e-P_g
σ^{+1}	$I_x + iI_y$	Upward transition $\Phi_g \rightarrow \Phi_e$
σ^{-1}	$I_x - iI_y$	Downward transition $\Phi_e \rightarrow \Phi_g$

For a single uncoupled spin the correspondence between the Cartesian and the spherical basis set is shown in Table 11.1. The density matrix, σ^o represents the equilibrium population difference between the ground and excited states and is referred to as zero-quantum coherence. Single quantum coherences are represented by σ^{+1}, which represents systems in transition from the ground to the excited state, and σ^{-1}, which represents transitions from the excited to the ground state.

The fact that the spherical basis set represents transitions in a single direction can be clearly seen from their corresponding density matrices:

$$\sigma^{+1} = \frac{\hbar}{2}\begin{bmatrix} 0 & 1 \\ 0 & 0 \end{bmatrix} \qquad \sigma^{-1} = \frac{\hbar}{2}\begin{bmatrix} 0 & 0 \\ 1 & 0 \end{bmatrix} \tag{11.1}$$

while the Cartesian representation generally represent both upward and downward transitions, for example:

$$I_x = \frac{\hbar}{2}\begin{bmatrix} 0 & 1 \\ 1 & 0 \end{bmatrix} \tag{11.2}$$

The Cartesian representation of the density matrix can be converted to the spherical representation using the following transformations:

$$I_x = \frac{1}{2}[\sigma^{+1} + \sigma^{-1}] \qquad I_y = \frac{1}{2i}[\sigma^{+1} - \sigma^{-1}] \tag{11.3}$$

11.1.1.1 Selection of Double Quantum Coherence in DQF-COSY Experiment

As an example of the selection of particular transition by phase cycling or pulsed-field gradients, consider the magnetization present in the DQF-COSY experiment during the delay Δ (See Fig. 9.10).

$$\begin{aligned} 2I_xS_y &= 2\frac{1}{2}(I^+ + I^-)\frac{1}{2i}(S^+ - S^-) \\ &= \frac{1}{2i}[I^+S^+ - I^+S^- + I^-S^+ - I^-S^-] \end{aligned} \tag{11.4}$$

where I^+ represents σ^{+1} for the I-spins, etc.

The density matrices I^+S^+ and I^-S^- are non-zero for only elements that represent double quantum transitions while the matrices I^+S^- and I^-S^+ have only non-zero elements that represent zero quantum transitions. Consequently, this Cartesian form of the density matrix ($2I_xS_y$) can be expressed using the spherical basis set as:

$$\rho = \sigma^{+2} + \sigma^0 + \sigma^{-2} \tag{11.5}$$

Coherence selection, either by pulsed-field gradients or phase cycling of pulses, would induce a phase shift in each of the separate coherences at this point in the experiment. The phase encoding depends on the coherence level of each element of the density matrix. After application of gradients, or phase alteration of the pulses, the above density matrix would have the form:

$$\rho = \sigma^{+2}e^{i\varphi_2} + \sigma^0 e^0 + \sigma^{-2}e^{-i\varphi_2} \tag{11.6}$$

where φ_2 is the phase shift associated with the σ^{+2} part of the density matrix. Note that the shift associated with the σ^{-2} term is exactly opposite to the phase shift of σ^{+2}.

If a second phase shift of $e^{-i\varphi_2}$ is induced in the signal, by the application of a gradient at some time later in the sequence, or by altering the receiver phase in the case of phase cycling, then the density matrix will become:

$$\begin{aligned} \rho &= e^{-i\varphi_2}[\sigma^{+2}e^{i\varphi_2} + \sigma^0 e^0 + \sigma^{-2}e^{-i\varphi_2}] \\ &= \sigma^{+2} + \sigma^0 e^{-2i\phi} + \sigma^{-2}e^{-i2\varphi_2} \end{aligned} \tag{11.7}$$

If we then assume that only signals with a phase shift of zero can be detected, then the final density matrix consists only of signals that were in a double-quantum state during the Δ period, all other signals will be absent from the spectrum.

The above example represents the general nature of coherence selection by phase cycling or pulsed-field gradients. In both cases the elements of the density matrix are encoded with a known phase shift, and this phase shift is exploited to reject all of the undesirable signals.

11.1.2 Coherence Changes in NMR Experiments

Before coherence selection methods can be applied to an NMR experiment it is necessary to know the coherences that are present during the experiment. Any NMR experiment can be considered to be a series of pulses that bracket periods of free evolution. Since periods of free evolution correspond to a rotation about the z-axis, there is no change in the coherence level during those periods, the density matrix simply develops a phase shift that is proportional to the frequency of the transition that is represented by each element of the density matrix, e.g. $\sigma^{+1} \rightarrow \sigma^{+1}e^{-i\omega t}$. In contrast, RF-pulses create new coherence levels in a system by virtue of the fact that they cause transitions between spin states.

11.1.2.1 Example - Coherence Order in a COSY Experiment

As an example, consider the COSY experiment. By writing the density matrix in both the Cartesian and spherical basis sets it is possible to follow the coherence changes during the experiment. The initial state of the system is:

$$\rho_0 = \sigma^0 = I_z + S_z = \begin{bmatrix} 1 & 0 & 0 & 0 \\ 0 & 0 & 0 & 0 \\ 0 & 0 & 0 & 0 \\ 0 & 0 & 0 & -1 \end{bmatrix} \tag{11.8}$$

Note that this density matrix is composed of only zero-quantum coherence, which correspond to populations in this case. After the first pulse:

$$\rho_1 = -I_y - S_y = \begin{bmatrix} 0 & -i & -i & 0 \\ i & 0 & 0 & -i \\ i & 0 & 0 & -i \\ 0 & i & i & 0 \end{bmatrix} \propto I^+ + I^- + S^+ + S^- = \sigma_I^1 + \sigma_I^{-1} + \sigma_S^1 + \sigma_S^{-1} \tag{11.9}$$

This density matrix is composed of an equal mixture of σ^{+1} and σ^{-1} coherences for both the I and S spins. During the t_1 evolution time the density matrix evolves under both J-coupling and chemical shift evolution. At the end of t_1 the part of the density matrix that will give rise to the crosspeak is:

$$\rho_2 = 2I_y S_z sin(\omega_I t_1) sin(\pi J t_1) \tag{11.10}$$

This is still a mixture of single quantum coherences, as indicated by the presence of only one transverse operator (I_y). Therefore, evolution of the system under either the chemical shift or the J-coupling Hamiltonian does not change coherence order, as discussed above.

The final pulse of the COSY experiment exchanges magnetization between the two coupled spins:

$$2I_y S_z \rightarrow -2I_z S_y \tag{11.11}$$

The detected signal is obtained in the usual fashion, by taking the trace of the density matrix and the operator that represents the observable. In the case of quadrature detection the observable for the I spin is $I^+ = I_x + iI_y$ (a similar expression could be written for the S spins), giving the following signal:

$$M^+(t) = M_x(t) + iM_y(t) = \text{Trace}\left[\,\rho(I_x + iI_y)\right] = \text{Trace}\left[\,\rho I^+\right] \tag{11.12}$$

Recall that;

$$I^+ = \begin{bmatrix} 0 & 0 & 1 & 0 \\ 0 & 0 & 0 & 1 \\ 0 & 0 & 0 & 0 \\ 0 & 0 & 0 & 0 \end{bmatrix} \quad \text{and} \quad I^- = \begin{bmatrix} 0 & 0 & 0 & 0 \\ 0 & 0 & 0 & 0 \\ 1 & 0 & 0 & 0 \\ 0 & 1 & 0 & 0 \end{bmatrix} \tag{11.13}$$

Therefore, the only density matrix that gives a non-zero trace when multiplied by I^+ is I^-. Consequently, the only coherence state that is detected with quadrature detection is σ^{-1}.

In summary, in the COSY experiment, the density matrix can be represented by zero quantum coherence prior to the first pulse, as single quantum coherence during t_1, and the detected signal is proportional to the contribution of -1 coherence to the density matrix.

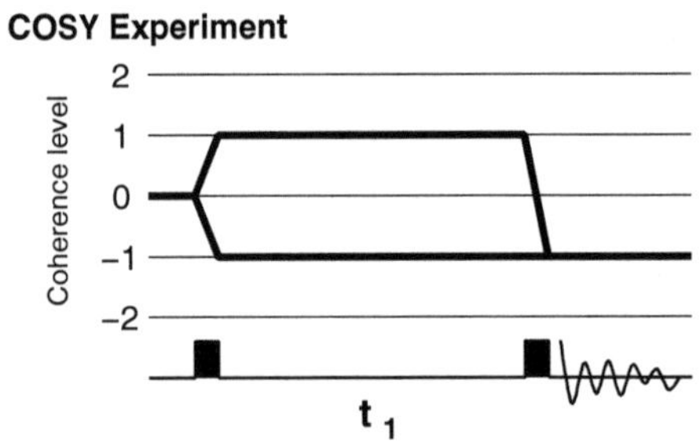

Figure 11.2 Coherence changes in the COSY experiment. The desired coherence changes in the COSY experiment are summarized with a coherence level diagram. The initial coherence level is zero (σ^0), the first pulse generates both σ^{+1} and σ^{-1} for both I and S spins. The second mixing pulse interchanges the label associated with each coherence (e.g. $\sigma_I^{+1} \rightarrow \sigma_S^{+1}$) and at the end of the experiment, only σ^{-1} is detected by quadrature detection.

11.1.3 Coherence Pathways

The coherence changes that occurred during the COSY experiment are neatly summarized by a coherence pathway diagram, as shown in Fig. 11.2. The system begins in a zero-quantum state, representing populations, is then transformed to single quantum states by the first pulse, followed by detection of -1 single quantum coherence after the second pulse. These transformations of the density matrix can also be represented by the following sets of *coherence paths*:

$$\begin{aligned} \sigma^0 &\rightarrow \sigma^{+1}(t_1) \rightarrow \sigma^{-1} \\ \sigma^0 &\rightarrow \sigma^{-1}(t_1) \rightarrow \sigma^{-1} \end{aligned} \tag{11.14}$$

The above coherence changes specify the *desired* changes in coherence during the COSY experiment. There are other coherence paths that are also present. These paths are undesired because they can lead to the presence of unwanted artifacts in the final spectra. For example, if the first RF-pulse was not exactly equal to 90°, then the first pulse will leave some magnetization along the z-axis, which is represented by zero quantum coherence. Therefore, the following pathway is one of a number of *undesirable* coherence pathways that can occur in the COSY experiment:

$$\sigma^0 \rightarrow \sigma^0 \xrightarrow{t_1} \sigma^0 \rightarrow \sigma^{-1} \tag{11.15}$$

Having defined the desired coherence changes in the COSY experiment, the following sections will illustrate how pulsed-field gradients (Section 11.2) or RF-pulse phase shifts (Section 11.3) can be used to select the desired coherence changes in an experiment while rejecting the undesired ones.

11.2 Phase Encoding With Pulsed-Field Gradients

11.2.1 Gradient Coils

To employ pulsed-field gradients in coherence editing it is necessary to use specially designed probes and electronics. In the case of the probe, one (z-gradient) or three (triple-axis: x, y, z-gradients) additional coils are placed next to the sample. Each coil will generate a magnetic field gradient within the sample when current is passed through the coil. The gradient coils are designed in such a manner that the generated B_z field depends on the *position* of the nuclear spin within the sample tube. For example, a triple axis gradient would generate the following magnetic field at a position (x, y, z) in the sample:

$$B_z = G_z \times z + G_x \times x + G_y \times y \tag{11.16}$$

where G_i represents the gradient strength in the i direction, or the change in the magnetic field in the z-direction due to a change in the coordinate in the ith dimension:

$$G_i = \partial G_i / \partial x_i \tag{11.17}$$

The applied gradients can be either positive or negative, are of a well defined amplitude, and can be activated for well defined periods of time using simple commands within the NMR pulse program.

11.2.1.1 Gradient Recovery

The pulsed-field gradients cause a perturbation of the static B_o field. Consequently it is necessary to provide a recovery period after application of the gradient to allow B_o field to regain its homogeneity. To shorten the recovery time, all commercial gradient coils are actively shielded, meaning that the magnetic field is driven back to homogeneity by an active process. Typical recovery times are on the order of $100-200$ μsec. The recovery time can be easily measured by simply applying a gradient pulse, followed by a delay, and then a 90° pulse; the peak-height will increase as the field homogeneity returns. If the recovery time is longer than 200 μsec it may be necessary to adjust the gradient hardware and/or software for optimal performance.

11.2.1.2 Effect of Position on Gradient Induced Phase Changes

During the application of a magnetic field gradient along the z-axis, the magnetic field will become $B_z = B_o + G_z \cdot z$. Therefore the precessional frequency of a spin will become:

$$\omega = \omega_s + \gamma G_z \times z \tag{11.18}$$

If the field gradient is applied for a period τ, then the evolution of the density matrix by the end of the gradient pulse is given by:

$$\rho(\tau) = e^{-i\omega\tau I_z} \rho e^{+i\omega\tau I_z} \tag{11.19}$$

corresponding to a rotation about the z-axis by an angle $\omega\tau$.

Different molecules in the NMR tube will experience different field strengths during the application of a magnetic field gradient because of their different locations within the sample. The effect of a simple z-gradient on the phase of spins is shown in Fig. 11.3. In this figure the nuclear spins labeled 'E' are at the center of the sample and do not precess in the rotating frame. Spins that are spatially above E experience a stronger magnetic field and therefore precesses more rapidly than the rotating frame, moving in a counter-clockwise direction. As spins become more distant from E (e.g. position H) the precessional rate increases because of the field gradient. In a similar manner, spins that are below 'E' in the sample experience a weaker magnetic field and

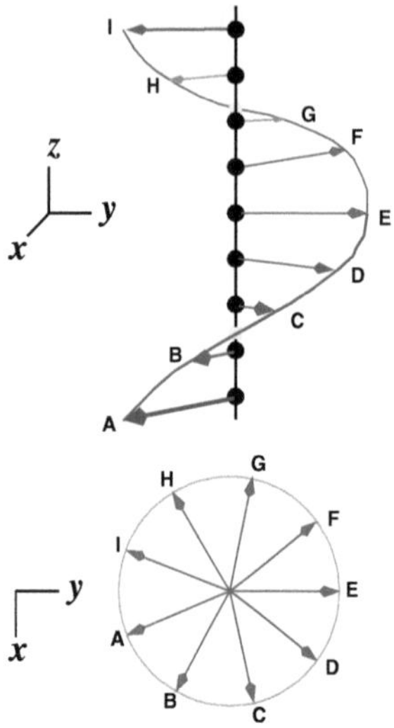

Figure 11.3 Effect of magnetic field gradients on the precessional rate of spins. This figure illustrates the effect of a field gradient along the z-axis. The top section of the figure is a side view of the sample and the bottom section provides a top view of the sample. The magnetization was initially along the y-axis in the rotating frame. The rate of rotation is such that spins at position 'E' are on resonance and therefore do not evolve. The middle of the sample, at $z = 0$ (E), experiences no change in field while the gradient is applied. The magnetic field increases as the displacement from the center increases. Spins above 'E' experience a larger field and spins below 'E' experience a smaller field. The direction of the arrows mark the position of the transverse magnetization after the application of the gradient.

will appear to precess in the clockwise direction. Since the pulsed-field gradient is applied for a defined period of time, the spatially varying precessional frequency will result in a phase shift of the density matrix. The size of this phase shift depends on the location of the spin within the sample.

11.2.2 Effect of Coherence Levels on Gradient Induced Phase Changes

The size of the phase shift induced by a gradient pulse also depends on the coherence level of the spin. It is easy to determine the relationship between the coherence state and the gradient induced phase change using the spherical basis set. Note that in the following calculations, we generalize the evolution during the presence of a gradient as a rotation about the z-axis.

For example, the effect of a z-rotation on σ^{+1} is as follows:

$$\begin{aligned} e^{-i\phi I_z}\sigma^{+1}e^{i\phi I_z} &= e^{-i\phi I_z}(I_x + iI_y)e^{i\phi I_z} \\ &= I_x cos\phi + I_y sin\phi + i\left[I_y cos\phi - I_x sin\phi\right] \\ &= I_x(cos\phi - isin\phi) + I_y(sin\phi + icos\phi) \\ &= I_x e^{-i\phi} + iI_y(cos\phi - isin\phi) \\ &= (I_x + iI_y)e^{-i\phi} \\ &= \sigma^{+1}e^{-i\phi} \end{aligned} \tag{11.20}$$

In a similar manner:

$$e^{-i\phi I_z}\sigma^{-1}e^{i\phi I_z} = \sigma^{-1}e^{+i\phi} \tag{11.21}$$

The effect of z-rotations on zero quantum coherence, σ^0 (e.g. $\sigma_I^{+1}\sigma_S^{-1} = I^+S^-$) is:

$$\begin{aligned} e^{-i\phi(I_z+S_z)}\sigma^0 e^{i\phi(I_z+S_z)} &= e^{-i\phi I_z}\sigma_I^{+1}e^{i\phi I_z}e^{-i\phi S_z}\sigma_S^{-1}e^{+i\phi S_z} \\ &= \sigma_I^{+1}e^{-i\phi}\sigma_S^{-1}e^{+i\phi} \\ &= \sigma_I^{+1}\sigma_S^{-1} \\ &= \sigma^0 \end{aligned} \tag{11.22}$$

The effect of z-rotations on double quantum coherence, for example σ^{+2} is:

$$\begin{aligned} e^{-i\phi(I_z+S_z)}\sigma^{+2} e^{i\phi(I_z+S_z)} &= e^{-i\phi I_z}\sigma_I^{+1}e^{i\phi I_z}e^{-i\phi S_z}\sigma_S^{+1}e^{+i\phi S_z} \\ &= \sigma_I^{+1}e^{-i\phi}\sigma_S^{+1}e^{-i\phi} \\ &= \sigma_I^{+1}\sigma_S^{-1}e^{-2i\phi} \\ &= \sigma^{+2}e^{-2i\phi} \end{aligned} \tag{11.23}$$

Generalizing from the above calculations, if a z-rotation of magnitude ϕ is applied to a coherence of level p, then the coherence will be phase shifted by $e^{-ip\phi}$, that is:

$$\sigma^p \xrightarrow{\phi} \sigma^p e^{-i\,p\phi} \tag{11.24}$$

Note that the phase shift depends on the product of the coherence level (p) and the phase shift induced by the magnetic field gradient (ϕ). If the molecules do not move during the experiment, then the induced phase shift will depend only on the coherence state of the spin at the time the gradient is applied.

11.2.2.1 Example: Coherence Selection in a DQF-COSY Using Gradients

The coherence level diagram for the DQF-COSY experiment is shown in Fig. 11.4. One of the desired coherence paths is:

$$\sigma^0 \rightarrow \sigma^{+1} \rightarrow \sigma^{+2} \rightarrow \sigma^{-1} \tag{11.25}$$

This pathway can be selected by utilizing a total of three gradient pulses, one during t_1, one during the Δ period and one during detection to refocus the desired coherence path, as illustrated in Fig. 11.4. The first gradient will label the single quantum coherence with a phase shift of:

$$\phi = -(+1)\gamma G_z \tau \tag{11.26}$$

This phase shift will be transferred to double quantum coherence by the second RF-pulse:

$$\sigma^{+1}e^{i\phi} \rightarrow \sigma^{+2}e^{i\phi} \tag{11.27}$$

The second gradient, of the same strength and length as the first, will induce an additional phase shift of $\phi = -2 \times \gamma(-G_z)\tau$ in the part of the density matrix that

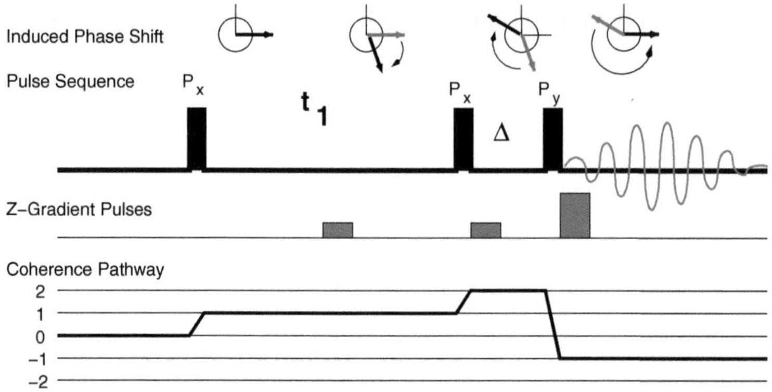

Figure 11.4. Coherence selection in a DQF-COSY experiment using gradients. One of the desired coherence pathways in a DQF-COSY experiment is highlighted by a thick line on the coherence scale shown in the lower part of the figure. Three field gradients are applied for coherence selection, one during t_1, one during the delay Δ, and one at the beginning of the acquisition time. The gradients are all positive in this example. The induced phase shifts for a particular position within the sample are shown on the top of the diagram. The first gradient induces a 70° phase shift. The second gradient increases the phase shift to 210° (70 + 140). The last gradient pulse reverses the phase shift that was induced by the first two gradients, resulting in detectable signal from this particular coherence path.

is represented by σ^{+2}. Therefore, the overall induced phase shift after the second gradient pulse is:

$$\phi = -[1+2]\,\gamma G_z \tau \tag{11.28}$$

After the application of the third RF-pulse it will be necessary to refocus this phase shift by applying a third gradient. At this point in the experiment the desired density matrix has a phase shift of $\phi = -[1+2]\,\gamma G_z \tau$ associated with it. Therefore a phase shift of opposite sign has to be induced in the density matrix. This can be attained by the application of a gradient that is either three times the length of the first gradient, or one that is three times as strong. This gradient induces the following change in σ^{-1}:

$$\sigma^{-1} e^{-[1+2]\gamma G_z \tau} \overset{G_3}{\rightarrow} \sigma^{-1} e^{+3\gamma G_z \tau} e^{-[1+2]\gamma G_z \tau} = \sigma^{-1} \tag{11.29}$$

Therefore, only magnetization that follows the coherence levels of +1, +2, and -1 will be refocused by the last gradient pulse and give rise to a detectable signal.

11.2.3 Coherence Selection by Gradients in Heteronuclear NMR Experiments

Pulsed-field gradients can also be used to select particular coherence paths in heteronuclear NMR experiments. For example, consider the HSQC pulse sequence shown in Fig. 11.5. The application of two gradient pulses, G_1 and G_2 will select coherence paths in which the magnetization is transverse on the X-spin during t_1 and then transverse on the proton during detection. Both gradient pulses are applied within a spin-echo sequence (e.g. $\tau - 180^\circ - \tau$) to prevent chemical shift evolution during

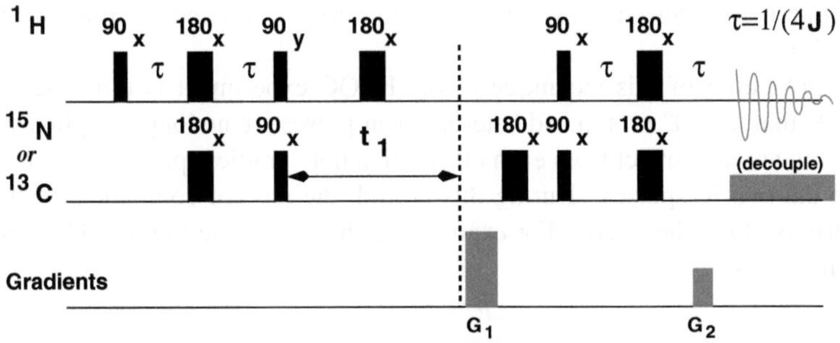

Figure 11.5. Use of pulsed-field gradients for coherence path selection in a HSQC experiment. The application of the first gradient (G_1) induces a phase shift in the various coherences that comprise the density matrix. If the ratio of the two gradients is γ_X/γ_H, then heteronuclear single quantum coherence that is present during t_1 is refocused as detectable proton magnetization by the second gradient.

the application of the gradient. In the case of the first gradient pulse, the spin-echo delay need only be as long as G_1 plus its recovery time. In the case of the second gradient pulse, the gradient is applied during the normal proton-heteronuclear refocusing period, τ.

During the t_1 evolution time the density matrix that represents the heteronuclear magnetization consists of σ^{+1} and σ^{-1}. Application of the first gradient, G_1, will induce a phase shift of $e^{-1\gamma_X G_1}$ and $e^{+1\gamma_X G_1}$, respectively, in each of these coherences. The factor γ_X is explicitly written to take into account the fact that the precessional frequency is proportional to γ (See Eq. 11.18). The second gradient will induce a phase shift of $e^{+1\gamma_H G_2}$ in the detectable proton magnetization, σ_H^{-1}. If the gradient strengths are adjusted such that

$$\frac{G_1}{G_2} = \frac{\gamma_H}{\gamma_X} \tag{11.30}$$

then the phase shift induced in σ_X^{+1} at the end of t_1 will be refocused by the gradient applied during the second INEPT period. Consequently, the only magnetization that will give rise to a detectable signal must have been transverse and associated with the X-spin during t_1 and then transfered to the proton for detection.

11.2.3.1 Coherence Rejection by Pulse-Field Gradients (z-filters)

Direct selection of individual coherences is very effective at suppressing unwanted signals. Unfortunately, direct selection of coherence leads to loss of signal because the phase shift that is induced in the σ^{-1} component of the density matrix by the first gradient pulse *cannot* be rephased by the second gradient. Consequently, one-half of the signal is lost by the gradient selection. Signal loss can be avoided by utilizing gradients to dephase unwanted coherences without affecting the desired coherences [15]. Recall that zero-quantum coherences (σ^o) are not affected by field gradients. Consequently, if the desired magnetization is stored along the z-axis during the application of a gradient pulse, it will not be affected by the gradient. In the case of a single spin, a gradient will have no effect when $\rho = I_z$. For two spins, a gradient will have no effect when $\rho = 2I_zS_z$. Gradient pulses that are applied to retain z-states of the magnetization are referred to as z-filters in the case $\rho = I_z$, and zz-filters in the case of $\rho = 2I_zS_z$.

The application of this technique to the HSQC experiment is illustrated in Fig. 11.6. Note that the 90° proton and heteronuclear pulses are no longer applied simultaneously, but are now offset from each other and a field gradient pulse has been applied between the two RF-pulses. During this period, the desired component of the density matrix is along the z-axis. For example, at the end of the first INEPT period the density matrix is:

$$\rho = 2I_xS_z \tag{11.31}$$

the 90° proton pulse along the y-axis converts this to:

$$\rho = 2I_zS_z \tag{11.32}$$

which is not affected by the applied pulsed-field gradient. The solvent signal, on the other hand, will be transverse during the application of the gradient and will thus receive a spatially dependent phase shift that will greatly attenuate its contribution to the final detected signal.

11.2.3.2 Reduction of Artifacts in 180° Pulses with Field Gradients.

Refocusing pulses in heteronuclear experiments can introduce artifacts into the spectrum if they do not completely invert the spins. Placement of a matched pair of gradient pulses on either of side of the 180° pulse will dephase artifacts that are generated from a non-ideal pulse. Examples of this technique are shown in Fig. 11.6, involving pairs of gradients G1 and G2 plus G5 and G6.

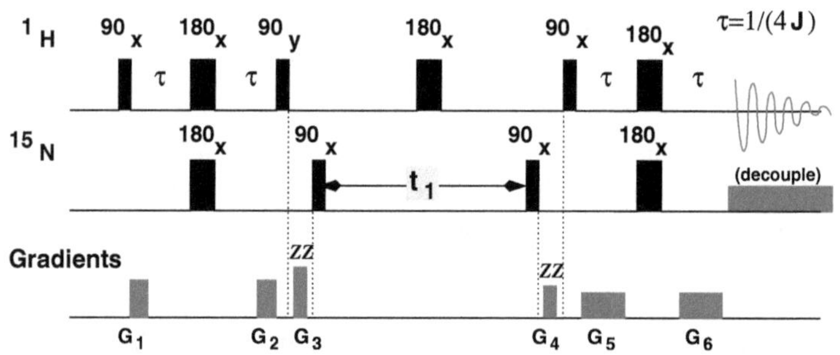

Figure 11.6. Artifact suppression with pulsed-field gradients in heteronuclear experiments. Pulsed-field gradients can be used to suppress artifacts by either the rejection of coherences (z-filter) or by the symmetrical placement of gradient pulses around 180° pulses. The two gradient pulses that act as zz-filters are G_3 and G_4. These are usually of different strengths such that G_4 does not accidentally refocus magnetization that was defocused by G_3. G_1 and G_2 serve to remove artifacts from the 180° proton and heteronuclear pulses during the first INEPT period. These two gradients *must* be the same amplitude. Gradient pulses G_5 and G_6 serve the same purpose during the second INEPT period. This pair of gradient pulses must also have the same amplitude, however the amplitudes of the G_1, G_2 pair need not be the same as the G_5, G_6 pair. Note that all of these gradient pulses have no effect on the normal magnetization transfers associated with the HSQC experiment, they simply remove artifacts and signals from protons that are not coupled to heteronuclear spins.

It is easy to show that the pair of gradients has no effect on the desired evolution of magnetization during these periods. Prior to the pulse, the density matrix associated with the proton magnetization is:

$$\rho = \sigma^{+1} + \sigma^{-1} \tag{11.33}$$

After the first gradient, but before the 180° pulse, a phase shift, G, has been induced in the magnetization by the gradient, giving:

$$\rho = \sigma^{+1}e^{-iG} + \sigma^{-1}e^{+iG} \tag{11.34}$$

The 180° pulse interchanges the values +1 and -1 coherences, as follows:

$$\begin{bmatrix} 0 & e^{-iG} \\ e^{+iG} & 0 \end{bmatrix} \xrightarrow{P_{180}} \begin{bmatrix} 0 & e^{+iG} \\ e^{-iG} & 0 \end{bmatrix} \tag{11.35}$$

giving the density matrix after the 180° pulse, but before the second gradient,

$$\rho = \sigma^{+1}e^{+iG} + \sigma^{-1}e^{-iG} \tag{11.36}$$

After the second gradient, these phase shifts are reversed, giving the original density matrix:

$$\rho = \sigma^{+1}e^{-iG}e^{+iG} + \sigma^{-1}e^{+iG}e^{-iG} = \sigma^{+1} + \sigma^{-1} \tag{11.37}$$

However, in the case of an imperfect pulse, some of the transverse magnetization may be converted to longitudinal magnetization. Since longitudinal magnetization is not affected by the second field gradient (σ^0) its phase will not be refocused by the second gradient and will not appear as observable magnetization.

11.3 Coherence Selection Using Phase Cycling

Phase cycling involves acquiring and co-adding the free induction decays from a number of experiments that are identical in *all* aspects, *except* the phases of one or more RF-pulses and the phase of the receiver. In contrast to field gradients, coherence editing is not obtained within a single scan, rather it is the summation of the data from all of the individual scans that results in the cancellation of unwanted signals. For example, if an experiment uses a four step phase cycle for one of the pulses then four individual FIDs will be acquired with pulse phases of 0, $\pi/2$, π, and $3\pi/2$. These four FIDs are added together such that the signal associated with the desired coherence path add constructively while signals associated with undesired paths are canceled by the summation.

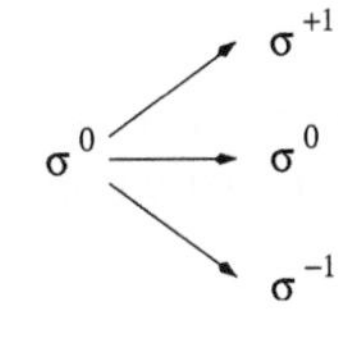

Figure 11.7. Coherence changes induced by RF-pulses. This figure shows the effect of a non-ideal 90° pulse on zero-quantum coherences. The pulse creates a mixture of zero and single quantum coherences.

11.3.1 Coherence Changes Induced by RF-Pulses

The following sections develop the theory of coherence selection by phase shifting of RF-pulses. The treatment follows Bodenhausen et al [20] and the reader is referred to the original publication for more details.

The application of RF-pulses to the spins generates new coherence levels from existing ones. For example, if the first 90° pulse in the COSY or DQF-COSY experiment is ideal, it will cause the following changes in coherences:

$$\sigma^0 \xrightarrow{P_1} \sigma^{+1} + \sigma^{-1} \tag{11.38}$$

If the 90° pulse is not ideal, a fraction of σ^0 will remain as σ^0, as indicated in Fig. 11.7.

In general, the effect of a pulse, P_i, on a single coherence is to generate a manifold of coherences:

$$P_i \sigma^p(t_i^-) P_i^{-1} = \sum_{p'} \sigma^{p'}(t_i^+) \tag{11.39}$$

where $\sigma^p(t_i^-)$ represents the density matrix before the pulse, and $\sum_{p'} \sigma^{p'}(t_i^+)$ represents ρ after the pulse.

The key to coherence editing using phase cycling is that a change in phase of the pulse will produce a phase shift of the coherences that are produced by the pulse, providing a means to label each coherence with a known phase shift.

11.3.1.1 Phase Shifts of Pulses are Z-rotations

To determine how changing the phase of the pulse generates a phase shift of coherence levels it is necessary to write a general expression for a phase shifted pulse and then evaluate the effect of this pulse on an arbitrary coherence level.

A change in the phase of a pulse is equivalent to the rotation about the z-axis of the operators that describe the pulse. For example, if the phase of a excitation pulse in an experiment is cycled in the following manner,

$$\mathrm{P}_x, \mathrm{P}_y, \mathrm{P}_{-x}, \mathrm{P}_{-y}$$

Then the pulse along the y-axis is related to the original pulse along the x-axis by a $90°(\pi/2)$ rotation about the z-axis, as follows:

$$P_y = e^{-iI_z\frac{\pi}{2}} P_x e^{iI_z\frac{\pi}{2}} \tag{11.40}$$

In a similar manner, the pulse along the minus x-axis is obtained by applying a $180°(\pi)$ rotation to the original pulse:

$$P_{-x} = e^{-iI_z\pi} P_x e^{iI_z\pi} \tag{11.41}$$

The effect of phase shifted pulses on the density matrix is obtained by applying the rotated pulse to the density matrix. For example, if an arbitrary density matrix, ρ, is transformed to ρ_x by an x-pulse with a flip angle of β as follows:

$$e^{-i\beta I_x} \rho e^{i\beta I_x} \rightarrow \rho_x \tag{11.42}$$

then the effect of a y-pulse on the same initial density matrix is given by:

$$\left[e^{-iI_z\pi/2}e^{-i\beta I_x}e^{iI_z\pi/2}\right]\rho\left[e^{-iI_z\pi/2}e^{i\beta I_x}e^{iI_z\pi/2}\right] \rightarrow \rho_y \tag{11.43}$$

This equation can be written for any arbitrary pulse, P_o, phase shifted by any arbitrary amount, ϕ:

$$e^{-i\phi I_z}P_o e^{+i\phi I_z}\rho e^{-i\phi I_z}P_o^{-1}e^{+i\phi I_z} \rightarrow \rho_\phi \tag{11.44}$$

Since any density matrix, ρ, can always be written as a linear sum of difference coherences (e.g. $\rho = \sum_{p=-n}^{+n} a_p\sigma^p$) it is possible to simplify the analysis by considering a single arbitrary coherence, σ^p.

Recalling that the effect of z-rotations on coherence levels is:

$$e^{-i\phi I_z}\sigma^p e^{+i\phi I_z} = \sigma^p e^{-ip\phi} \tag{11.45}$$

then, the following is true:

$$e^{+i\phi I_z}\sigma^p e^{-i\phi I_z} = \sigma^p e^{+ip\phi} \tag{11.46}$$

The following shows the effect of a phase shifted pulse when applied to a coherence level of p, to give ρ_ϕ:

$$\begin{aligned}\rho_\phi &= \left[e^{-i\phi I_z}P_o e^{+i\phi I_z}\right]\sigma^p\left[e^{-i\phi I_z}P_o^{-1}e^{+i\phi I_z}\right]\\ &= e^{-i\phi I_z}P_o\left[e^{+i\phi I_z}\sigma^p e^{-i\phi I_z}\right]P_o^{-1}e^{+i\phi I_z}\\ &= e^{-i\phi I_z}P_o\left[\sigma^p e^{+ip\phi}\right]P_o^{-1}e^{+i\phi I_z}\\ &= e^{-i\phi I_z}\left[P_o\sigma^p P_o^{-1}\right]e^{+i\phi I_z}e^{+ip\phi}\end{aligned} \tag{11.47}$$

Assuming that the application of a pulse to a single coherence generates a manifold of different coherences after the pulse:

$$P_o\sigma^p P_o^{-1} \rightarrow \sum_{p'}\sigma^{p'}$$

then,

$$\begin{aligned}\rho_\phi &= e^{-i\phi I_z}\sum_{p'}\sigma^{p'}e^{i\phi I_z}e^{+ip\phi}\\ &= \sum_{p'}\left[e^{-i\phi I_z}\sigma^{p'}e^{+i\phi I_z}\right]e^{+ip\phi}\\ &= \sum_{p'}\sigma^{p'}e^{-ip'\phi}e^{+ip\phi}\\ &= \sum_{p'}\sigma^{p'}e^{-i\Delta p\phi}\end{aligned} \tag{11.48}$$

where ϕ is the phase shift of the pulse relative to P_o and Δp is the change in coherence level caused by the pulse ($\Delta p = p(t_i^+) - p(t_i^-)$).

Equation 11.48 shows that if a pulse produces a coherence jump of Δp, from σ^p to $\sigma^{p'}$, and if that pulse is phase shifted by an angle ϕ, then the coherence $\sigma^{p'}$ will be shifted in phase by $-i\Delta p\phi$ due to phase shifting of the pulse. This is the key step in phase encoding the magnetization for coherence selection.

11.3.1.2 Description of Coherence Pathways Using Coherence Jumps

Each coherence path in an experiment can be defined by either the coherences present during various time periods between pulses or *equivalently* by RF-pulse induced coherence jumps. The former description is useful for coherence selection with pulsed-field gradients while the latter is more useful for coherence selection by phase cycling of RF-pulses.

For example, in the COSY experiment there are three coherence paths that connect the initial state ($\rho = \sigma^0$) to the final detected coherence level, σ^{-1}. The location of these coherence jumps in the pulse sequence are shown in Fig. 11.8. The first two paths are desirable, while the third path will give rise to an artifact in the spectrum.

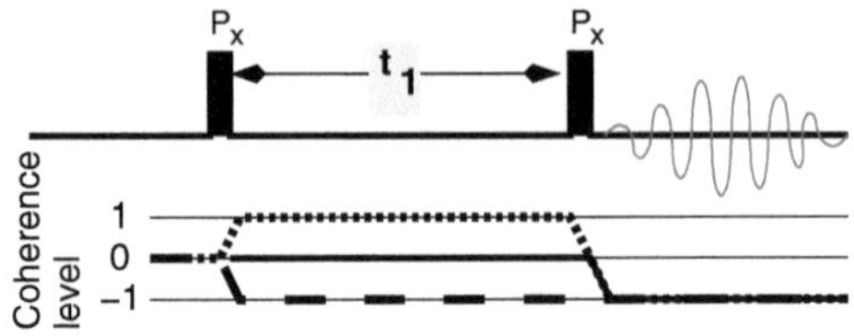

Figure 11.8. Coherence jumps in the COSY experiment. Three possible paths of coherence changes in the COSY experiment are illustrated. Note that only paths that end at a coherence level of -1 will give a detectable signal. The first path is represented by the dotted line and is described by coherence jumps of $\Delta p_1 = +1,\ \Delta p_2 = -2$. The second path is represented by the dashed line and consists of coherence jumps of $\Delta p_1 = -1,\ \Delta p_2 = 0$. Both of these paths will give rise to a normal COSY spectrum. The solid line shows the third undesired coherence path, with $\Delta p_1 = 0, \Delta p_2 = -1$.

Phase encoding of the different coherences can be achieved by changing the phase of the RF-pulses. The resultant phase shift of the coherence is the product of the phase shift of the RF-pulse, ϕ, and the change in coherence, Δp. Since a single coherence pathway is defined by a unique set of coherence jumps, it is possible to select out specific pathways by the appropriate phase cycling of pulses in much the same way gradients were used to select coherence paths. The phase shift that is associated with the desired pathway is refocused at the end of the experiment such that the desired coherence paths will add constructively to the overall signal.

The aggregate phase of the signal that is associated with a particular coherence path depends on the coherence jumps associated with that pathway and the phase shifts of the pulses at each coherence change. For every coherence path (j), the *change* in the coherence levels can be represented as a vector:

$$\vec{\Delta p_j} = (\Delta p_1, \Delta p_2, \ldots) \tag{11.49}$$

For the first path listed above for the COSY experiment (Fig. 11.8): $\vec{\Delta p} = (+1,\ -2)$. The phase shifts associated with each RF-pulse can also be written as a vector:

$$\vec{\phi} = (\phi_1, \phi_2, \ldots) \tag{11.50}$$

The total phase shift that is accumulated over any particular coherence path, j, is just the dot product of these two vectors:

$$\Phi_j = \vec{\Delta P} \cdot \vec{\phi} \tag{11.51}$$

For example, the total phase associated with the first coherence path in the COSY experiment is:

$$\Phi_1 = \Delta P_1 \phi_1 + \Delta P_2 \phi_2 = (+1) \times \phi_1 + (-2) \times \phi_2 \tag{11.52}$$

In general, the signal associated with the jth path is:

$$S_j = S_j^0 e^{-i\Phi_j} \tag{11.53}$$

where S_j^o represents the signal that would be obtained with no phase shifting of pulses (e.g. $\vec{\phi} = 0$).

11.3.2 Selection of Coherence Pathways

The detected signal will contain a contribution for each coherence path in the experiment:

$$S = \sum_j S_j = \sum_j S_j^o e^{-i\Phi_j} \tag{11.54}$$

Using the COSY experiment discussed in Fig. 11.8 as an example, the final signal from any given scan is:

$$S = S_1^o e^{-i\Phi_1} + S_2^o e^{-i\Phi_2} + S_3^o e^{-i\Phi_3} \tag{11.55}$$

The first two terms represent signals from desirable coherence paths, while the third term represents the signal from an undesirable path that will be eliminated by the phase cycle.

In contrast to coherence selection by pulsed-field gradients, the signals with different phases, S_j, do not cancel. Therefore, the selection of a particular coherence path is accomplished by the selection of pulse phases and receiver phase such that the desired signal always adds for each scan of the phase cycle while the signals from undesired paths sum to zero at the completion of the phase cycle.

11.3.2.1 Defining Phase cycles

The required phase shift associated with one pulse, ϕ, will depend on how many coherence jumps have to be filtered out by the particular pulse. A phase shift of:

$$\phi = \frac{k2\pi}{N} \quad k = 0,\ 1....(N-1) \tag{11.56}$$

will select out coherence orders: $\Delta p \pm nN$; $n = ..., -2, -1, 0, 1, 2, ...$, provided that the receiver phase is set to $-i\Delta p \times \phi$. This setting of the receiver phase will insure that desired signal will always appear as the real signal to the receiver, resulting in the co-addition of this signal over the entire phase cycle. In contrast, all other signals will sum to zero.

As an example, consider phase cycling of a single pulse with the intent of selecting a coherence change, Δp, of -2. If N is set to 3, corresponding to a three step phase cycle (pulse phases of 0, $2\pi/3$, and $4\pi/3$), and the receiver phase is set to 0, $-4\pi/3$, and $-8\pi/3$, then coherence jumps of -2 $\pm 3n$ will be retained, as shown in Fig. 11.9.

The result illustrated in Fig. 11.9 can be easily proven in general. First assume that an N-step phase cycle results in the positive selection of coherence jumps, $\Delta p \pm nN$. As an example let $\Delta p' = \Delta p + N$. To select this coherence jump it would be necessary to set the receiver phase to $e^{-i\Delta p\phi}$. This receiver phase is represented below by phase shifting the signal in the opposite direction, e.g. multiplying it by $e^{+i\Delta p\phi}$ (this term is underlined in Eq. 11.58). The detected signal will be the sum of all three signals, one for each phase of the pulse. The phase shift introduced in the signal by each RF-pulse is:

$$e^{-i\Delta p \frac{2\pi}{N} k} \tag{11.57}$$

where $k = 0, 1, 2$ and $N = 3$.

The net signal is given by the sum over the phase cycle:

$$\begin{aligned} S &= \sum_{k=0}^{N-1} \underline{e^{i\Delta p' \frac{2\pi}{N} k}} S_0 e^{-i\Delta p \frac{2\pi}{N} k} \\ &= \sum_{k=0}^{N-1} e^{i(\Delta p + N) \frac{2\pi}{N} k} S_0 e^{-i\Delta p \frac{2\pi}{N} k} \\ &= \sum_{k=0}^{N-1} e^{i\Delta p \frac{2\pi}{N} k} e^{iN \frac{2\pi}{N} k} S_0 e^{-i\Delta p \frac{2\pi}{N} k} \end{aligned} \tag{11.58}$$

$e^{iN\frac{2\pi}{N}k} = e^{i2\pi k} = 1$ since $k \in I$, therefore,

$$\begin{aligned} S &= \sum_{k=0}^{N-1} e^{i\Delta p \frac{2\pi}{N} k} S_0 e^{-i\Delta p \frac{2\pi}{N} k} \\ &= N S_0 \end{aligned} \tag{11.59}$$

Now consider the case of rejection of the signal: $\Delta p' = \Delta p + m \quad m \neq N$

$$\begin{aligned} S &= \sum_{k=0}^{N-1} S_0 e^{i\Delta p' \frac{2\pi}{N} k} e^{-i\Delta p \frac{2\pi}{N} k} \\ &= S_0 \sum_{k=0}^{N-1} e^{i(\Delta p + m) \frac{2\pi}{N} k} e^{-i\Delta p \frac{2\pi}{N} k} \\ &= S_0 \sum_{k=0}^{N-1} e^{i\Delta p \frac{2\pi}{N} k} e^{im \frac{2\pi}{N} k} e^{-i\Delta p \frac{2\pi}{N} k} \end{aligned} \tag{11.60}$$

$$= S_0 \sum_{k=0}^{N-1} e^{im \frac{2\pi}{N} k} \tag{11.61}$$

To show that this sum is zero, consider the following:

$$\begin{aligned} S &= 1 + e^{im\frac{2\pi}{N}} + e^{im\frac{2\pi}{N}2} + ... + e^{im\frac{2\pi}{N}(N-1)} \\ e^{im\frac{2\pi}{N}} S &= e^{im\frac{2\pi}{N}} + e^{im\frac{2\pi}{N}2} + ... + e^{im2\pi} \\ S - e^{im\frac{2\pi}{N}} S &= (1 - e^{im2\pi}) \\ S(1 - e^{im\frac{2\pi}{N}}) &= 0 \end{aligned} \tag{11.62}$$

If $m = N$ then $S(1 - e^{im\frac{2\pi}{N}}) = S \cdot 0$ and S can have any value. If $m \neq N$ then $S = 0$ since $(1 - e^{im\frac{2\pi}{N}}) \neq 0$.

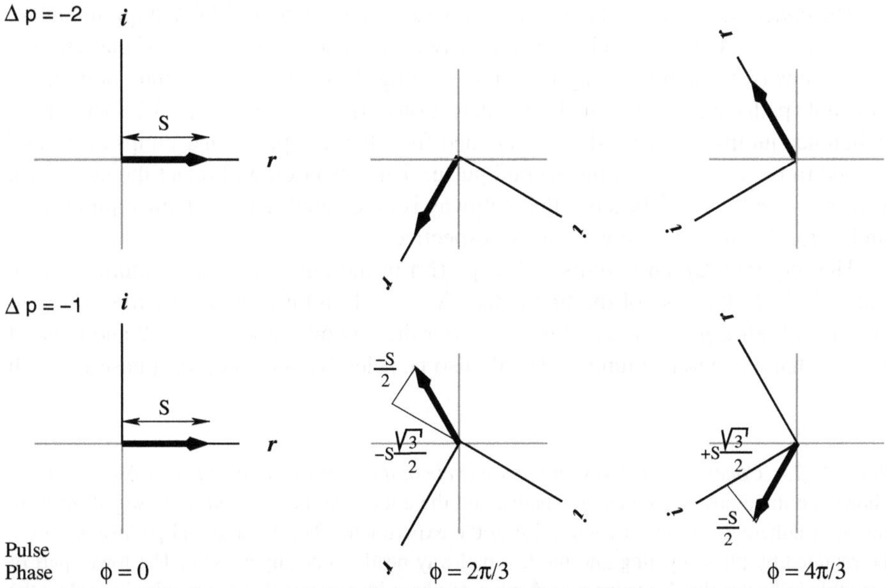

Pulse phase	*Signal*	*Phase of Receiver*	*Real Channel*	*Imaginary Channel*
$\mathbf{\Delta p = -2}$				
0	$Se^{-i(-2)\cdot 0}$	0	S	0
$2\pi/3$	$Se^{-i(-2)\cdot 2\pi/3}$	$-(-2)\cdot 2\pi/3$	S	0
$4\pi/3$	$Se^{-i(-2)\cdot 4\pi/3}$	$-(-2)\cdot 4\pi/3$	S	0
Net sum			$3S$	0
$\mathbf{\Delta p = -1}$				
0	$Se^{-i(-1)\cdot 0}$	0	S	0
$2\pi/3$	$Se^{-i(-1)\cdot 2\pi/3}$	$-(-2)\cdot 2\pi/3$	$-S/2$	$-S\frac{\sqrt{3}}{2}$
$4\pi/3$	$Se^{-i(-1)\cdot 4\pi/3}$	$-(-2)\cdot 4\pi/3$	$-S/2$	$+S\frac{\sqrt{3}}{2}$
Net sum			0	0

Figure 11.9. Selection and rejection of coherences in the COSY experiment by phase cycling. The upper series of diagrams show the signal for pulse phases of 0, $2\pi/3$, and $4\pi/3$ for a coherence jump of $\Delta p = -2$. The lower series of diagrams show the signal for a coherence jump of -1. The real and imaginary axis of the receiver are also indicated. The receiver phase is set to: $-\Delta p\phi$, where $-\Delta p$ is the desired coherence jump and ϕ is the phase of the RF-pulse. In this example, the receiver phase has been set to select the -2 coherence jump. Consequently, the signal associated with that coherence is always detected as the real signal by the receiver. In the case of a coherence jump of -1, the signal is not in phase with the receiver and the net sum of the real and imaginary channels is zero after completing all three scans of the phase cycle. The first column of the table shows the phase of the pulse, the second column shows the signal, including the phase shift introduced by the coherence jump and the change in the phase of the pulse. The third column shows the receiver phase setting for each scan. The last two columns show the signal for the real and imaginary channels, respectively. Note that the signals add in the case of $\Delta p = -2$ but cancel in the case of $\Delta p = -1$.

11.3.2.2 Example - Coherence Selection in the COSY Experiment

The coherence paths and associated coherence jumps in the COSY experiment are shown in Part A of Table 11.2. The first two give rise to the desired signal and the third needs to be suppressed by the phase cycling. It is necessary to retain both of the desirable paths to allow for quadrature detection during t_1 (see Chapter 12). Therefore, coherence jumps of ± 1 need to be retained from the first pulse, and jumps of -2, and 0 need to be retained from the second pulse. This selection will reject the coherence path $\sigma^o \rightarrow \sigma^0 \rightarrow \sigma^{-1}$ because that pathway is associated with coherence jumps of 0 and -1 for the first and second pulses, respectively.

The required Δp and number of steps (N) in the phase cycle are summarized in Table 11.2. In the case of the first pulse, $\Delta p = +1$ and it is necessary to set N=2 to select for both $\Delta p = +1$ and $\Delta p = -1$. For the second pulse $\Delta p = -2$ and N is set to 2 so that a coherence jump of 0 will also be selected. The receiver phase for each

Table 11.2. Coherence changes and phase cycle in the COSY experiment. Part A of this table shows the three possible coherence paths, and the equivalent description of these pathways in terms of coherence jumps at each pulse in the experiment. The 1st and 2nd pathways should be retained by phase cycling and the last pathway needs to be suppressed. The lower part of the table indicates the Δp value and number of phase increments that are required to select the coherence jumps for *both* desired pathways. The coherence jumps that would be selected by these values of Δp and N are also shown.
Part B shows the four-step phase cycle of this experiment. The phases for the RF-pulses, and the associated receiver phases, are shown. The short-hand notation for pulse and receiver phases is: $0 = 0$, $\pi/2 = 1$, $\pi = 2$, $3\pi/2 = 3$. The receiver phase is calculated using Eq. 11.63, using $\Delta p_1 = +1$ and $\Delta p_2 = -2$. For example, the third step of the phase cycle has pulse phases of $\phi_1 = \pi$ and $\phi_2 = 0$, therefore the receiver phase is: $\varphi_{rec} = -[(+1)(\pi) + (-2)(0)] = -\pi = \pi$. This phase is represented by a 2 in the table.

A:

Coherence Path	*Coherence Jumps* (Δp) *1st Pulse*	*2nd Pulse*
$\sigma^0 \rightarrow \sigma^{+1} \rightarrow \sigma^{-1}$	+1	- 2
$\sigma^0 \rightarrow \sigma^{-1} \rightarrow \sigma^{-1}$	-1	0
$\sigma^0 \rightarrow \sigma^{\ 0} \rightarrow \sigma^{-1}$	0	- 1
Δp	+1	- 2
Number of Steps (N)	2	2
Coherence jumps selected	$\Delta p = ... -3, -1, +1, +3...$	$\Delta p = ... -4, -2, 0, +2...$

B:

	Phase Cycle Step 1	2	3	4
First Pulse	0	0	2	2
Second Pulse	0	2	0	2
Receiver Phase (φ_{rec})	0	0	2	2

set of phase values is calculated from:

$$\begin{aligned} \varphi_{rec} &= -i\vec{\Delta p_j} \cdot \vec{\phi} \\ &= -i(\Delta p_1 \phi_1 + \Delta p_2 \phi_2) \end{aligned} \tag{11.63}$$

The pulse phases and associated receiver phases are given in part B of Table 11.2. Since the phase of each RF-pulse has to be cycled independently of the other, it is necessary to perform a four step phase cycle to obtain all possible permutations of the pulse phases (e.g. 2 × 2). The first two steps in the cycle serve to select the desired coherence jumps at the second pulse when the phase of the *first* pulse is 0. After the phase of the first pulse is changed to 2 (π), it is necessary to repeat the two-step phase cycle of the second pulse to select the desired coherences at the second pulse. After all four steps of the phase cycle are summed the contribution of the undesired $\sigma^o \rightarrow \sigma^o \rightarrow \sigma^{-1}$ coherence path would be completely suppressed.

11.3.3 Phase Cycling in the HMQC Pulse Sequence

Phase cycling can also be used for coherence editing in heteronuclear pulse sequences. To simplify the analysis, the coherence order changes are followed independently for each type of spin. In contrast to coherence selection with gradients, the phase shifts in the coherences that are introduced by phase cycling are *independent* of the type of nucleus. Consequently, the overall receiver phase that is required to select the desired coherence pathways is simply $-i\vec{\Delta p} \cdot \vec{\phi}$.

Coherence levels in the HMQC Experiment

The HMQC pulse sequence is shown in Fig. 11.10. The coherence levels that are present at various locations in the sequence can be inferred from the Cartesian representation of the density matrix:

$$\begin{array}{ccccccccccccc} I_z & \stackrel{P^H_{90}}{\rightarrow} & -I_y & \stackrel{\Delta}{\rightarrow} & I_x S_z & \stackrel{P^S_{90}}{\rightarrow} & & \stackrel{\frac{t_1}{2}\ 180^o_H\ \frac{t_1}{2}}{\rightarrow} & & \stackrel{P^S_{90}}{\rightarrow} & I_x S_z & \stackrel{\Delta}{\rightarrow} & I_y \\ \sigma^o & & \sigma^{\pm1}_H & & & & & \sigma^{\pm1}_H + \sigma^{\pm1}_S & & & \sigma^{\pm1}_H & & \end{array}$$

The coherence changes in the HMQC experiment are illustrate in Fig. 11.10. The first proton pulse creates single quantum proton coherence. The next heteronuclear pulse generates single-quantum X-nucleus coherence that persists throughout the t_1 labeling period. During the t_1 period the 180° pulse on the proton converts the σ^{+1}_H

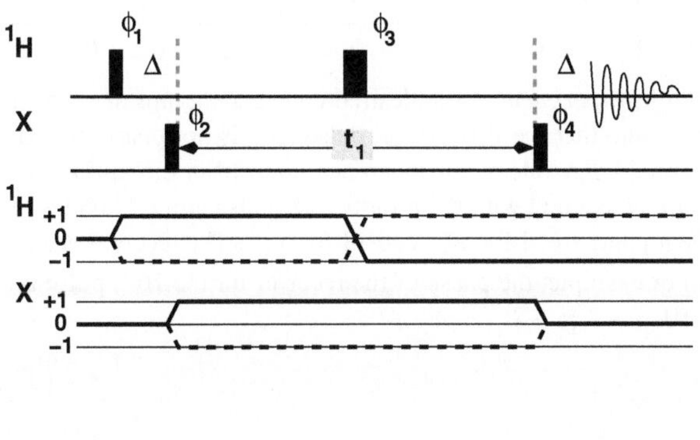

Figure 11.10 Coherence jumps in the HMQC experiment. The separate coherence jumps for the proton and heteronuclear spin are shown for an HMQC experiment. The solid line and the dashed line show the coherence paths that are to be retained by phase cycling.

coherence to σ_H^{-1} coherence, a net change of ± 2. This two level change in the coherence can be understood by writing I_x as the sum of the raising and lowering operators:

$$I_x = \frac{1}{2}(I^+ + I^-) \tag{11.64}$$

The effect of the 180° pulse on I^+ is:

$$I^+ = I_x + iI_y \rightarrow I_x - iI_y = I^- \tag{11.65}$$

or a net change in coherence of -2.

The single quantum coherence associated with the X-nucleus is returned to zero quantum coherence by the second heteronuclear pulse, and the density matrix refocuses to single quantum coherence by the end of the Δ delay.

Defining the Phase Cycle

The first proton pulse is generally not phase cycled since imperfections in this pulse cannot give rise to detectable magnetization. The phases of the three remaining pulses are cycled to remove artifacts. The desired coherence jumps and number of phase shifts that are associated with each pulse are shown in Table 11.3.

The coherence jump associated with the first X-nucleus pulse is +1. However, since a jump of -1 must also be preserved, a two step phase cycle is required. In a similar manner, the desired coherence jumps associated with the last X-nucleus pulse is either -1 (solid line in Fig. 11.10) or +1 (dotted line), required a two step phase cycle. A total of four steps ($N = 4$) are required for the proton 180° pulse since the desired coherence changes are either -2 or +2. Choosing a Δp of +2, and $N = 4$, insures that only Δp values of -2 and +2 will be retained. Note that coherence jumps of -6 and +6 are also retained by this selection since $\Delta p = \Delta p \pm N$, but such high coherence levels cannot be attained with two coupled spins.

Since each phase cycle operates independently of the other phase cycles, a total of 16 scans are required ($2 \times 4 \times 2$) to suppress all of the artifacts. The complete phase cycle, including the receiver phase, is shown in Table 11.3.

		Experiment Number															
Pulse	Δp	1	2	3	4	5	6	7	8	9	1	011	12	13	14	15	16
2	1	0	2	0	2	0	2	0	2	0	2	0	2	0	2	0	2
3	2	0	0	1	1	2	2	3	3	0	0	1	1	2	2	3	3
4	1	0	0	0	0	0	0	0	0	2	2	2	2	2	2	2	2
φ_{rec}	-	0	2	2	0	0	2	2	0	2	0	0	2	2	0	0	2

Table 11.3. Phase cycle of the HMQC experiment. The desired coherence jumps are shown in the second column from the left. Note that the first proton pulse (ϕ_1) is not phase cycled. The first heteronuclear pulse (ϕ_2) should generate a ± 1 coherence jump from the initial σ_X^0. The allowed coherence jumps that are associated with the proton 180° pulse are ± 2 and those associated with the final heteronuclear pulse are -1 for σ_X^{+1} and +1 for σ_X^{-1}. The receiver phase, φ_{rec}, is calculated from $-\vec{\Delta p} \cdot \vec{\phi}$. For example, the phase of the receiver for the 16th pulse is: $-[(1)(2) + (2)(3) + (1)(2)] = -10 = -2 = +2$.

11.4 Exercises

1. Draw the coherence level changes for a triple-quantum filtered COSY (TQF-COSY) experiment. The pulse sequence for this experiment is identical to that of a DQF-COSY. However, only magnetization that can enter a triple-quantum state is retained in the experiment. This experiment, along with higher quantum filtered experiments, are described by [115] and are useful for filtering the DQF-COSY according to amino acid residue type.

2. Determine the phase cycle for a triple-quantum filtered COSY experiment.

3. How would you accomplish the same coherence editing with pulsed-field gradients?

4. A TQF-COSY spectra is acquired on a sample containing glycine and serine in D_2O, what crosspeaks are present in the spectrum? How do they differ from the DQF-COSY?

11.5 Solutions

1. The coherence diagram is shown in Fig. 11.11. There are four desired coherence paths. The coherence changes associated with each of these paths is given in the table below.

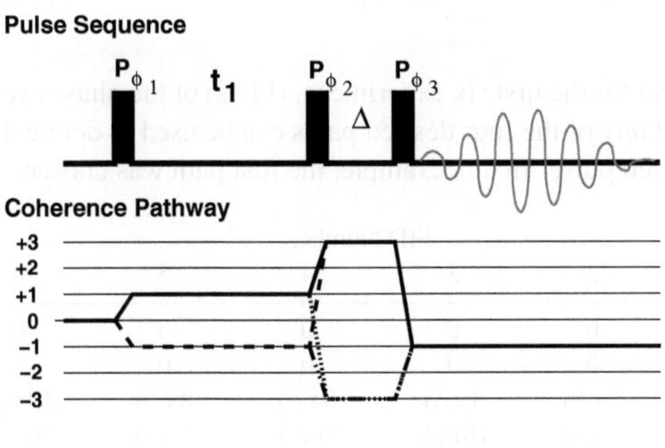

Figure 11.11 Coherence levels in the TQF-COSY Experiment.

2. The four desired coherence paths, and the associated coherence jumps are:

Path	Coherence Jumps 1st Pulse	 2nd Pulse	 3rd Pulse
Path 1 ($\sigma_o \rightarrow \sigma_{+1} \rightarrow \sigma_{+3} \rightarrow \sigma_{-1}$)	+ 1	+2	-4
Path 2 ($\sigma_o \rightarrow \sigma_{+1} \rightarrow \sigma_{-3} \rightarrow \sigma_{-1}$)	+ 1	-4	+2
Path 3 ($\sigma_o \rightarrow \sigma_{-1} \rightarrow \sigma_{-3} \rightarrow \sigma_{-1}$)	- 1	-2	+2
Path 4 ($\sigma_o \rightarrow \sigma_{-1} \rightarrow \sigma_{+3} \rightarrow \sigma_{-1}$)	- 1	+4	-4

The desired coherence jumps associated with the first pulse are ± 1, while the undesired change is zero. Thus a 2 step phase cycle will suffice, with phase shifts of 0 and π. If Δp is set to -1, the two step phase cycle will retain all other coherence changes of $-1 \pm 2n, n \in I$, which will include the jump of $+1$.

The second pulse has to retain the coherence changes of -4, -2, +2, and +4. Since it is necessary to retain both +2 and +4 (or -2 and -4), it is only possible to use a two-step phase cycle, again of 0 and π. If the coherence jump, Δp, is set to +4, then this phase cycle will retain all coherence changes of $+4 \pm 2n, n \in I$, which will include +2, 0, -2, and -4. Note that it is not possible to exclude a coherence jump of 0. The coherence selection by this phase cycle is summarized below, where the desired coherence jumps are set in bold type face, and the allowed coherence jumps are underlined.

<u>-4</u> -3 **<u>-2</u>** -1 <u>0</u> +1 **<u>+2</u>** +3 **<u>+4</u>**

The third pulse should only pass coherence jumps of +2 or -4. Therefore, an N value of 6 is appropriate, giving a six step phase cycle of:

$$0,\ 1\frac{2\pi}{6},\ 2\frac{2\pi}{6},\ 3\frac{2\pi}{6},\ 4\frac{2\pi}{6},\ 5\frac{2\pi}{6} \tag{11.66}$$

The total number of experiments to complete the phase cycling for all three pulses is 24 ($2 \times 2 \times 6$). The receiver phase for each phase combination is given by:

$$\phi_{rec} = -\sum_{l=1}^{3} \Delta p_l \phi_l \tag{11.67}$$

The pulse and receiver phase for the first six experiments (FIDs) of the phase cycle are shown below. Note that *any* of the four desired paths can be used to define the coherence jump, Δp, for each pulse. In this example, the first path was chosen.

		FID Number					
	Δp	1	2	3	4	5	6
Pulse 1 phase	+1	0	0	0	0	0	0
Pulse 2 phase	+2	0	0	0	0	0	0
Pulse 3 phase	-4	0	$2\pi/6$	$4\pi/6$	$6\pi/6$	$8\pi/6$	$10\pi/6$
ϕ_{rec}		0	$8\pi/6$	$16\pi/6$	$24\pi/6$	$32\pi/6$	$40\pi/6$

The next 6 FIDs would have a phase of π for the 1st pulse, resulting in an additional receiver phase of $-\pi$ ($\Delta p = +1$). The subsequent 6 FIDs would have a phase of 0 for the 1st pulse and a phase of π for the 2nd pulse. Since the coherence jump for the second pulse is $+2$, no change in the receive phase is required. The last 6 FIDS would have a phase of π for the 1st pulse and a phase of π for the 2nd pulse, the receiver would be shifted by an additional $-\pi$.

3. Three gradients would be required, as with the DQF-COSY experiment shown in Fig. 11.4. The desired coherence path would have obtain a phase shift of $\mathbf{3}\gamma G_z \tau$ from the second gradient pulse. To refocus the desired signal, the third gradient would have to have a strength that is *four* times that of the previous gradients.

4. The TQF-COSY experiment will filter out any pairs of coupled spins that are not coupled to another spin, as the spin cannot enter a triple quantum state of the form: $I_x S_x K_x$. Since glycine has only two α-protons, its peaks would not be present in the TQF-COSY spectrum. In contrast, the α-proton of serine is coupled to the two β-protons, and can enter a triple quantum state and thus the TQF-COSY spectrum of serine is identical to its DQF-COSY spectrum.

Chapter 12

QUADRATURE DETECTION IN MULTI-DIMENSIONAL NMR SPECTROSCOPY

Quadrature detection provides a means to discriminate between positive and negative frequencies and permits the phasing of the spectrum to obtain pure absorption mode lineshapes. In the case of one-dimensional NMR experiments quadrature detection is attained by routing the signal from the receiver through two different circuits to produce a $cos(\omega t)$ modulated signal, commonly called the real signal, and a $sin(\omega t)$ modulated signal, which is commonly called the imaginary signal. Linear combination of these two signals produces a complex signal, the Fourier transform of which produces a resonance line at a single frequency that can be phased to give a pure absorption mode lineshape.

Quadrature detection in multi-dimensional NMR experiments is equally important for exactly the same reasons as with one-dimensional spectroscopy. Sign discrimination of the frequencies permits placing the transmitter in the center of the indirectly detected spectrum for optimal excitation of the spins. In addition, the observed linewidth will be narrowest if a pure absorption lineshape can be generated.

There are three strategies that have been developed to obtain quadrature detection in the indirectly detected domain. The first strategy, time-proportional-phase-increments (TPPI) was one of the earliest methods of obtaining quadrature detection in multidimensional experiments and is still used in some applications [106]. The TPPI method shifts the apparent frequency of the transmitter to the edge of the spectrum, which allows for discrimination between positive and negative frequencies.

The remaining two techniques of quadrature detection, the hypercomplex method and the echo-antiecho method, are similar in the sense that they both result in the production of cosine and sine modulated data, which can be combined to yield a complex signal in the indirectly detected domain. The hypercomplex method is generally utilized when phase cycling is employed to select coherence pathways. The echo-antiecho method is utilized when pulsed field gradients are used for coherence selection.

The following discussion will focus on heteronuclear two-dimensional NMR experiments, such as the HSQC experiment. However, the discussion is equally applicable to homonuclear experiments, such as the COSY and TOCSY experiments.

12.1 Quadrature Detection Using TPPI

TPPI stands for **T**ime **P**roportional **P**hase **I**ncrements. This method generates an *apparent* shift of the transmitter to the edge of the spectrum by altering the apparent frequencies of the peaks. It is implemented by shifting the phase of *one* of the 90° heteronuclear pulse that bracket the t_1 evolution period. The pulse whose phase is shifted is often referred to as the *quadrature pulse*, as illustrated in Fig. 12.1. The phase of the quadrature pulse is incremented by $\pi/2$ for *each* t_1 point, i.e., the phase of the quadrature pulse is set to x for t_1=0, y for $t_1 = \Delta t_1$, $-x$ for $t_1 = 2\Delta t_1$, $-y$ for $t_1 = 3\Delta t_1$, etc. Note that this phase increment is independent of the phase cycling that would be used for coherence selection.

The effect of this phase shifting on the detectable magnetization can be easily evaluated using the product operator representation of the density matrix. In the HSQC experiment, the magnetization at the end of the t_1 period is:

$$2I_zS_ycos(\omega_S t_1) - 2I_zS_xsin(\omega_S t_1) \tag{12.1}$$

If the phase of the second 90° heteronuclear pulse is shifted for each t_1 point then the following density matrices occur after the pair of proton and heteronuclear 90° pulses that begin the second INEPT period:

$$\begin{aligned}
2I_zS_ycos(\omega_S t_1) - 2I_zS_xsin(\omega_S t_1) &\xrightarrow{H_{90(X)}N_{90(X)}} -2I_yS_zcos(\omega_S t_1)\\
2I_zS_ycos(\omega_S t_1) - 2I_zS_xsin(\omega_S t_1) &\xrightarrow{H_{90(X)}N_{90(Y)}} -2I_yS_zsin(\omega_S t_1)\\
2I_zS_ycos(\omega_S t_1) - 2I_zS_xsin(\omega_S t_1) &\xrightarrow{H_{90(X)}N_{90(-X)}} +2I_yS_zcos(\omega_S t_1)\\
2I_zS_ycos(\omega_S t_1) - 2I_zS_xsin(\omega_S t_1) &\xrightarrow{H_{90(X)}N_{90(-Y)}} +2I_yS_zsin(\omega_S t_1)
\end{aligned} \tag{12.2}$$

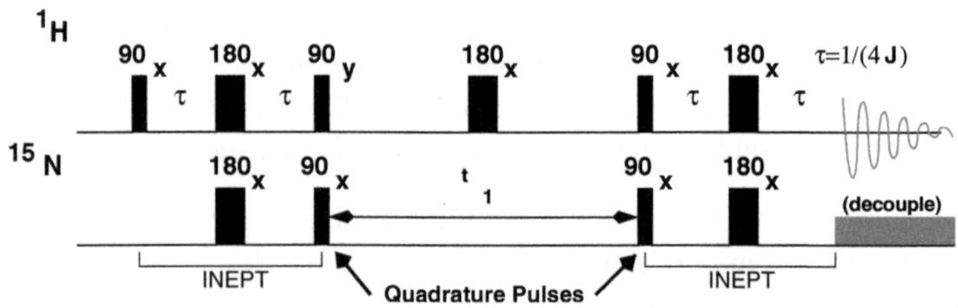

Figure 12.1. Phase shifting of pulses for quadrature detection. The pulses that bracket the t_1 evolution period are indicated as *quadrature pulses* . In this particular experiment *either* one, but not both, of these pulses can be phase shifted to produce quadrature detection. Note, that some experiments may require phase shifting of more than one pulse prior to the t_1 period.

where only the detectable single quantum term, $2I_yS_z$ has been kept. The general expression for the $2I_yS_z$ component of the density matrix for any t_1 time point is:

$$\rho_n \propto cos(\omega_S n\Delta t_1 + n\frac{\pi}{2}) \tag{12.3}$$

This can be rearranged as follows:

$$\begin{aligned} \rho_n &\propto cos(n(\omega\Delta t + \frac{\pi}{2})) \\ &\propto cos\left[n\Delta t\left[\omega + \frac{\pi}{2\Delta t}\right]\right] \\ &\propto cos\left[n\Delta t\left[\omega + \omega_{TPPI}\right]\right] \end{aligned} \tag{12.4}$$

The second term in the argument of the cosine function shows that the frequency of the signal has by shifted by ω_{TPPI} because of the phase shifting of the quadrature pulse.

After Fourier transformation in t_1, the spectra will contain peaks at $-\omega$ and $+\omega$ due to the Fourier transform of $cos(\omega t)$. The peaks that occur at $-\omega$ are an artifact of the Fourier transform and are often referred to as *folded* peaks. To prevent the folded peaks from overlapping with the actual peaks the frequency shift, ω_{TPPI}, is selected to be at least 1/2 the spectral width. This change in the apparent resonance frequencies will place all of the peaks to one side of the transmitter, as illustrated in Panel C of Fig. 12.2 and none of the folded peaks will overlap with the true peaks.

To generate an apparent frequency shift, ω_{TPPI} that is equal to one-half of the width, f_{sw}, requires that

$$\omega_{TPPI} = \frac{2\pi f_{sw}}{2} = \frac{\pi}{2\Delta t} \tag{12.5}$$

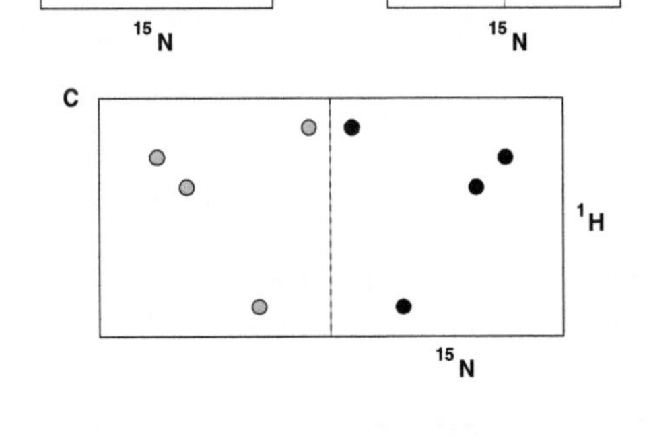

Figure 12.2 Quadrature detection in 2D with TPPI. Panel A shows a hypothetical two-dimensional ^{1}H-^{15}N HSQC spectrum that contains four resonance lines. Panel B shows the spectrum after Fourier transform of $cos(\omega_N t_1)\ e^{i\omega_H t_2}$ in the t_2 and the t_1 dimensions. The folded peaks are shaded light gray. In practice the folded and true peaks cannot be distinguished. Panel C shows the spectrum obtained using the TPPI method of quadrature detection. The peaks have been shifted 1/2 a sweepwidth to the right. Fourier transform of this signal also yields folded peaks, but they do not overlap with the true peaks. The left part of the spectrum would be discarded after processing.

Therefore, the t_1 dwell time must be equal to:

$$\Delta t = \frac{1}{2f_{sw}} \tag{12.6}$$

This is half of the normal dwell time and thus if TPPI is used for quadrature detection the spectra have to be sampled twice as frequently in the t_1 domain, doubling the number of FIDs that have to be acquired for digitization of the indirectly detected domain.

12.2 Hypercomplex Method of Quadrature Detection

The hypercomplex method of quadrature detection was originally described by States et al. [154] and is ofter referred to as the States method of quadrature detection. The hypercomplex method uses the same strategy for the generation of quadrature detection as the directly detected signal. Two separate signals, one that is amplitude modulated as $cos(\omega t_1)$, and one that is modulated as $sin(\omega t_1)$, are collected. The separate $cos(\omega_S t_1)$ and $sin(\omega_S t_1)$ signals provide orthogonal functions that can be combined to generate a single peak at ω_S. Furthermore, the availability of both components permits the phasing of the spectrum in t_1 by linear combination of the two components, as discussed in Chapter 2 for the directly detected domain.

The separate *cosine* and *sine* modulated signals are obtained by shifting the phase of one of the RF-pulses that bracket the indirect (e.g. t_1) evolution period by 90° and collecting a separate FID with the *same* evolution delay time; the only change in the experiment is a 90° shift of the quadrature pulse. The first phase setting generates the $cos(\omega_S t_1)$ signal and the second phase setting generates the $sin(\omega_S t_1)$ signal. The generation of each term can be understood by considering the effect of shifting the phase of the second heteronuclear pulse on the component of the density matrix that gives rise to the observable signal. The density matrix at the end of the t_1 evolution period is:

$$\rho = 2I_zS_y cos(\omega_S t_1) - 2I_zS_x sin(\omega_S t_1) \tag{12.7}$$

After application of the proton 90° pulse along the x-axis, this density matrix becomes:

$$\rho = -2I_yS_y cos(\omega_S t_1) + 2I_yS_x sin(\omega_S t_1) \tag{12.8}$$

Application of the heteronuclear pulse with a phase of x gives:

$$-2I_yS_y cos(\omega_S t_1) + 2I_yS_x sin(\omega_S t_1) \stackrel{90_{N(X)}}{\longrightarrow} -2I_yS_z cos(\omega_S t_1) + 2I_yS_x sin(\omega_S t_1) \tag{12.9}$$

while application of a heteronuclear pulse with a phase of y gives:

$$-2I_yS_y cos(\omega_S t_1) + 2I_yS_x sin(\omega_S t_1) \stackrel{90_{N(Y)}}{\longrightarrow} -2I_yS_y cos(\omega_S t_1) - 2I_yS_z sin(\omega_S t_1) \tag{12.10}$$

The double-quantum terms (e.g. $2I_yS_y$) do not evolve into detectable signal while the single quantum terms refocus into in-phase proton magnetization during the second INEPT period to give:

$$\begin{aligned} -2I_yS_z cos(\omega_S t_1) &\rightarrow +I_x cos(\omega_S t_1) \\ -2I_yS_z sin(\omega_S t_1) &\rightarrow +I_x sin(\omega_S t_1) \end{aligned} \tag{12.11}$$

	$t_2 \longrightarrow$									
$t_1 \downarrow$	0	1	2	3	4	5	6	7	.	*n*
t_1=0:cos	x	x	x	x	x	x	x	x	.	.
t_1=0:sin	x	x	x	x	x	x	x	x	.	.
$t_1=\Delta t_1$:cos	x	x	x	x	x	x	x	x	.	.
$t_1=\Delta t_1$:sin	x	x	x	x	x	x	x	x	.	.
$t_1=2\Delta t_1$:cos	x	x	x	x	x	x	x	x	.	.
$t_1=2\Delta t_1$:sin	x	x	x	x	x	x	x	x	.	.
.	.	.	.	.	.	.	.	.	.	.
2p	.	.	.	.	.	.	.	.	.	.

Figure 12.3 Data matrix structure in hypercomplex quadrature detection. Raw data matrix for hypercomplex quadrature detection. Two spectra are acquired per time point. One that is amplitude modulated as $cos(\omega_S t_1)$ and the other as $sin(\omega_S t_1)$. Therefore a total of $2p$ points have to acquired.

thus the experiment performed with a pulse phase of x gives a proton signal that is amplitude modulated by $cos(\omega_S t_1)$ while the experiment performed with a pulse phase of y gives a signal that is amplitude modulated by $sin(\omega_S t_1)$.

The data matrix from a hypercomplex experiment is shown in Fig. 12.3. Note that each t_1 time point has two FIDs associated with it, one cosine modulated and the second sine modulated. This matrix would be processed by first performing the Fourier transform in t_2, phasing of the spectrum and discarding the imaginary part. Then each column would be subject to transformation in t_1. The *cosine* and *sine* modulated points can be transformed separately and then the addition of the two spectra will cancel the folded peaks. Alternatively, the signals can be combined to produce a complex signal, $e^{i\omega_S t_1}$, that can be subjected to a complex Fourier transform, to yield a single peak at ω_S.

Although it appears that the hypercomplex method requires the collection of twice as many FIDs as the TPPI method, recall that the indirect time domain has to be sampled twice as frequently with the TPPI method, thus the number of FIDs is identical for both methods.

12.2.1 States-TPPI - Removal of Axial Peaks

Axial peaks are artifactual peaks in a multi-dimensional NMR spectrum. These peaks appear in the center of an indirectly detected domain, i.e. at $\omega = 0$. Axial peaks arise from any magnetization that appears in the directly detected time domain, but is not frequency labeled during the indirectly detected time domain, giving an artifact peak located at zero frequency in ω_1. For example, in the COSY experiment, T_1 relaxation during the first time evolution period will create proton longitudinal magnetization (I_z) that is not labeled with any frequency in t_1. The 90° pulse at the end of t_1 will convert this magnetization into an observable signal. Fourier transformation of this signal in both time domains will produce an axial peak at zero frequency in ω_1. The axial peaks can be quite large and will distort the spectrum in the region of zero frequency in ω_1.

A simple way to remove axial peaks is to cycle the phase of the quadrature pulse and the receiver phase by π as part of the phase cycle associated with the pulse sequence. The desired signals will be inverted by the phase shift of the quadrature pulse, while the axial signals will not be inverted since they are not affected by the first pulse. Since the receiver phase is also inverted, the desired signals will add while the axial signals will cancel.

The problem with the above approach is that it doubles the number of transients that have to be acquired in order to remove axial peaks. Although this is seldom a problem in a two-dimensional experiment, it can cause a three-dimensional experiment to take twice as long as required to obtain adequate signal.

An alternative solution to axial peak removal was introduced by Marion et al [105] and is ofter referred to as the States-TPPI method. This method works by changing the frequency of the axial peaks such that they appear on the edge of the two-, or multi-dimensional spectrum. This is accomplished by shifting *both* the phase of the quadrature pulse *and* the phase of the receiver by π for each new value of t_1. This 180° phase shift is applied to both quadrature components. For example, if the two quadrature pulses in t_1 were x and y, with a receiver phase of x, then the next t_1 point would have phases of $-x$ and $-y$, with a receiver phase of $-x$.

Since both the quadrature pulse and the receiver have the same phase relationship the coherence paths that lead to the normal crosspeaks are not changed. However, the axial peaks will experience an effective phase change of π for each t_1 value due to the change in the receiver phase. The signal associated with the axial peak in this case is:

$$S_{axial}(t_1) = e^{i(\omega n\Delta t + n\pi)} \tag{12.12}$$

This can be rearranged to give,

$$S_{axial}(t_1) = e^{i[n\Delta t(\omega + \pi/\Delta t)]} \tag{12.13}$$

The apparent frequency shift of the axial peaks is:

$$\Delta\omega = 2\pi\Delta f = \frac{\pi}{\Delta t} \tag{12.14}$$

Since $1/\Delta t = sw$, the shift in frequency is:

$$\Delta f = sw/2 \tag{12.15}$$

The axial peaks are now shifted to the edge of the spectrum, as illustrated in Fig. 12.4. The similarity to the TPPI method of quadrature detection is apparent, in both cases the phase of the quadrature pulse is incremented by a constant amount for each new t_1 value and in both cases the frequencies of the peaks are shifted by one-half of the sweepwidth.

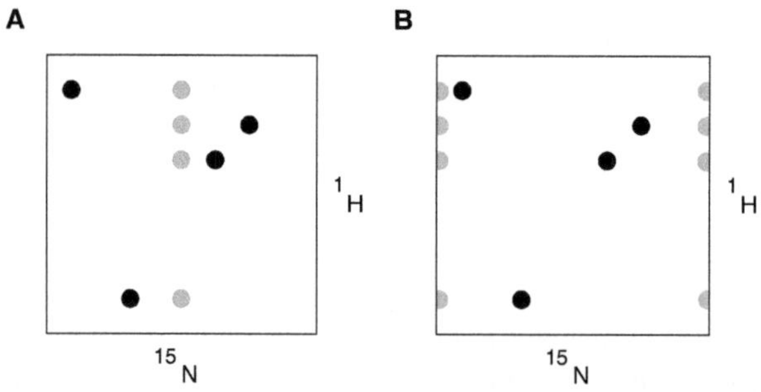

Figure 12.4 Axial peak suppression using States-TPPI. The left spectrum was acquired without phase shifting the first quadrature pulse and the receiver by π. The axial peaks are shown in light gray. The right spectrum was acquired with phase shifting, and the axial peaks are now on the edge of the spectrum.

12.3 Sensitivity Enhancement

During the course of a multi-dimensional NMR experiment it is common to convert a portion of the starting magnetization into undetectable signal. For example, in the case of the HSQC experiment, the double-quantum terms that are created by the pair of proton and heteronuclear pulses at the end of t_1 ($2I_yS_y$, $2I_yS_x$) do not result in detectable signal. The loss of detectable magnetization reduces the intensity of the observed resonance lines.

In the case of HSQC experiment this loss of signal can be prevented by the inclusion of an additional refocusing period that will convert these double-quantum terms to detectable single quantum terms [31]. The additional refocusing period is incorporated into the HSQC experiment as illustrated in Fig. 12.5. During this second refocusing period the single quantum magnetization that was refocused by the first INEPT period is stored along the z-axis and then subsequently returned to detectable magnetization. This technique is referred to as sensitivity enhancement (SE).

We begin analysis of this sequence at the end of t_1 period. The density matrix at that point is represented by the following product operators

$$\rho = 2I_zS_ycos(\omega_St_1) - 2I_zS_xsin(\omega_St_1) \tag{12.16}$$

In this experiment two FIDs are acquired for each t_1 time point. One with the phase ϕ set to x and one with the phase set to $-x$. The two spectra are stored separately and then linear combinations of the two spectra are used to generate $cosine$ and $sine$ modulated spectra.

After application of the proton 90° pulse and the 90° heteronuclear pulse with a phase (ϕ) equal to 0 or π(-x), the density matrix is:

$$\rho = \eta 2I_yS_zcos(\omega_St_1) + 2I_yS_xsin(\omega_St_1) \tag{12.17}$$

η is -1 if the phase of the heteronuclear pulse is 0 and +1 if the phase is π. This density matrix evolves due to J-coupling during the 1st INEPT period as:

$$\eta 2I_yS_zcos(\omega_St_1)+2I_yS_xsin(\omega_St_1) \rightarrow -\eta I_xcos(\omega_St_1)+2I_yS_xsin(\omega_St_1) \tag{12.18}$$

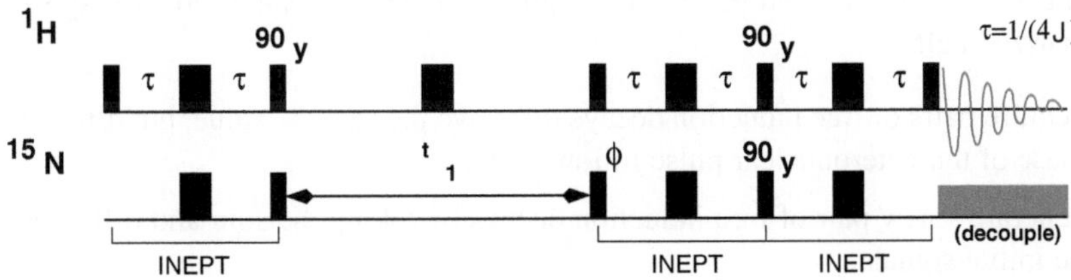

Figure 12.5. *Sensitivity enhanced HSQC.* Narrow bars correspond to 90° pulses while the wide bars are 180° pulses. All pulses are along the x-axis unless otherwise noted. The phase of the heteronuclear pulse marked ϕ is first set to zero (x) and then to π ($-x$) while the t_1 delay is keep constant. These two FIDs are stored in separate locations and $cosine$ and $sine$ modulated signals are generated from these two FIDs as described in the text.

Note that double-quantum term $2I_yS_x$, does not evolve due to J-coupling. Furthermore, chemical shift evolution is suppressed by the pair of 180° pulses that are applied to both the proton and heteronuclear spin in the middle of the INEPT period.

The pair of y-pulses convert the density matrix as follows:

$$-\eta I_x cos(\omega_S t_1) + 2I_yS_x sin(\omega_S t_1) \rightarrow \eta I_z cos(\omega_S t_1) - 2I_yS_z sin(\omega_S t_1) \quad (12.19)$$

During the 2nd INEPT period the first term (I_z) does not evolve while the second term is refocused:

$$\eta I_z cos(\omega_S t_1) - 2I_yS_z sin(\omega_S t_1) \rightarrow \eta I_z cos(\omega_S t_1) + I_x sin(\omega_S t_1) \quad (12.20)$$

The last proton pulse, along x, converts the density matrix to:

$$\eta I_z cos(\omega_S t_1) + I_x sin(\omega_S t_1) \rightarrow -\eta I_y cos(\omega_S t_1) + I_x sin(\omega_S t_1) \quad (12.21)$$

The two density matrices obtained for $\phi = 0$ and $\phi = \pi$ are:

$$\begin{array}{ll} \phi = 0 & \rho = I_y cos(\omega_S t_1) + I_x sin(\omega_S t_1) \\ \phi = \pi & \rho = -I_y cos(\omega_S t_1) + I_x sin(\omega_S t_1) \end{array} \quad (12.22)$$

Quadrature detection is obtained by taking the sum and the difference of these two density matrices:

$$\begin{aligned} \Sigma &= [I_y cos(\omega_S t_1) + I_x sin(\omega_S t_1)] + [-I_y cos(\omega_S t_1) + I_x sin(\omega_S t_1)] \\ &= 2I_x sin\omega_S t_1 \quad (12.23) \\ \Delta &= [I_y cos(\omega_S t_1) + I_x sin(\omega_S t_1)] - [-I_y cos(\omega_S t_1) + I_x sin(\omega_S t_1)] \\ &= 2I_y cos\omega_S t_1 \quad (12.24) \end{aligned}$$

Before constructing the complex signal from the sum and difference signals it is necessary to shift the phase of one of them by applying a zero-order 90° phase shift. This operation is required because the sum signal is initially present in the real channel of the FID (I_x) while the difference signal is initially present in the imaginary channel of the FID (I_y). The overall process of acquiring sensitivity enhanced (SE) spectra is summarized below:

1. Acquire pairs of free induction decays that have the same t_1 value, but differ in the phase of the heteronuclear pulse following t_1.
2. Generate a new pair of free induction decays by taking the sum and difference of the initial signals.
3. Apply a 90° zero-order phase correction to the difference signal, converting $I_y \rightarrow I_x$.
4. The resultant data matrix can now be processed as if it were acquired with the hypercomplex method.

12.4 Echo-AntiEcho Quadrature Detection: N-P Selection

In our discussion of coherence selection using pulsed field gradients in Chapter 11 it was determined that it was only possible to select one of the two coherences present during t_1, either σ^{+1} or σ^{-1}. Therefore the detected magnetization is of the form:

$$S(t_1, t_2) = e^{-t_1/T_2} e^{+i\omega_S t_1} e^{-t_2/T_2} e^{+i\omega_I t_2} \tag{12.25}$$

Since the selected coherence corresponds to precession at positive frequencies or negative frequencies, these coherences are often referred to as P- and N-coherence levels and the selection of one of the two pathways is referred to as N-P selection.

The selection of either the N or P coherence completely defines the frequency of the signal, because the Fourier transform of $e^{+i\omega_S t}$ is a single delta function at ω_S. The lineshape of this peak, however, is non-Lorentzian. The lineshape can be determined by considering the Fourier transform of the exponentially decaying function, e^{-t/T_2}, as follows. Transformation in the directly detected dimension yields:

$$S(t_1, \omega_2) = e^{-t_1/T_2} e^{+i\omega_S t_1} [A(\omega_2) + iD(\omega_2)] \tag{12.26}$$

where $A(\omega_2)$ and $D(\omega_2)$ represent absorption and dispersion lineshapes in the directly detected frequency domain, ω_2. The position of these lines in ω_1 is defined by ω_I.

The transform in t_1 gives the final two-dimensional lineshape.

$$\begin{aligned} S(\omega_1, \omega_2) &= [A(\omega_1) + iD(\omega_1)][A(\omega_2) + iD(\omega_2)] \qquad (12.27) \\ &= [A(\omega_1)A(\omega_2) - D(\omega_1)D(\omega_2)] + i[A(\omega_1)D(\omega_2) + D(\omega_1)A(\omega_2)] \end{aligned}$$

where $A(\omega_1)$ and $D(\omega_1)$ represent lineshapes in the indirectly detected frequency domain, ω_1. The position of these lines in ω_2 is defined by ω_S.

Equation 12.27 indicates that the real component of the lineshape is a mixture of an absorption mode lineshape and a dispersion mode lineshape and is often referred to as a phase-twisted lineshape. This mixed lineshape has both positive and negative lobes and unacceptably broad dispersion tails.

12.4.1 Absorption Mode Lineshapes with N-P Selection

An HSQC pulse sequence that produces pure absorption mode lineshapes with N-P selection was introduced by Kay et al in 1992 [82] and is shown in Fig. 12.6. In this experiment, absorption mode lineshapes are obtained by collecting two independent FIDs for each t_1 time point. As shown below, these FIDs differ in the weighting of N- and P-coherences. A linear combination of these two FIDs generates *cosine* and *sine* modulated signals that give pure absorption mode lineshapes with frequency discrimination. The signal intensity that is lost by coherence path selection is recovered by a sensitivity enhanced refocusing period.

Analysis of the density matrix for this sequence is more involved than the hypercomplex sensitivity enhanced HSQC discussed in the previous section due to the coherence selection using field gradients. However, the principle is the same, one half of the magnetization is refocused in the first INEPT period and then stored along the z-axis. The second half of the magnetization is refocused during the second INEPT period.

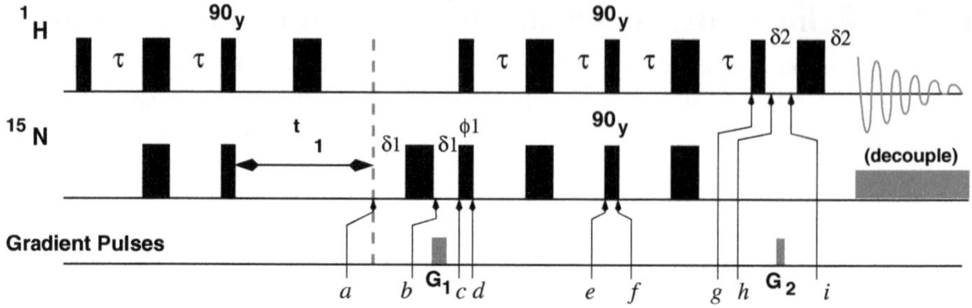

Figure 12.6. Sensitivity enhanced gradient HSQC pulse sequence. The narrow bars represent 90° pulses and the wider bars represent 180° pulses. The letters *a-i* mark the positions the pulse sequence that are discussed in Table 12.1. The delay τ is set to 1/(4J). The delay $\delta 1$ is equal in length to the first gradient plus an additional 100 μsec for recovery of the homogeneous field after the gradient pulse. The delay $\delta 2$ is equal to the length of the second gradient plus 100 μsec for recovery of the field . Note that both of the gradients are applied during a spin-echo sequence in order to refocus any evolution due to chemical shift. The intensity of the gradient pulses is equal to the ratio of the γ's of each spin: $\frac{G_1}{G_2} = \frac{\gamma_H}{\gamma_X}$. Typical values for X=^{15}N are G_1=30 G/cm, applied for 2.7 msec, and G_2=29.05 G/cm, applied for 0.25 msec. One scan is acquired with with the phase of the heteronuclear pulse marked with ϕ_1 set to 0, a second scan is acquired with the same value of t_1, but with ϕ_1 set to π and the amplitude of the second gradient is inverted. These two scans are combined to generate cosine and sine modulated signals, as discussed in the text. Note that additional pulse field gradients can be applied as zz-filters and π-clean filters, see Chapter 11.

The analysis is easiest if the density matrix is expressed in a Cartesian representation when evaluating the effect of pulses and in the spherical basis when considering the effect of gradients. The following conversions will be useful:

$$\begin{aligned} \sigma^+ &= S_x + iS_y \qquad & S_x &= \frac{1}{2}(\sigma^+ + \sigma^-) \\ \sigma^- &= S_x - iS_y \qquad & S_y &= \frac{1}{2i}(\sigma^+ - \sigma^-) \end{aligned} \tag{12.28}$$

After the end of the t_1 evolution time, indicated by point a in Fig. 12.6, the magnetization is given by:

$$\rho_a = 2I_z[S_y cos(\omega_S t_1) - S_x sin(\omega_S t_1)] \tag{12.29}$$

Casting this density matrix in term of spherical basis (coherences):

$$\begin{aligned} \rho_a &= 2I_z[\frac{1}{2i}(\sigma^{+1} - \sigma^{-1})]cos(\omega_S t_1) - [\frac{1}{2}(\sigma^{+1} + \sigma^{-1})]sin(\omega_S t_1) \\ &= 2I_z\left[-\frac{i\sigma^+}{2}[cos(\omega_S t_1) - isin(\omega_S t_1)] + \frac{i\sigma^-}{2}[cos(\omega_S t_1) + isin(\omega_S t_1)]\right] \\ &= 2I_z\left[-\frac{i\sigma^+}{2}e^{-i\omega_S t_1} + \frac{i\sigma^-}{2}e^{+i\omega_S t_1}\right] \end{aligned} \tag{12.30}$$

The next segment of the experiment is a spin-echo sequence. Ignoring chemical shift evolution during the δ_1 period, the 180° heteronuclear pulse will interchange the values associated with each coherence, giving the following value for the density matrix at point b in the pulse sequence:

$$\rho_b = 2I_z \left[\frac{i\sigma^+}{2} e^{+i\omega_S t_1} - \frac{i\sigma^-}{2} e^{-i\omega_S t_1} \right] \tag{12.31}$$

The first gradient pulse, $G1$, will induce the following phase shifts:

$$\rho_c = 2I_z \left[\frac{i\sigma^+}{2} e^{+i\omega_S t_1} e^{-i\gamma_S G_1} - \frac{i\sigma^-}{2} e^{-i\omega_S t_1} e^{+i\gamma_S G_1} \right] \tag{12.32}$$

Simplifying the above by combining the chemical shift and gradient induced phase shifts as $\varphi = \omega t_1 - \gamma_S G_1$:

$$\rho_c = 2I_z \frac{i}{2} \left[\sigma^+ e^{+i\varphi} - \sigma^- e^{-i\varphi} \right] \tag{12.33}$$

This density matrix is converted back to the Cartesian representation to give:

$$\begin{aligned} \rho_c &= 2I_z \frac{i}{2} \left[(S_x + iS_y) e^{+i\varphi} - (S_x - iS_y) e^{-i\varphi} \right] \\ &= -2I_z \left[S_x sin\varphi + S_y cos\varphi \right] \end{aligned} \tag{12.34}$$

The analysis of the remaining section of the pulse sequence is shown in Table 12.1. The final proton signals obtained with $\phi_1 = 0$ and $\phi_1 = \pi$ are:

$$\begin{aligned} \phi_1 &= 0 \qquad \rho_0 = I_x cos\omega_S t_1 - I_y sin\omega_S t_1 \\ \phi_1 &= \pi \qquad \rho_\pi = I_x cos\omega_S t_1 + I_y sin\omega_S t_1 \end{aligned} \tag{12.35}$$

As with the sensitivity enhanced HSQC experiment discussed previously, these signals are added and subtracted to give:

$$\sum = \rho_o + \rho_\pi = I_x cos\omega_S t_1 \qquad \Delta = \rho_o - \rho_\pi = I_y sin\omega_S t_1 \tag{12.36}$$

A 90° zero phase shift is applied to the Δ signal, converting the I_y to I_x, thus generating the desired *cosine* and *sine* modulated signals in the indirectly detected domain.

Event	$\phi_1 = 0$, G_2 Positive	$\phi_1 = \pi$, G_2 Negative
d	$2I_yS_xsin\varphi + 2I_yS_zcos\varphi$	$2I_yS_xsin\varphi - 2I_yS_zcos\varphi$
	$\downarrow$ INEPT	$\downarrow$ INEPT
e	$2I_yS_xsin\varphi - I_xcos\varphi$	$2I_yS_xsin\varphi + I_xcos\varphi$
	$\downarrow 90^H_y, 90^X_y$	$\downarrow 90^H_y, 90^X_y$
f	$-2I_yS_zsin\varphi + I_zcos\varphi$	$-2I_yS_zsin\varphi - I_zcos\varphi$
	$\downarrow$ INEPT	$\downarrow$ INEPT
g	$I_xsin\varphi + I_zcos\varphi$	$I_xsin\varphi - I_zcos\varphi$
	$\downarrow 90^H_x$	$\downarrow 90^H_x$
h	$I_xsin\varphi - I_ycos\varphi$	$I_xsin\varphi + I_ycos\varphi$
	$\downarrow$ Conversion to Spherical Representation	
	$\frac{1}{2}(\sigma^+ + \sigma^-)sin\varphi - \frac{1}{2i}(\sigma^+ - \sigma^-)cos\varphi$	$\frac{1}{2}(\sigma^+ + \sigma^-)sin\varphi + \frac{1}{2i}(\sigma^+ - \sigma^-)cos\varphi$
	$=\frac{1}{2i}\left[-\sigma^+e^{-i\varphi} + \sigma^-e^{+i\varphi}\right]$	$=\frac{1}{2i}\left[\sigma^+e^{+i\varphi} - \sigma^-e^{-i\varphi}\right]$
	$\downarrow$ Application of G_2	
i	$\frac{1}{2i}\left[-\sigma^+e^{-i\omega_St_1} + \sigma^-e^{+i\omega_St_1}\right]$	$\frac{1}{2i}\left[\sigma^+e^{+i\omega_St_1} - \sigma^-e^{-i\omega_St_1}\right]$
	$\downarrow$ Conversion to Cartesian Representation	
	$= \frac{1}{2i}[-(I_x + iI_y)e^{-i\omega_St_1}$	$= \frac{1}{2i}[+(I_x + iI_y)e^{+i\omega_St_1}$
	$+(I_x - iI_y)e^{+i\omega_St_1}]$	$-(I_x - iI_y)e^{-i\omega_St_1}]$
	$= \frac{1}{2i}[I_x(e^{i\omega_St_1} - e^{-i\omega_St_1})$	$= \frac{1}{2i}[I_x(e^{i\omega_St_1} - e^{-i\omega_St_1})$
	$-iI_y(e^{i\omega_St_1} + e^{-i\omega_St_1})]$	$+iI_y(e^{i\omega_St_1} + e^{-i\omega_St_1})]$
	$= I_xsin(\omega_St_1) - I_ycos(\omega_St_1)$	$= I_xsin(\omega_St_1) + I_ycos(\omega_St_1)$

Table 12.1. Evolution of the density matrix in the gradient selected sensitivity enhanced HSQC. This analysis begins after the heteronuclear pulse of phase ϕ_1. The letters *d-i* on the left refer to the position in the pulse sequence, see Fig. 12.6. The density matrix that is created by a phase of $\phi_1 = 0°$ is shown on the left and that for a phase of π is shown on the right. Note that the second gradient pulse is positive for ϕ_1=0 and negative for $\phi_1 = \pi$. In the case of coherence selection by G_2 (position i in the pulse sequence), a positive value for this gradient will induce the following phase shifts in the coherences: $\sigma^+ \rightarrow \sigma^+e^{-i\gamma_H G_2}$. If $\gamma_X G_1 = \gamma_H G_2$, then this phase shift will cancel the phase introduced by the first gradient, leaving only the phase induced by chemical shift evolution during t_1.

Chapter 13

RESONANCE ASSIGNMENTS: HOMONUCLEAR METHODS

13.1 Overview of the Assignment Process

NMR spectroscopy can be used to determine the structure of a protein (see Chapter 17), and to provide detailed information on the dynamics of the protein over a wide range of time-scales (see Chapter 19). However, before NMR can be used to investigate the structure and dynamics of a protein it is necessary to assign resonance lines in the spectrum to specific atoms in the protein. There are four main steps that are used in traditional assignment strategies for proteins. These steps are described below and illustrated in Fig. 13.1.

1. Collect all resonance frequencies that are associated with spins on the same residue. This collection of resonances is often referred to as a *spin-system*. Ultimately, the resonances within a spin-system will be assigned to atoms within a particular residue. The set of frequencies in a spin-system is often divided into main-chain atoms (H_N, N_H, H_α, C_α, CO) and side-chain atoms.
2. Spin-systems are grouped based on their most likely amino acid type.
3. Determine which pairs of spin-systems arise from *adjacent* residues in the protein. Extend this pairwise association to connect as many spin-systems as possible into a linear chain that represents a segment of the poly-peptide chain. Generally, a number of disconnected segments are obtained due to missing signals, chemical shift overlap, and the presence of prolines, which lack an amide proton.
4. Associate the connected segment of spin-systems with the segment of primary sequence that best matches the likely amino acid type of the spin-systems.

In this chapter we will discuss homonuclear and ^{15}N-separated homonuclear assignment methods. The details of heteronuclear triple resonance methods of assignment are discussed in the following chapter. In these two chapters several pulse sequences are presented; it may be useful to refer to Appendix D for a review of the basic building blocks of these pulse sequences before continuing.

Both the homonuclear and heteronuclear approaches to chemical shift assignments are limited by the molecular weight of the protein. These limitations arise from the

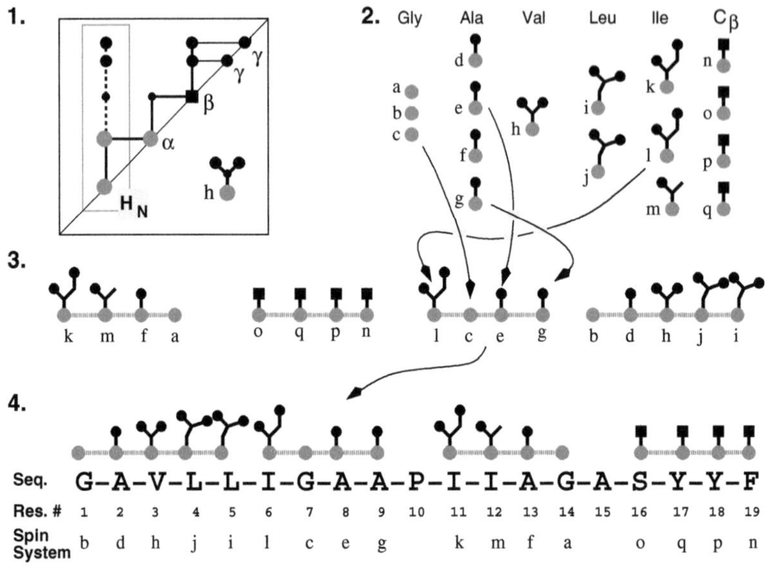

Figure 13.1. Overview of the resonance assignment strategy.

Step 1: Resonances are collected into spin-systems. In this example, a spin-system is assembled based proton-proton scalar coupling, either in a COSY experiment (resonances connected by solid line) or in a TOCSY experiment (resonances boxed and connected by dotted line.). The large gray dots represent main-chain atoms, the large black dots represent methyl groups, and the square black boxes represent β-carbons. The spin-systems are given *arbitrary* labels, e.g. $a - q$ at this point.

Step 2: Spin-systems are then grouped based on their most likely amino acid type. For example, the spin-system defined in panel 1 (spin-system h) is likely a Val residue because of its two methyl groups. Many amino acid types, for example Gly, have characteristic chemical shifts and are readily grouped together. The last grouping shown in this illustration, labeled C_β, includes all residues that possess a β-carbon/proton that is not scalar coupled to another atom besides the α-carbon/proton. This group includes Ser, Cys, Asp, Asn, as well as the aromatic residues. In the case of Tyr, Phe, Trp, and His, the aromatic protons are not coupled to the H_β protons, making it difficult to correlate the aromatic protons to the remainder of the resonances in the spin-system. Not illustrated are non-methyl containing spin-systems with C_γ carbons, such as Glu, and Gln. Met is also contained in this group since the scalar coupling of its methyl group to the rest of the sidechain is quite weak. spin-systems that contain long side-chains, such as Lys and Arg, are also not shown.

Step 3: Adjacent spin-systems are connected either through space *via* dipolar coupling (homonuclear methods), or through-bond *via* scalar coupling (heteronuclear methods). For example, spin-systems l and c would be connected based on the presence of a NOESY crosspeak between their amide protons, indicating that these two protons are within 5 Å of each other. If the HNCA experiment was used to connect spin-systems then the inter-residue C_α shift of spin-system c would be equal to the intra-residue C_α shift of l. This process is extended to include spin-systems e and then g into the sequence of connected spin systems.

Step 4: The likely amino acid assignment of the collected spin-systems is used to guide the placement of the connected segment on to the primary sequence. For example, the segment $l-c-e-g$ consists of spin-systems whose amino acid type match residues 6-9. Note that no assignments are possible for Pro10 since it lacks an amide proton. Ala15 remains also unassigned because no corresponding spin-system was detected, possibly due to exchange broadening of its amide resonance (see Chapter18).

Table 13.1. Molecular weight limitations for chemical shift assignments.

Mol. Weight	*Technique*	*Observed Spins*	*Dimensionality*
<10 kDa	Homonuclear	^{1}H	2D
10-15 kDa	^{15}N-homonuclear†	^{1}H, ^{15}N	3D, 4D
15-30 kDa	Triple Resonance‡	^{1}H, ^{15}N, ^{13}C	3D, 4D
30-60 kDa	Triple Resonance/deuterated‖	^{1}H, ^{15}N, ^{13}C	3D, 4D
60-100 kDa	Triple Resonance/deuterated/TROSY‖	^{1}H, ^{15}N, ^{13}C	3D, 4D

†Requires uniform labeling of protein with ^{15}N.

‡Requires uniform labeling with ^{15}N and ^{13}C.

‖Requires uniform labeling with ^{15}N, ^{13}C and replacement of CH groups with CD.

increased spin-spin relaxation rate of larger proteins, which reduces the sensitivity of multi-dimensional NMR experiments, and the larger number of resonance lines, which leads to signal overlap and degeneracies[1]. Approximate molecular weight limitations are presented in Table 13.1 and the origin of these limitations are discussed in more detail below.

Homonuclear techniques, which observe only protons, are the most limited with respect to the size of the protein that can be assigned. This limitation is due to resonance overlap that can only be marginally resolved using two-dimensional methods. In addition, the sequential linking of adjacent spin-systems relies on detecting through-space dipolar coupling between protons on adjacent residues. This approach is far less reliable than the use of scalar couplings because it depends on the conformation of the peptide chain. In addition, a significant fraction of through space dipolar couplings are between *non*-sequential residues, confounding the assignment process.

^{15}N-homonuclear techniques, which utilize the ^{15}N shift to provide an additional dimension to homonuclear experiments, were first introduced around 1989 [104]. This approach increases the molecular weight of proteins that can be studied to some degree. The increased performance is due solely to the increased resolution in three- or four-dimensional experiments. The method of connecting spin-systems still relies on through-space dipolar couplings.

Heteronuclear triple resonance experiments observe signals from ^{1}H, ^{13}C, and ^{15}N and offer distinct advantages over homonuclear techniques, permitting the assignment of proteins in 15-30 kDa range. The principal difference between homonuclear and heteronuclear approaches is the method used to connect adjacent spin-systems. In the case of heteronuclear methods, the scalar coupling between ^{15}N and ^{13}C *across* the peptide bond permits the sequential assignment of a protein's mainchain atoms. This process relies solely on chemical bonding between atoms, is completely independent of the structure of the protein, and provides unambiguous inter-residue connectivities, except in the presence of chemical shift degeneracies.

[1]Two atoms are considered degenerate if they have the same chemical shift.

Another favorable attribute of triple resonance experiments is the larger J-coupling between heteronuclear spins. The larger J-couplings permits a more rapid transfer of magnetization between spins, making triple resonance experiments more sensitive in larger proteins that have shorter spin-spin relaxation times. The presence of additional NMR active nuclei, such as ^{15}N and ^{13}C, permit the separation of the resonance signals over additional chemical shift scales, increasing the resolution of experiments. Finally, the carbon frequencies can also be used to readily identify the likely amino acid type of the spin-system.

The molecular weight limit for triple resonance assignment methodologies is mostly due to the rapid spin-spin relaxation time of larger, slowly tumbling, proteins. The principal mechanism of relaxation is due to proton-proton and proton-carbon dipolar coupling. Deuteration of a protein, by replacing the aliphatic and aromatic protons with deuterons, decreases the spin-spin relaxation rates because of the smaller gyromagnetic ratio of the deuteron. This extends the molecular weight limit of triple-resonance techniques, permitting the assignment of proteins as large as 60 kDa.

The overall sensitivity of ^{15}N separated and triple resonance experiments and can be increased further by utilizing the interference between dipolar and CSA relaxation of the amide group to decrease the spin-spin relaxation rate of both the amide proton and nitrogen. This technique is usually referred to as TROSY, or **T**ransverse **r**elaxation **o**ptimized **s**pectroscop**y**. With TROSY techniques it has been possible to assign systems as large as 100 kDa. An introductory description of the TROSY technique is presented in Chapter 15.

13.2 Homonuclear Methods of Assignment

If the protein or peptide is unlabeled, i.e. the levels of ^{13}C and ^{15}N are at natural abundance, then it is only possible to use homonuclear methods to obtain resonance assignments. It is generally difficult to study protein that are greater than 7-10 kDa using these methods.

Step 1 of the assignment process uses DQF-COSY and TOCSY experiments to define the spin-systems. The theory of these experiments were discussed in Chapter 9 and practical pulse sequences for the DQF-COSY and TOCSY experiments can be found in Chapter 15. Step two in the assignment process utilizes characteristic proton shifts, which were provided in Chapter 1, to determine the most likely amino acid type of each spin-system.

The third step in the assignment process will utilize dipolar coupling between protons on adjacent residues to determine which spin-systems are adjacent to each other. Dipolar coupling is discussed in considerable detail in Chapter 16. The NMR experiment that elucidates dipolar couplings between protons is called a NOESY experiment, which stands for Nuclear Overhauser Spectroscopy. The effect of dipolar coupling was first observed for nuclear-electron magnetic dipoles by Overhauser [122].

The intensity of crosspeaks in the NOESY experiment is proportional to $1/d^6$, where d is the inter-proton distance. Typically, distances of 5 Å or less can be detected with this experiment. The measured inter-proton distances are usually referred to as NOEs. With respect to resonance assignments, NOEs involving the amide proton and/or the H_α proton are typically utilized because these resonances are in a less crowded region of the spectrum.

A number of inter-proton distances that are typically observed for mainchain atoms are shown in Fig. 13.2 and listed in Table 13.2. In the case of a β-sheet, the strongest NOE between adjacent residues is the NOE between the amide proton and the H_α proton of the *preceding* residue. The NOE between amide protons on adjacent residues in a β-strand conformation is comparably weak and is therefore not generally useful for confirming adjacent spin-systems. Note that the amide proton also shows two moderately intense NOEs to protons on the adjacent strand. These could be interpreted, incorrectly, as NOEs between adjacent residues, causing errors in the assignments.

In the case of the α-helix, the NOE between adjacent amide protons is strong due to the short inter-proton distance of 2.8 Å. The NOE between the H_α proton and the amide proton of the following residue is somewhat weaker (d = 3.5 Å), but still useful for verifying the inter-spin-system connection. An amide proton is also close to the amide proton of the *i+2* (4.2 Å) and the *i+3* residue (4.7 Å), which provides redundant information regarding the sequential ordering of spin-systems.

Several inter-residue NOEs involving H_N and H_α protons are generally detectable in NOESY spectra, regardless of the secondary structure of the residue. Consequently, under favorable conditions it should be possible to sequentially link spin-systems on

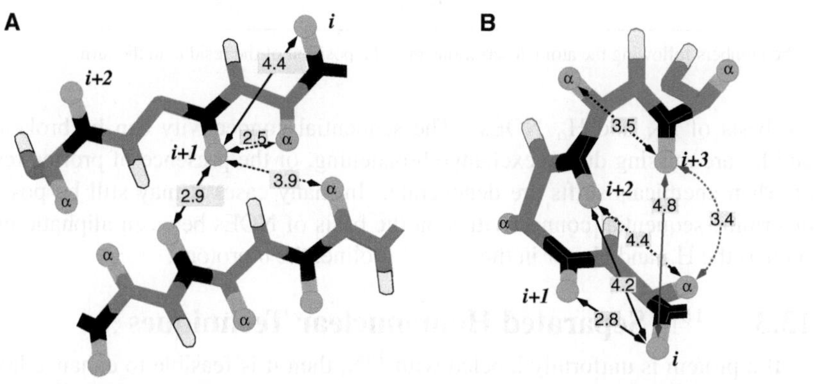

Figure 13.2. Inter-proton distances in regular secondary structures. Inter-residue distances in an β-sheet (Panel A) and an α-helix (Panel B). Inter-proton distances are shown in Angstroms. All of these distances should be detectable in a NOESY experiment, however longer distances will give weaker crosspeaks. Light gray atoms represent amide and H_α protons, dark gray atoms are carbon, white atoms with a black outline are carbonyl oxygens, black atoms are amide nitrogens.

Panel A: An amide proton will typically show four NOEs, two to the preceding residue and two to non-sequential residues across the strand. The sequential NOE between the amide and the H_α-proton of the preceding residue is intense because of the short (2.5 Å) distance between the two protons.

Panel B: The first amide in the helix is labeled i and distances to the amide protons of the next three residues are indicated with solid arrows. The H_α proton is relatively close to the H_N proton of the next residue (3.5 Å, dashed line, shown for H_α^{i+2} to H_N^{i+3}) as well as to the amide proton of the $i + 3$ residue.

Table 13.2. Selected inter-residue distances for sequential assignments. Proton-proton inter-residue distances, given in Angstroms, were obtained from [170]. The first index of d_{ij} gives the atom on the *i-th* residue, the second index gives the atom on the *i+n* residue, where n varies from 1 to 4. Using the first entry of the table as an example: In an α-helix the H_α proton on the ith residue is 3.5 Å from the amide proton (N) of the $i + 1$ residue.

Secondary Structure	*Interacting Atoms*	*i+1*	*i+2*	*i+3*	*i+4*
α-helix	$d_{\alpha N}$	3.5	4.4	3.4	4.2
	d_{NN}	2.8	4.2	4.8	6.1
3_{10}-helix	$d_{\alpha N}$	3.4	3.8	3.3	
	d_{NN}	2.6	4.1	5.2	
β-strand, parallel(||))	$d_{\alpha N}$	2.2			
	d_{NN}	4.3			
β-strand, anti-parallel (anti-||)	$d_{\alpha N}$	2.2			
	d_{NN}	4.2			
Type I and Type II		Type I	Type II		
Turns†	$d_{\alpha N}(2,3)$	3.4	2.2		
	$d_{\alpha N}(3,4)$	3.2	3.2		
	$d_{\alpha N}(2,4)$	3.6	3.3		
	$d_{NN}(2,3)$	2.6	4.5		
	$d_{NN}(3,4)$	2.4	2.4		
	$d_{NN}(2,4)$	3.8	4.3		

†The numbers following the atom descriptions give the position of the residue in the turn.

the basis of H_N and H_α NOEs. The sequential connectivity can be broken when amides are missing due to exchange broadening, or the presence of proline residues, or when chemicals shifts are degenerate. In many cases it may still be possible to determine sequential connectivities on the basis of NOEs between aliphatic protons, such as the H_α and H_β, or in the case of proline, the δ protons.

13.3 ^{15}N Separated Homonuclear Techniques

If a protein is uniformly labeled with ^{15}N, then it is feasible to enhance homonuclear 2D experiments by the addition of a third frequency dimension, the ^{15}N chemical shift. This additional dimension increases the resolution by separating spin-systems (residues) by the chemical shift of the amide nitrogen, greatly increasing the ability to resolve resonance peaks. Fig. 13.3 shows a 2D-TOCSY (lower left) and a single slice from the corresponding 3D experiment (lower right). The 2D-TOCSY is highly overlapped and only a few spin-systems, such at the amide with a proton chemical shift of 6.25 ppm, are cleanly resolved. In contrast, the 3D experiment is capable of resolving almost all spin-systems. In the example shown in Fig. 13.3, the three spin-systems with an amide proton shift of 9.6 ppm are unresolved in the 2D-TOCSY spectrum, but become resolved in the 3D-TOCSY by virtue of their different nitrogen shifts. The TOCSY signals from one of these three residue, with $\delta_N = 130.3$ ppm and $\delta_H = 9.6$ ppm, are boxed in the 3D-TOCSY spectrum.

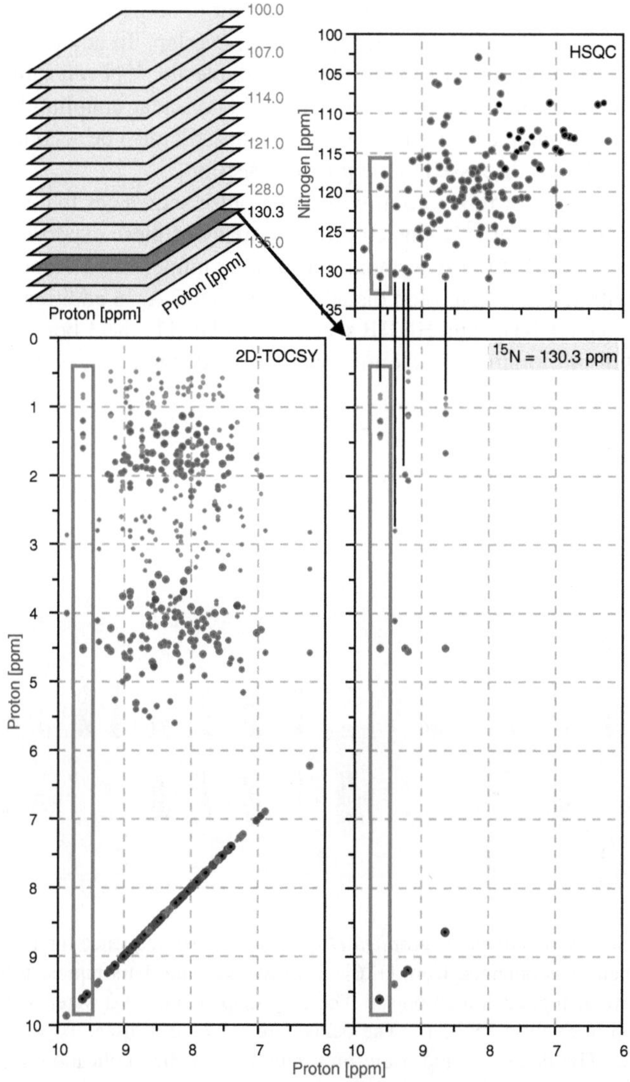

Figure 13.3. Increased resolution in a 3D-^{15}N separated TOCSY. The amide-aliphatic region of a 2D-TOCSY spectrum of a 130 residue protein is shown in the lower left. The 3D spectrum, represented as a stack of individual 2D-spectra at various ^{15}N frequencies is shown in the upper left. One of these spectra, or slices, at a nitrogen frequency of 130.3 ppm, is shown on the lower right. For reference, a 2D proton-nitrogen HSQC spectrum is shown in the upper right. The slice from the 3D experiment contains five residues with ^{15}N shifts of $\approx$130.3 ppm. These residues are highlighted by the vertical lines in the slice from the 3D-TOCSY spectrum. Three residues, highlighted with boxes in the 2D-TOCSY and HSQC spectra, have a proton chemical shift of $\approx$9.6 ppm and are degenerate in the 2D-TOCSY, but become resolved in the 3D-TOCSY spectrum. The boxed peaks in the slice from the 3D-TOCSY arise from a single residue.

The increase in resolution permits the assignment of somewhat larger proteins by homonuclear methods because of reduced spectral overlap. In addition, the incorporation of ^{15}N also generates a sample that is suitable for ^{15}N relaxation studies to probe backbone dynamics. However, since there is no scalar coupling between adjacent residues, the sequential ordering of spin-systems still must be accomplished using inter-residue NOEs, as discussed in the previous section.

Although the focus of this section will be on pulse sequences that are commonly used for the assignment of ^{15}N labeled proteins, many of these experiments also provide constraints for structure determination. For example, the 3D and 4D NOESY experiments will detect inter-proton distances less than 5 Å. In addition, the intensity of crosspeaks in the HNHA and HNHB spectra are related to the 3-bond coupling constants, providing information on the torsional angles of the N-C$_\alpha$ and C$_\alpha$-C$_\beta$ bonds, respectively.

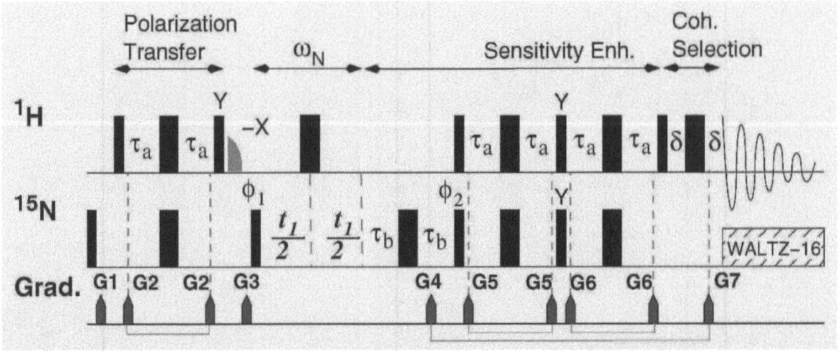

Figure 13.4. Sensitivity enhanced gradient HSQC. This pulse sequence for a proton-nitrogen HSQC experiment was obtained from [176]. Narrow and wide bars represent 90° and 180° pulses, respectively, applied at high power. The gray shaped pulse is a water selective flipback pulse, discussed in Section 15.2.2.6. The delays are $\tau_a = 2.3$ msec, $\tau_b = 1.5$ msec, and $\delta = 0.5$ msec. The delays τ_b and δ are only required for the application and recovery of gradients G4 and G7, respectively, and should be made as short as possible. Pulse phases are $\phi_1 = x, -x; \phi_2 = x; \phi_{rec} = x, -x$. ϕ_1 is incremented by 180° along with the receiver to shift axial peaks to the edge of the spectrum (TPPI). Quadrature detection of nitrogen is achieved by N- and P-type data collection. Specifically, a second scan is acquired at the same t_1 value, but ϕ_2 is incremented by 180° and the sign of gradient G7 is inverted. The two FIDs are added to give a signal proportional to $cos(\omega_N t_1)$ or subtracted to give a signal proportional to $sin(\omega_N t_1)$, see Section 12.4 for details. Gradient strength used by Zhang et al. [176] were G1(4 G/cm, 1 msec), G2 (5 G/cm, 750 μsec), G3 (-15 G/cm, 1.5 msec), G4 (30 G/cm, 1.25 msec), G5=G6 (4 G/cm, 500 μsec), G7 (27.8 G/cm, 125 μsec). Gradient G1 insures that only magnetization originating on the proton is detected. Gradient pairs G2, G5 and G6 remove artifacts associated with the 180° pulses. Gradient G3 is a zz-filter and gradients G4 and G7 are used for N-P selection. Note that a slight adjustment of the ratio of these two gradients may be required to obtain maximum signal.

13.3.1 2D ^{15}N HSQC Experiment

The 2D ^{15}N HSQC experiment was discussed in some detail in Chapter 10 as an illustration of many of the central features of 2D heteronuclear methods. The version presented here, in Fig. 13.4, is commonly used in many NMR laboratories and sufficient information has been provided to allow the user to implement the sequence. It is included in this chapter to facilitate the analysis of the other ^{15}N separated 3D sequences.

The overall path of magnetization transfer in the HSQC experiment is as follows:

$$H_N \stackrel{INEPT}{\longrightarrow} \omega_N(t_1) \stackrel{INEPT-SE}{\longrightarrow} H_N(t_2).$$

A brief review of this sequence using product operators is as follows. The first INEPT, or polarization transfer period transfers the proton polarization to the nitrogen. This magnetization is labeled with the nitrogen frequency during the t_1 period and becomes encoded with the nitrogen coherence by gradient G4:

$$I_z \rightarrow -I_y \rightarrow 2I_xN_z \rightarrow 2I_zN_y \rightarrow 2I_zN_ye^{i\omega_Nt_1} \rightarrow 2I_zN_ye^{i\omega_Nt_1}e^{i\gamma_NG4} \quad (13.1)$$

The remaining two INEPT transfers refocus the anti-phase magnetization in a manner that increases the sensitivity of the experiment (see Section 12.3 for details):

$$2I_zN_ye^{i\omega_Nt_1}e^{i\gamma_NG4} \rightarrow [I_xcos(\omega_Nt_1) - I_ysin(\omega_Nt_1)]\,e^{i\gamma_NG4} \quad (13.2)$$

The final gradient, G7, refocuses the phase shift introduced by gradient G4, giving:

$$\begin{aligned}[I_xcos(\omega_Nt_1) \quad - \quad & I_ysin(\omega_Nt_1)]e^{i\gamma_NG4} \rightarrow \\ & [I_xcos(\omega_Nt_1) - I_ysin(\omega_Nt_1)]\,e^{i\gamma_NG4\gamma_HG7} \\ & = [I_xcos(\omega_Nt_1) - I_ysin(\omega_Nt_1)] \quad (13.3)\end{aligned}$$

13.3.2 3D ^{15}N Separated TOCSY Experiment

This experiment is often referred to as the HSQC-TOCSY experiment. In this experiment the overall path of magnetization transfer is as follows (see Fig. 13.5):

$$H_{aliphatic}(t_1) \stackrel{TOCSY}{\longrightarrow} H_N \stackrel{INEPT}{\longrightarrow} \omega_N(t_2) \stackrel{INEPT-SE}{\longrightarrow} H_N(t_3)$$

The 3D HSQC-TOCSY experiment can correlate all of the aliphatic proton resonances with the amide group of a residue. In the case of small proteins, practically all of the members of a spin-system can be identified in this experiment.

The pulse sequence for the 3D ^{15}N separated TOCSY experiment is given in Fig. 13.6 and an illustration of the resultant spectra was shown in Fig. 13.3. This sequence has many of the same features as the HSQC experiment discussed in a previous chapter, namely coherence selection by gradients and sensitivity enhancement of the nitrogen magnetization. In addition, the sequence is also designed to minimize the saturation of the water protons by the use of selective water flip-back pulse (gray shaped pulse in Fig. 13.6). The correct handling of the water magnetization requires a slight modification to the phase cycle, as discussed below.

13.3.2.1 Product Operator Analysis of the 3D TOCSY Experiment

The first proton pulse (ϕ_1), followed by the t_1 frequency labeling period generates:

$$I_z \stackrel{P_90}{\rightarrow} I_x \rightarrow I_x cos(\omega_H t_1) + I_y sin(\omega_H t_1) \quad (13.4)$$

The proton y-pulse that precedes the DIPSI-2 mixing sequence will place the I_x magnetization on the z-axis. The I_y term will be dephased by the DIPSI decoupling scheme and therefore it will not give rise to detectable signal and can be ignored. During the mixing period, the magnetization will be passed to the amide proton, using the H_β proton as an example:

$$I_x^\beta cos(\omega_{H\beta} t_1) \stackrel{P_{90y}}{\rightarrow} I_z^\beta cos(\omega_{H\beta} t_1) \stackrel{TOCSY}{\rightarrow} I_z^\alpha \stackrel{TOCSY}{\rightarrow} I_z^{HN} cos(\omega_{H\beta} t_1) \quad (13.5)$$

Note that the magnetization does not pass directly from the H_β proton to the H_N because the coupling between these two protons is essentially zero. Rather, it is first passed to H_α and then to the amide proton.

During the first INEPT period, the magnetization is sent to the amide nitrogen:

$$I_z^{HN} \stackrel{P_{90x}}{\rightarrow} -I_y \stackrel{\tau_a - 180 - \tau_a}{\longrightarrow} 2I_x N_z \stackrel{P_{90y}}{\rightarrow} 2I_z N_z \quad (13.6)$$

This magnetization is labeled with the nitrogen frequency during the t_2 evolution period:

$$2I_z N_z \stackrel{P_{90\phi 3}}{\rightarrow} -2I_z N_y \stackrel{t2/2 - \pi - t2/2}{\longrightarrow} -2I_z[N_y cos(\omega_N t_2) - N_x sin(\omega_N t_2)] \quad (13.7)$$

The subsequent τ_b periods permit the application of Gradient 4 without evolution of the nitrogen chemical shift. The subsequent two INEPT periods will refocus both the N_x and N_y terms, as discussed in Chapter 12, to give the final detected signal for the selfpeak:

$$cos(\omega_{HN} t_1) cos(\omega_N t_2) cos(\omega_{HN} t_3) \quad (13.8)$$

which represents magnetization that remained on the amide proton after the TOCSY transfer period.

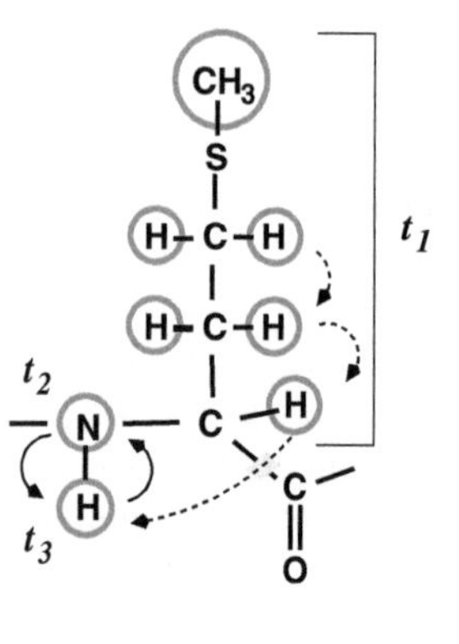

Figure 13.5 Magnetization transfer pathway in a proton TOCSY experiment. The magnetization pathway in the 3D TOCSY experiment begins on the side-chain protons. These are frequency labeled during t_1 and the magnetization is passed, in a relay fashion by the TOCSY mixing sequence (dotted arrows), to the amide proton. The magnetization is transferred to the amide proton via an INEPT transfer where the frequency of the nitrogen is recorded. Another INEPT transfer occurs, leaving the magnetization on the amide proton for detection. The amino acid Met is used in this example to illustrate that the methyl group will not yield crosspeaks because the scalar coupling between the methyl protons and the Hγ protons is small. Note that all of the magnetization transfers are within a residue.

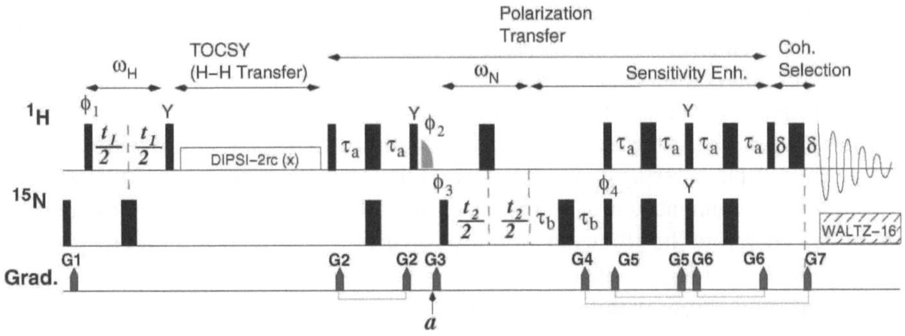

Figure 13.6. Pulse sequence for a 3D ^{15}N separated TOCSY experiment. Experiment and parameters are from [176]. Narrow and wide rectangular bars represent 90° and 180° pulses, respectively. The gray shaped pulse, with phase ϕ_2 is a water selective flipback pulse, see Section 15.2.2.6. The proton 90° pulse lengths were 10.6 μsec, corresponding to a field strength of 23.5 kHz. Isotropic mixing is accomplished using the relaxation compensated DIPSI-2 sequence [32] (total time of 55 msec) using a field strength of 9.5 kHz, which is equivalent to a proton 90° pulse of 26.3 μsec. Nitrogen 90° pulse length was 45 μsec, corresponding to a 5.5 kHz field strength. Nitrogen decoupling during detection (t_3) was accomplished using a 1 kHz field strength (250 μsec 90° pulse).

The delays τ_a, τ_b and δ are set to 2.3, 1.5 and 0.5 msec, respectively. τ_a is $1/(4J_{NH})$. The delays τ_b and δ, need only be long enough for the application and recovery of gradients G4 and G7, respectively.

Pulse phases are $\phi_1 = y, -y$; $\phi_2 = x, -x$; $\phi_3 = x, x, -x, -x$; $\phi_4 = x$; $\phi_{rec} = x, -x, -x, x$. When ϕ_1 is incremented by 90° for hypercomplex quadrature detection in t_1, the phase cycle for ϕ_1 becomes $-x, x$. In this case it is necessary to change ϕ_2 to $-y, y$ for proper management of the water magnetization, see text. In the case of nitrogen evolution, N- and P-type data were acquired in separate FIDs by incrementing the phase of the nitrogen pulse with phase ϕ_4 by 180° and by inverting the sign of gradient G7. Axial peaks were shifted to the edge of the spectrum by incrementing the phase of the same nitrogen pulse and the receiver phase by 180°. Gradient strengths used by Zhang et al were: G1 (1 msec, 4 G/cm), G2 (0.5 msec, 8 G/cm), G3 (1.5 msec, -15 G/cm), G4 (1.25 msec, 30 G/cm), G5=G6 (0.5 msec, 4 G/cm), G7 (1125 μsec, 27.8 G/cm). Note that gradient G3 is negative. Gradient G1 removes any nitrogen magnetization, insuring that the detected signal arises from protons. Gradient pairs G2, G5, and G6 remove artifacts associated with the 180° pulses. G3 is a zz-filter, it also dephases any transverse water magnetization. Gradients G4 and G7 used for nitrogen quadrature detection, as discussed above.

In addition, crosspeaks will arise from the following signal:

$$S(t_1, t_2, t_3) = \sum_{i=1}^{N} \gamma cos(\omega_i t_1) cos(\omega_N t_2) cos(\omega_{HN} t_3) \tag{13.9}$$

where the sum is over all sidechain protons that are coupled, either directly (e.g. H_α) or indirectly (e.g. H_β) to the amide proton. The efficiency of transfer from the sidechains protons to the amide proton is represented by γ.

A comprehensive summary of transfer efficiencies has been presented by Cavanagh et al [29] and a smaller sampling is presented in Chapter 9. In general, the further the side-chain proton is from the amide proton, the larger the number of intervening spins and the weaker the signal. For example, the methyl protons of Val will give more intense crosspeaks than the methyls of Leu. In addition to residue specific differences, the intensity of the crosspeak also depends on the secondary structure of the residue because all of the magnetization must funnel through the H_α proton to be detected on the amide. Therefore the efficiency of transfer depends on $J_{H\alpha H_N}$ for *all* crosspeaks. This dependence on $J_{H\alpha H_N}$ generates weak crosspeaks for residues in α-helices, due to the small value of this coupling constant.

13.3.2.2 Water Saturation in 3D-TOCSY Experiment

As with all experiments that detect amide protons it is important that the water magnetization is restored to the $+z$-axis at the end of each scan. Otherwise, the non-equilibrium state of the water magnetization can be transferred to the amide protons by chemical exchange, leading to a loss of signal. The correct handling of the water requires that the phase of the water selective pulse, ϕ_2, is altered when ϕ_1 is changed for quadrature detection. When $\phi_1 = y$, the density matrix representing the water magnetization, to the point marked a in the sequence, evolves as follows:

$$I_z \overset{P_{90}(\phi_1=y)}{\longrightarrow} I_x \overset{t1}{\rightarrow} I_x \overset{P_{90y}}{\rightarrow} -I_z \overset{DIPSI}{\longrightarrow} -I_z \overset{P_{90}}{\rightarrow} I_y \overset{2\tau_a}{\rightarrow} -I_y \overset{P_{90y}}{\rightarrow} -I_y \overset{P(\phi 2=x)}{\longrightarrow} -I_z \tag{13.10}$$

When the phase of the first proton pulse is changed to $-x$, for quadrature detection, the water evolves to point a as follows:

$$I_z \overset{P(\phi_1=-x)}{\longrightarrow} I_y \overset{t1}{\rightarrow} I_y \overset{P_{90y}}{\rightarrow} I_y \overset{DIPSI}{\longrightarrow} I_y \overset{P_{90}}{\rightarrow} I_z \overset{2\tau_a}{\rightarrow} -I_z \overset{P_{90y}}{\rightarrow} -I_x \overset{P(\phi 2=-y)}{\longrightarrow} -I_z \tag{13.11}$$

In the subsequent part of the sequence, from point a to detection, the water magnetization follows the following path:

$$-I_z \overset{P_{180}\phi_1}{\rightarrow} +I_z \overset{P_{90x}}{\rightarrow} -I_y \overset{2\tau_a}{\rightarrow} -I_y \overset{P_{90y}}{\rightarrow} -I_y \overset{2\tau_a}{\rightarrow} -I_y \overset{P_{90x}}{\rightarrow} -I_z \overset{P_{\delta-180-\delta}}{\rightarrow} +I_z \tag{13.12}$$

Therefore the water magnetization will be placed along the $+z$-axis, regardless of the quadrature phase of ϕ_1.

13.3.3 The HNHA Experiment - Identifying H_α Protons

Although the TOCSY experiment can identify all of the protons of a spin-system, it cannot automatically differentiate between the type of proton, an important consideration for identifying the most probable amino acid type of the spin-system. Although it is generally true that H_α protons will give stronger crosspeaks in the TOCSY spectrum than other side-chain protons, the actual intensity of the crosspeaks will depend on the individual J-couplings throughout the residue.

H_α protons can be unambiguously identified in the HNHA experiment [162, 90]. This is a three dimensional ^{15}N separated experiment with the following pathway of magnetization transfer:

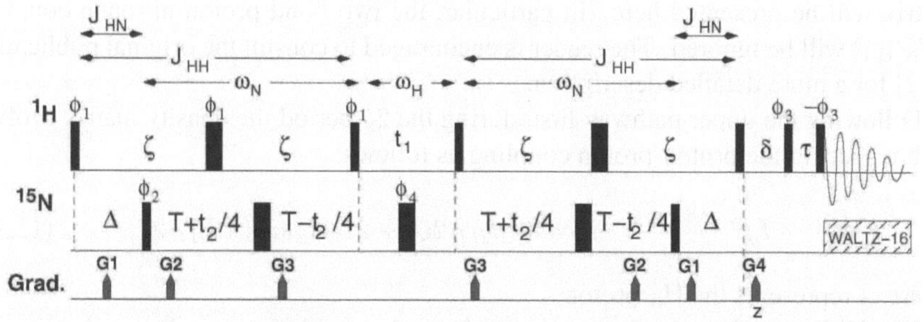

Figure 13.7. *HNHA pulse sequence.* This experiment was obtained from [90] and correlates the amide proton to the H_α proton within the same residue. Periods of evolution under J-coupling or chemical shift are indicated above the sequence. Typical proton and nitrogen power levels can be found in the legend to Fig. 13.6. The proton transmitter should be placed on water. Delays are ζ= 13.4 msec, Δ= 5.3 msec $\approx 1/(2J)$, δ= 2 msec, τ= 90 μsec. T is defined by Δ and ζ: 2T= $2\zeta - \Delta$. The length of τ depends on the frequency separation between the water and the center of the amide region, see text. All gradients are 25 G/cm sine-bell shaped, except for G4, which is -24 G/cm. G4 is a z-filter. Durations were G1= 0.3 msec, G2= 0.45 msec, G3= 0.75 msec, and G4= 0.5 msec. Phases are $\phi_1 = 2x, 2(-x); \phi_2 = x; \phi_3 = 4x, 4y; \phi_4 = x, y; \phi_{rec} = x, 2(-x), x, y, 2(-y), y$. Quadrature in t_1 and t_2 is generated using the States-TPPI method, by shifting the phase of ϕ_1 and ϕ_2, respectively.

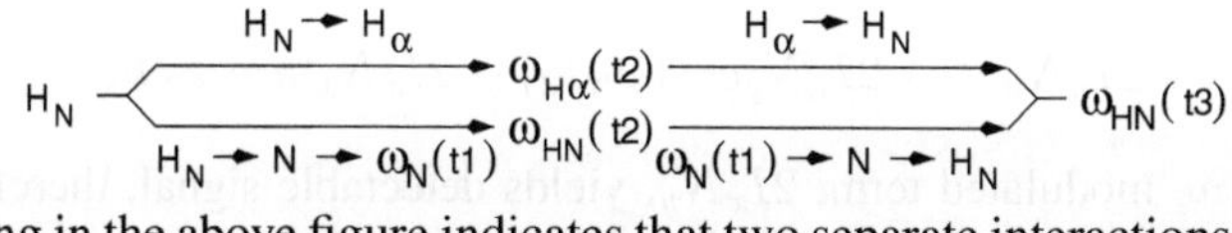

The branching in the above figure indicates that two separate interactions involving the amide proton occur simultaneously: coupling to the H_α proton, and coupling to the amide nitrogen. The experiment yields selfpeaks at $(\omega_N, \omega_{HN}, \omega_{HN})$ and crosspeaks at $(\omega_N, \omega_{H\alpha}, \omega_{HN})$. The latter provide the chemical shift of the H_α proton within a spin-system.

The intensity of the H_α crosspeaks is related to the size of the J-coupling between the amide and the H_α protons:

$$\frac{I_{cross}}{I_{self}} = -tan^2(\pi J 2\zeta) \tag{13.13}$$

where I_{cross} and I_{self} are the intensities of the cross- and selfpeaks, respectively. These intensities have to be corrected for different relaxation rates of the magnetization that occurs during the sequence. The time allowed for the transfer of magnetization between the two protons is represented by ζ. The pulse sequence for this experiment is presented in Fig. 13.7 and a product operator analysis is given below.

13.3.3.1 Product Operator Analysis of HNHA Experiment

The analysis of this experiment is rather involved and requires a consideration of the three bond proton-proton coupling between the amide and H_α protons, as well as the one and two bond couplings between the nitrogen and the amide and H_α protons, respectively. Consequently, only the major features of the evolution of the density matrix will be presented here. In particular, the two bond proton-nitrogen coupling ($^2J_{\mathrm{N\,H}\alpha}$) will be ignored. The reader is encouraged to consult the original publication [162] for a more detailed description.

Following the upper pathway first, during the 2ζ period the density matrix evolves with respect to the proton-proton coupling as follows:

$$I_z \stackrel{P_{90x}}{\rightarrow} -I_y \stackrel{\zeta-180^\circ-\zeta}{\longrightarrow} -I_y cos(\pi J_{HH}2\zeta) + 2I_x A_z sin(\pi J_{HH}2\zeta) \qquad (13.14)$$

where A represents the H_α proton.

In addition to the proton-proton coupling, the amide proton also evolves under coupling its attached nitrogen, the period Δ is a HMQC-type polarization transfer, followed by a 90° nitrogen pulse:

$$-I_y \stackrel{\Delta}{\rightarrow} 2I_x N_z \stackrel{P_{90}}{\rightarrow} -2I_x N_y \qquad (13.15)$$

The subsequent $T + t_2/4 - 180^\circ - T - t_2/4$ interval is a constant time segment, causing evolution of the nitrogen chemical shift for a period of $t_2/2$. Note that the second half of the nitrogen evolution occurs during the next 2ζ period, giving a total evolution of t_2. During these constant time periods evolution by $\mathrm{J_{HN}}$ coupling does not occur since both the coupled proton and nitrogen spins are transverse[2]. Therefore the $-2I_x N_y$ term evolves as:

$$-2I_x N_y \rightarrow -2I_x N_y cos(\omega_N t_2) + 2I_x N_x sin(\omega_N t_2) \qquad (13.16)$$

Only the *cosine* modulated term, $2I_x N_y$, yields detectable signal, therefore the *sin* term will be discarded at this point in the analysis.

Combining this chemical shift evolution with the proton to nitrogen polarization transfer and the evolution under proton-proton coupling (Eq. 13.14) gives the following for the density matrix just prior to the t_1 period:

$$[-cos(\pi J_{HH}2\zeta)I_x + sin(\pi J_{HH}2\zeta)2I_y A_z]\, 2N_y cos(\omega_N t_2) \qquad (13.17)$$

The 90° proton pulse converts the above to:

$$[-cos(\pi J_{HH}2\zeta)I_x + sin(\pi J_{HH}2\zeta)2I_z A_y]\, 2N_y cos(\omega_N t_2) \qquad (13.18)$$

and subsequent chemical shift evolution during t_1, where only the cosine terms are kept, produces:

$$[-cos(\pi J_{HH}2\zeta)cos(\omega_{HN}t_1)I_x + sin(\pi J_{HH}2\zeta)cos(\omega_{H\alpha}t_1)2I_z A_y]\, 2N_y cos(\omega_N t_2) \qquad (13.19)$$

The remaining part of the sequence reverses all of the above, such that at the end of the last ζ period the density matrix is:

$$I_y\, [-cos(\pi J_{HH}2\zeta)cos(\omega_{HN}t_1) + sin(\pi J_{HH}2\zeta)cos(\omega_{H\alpha}t_1)]\, cos(\omega_N t_2) \qquad (13.20)$$

[2]Terms such as $2I_x N_y$ or $2I_y N_y$ do not evolve under J-coupling, see Section 10.2.1.1.

the 90° proton pulse places this magnetization along the z-axis. Application of gradient G4 removes any coherent transverse magnetization leaving the desired signal along the z-axis.

The final segment of the sequence, $P_{90}^{\phi 3} - \tau - P_{90}^{-\phi 3}$, is jump-and-return selective excitation sequence that will place the magnetization associated with the amides in the x-y plane, but will leave the water along the z-axis (see Section 15.2.2.5 for a discussion of water suppression by the jump-return sequence). The final detected signal is:

$$\left[-cos^2(\pi J_{HH}2\zeta)cos(\omega_{HN}t_1) + sin^2(\pi J_{HH}2\zeta)cos(\omega_{H\alpha}t_1)\right] cos(\omega_N t_2)e^{i\omega_{HN}t_3} \tag{13.21}$$

Selfpeaks will have an intensity proportional to:$-cos^2(\pi J_{HH}2\zeta)$ while crosspeaks will have an intensity proportional to: $sin^2(\pi J_{HH}2\zeta)$.

These intensities have to be corrected for differential relaxation of the magnetization during the ζ periods. During this period the magnetization associated with the selfpeaks was represented by the product operator $2I_xN_y$ while the magnetization associated with the crosspeak is of the form $4I_yN_yA_z$. Due to the presence of the A_z term the relaxation rate of the magnetization is increased by the spin-lattice relaxation rate, R_1, of the H_α proton:

$$R_2^{2I_yN_yA_z} \approx R_2^{2I_xN_y} + R_1^{A_z} \tag{13.22}$$

Therefore the magnetization associated with the selfpeak decays approximately as:

$$I_{self} = I_o e^{-R_2 4\zeta} \tag{13.23}$$

while the magnetization associated with the crosspeak decays approximately as:

$$I_{cross} = I_o e^{-R_2 4\zeta} e^{-R_1 4\zeta} \tag{13.24}$$

The coupling constant is calculated as:

$$J = \frac{1}{\pi 2\zeta} arctan\left[\sqrt{\frac{|I_{cross}e^{+R_1 4\zeta}|}{|I_{self}|}}\right] \tag{13.25}$$

In order to correct the intensities it is necessary to know R_1 for each H_α proton. An R_1 value of 10 sec^{-1} was used by Vuister and Bax [162] to correct for this effect in Staphylococcal nuclease, a 15 kDa protein. Values for other proteins can be obtained by assuming that R_1 is proportional to the rotational correlation time, as discussed in Chapter 19.

13.3.4 The HNHB Experiment- Identifying H_β Protons

The three-bond coupling between the amide nitrogen and the H_β protons can be used to correlate the latter with the amide resonances of a spin-system. The pulse sequence, modified from the original version [6] to include pulsed field gradients, is shown in Fig. 13.8. The overall pathway of magnetization transfer can be summarized as follows:

$$H_N \overset{J_{HN}}{\longrightarrow} N_H \overset{\omega_N+J}{\longrightarrow} H_\beta cos(\omega_N t_1) \overset{\omega_{H\beta}}{\longrightarrow} H_\beta cos(\omega_{H\beta}t_2)cos(\omega_N t_1) \quad (13.26)$$
$$\overset{J_{NH\beta}}{\longrightarrow} N_H cos(\omega_{H\beta}t_2)cos(\omega_N t_1) \overset{J_{NH}}{\longrightarrow} H_N cos(\omega_{H\beta}t_2)cos(\omega_N t_1)$$

The intensities of the crosspeaks are:

$$I_{cross} = sin^2(\pi J_{NH_\beta}\Delta)sin^2(\pi J_{NH}\Delta)\Pi_k cos^2(\pi J_{Nk}\Delta) \quad (13.27)$$

where k is the sum over all protons, besides the H_β and H_N protons, that are coupled to the amide nitrogen.

The evolution of the density matrix that gives rise to the above signal will be briefly discussed. Beginning with proton magnetization along the z-axis, $\rho = I_z$, the proton pulse and subsequent INEPT period produce the following:

$$I_z \overset{P90x}{\rightarrow} -I_y \overset{INEPT}{\longrightarrow} 2I_xN_z \quad (13.28)$$

The subsequent proton and nitrogen 90° pulses generate $\rho = 2I_zN_y$. This term evolves in the subsequent constant time period under nitrogen chemical shift as well as J-coupling to the amide proton, the H_β proton, as well as to any other protons coupled to the nitrogen, such as the H_α proton. Coupling of the nitrogen to the amide proton is refocused by setting the delay, Δ to an odd multiple of $1/(2J_{NH})$, as described below:

$$\begin{aligned} 2I_zN_y &\rightarrow 2I_zN_ycos(\pi J\Delta) - N_xsin(\pi J\Delta) \\ &= 2I_zN_ycos(\pi\frac{n}{2}) - N_xsin(\pi\frac{n}{2}) \\ &= -N_x \quad n = 1, 3, 5... \end{aligned} \quad (13.29)$$

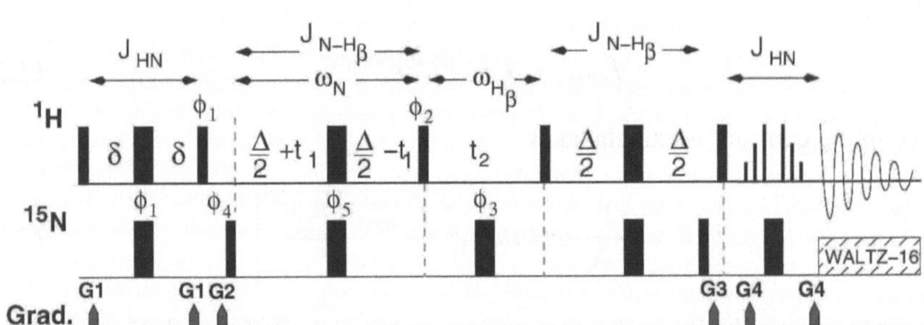

Figure 13.8. *HNHB pulse sequence.* This experiment has been modified from the original sequence [6] to include pulse field gradients and WATERGATE for suppression of the solvent (see Section 15.2.2.4). The delay, δ is set to $1/(4J_{NH})$, or shorter depending on the relaxation rate of amide proton. The delay Δ is set to odd multiples of $1/(2J_{NH})$, 38 msec is a typical value. Pulse phases are $\phi_1 = y, -y; \phi_2 = 2(x), 2(-x); \phi_3 = 4(x), 4(y), 4(-x), 4(-y); \phi_5 = 16(x), 16(-x); \phi_{rec} = x, 2(-x), x, -x, 2(x), -x$. Quadrature in nitrogen (t_1) is States-TPPI, attained by incrementing ϕ_4 from x to y. Quadrature in t_2 is obtained using the same method, except that ϕ_2 is incremented by 90°. Typical Gradient levels would be, G1 (8G/cm, 500 μsec), G2 (16 G/cm, 200 μsec), G3 (12 G/cm, 200 μsec, G4 (24 G/cm, 250 μsec). The gradient pair labeled G1 removes artifacts associated with the 180° pulses, G2 and G3 are zz-filters and G4 dephases the water magnetization.

At the same time, the magnetization also evolves under the nitrogen chemical shift:

$$-N_x \rightarrow -N_x cos(\omega_N t_1) - N_y sin(\omega_N t_1) \tag{13.30}$$

as well as due to J-coupling to the H_β protons, represented by B here:

$$-N_x cos(\omega_N t_1) \stackrel{J_{NH\beta}}{\rightarrow} [-N_x cos(\pi J_{NH\beta}) - 2N_y B_z sin(\pi J_{NH\beta})] \, cos(\omega_N t_1) \tag{13.31}$$

where only the $cos(\omega_N t_1)$ term has been kept. The proton pulse just prior to t_2 makes the product operator associated with the H_β proton transverse, allowing it to become modulated with its chemical shift:

$$2N_y B_z \stackrel{P_{90,H}}{\rightarrow} 2N_y B_y \rightarrow 2N_y \left[B_y cos(\omega_{H\beta} t_2) - B_x sin(\omega_{H\beta} t_2)\right] \tag{13.32}$$

Note that evolution of the ^{15}N chemical shift is suppressed by the nitrogen 180° pulse in the middle of the t_2 period.

The 90° proton pulse after the t_2 period converts B_y back to B_z, permitting refocusing of the coupling between the nitrogen and the H_β proton by the INEPT sequence:

$$2N_y B_z \stackrel{J_{NH\beta}}{\longrightarrow} 2N_y B_z cos(\pi J_{NH\beta}\Delta) - N_x sin(\pi J_{NH\beta}\Delta) \tag{13.33}$$

Evolution due to coupling between the amide nitrogen and its proton also occur during this period. Since only the N_x term will give rise to detectable signal, only the N_x term will be followed:

$$\begin{aligned} -N_x sin(\pi J_{NH\beta}\Delta) \stackrel{J_{NH_N}}{\rightarrow} & -sin(\pi J_{NH\beta}\Delta)[N_x cos(\pi J_{NH_N}\Delta) \\ & + 2N_y I_z sin(\pi J_{NH_N}\Delta)] \end{aligned} \tag{13.34}$$

Since $\Delta = n/(2J_{NH_N})$, $cos(\pi J_{NH_N}\Delta) = 0$, and the above gives:

$$-N_x sin(\pi J_{NH_\beta}\Delta) \stackrel{J_{NH_N}}{\rightarrow} -sin(\pi J_{NH_\beta}\Delta) 2N_y I_z sin(\pi J_{NH_N}\Delta) \tag{13.35}$$

The subsequent pair of 90° pulses, followed by the final INEPT transfer, which is a WATERGATE segment for water suppression, produces:

$$-2N_y I_z \stackrel{P90}{\rightarrow} -2N_z I_y \stackrel{INEPT}{\rightarrow} I_x \tag{13.36}$$

The final signal is:

$$S(t_1, t_2, t_3) = I_{cross} cos(\omega_N t_1) cos(\omega_{H\beta} t_2) e^{i\omega_{HN} t_3} \tag{13.37}$$

13.3.5 Establishing Spin-system Connectivities With Dipolar Coupling

The TOCSY, HNHA, and HNHB experiments only provide information on proton connectivities *within* a residue or spin-system. In order to sequentially connect spin-systems it is necessary to identify inter-residue connectivities. These cannot be established using experiments that utilize proton-proton J-coupling because these couplings are very weak across the peptide bond. Consequently it is necessary to use dipolar, or through space, interactions to detect inter-residue couplings. These experiments are usually referred to as NOESY experiments and routinely give rise to crosspeaks if the coupled protons are within 5 Å of each other. Although the emphasis in this section is on establishing inter-residue connectivities, NOESY spectra can also aid in defining spin-systems because many of the protons within an amino acid sidechain are close enough to each other to give rise to intra-residue crosspeaks in the NOESY spectrum. Dipolar coupling is discussed in detail in Chapter 16.

A three dimensional ^{15}N-separated NOESY experiment is shown in Fig. 13.9. The experiment begins by frequency labeling all of the protons, this information is transferred, via dipolar coupling, to any amide protons within approximately 5Å of the original proton during the mixing time (τ_m). The magnetization is then transferred to the amide nitrogen where it becomes labeled with the nitrogen chemical shift. The magnetization is transfered back to the amide proton for detection. The final signal is:

$$\eta cos(\omega_{Hk}t_1)cos(\omega_N t_2)e^{i\omega_{HN}t_3} \qquad (13.38)$$

where η is inversely proportional to the sixth power of the distance between proton k and the amide proton ($\eta \propto 1/d^6$).

The sequence presented in Fig. 13.9 is similar in style to that of the 3D-TOCSY sequence, consequently the following discussion will be brief and will focus primarily on how the water magnetization is restored to the z-axis at the end of each scan.

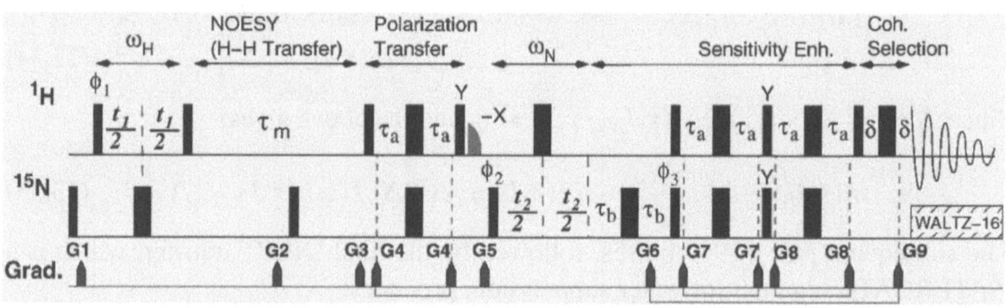

Figure 13.9. Pulse sequence for a 3D-^{15}N Separated NOESY experiment (HSQC-NOESY). This experiment was obtained from Zhang et al. [176]. Most of the experimental parameters are described in Fig. 13.6. Parameters that are specific to this experiment are described here. The NOESY mixing time, τ_m is typically set to 150 msec. The phase cycle is $\phi_1 = 135°, 315°; \phi_2 = 2(x), 2(-x); \phi_3 = x, \phi_{rec} = x, 2(-x), x$. Quadrature in ω_1 is obtained with States-TPPI by shifting the phase of ϕ_1. N- and P-coherences in ω_N are selected by inverting the sign of gradient G9 and inverting the phase of ϕ_3 to $-x$ from x. See Chapter 12 for processing. Gradient strengths are: G1 (4 G/cm, 1 msec), G2 (10 G/cm, 1 msec), G3 (8 G/cm, 0.5 msec), G4 (5 G/cm, 750 μsec), G5 (-15 G/cm, 1.5 msec), G6 (30 G/cm, 1.25 msec), G7=G8 (4 G/cm, 0.5 msec), G9 (27.8 G/cm, 125 μsec). Gradients G1 and G3 insure that only magnetization originating of protons is detected. Gradient G2 dephases any residual transfer water magnetization, see text. Gradient G5 is a zz-filter. Gradient pairs G4, G7, and G8 remove artifacts associated with the 180° pulses. Gradients G6 and G9 select N- or P-nitrogen coherences.

The key feature of management of the water magnetization is the use of radiation damping to place the water along the $+z$-axis by the end of the mixing period, τ_m. The subsequent INEPT, or polarization transfer period, will place the water magnetization along the y axis. The water selective flip-back pulse, with phase $-x$, will place the water along the $-z$-axis. The application of gradient G5 immediately after this pulse will inhibit radiation damping by dephasing any water magnetization in the x-y plane. The proton 180° pulse in the middle of the t_2 evolution period brings the water back to $+z$. During the remaining two INEPT periods, and the coherence selection spin echo, the water magnetization is transformed as follows:

$$I_z \xrightarrow{90_X} -I_y \xrightarrow{180_X} +I_y \xrightarrow{P90_Y} +I_y \xrightarrow{180_X} -I_y \xrightarrow{90_X} -I_z \xrightarrow{180_X} +I_z \qquad (13.39)$$

To insure that there is sufficient transverse water magnetization at the beginning of τ_m to cause radiation damping, regardless of the quadrature phase of ϕ_1, it is necessary to set the initial phase of ϕ_1 to 135°. In this case the water magnetization will be:

$$M_x = I_o cos(45°) \qquad M_y = -I_o sin(45°) \qquad (13.40)$$

The 90° proton pulse at the end of the t_1 labeling period will leave the x-component, or $I_o/\sqrt{2}$, of the water magnetization in the transverse plane, leading to effective radiation damping.

When the *sine* modulated proton magnetization is selected in the second step of the States-TPPI protocol, ϕ_1 will become 225° $(= 130° + 90°)$. The resultant 90° pulse will produce the following state for the water magnetization:

$$M_x = -I_o cos(45°) \qquad M_y = -I_o sin(45°) \qquad (13.41)$$

The 90° pulse at the end of the t_1 period will leave the x-component, or $I_o/\sqrt{2}$ in the transverse plane, again leading to effective radiation damping.

If the phase of the first proton pulse had been set to $+x$ (0°), then the entire water magnetization would have placed along the minus z-axis at the beginning of the mixing period, which would be relatively inefficient for radiation damping. When the *sine* component was acquired, by shifting the phase of the ϕ_1 pulse to $+y$, the entire water magnetization would be left in the transverse plane, leading to very efficient radiation damping. Consequently, the $cos(\omega_H t_1)$ and $sin(\omega_H t_1)$ quadrature signals would show significant differences in the water signal at the end of each scan. By setting ϕ_1 to 135°, efficient radiation damping is insured in both cases.

13.3.5.1 4D-^{15}N Separated NOESY

The three-dimensional experiment can be easily extended to four-dimensions. The pulse sequence is shown in Fig. 13.10 and a simulated spectrum is shown in Fig. 13.11.

The overall experiment can be summarized as follows:

$$\omega_N^j t_1 \xrightarrow{J_{NH}} \omega_H^j t_2 \xrightarrow{Dipolar} \omega_N^k t_3 \xrightarrow{J_{NH}} \omega_H^k t_4$$

and the overall signal for a crosspeak is:

$$\eta cos(\omega_N^j t_1) cos(\omega_H^j t_2) cos(\omega_N^k t_3) e^{i\omega_H^k t_4} \qquad (13.42)$$

The amide nitrogen and proton frequencies of one spin-system, e.g. j, are correlated with the amide nitrogen and proton frequencies of any other amide group, e.g. k, within approximately 5 Å.

The experiment can be readily described using the pulse-sequence elements discussed in Appendix D. The initial polarization of the amide proton is transferred to the amide nitrogen by an INEPT sequence. Following frequency labeling by ω_N the magnetization is transferred back to the same amide proton using a semi-constant time period. Frequency labeling by the amide proton also occurs during this period. The magnetization is transferred from one amide proton to another during the mixing time, τ_m. The remainder of the sequence consists of a INEPT transfer to the amide nitrogen, frequency labeling by ω_N, and transfer back to the amide proton by an INEPT/WATERGATE sequence, whose frequency is detected in t_4.

The principal advantage of the 4D NOESY experiment over the corresponding three-dimensional experiment is that both the proton and nitrogen shifts of each amide group that are within 5 Å of each other are obtained, as illustrated in Fig. 13.11. Having both the amide nitrogen and proton shifts of each amide group facilitates the unique assignment of crosspeaks in the NOESY spectra to individual amide groups.

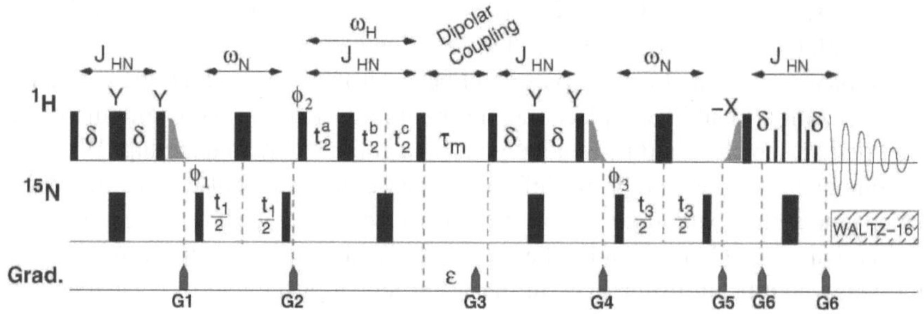

Figure 13.10. Pulse sequence for a 4D amide-amide NOESY experiment. This experiment was obtained from [71]. The gray pulses represent 1 msec half-Gaussian water selective pulses. All pulses are along the x-axis unless otherwise noted. The delay δ is set to $1/(4J_{NH})$, or slightly shorter depending on the R_2 of the amide proton. The delay ϵ is set to 170 msec and allows for radiation damping to return the water to the z-axis. t_2^a, t_2^b, and t_2^c are associated with a semi-constant time evolution, see Table D.1 for details. Phase cycling is: $\phi_1 = x, -x; \phi_2 = 45°; \phi_{rec} = x, -x$. ϕ_2 is set to 45° to insure equal treatment of the water magnetization for both quadrature phases, see discussion with 3D-NOESY experiment. Quadrature detection in t_1, t_2, and t_3 is accomplished using the States-TPPI method, by incrementing ϕ_1, ϕ_2, and ϕ_3, respectively. Typical gradients are sine squared in shape with amplitudes and durations of: G1 (2.15 msec, 20 G/cm), G2 (1.35 msec, 20 G/cm), G3 (5.0 msec, 12 G/cm), G4 (3.0 msec, 16 G/cm), G5 (1.0 msec, 20 G/cm), and G6 (0.30 msec, 20 G/cm).

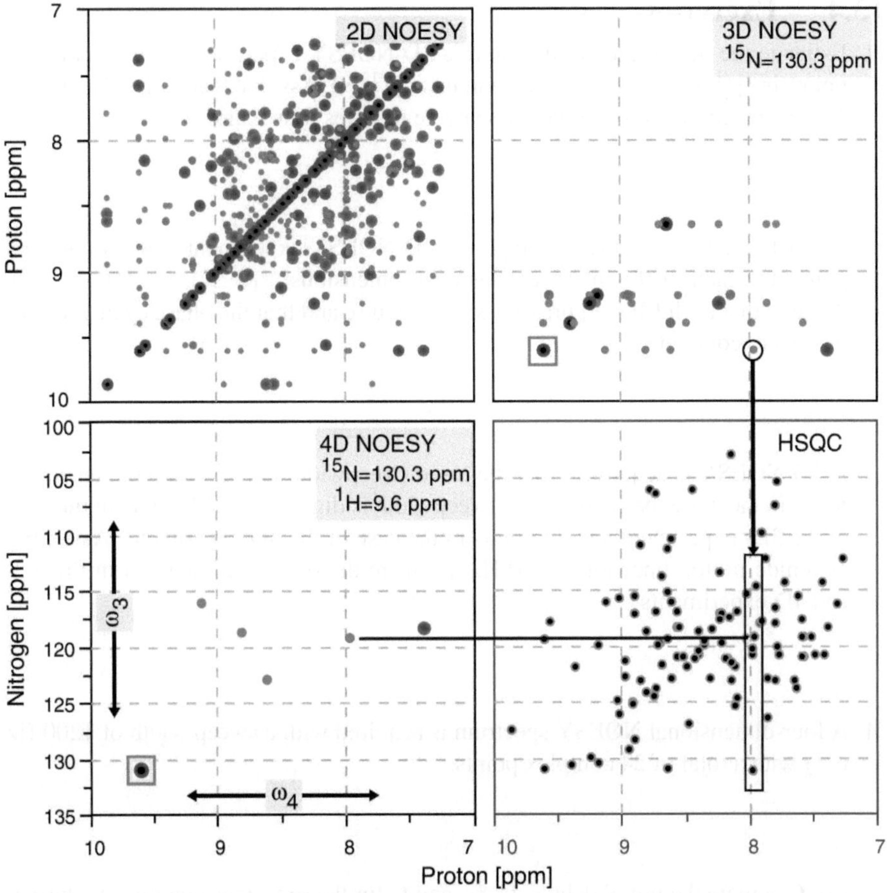

Figure 13.11. Increased resolution in 4D-NOESY spectra. The 4D-NOESY experiment allows the unique assignment of crosspeaks to amide groups. The top left panel shows the amide region of a 2D-NOESY of a 130 residue protein. Larger spots correspond to more intense peaks, which arise from amide protons that are close in space. The top right panel show a slice from the 3D-experiment, at an amide nitrogen frequency of 130.3 ppm. The lower right panel is a 2D-HSQC experiment and the lower left panel shows a 2D plane that was extracted from the 4D-NOESY spectrum at the indicated nitrogen and protons shifts. The boxed peak indicates the self-peak in both NOESY spectra. The considerable overlap in the 2D-spectra makes it difficult to obtain unique assignments of the NOESY crosspeaks. Separation of the 2D-spectrum by the ^{15}N chemical shift, to generate the 3D-NOESY spectrum, resolves most of the crosspeaks. However, there are many possible assignments for the crosspeaks. For example, the circled crosspeak at 9.6, 8.0 ppm represents an NOE between an amide group with a chemical shift $\omega_N = 130.3$ ppm, $\omega_H = 9.6$ ppm with another amide whose proton shift is 8.0 ppm. The area of the HSQC spectrum that is enclosed by a rectangle (lower right) shows that there are 5 or 6 possible assignments for the other amide that participates in this NOESY crosspeak. This ambiguity arises because the *nitrogen* shift of the crosspeak is not known. In the 4D-NOESY spectrum (lower left), both the nitrogen and proton frequencies of the amides are obtained, allowing an unambiguous assignment of this peak to a single amide group. The horizontal line indicates the position of this peak in both the 4D-NOESY and the HSQC spectrum.

13.4 Exercises

1 Estimate the time required to acquire a 3D-NOESY experiment with 64 complex points in $t_1(^1\text{H})$, and 32 complex points in $t_2(^{15}\text{N})$. Assume that each FID requires 1.1 sec to acquire and that the phase cycle requires four scans.

2 Estimate the time required to acquire a 4D-NOESY experiment with 24 complex points for each of the indirectly detected dimensions: $t_1(^{15}\text{N})$, $t_2(^1\text{H})$, $t_3(^{15}\text{N})$. Assume that each FID requires 1.1 sec to acquire and that the phase cycle requires 2 scans to complete.

3 A 3D-NOESY is acquired using the same sweepwidth in the proton dimension as for t_2 in the 4D-experiment. This sweepwidth is digitized by 64 points in the case of the 3D- experiment and 24 points in the case of the 4D-experiment. Assuming an amide proton linewidth of 10 Hz, estimate the observed linewidth in the 3D- and 4D-experiments.

4 A four-dimensional NOESY spectrum is acquired with a sweep width of 1200 Hz in t_2 with a total of 24 complex points.

(a) Calculate the initial delays, t_a, t_b, and t_c for the semi-constant time evolution, assuming $J_{NH} = 92$ Hz.

(b) Determine the values for the increments for each period.

5 The following figure gives a TOCSY spectrum and a NOESY spectrum for a 4-residue segment that is contained within a larger protein. The partial sequence for this protein is:

Gly_{46}-Glu_{47}-Asp_{48}-Ile_{49}-Cys_{50}-Gly_{51}-Asp_{52}-Gly_{53}-Val_{54}-Leu_{55}-Glu_{56}-Ile_{57}

Only NOE peaks between protons on this segment are shown. On the basis of these spectra, determine the assignments for this segment of the protein and suggest a secondary structure of this segment.

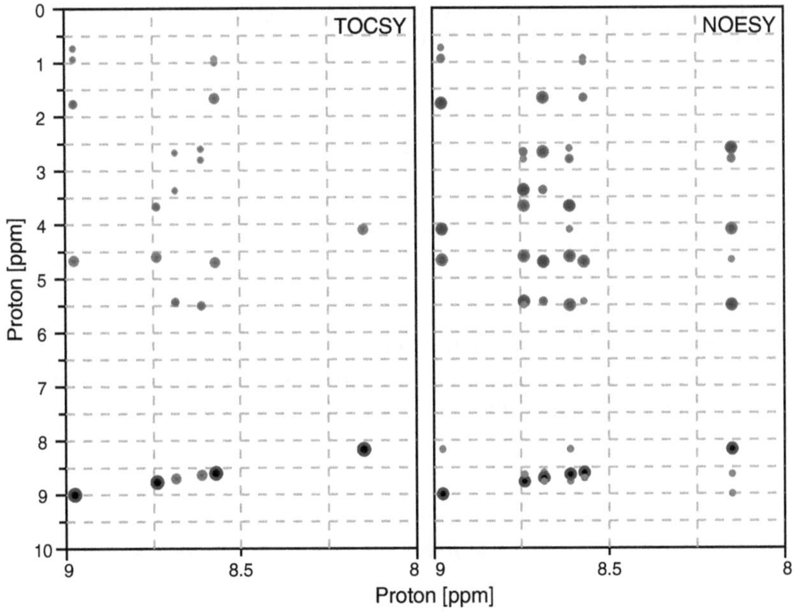

13.5 Solutions

1 The number of points that need to be acquired for each dimension are doubled due to quadrature detection. Each point requires 4.4 sec to acquire due to the requirement to complete the phase cycle. The total time is:
$64 \times 2 \times 32 \times 2 \times 4.4$ sec = 36044 sec = 10 hours.

2 Each point requires 2.2 sec to acquire. The total time is:
$(24 \times 2)^3 \times 2.2$ = 243,302 sec = 67 hours (2.8 days).
A 4D-experiment, even with significantly lower digital resolution than a similar 3D-experiment, can take 6-10 times longer.

3 The total acquisition time is given by: $T_{ACQ} = (N - 1) \times t_{dw}$. This defines the length of the square wave that the true FID was multiplied by to generate the observed FID. The Fourier transform of the product of the square wave and the FID is the convolution of their individual transforms. A reasonable estimate of the final linewidth is simply the sum of the intrinsic linewidth (10 Hz) and the width of the sinc function that was generated from the square wave. The Fourier transform of the sinc function is given in Appendix A, and the width is approximately: $\Delta\nu = 1/\tau_{ACQ}$. In the case of the 3D-experiment: $\Delta\nu = 1/(833\ \mu\text{sec} \times 63) = 19$ Hz, or the digital resolution contributes an additional 19 Hz to the linewidth. In the case of the 4D-experiment: $\Delta\nu = 1(833\ \mu\text{sec} \times 23) = 52$ Hz. The lines in the 4D-NOESY will be twice as broad as those in the 3D-NOESY, reducing the resolution in the spectrum.

4 (a) The pulse sequence given in Fig. 13.10 uses version *ii* of the semi-constant time sequences. Using Table D.1:

$$\begin{aligned} t_a^o &= 1/4J = 2.71 \text{ msec} \\ t_b^o &= 0 \\ t_c^o &= 1/4J = 2.71 \text{ msec} \end{aligned}$$

(b) The dwell time, $\Delta t = 1/1200 = 833\ \mu\text{sec}$ and the index of the last point, N is 23. However, $N = 25$ will be used to insure that the shortest delay for t_a will still allow room for the 180° pulse. This gives the following:

$$\begin{aligned} \delta t_a &= -2.71 \text{ msec}/25 = -108\ \mu\text{sec} \\ \delta t_b &= 833\ \mu\text{sec}/2 - 108\ \mu\text{sec} = +308\ \mu\text{sec} \\ \delta t_c &= 833\ \mu\text{sec}/2 = +416.5\ \mu\text{sec} \end{aligned}$$

5 Annotated spectra are shown below. The first step is to characterize the 6 spin-systems that are present in the TOCSY. In this example the H_α protons all have chemical shifts higher then the H_β shift. If the sample was labeled with ^{15}N the identity of the H_α protons could be confirmed with a HNHA experiment. The chemical shifts associated with each spin-system, and the most likely residue type are as follows:

Spin Sys.	HN	H_α	H_β	Others	Likely Amino Acid
A	8.90	4.6	1.70	0.7, 0.9	Val, Leu, or Ile based on two methyls.
B	8.75	4.6, 3.6	-	-	Gly if these are two H_α protons. Could also be Ser if one proton is actual H_β. Could also be a Thr, with the methyl resonance missing.
C	8.70	5.4	3.4, 2.7		Typical β-residue, likely Asp, Asn, Cys. Phe, Trp, Tyr are also possible, but no aromatics protons are present.
D	8.60	5.5	2.8, 2.6		Same discussion as C.
E	8.57	4.7	1.7	0.97, 0.92	Same discussion as A.
F	8.15	4.08	-	-	Same discussion as B, except two H_α shifts are degenerate.

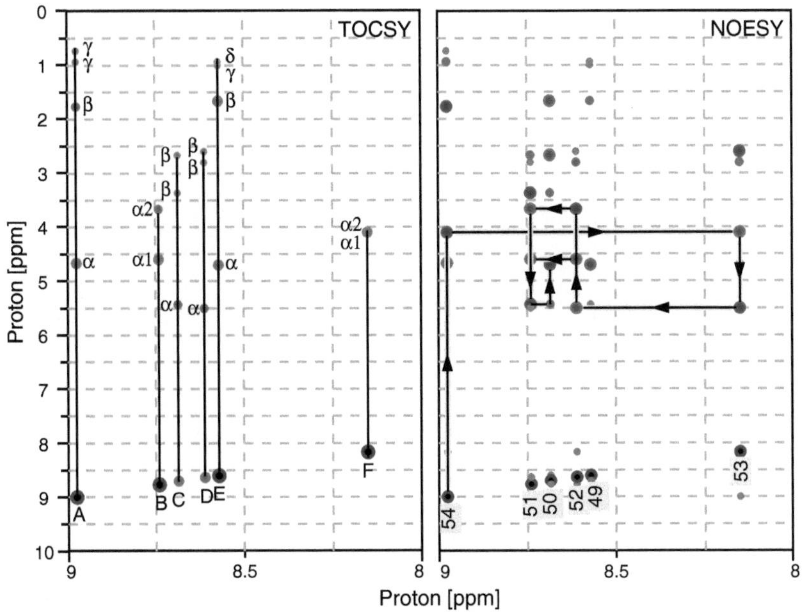

The NOESY spectrum is used to determine adjacent spin-systems. The amide-amide peaks are relatively weak in the NOESY, but there are a number of strong NOEs between amides and H_α protons, suggesting β-sheet structure. To follow the connectivities, one starts at an amide and finds a strong NOE to an H_α proton that is not on the same residue, this is likely the H_α proton of the preceding residue. In this example, it is possible to begin at spin-system A, find an NOE to the H_α proton of spin-system F. The amide of spin-system F has a strong NOE to the H_α proton of spin-system D. This type of linkage can be followed all the way to spin-system E, as indicated by the arrows. Therefore, the order of the spin-systems, written from the N-Terminus, is:

E C B D F A

This order can be combined with the residue type information presented above to give the following representations of the sequence, with the most likely amino acids in bold and the correct amino acid underlined:

$$\begin{bmatrix} Val \\ Leu \\ \underline{Ile} \end{bmatrix} - \begin{bmatrix} Asp \\ Asn \\ \underline{Cys} \end{bmatrix} - \begin{bmatrix} \underline{\textbf{Gly}} \\ Ser \\ Thr \end{bmatrix} - \begin{bmatrix} \underline{Asp} \\ Asn \\ Cys \end{bmatrix} - \begin{bmatrix} \underline{\textbf{Gly}} \\ Ser \\ Thr \end{bmatrix} - \begin{bmatrix} \underline{Val} \\ Leu \\ Ile \end{bmatrix}$$

Comparing this ordering to the known amino-acid sequence:

Gly_{46}-Glu_{47}-Asp_{48}-Ile_{49}-Cys_{50}-Gly_{51}-Asp_{52}-Gly_{53}-Val_{54}-Leu_{55}-Glu_{56}-Ile_{57}

indicates that spin-system E is Ile_{49}, C is Cys_{50}, etc. Note the importance of Gly and methyl containing residues in positioning the connected spin-systems within the primary sequence.

Chapter 14

RESONANCE ASSIGNMENTS: HETERONUCLEAR METHODS

Triple-resonance experiments, involving ^{15}N, ^{13}C and ^{1}H spins, provide a much more reliable and robust method of obtaining resonance assignments. In the case of main chain assignments, the direct and relatively large scalar coupling between carbon and nitrogen across the peptide bond can be used to directly link spin systems. In the case of sidechain assignments, the large carbon-carbon coupling can be used to efficiently pass magnetization throughout the sidechain using isotropic (TOCSY) mixing. Alternatively, INEPT-like transfers between carbon atoms can be used to transfer magnetization in a well controlled fashion. Fig. 14.1 provides a summary of the heteronuclear coupling constants that can be exploited for the transfer of magnetization between spins.

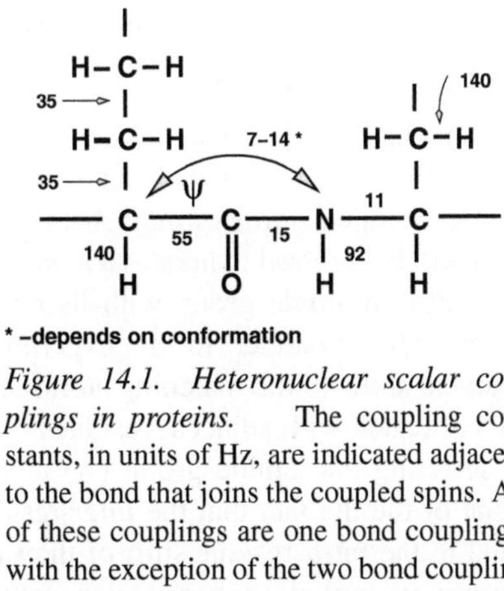

Figure 14.1. Heteronuclear scalar couplings in proteins. The coupling constants, in units of Hz, are indicated adjacent to the bond that joins the coupled spins. All of these couplings are one bond couplings, with the exception of the two bond coupling between the amide nitrogen and the C_α carbon of the preceding residue. All of the one-bond couplings are essentially independent of the secondary structure. In contrast, the two-bond coupling between the nitrogen and the C_α carbon, which depends on the ψ angle. Note that the two-bond coupling between the amide nitrogen and its own carbonyl carbon is essentially zero, thus it is only practical to directly correlate the amide nitrogen shift with the carbonyl shift of the preceding residue.

The remainder of this chapter will be divided into two sections. The first will discuss the general strategy for obtaining mainchain assignments by triple-resonance techniques. Three pulse sequences will then be described in some detail, providing the reader with sufficient information to analyze the large number of triple-resonance sequences that are currently described in the literature. The second section of this chapter will briefly discuss approaches to obtaining sidechain assignments, primarily by the use of carbon-TOCSY experiments.

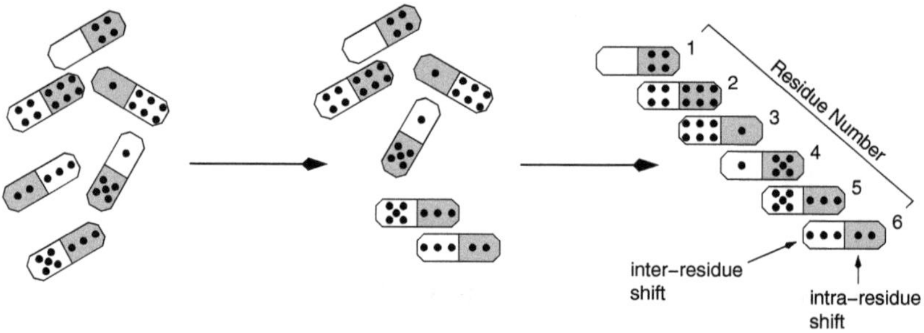

Figure 14.2. Determining sequential assignments using inter-residue chemical shifts. Each domino represents an amide group in a six residue protein. The number of dots on each half of each domino represent the chemical shift of the matching atom, such as the C_α shift. The shaded half of the domino represents the intra-residue chemical shift while the unshaded region represents the inter-residue chemical shift. Initially, the dominoes are unordered (left) but can be arranged in order such that the inter-residue shift of one amide group is matched to the intra-residue shift of the previous amide group in the polypeptide chain, as shown on the right.

14.1 Mainchain Assignments

14.1.1 Strategy

Mainchain, or backbone[1] assignments are obtained by first correlating the chemical shifts of the amide group (N_H,H_N) of a spin-system with *both* the inter- and intra-residue chemical shifts of mainchain or sidechain atoms. The sidechain or mainchain atom will be referred to here as a *matching atom.* For example, the HNCA experiment correlates an amide group with its own α-carbon as well as the α-carbon shift of the preceding residue. In this experiment the α-carbon is the matching atom. The chemical shifts of the matching atoms are then used to determine which amide groups are associated with adjacent residues on the polypeptide chain. This is accomplished by selecting one amide group (A) and then identifying the adjacent amide (B) by virtue of the the fact that the *inter*-residue shift of the matching atom of amide A is equal to the *intra*-residue shift of the same matching atom of amide B. This process is repeated until all sequential connectivities are obtained, as illustrated in Fig. 14.2. Under favorable conditions it is possible to determine the complete ordering of the spin-systems in the protein by this method.

The above process can only lead to complete sequential connectivities of the residues if the chemical shift of the matching atoms are unique for all residues in the protein. Otherwise an ambiguity will eventually occur, in that more than one amide group can be the sequential residue. This ambiguity can often be resolved by using a second, or third, matching atom. For example, the carbonyl- and β-carbon shifts are routinely used for this process. In non-deuterated protons, the α- and β-protons are also commonly used a matching atoms. The use of an additional chemical shift to resolve degeneracies is illustrated in Fig. 14.3.

[1]Mainchain, or backbone, atoms are generally considered to be the H_N, N, C_α, H_α, and C' (carbonyl carbon) atoms.

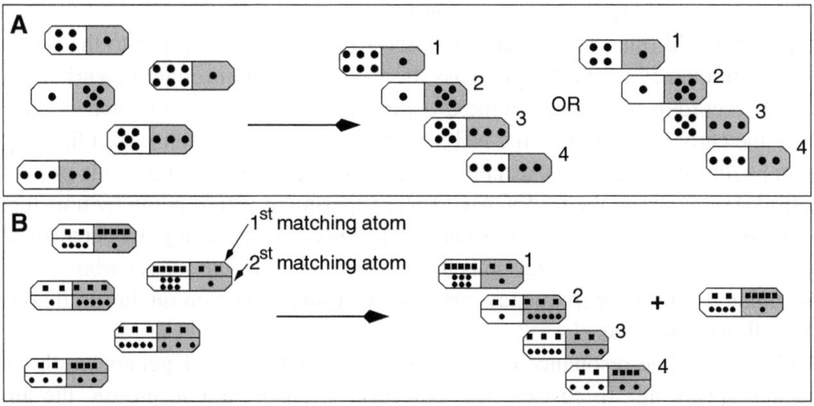

Figure 14.3. Resolving degeneracies in the assignment process. The use of an additional matching atom to resolve degeneracies is illustrated. *Panel A:* One matching atom has been used and two of the five residues have the same intra-residue chemical shift (δ = "one"). When the residues are ordered, there are two possible choices for the first residue in the chain and thus the assignment of this residue is ambiguous. *Panel B:* Inclusion of an additional matching atom (e.g. β-carbon) resolves this degeneracy by providing an additional chemical shift to distinguish between the two choices.

In practice, due to missing data, or unresolvable ambiguities, or the presence of proline residues that lack an amide proton, it will not be possible to unambiguously order all of the spin-systems. Instead, one obtains segments of linked residues of varying lengths. These segments have to be correctly located in the known primary sequence of the protein in a manner similar to that presented in Fig. 13.1. Therefore, the amino acid type of each residue (spin-system) in the segment has to be determined.

As with the case of homonuclear assignment methods, determination of the type of the amino acid of the spin-system is based on the characteristic chemical shifts of the atoms within an amino acid residue. The assignment of residue type is greatly facilitated by knowledge of the carbon chemical shifts. For example, a residue with a β-carbon shift of 20 ppm and a β-proton shift of 1.0 ppm is very likely to be an alanine (see Fig. 1.16 for chemical shifts). Although the process of residue identification does not always lead to a definitive assignment of residue type, usually there is sufficient information to correctly associate a connected segment of spin-systems with the corresponding region of the primary sequence of the protein.

14.1.2 Methods for Mainchain Assignments

A large number of triple-resonance pulse sequences have been devised for mainchain assignments and a number of these are listed in Table 14.1. The pathway of magnetization transfer in a small number of these is shown in Fig. 14.4. The usual nomenclature for these experiments is to list the nuclei in the order that frequency labeling occurs in the pulse sequence. If magnetization is passed through a spin with no frequency labeling then that spin is enclosed in parenthesis. In addition, an experiment that begins with the magnetization on the amide proton, indicated as 'HN' in the experiment name, generally implies an 'out-and-back' experiment, i.e. the magnetization is returned to the amide proton for detection. For example the HN(CO)CA experiment would transfer the magnetization from the amide proton to the nitrogen, record the frequency of the N spin, pass the magnetization through the carbonyl carbon (CO), record the frequency of the C_α spin, and then return the magnetization back through the carbonyl carbon to the NH proton for detection. In contrast, if the magnetization begins elsewhere, it is usually passed in one direction to the amide proton for detection. For example the (HA)CA(CO)NH experiment would begin by transferring the α-proton magnetization to the α-carbon, followed by recording the chemical shift of the α-carbon. The magnetization is then transfered to the carbonyl carbon and then to the amide nitrogen for frequency labeling, with final detection on the amide proton on the following residue.

Each of the triple-resonance experiments listed in Table 14.1 generate a three dimensional spectrum. In cases where detection is on the amide proton, the amide nitrogen and proton frequencies generally comprise the second and third frequency dimension and the remaining dimension corresponds to the matching atom. A slice from this three dimensional spectrum at the nitrogen frequency of an amide group will show crosspeaks at the intersection of the amide proton chemical shift and the shift of the third spin (matching atom). Some experiments, such as the HNCA ex-

Table 14.1. Triple-resonance experiments for assignments. A comprehensive description of a number of these experiments can be found in Sattler et al [142]. An earlier review by Bax and Grzesiek [11] should also be consulted for more information.

Experiment	*Correlated Atoms*	*References*
	Mainchain Carbon Shifts†	
HNCO	CO^{i-1} with NH‡	[75, 68, 86, 116, 174]
HN(CA)CO	CO with NH	[39, 173]
HNCA	$C_\alpha^{i,\,i-1}$ with NH	[68, 75, 173, 172]
HN(CO)CA	C_α^{i-1} with NN	[81, 173]
HNCACB	$[C_\beta C_\alpha]^{i,\,i-1}$ with NH	[168]
CBCANH	$[C_\beta C_\alpha]^{i,\,i-1}$ with NH	[67]
HN(CO)CACB	$[C_\beta C_\alpha]^{i-1}$ with NH	[66]
	Mainchain Proton Shifts	
HN(CA)HA	$H_\alpha^{i,\,i-1}$ with NH	[40]
HN(COCA)HA	H_α^{i-1} with NH	[41]
HAHB(CO)NH	$[H_\beta H_\alpha]^{i-1}$ with NH	[67]
HAHBNH	$[H_\beta H_\alpha]^i$ & $[H_\beta H_\alpha]^{i-1}$ with NH	[163]
HCACO	C_α, CO, H_α	[69]
	Other Sidechain Shifts	
HCCH TOCSY	All protons with H_α & C_α	[9, 35, 55, 85]
C-C TOCSY (CO)NH	All C^{i-1} with NH	[65]
H-H TOCSY (CO)NH	All H^{i-1} with NH	[65]

† The mainchain includes the β-proton and carbon. ‡ "NH" implies correlation to both amide proton and nitrogen.

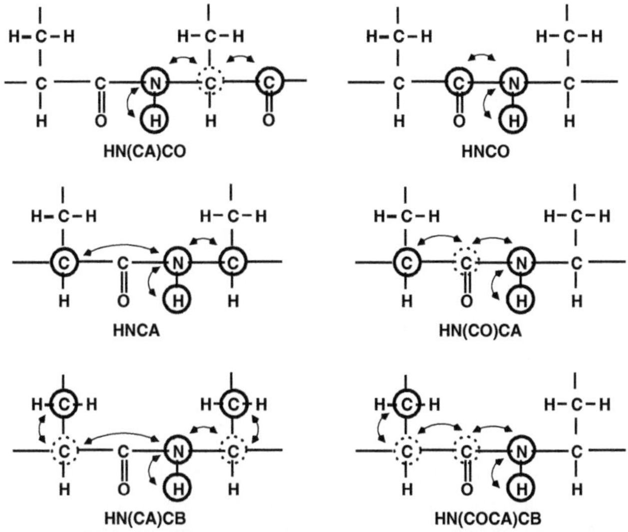

Figure 14.4. Triple-resonance experiments for main chain assignments. The magnetization transfer pathways in selected triple-resonance experiments are shown. In these diagrams the atoms that are circled define the recorded chemical shifts. The atoms that are circled with a dotted line serve to transfer magnetization between atoms, but their chemical shift is not recorded. The arrows indicate the direction of magnetization transfer. In all cases these experiments begin by transferring the amide proton polarization to the amide nitrogen. Experiments that give exclusively inter-residue shifts are on the right. These experiments generate one crosspeak per amide group. For the two experiments in the first row, the HN(CA)CO and HNCO, the matching atom is the carbonyl carbon. The HN(CA)CO can in principal give both inter- and intra-residue carbonyl shifts. However, the inter-residue peak is generally of low intensity and often not observable. The HNCO experiment gives the chemical shift of the carbonyl of the preceding residue. The HNCA and HN(CA)CB experiments give both the intra- and inter-residue α-carbon (HNCA) or β-carbon (HN(CA)CB) shifts. Thus two crosspeaks are observed per amide group. The HN(CO)CA and HN(COCA)CB experiments only correlate the inter-residue α-carbon (HN(CO)CA) or β-carbon (HN(COCA)CB) shift to the amide group, giving one crosspeak per residue.

periment, give both the inter- and intra-residue shifts for the matching atom, and thus will show two crosspeaks for each amide group. In order to unambiguously identify whether the peak originates from the inter- or intra-residue spin requires data from a complementary experiment that generates signals from only one of the two matching atoms. For example, the HN(CO)CA experiment complements the HNCA experiment by providing only inter-residue α-carbon shifts.

The general appearance of these spectra is illustrated in Fig. 14.5, using the HNCA and the HN(CO)CA experiments as an example. Other triple-resonance spectra would be very similar in appearance, with the significant difference being the carbon frequency axis. In the HNCO and HN(CA)CO experiments carbonyl shifts would be observed while in the HN(CA)CB and HN(COCA)CB experiments the β-carbon shifts would be observed.

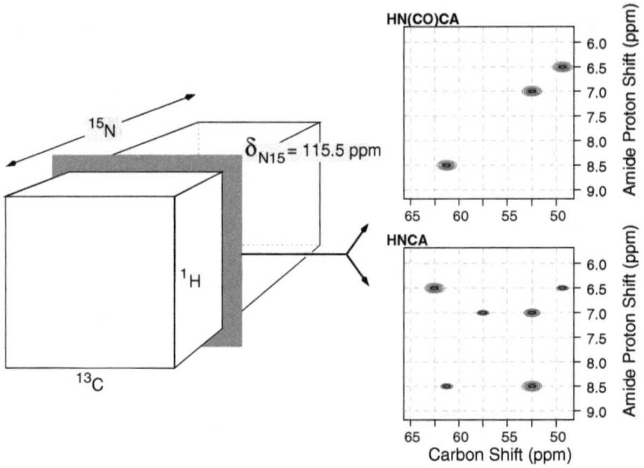

Figure 14.5. Illustration of HNCA and HN(CO)CA Spectra. The left section of the diagram shows the three-dimensional spectrum with the three frequency axis labeled. A single slice from this spectrum is taken at a nitrogen frequency of 115.5 ppm and the resultant two dimensional slices are shown on the right. The HN(CO)CA spectrum is shown on the top and the HNCA spectrum is on the bottom. This protein contains three residues whose amide nitrogen has a chemical shift of 115.5 ppm. The amide proton frequencies of these residues are 6.5, 7.0, and 8.5 ppm. The HN(CO)CA spectrum gives the chemical shifts of the α-carbon that precedes each residue. In this case the shifts are 48.75, 52.5, and 60.25 ppm for the first, second, and third amide proton shifts, respectively. The HNCA spectrum gives both the inter- and intra-residue shifts for each residue. The intra-residue crosspeak is usually more intense than the inter-residue peak because the one bond intra-residue coupling is generally usually larger than the two-bond inter-residue coupling. In this example, the opposite is true for the residue with H_N = 7.0 ppm. Using the HN(CO)CA spectrum to unambiguously identify the intra-residue peak in the HNCA spectrum gives intra-residue α-carbon shifts of 62.5, 57.5, and 52.5 ppm for the first, second, and third amide proton shifts, respectively.

14.2 Description of Triple-resonance Experiments

In this section, a detailed description of the HNCO, HNCA, and HN(CO)CA experiments will be given. The reader should review the elements of pulse sequences that are presented Appendix D before reading this section. Excitation of distinct types of carbons, e.g. CO and C_α are required in these experiments. In practice it is necessary to utilize semi-selective carbon pulses for both excitation and decoupling. A discussion of these two important pulse sequence elements is presented after the description of the pulse sequences, in Section 14.3.

14.2.1 HNCO Experiment

The HNCO experiment is one of the simplest triple-resonance experiments. It correlates the amide group with the carbonyl carbon (C') shift of the preceding residue. The overall pathway of the magnetization is as follows:

$$H_N \rightarrow N \rightarrow C'^{(i-1)}(t_1) \rightarrow N(t_2) \rightarrow H_N(t_3) \qquad (14.1)$$

The pulse sequence for the HNCO experiment is shown on the following two pages.

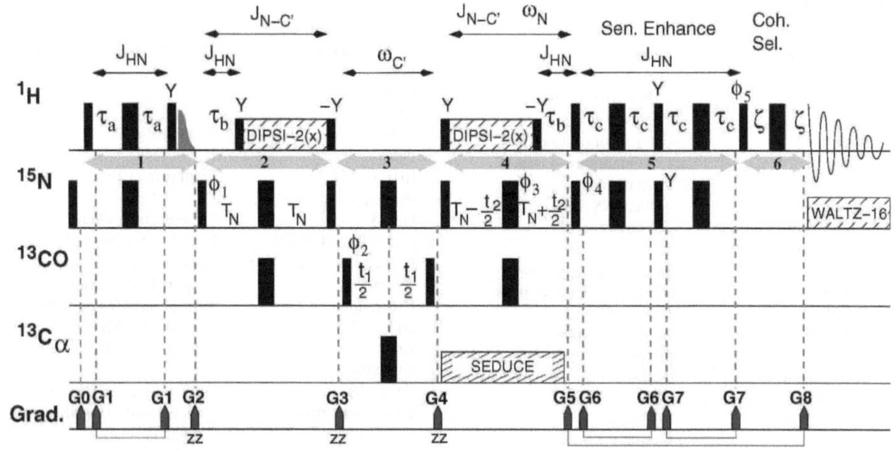

Figure 14.6. Pulse sequence for the HNCO experiment. This experiment is from [86]. Full height narrow bars and wide bars on the proton and nitrogen channels represent 90° and 180° high power pulses, respectively. The shaped gray colored proton pulse is a water flipback pulse. The DIPSI-2 decoupling scheme would be applied using a power level that yields a 90° ^{1}H pulse of 50 μsec ($\gamma B_1 = 5$ kHz). The proton pulses on either side of the DIPSI sequence would be applied at the same power level to insure the correct phase relationship between these pulses and the DIPSI decoupling. Note that both proton and carbon decoupling are turned off before the application of a gradient pulse. The carbon pulses are applied as semi-selective pulses , as discussed in the text. The transmitters for the carbonyl and C_α carbon channels are set to 175 and 55 ppm, respectively. If only one carbon channel is available, the carbon transmitter would be placed in the center of the C_α carbons and the carbonyl pulses would be generated by frequency shifted pulses, as described in Section 14.3.4. The coupling between the C_α and the nitrogen during Period 4 is removed by application of a SEDUCE decoupling sequence to the C_α carbons, as discussed in the text, using a field strength of: $\gamma B_1 = 1.7$ kHz. On spectrometers with a single carbon channel, it is more convenient to use a single 180° pulse, as discussed in Section 14.2.1.1 on page 287.

The delay T_N is 12.4 msec, allowing the evolution of nitrogen-carbonyl coupling. The delays τ_a and τ_c are set to 2.3 msec (<1/(4J_{NH})) and τ_b is set to 5.5 msec (=1/(2J_{NH})). The delay ζ is required for the application and recovery of gradient G8 and is set to 0.5 msec. The phase cycle is $\phi_1 = x, -x; \phi_2 = 4(x), 4(-x); \phi_3 = 2(x), 2(-x); \phi_4 = x; \phi_5 = x, \phi_{rec} = 2(x, -x), 2(-x, x)$. Quadrature detection in t_1 (carbonyl) is obtained using States-TPPI, implemented by phase shifting the carbonyl pulse labeled ϕ_2 by 90°. Quadrature detection in nitrogen is obtained using N-, P-selection. Two FIDs are collected for each t_2 value; for the second, the phase of ϕ_4 is inverted as is the sign of gradient G5. Processing is as described in Chapter 12, Section 12.4.1. Gradient are usually sine-shaped with the following strengths: G0(8 G/cm, 0.5 msec), G1(5 G/cm, 0.5 msec), G2(15 G/cm, 2 msec), G3(20 G/cm, 0.75 msec), G4(5 G/cm, 0.2 msec), G5 (30 G/cm, 1.25 msec), G6(5 G/cm, 0.3 msec), G7(10 G/cm, 0.2 msec), G8(27.8 G/cm, 0.125 msec). Gradient G0 insures that the detectable magnetization arises only from the amide proton. Gradient pairs G1, G6, and G7 remove artifacts associated with imperfect 180°pulses. Gradients G2, G3, and G4 are zz-filters.

HNCO Pulse Sequence

Pulse Sequence	*Comments*
6 Loop	^{15}N Evolution Loop
5 Loop	^{15}N N-P selection Loop
4 Loop	CO Evolution Loop
3 Loop	CO Quadrature Loop
zero	Zero memory
2 Loop	Collect NS scans
power level[High]:N	Set power levels
power level[High]:Cα	
delay d1	Inter-scan delay (1sec)
(p90 x):N	90° Nitrogen purge pulse
Grad 0	Gradient 0
(p90 x):H	90° Proton pulse
Grad1	Gradient 1
delay τ_a-Grad1	1/2J$_{NH}$
(p180 x):N (p180 x):H	180° on H and N
delay τ_a-Grad1	1/2J$_{NH}$
Grad1	Gradient 1
(p90 y):H	$2I_xN_z \rightarrow 2I_zN_z$
power level[flipback]:H	
(p90$_{Shaped}$ x):H	Water flipback
Grad2	Gradient 2
Refocus NH and Transfer to CO	
power level[dipsi]:H	Reduce proton power
(p90 ϕ_1):N	$2I_zN_z \rightarrow 2I_zN_y$
delay τ_b	$2I_zN_y \rightarrow N_x$
(p90$_{lp}$ y):H	Place H along x for DIPSI
H-decouple: DIPSI ON	DIPSI turned on
delayT$_N - \tau_b$	
(p180 x):N (p180 x):CO	Dual 180° on N and CO
delay T$_N$	
(p90 x):N	$2N_yC_z \rightarrow 2N_zC_z$
H-decouple: OFF	Turn off DIPSI decoup.
(p90$_{lp}$ $-y$):H	Proton back to z-axis
Grad3	zz-filter
(p90 ϕ_2):CO	$2N_zC_z \rightarrow 2N_zC_y$
CO frequency label	
delay t$_1$/2	First half of t$_1$ period
(p180 x):N (p180 x):C$_\alpha$	Pulses for N and C$_\alpha$ decoup.
delay t$_1$/2	Second half of t$_1$ period
Return to NH	
(p90 x):CO	$2N_zC_y \rightarrow 2N_zC_z$
Grad4	zz-filter
(p90$_{lp}$ y):H	Place H on x for DIPSI
H-decouple: DIPSI ON	DIPSI turned on
power level[Seduce]:Cα	Reduce C$_\alpha$ power for Seduce
(p90 x):N	$2N_zC_z \rightarrow 2N_yC_z$
C$_\alpha$-dec.: SEDUCE ON	C$_\alpha$ Seduce Decoupling
delay T$_N$-t$_2$/2	1st half of constant time
(p180 ϕ_3):N (p180 x):CO	180° pulses on N and CO
delay T$_N + t_2/2 - \tau_b$	2nd half of T$_N$, minus τ_b
H-decouple: OFF	Turn off H-decoup.
(p90$_{lp}$ $-y$):H	Return proton to z-axis
delay τ_b-Grad5	$N_x \rightarrow 2I_zN_y$
C$_\alpha$-decouple: OFF	Turn off Seduce Decoup.
Grad5	Coherence Gradient
power level[High]:H	Proton power to high
Sensitivity Enhancement	
(p90 x):H (p90 ϕ_4):N	
Grad6	
delay τ_c-Grad6	
(p180 x):N (p180 x):H	Dual 180° pulse on H and N
delay τ_c-Grad6	
Grad6	
(p90 y):N (p90 y):H	Dual 90° pulse on H and N
Grad7	
delay τ_c-Grad7	
(p180 x):N (p180 x):H	Dual 180° pulse on H and N
delay τ_c-Grad7	
Grad7	
Coherence Selection	
(p90 ϕ_5):H	Final proton 90
delay ζ	
(p180 x):H	Final proton 180
delay ζ-Grad8	
Grad8	Coherence Gradient
power level[low]:N	Lower ^{15}N power (decoup.)
N-decouple WALTZ ON	Turn on nitrogen decoup.
acq ϕ_{rec}	Acquire FID, phase=ϕ_{rec}
N-decouple OFF	Turn of nitrogen decoup.
go to 2, NS times	loop to 2 for NS scans
write	write FID to DISK
CO Evolution	
$\phi_2 + 90$	Add 90 to ϕ_2 (STATES)
go to 3, one time	Collect quad FID, NS scans
$t1/2 = t1/2 + \Delta t1/2$	Change t1
ϕ_{rec}+180	Add 180° to ϕ_{rec} (TPPI)
go to 4, p times	Collect p pairs of FIDS
t1=0	CO evolution done, t$_1$ reset to 0
^{15}*N Evolution*	
increment ph4+180°	N-P selection
invert Grad5	N-P selection
go to 5, one time	Collect the second N-P FID
$t_2/2 = t_2/2 + \Delta t_2/2$	Change t2
go to 6, q times	Collect q pairs of FIDS
exit	End experiment

Phases

ϕ_1= 0 2	ϕ_4= 0
ϕ_2= 0 0 0 0 2 2 2 2	ϕ_5= 0
ϕ_3= 0 0 2 2	ϕ_{rec}= 0 2 0 2 2 0 2

Figure 14.7. HNCO pulse sequence code. Pulses are indicated by: (pθ *phase*)*:nuclei*, where θ is the flip-angle, *phase* is the phase of the pulse, and $nuclei$ indicates the RF-channel. For example, (p180 y):C$_\alpha$ represents a 180° pulse along the y-axis, applied to the C$_\alpha$ carbons. 'power level' indicates that the transmitter power on the indicated RF-channel is changed to the level indicated within the square brackets. For example *power level[dipsi]:H* changes the RF-power of the proton channel to the level required for the application of the DIPSI decoupling sequence. Note that decoupling must be actively turned off. Grad n indicates the application of a gradient pulse, including the recovery time after the gradient. The length of gradients are subtracted from evolution delays. For example, the $\zeta - P180 - [G8] - \zeta$ segment at the end of the sequence. See Fig. 14.8 for a discussion of the looping and order of data collection.

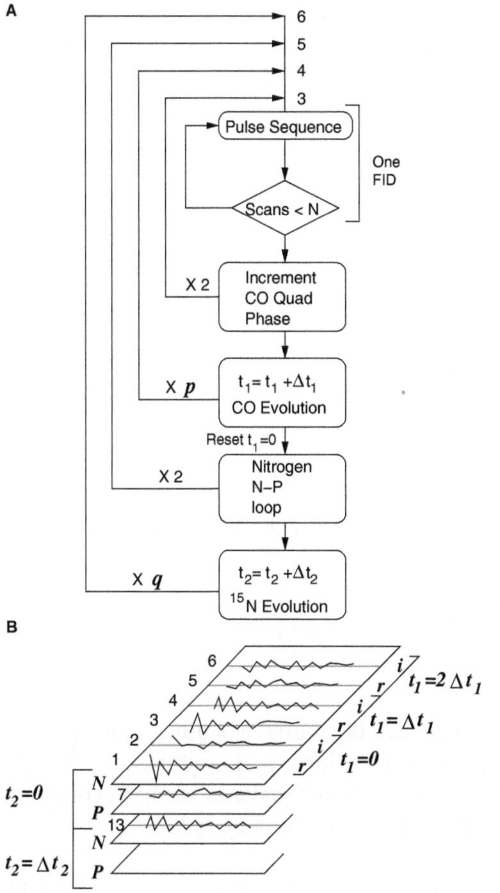

Figure 14.8 Data collection in HNCO experiment. Part A illustrates the looping performed by the pulse sequence code (Fig. 14.7). The *cosine*(real) and *sine*(imaginary) modulated FIDs are collected for carbonyl evolution. The carbonyl evolution time is then incremented, and another pair of FIDs are collected. After a complete set of $2p$ FIDs are collected, corresponding to the N-type dataset for ^{15}N evolution, a second set of $2p$ FIDs are collected for the P-type dataset. The ^{15}N evolution time is then incremented, and collection of the carbonyl evolution is then repeated. Part B shows the time-domain data assuming three carbonyl evolution time points ($p = 3$) and two nitrogen evolution points ($q = 2$). The first six FIDs are associated with $t_2 = 0$ and the N-dataset. The seventh FID is the first FID of the P-dataset, t_2 is still zero and will not be incremented until the 13th FID.

The version of the HNCO presented in Fig. 14.6 utilizes gradients for coherence selection (quadrature detection) of nitrogen, is of enhanced sensitivity (see Section 12.3) and restores the water magnetization to the z-axis at the end of the pulse sequences. The only new feature in this sequence is the use of selective pulses for carbon excitation and decoupling, such that the magnetization of the α-carbons can be manipulated independently from that of the carbonyl carbon. This aspect of the pulse sequence will be discussed in more detail in Section 14.3.

14.2.1.1 HNCO Experiment: Product Operator Analysis

A detailed analysis of this sequence using product operators is provided below, the section labels correspond to the labeled periods in the pulse sequence (Fig. 14.6).

Segment 1: The first segment transfers the proton polarization to the nitrogen with a standard INEPT sequence:

$$I_z \rightarrow -I_y \stackrel{INEPT}{\longrightarrow} 2I_xN_z \stackrel{P90_H P90_N}{\longrightarrow} 2I_zN_y \qquad (14.2)$$

Segment 2: In this phase of the sequence the density matrix evolves under a Hamiltonian that contains evolution due to three different couplings: J_{HN}, $J_{NC'}$, and $J_{NC\alpha}$, plus the nitrogen chemical shift. The experiment is designed in such a way that only J_{HN} and $J_{NC'}$ coupling occurs. Evolution of the system by $J_{NC\alpha}$, as well as the nitrogen chemical shift, are suppressed during this segment. The net result is that the anti-phase proton-nitrogen magnetization is refocused to in-phase nitrogen magnetization ($2I_zN_y \rightarrow N_x$). Simultaneously, anti-phase nitrogen-carbonyl(C') carbon magnetization is generated ($N_x \rightarrow 2N_yC'_z$).

The evolution of the density matrix during this segment will be evaluated using product operators. In theory, it is necessary to consider evolution under all four terms, i.e.:

$$\begin{aligned}\rho(\tau) &= e^{i\pi J_{(HN)}\tau I_z N_z} e^{i\pi J_{(NC')}\tau N_z C'_z} e^{i\pi J_{(NC\alpha)}\tau N_z C^\alpha_z} e^{i\omega_N \tau N_z} \rho_o \\ &\quad e^{-i\pi J_{(HN)}\tau I_z N_z} e^{-i\pi J_{(NC')}\tau N_z C'_z} e^{-i\pi J_{(NC\alpha)}\tau N_z C^\alpha_z} e^{-i\omega_N \tau N_z} \quad (14.3)\end{aligned}$$

during all periods of free precession of the spins. This analysis can be quite tedious. Therefore, we will evaluate the individual evolution of each of these terms over the entire segment, using the concepts developed in Appendix D, and then determine the net effect of all four terms on the density matrix.

Chemical Shift Evolution: It is clear that chemical shift evolution of the transverse nitrogen magnetization does not occur because the 180° nitrogen pulse is symmetrically placed in this time interval.

Proton-Nitrogen Coupling: The coupling between the nitrogen and the amide proton is active during this period. It is allowed to evolve for a period of τ_b, after which the application of the DIPSI decoupling on the protons prevents further evolution. Therefore the evolution of the density matrix due to this coupling is:

$$2I_zN_y \rightarrow 2I_zN_y cos(\pi J_{NH}\tau_b) - N_x sin(\pi J_{NH}\tau_b) \quad (14.4)$$

If τ_b is set to exactly $1/(2J_{NH})$ then the evolution can be simplified:

$$2I_zN_y \rightarrow N_x \quad (14.5)$$

Nitrogen-Carbonyl Coupling: The coupling between the nitrogen and the carbonyl carbon occurs during this entire period because a 180° pulse was applied to *both* spins. Consequently,

$$N_x \rightarrow N_x cos(\pi J_{NC} 2T_N) + 2N_yC'_z sin(\pi J_{NC} 2T_N) \quad (14.6)$$

If $T_N = 1/(4J_{NC})$ then only the anti-phase term remains: $2N_yC'_z sin(\pi J_{NC} 2T_N)$. In practice, T_N is usually set to a shorter time, e.g. 12.4 msec versus 16.6 msec, to reduce the effects of signal loss due to relaxation during the $2T_N$ period. In this case the N_x term is removed by phase cycling of ϕ_2.

Nitrogen-C_α Coupling: Since a 180° pulse is applied to the nitrogen, but not to the C_α carbons, this coupling is refocused over the $2T_N$ period.

Segment 3: During this period, the carbonyl magnetization is brought into the transverse plane, its chemical shift is recorded, and then it is returned to the z-axis. The transverse carbonyl carbon can also evolve from coupling to both the nitrogen and the α-carbon. Both of these interactions are suppressed by the 180° pulses that are applied to the nitrogen and α-carbon spins. Therefore the net evolution is:

$$2N_yC'_z \xrightarrow{P90_N P90_C} -2N_zC'_y \xrightarrow{t_1} -2N_zC'_y cos(\omega_{C'}t_1) + 2N_zC'_x sin(\omega_{C'}t_1)$$
$$\xrightarrow{P90_N P90_C} 2N_yC'_z cos(\omega_{C'}t_1) \qquad (14.7)$$

where the $sin(\omega_C t_1)$ term has been dropped since it will not give rise to detectable magnetization.

Segment 4: The evolution of the system during this segment is similar to that in segment 2, but the nitrogen chemical shift is recorded in a constant time manner (see Section D.2.3). This is accomplished by moving the 180° nitrogen and carbonyl pulses to the left for each subsequent t_2 increment (see Fig. 14.6). Coupling between the nitrogen and the carbonyl carbon occurs during the entire $2T_N$ period since 180° pulses are applied to both spins simultaneously. Because the 180° nitrogen pulse is no longer centered in the interval, coupling between the nitrogen and the C_α carbon can now occur. This coupling is prevented by the application of the SEDUCE sequence to the C_α carbons. Note that the SEDUCE decoupling has to be turned off during application of G5, however, the degree of evolution due to $J_{NC\alpha}$ is small during the time period.

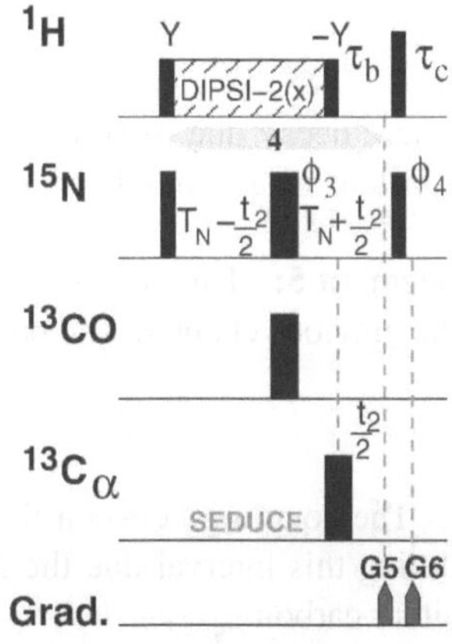

Figure 14.9. Refocusing N-Cα coupling in the HNCO experiment. The N-$C\alpha$ coupling during evolution of the nitrogen chemical shift in the HNCO experiment can be refocused using a 180° C_α carbon pulse. Segment 4 of the HNCO sequence is shown with the SEDUCE decoupling sequence (gray rectangle) replaced by a 180° C_α pulse. The 180° carbon pulse is easier to implement than the SEDUCE sequence on three-channel spectrometers.

The use of SEDUCE decoupling to prevent the evolution of the N-C_α coupling is convenient if a separate RF-channel is available for the α-carbon frequency. However, on three-channel spectrometers only a single RF-channel is available for carbon excitation. In this case implementation of SEDUCE decoupling is cumbersome because the decoupling would have to be interrupted during the application of the carbonyl 180° pulse. An alternative strategy for removal of the N-C_α coupling involves replacement of the SEDUCE sequence with a single 180° pulse, as shown in Fig. 14.9. The net evolution of the density matrix due to N-$C\alpha$ coupling can be easily shown to be zero for this segment by calculating the net phase change in the density matrix over the interval. Recall that a 180° pulse applied to *either* of the two coupled spins, N or C_α, will reverse the evolution of the density matrix. Therefore the net phase angle is:

$$\Theta_{J_{NC\alpha}} = [T_N - \frac{t_2}{2}] - [T_N] + [\frac{t_2}{2}] \quad = 0 \qquad (14.8)$$

The proton-nitrogen coupling is suppressed by the DIPSI decoupling scheme, except for the period τ_b at the end of this segment. Therefore the evolution of the system can be described as follows:

$$2N_yC'_z \stackrel{2T_N J_{NC}}{\longrightarrow} -N_x \stackrel{\tau_b J_{NH}}{\longrightarrow} -2N_yI_z \tag{14.9}$$

The pair of proton and nitrogen pulses convert this density matrix to:

$$-2N_yI_z \rightarrow 2N_zI_y \tag{14.10}$$

Coherence selection is also applied in this segment, by gradient G5. Therefore the overall density matrix at the end of segment 4 is:

$$2N_zI_y cos(\omega_C t_1) e^{i\omega_N t_2} e^{i\gamma_N G5} \tag{14.11}$$

where decay due to spin-spin relaxation of the nitrogen magnetization, with a time constant of T_2, has been ignored. This would decrease the signal by a factor of e^{-2T_N/T_2}.

Segment 5: This segment is exactly the same as the HSQC experiment presented in the previous chapter. The pair of INEPT sequences refocus the $2N_zI_y$ term to:

$$2N_zI_y e^{i\omega_N t_2} \rightarrow I_x cos(\omega_N t_2) - I_y sin(\omega_N t_2) \tag{14.12}$$

The coupling between the nitrogen and the C_α and carbonyl carbon is refocused during this interval due the application of the 180° pulse to the nitrogen, but not to either carbon.

Segment 6: The last segment selects the appropriate nitrogen coherence:

$$e^{i\gamma_N G5} \rightarrow e^{i\gamma_N G5} e^{i\gamma_H G8} = 1 \tag{14.13}$$

The final signal, after processing the data for nitrogen N-P selection is:

$$cos(\omega_C t_1) cos(\omega_N t_2) e^{i\omega_H t_3} \tag{14.14}$$

14.2.1.2 Water Management in the HNCO Experiment

Restoring the water magnetization to the z-axis is more complicated in this experiment because of the proton decoupling that is applied during segments 2 and 4. In this case the y-pulses that bracket the DIPSI sequences will place the water magnetization along the x-axis. Since the DIPSI sequence is applied along the x-axis the water will remained aligned along the B_1 field during the entire decoupling sequence. After the decoupling is turned off, the water is restored to the z-axis by the $-y$ pulse.

The overall path of the water magnetization in the experiment is as follows:

$$I_z \stackrel{INEPT}{\longrightarrow} I_y \stackrel{Flipback}{\longrightarrow} I_z \stackrel{y\ DIPSI\ -y}{\longrightarrow} I_z \stackrel{y\ DIPSI\ -y}{\longrightarrow} I_z \stackrel{P90_x}{\longrightarrow} -I_y \stackrel{P180_x}{\longrightarrow} I_y$$
$$\stackrel{P90_y}{\longrightarrow} I_y \stackrel{P180_x}{\longrightarrow} -I_y \stackrel{P90_{\phi 5}}{\longrightarrow} -I_z \stackrel{P180_x}{\longrightarrow} I_z \tag{14.15}$$

14.2.1.3 Phase Cycle of HNCO Experiment

The phase cycle of this experiment is quite simple, and it is possible to analyze the phase cycle using coherence levels that were discussed in Chapter 12. As a reminder, recall that a coherence jump of $\Delta p \pm N$ can be selected from other changes in coherence levels by setting the phase increment of a pulse to $2\pi/N$ and the receiver phase to $-\Delta p \cdot \phi$.

The complete phase cycle in the HNCO experiment is summarized in Table 14.2. The nitrogen pulse, with phase ϕ_1, creates single quantum nitrogen coherence from a zero quantum state: $\sigma_N^0 \rightarrow \sigma_N^{+1} + \sigma_N^{-1}$. The change in coherence order, Δp, is +1 at this step of the pulse sequence. However, it is also necessary to retain the -1 change, therefore a two step phase cycle is dictated such that jumps of +1 and -1 are both retained. Therefore the cycle is simply x, and -x. Cycling this pulse insures that only magnetization that is associated with the N_z at this point in the experiment is detected.

The 90° pulse applied to the carbonyl carbon channel, with phase ϕ_2 generates exactly the same coherence changes as the nitrogen pulse discussed above, therefore it must also be a two step phase cycle, with pulse phases of x or -x. Cycling this pulse insures that only magnetization that is frequency labeled with the carbonyl chemical shift will be detected.

The final pulse that is phase cycled is the 180° nitrogen pulse during the frequency labeling period. The nitrogen magnetization is transverse during this period, therefore the coherence jump that is generated by this pulse is ± 2, i.e. $\sigma^{-1} \rightarrow \sigma^{+1}$ and $\sigma^{+1} \rightarrow \sigma^{-1}$. Therefore, this pulse is also subject to a two step phase cycle. Cycling this pulse removes artifacts associated with an imperfect 180° pulse, which would generate coherence changes such as $\sigma^{+1} \rightarrow \sigma^{o}$. Note that it is difficult to accomplish this coherence editing using pulsed field gradients because the constant time evolution does not provide sufficient time for the application of a gradient pulse.

The receiver phase is given by the following:$\phi_{rec} = -\sum \Delta p_i \phi_i$, for example:

$$\begin{aligned} Scan\ 2: \quad \phi_{rec} &= -[(+1)(\pi)) + (+1)(0\pi) + (+2)(0\pi)] \\ &= -\pi \equiv -x \end{aligned} \tag{14.16}$$

Note that the phase of ϕ_3 has no effect on the receiver phase since Δp is +2 and the phase shift is π, inducing a net phase change of 2π on the signal.

Table 14.2. HNCO phase cycle. ϕ_1 is the phase of the nitrogen pulse prior to T_N, ϕ_2 is the phase of the carbonyl pulse just prior to the t_1 period, and ϕ_3 is the phase of the 180° nitrogen pulse during the constant time evolution.

Pulse/ Receiver	Δp	Scan Number 1	2	3	4	5	6	7	8
$\phi_1^\dagger$	+1	x	-x	x	-x	x	-x	x	-x
ϕ_2	+1	x	x	x	x	-x	-x	-x	-x
ϕ_3	+2	x	x	-x	-x	x	x	-x	-x
ϕ_{rec}	$-\sum \Delta p_i \phi_i$	x	-x	x	-x	-x	x	-x	x

† $x = 0, y = \pi/2, -x = \pi, -y = 3\pi/2$. This is often encoded in pulse programs as: $x = 0, y = 1, -x = 2, -y = 3$. See Fig. 14.7.

14.2.2 HNCA Experiment

The HNCA experiment correlates the amide group with the chemical shift of its own C_α (i) and the C_α of the preceding residue $(i-1)$. The pathway of magnetization transfer is as follows:

$$H_N \rightarrow N \rightarrow \begin{matrix} C_\alpha^i(t_1) \\ \\ C_\alpha^{i-1}(t_1) \end{matrix} \rightarrow N(t_2) \rightarrow H_N(t_3) \tag{14.17}$$

The most significant difference between the HNCA experiment and the HNCO experiment is that the magnetization is transferred from the nitrogen to the C_α carbon in the HNCA experiment while it is transferred to the carbonyl carbon in the HNCO experiment. Consequently, the pulse sequence of the two experiments are identical, with the following two changes:

1. The pulses on the two carbon types are switched, i.e. pulses that were applied to the carbonyl carbon are now applied to the C_α carbon.
2. In the HNCA experiment the crosspeak intensity depends on the product of two trigonometric functions: $I \propto sin(\pi J_1 \tau) \times cos(\pi J_2 \tau)$, therefore the maximum intensity is obtained when the delay T_N is shorter than 1/(4J) (see Fig. 14.10).

The intensity of the crosspeaks in the HNCA experiment depends on the size of the coupling between the nitrogen and the C_α carbon. During the T_N period, the density matrix evolves due to coupling to the intra-residue carbon as follows:

$$N_x \overset{J_{NC_\alpha^i}}{\rightarrow} N_x cos\Theta_i + 2N_y C_z^i sin\Theta_i \tag{14.18}$$

Evolution also occurs due to the two bond coupling to C_α of the preceding residue. Therefore the density matrix given in Eq. 14.18 evolves further, giving:

$$\begin{aligned} N_x cos\Theta_i + 2N_y C_z^i sin\Theta_i &\overset{J_{NC_\alpha^{i-1}}}{\rightarrow} cos\Theta_i[N_x cos\Theta_{i-1} + 2N_y C_z^{i-1} sin\Theta_{i-1}] \\ &+ 2C_z^i sin\Theta_i[N_y cos\Theta_{i-1} - 2N_x C_z^{i-1} sin\Theta_{i-1}] \\ &= N_x cos\Theta_i cos\Theta_{i-1} + 2N_y C_z^{i-1} cos\Theta_i sin\Theta_{i-1} \\ &+ 2N_y C_z^i sin\Theta_i cos\Theta_{i-1} - 4N_x C_z^i C_z^{i-1} sin\Theta_i sin\Theta_{i-1} \end{aligned} \tag{14.19}$$

where $\Theta_i = \pi J_{NC_\alpha^i} 2T_N$, and, $\Theta_{i-1} = \pi J_{NC_\alpha^{i-1}} 2T_N$.

Of these four terms, only the middle two give rise to detectable magnetization:

$$2N_y C_z^{i-1} cos\Theta_i sin\Theta_{i-1} \qquad 2N_y C_z^i sin\Theta_i cos\Theta_{i-1} \tag{14.20}$$

generating the inter- and intra-residue peaks, respectively.

Since there are two transfer periods, the trigonometric terms become squared, giving the following for the intensities of the two peaks:

$$I_{Inter} = [cos\Theta_i sin\Theta_{i-1}]^2 \qquad I_{Intra} = [sin\Theta_i cos\Theta_{i-1}]^2 \tag{14.21}$$

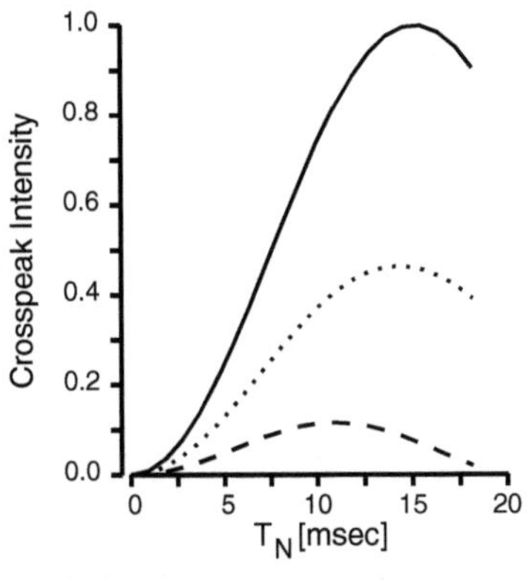

Figure 14.10 Relative sensitivity of the the HNCO and HNCA experiments. Crosspeak intensities are plotted for the HNCO experiment (solid curve), the inter- (dashed) and intra-residue (dotted) crosspeaks in the HNCA experiment as a function of T_N. The intensity of the HNCO experiment is given by $sin^2(\pi J_{NC} 2T_N)$. See text for the equivalent formula for the HNCA crosspeaks. The following coupling constants were used: ${}^1J_{NC'} = 15$ Hz, ${}^1J_{NC^i_\alpha} = 11$ Hz, ${}^2J_{NC^{i-1}_\alpha} = 7$ Hz.

Intensities for the inter- and intra-residue crosspeaks are shown in Fig. 14.10, along with the crosspeak intensity in the HNCO experiment. This figure shows that the intensity of the peaks in the HNCA experiment are smaller than in the HNCO experiment. In addition, the inter-residue crosspeak in the HNCA experiment is generally less intense than the intra-residue peak because the two-bond inter-residue coupling is smaller than the one-bond intra-residue coupling. However, for a small number of residues, the inter-residue coupling constant is larger than the intra-residue one, which reverses the intensities of the two peaks.

14.2.2.1 HN(CO)CA Experiment

The HN(CO)CA experiment correlates the amide nitrogen with *only* the C_α of the preceding residue. It is a valuable experiment in that it distinguishes between inter- and intra-residue peaks on the HNCA experiment. The path of magnetization transfer is as follows:

$$H_N \rightarrow N \rightarrow C'^{(i-1)} \rightarrow C_\alpha^{i-1}(t_1) \rightarrow C'^{(i-1)} \rightarrow N(t_2) \rightarrow H_N(t_3) \qquad (14.22)$$

The pulse sequence for this experiment is given in Fig. 14.11. As anticipated from the magnetization transfer path, it is very similar to the HNCO experiment, with the addition of short INEPT segment to transfer the magnetization from the carbonyl carbon to the C_α carbon. These segments are indicated on the pulse sequence by $C \rightarrow C$. Due to the similarity to the HNCO only this section of the sequence will be discussed in detail. The reader is referred to the HNCO experiment for a description of the events that occur during segments 1-6. Note, however, that the gradient labels have changed. In particular gradients G9 and G12 are utilized for N-P selection in the HN(CO)CA experiment. Transfer of magnetization from the carbonyl carbon to the C_α is accomplished using a standard INEPT segment, with the delays set to somewhat less that $1/(4J_{C\alpha C'})$ to reduce signal loss from relaxation. Ignoring the hashed C_α pulses for the moment, the density matrix evolves as:

$$2N_zC'_y \rightarrow 2N_z[C'_y cos(\pi J_{C\alpha C}\tau_d) - 2C'_x C^\alpha_z sin(\pi J_{C\alpha C}\tau_d)] \qquad (14.23)$$

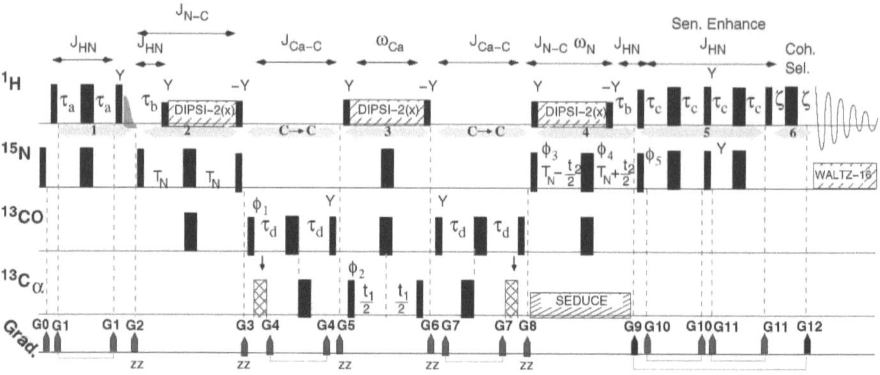

Figure 14.11. The HN(CO)CA experiment. The pulse sequence for an HN(CO)CA experiment is presented. This was modified from Yamazaki et al. by removal of constant time evolution in t_1 and removal of ^{2}H decoupling [173]. Proton and nitrogen pulses are as described for the HNCO sequence. Note that decoupling is turned off when gradients are applied. The pulses applied to the carbons are semi-selective as described in Section 14.3. The two hashed C_α pulses, marked with small arrows compensate for Bloch-Siegert shifts, see text. The delays are the same as in the HNCO experiment, except that τ_d is set to 4.1 msec, which is slightly less than $1/(4J_{C\alpha C'})$ (4.55 msec). The phase cycle is: $\phi_1 = x, -x; \phi_2 = 2(x), 2(-x); \phi_3 = x; \phi_4 = 4(x), 4(-x); \phi_5 = x; \phi_{rec} = x, -x, -x, x$. Quadrature detection in t_1 is attained by phase shifting ϕ_2 in a States-TPPI manner. Quadrature in nitrogen is by N-P selection. For each t_2 value two FIDs are collected, the second with gradient G9 inverted as well as the phase of the pulse labeled ϕ_5. Axial peaks in the nitrogen dimension are shifted to the edge of the spectrum by incrementing ϕ_3 and the receiver by 180° for each new t_2 values. Gradients are sine-shaped and the levels used by Yamazaki et al. [173] were: G0(0.5 msec, 8 G/cm), G1(0.5 msec, 4 G/cm), G2(1 msec, 10 G/cm), G3(1 msec, 1 G/cm), G4(1 msec, 7 G/cm), G5(1 msec, -15 G/cm), G6(1 msec, 8 G/cm), G7(1 msec, 0.4 G/cm), G8(0.6 msec, 10 G/cm), G9(1.25 msec, 30 G/cm), G10(0.5 msec, 8 G/cm), G11(0.3 msec, 2 G/cm), G12(0.125 msec, 27.8 G/cm).

The subsequent 90° carbonyl pulse, along the y-axis, converts this to:

$$2N_z[C'_y cos(\pi J_{C\alpha C'}\tau_d) - 2C'_z C^\alpha_z sin(\pi J_{C\alpha C'}\tau_d)] \tag{14.24}$$

and gradient G5 acts as a zz-filter, destroying the $2N_zC'_y$ term. The subsequent Cα pulse (phase=ϕ_2) converts $N_zC'_zC^\alpha_z$ to:

$$-N_zC_zC^\alpha_y \tag{14.25}$$

which evolves due to the C_α chemical shift in t_1. The application of 180° pulses on the nitrogen and carbonyl carbon prevent evolution of the N-C_α and the C'-C_α coupling during t_1.

14.2.2.2 Bloch-Siegert Shift Compensation

Bloch-Siegert shifts are changes in the chemical shift of a resonance due to the presence of an oscillating magnetic field while the spin is precessing in the transverse plane [19]. In this case the total magnetic field felt by the spin is the vector sum of the two fields. For a spin that is off-resonance by $\Delta\omega$ rad/sec, the effective field is:

$$B_{eff} = \sqrt{[\Delta\omega/\gamma]^2 + [B_1]^2} \tag{14.26}$$

Block-Siegert shifts occur in many triple-resonance experiments because the INEPT segments used to transfer magnetization between the C_α and carbonyl carbons involve the transfer of magnetization between *homonuclear* spins. Consequently, the resonance frequencies of the two coupled spins are relatively close to each other, within tens of kHz, in this case. As a result, the magnetic field associated with the 180° pulse that is applied to one carbon can change the effective magnetic field, B_{eff} of the other carbon.

Selective RF-pulses cause a *pseudo*-Bloch-Siegert phase shift. Consider the first carbon-carbon INEPT period in the HN(CO)CA experiment (labeled $C \rightarrow C$ in Fig. 14.11). In this case the magnetization associated with the carbonyl carbon is transverse while the 180° inversion pulse is applied to the α-carbon. This inversion pulse is selective for the α-carbon and will therefore cause a 2π rotation of carbonyl magnetization (see following section on selective pulses). Since the carbonyl carbon has not evolved during the C_α pulse, a phase shift of $\omega_C\tau_{180}$ will be associated with the carbonyl carbon magnetization, as illustrated in Fig. 14.12.

If this phase shift is uncorrected then the density matrix, at the end of the INEPT period will be (assuming $\tau_d = 1/(4J_{C'C\alpha})$):

$$-2N_zC'_xC^\alpha_z cos(\theta) - 2N_zC'_yC^\alpha_z sin(\theta) \tag{14.27}$$

and the subsequent 90_y carbonyl pulse will only convert a fraction, $cos(\theta)$, of this density matrix to the desired form: $N_zC'_zC^\alpha_z$.

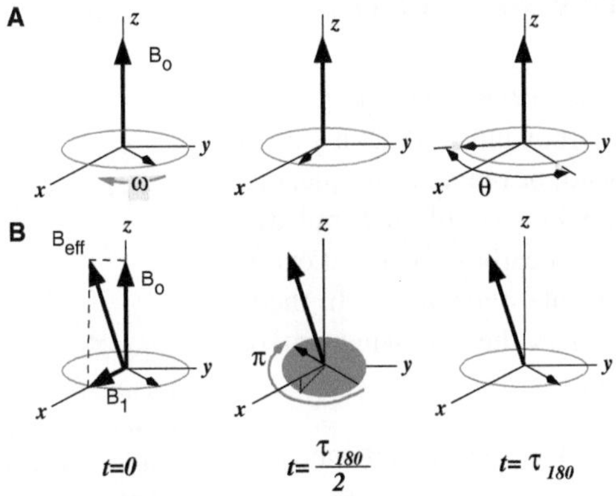

Figure 14.12. The origin of Bloch-Siegert phase shifts. Part A shows the evolution of the transverse carbonyl carbon magnetization due to its chemical shift, in the absence of the 180° C_α pulse. During a time τ_{180} (the length of the C_α pulse) the carbonyl magnetization precesses an angle: $\theta = \omega_C\tau_{180}$. Panel B shows the effect of the selective 180° C_α pulse on the carbonyl magnetization. When the pulse is applied, precession about B_o is interrupted and the carbonyl magnetization precesses about the effective magnetic field, B_{eff}, returning to its starting position at the end of the pulse. Therefore the overall phase shift introduced by the 180° pulse is $\omega_C\tau_{180}$ (see [142] for more details).

This type of Bloch-Siegert phase shift can be corrected by applying a compensating pulse. The compensating pulse and the refocusing pulse have to be on opposite sides of the 180° carbonyl pulse, as indicated in Fig. 14.11. Under these conditions, the compensating pulse will introduce a phase shift of θ in the transverse carbonyl magnetization. This phase shift will be negated, to $-\theta$, by the 180° *carbonyl* pulse. The second 180° C_α pulse will introduce a phase shift of θ, which will now cancel the phase shift introduced by the first C_α pulse. To insure that the compensating pulse does not interfere with evolution of the scalar coupling, it is placed adjacent to the 90°pulses, as indicated in Fig. 14.11.

The 180° carbonyl pulse in the middle of the t_1 period also causes a phase shift of the precessing C_α magnetization. However, in this case there is no real loss of signal, since this Bloch-Siegert phase shift simply adds to the phase introduced by precession and can be easily removed by phasing the spectrum after processing.

The application of selective decoupling during an evolution period will cause a true Bloch-Siegert shift in the transverse magnetization. The decoupling pulses will cause a small change in the effective field, leading to a change in the precessional rate of the transverse spin, causing a change in the observed resonance frequency. The change in chemical shift would be largest for those spins that are closest to the decoupling frequency. For example, during the t_1 period of the HN(CO)CA experiment, if the scalar coupling between the transverse C_α and the carbonyls was removed by the application of a selective decoupling scheme, a small change in the chemical shift of the α carbons would be observed. Consequently, it is preferable to use 180° pulses for decoupling.

14.3 Selective Excitation and Decoupling of ^{13}C

14.3.1 Selective 90° Pulses

Most triple-resonance experiments require selective 90° and 180° carbon pulses that only excite a relatively narrow band, or range, of frequencies. The most common application of such pulses is excitation of the carbonyl carbons, without excitation of the C_α carbons and vice versa. In theory such pulses should satisfy the following criteria, in order of importance.

1. Uniform excitation over the desired bandwidth.
2. No excitation of magnetization in other regions of the spectrum.
3. Constant phase of the excited magnetization within the bandwidth.
4. Short duration.

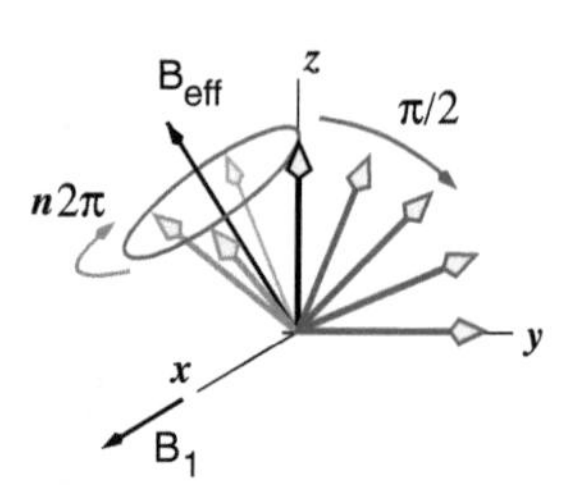

Figure 14.13. Excitation of spins by selective pulses. A selective 90° pulse. The applied B_1 field rotates on-resonance spins (black arrows) by 90° ($\pi/2$) while off-resonance spins (light gray arrows) precess about B_{eff} by an angle of $n2\pi$, giving a null excitation.

In many cases, these selective pulses are non-rectangular in shape. Non-rectangular pulses are produced from a waveform generator that breaks the pulse into a number of segments, each with its own amplitude and phase. During the experiment, this waveform is used to control the transmitter, allowing the generation of shaped pulses.

There are three general classes of selective pulses, 90° pulses, 180° inversion pulses (i.e. $C_z \rightarrow -C_z$) and 180° refocusing pulses (i.e. $C_y \stackrel{\tau-P180-\ tau}{\rightarrow} -C_y$). The first two categories are described here, the reader is referred to Geen and Freeman [61] and Freeman [56] for a more detailed discussion of 180° refocusing pulses.

Most spectrometer software packages provide a pulse simulator that can be used to determine the excitation properties of selective pulses. It is well worth the effort to optimize these pulses because non-optimal pulses will lead to sensitivity losses and in some cases produce additional artifacts in the spectra.

The underlying theory of selective pulses will be illustrated with 90° pulses. The desired outcome is to rotate the magnetization of on-resonance spins by $\pi/2$ and the off-resonance spins, at the location of null excitation, by a multiple of 2π, as illustrated in Fig. 14.13.

In the rotating frame the magnetic field felt by the on-resonance spins is simply B_1, which can be written in angular frequency units as $\omega_1 = \gamma B_1$. Off-resonance, the effective field is larger, and given by:

$$\omega_{eff} = \sqrt{\omega_1^2 + (\Delta\omega)^2} \tag{14.28}$$

where $\Delta\omega$ is the angular frequency difference between the transmitter frequency and the location of null excitation. A pulse of duration τ_{90} generates a $\pi/2$ rotation for the on-resonance spins:

$$\frac{\pi}{2} = \omega_1 \tau_{90} \tag{14.29}$$

and a rotation of $n2\pi$ at the frequency of null excitation:

$$n2\pi = \omega_{eff}\tau_{90} \tag{14.30}$$

Eliminating τ_{90} and solving for the RF-field strength, ω_1, gives:

$$\begin{aligned} 4n\omega_1 &= \sqrt{\omega_1^2 + \Delta\omega^2} \\ 16n^2\omega_1^2 &= \omega_1^2 + \Delta\omega^2 \\ \omega_1 &= \frac{\Delta\omega}{\sqrt{16n^2 - 1}} \end{aligned} \tag{14.31}$$

The pulse length is therefore:

$$\tau_{90} = \left[\frac{\pi}{2}\right]\frac{1}{\omega_1} = \pi\frac{\sqrt{16n^2-1}}{2\Delta\omega} = \frac{\sqrt{16n^2-1}}{4\Delta\nu}$$

where $\Delta\nu = \Delta\omega/(2\pi)$, or the location of null excitation in Hz.

The shortest selective pulse is obtained for $n = 1$, giving a length of:

$$\tau_{90} = \frac{\sqrt{15}}{4\Delta\nu} \tag{14.32}$$

Longer selective pulses, e.g. $n = 2$, can have better excitation properties.

An assortment of selective 90° pulses are shown in Fig. 14.14. This figure illustrates pulses for selective excitation of carbon on a 500 MHz instrument. At higher static magnetic field strengths these pulses will be more selective due to the increased the separation between the C_α an carbonyl carbons.

The two simple rectangular pulses, corresponding to $n = 1$ and $n = 2$ perform remarkably well for excitation of α-carbon with a null at the carbonyl frequency, as indicated in Panels A and B of Fig. 14.14. In the reverse situation, excitation of the

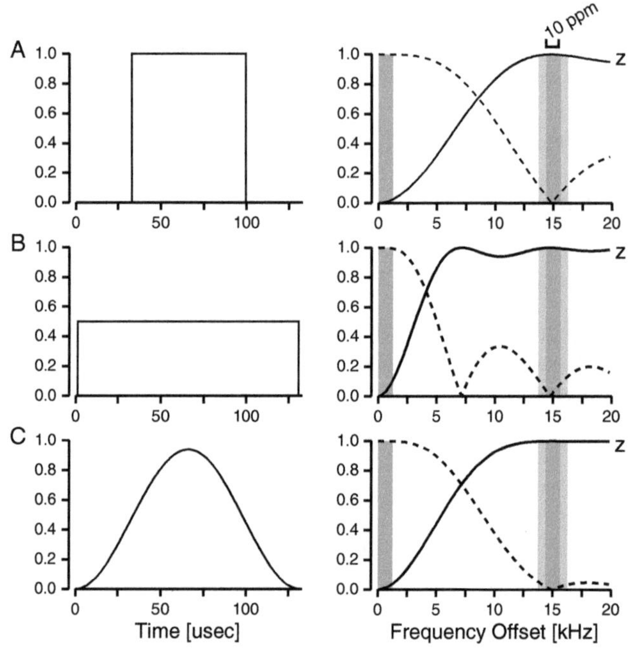

Figure 14.14. Selective 90° carbon pulses. In this example the desired region of null excitation is defined to be 15 kHz from the transmitter. This is the frequency separation between the C_α and carbonyl carbons at ν_C=125 MHz (ν_H=500 MHz). The left side of the diagram shows the pulse shape. A and B are rectangular pulse with lengths of 64.5 μsec ($n = 1$) and 132.3 μsec ($n = 2$). C is a $sine^2$ shape of length 130 μsec. The right side of the figure shows the excitation profile of each pulse. The solid line shows the z-magnetization and the dotted line shows the magnetization present in the transverse plane ($\sqrt{M_x^2 + M_y^2}$). Note that these excitation profiles are symmetric about the origin and the profile for negative frequencies have not been shown. The dark shaded vertical bars represent a bandwidth of 10 ppm on a 500 MHz spectrometer. This is one-half of the range of C_α frequencies and the full range for the carbonyls. The lighter, and wider, shaded bar at 15 kHz represents the full carbon bandwidth associated with the C_α carbons at this spectrometer frequency. All three of these pulses are appropriate for excitation of the C_α carbons with a null excitation at the carbonyl carbon 15 kHz away. In contrast, if a selective carbonyl pulse was generated by placing the transmitter on the carbonyl frequency, then the $sine^2$ pulse is the only pulse with a sufficiently wide region of null excitation 15 kHz away to include all of the C_α carbon frequencies.

carbonyl with a null at the α-carbon, neither of these two pulse generates a region of null excitation that is sufficiently wide to cover the entire α-carbon range. However, a pulse that is shaped in the form of a sin^2 function (Fig. 14.14, Panel C) has very good performance, both in the excitation bandwidth and the lack of excitation in the region of null excitation. Consequently, the sin^2 pulse is preferable over either rectangular pulse for elective excitation of carbonyl carbons. However, it is somewhat longer than the short rectangular pulse, shown in part A of Fig. 14.14, which may case timing problems in some experiments.

14.3.2 Selective 180° Pulses

The RF-pulse field strength, and corresponding pulse length, to generate a selective 180° pulse is:

$$\omega_1 = \frac{\Delta\omega}{\sqrt{4n^2-1}} \qquad \tau_{180} = \frac{\sqrt{4n^2-1}}{2\Delta\nu} \qquad n = 1, 2, ... \tag{14.33}$$

Selective 180° inversion pulses are generally used to convert z-magnetization to minus z-magnetization for decoupling purposes. For example, a selective 180° pulse would be applied to the C_α carbon in the HN(CO)CA experiment (Sections labeled

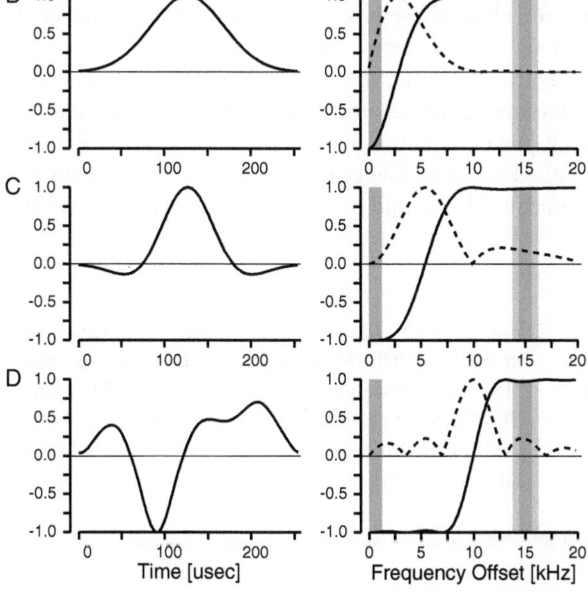

Figure 14.15 Selective 180° pulses. Panel A is a 115.5 μsec rectangular pulse, B is Gaussian in shape, C is a Hermite polynomial and D is an I-Burp1 pulse [61]. Pulses B-D are 256 μsec in length. The excitation profiles are shown on the right. The solid line is the z-magnetization and the dotted line is the transverse magnetization. The shaded vertical bars are described in Fig. 14.14. At a spectrometer frequency of 500 MHz, the Hermite polynomial pulse shape provides sufficient inversion bandwidth for the C_α carbons with minor excitation of the carbonyl carbons. At higher spectrometer frequencies, such as 900 MHz, the I-Burp1 pulse would be more suitable to generate inversion over a larger frequency range.

$C \rightarrow C$) when the magnetization is transferred from the carbonyl to the C_α carbon during the INEPT period.

A number of selective 180° inversion pulses are shown in Fig. 14.15. The rectangular, as well as the Gaussian, shaped pulses perform adequately at this particular proton frequency (500 MHz). However, the excitation bandwidth for both of these pulses is narrower than the Hermite or I-Burp pulse. Consequently the Hermite or I-Burp pulses are more suitable as selective inversion pulses, especially at higher magnetic field strengths where the desired frequency range for excitation is larger. Unfortunately, both the Hermite and I-burp pulses generate a small amount of transverse magnetization at the null frequency. The transverse magnetization leads to signal loss during the INEPT period because the two-spin term, $C_x C_x^\alpha$ no longer evolves under scalar coupling. The amount of excitation at the null frequency decreases as the pulse length increases. Consequently, pulse lengths that are longer than those shown in Fig. 14.15 would be more appropriate at lower fields.

14.3.3 Selective Decoupling: SEDUCE

Shaped pulses can also be used to selectively decouple carbons. For example, in both the HNCO and the HN(CO)CA experiments, the C_α carbons are decoupled during the evolution of the nitrogen chemical shift. One of the most widely used selective decoupling schemes is the SEDUCE sequence introduced by McCoy and Mueller [110]. In this sequence a shaped pulse, which is similar in shape to a sin^2 pulse, is applied using the phases present in the WALTZ-16 decoupling sequence. The fundamental rotation element of the the WALTZ-16 sequence is:

$$R = \left[\frac{\pi}{2}\right]_x \pi_{-x} \left[\frac{3\pi}{2}\right]_x = 1\bar{2}3 \qquad (14.34)$$

This rotation element is combined with its phase inverted replica to generate the complete WALTZ-16 decoupling scheme (see Chapter 7 for additional details).

The WALTZ-16 rotation element, containing the sin^2 shaped pulse, is shown in Fig. 14.16. The bandwidth of this decoupling scheme is very similar to that found using WALTZ decouping with high-power rectangular pulses, approximate $\pm 1.75\gamma B_1$. For example, a 825 Hz field, applied to the middle of the C_α carbons would decouple ± 11 ppm on a 500 MHz spectrometer. This would be more than adequate to cover the 20 ppm C_α range.

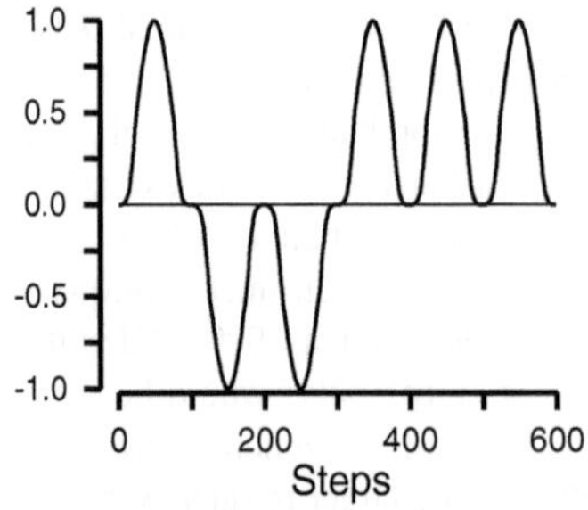

Figure 14.16 SEDUCE decoupling. The pulses used in the SEDUCE decoupling scheme are shown. Each 100 step shaped-pulse acts as a selective 90° pulse. The fundamental rotation element of the WALTZ decoupling scheme, $1 - \bar{2} - 3$, is illustrated here as a total of 6 selective pulses, giving a total rotation angle of 180°.

14.3.4 Frequency Shifted Pulses

Most triple-resonance sequences require at least four separate RF-channels. For example the HNCO and HN(CO)CA experiments require ^{1}H, ^{15}N, C_α, and carbonyl carbon pulses. Many spectrometers have only three channels, therefore both carbon frequencies have to be generated from the same channel. Changing the carbon transmitter frequency can be accomplished in several ways.

A straight-forward method is to simply change the frequency of the transmitter when required. For example, in the HNCO experiment the frequency of the carbon transmitter would initially be placed at 175 ppm (carbonyl) and then it would be switched to 55 pm for the first 180° pulse on the α-carbons. Using this approach it would be necessary to change the carbon transmitter frequency 5 times in the HNCO experiment. The principle problem with this approach is that a change in the transmitter frequency usually causes an unpredictable change in the phase of the pulse when the frequency is returned to its original value. Often it is imperative that the phases of pulses that are applied to one type of spin have a fixed relationship to each other. For example, in the t_1 evolution period of the HNCO experiment the pair of 90° carbonyl pulses must have a well defined phase relationship in order to record the chemical shift evolution of the carbonyl carbons. Changing the frequency of the carbon transmitter in the middle of the t_1 period to generate the 180° α-carbon pulse would destroy this phase coherence.

The other method of generating pulses to excite different types of carbon spins is to set the transmitter at a single frequency and then change the *apparent* frequency of the pulse. The change in the apparent frequency of the pulse is generated by changing the phase of the pulse *while* it is being applied. Currently, this is implemented on spectrometers by dividing the pulse into short segments, or lamina, and then incrementing the phase for each segment. This approach can be applied to both rectangular as well as non-rectangular shaped pulses and multiple frequencies can be encoded in the phase shifts. Pulses of this type are commonly referred to as shifted laminar pulses and were introduced by Patt [125].

For example, a 50 μsec rectangular pulse might be divided into 100 segments, each of which is 500 nsec in length (Δt= 500 nsec). The phase of the first segment would be defined by the desired phase of the pulse, e.g. along the x-axis. The RF-pulse that is applied to the sample for the nth segment would be:

$$e^{i\omega n\Delta t}e^{in\phi} \tag{14.35}$$

where the first term, $e^{i\omega n\Delta t}$, represent the normal B_1 field from the RF-pulse and the second term indicates the phase of the nth segment of the laminar pulse.

The above equation can be re-arranged as follows to show how the frequency of the pulse is changed:

$$\begin{aligned} &= e^{i(\omega n\Delta t+n\phi)} \\ &= e^{i(\omega n+n\phi/\Delta t)\Delta t} \\ &= e^{i(\omega+\phi/\Delta t)n\Delta t} \\ &= e^{i(\omega+\Delta\omega)n\Delta t} \end{aligned} \tag{14.36}$$

where $\Delta\omega = \phi/\Delta t$. The phase shift associated with each segment, ϕ, causes a shift in the frequency of the applied pulse, $\Delta\omega$.

Given the length of each segment of the pulse, it is possible to calculate the phase shift per segment to give the desired frequency shift. Most NMR spectrometers perform these calculations in an automatic fashion for the user. It is only necessary to indicate the desired shape (e.g. Gaussian), the number of segments in the pulse (e.g. 400), the overall length of the pulse (e.g. 256 μsec) and the desired frequency shift (e.g. 15 kHz).

14.4 Sidechain Assignments

The assignment of sidechain atoms is a necessary prerequisite for structure determination. In addition, these assignments are necessary to study chemical exchange or the relaxation properties of sidechain groups, such as methyls. However, sidechain assignments are clearly not required for studies on the relaxation/chemical exchange properties of mainchain amide groups.

The assignments of sidechain atoms are generally obtained by using scalar coupling to correlate the shifts of the unknown sidechain atoms to mainchain or sidechain atoms that have already been assigned. For smaller proteins, less than 20 kDa, near complete sidechain assignments for both carbon and protons can be obtained using TOCSY and DQF-COSY experiments. As the protein size increases it becomes increasingly more difficult to obtain complete assignments because of the relaxation losses during the magnetization transfer periods. For proteins in the range of 20-30 kDa it is generally possible to assign the sidechain resonances of methyl containing residues, such as Val, Ile, Leu, Thr, and Ala by correlating the intense methyl resonance to the amide pro-

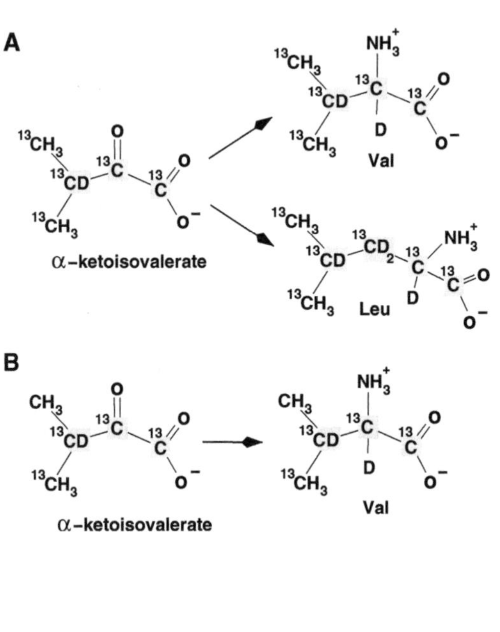

Figure 14.17 Compounds for the production of proteins uniformly labeled with ^{13}C and methyl-protonated at Val and Leu. In panel A uniformly ^{13}C labeled α-isoketovalerate is used to biosynthetically produce labeled Val and Leu. Note that the bacterial cells would have to be cultured in the presence of D_2O and uniformly labeled ^{13}C glucose, as well as the labeled α-isoketovalerate, to achieve the labeling pattern shown for Val and Leu. Panel B shows the labeling pattern for hemi-methyl labeled α-isoketovalerate. Biosynthetic incorporation of this compound will produce Leu and Val residues that are ^{13}C labeled at only one methyl carbon. Labeling a single methyl carbon removes a branch point in the network of ^{13}C coupled spins, increasing the sensitivity of ^{13}C-TOCSY and COSY experiments.

ton. In the range of 30-40 kDa deuteration of the aliphatic protons may be necessary to prevent relaxation losses. Deuteration reduces the contribution of the attached hydrogen to the carbon relaxation, making carbon TOCSY sequences more sensitive. Since the aliphatic protons have been replaced by deuterons most of the assignments have to be obtained by 'out-and-back' methods. Selective labeling with protonated and uniformly ^{13}C-labeled pyruvate [137], α-ketobutyrate , and α-ketoisovalerate [64] increases the protonation level of the methyl resonances significantly, allow the use of TOCSY-like experiments for the assignment of methyl resonances in large proteins. For proteins with molecular weights in the excess of 50 kDa it is generally necessary to label the methyl-containing residues in such a way as to remove branch points in the magnetization transfer pathways; increasing the sensitivity of the experiments [161]. For example, biosynthetically labeling with ^{13}C-^{12}C α-ketoisovalerate will result in the production of proteins in which the ^{13}C-spins in the Val and Leu residues form a linear systems of coupled atoms (see Fig. 14.17).

14.4.1 Triple-resonance Methods for Sidechain Assignments

The magnetization transfer paths for several triple-resonance experiments that are directed at sidechain assignments are shown in Fig. 14.18. These experiments can be divided into two classes. The four experiments shown on the right side of Fig. 14.18 use INEPT segments to transfer magnetization between spins. In contrast, the experiments shown on the left side of the figure use isotropic mixing (TOCSY) to exchange magnetization between all spins, permitting the assignment of practically all of the aliphatic protons and carbons.

The experiments that use INEPT segments, such as the CBCA(CO)NH are generally more sensitive than the TOCSY experiment and give less crowded spectra. They also provide unambiguous identification of the type of atom. For example, the CBCA(CO)NH experiment will correlate the β- and α-carbon shifts of one residue to the amide group of the following residue. The crosspeak associated with the α-carbon can be readily identified from the HN(CO)CA experiment, permitting unambiguous identification of the β-carbon shift.

The TOCSY based experiments utilize isotropic mixing of the carbon magnetization to relay the magnetization to either the α-carbon and proton, or through the carbonyl or C_α carbon to the amide for detection. The isotropic mixing of carbon magnetization is advantageous because of the larger one bond J-coupling between carbons of approximately 35 Hz. The larger J-coupling permits more rapid magnetization transfer among the spins, consequently signal loss due to relaxation is not as severe as in the proton-based TOCSY experiments.

In the TOCSY-based experiments it is usually more convenient to relay the sidechain chemical shift information to the amide group for detection because the amide region of the spectrum is more resolved than the α-carbon/proton region. Generally, the magnetization is passed from the α-carbon to the carbonyl carbon, and then across the peptide bond to the amide nitrogen of the following residue. Thus the sidechain resonances of one residue $(i-1)$ are correlated with the amide group of the following residue, (i).

Although it is also possible to pass the magnetization from the α-carbon to the amide nitrogen on the same residue, utilizing the one-bond C_α-N coupling, the spin-spin relaxation time (T_2) of the C_α carbon is generally very short, resulting in unacceptable signal loss in larger proteins (>20 kDa) during the transfer period. In addition, both inter- and intra-residue correlations are observed in this case, which leads to very crowded TOCSY spectra. Consequently, experiments that transfer the magnetization through the carbonyl to the amide group on the following residue are generally preferred.

14.4.2 The HCCH Experiment

Experiments that use TOCSY and INEPT segments to transfer magnetization from the sidechains to the amide group, such as the CBCA(CO)NH experiment, conveniently associate the sidechain resonances with an assigned amide group. However, experiments of this type are often of low sensitivity due to loss of signal during the transfer of magnetization to the amide nitrogen. Consequently, higher sensitivity is obtained if a sidechain proton is utilized for the final detection of the signal. An ex-

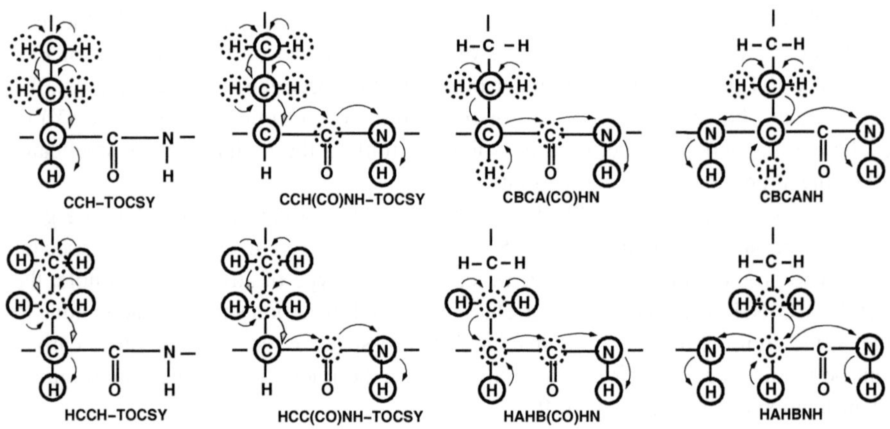

Figure 14.18. Triple-resonance experiments for sidechain assignments. Magnetization transfer pathways are shown for triple-resonance experiments that are directed at sidechain assignments. Arrows indicate direction of magnetization transfer. Atoms that are circled with dotted lines serve as the initial source of magnetization (e.g. H) or are involved in the transfer of magnetization between two atoms (e.g. CO, carbonyl). Atoms that are circled with solid lines have their chemical shift recorded in the experiment. All of these experiments are acquired as three-dimensional spectra. The four experiments on the left utilize TOCSY transfer, as indicated by the open-headed arrowheads, to correlate the sidechain chemical shifts to either the α-carbon and proton (far left) or the amide group (middle pair). Therefore, these sequences can, at least in theory, provide all of the sidechain chemical shifts with the exception of aromatic protons and carbons. The aromatic carbon shifts are generally too distant from the C_β shifts to permit effective transfer by isotropic mixing schemes. The sequences on the right side of the figure use INEPT segments for transfers and will only provide the β-carbon (top right) or β-proton (bottom right) chemical shifts, correlated with the amide chemical shifts.

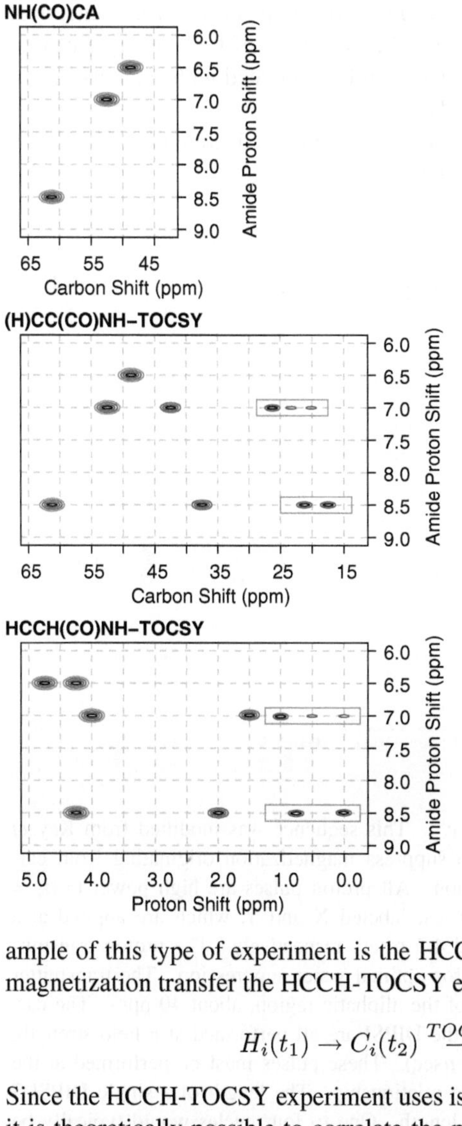

Figure 14.19 Sidechain assignments using TOCSY experiments. These two-dimensional spectra are slices at a particular amide frequency from the original three-dimensional spectra. The top spectrum, an HN(CO)CA, is provided to indicate the location of the C_α^{i-1} resonance, thus allowing unambiguous identification of the α-carbon. An HN(COCA)HA experiment would be used to unambiguously identify α-protons. TOCSY experiments, correlating the carbon (middle) and proton (bottom) shifts with the amide proton of the following residue are shown. The boxed peaks represent proton and carbon resonance that would not be obtained from triple-resonance experiments used for mainchain assignments, such as the γ and δ carbons and protons.

ample of this type of experiment is the HCCH-TOCSY experiment. The pathway of magnetization transfer the HCCH-TOCSY experiment is as follows:

$$H_i(t_1) \rightarrow C_i(t_2) \overset{TOCSY}{\longrightarrow} C_j \rightarrow H_j(t_3) \tag{14.37}$$

Since the HCCH-TOCSY experiment uses isotropic mixing to transfer magnetization, it is theoretically possible to correlate the proton detected in t_3 to all other ^{13}C and ^{1}H spins in the sidechain. However, the intensity of the crosspeaks will depend on the transfer efficiency as discussed in Section 9.5, consequently peaks that arise from multiple carbon-carbon transfers can be quite weak. Furthermore, it is usually impossible to observe crosspeaks that arise from the transfer of magnetization from aromatic to aliphatic carbons because of the large frequency difference between the two types of carbons. Therefore, in practice, not all expected correlations are observed in this experiment.

The pulse sequence for the HCCH-TOCSY experiment is shown in Fig. 14.20 and a brief discussion of the more important features of each segment of the sequence follows.

Segment 1: During this period the magnetization evolves under the proton chemical shift as well as the proton-carbon coupling. The net effect of both of these can be readily calculated by determining the net phase angle associated with each evolution, as discussed in Section D.2.5 of Appendix D.

The first 90° pulse generates $-I_y$ from the proton z-magnetization (I_z). The phase angle induced by proton chemical shift evolution is:

$$\Theta_{\omega_H} = \omega_H[\tau_a + \frac{t_1}{2} + \frac{t_1}{2} - \tau_a] = \omega_H t_1 \tag{14.38}$$

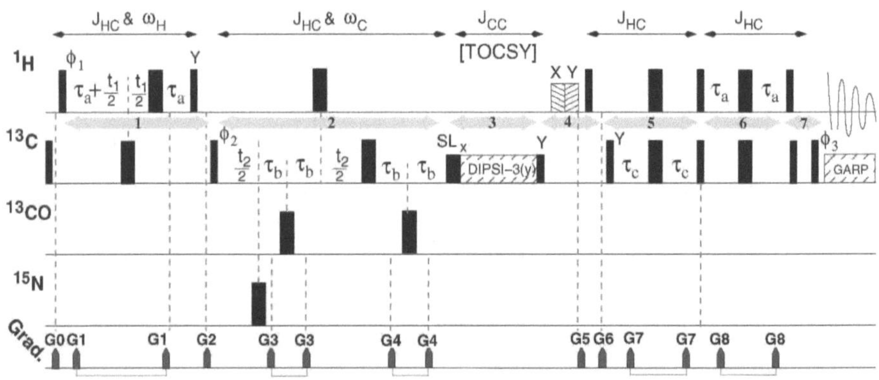

Figure 14.20. HCCH-TOCSY pulse sequence. This sequence was modified from Kay *et al.* [85] by the addition of gradient G0, to suppress magnetization originating from carbon. Parameters are from the same publication. All proton pulses are high power (γB_1 = 25 kHz) except the two proton spin-lock pulses, labeled X and Y, which are applied at a lower power level of γB_1=10 kHz for 7 and 4.3 msec, respectively. The proton transmitter would be placed on the water resonance for optimal water suppression. The transmitter for the ^{13}C channel is placed in the middle of the aliphatic region, about 40 ppm. The carbon spin-lock, DIPSI, and y-pulse following the DIPSI are all performed at a field strength, $\gamma B_1 = 8.2$ kHz (equivalent to $P_{90} = 30.5$ μsec). These pulses must be performed at the same power level to insure a consistent phase relationship. The length of a single DIPSI-3 mixing sequence is 6.63 msec for this pulse length. One to four cycles would typically be used, giving total mixing times from 6.63 msec to 26.52 msec. The power level for GARP decoupling during acquisition was 3.5 kHz. The carbonyl pulses were applied as 250 μsec SEDUCE-1 shaped pulses. The high power nitrogen 180° pulse is applied in the middle of the amide region, about 115 ppm. The delays are as described in the text. The phase cycle is: $\phi_1 = x, -x; \phi_2 = 2(x), 2(-x); \phi_3 = 4(x), 4(-x); \phi_{rec} = x, -x, -x, x$. Quadrature detection in t_1 and t_2 is accomplished by altering ϕ_1 and ϕ_2, respectively, according to States-TPPI. Although rectangular gradients were used in the original version, *sine*-shaped gradients are preferred. Gradients are applied as follows: G0 (10 G/cm, 1 msec), G1 (8 G/cm, 0.5 msec), G2=G3 (8 G/cm,2 msec), G4 (8 G/cm, 0.3 msec), G5 (30 G/cm, 7 msec), G6 (30 G/cm, 4.4 msec), G7=G8 (8 G/cm, 0.5 msec). Gradient pulse G2 is a zz-filter, gradient pulses G1, G3, G4, G7, and G8 remove imperfections associated with the 180° pulses.

The evolution due to CH coupling is evaluated in the same way, reversing the direction of evolution if a 180° pulse is applied to either spin:

$$\Theta_J = \pi J[\tau_a + \frac{t_1}{2} - \frac{t_1}{2} + \tau_a] = \pi J 2\tau_a \tag{14.39}$$

Assuming that $\tau_a = 1/(4J)$ the evolution of the density matrix due to J-coupling and proton chemical shift evolution is:

$$-I_y \rightarrow 2C_z[I_x cos(\omega_H t_1) + I_y sin(\omega_H t_1)] \tag{14.40}$$

The proton and carbon 90° pulses at the end of this segment places the proton magnetization along the z-axis and brings the carbon magnetization to the y-axis, e.g. $2C_zI_x \rightarrow 2C_yI_z$.

The gradient pair, G1, removes artifacts associated with the 180° pulse. This is especially important for the carbon pulse, since the chemical shift range is large, leading to off-resonance effects. Gradient G2 is a zz-filter.

Segment 2: This segment accomplishes two goals, refocusing of the antiphase carbon proton magnetization ($2C_yI_z$) to pure in-phase magnetization (C_x) and recording the carbon chemical shift. The density matrix evolves under the following Hamiltonian during this time:

$$\mathcal{H} = J_{C_\alpha N} + J_{CN} + J_{CC} + \omega_C + J_{HC} \tag{14.41}$$

The desired evolution is from the last two terms.

The evolution of the density matrix under all five of these interactions can be calculated readily using the concept of phase angles, beginning with the first two couplings:

$$\begin{aligned} \Theta_{J_{C\alpha N}} &= \pi J[(\frac{t_2}{2}) - (\tau_b + \tau_b + \frac{t_2}{2}) + (\tau_b + \tau_b)] \\ &= 0 \end{aligned} \tag{14.42}$$

$$\begin{aligned} \Theta_{J_{C\alpha C}} &= \pi J[(\frac{t_2}{2} + \tau_b) - (\tau_b + \frac{t_2}{2}) + (\tau_b - \tau_b)] \\ &= 0 \end{aligned} \tag{14.43}$$

Evolution due to the first two couplings has been removed by the application of 180° pulses on both the carbonyl and nitrogen during this period.

Evolution due to coupling between aliphatic carbons cannot be removed by a 180° pulse because of the similar chemical shifts of aliphatic carbons. The net phase angle due to evolution of the density matrix from carbon-carbon coupling is:

$$\begin{aligned} \Theta_{J_{CC}} &= \pi J[\frac{t_2}{2} + \tau_b + \tau_b + \frac{t_2}{2} + \tau_b + \tau_b] \\ &= \pi J[t_2 + 4\tau_b] \end{aligned} \tag{14.44}$$

Therefore the changes in the density matrix are:

$$2I_zC_y \rightarrow 2I_z\left[C_y cos[\pi J(t_2 + 4\tau_b)] - C_x sin[\pi J(t_2 + 4\tau_b)]\right] \tag{14.45}$$

note that only the $2I_zC_y$ term will give rise to detectable magnetization after it is converted to C_z under the influence of carbon-proton coupling. The $cos[\pi J(t_2+4\tau_b)]$ term will cause in-phase splitting of the resonance in carbon dimension.

The carbon chemical shift evolution is:

$$\begin{aligned}\Theta_{\omega C} &= \omega_C[(\frac{t_2}{2}+\tau_b+\tau_b+\frac{t_2}{2})-(\tau_b+\tau_b)] \\ &= \omega_C t_2\end{aligned} \qquad (14.46)$$

The proton-carbon coupling evolves as:

$$\begin{aligned}\Theta_{J_{HC}} &= \pi J[(\frac{t}{2}+\tau_b+\tau_b)-(\frac{t}{2})+(\tau_b+\tau_b)] \\ &= \pi J 4\tau_b\end{aligned} \qquad (14.47)$$

If τ_b is set to 1/(8J$_{CH}$), then CH groups will be completely refocused during the delay, $2C_yI_z \rightarrow C_x$. However, CH_2 and CH_3 groups are not refocused using this delay time because they evolve under coupling to one or two of the other protons, respectively. A more detailed analysis has been presented in Section 10.2.6. Consequently, τ_b is generally set to approximately 1/(16J$_{CH}$), however a value of 1/(8J$_{CH}$) will be used in this analysis. Therefore, the net result at the end of segment 2, ignoring carbon-carbon coupling, is:

$$2I_zC_y \rightarrow C_x cos(\omega_C t_2) + C_y sin(\omega_C t_2) \qquad (14.48)$$

Segment 3: The spin-lock pulse (SL$_x$) at the beginning of this segment will dephase any magnetization that is not along the x-axis. Dephasing occurs because the B_1 field is inhomogeneous, thus magnetization that is aligned along the y-axis will precess at different rates about the x-axis, depending on the location of the spin within the sample.

During the DIPSI-3 mixing period the magnetization is passed from one carbon (C_x^i) to another (C_x^j) by virtue of the relatively strong one-bond coupling:

$$C_x^i \stackrel{TOCSY}{\longrightarrow} C_x^j \qquad (14.49)$$

this magnetization is then returned to the z-axis by the y-pulse at the end of the mixing period.

Segment 4: This segment uses proton spin-lock pulses, in combination with gradients G5 and G6, to saturate the water. Since the observed protons are not readily exchangeable, saturation of the water has only a small effect on the sensitivity of this experiment. The spin-lock pulses will dephase any water magnetization that is not along the x- or y-axis, respectively. The gradient, G5, will dephase any remaining transverse magnetization. The proton 90° pulse following Gradient 5 will bring the z-component of the water magnetization into the x-y plane where it will be dephased by gradient G6.

Segment 5: This is the first of a series of two INEPT segments, the first converts the in-phase carbon magnetization to anti-phase carbon-proton magnetization. The pair of 90° pulses places the carbon on the z-axis and brings the proton into the transverse plane. The delay, τ_c is set to approximately $1/(8J_{CH})$ such that magnetization from CH, CH_2 and CH_3 groups can be converted to $2C_yI_z$:

$$C_z \overset{P90^Y_C}{\longrightarrow} C_x \overset{INEPT}{\longrightarrow} 2C_yI_z \quad (14.50)$$

Segment 6: The second INEPT refocuses the magnetization to in-phase proton magnetization:

$$2C_yI_z \overset{P90_H P90_C}{\longrightarrow} 2C_zI_y \overset{INEPT}{\longrightarrow} I_x \quad (14.51)$$

The delay, τ_a is ideally set to $1/(4J_{CH})$, however shorter times are generally used to compensate for relaxation during the transfer of magnetization.
Segment 7: This segment purges product operators that would give rise to artifacts in the spectrum. Since the focusing delay in segment 5, τ_c, is much shorter than $1/(4J_{CH})$, a significant portion number of undesirable terms arise that can cause artifacts in the spectrum. The two carbon pulses at the end of the pulse sequence remove these artifacts. The final proton pulse does not affect the desired signal, but rotates any remaining water magnetization to the z-axis, where it is undetectable.

Taking into account the evolution of the system due to chemical shift and carbon-carbon couplings gives the following for the final detected signal for the crosspeaks:

$$\sum_{ij} \eta_{ij} cos(\omega^i_H t_1) cos(\pi J_{CC}(t_2 + 4\tau_b)) cos(\omega^i_C t_2) e^{i\omega^j_H t_3} \quad (14.52)$$

where η_{ij} is the transfer efficiency of the isotropic mixing between spin i and j and $\sum_{ij}$ represents all coupled carbons within an amino acid residue. The spectrum also contains diagonal peaks arise from magnetization that is not transferred by the isotropic mixing, and therefore has the same proton frequency in ω_1 and ω_3.

Two-dimensional planes from the three-dimensional HCCH-TOCSY spectra are shown in Fig. 14.21. The particular carbon frequency that was selected is at the C_α chemical shift for three residues, one Ile and two Thr. Note that the intensity of the crosspeaks depends on the mixing time and that carbon atoms that are more distant from the C_α do not become apparent until longer mixing times are used. Since spin-spin relaxation during the mixing period will attenuate the crosspeak intensity, it may be difficult to detect crosspeaks that arise from transfer between distant carbons in proteins that are larger than 20 kDa.

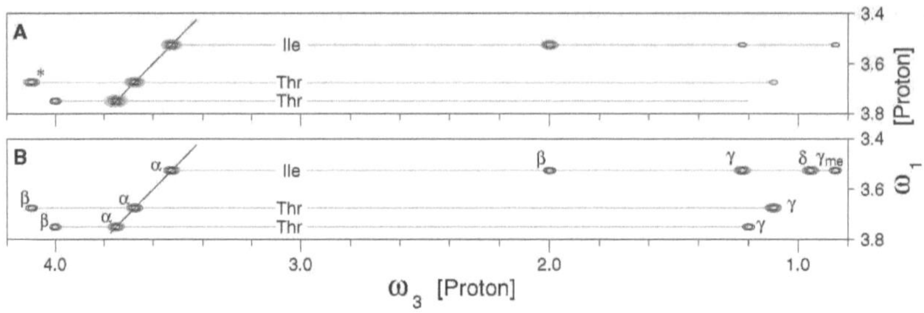

Figure 14.21. HCCH-TOCSY experiment. This figure shows a slice from a HCCH-TOCSY experiment at $\omega_2 = 66.5$ ppm. Panels A and B corresponds to DIPSI-3 mixing times of 7.6 and 22.82 msec, respectively. This assumes a 90° pulse length of 35 μsec, corresponding to a B_1 field strength of 7.142 kHz. At $\nu_C = 125$ MHz ($\nu_H = 500$ MHz) this would correspond to a $\sim$ 60 ppm range for effective magnetization transfer during the DIPSI-3 period. This particular slice contains three residues, two Thr and one Ile. The chemical shift of $\omega_2 = 66.5$ ppm corresponds to the α-carbon shift of all of these residues. Note that the crosspeak for the δ methyl of Ile and for the γ methyl of one of the Thr residues is too weak to observe at the shorter mixing time. The magnetization transfer path for the peak marked with an '*' in panel A can be summarized as follows:

$$\omega_{H\alpha}(t_1) \rightarrow \omega_{C\alpha}(t_2) \stackrel{TOCSY}{\rightarrow} C_\beta \stackrel{INEPT}{\rightarrow} \omega_{H\beta}(t_3)$$

14.5 Exercises

1. Indicate, on a chemical diagram of a dipeptide, the magnetization transfer path in the HN(COCA)HA experiment. Describe how the path would differ in the HA(CACO)NH experiment.

2. How would omitting the 180° C_α pulse in the HNCO experiment affect the appearance of the spectrum?

3. The following is an simplified version of the HNCBCA experiment [168]. This experiment is an extension of the HNCA experiment and gives inter- and intra-residue crosspeaks between both the C_α and C_β carbons and the amide nitrogen and proton.

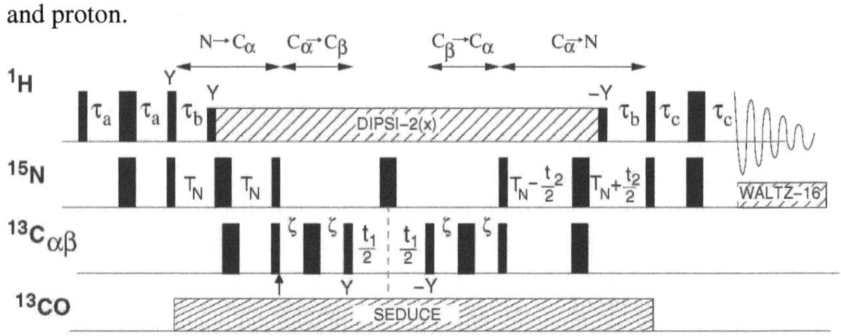

The product operator representation of the density matrix immediately after the application of simultaneous nitrogen and $C_{\alpha\beta}$ pulses (marked with an arrow) is: $-N_z A_y$, where N and A represents the nitrogen and C_α carbon, respectively.

Show that if the delay ζ is set to $1/(8J_{\alpha\beta})$ that the sign of the crosspeaks from the C_α carbons are inverted relative to those from the C_β carbons and that both are of equal intensity.

4. Calculate the length of a selective 90° rectangular pulse that would excite carbonyl carbons, but not C_α carbons on a 800 MHz (ν_H) spectrometer.

5. What is the approximate pulse length would you use for SEDUCE decoupling of the C_α carbons in the HN(CO)CA experiment, assuming the same conditions in Question 4.

6. The following two-dimensional spectra are slices from a HNCA experiment at the indicated nitrogen frequencies. Determine the order of these four spin-systems and give the assignments for the C_α carbon, amide nitrogen, and amide proton. What is the most likely amino acid type of the residue that precedes the first of these four residues?

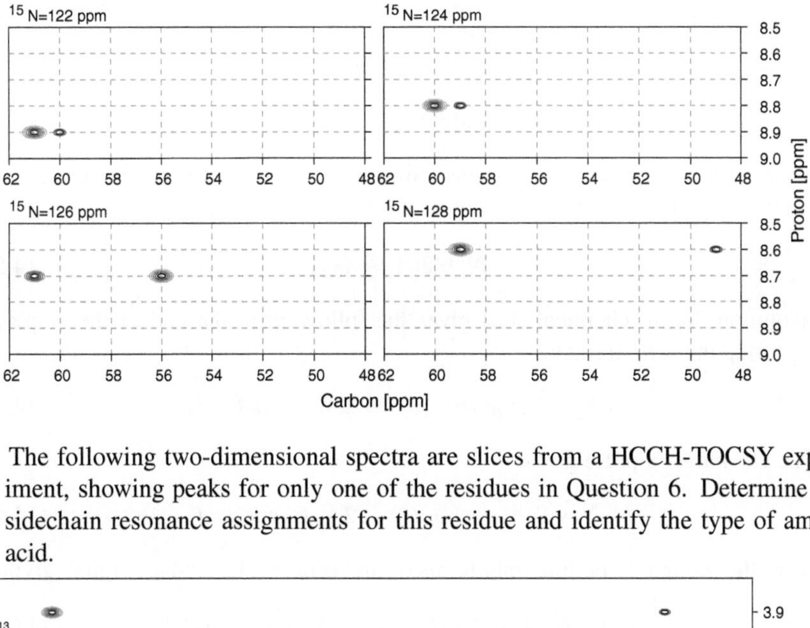

7. The following two-dimensional spectra are slices from a HCCH-TOCSY experiment, showing peaks for only one of the residues in Question 6. Determine the sidechain resonance assignments for this residue and identify the type of amino acid.

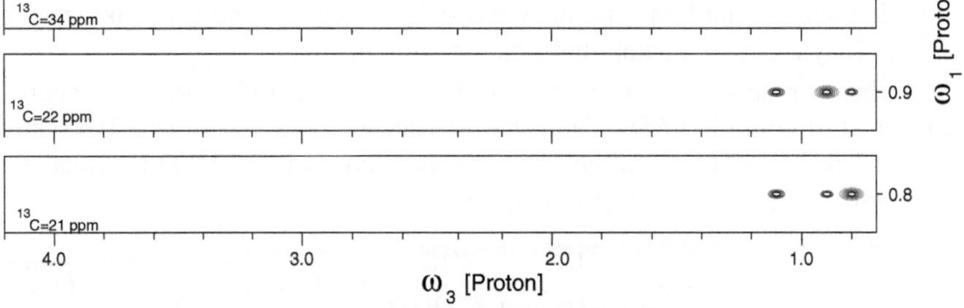

8. Sketch the two-dimensional carbon-proton plane from CBCA(CO)NH spectrum that would be obtained from the same residue in Question 7.

14.6 Solutions

1. In the HN(COCA)HA experiment, the magnetization goes from the amide group to the H_α of the preceding residue, *via* the carbonyl carbon and then the α-Carbon. Since it is an out-and-back type experiment, the magnetization returns by the same path. In the HA(CACO)NH the magnetization begins on the HA and is transferred to the amide group of the following residue. Thus both experiments give the same information.
2. This C_α pulse would refocus the evolution due to coupling between the C_α and carbonyl spins. If it were to be omitted, this coupling would evolve during t_1, giving rise to a 55 Hz splitting of the resonances in the carbonyl-dimension.
3. During the period $\zeta - P180 - \zeta$ the coupling between the C_α and C_β carbons is active because the 180° pulse is non-selective for these spins, therefore both are inverted. However, the evolution due to coupling between the Nitrogen and the C_α carbon is suppressed by this pulse. Therefore:

$$-N_zA_y \rightarrow -N_z[A_ycos(\pi J2\zeta) - 2A_xB_zsin(\pi J2\zeta)] \tag{14.53}$$

where B represents the C_β carbons. Since $\zeta = 1/(8J)$, the trigonometric terms are both equal to $\sqrt{2}$. The $90°_y$ pulse causes:

$$-N_z\sqrt{2}[A_y + A_zB_x] \tag{14.54}$$

Evolution due to chemical shift gives the following at the end of the t_1 period (ignoring the $sine$ terms):

$$-N_z\sqrt{2}[A_ycos(\omega_\alpha t_1) + 2A_zB_xcos(\omega_\beta t_1)] \tag{14.55}$$

The 90° *minus* y pulse gives:

$$-N_z\sqrt{2}[A_ycos(\omega_\alpha t_1) - 2A_xB_zcos(\omega_\beta t_1)] \tag{14.56}$$

After the second 2ζ period, only terms of the form N_zA_y produce signal, giving:

$$-N_z\sqrt{2}[A_y\sqrt{2}cos(\omega_\alpha t_1) - A_y\sqrt{2}cos(\omega_\beta t_1)] \tag{14.57}$$

where the additional $\sqrt{2}$ arises from the $cos(\pi J2\zeta)$ term.

The first and second terms are the crosspeaks for the C_α and C_β peaks, respectively. They are equal in amplitude, but opposite in sign.

4. The chemical shift difference between the carbonyl and C_α carbons is approximately 120 ppm (175-55). The carbon frequency on a $\nu_H = 800$ MHz spectrometer is 200 MHz, therefore 120 ppm corresponds to a 24,000 Hz frequency separation. The shortest selective pulse is:

$$\tau_{90} = \frac{\sqrt{15}}{4\Delta\nu} = \frac{\sqrt{15}}{4 \times 24000} = 40\ \mu\text{sec}. \tag{14.58}$$

5. The C_α carbons range from approximately 45 ppm to 65 ppm, a span of 20 ppm, or 4,000 Hz, on a ν_H=800 MHz spectrometer. Assuming that the bandwidth of the decoupling scheme is 1.75 times γB_1, a field strength of 2286 Hz is required, corresponding to a pulse length of 109 μsec ($\gamma B_1 = 1/(4\tau_{90})$).
6. Each pair of peaks corresponds to a amide group. Generally the more intense peaks is the intra-residue C_α carbon. Writing the four spin-systems in order of the ^{15}N Shift:

Spin-system	^{15}N	^1H_N	C^i_α	C^{i-1}_α
A	122 ppm	8.9	61	60
B	124 ppm	8.8	60	59
C	126 ppm	8.7	56	61
D	128 ppm	8.6	59	49

Re-ordering these, such that the C^{i-1}_α of one amide is the same as the C_α of the preceding residue gives the assignment:

Spin-system	^{15}N	^1H_N	C^i_α	C^{i-1}_α	*Assignment*
D	128 ppm	8.6	59	49	1
B	124 ppm	8.8	60	59	2
A	122 ppm	8.9	61	60	3
C	126 ppm	8.7	56	61	4

The preceding residue is likely glycine, given that the chemical shift of its C_α is 49 ppm.
7. The upper-most slice is at a carbon shift of 61 ppm, corresponding to residue 3 from the backbone assignments. This slice gives the H_α shift of 3.9 ppm. The additional peaks in the this slice is likely the H_β at 1.1 ppm. The second slice give the carbon shift associated with this peak, 34 ppm. Supporting its assignment as a H_β proton. The two other protons resonance at 0.8 and 0.9 ppm. The third slice shows that the carbon frequency of the 0.9 ppm resonance is 22 ppm and the fourth slice provides the carbon frequency of the 0.8 ppm peak, at 21 ppm. The carbon and proton frequencies associated with these pairs of carbon and proton shifts suggest that they are methyls. Therefore the residue is likely Val, with the following resonance assignments:

^{15}N	122 ppm	^1H	8.9 ppm
$^{13}C_\alpha$	61 ppm	$^1H_\alpha$	3.9 ppm
$^{13}C_\beta$	34 ppm	$^1H_\beta$	1.1 ppm
$^{13}C_{Methyl}$	22 ppm	$^1H_{Methyl}$	0.9 ppm
$^{13}C_{Methyl}$	21 ppm	$^1H_{Methyl}$	0.8 ppm

8. The CBCA(CO)HN experiment will transfer the C_α and C_β chemical shift information of $(i-1)^{th}$ residue to the amide group of the ith residue. In this case the carbon shifts of residue 3 will be associated with amide shifts of residue 4, giving the following two dimensional slice from the 3D-experiment.

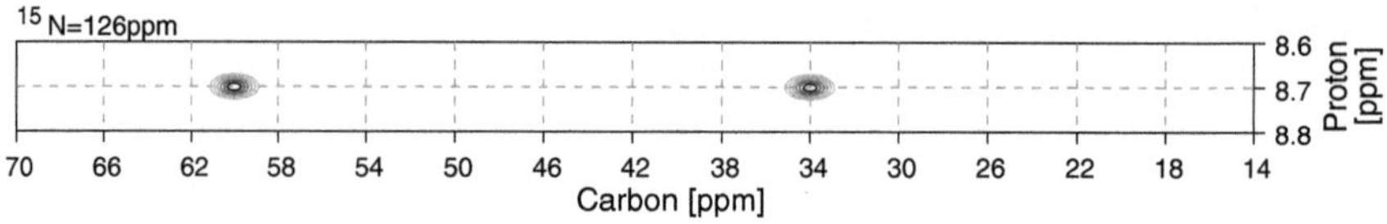

Note that the carbon chemical shift scale is much larger than in the HNCA experiment in order to accommodate the wide range of potential C_β shifts.

Chapter 15

PRACTICAL ASPECTS OF N-DIMENSIONAL DATA ACQUISITION AND PROCESSING

This chapter describes some of the more practical aspects of acquiring and processing multi-dimensional NMR data. It is divided into the following sections:

- Sample Preparation.
- Solvent Suppression.
- Instrument Configuration.
- Calibrating Shaped Pulses.
- Defining Indirect Acquisition Parameters.
- Processing 3-Dimensional Experiments.

15.1 Sample Preparation

15.1.1 NMR Sample Tubes

High-quality NMR tubes are the only tubes that should be used in high resolution spectrometers. The slight bend and non-circular characteristics of less expensive tubes can damage a high resolution probe because of the close fit between the probe and the sample tube.

New sample tubes should be cleaned with a dilute detergent solution. Solid protein residue can be removed from previously used tubes by soaking the tube in 4M HCl overnight or in household bleach for no more than 30 min.

15.1.2 Sample Requirements

Most of the experiments described in chapters 13 and 14 will require a protein concentration ranging from 0.5 to 1 mM. It is possible to acquire usable two-dimensional spectra with concentrations on the order of 0.1 mM and signal averaging for a total acquisition time of 4-12 hours. Such data may be useful for ligand titrations and evaluating the effect of sample conditions, such as salt concentration, pH, or temperature, on the protein and quality of the spectrum. The recent introduction of cryogenic probes lowers the required concentration by a factor of 2 to 4, however the increase in sensitivity is reduced by the presence of mobile ions in the buffer solution (see [87]).

Sample volumes of 0.4 to 0.5 ml are necessary in standard 5 mm NMR tubes to insure that the sample is sufficiently long to place the solvent/air interface away from the transmit/receive coils. Otherwise the large difference in the magnetic susceptibility between the air and the solvent can lead to inhomogeneities in the magnetic field that are difficult, if not impossible, to remove by shimming. Microcells can be used with smaller volumes, on the order of 0.3 ml, thus only about 7 mg of a 20 kDa protein is required for a sample. One commonly used microcell has a solid glass plug at the bottom of the tube and a solid glass insert that is placed above the sample, displacing the air. The magnetic susceptibility of the glass is formulated to be similar to that of the solvent, minimizing changes in the magnetic field at the ends of the sample (see Fig. 15.1). The sample should be free of macroscopic particles as these can cause difficulties in shimming, again due to susceptibility differences between the solvent and the particulate matter. Centrifugation for 10 min at 10,000 x g is generally sufficient. Air bubbles have the same effect as particles, hence they should also be absent from the sample.

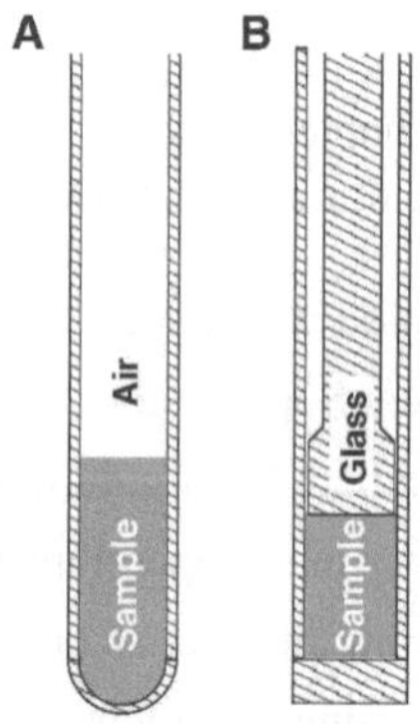

Figure 15.1. NMR sample microcells. A standard (A) NMR tube is compared to a microcell (B). The solvent-water interface in the normal NMR tube is replaced by a glass insert, placed immediately above the sample.

Ideally, samples of low molecular weight proteins should also be degassed to remove dioxygen. O_2 is paramagnetic and its unpaired electron is effective at relaxing nuclear spins because of the large gyromagnetic ratio of the electron. Contributions of this relaxation process to the linewidth can be on the order of 1 Hz. This additional broadening is less significant for larger proteins that have linewidths on the order of 10-20 Hz and degassing may not be necessary.

A salt concentration larger that 0.20 M (monovalent ions) should be avoided if possible. The high salt can cause difficulty in tuning of the probe and will reduce the sensitivity of both traditional and cryo-probes. The effect with cryo-probes is more pronounced due to their higher Q. In addition, the high salt absorbs the radio-frequency field applied to the sample during pulses and decoupling. This has three consequences. First, longer pulse lengths are required for excitation, leading to a smaller bandwidth of excitation. Second, the homogeneity of the applied pulses is reduced; the molecules in the center of the sample will experience a somewhat weaker RF-field. Third, significant amounts of heat can be deposited in the sample during decoupling, causing temperature changes in the sample during the experiment, leading to protein denaturation under extreme conditions.

The pH of the solvent should not exceed approximately 7.0. At higher pH values, the large exchange rate of the amide proton can cause broadening of the amide proton linewidth due to chemical exchange, as discussed in more detail in the following section.

The solvent should be buffered against pH changes. Buffers should be chosen that either do not contain protons, such as phosphate, or are readily available in deuterated form. It is usually prudent to include sodium azide at a concentration of 0.02% to prevent microbial growth in the sample. Finally, it is also necessary to include 5-10% D_2O in the solvent for proper operation of the spectrometer lock.

It is frequently necessary to optimize the solution conditions of a sample prior to extended data collection. The solvent conditions should be selected to give monomeric protein, as indicated by the observed linewidths in the spectra. The large effect of temperature on the linewidth, as discussed in the Chapter 19, should be taken into account in the assessment of linewidth.

15.2 Solvent Considerations - Water Suppression

15.2.1 Amide Exchange Rates

Resonance signals from the amide protons provide the cornerstone for practically all NMR experiments that are discussed in this text. Unfortunately, the amide proton readily exchanges with solvent. This exchange rate is catalyzed by both acid and base and the effect of pH on the rate of exchange is shown in Fig. 15.2. These rates correspond to groups that are fully exposed to the solvent. Rates that are several orders of magnitude slower can be observed for protons that are buried and not exposed to solvent. Since these buried amides eventually exchange with solvent, they must become transiently exposed to solvent as the protein samples multiple conformations.

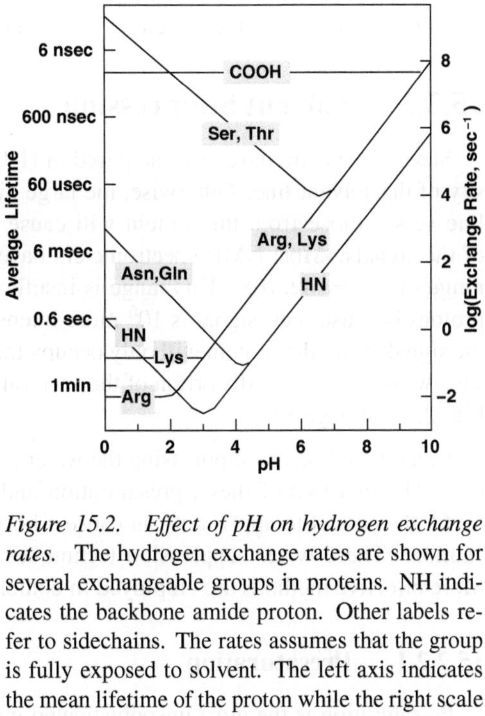

Figure 15.2. Effect of pH on hydrogen exchange rates. The hydrogen exchange rates are shown for several exchangeable groups in proteins. NH indicates the backbone amide proton. Other labels refer to sidechains. The rates assumes that the group is fully exposed to solvent. The left axis indicates the mean lifetime of the proton while the right scale gives the exchange rate (Adapted from [171]).

If the rate at which the amide hydrogen exchanges with water is sufficiently fast, signals from the amide protons can disappear from the spectrum. For example, an amide proton resonance at 5.5 ppm is ≈500 Hz from the water resonance line on a 600 MHz (ν_H) spectrometer. If the rate of exchange is much faster than 500 sec^{-1}, the system will be in fast exchange and the observed chemical shift of the amide will be heavily weighted to that of the water due to the high concentration of solvent in the sample, i.e. the amide resonance would essentially appear at the solvent frequency (see Chapter 18).

Inspection of Fig. 15.2 shows that pH values less than approximately 6.0 are required to give an exchange rate that is slower than 500 sec^{-1} for fully exposed amides. At this pH the sidechain NH groups of arginine and lysine residues will be in fast exchange with the water resonance since their exchange rates are 100 times that of the NH proton. Consequently, their NH protons cannot be observed. The amide exchange rate of the amino terminus is similar to ϵ-amino group of lysine, therefore it is often difficult to observe resonance signals from this group as well. In a similar fashion, it would also be impossible to observe resonance signals from the hydroxyl group of serine and threonine at any pH, unless the group participates in hydrogen bonding or is buried in the protein, causing a reduction in the exchange rate.

Amide hydrogen exchange can also cause broadening of resonance lines, even if the observed resonance signal arises from a proton under that exchanges slowly with the solvent. An exchange rate of 50 sec^{-1} would give an increase the linewidth by 50 Hz, since the linewidth is increased by the exchange rate ($\Delta\nu = k_{ex} + 1/(\pi T_2)$), as discussed in Chapter 18. Therefore, pH values somewhat less than 6.0 will generally give narrower lines for solvent exposed amide protons. Note that this source of line-broadening occurs irrespective of the chemical shift difference between the water and the amide proton since the system is in slow exchange.

15.2.2 Solvent Suppression

Since the spectra have to be acquired in H_2O, it is necessary to attenuate the intensity of the solvent line. Otherwise, the large difference in the intensity of the solvent line versus those from the protein will cause problems with the proper digitization of the signals. Most NMR spectrometers have 16-bit digitizers, allowing a dynamic range of $2^{15} = 32,768$. This range is insufficient to digitize the FID from the H_2O protons because their signal is 10^5 more intense than that of the protein. As a result, the signals from the protein will only occupy the lowest one or two bits of the digitizer. This will cause severe distortion of the spectral lines from the protein, as illustrated in Fig. 2.17 on page 53.

Several methods of suppressing the water signal will be presented in the following text. The first two of these, presaturation and heteronuclear spin-lock methods, are by far the easiest to apply, but can cause a host of problems and should be used with caution. These water suppression techniques are found in earlier pulse sequences. More effective methods are employed in contemporary experiments.

15.2.2.1 Presaturation

Presaturation is the most unsophisticated method of solvent saturation. The water resonance is simply irradiated with a low power radio-frequency field, as illustrated in Fig. 15.3. This field causes transitions from the ground state to the excited state over a bandwidth of approximately 1 Hz. If the rate of excitation exceeds the rate of spin-lattice relaxation (R_1) then the two population levels become equal and no water signal is observed. Although pre-saturation is a very reliable method of solvent suppression, it causes a number of problems (see [95]).

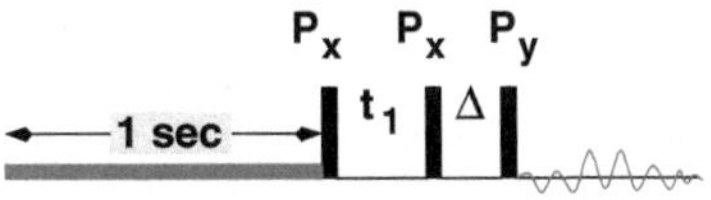

Figure 15.3 Solvent presaturation. The application of solvent presaturation prior to the collection of a DQF-COSY experiment is illustrated. The transmitter frequency would be set to the resonance frequency of the water and the presaturation pulse would be applied at a power that is approximately 40 dB lower than the high power used for pulses.

First, the efficiency of presaturation is affected by the width of the water resonance line. Higher power levels are required to equalize the ground and the excited state over a wider frequency range.

Second, presaturation of the solvent requires approximately 1 second. Consequently, any exchangeable protons with an exchange rate of greater than $\approx$ 1 sec^{-1} will also become saturated because the saturated water protons will replace the exchangeable proton. This includes amide, hydroxyl, and carboxyl protons. Furthermore, this saturation can be effectively transfered to nearby non-exchangeable protons by dipolar coupling. In this case dipolar coupling can be thought of as a mechanism for the system to approach thermal equilibrium. Saturation of the exchangeable proton increases its spin temperature [1]. This 'thermal energy' is then transferred from the exchangeable proton to any other protons within $\approx$ 5 Å of the exchangeable proton, reducing the population difference between their ground and excited states.

Third, the application of the pre-saturation pulse to the water resonance will also saturate any H_α protons whose resonance signals are close to the water resonance, causing the resonance line from the α-proton to disappear. The intensity of any non-exchangeable protons that are within 5 Å of an α-proton will also become reduced.

The rate of transfer of saturation by dipolar coupling depends on the spectral density function, $J(\omega)$ at $\omega = 0$, therefore, the transfer of saturation become more efficient as the size of the protein increases. Consequently, pre-saturation should not be used on proteins whose molecular weight is greater than $\approx$ 10 kDa.

15.2.2.2 Heteronuclear Spin-Lock for Water Saturation

This method is frequently used in heteronuclear experiments and takes advantage of the *lack* of coupling between the water proton and the nitrogen to place the water protons in a state that can be selectively saturated, while preserving the intensity of the amide protons [113].

As an example, consider the initial portion of an HMQC experiment, as shown in Fig. 15.4. The product operator associated with HN groups at the end of the polarization transfer step is antiphase, represented by the product operator $2I_xN_z$. In contrast, the water magnetization is found along the y-axis and is represented by the product operator I_y. If a 1-2 msec high power proton pulse, commonly called a spin-lock pulse, is

[1]Recall that the population difference between two energy levels is given by the Boltzmann equation:

$$\frac{n_e}{n_g} = e^{-\Delta E/kT}$$

If the protons are saturated, $n_e = n_g$, therefore the spin temperature (T) $= \infty$.

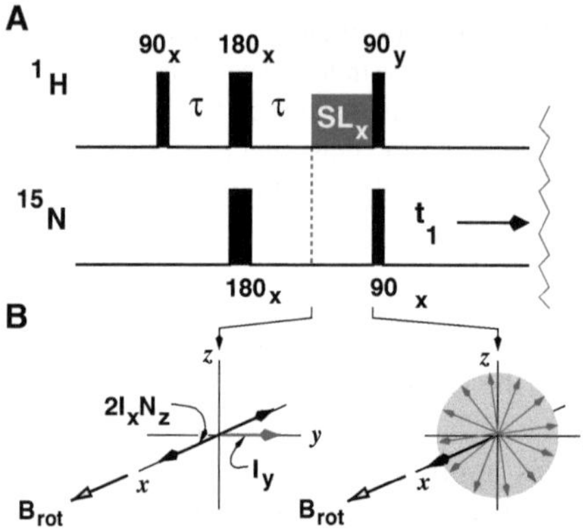

Figure 15.4 Spin-lock water suppression. Panel A shows the initial section of an HMQC experiment, with the inclusion of the spin-lock pulse (SL_x) at the end of the INEPT period. Panel B shows the magnetization just prior to application of the spinlock (left). The effective field in the rotating frame is B_{rot}. The magnetization associated with the water, I_y, is illustrated as a dark gray vector with a single arrowhead. The antiphase magnetization associated with the NH group is shown as a vector with an arrowhead at each end. The spin-lock pulse causes the water magnetization to become random in the y-z plane (right).

applied along the x-axis, the effective magnetic field in the rotating frame will be along the x-axis. The magnetization associated with the amide proton is aligned with this field and will remain 'locked' along that axis during the application of the spin-lock pulse. The water magnetization, on the other hand, will precess around the effective field. Since the B_1 field generated by the transmitter coils is relatively *in*homogeneous, the water magnetization will dephase rapidly and essentially become saturated. Since the saturation occurs in a few msec, this technique of water suppression generally has a smaller impact on the intensity of the exchangeable amides because there is insufficient time for chemical exchange to occur. However, the saturated state of the water can eventually be transferred to the amide and aliphatic protons by the processes described in the above section, especially if the relaxation delay period between scans is too short to allow the recovery of the water magnetization prior to the next scan.

15.2.2.3 Coherence Selection

The coherence state of the water spins during the pulse sequence can be used to remove the solvent line with gradients. For example, in the DQF-COSY experiment the solvent line cannot participate in double quantum coherence. Therefore, the gradients that are applied to select for double quantum magnetization during the Δ period dephase the water and eliminate it from the spectrum.

In the case of heteronuclear experiments, it is possible to also suppress the water using a similar strategy, as illustrated in Fig. 15.5. The application of a gradient during the t_1 evolution period, when the nitrogen magnetization is transverse, will introduce a phase shift into the signal:

$$\phi = e^{iG_1\gamma_N\tau_1}$$

Application a refocusing gradient during the INEPT period, will introduce an additional phase shift:

$$\phi = e^{-iG_2\gamma_H\tau_2}$$

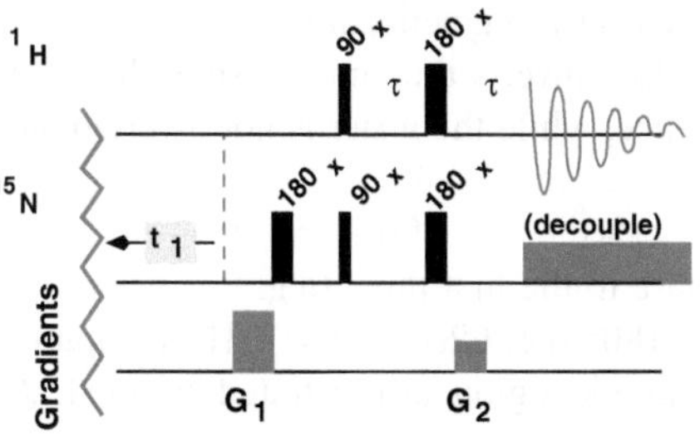

Figure 15.5 Water suppression by coherence selection. The latter portion of an HSQC experiment is shown. Gradients G1 and G2 are designed to refocus the magnetization associated with the NH group. Consequently they will dephase the magnetization associated with water, reducing its signal intensity.

G_1, G_2, τ_1, and τ_2 are adjusted such that these two phase shifts cancel, therefore the signal associated with the NH group is observed as in a normal HSQC. The water magnetization, on the other hand, dephases during the first gradient as:

$$\phi = e^{iG_1\gamma_H\tau_1}$$

and therefore cannot be refocused by the second gradient and remains undetectable. Since it is only possible to refocus one of the coherences, the above approach will cause a loss of signal intensity. Sensitivity enhanced implementations of this method of solvent saturation do not suffer this loss, as discussed in Chapter 14.

Note that this strategy relies on the water being transverse during the application of each gradient. The water magnetization can undergo a process called radiation damping whereby the intense transverse magnetization of the water protons induces an oscillating magnetic field that stimulates relaxation of the water spins back to the ground state. Consequently, care must be taken to insure that the water magnetization is in the correct state during the application of gradients, as discussed in chapters 13 and 14.

15.2.2.4 WATERGATE: Selective Dephasing of Water by Gradients

A extremely versatile water suppression technique, called WATERGATE (Water suppression by gradient-tailored excitation) [132, 150], is shown in Fig. 15.6. The WATERGATE sequence consists of a dephasing gradient, followed by a *selective* 180°

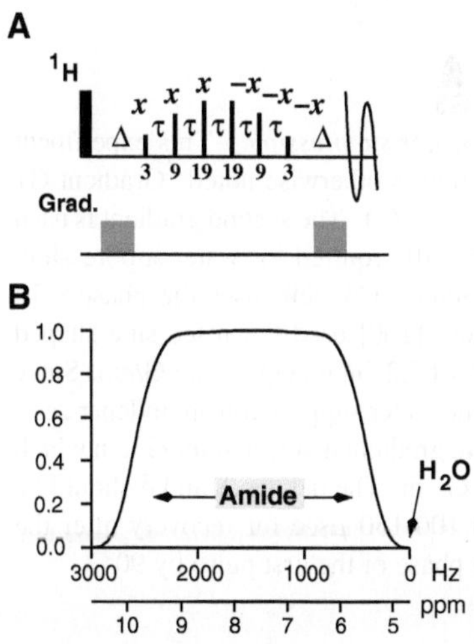

Figure 15.6 WATERGATE element for water suppression. Panel A shows the WATERGATE sequence, consisting of a 6-pulse element that is flanked by two identical gradients. The 90° proton pulse would be the last pulse in a NOESY experiment, for example. The delays Δ are as short as possible to allow application and recovery of the gradient pulses. Panel B shows the excitation profile for the 180° pulse, assuming $\tau = 333$ μsec. The ppm scale is appropriate for a spectrometer frequency of 500 MHz.

pulse, and then a second gradient that will refocus any magnetization that was inverted by the selective 180° pulse. The selective pulses inverts the amides, such that they refocus at the end of the WATERGATE sequence, while the water is not inverted and is further dephased by the second gradient.

The selective pulse is: $3\alpha - \tau - 9\alpha - \tau - 19\alpha - \tau - 19\alpha - \tau - 9\alpha - \tau - 3\alpha$, where the last three pulses are opposite in phase to the first three (e.g. $-x$ versus x). The total rotation angle of all six pulses equals 180° (i.e. $62\alpha = 180°$). The excitation profile of this pulse is remarkably flat over the amide region, as illustrated in Fig. 15.6, showing a null excitation at the following frequencies:

$$\pm k/\tau, \quad k = 0, 1, 2, 3 \ldots \tag{15.1}$$

where τ is the spacing between the pulses.

Improved WATERGATE sequences, which utilize 8 and 10 pulses for the selective 180° pulses, have been published [99]. These provide more uniform excitation of the region between the points of null excitation.

The WATERGATE pulse element can be appended to the end of a homonuclear experiment, such as COSY, or NOESY experiment. The inclusion of the WATERGATE sequence in a homonuclear COSY experiment is shown in Fig. 15.7. The WATERGATE sequence can also be included in heteronuclear experiments as part of the last INEPT transfer segment. In which case, the delay δ, inclusive of τ and pulsewidths, is set to 1/(4J).

The WATERGATE sequence can still perturb the intensity of amide proton resonances (see [95]). Since it dephases the water while in the transverse plane, the sequence will generate a non-equilibrium population of the ground and excited state of the water magnetization. This non-equilibrium state can be transferred to the amide protons during the recycle delay period, reducing their signal intensity. This signal loss can be avoided by either by utilizing long recycle delay periods or by using water

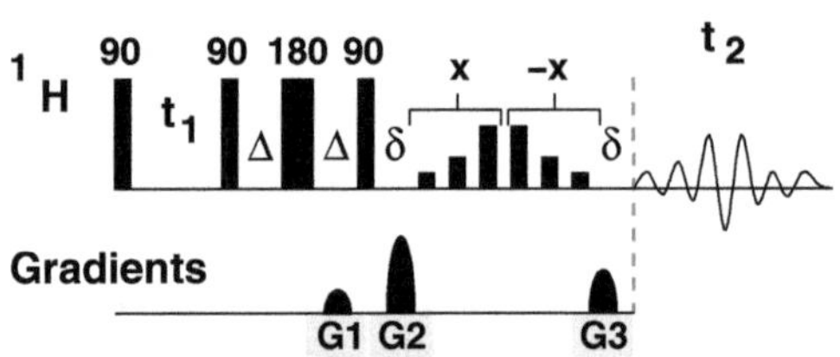

Figure 15.7. DQF-COSY sequence with WATERGATE water suppression. This experiment was described by [160]. All pulses are along the x-axis unless otherwise noted. Gradient G1 encodes the double quantum magnetization with phase $\phi = 2 \times G1$. The second gradient is used to refocus this phase shift as well as to induce the phase shift required for water suppression. Therefore, its intensity is $2 \times G1 + G3$. The final gradient, G3, refocuses the phase shift associated with water suppression. Trimble and Bernstein [160] used 0.5 msec sine shaped gradients with the following values: G1 = 24.5 G/cm, G2 = 65.2 G/cm, G3 = 16.3 G/cm. Since the gradients for selection of double quantum coherence and water suppression are independent, the size of middle gradient (G2) can be reduced if a negative gradient is selected for G3, in which case G2 = 32.6 G/cm if G1 = 24.5 G/cm and G3 = -16.3 G/cm. The delays Δ and δ should be as short as possible, allowing time for the gradient and 100-150 μsec for recovery after the gradient. Quadrature detection is obtained by shifting the phase of the first pulse by 90°.

selective pulses to return the magnetization to the z-axis prior to the application of the WATERGATE sequence. The water selective pulses are referred to as water flip-back pulses and are discussed below.

15.2.2.5 Jump-and-Return

The jump-and-return method of water suppression is a simpler version of the WATERGATE sequence. It was introduced by Plateau and Gueron in 1982, primarily for use in acquiring NMR spectra of nucleic acids in H_2O [133]. It has also been used in a number of multi-dimensional NMR experiments, such as the HNHA experiment discussed in Chapter 13.

As indicated in Fig. 15.8 This method uses two high-power proton pulses, separated by a delay τ. Assuming that the transmitter is on water, the second pulse in the jump-and-return sequence will return the water to the z-axis because the water magnetization does not precess during the τ period. In contrast, the magnetization associated with the amide resonances will precess 90° during the τ period such that when the second pulse is applied, it will be aligned with the RF-pulse. Consequently, the RF-pulse will be ineffective at rotating the amides, leaving their magnetization in the transverse plane for detection. The delay, τ should be set to $1/(4\nu)$, where ν is the frequency separation between the water resonance and the *center* of the amide proton resonances, 1500 Hz for a 500 MHz spectrometer, equivalent to 3 ppm.

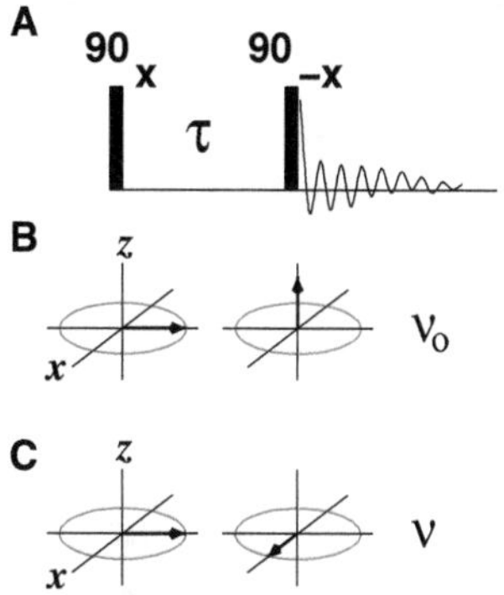

Figure 15.8. Jump-and-return sequence for water suppression. The jump-and-return sequence for water suppression is shown in part A. Part B and C show the effect of this pulse sequence for on-resonance (B) and off-resonance (C) spins.

Efficient water suppression using the jump-and-return sequence requires a narrow linewidth for the water resonance. Otherwise the water magnetization that is slightly off-resonance distorts the baseline of the spectrum. In addition, the excitation profile for the jump-and-return is narrower than that of the WATERGATE sequence. Consequently most current NMR experiments employ the WATERGATE sequence for water suppression.

15.2.2.6 Water Flip-back Pulses

The influence of the state of water magnetization on the intensity of the amide protons can be reduced substantially if the water magnetization is returned to the z-axis at the end of the pulse sequence [5]. Water selective 'flip-back' pulses can be used to rotate the water magnetization back to the z-axis, restoring thermal equilibrium of the water magnetization. In the case of the WATERGATE sequence the water selective flip-back pulse is applied prior to the 90° pulse. If the phase of the flip-back pulse is opposite to that of the 90° pulse then the water magnetization will be returned to the z-axis by the non-selective 90° pulse, i.e.:

$$I_z \stackrel{P90_{-x}}{\longrightarrow} I_y \stackrel{P90_x}{\longrightarrow} I_z$$

The inclusion of water flip-back pulses and WATERGATE water suppression in a 2-dimensional homonuclear NOESY [98] and TOCSY experiment [57] is shown in Fig. 15.9. The inclusion of the water flip-back pulse increases the average signal intensity of the amide protons by approximately 20%.

The TOCSY sequence is nearly identical to the NOESY experiment since the isotropic mixing is performed while the magnetization is along the z-axis. A significant difference between the two experiments is that the NOESY experiment relies

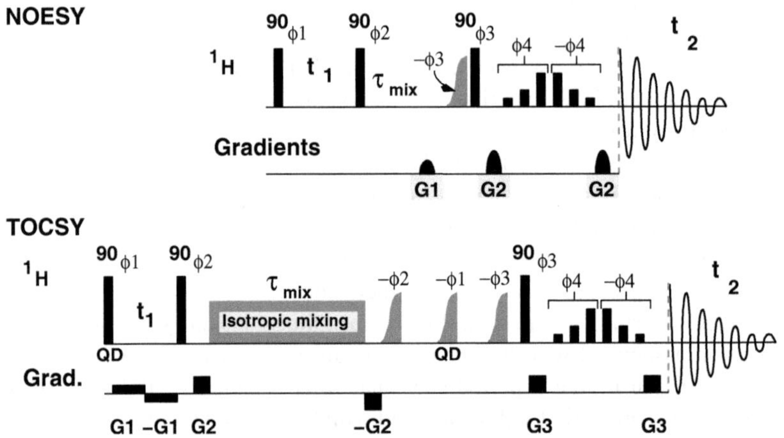

Figure 15.9. 2D-NOESY & TOCSY sequences with water flip-back pulses.
TOP: A two-dimensional NOESY experiment that incorporates WATERGATE water suppression and a water flip-back pulse is shown. This was adapted from Lippens et al. [98] by replacing the original WATERGATE pulse element with the 3-9-19-19-9-3 element. Pulse phases are $\phi_1 = x, -x, y, -y, -x, x, -y, y$; $\phi_2 = x, x, y, y, -x, -x, -y, -y$; $\phi_3 = \phi_2$; $\phi_4 = -x, -x, -y, -y, x, x, y, y$; $\psi_{rec} = x, -x, y, -y, -x, x, -y, y$. Gradients are sine-shaped and are 5 G/cm for G1, and 15 G/cm for G2. All three gradients would be applied for 1 msec. The sequence relies on radiation damping to restore the water magnetization to the z-axis during the mixing time, therefore mixing times of less than 100 msec should be avoided. Gradient G1 is applied at the end of the mixing time to dephase residual transverse water magnetization. The water selective flip-back pulse is shaded gray and has a phase of $-\phi_3$. It was applied with a transmitter field strength of 160 Hz, corresponding to the power required for a 1.56 msec pulse. Quadrature detection would be obtained by incrementing ϕ_2 by 90° and collecting a second FID with the same value for t_1.
BOTTOM: WATERGATE-TOCSY. This pulse sequence was adapted from [57]. The tall black bars refer to high power 90° pulses and the shaped gray pulses are selective water flip-back pulses. Isotropic mixing occurs while magnetization is along the z-axis and can utilize any of the mixing schemes discussed in Chapter 9. The pulse phases are: $\phi_1 = 4(x), 4(y), 4(x), 4(y), 4(-x), 4(-y), 4(-x), 4(-y)$; $\phi_2 = 4(x), 4(y), 4(-x), 4(-y)$; $\phi_3 = x, y, -x, -y$; $\phi_4 = x, y, -x, -y, -x, -y, x, y$; $\psi_{rec} = 4(x, y, -x, -y), 4(-x, -y, x, y)$. Quadrature detection is obtained by incrementing the phases of both pulses that are labeled with 'QD': the first 90° pulse (ϕ_1) and the middle flip-back pulse. Gradients are applied as rectangular shapes, G1 = 0.25 G/cm, G2 = 1.3 G/cm, and G3 = 13 G/cm. Gradients G1 and G2 are applied at low levels such that it is not necessary to include gradient recovery periods. The length of G1 is equal to $t_1/2$, G2 and G3 are 1 msec.

on radiation damping during the mixing time to return the water magnetization to the z-axis at the end of the sequence whereas the TOCSY sequence suppresses radiation damping during the t_1 evolution period by the application of the bipolar gradients (G1) that dephase the transverse water magnetization and then refocus it at the end of the t_1 period. The pulses phases are designed to restore the water magnetization to the z-axis when the FID is acquired.

When the data is collected for quadrature detection in the TOCSY experiment it is necessary to shift the pulse of the first 90° pulse by 90°, from x to y. To insure that the water magnetization is returned to the z-axis, it is also necessary to modify the phase of the middle water flip-back pulse, giving the following for the transformation of the water magnetization during the experiment:

$$I_z \stackrel{\phi_1=x+\pi/2}{\longrightarrow} I_x \stackrel{t_1}{\rightarrow} I_x \stackrel{\phi_2}{\rightarrow} I_x \stackrel{TOCSY}{\longrightarrow} I_x \stackrel{-\phi_2}{\rightarrow} I_x \stackrel{-\phi_1}{\rightarrow} I_z \stackrel{-\phi_3}{\rightarrow} I_y \stackrel{\phi_3}{\rightarrow} I_z \qquad (15.2)$$

15.2.2.7 Selection of Flip-back Pulses

A selection of flip-back pulses that have been used in the literature is shown in Fig. 15.10, along with their excitation profiles. It is clear from the excitation profiles presented in Fig. 15.10 that all of these pulses have sufficient bandwidth to uniformly rotate the water magnetization by 90°. However, they differ considerably in the degree of perturbation of the amide proton region. Ideally, this region of the spectrum

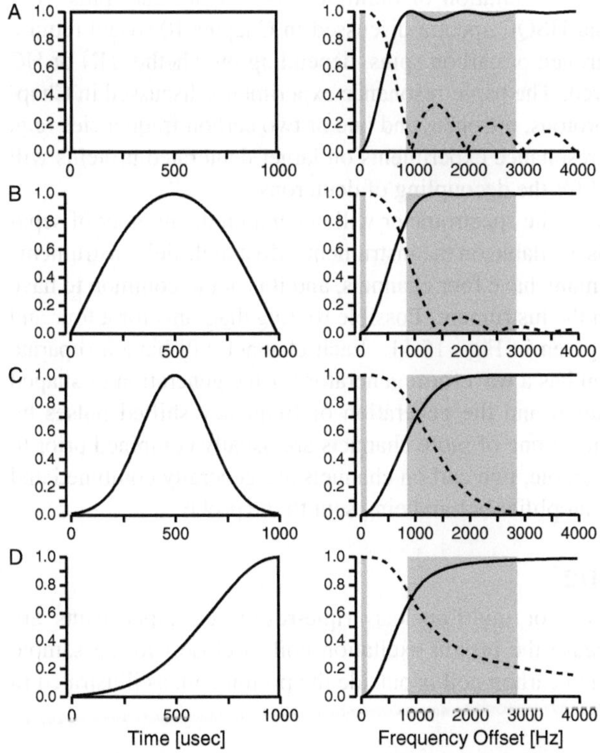

Figure 15.10 Water flip-back pulses. A rectangular (A), sine-shaped (B), Gaussian (C), and half-Gaussian (D) water flip-back pulses are shown (left panels) along with their corresponding excitation profiles (right panels). All pulses are 1 msec in length. The solid line in the right panel gives the z-component of the magnetization and the dotted line gives the amount of magnetization in the xy-plane ($\sqrt{M_x^2 + M_y^2}$). The narrow shaded area at $\nu = 0$ in the right panel indicates the typical width of the water resonance. The wide shaded area represents the chemical shift range of the amide protons (6.5 to 10.5 ppm) on a 500 MHz spectrometer. The chemical shift of the water protons is set at 4.8 ppm.

should not be affected by the selective water pulse, i.e. M_z should be 1.0 and $M_{x,y}$ should be zero in the amide proton region. In practice, both the rectangular and sine shaped pulses lead to lobes of excitation throughout the amide region, as shown by the non-zero transverse magnetization. The excitation of the amide protons by a rectangular pulse is substantial and this type of selective pulse should not be used, except perhaps at very high magnetic field strengths (see below). The excitation profile of the Gaussian shaped pulses are somewhat better, in that the excitation profile decays smoothly to zero.

All four of the flip-back pulses shown in Fig. 15.10 are unacceptable at the illustrated field strength ($\nu_H = 500$ MHz) because of the excitation of the amide region. The sine (B) and Gaussian (C) shape would be functional at 900 MHz, because the frequency range of the amide protons would be shifted further from the H_2O resonance. The sine and Gaussian pulses can be used at 500 MHz if the pulse length is increased. The excitation profile of the pulse is approximately equal to the inverse of pulse width. For example, a 2 msec sine or Gaussian pulse would show adequate selectivity at a proton resonance frequency of 500 MHz. The actual excitation profile of the pulse should be calculated prior to use with numerical simulations. Most major spectrometers have a pulse simulator included with the instrumentation software.

15.3 Instrument Configuration

Most multi-dimensional NMR experiments, with the exception of homonuclear proton experiments, require the excitation of multiple types of nuclear spins. For example, the two-dimensional HSQC spectra discussed in Chapter 10 would require excitation of protons and nitrogen or carbon spins, depending on whether HN or HC correlations are being observed. The triple-resonance experiments discussed in Chapter 14 require excitation of protons, nitrogen, and one or two carbon frequencies (e.g. CO and C_α). Finally, triple-resonance experiments on larger deuterated proteins will require an additional channel for the decoupling of deuterons.

The specific configuration of the spectrometer will depend on the number of separate radio-frequency channels available on the instrument. Most high-field instruments have at least three channels, many have four channels, and it is not uncommon to have five independent channels on the instrument. Possible routing diagrams for a four and five channel instrument are given in Fig. 15.11. Each channel will have a separate frequency generator and often has a waveform generator for the generation of shaped pulses by amplitude modulation and the generation of frequency shifted pulses by phase modulation. The output of one of more channels are usually combined prior to input to the amplifier. For example, two carbon channels are generally combined and routed to a single broadband amplifier before being sent to the probe.

15.3.1 Probe Tuning

Most probes that are used for multi-nuclear triple-resonance experiments are termed 'inverse probes' because the proton excitation coil in closest to the sample and the dual tuned nitrogen and carbon coil is outside the proton coil, as illustrated in Fig. 15.12.

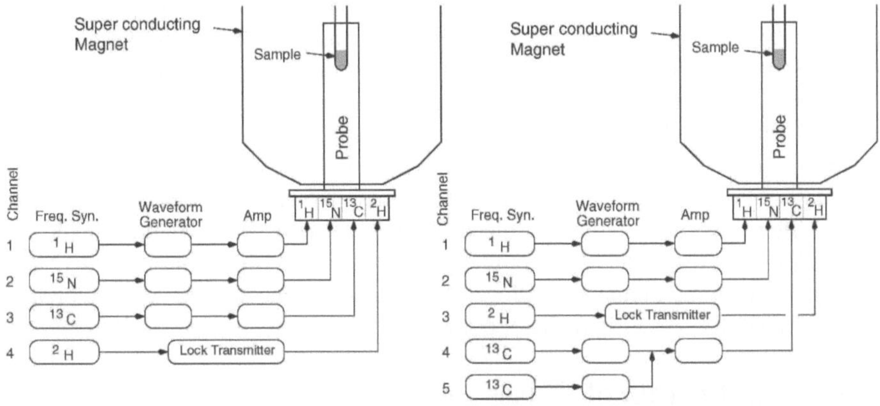

Figure 15.11. Configuration of spectrometer channels. Possible layout of channels in a three channel (left) or a four channel instrument (right) are shown. In general, the deuterium channel is not counted towards the number of channels in the spectrometer. The sample and probe are labeled. In the three channel instrument, carbon pulses with different frequencies, such as CO and C_α would be generated by either changing the frequency of the synthesizer or using phase modulation to shift the apparent frequency. If deuterium decoupling was used, the lock transmitter would be controlled as if it were a fourth channel, otherwise it would be independent from the other channels. In a four channel instrument two carbon frequencies can be independently specified. However, the output from the waveform generators would be combined and then used as input to the amplifier. Deuterium decoupling would utilize the deuterium channel as if it were a fifth channel. Frequency filters are not illustrated in this diagram, but would be placed between the output of the amplifier and the input to the probe.

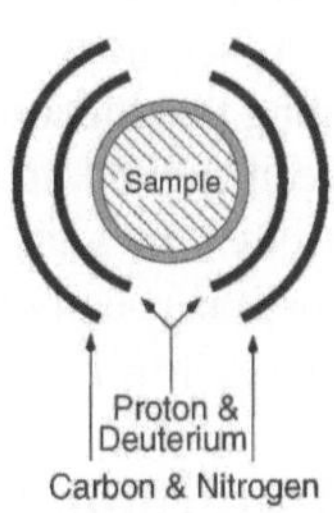

Figure 15.12. Excitation coils in an inverse probe. The position of transmitter coils in a triple-resonance inverse probe are shown, viewed looking down on the sample. The inner coil is dual tuned for proton and deuterium, while the outer coil is dual tuned for nitrogen or carbon.

Each channel on the probe should be tuned for new samples or if the temperature of the sample is changed more than $\approx 5°$ C. Probe tuning was discussed in some detail in Chapter 2. In all instruments, the 2nd channel of the preamplifier is used for tuning of the heteronuclear (^{15}N, ^{13}C) coils. In most instruments it is necessary to physically change the cabling when tuning a channel on the probe. For example, if the carbon channel is to be tuned then a cable will have to be connected from the 2nd channel to the carbon channel on the probe. Newer preamplifiers can automatically switch the signal routing during tuning, such that the tuning signal is sent to the correct channel on the probe.

Recall that tuning involves the adjustment of the complex impedance of the transmit/receive coil to account for changes in the magnetic susceptibility of the sample, as discussed in Chapter 2, Fig. 2.8. Nuclei that resonate at lower frequencies, such as

^{15}N, are less sensitive to these changes. In addition, nuclei that are excited by the outer coil of an inverse probe are also less sensitive to the changes in the sample because less of the coil is occupied by the sample. Consequently, the nitrogen channel is the least sensitive to changes in tuning, followed by carbon and then the proton channel. Consequently, the channels are generally tuned in the following order: nitrogen, carbon, and then proton. Since the nitrogen and carbon channel share the same coil, it may be necessary to re-tune the nitrogen channel if there are large changes in the carbon tuning. After the probe is correctly tuned, the pulse lengths for excitation should be within a 5-10% of the levels found for previous samples of similar ionic strength. Note that samples with high ionic strength can be difficult to tune and will require higher power to obtain the same flip-angle of pulses.

15.4 Calibration of Pulses

15.4.1 Proton Pulses

High power proton pulses can be calibrated directly, using the techniques discussed in Chapter 2. Shaped proton pulses, such as the water flip-back pulses discussed above, are best calibrated in the context that they are to be used. Fig. 15.13 shows one possible way to calibrate flip-back pulses. The transmitter is placed on water and the power required for a high power 90° pulse is obtained by calibrating a 360° pulse. Then, both pulses are used and the power level of the flip-back pulse is adjusted such that the water signal is nulled, indicating that the water magnetization has been excited and returned to the z-axis by the pair of pulses. Note that in most spectrometers, a change in the power level leads to small phase shift in the RF-pulse. Thus it is usually necessary to calibrate both the power level as well as the phase shift, φ.

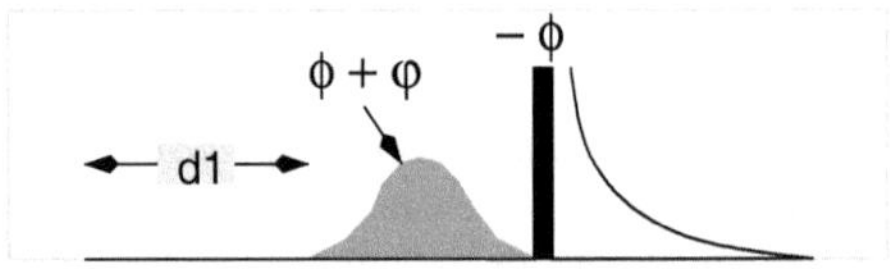

Figure 15.13. Calibration of flip-back pulses. A pulse sequence to calibrate proton flip-back pulses is shown. The relaxation time, d1, is on the order of 2 sec. The gray Gaussian shaped flip-back pulse is of duration of 1-1.5 msec. Its phase is $\phi+\varphi$, where ϕ is x, y, $-x$, or $-y$ and φ is a small correction to the phase due to power level differences, typically $|\varphi| \leq 5°$. The tall black rectangle is a high power 90° pulse, typically of 10 μsec in duration. Its phase is opposite to that of the flip-back pulse, thus it will return the water to the z-axis, giving a zero signal if the flip-back pulse is properly calibrated.

15.4.2 Heteronuclear Pulses

Since each heteronuclear channel has different electronic components, which may have different power losses, it is necessary to calibrate pulses using *exactly* the same hardware set-up as will be run in the actual experiment. Therefore, heteronuclear pulses, such as nitrogen and carbon, will have to be calibrated indirectly. Ideally, these pulses should be calibrated on the actual sample.

An HSQC experiment can be used to calibrate pulses with the actual sample. Although this experiment has four heteronuclear pulses (two 180° and two 90° pulses), the maximum signal is obtained when the pulses are of the correct length. This approach works because, to a first approximation, the signal intensity in the HSQC exper-

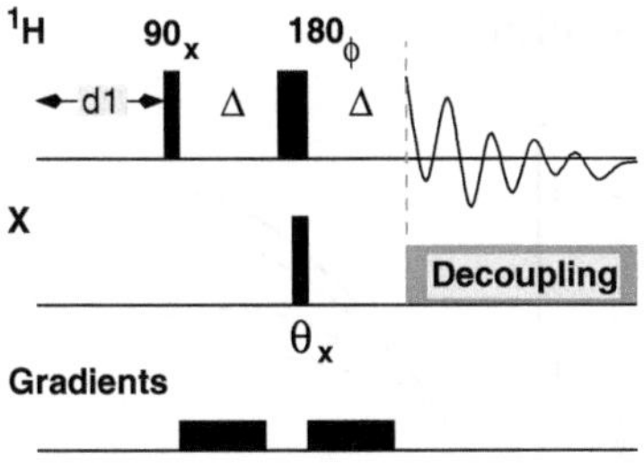

Figure 15.14 Pulse sequence for calibration of heteronuclear pulses. The heteronuclear pulse labeled θ_x is applied along the x-axis and has flip-angle of θ. Phase cycling is: $\phi = x, y, -x, -y$; $\psi_{rec} = x, -x, x, -x$. The delay Δ should be set to 1/(2J). The gradients are optional, but can be applied at 5 G/cm for most of the Δ period, followed by 100 μsec recovery period before the application of pulses or data acquisition. Decoupling is also optional, but can be applied if approximate power levels are known. The delay, d1, should be several seconds because of the long T_1 values for small molecules.

iment is given by $\left[(2cos^2\delta - 1)(cos\delta)\right]^2$, where δ is the deviation of the pulse length from ideality. For calibrating carbon pulses, it is best to place the carbon transmitter on the methyl carbon ($\approx$ 18 ppm) or in the aromatic ($\approx$ 130 ppm) region to minimize the effect of resonance offset on the efficiency of the pulses.

If the actual sample cannot be used because of low sensitivity, a 100 mM solution of ^{15}N labeled formamide (in H_2O) or 10-20 mM solution ^{13}C labeled glycine (in D_20) can be used for calibration using the simple pulse sequence described in this section. To minimize changes in the tuning between the calibration samples and the actual sample the calibration samples should be in the same buffer that is used for the protein. Nevertheless, the probe should be tuned prior to pulse calibration and then retuned after the sample is inserted.

A pulse sequence that can be used to calibrate heteronuclear pulses is shown in Fig. 15.14. This sequence can also be used to calibrate shaped pulses by replacement of the heteronuclear rectangular pulse with the shaped pulse. Analysis of this sequence using product operators is straight-forward. Proton chemical shift evolution can be neglected in this analysis because of the 180° proton refocusing pulse. Beginning with equilibrium proton magnetization:

$$I_z \overset{90^H_x}{\rightarrow} -I_y \overset{\Delta}{\rightarrow} 2I_xS_z \overset{180^H_\phi}{\rightarrow} 2I_xS_z \overset{\theta_x}{\rightarrow} 2I_x\left[S_zcos\theta - S_ysin\theta\right] \overset{\Delta}{\rightarrow} I_ycos\theta - 2I_xS_ysin\theta \tag{15.3}$$

The term $-2I_xS_y$ is not detectable, therefore the intensity of the observed signal, as a function of the heteronuclear flip-angle, is:

$$I(\theta) = cos\theta \tag{15.4}$$

A 90° heteronuclear pulse will give a null signal and a 180° pulse will give an inverted signal. In practice, a spectrum is acquired with $\theta = 0°$ and phased to give a positive absorptive spectrum. The same processing parameters are applied for non-zero values of θ.

The 90° pulse lengths for carbon and nitrogen at full transmitter power depend on a number of factors, such as the probe and sample components, however, typical values range from 10-15 μsec for carbon and 35-55 μsec for nitrogen.

15.5 T_1, T_2 and Experimental Parameters

Knowledge of the relaxation rates for protons an other spins in the protein are essential for the optimal set-up of most NMR experiments. These rates are sensitive to both the rotational correlation time of the protein *and* the magnetic field strength (see Figs. 15.16 and 15.17). Therefore, the effect of T_1 and T_2 on the data acquisition parameters should be considered for each protein and magnetic field strength.

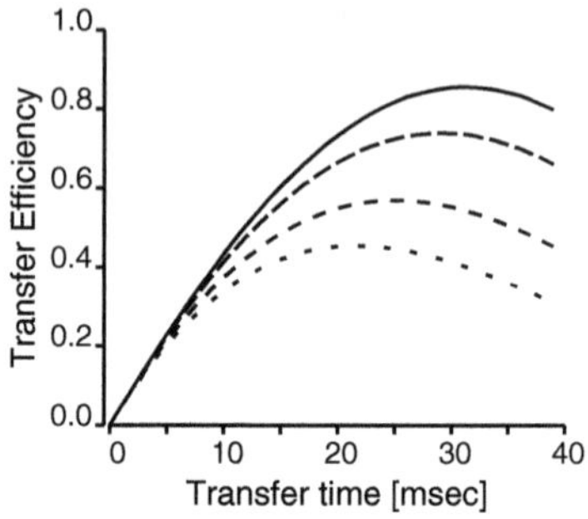

Figure 15.15. Effect of T_2 on nitrogen-carbonyl polarization transfer. The time dependence for polarization transfer from the nitrogen to the carbonyl carbon, such as in an HNCO experiment, is shown for proteins of different sizes: 10 kDa (solid line), 25 kDa (long dashed line), 50 kDa (short dashed line), and 80 kDa (dotted line). The nitrogen-carbonyl carbon coupling constant was 15 Hz, giving an optimal transfer time in the absence of relaxation of 33.3 msec ($= 1/(2J_{NC})$). Note that in larger proteins it is necessary to shorten the transfer time because of the shorter T_2 of the nitrogen.

T_1 Considerations: Since most NMR experiments begin with proton magnetization, the amount of proton longitudinal magnetization at the beginning of each scan affects the intensity of the signal. In the steady-state, the level of longitudinal magnetization is given by:

$$I_z = 1 - e^{-T/T_1} \quad (15.5)$$

where T is the sum of the delay before the scan and the acquisition time. A delay of at least 1.5 times T_1 should be used between each scan, giving a steady-state magnetization of 78% of what would be measured for an infinitely long inter-scan delay.

T_2 Considerations: The T_2 is the lifetime of coherent transverse magnetization. Consequently, the T_2 indicates how long the FID *should* be sampled. Since the signal decays to 37% $(1/e)$ during a single T_2 period, there is seldom any benefit to acquire data for longer than ≈ 2.5 times T_2, because subsequent data points contain less than 10% of the signal.

The T_2 also affects the optimal length of polarization transfer periods. For example, in the HNCO experiment, the amount of in-phase nitrogen magnetization that is transfered to anti-phase carbonyl magnetization is:

$$N_x \rightarrow 2N_yC_z sin(\pi J_{CN}\tau)e^{-\tau/T_2} \quad (15.6)$$

where τ is the polarization transfer period and T_2 is the nitrogen spin-spin relaxation rate. The maximum for this function does *not* occur when $\tau = 1/(2J_{CN})$, but at a shorter time, depending on the rate of decay of the transverse nitrogen magnetization, as illustrated in Fig. 15.15.

15.5.1 Fundamentals of Nuclear Spin Relaxation

Nuclear spin relaxation is caused by fluctuations in the local magnetic field due to molecular reorientation. In the case of protons, the field fluctuations are largely due to dipolar coupling between the spins (see Chapter 16). These fluctuations, if of the appropriate frequency, induce transitions between quantum states. In the case of heteronuclear spins, the anisotropic shielding (chemical shift anisotropy, CSA) of the nucleus will also cause field fluctuations as the orientation of the molecule changes with respect to the external magnetic field. The relationship between molecular motion and relaxation is explored in considerable detail in Chapter 19 and only a brief summary will be given here.

The spectral density function, $J(\omega)$ describes the intensity of the magnetic field fluctuations at any given frequency. $J(\omega)$ is related to the rotational correlation time of the protein, τ_c, as follows:

$$J(\omega) = \frac{\tau_c}{1 + \omega^2 \tau_c^2} \tag{15.7}$$

The rotational correlation time is given by the Stokes-Einstein Equation:

$$\tau_c = \frac{4\pi\eta r^3}{3kT} \tag{15.8}$$

where η is the viscosity and r is the radius of the protein. The rotational correlation time is the average time required for a protein to rotate 1 radian and is approximately 1 nsec per 2.6 kDa of molecular mass. Spectral density functions for small, intermediate, and large proteins are shown in Panel B of Fig. 15.16. The spectral density function for large proteins, which tumble slowly and therefore generate low frequency field fluctuations, is large at low frequencies.

Spin-lattice relaxation, which occurs with a time constant of T_1, is due to the stimulation of single- and double-quantum transitions and is thus sensitive to field fluctuations at ω_s and $2\omega_s$, where ω_s is the resonance frequency of the spin. The dependence of the proton T_1 on the spectral density function is:

$$\frac{1}{T_1} = \frac{6}{20} d^2 \left[J(\omega_H) + 4J(2\omega_H) \right] \tag{15.9}$$

where $d^2 = \hbar^2 \gamma_H^4 / r^6$ and r is the inter-proton distance.

Spin-spin relaxation, which occurs with a time constant of T_2, is due to zero-, single-, and double-quantum transitions and is thus sensitive to field fluctuations that occur at $\omega = 0$ as well as at higher frequencies. The dependence of the proton T_2 on the spectral density function is:

$$\frac{1}{T_2} = \frac{3}{20} d^2 \left[3J(0) + 5J(\omega_H) + 2J(2\omega_H) \right] \tag{15.10}$$

The T_2 of a heteronuclear spin, such as ^{15}N, is caused by both dipolar coupling and the chemical shift anisotropy:

$$\begin{aligned} \frac{1}{T_2} &= \frac{d^2}{20} \left[4J(0) + J(\omega_H - \omega_N) + 3J(\omega_N) + 6J(\omega_H) + 6J(\omega_H + \omega_N) \right] \\ &\quad + \frac{1}{45} \omega_N^2 \Delta\sigma^2 \left[4J(0) + 3J(\omega_N) \right] \end{aligned} \tag{15.11}$$

where $d^2 = \gamma_H^2 \gamma_N^2 \hbar^2 / r^6$ and $\Delta\sigma$ represents the anisotropic chemical shift.

15.5.2 Effect of Molecular Weight and Magnetic Field Strength on T_1 and T_2

The effect of molecular weight (τ_c) and the magnetic field strength on proton relaxation times is illustrated in Fig. 15.16. Figure 15.17 shows the spin-spin relaxation time (T_2) for the amide nitrogen, carbonyl carbon, and C_α carbon as a function of these parameters. Note that at lower magnetic field strengths the amide nitrogen and carbonyl carbon have similar spin-spin relaxation times. In contrast, the C_α carbon is efficiently relaxed by its attached proton. However, at high magnetic field strengths, the contribution of the CSA to the carbonyl relaxation increases significantly because of the ω^2 dependence, leading to very short relaxation times for the carbonyl carbon at 900 MHz.

15.5.2.1 Effects of Molecular Weight on the Proton T_1

Since T_1 depends on $J(\omega)$ and $J(2\omega)$ the effect of the rotational correlation time on the observed T_1 can be explained by considering the intensity of the spectral density at ω and 2ω for various values of τ_c. Beginning with fast motions, or short τ_c values, the spectral density is of low intensity at these two frequencies (solid curve in

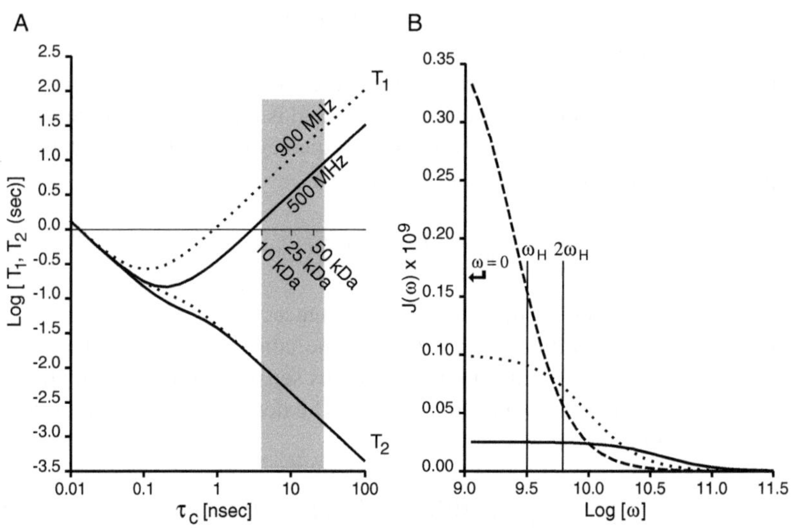

Figure 15.16. Effect of molecular weight on proton relaxation. Panel A shows proton T_1 and T_2 values as a function of the rotational correlation time, τ_c (note the *log* scale). These were computed assuming a spherical protein and an inter-proton distance of 1.7 Å. This value approximates the average proton density surrounding an amide proton. The individual curves for T_1 and T_2 are labeled, the solid line represents relaxation at a proton frequency of 500 MHz while the dotted line represents relaxation at 900 MHz. The gray rectangle indicates the range of molecular sizes, from 10 kDa to 60 kDa, that are routinely studied. Panel B shows spectral density functions for rotational correlation times of 0.025 (solid), 0.10 (dotted), and 0.40 nsec (dashed). These times are near the T_1 minimum. The ω_H and $2\omega_H$ values for a proton frequency of 500 MHz are indicated.

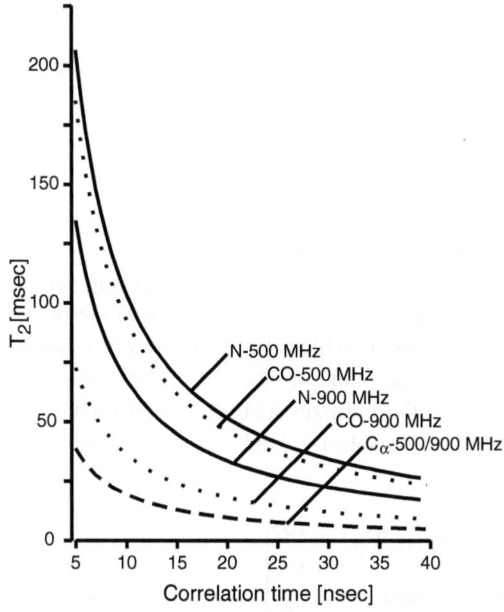

Figure 15.17 Effect of molecular weight and field strength on T_2 relaxation of heteronuclear spins. The spin-spin relaxation time (T_2) is shown for the amide nitrogen (solid line), carbonyl carbon (dotted), and C_α carbon (dashed line) as a function of rotational correlation time, τ_c. A 50 kDa protein will have a $\tau_c \approx 20$ nsec. These calculations were performed for two different spectrometer frequencies, 500 MHz and 900 MHz, as indicated on the figure. The nitrogen and carbonyl relaxation times were estimated by considering the contribution from chemical shift anisotropy (CSA) as well as dipolar coupling. In the case of the C_α carbon, the relaxation time was estimated by only considering dipolar coupling, hence there is no dependence on spectrometer frequency.

Fig. 15.16, Panel B). Therefore, spin-lattice relaxation is inefficient and the T_1 is relatively long. As the rotational motion slows, the profile of the spectral density changes becoming more intense at lower frequencies. When $\omega\tau_c$ is approximately one, the spectral density at ω and 2ω is large (dotted line in Panel B Fig. 15.16), and efficient spin-lattice relaxation occurs, giving a minimum in the T_1. As the motion becomes even slower, the intensity of $J(\omega)$ first drops at 2ω (dashed curve) and then eventually at ω. The small values of $J(\omega)$ at these two frequencies will cause inefficient spin-lattice relaxation, hence a longer T_1 for large molecules.

The magnetic field dependence on the T_1 is explained in a similar manner. For any values of τ_c corresponding to a molecular weight greater than 10 kDa, a higher magnetic field strength will cause $J(\omega)$ to be sampled at higher frequencies. Since $J(\omega)$ always decreases as ω increases, the spin-lattice relaxation will be less efficient and T_1 will become longer. Consequently, longer recycle delays need to be used at higher field strengths if the same level of steady-state magnetization is desired.

15.5.2.2 Effects of Molecular Weight on the Proton T_2

In contrast to the parabolic-like behavior of T_1, the T_2 for the proton (Fig. 15.16), as well as for the heteronuclear spin (Fig. 15.17), decreases as the rotational correlation time increases. This relationship is a result of the contribution of $J(0)$ to the spin-spin relaxation rate. As the rotational correlation time increases, the spectral density at $\omega = 0$ also increases. In fact, for most proteins the spectral density at zero frequency dominates the spin-spin relaxation. For larger proteins, the proton T_2 is:

$$\frac{1}{T_2} \approx \frac{3d^2}{20} 3J(0) = \frac{9d^2}{20}\tau_c \qquad (15.12)$$

and for the nitrogen T_2:

$$\frac{1}{T_2} \approx \frac{d^2}{20} 4\tau_c + \frac{1}{45}\omega_N^2 \Delta\sigma^2 4\tau_c \tag{15.13}$$

Since the rotational correlation time, τ_c is proportional to the molecular weight, the T_2 is inversely proportional to the molecular weight:

$$T_2 \propto 1/MW \tag{15.14}$$

as the size of the molecule increases, the T_2 becomes shorter.

In larger molecules, it becomes increasingly more difficult to transfer magnetization between spins because of the relaxation losses that are incurred during the polarization transfer periods. There are two methods by which the T_2 of the spins can be lengthened. The first involves changing the rotational correlation time of the sample by changing the temperature. The second method utilizes relaxation interference between dipolar coupling and the chemical shift anisotropy. Each of these methods will be briefly discussed below.

15.5.3 Effect of Temperature on T_2

As the temperature of the sample is increased, the rotational correlation time of the sample decreases. This effect is due to two factors. First, τ_c is inversely proportional to the temperature (see Eq. 15.8). Second, the temperature also affects the viscosity of water, η, causing a steep dependence of the linewidth on the temperature, as illustrated in Fig. 15.18. A modest 10° increase in temperature increases the T_2 by 24%.

In general, raising the temperature of the sample increases T_2. However, the rate of exchange processes are also increased at higher temperature, which can lead to shorter T_2 times (see Chapter 18) as well as increasing the rate of amide proton-deuterium exchange.

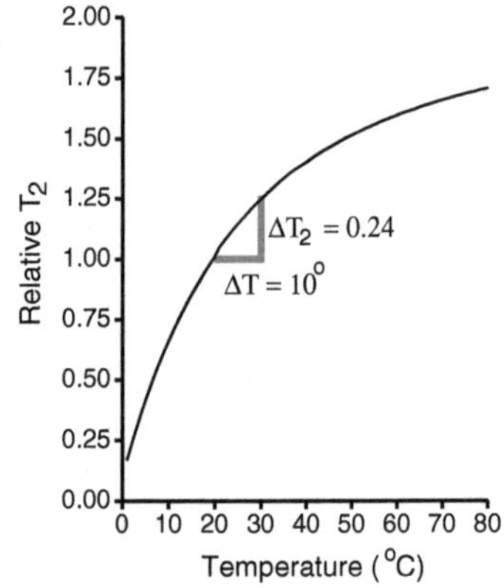

Figure 15.18. Temperature effects on T_2 relaxation. The T_2 was calculated assuming a 50 kDa protein. The T_2 at 20°C was set to 1.

15.5.4 Relaxation Interference: TROSY

TROSY (Transverse relaxation-optimized spectroscopy) is a technique that increases the resolution and sensitivity of heteronuclear NMR experiments on larger proteins at high magnetic field strengths. This technique has been applied to two-dimensional proton-nitrogen [129] and proton-carbon [130] correlated spectroscopy, as well as to a large number of triple resonance experiments, such as the HNCA experiment [140]. Several comprehensive reviews of this technique are available, see [128].

The TROSY technique works by taking advantage of the interference between dipolar coupling and the chemical shift anisotropy in the relaxation of coupled heteronuclear spins to produce narrow resonance lines. The magnetic field fluctuations that drive the relaxation of the proton or heteronuclear spin are generated by the chemical shift anisotropy of the relaxing spin and the field generated from the dipolar coupled spin. As illustrated in Fig. 15.19, the *correlated* motion of the spin and its dipolar coupled partner cause these two fields to either cancel or add, depending on the spin state (α or β) of the coupled spin. When there is destructive interference between these two relaxation mechanisms (Panel A in Fig. 15.19), the oscillating field becomes small and the relaxation rate of the spin decreases, leading to longer spin-spin relaxation times and narrower resonance lines for that component of the doublet. In contrast, as shown in Panel B of Fig. 15.19, the other component of the doublet will experience enhanced field fluctuations, leading to a shortened spin-spin relaxation time.

Since the field fluctuations due to CSA depend on the square of the magnetic field, the degree of cancellation depends on the field strength. Figure 15.20 shows the effect of spectrometer frequency on the cancellation of CSA and dipolar fields for the nitrogen spin in a proton-nitrogen group. The contributions of the nitrogen CSA to relaxation of the nitrogen spin becomes equal to the field from dipolar coupling at a spectrometer frequency of approximately 1100 MHz. Note that significant narrowing of the resonance line is observed at spectrometer frequencies that are in common use, i.e. 700-800 MHz.

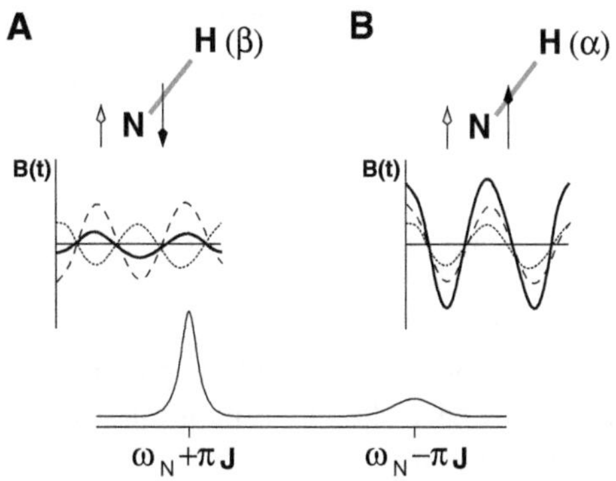

Figure 15.19. Origin of the TROSY effect. Panel A shows a nitrogen spin coupled to a proton in the β spin state while Panel B shows the nitrogen coupled to a proton in the α spin state. The open-headed arrows indicate magnetic field fluctuations from chemical shift anisotropy (CSA) while the closed-headed arrow indicates the dipolar field produced by the proton at the nitrogen nucleus. Note that the direction of the dipolar field depends on the spin state of the proton. The magnetic field fluctuations as a function of time, B(t), are shown in the middle section of the figure. In the case of Panel A, the fluctuations due to CSA (dotted line) partially cancel those from the dipolar field (dashed line) to give relatively small field fluctuations (heavy solid line). This nitrogen atom experiences smaller field fluctuations, and is therefore relaxed less efficiently, thus giving a narrower line. In Panel B, the opposite occurs, and the two fields add to give a larger oscillating field, enhancing the relaxation of the nitrogen coupled to a proton in the α spin-state. The resultant NMR spectrum is shown in the bottom of the figure. The nitrogen atom that is coupled to a proton in the β spin state experiences smaller field fluctuations and has the narrowest line in the doublet.

To illustrate the increased resolution that is provided by the TROSY technique we will consider a simple two-dimensional nitrogen-proton coupled experiment, i.e. a proton-nitrogen HSQC experiment. If decoupling of the spectrum during t_1 and t_2 is *not* performed, then a quartet will be found for each NH group, as illustrated in Panel A of Fig. 15.21. Each peak in the quartet arises from a distinct pair of single-quantum nitrogen and proton transitions. The frequency and relaxation rate of each of these transitions is described in Table 15.1.

All four components of the quartet in the coupled HSQC spectrum show different linewidths due to different

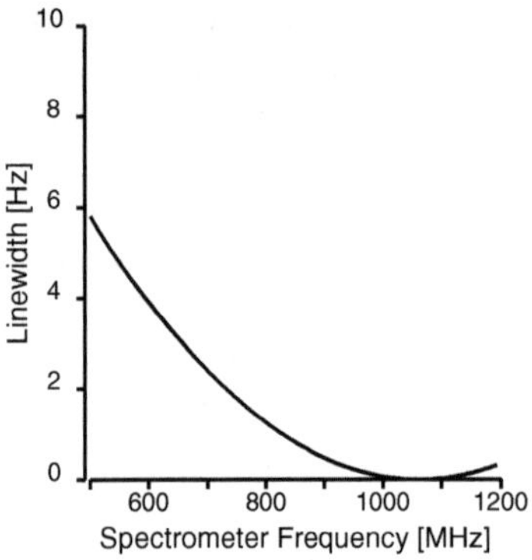

Figure 15.20 Field dependence of the TROSY effect for a NH spin-pair. The linewidth of an NH cross-peak in the ^{15}N dimension of a TROSY spectrum is shown. At 500 MHz, the field fluctuations due to dipolar coupling are not completely canceled, giving a residual linewidth of 6 Hz. As the magnetic field is increased, the contribution of the CSA to the field fluctuations increases and eventually cancels the contribution from dipolar coupling at a spectrometer frequency of 1050 MHz. An inter-nuclear distance of 1.08 Å and a CSA tensor, $\Delta\sigma_N$, of -160 ppm was used for this calculation. Other relaxation mechanisms, such as long range nitrogen-proton coupling, have been ignored in this calculation. Therefore, the observed linewidth at 1050 MHz would be greater than zero.

relaxation rates of each transition. Since the CSA for the amide proton is negative ($\Delta\sigma_H \approx -16$ ppm), the $2 \leftrightarrow 4$ proton transition will have the smallest relaxation rate for the two proton transitions. The nitrogen CSA is also negative ($\Delta\sigma_N \approx -160$ ppm), and therefore the $3 \leftrightarrow 4$ nitrogen transition will have the narrowest line. The peak in upper left of Panel A of Fig. 15.21 measures the nitrogen 3-4 transition during t_1 and the proton 2-4 transition during t_2. Since both of these transitions have slow relaxation rates, this component of the multiplet is narrow in both dimensions.

If proton decoupling is applied during t_1, and nitrogen decoupling during t_2, as in a normal HSQC experiment, then the spin states of the coupled spins are exchanged,

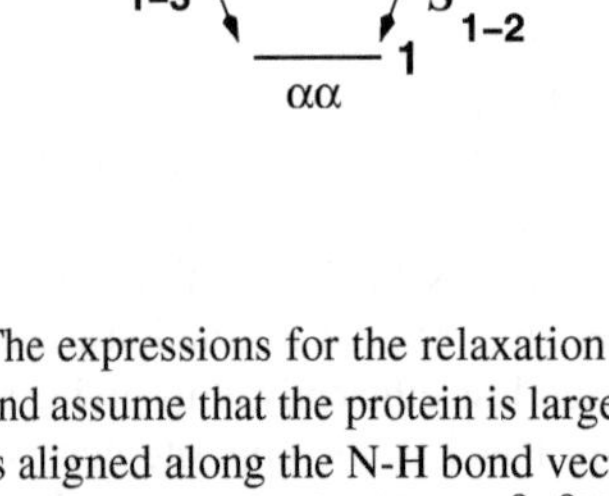

Table 15.1 TROSY: Relaxation rates. The energy level diagram for two coupled spins is shown to the left. Here, I refers to the proton and S refers to the nitrogen. The relaxation rates and frequencies for each of the single quantum transitions are given in the table below. The thicker arrows in the left diagram represent transitions that have narrow resonance lines.

Transition	*Frequency*	*Rate*
$1 \leftrightarrow 3$	$\omega_H + \pi J$	$4J(0)[(p - \delta_H)^2]$
$2 \leftrightarrow 4$	$\omega_H - \pi J$	$4J(0)[(p + \delta_H)^2]$
$1 \leftrightarrow 2$	$\omega_N - \pi J$	$4J(0)[(p - \delta_N)^2]$
$3 \leftrightarrow 4$	$\omega_N + \pi J$	$4J(0)[(p + \delta_N)^2]$

The expressions for the relaxation rates presented in the above table were obtained from [129] and assume that the protein is large such that $J(0) >> J(\omega)$, and that the chemical shift tensor is aligned along the N-H bond vector. $J(0)$ is the spectral density at $\omega = 0$ and is defined here as: $J(\omega) = (2/5)[\tau_c/(1+\omega^2\tau_c^2)]$. p and δ_X represent the contribution of dipolar coupling and chemical shift anisotropy to relaxation, respectively:

$$p = \frac{1}{2\sqrt{2}}\frac{\gamma_H\gamma_N\hbar}{r^3} \qquad \delta_X = \frac{1}{3\sqrt{2}}\gamma_X B_o \Delta\sigma_X$$

where B_o is the static magnetic field, and $\Delta\sigma_X$ is the chemical shift anisotropy for spin X.

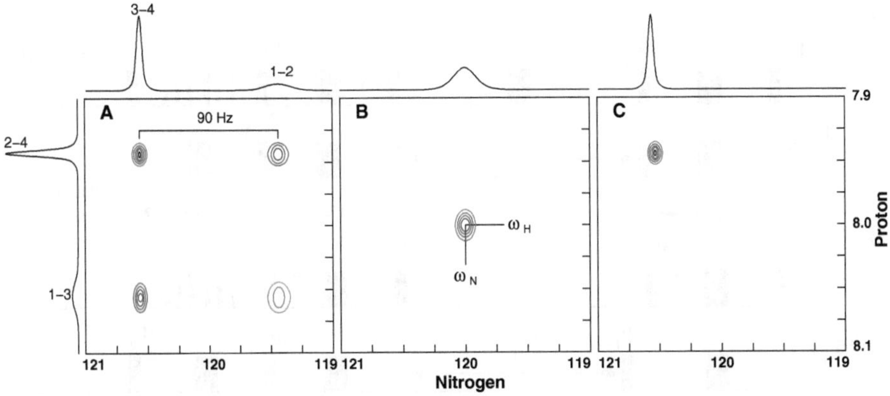

Figure 15.21. Decreased linewidth in two-dimensional ^1H-^{15}N *TROSY spectra.* Panel A shows a region of a coupled HSQC spectrum that contains the quartet of peaks associated with a single NH group. In this experiment the 180° refocusing pulse was omitted during t_1 and nitrogen decoupling was omitted during t_2. Therefore, a 90 Hz splitting, equal to J_{NH}, is observed in each dimension. Note that the linewidth of each member of the quartet is different, depending on the transition that generated the crosspeak. For example, the peak in the lower right is broad in both the nitrogen and proton dimensions, and arises from measuring the $1 \leftrightarrow 2$ transition in t_1 and the $1 \leftrightarrow 3$ in t_2. The one-dimensional spectra that are shown in Panel A are cross sections through the doublet in either the nitrogen dimension (top) or proton dimension (side). The transition associated with each peak is indicated above the peak. Panel B shows the effect of decoupling in both t_1 and t_2 dimensions on the spectrum (i.e. a traditional HSQC experiment). A single peak is obtained, and the observed proton and nitrogen linewidths are the average of the relaxation rates for each transition. The cross section in the nitrogen dimension is shown above the two-dimensional spectrum. Note that this peak is broader than the narrowest component (upper left) in Panel A. Panel C shows the TROSY spectrum of the same NH group. Only the most slowly relaxing peak, corresponding to the $3 \leftrightarrow 4$ transition during t_1 and the $2 \leftrightarrow 4$ transition in t_2, is retained in this spectrum.

or mixed, during t_1 and t_2 evolution, leading to an *averaging* of relaxation rates. For example, the proton relaxation rate during t_2 would be the average of the rates for the 1-3 and 2-4 transition, or $p^2 + \delta_H^2$. The averaging of the relaxation rates causes the observed line to become broader than the narrowest component of the original quartet, as illustrated in Panel B of Fig. 15.21. For larger proteins, the more rapidly relaxing transition will dominate the relaxation properties of the observed peak, leading to very broad lines that will be undetectable in large systems.

The spectrum that is obtained from the TROSY experiment is shown in Panel C of Fig. 15.21. The single peak in the spectrum is the narrowest component of the quartet and corresponds to magnetization that was associated with the 3-4 transition during nitrogen evolution and with the 2-4 transition during proton evolution. The remaining three components of the quartet been removed by phase cycling in the pulse sequence.

Since the TROSY technique only records one of the four peaks in the original multiplet, its intrinsic sensitivity is one-fourth that of a normal HSQC. However, in the case of proteins that are larger than approximately 40-50 kDa, the observed signal-

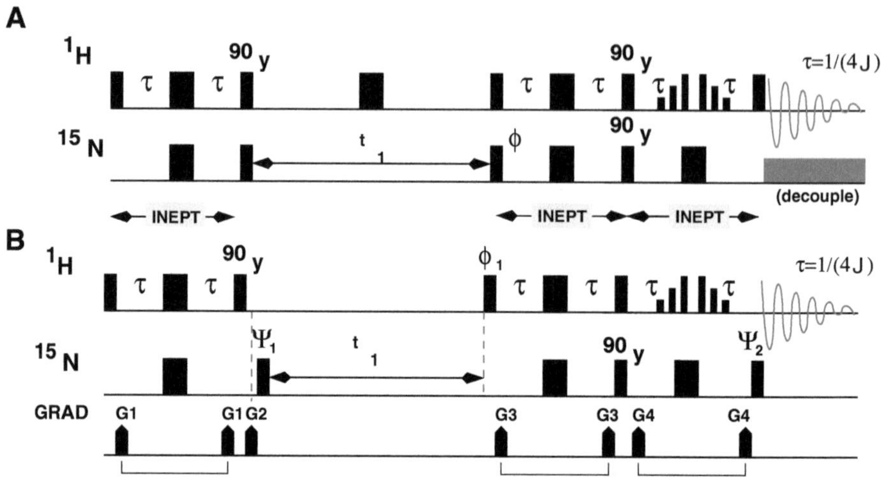

Figure 15.22. Comparison of TROSY versus HSQC pulse sequences. Panel A shows a simple sensitivity enhanced HSQC sequence (see Chapter 10) while Panel B shows the TROSY sequence from [129]. Narrow and wide bars indicate high-power 90° and 180° pulses, respectively. The series of 6 proton pulses in the last INEPT segment of both sequences is a selective WATERGATE pulse. The delay τ is equal to 1/(4J). The pulse phases are: $\Psi_1 = y, -y, -x, x, y, -y, -x, x$; $\Psi_2 = 4(x), 4(-x)$; $\phi_1 = 4(y), 4(-y)$; $\phi_{rec} = x, -x, -y, y, x, -x, y, -x$. Note that the y and $-y$ phases will have to be interchanged for Varian spectrometers. The gradient durations and strengths are: G1 (0.4 msec, 30 G/cm), G2 (1 msec, -60 G/cm), G3 (0.4 msec, 50 G/cm), G4 (0.6 msec, 48 G/cm). Quadrature detection in t_1 is obtained by acquiring an additional FID with the phase of Ψ_1 shifted by 90°.

to-noise in a TROSY spectrum will be *higher* than that found in a standard HSQC spectrum because of the narrower linewidth of the observed peak and the fact that the most slowly decaying transition is used in the polarization transfer step.

The experimental pulse sequence that gave rise to the TROSY spectrum is shown in Fig. 15.22. Superficially, the TROSY sequence resembles a sensitivity enhanced HSQC experiment. In both the experiments the first INEPT period serves to transfer the proton polarization to the nitrogen spin. Following chemical shift evolution during t_1, the next two INEPT periods transfer the magnetization to the amide proton for detection. The TROSY experiment differs from an HSQC experiment in the following three important respects:

1. There is no 180° proton pulse during the t_1 period, thus the spin state of the proton is not inverted during this period and the nitrogen magnetization follows either the 1-2 or 3-4 transition. The two nitrogen transitions do not mix.
2. There is no nitrogen decoupling during the t_2 evolution period, thus the spin state of the coupled nitrogen is not inverted during acquisition. Consequently, the proton magnetization follows either the 1-3 or the 2-4 transition.
3. The phase cycle is designed to keep only one single quantum coherence path, namely 3-4 during t_1 and 2-4 during t_2. Signals from the other three paths are

canceled by the phase cycle. This use of phase cycling is different than that discussed in Chapter 11. Specifically, the phase cycle in the TROSY experiment retains only a one of the two single quantum transition during each evolution period. The phase cycle given in Fig. 15.22 is designed for Bruker spectrometers. It is important to note that Varian spectrometers utilize a different convention for the y and -y phases of pulses. Therefore, implementation of this TROSY sequence on a Varian spectrometer would require interchanging the y and -y phases.

15.5.5 Determination of T_1 and T_2

Heteronuclear relaxation times are generally used to investigate protein dynamics. Methods to measure these times, as well as how to characterize dynamics, are presented in Chapter 19. Methods to measure the proton T_1 and T_2 are presented here.

Measuring Proton T_1: The proton T_1 can be obtained from an inversion recovery experiment, shown in Fig. 15.23. This experiment begins by inverting the proton magnetization with a 180° pulse, waiting a delay T, and then measuring the amount of z-magnetization with a 90° read pulse.

If the first 180° pulse is not ideal, either due to mis-setting of power levels, or to resonance off-set effects, then some of the initial magnetization will be found in the transverse plane after the pulse. This magnetization can contribute to the final signal, leading to an oscillatory behavior of the measured signal. The unwanted transverse magnetization is removed by the application of a pulsed-field gradient. Alternatively, a two step phase cycle can be used with the phase of the 90° pulse set to $x, -x$ with a receiver phase of x and $-x$ [49].

The dependence of the z-magnetization on T is given by:

$$M_z(T) = M_o(1 - 2e^{-T/T_1}) \tag{15.15}$$

where M_o is the equilibrium value of the z-magnetization. Spectra are acquired for different values of T and the spin-lattice relaxation time is obtained by fitting the data to Eq. 15.15. Alternatively, the relaxation delay time T can be varied to find the value, T_{null}, that gives a null signal, from which the T_1 can be obtained from the following expression: $T_1 = T_{null}/ln2$.

Measuring Proton T_2: The easiest way to obtain an estimate of the proton T_2 is from the linewidth: $T_2 = 1/(\pi\Delta\nu)$. Alternatively, the T_2 can be obtained from the decay of the FID.

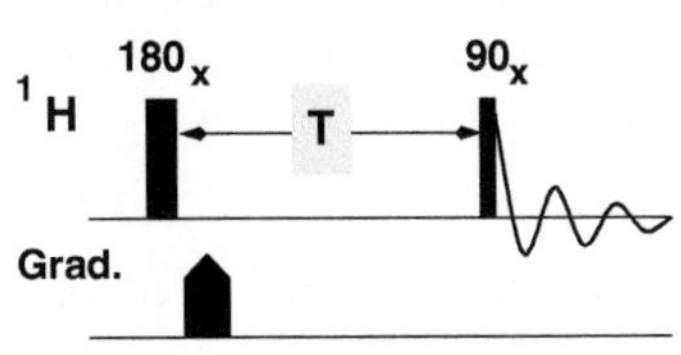

Figure 15.23 Pulse sequence for measuring the proton T_1. A 150 μsec 20 G/cm pulsed field gradient, will dephase any transverse magnetization after the 180° pulse. Residual water signal can be suppressed by presaturation. The delay between each scan should be 5 times T_1 to insure recovery of z-magnetization to equilibrium. The relaxation delay time, T, which includes the time for application and recovery of the field gradient, is varied from 0 to $\approx 4 - 5 \times T_1$.

15.6 Acquisition of Multi-Dimensional Spectra

The following sections discuss general aspects of acquiring and processing multi-dimensional NMR spectra. One of the most important tasks in setting up experiments of this type is defining the transmitter offset and sweepwidth in each dimension of the spectra. If possible, one should employ *exactly* the same parameters for each dimension in each experiment. For example, the transmitter offset, sweepwidth *and* number of data points should be identical in the ^{15}N dimension of every experiment. This concept is particularly important in complementary triple resonance experiments (e.g.HNCA and HN(CO)CA), such that the *positions* of common peaks (e.g. inter-residue correlations) are the same in both spectra.

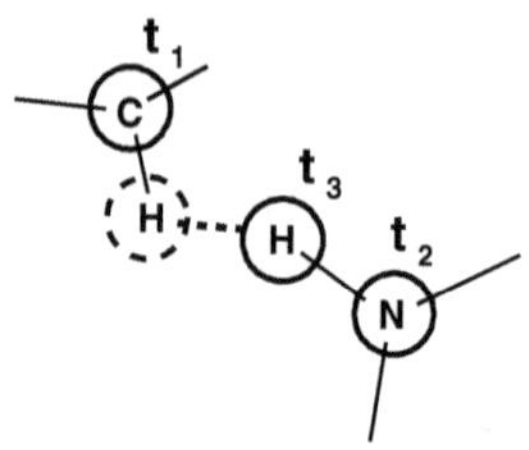

Figure 15.24. Magnetization transfer in the CN-NOESY experiment. Solid circles indicate that the chemical shift of the atom is recorded during the experiment during the indicated time domain. Dotted circles indicate that the magnetization is passed through these atoms, but their frequencies are not measured.

The 3D carbon-nitrogen NOESY (CN-NOESY) experiment will be used to illustrate how to define a number of experimental parameters. The magnetization transfer and frequency labeling that occurs in this experiment are illustrated in Fig. 15.24. The chemical shift of the carbon is recorded in t_1 and this information is passed to the attached proton by scalar coupling and then to a near-by amide proton by dipolar coupling, as illustrated by the dotted line in Fig. 15.24. The chemical shift of the amide nitrogen is recorded in t_2 and the magnetization is returned to the amide protons for final detection in t_3. Consequently, this experiment generates crosspeaks whose three frequencies are: ω_1, carbon shifts for those carbon-attached protons that are within ≈ 5 Å of an amide proton; ω_2, the nitrogen shift of the amide group; ω_3 the proton shift of the amide group. The pulse sequence for the CN-NOESY experiment is presented in Fig. 15.25.

15.6.1 Setting Polarization Transfer Delays

Most heteronuclear experiments have segments that transfer polarization between nuclei. For example, in the CN-NOESY experiment, the initial part of the experiment is an INEPT segment that transfers the proton polarization to the carbons. The density matrix evolves during this period as:

$$I_y \rightarrow [I_y cos(\pi J 2\Delta) - 2I_x S_z sin(\pi J 2\Delta)]\, e^{-2\Delta/T_2} \qquad (15.16)$$

where the factor $e^{-2\Delta/T_2}$ represents losses due to spin-spin relaxation of the transverse proton. Setting $\Delta = 1/(4J)$ leads to complete conversion of in-phase proton magnetization to anti-phase proton-carbon magnetization:

$$I_y \rightarrow -2I_x S_z e^{-2\Delta/T_2} \qquad \Delta = 1/(4J) \qquad (15.17)$$

However, due to proton relaxation, the amount of magnetization transferred is reduced by $e^{-2\Delta/T_2}$. The delay, Δ, that gives the maximum amount of transferred magnetization is obtain by finding the maximum of:

$$sin(\pi J 2\Delta) e^{-2\Delta/T_2} \qquad (15.18)$$

This maximum can be found experimentally by varying Δ and observing the signal intensity. Alternatively, if a good estimate of the T_2 is known, then the value of Δ can be computed. If an experiment contains multiple polarization transfer segments, such as the HNCO experiment shown in Section 14.2.1 on page 282, then each of these should be optimized to maximize the transfer of magnetization.

15.6.2 Defining the Directly Detected Dimension: t_3

The detected signal in the CN-NOESY, as with most other triple-resonance experiments, consists solely of resonance signals from the amide protons; the region of the spectrum for 0 to 5 ppm is usually void of resonance peaks. Therefore, in theory, it would be possible to place the proton transmitter in the middle of the amide region (e.g. 8 ppm) and set the dwell time to give a sweepwidth of ± 3 ppm. However, under these conditions the residual solvent line may give rise to artifacts in the spectrum because the solvent is no longer on resonance. Consequently it is advisable to place the transmitter on the solvent and collect a spectrum with a sweepwidth that is sufficient to cover the entire proton spectrum without aliasing any of the resonance lines. There is absolutely no harm in collecting such a wide sweepwidth in the directly detected domain because there is no increase the experiment time and the additional points in the spectrum that represent the aliphatic region can be discarded during processing to reduce the data storage requirements for the final spectrum.

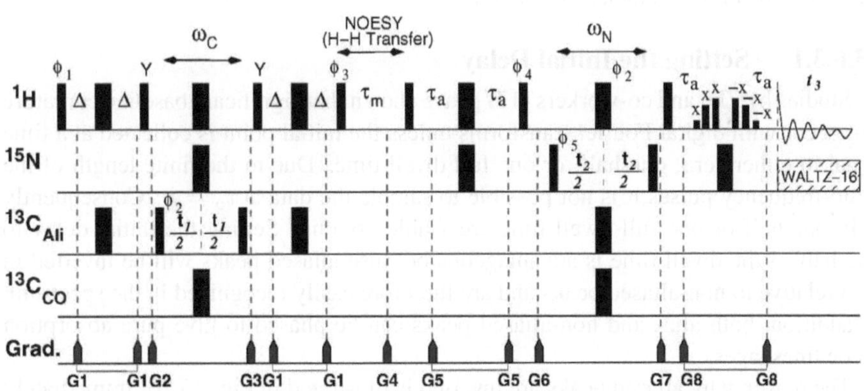

Figure 15.25. Pulse sequence for a carbon (t_1), nitrogen (t_2) separated NOESY. Pulses on proton and nitrogen are high power, except for the nitrogen decoupling sequence at the end of experiment. Carbon pulses are semi-selective pulses designed to excite the aliphatic region ($^{13}C_{ali}$) or the carbonyl region ($^{13}C_{CO}$). Δ is $1/(4J_{CH})$ and $\tau_a = 1/(4J_{HN})$. Gradients are: G1(7 G/cm, 400 μsec), G2(15 G/cm, 700 μsec), G3(17 G/cm, 700 μsec), G4(5 G/cm, 1 msec), G5(4 G/cm, 500 μsec), G6(5 G/cm, 800 μsec), G7(8 G/cm, 800 μsec), G8(17 G/cm, 1 msec). The gradient G8 is applied along the x, y, z axis, all other gradients are z gradients. Gradients G2, G3, G6, and G7 are zz-filters. The phase cycle is: $\phi_1 = x$; $\phi_2 = x, -x$; $\phi_3 = (x)_8, (-x)_8$; $\phi_4 = (y)_4, (-y)_4$; $\phi_5 = x, x, -x, -x$; $\psi_{rec} = x, -x, -x, x, -x, x, x, -x, -x, x, x, -x, x, -x, -x, x$. Quadrature detection using the States-TPPI method is obtained by shifting the phase of ϕ_2 for carbon and ϕ_5 for nitrogen.

15.6.3 Defining Indirectly Detected Dimensions

The indirectly detected dimensions in the experiment (^{13}C and ^{15}N) have to be acquired by systematically incrementing the respective evolution delays. Consequently it is necessary to define four parameters for *each* indirectly detected dimension:

1. The duration of the delay prior to acquiring the first point, t_o.
2. The time increment, or dwell time (τ_{dw}), between points. Recall from chapter 2 that the dwell time defines the spectral width in the frequency domain: $SW = 1/\tau_{dw}$. Sampling a larger spectral width than necessary will result in a lower resolution spectrum if the number of points is kept constant. Alternatively, if the spectral width is too wide, then more points will have to be acquired to attain the same resolution.
3. The total number of points, n, to collect. The number of points and the dwell time define the total acquisition time for a given dimension, $T_{acq} = n\tau_{dw}$. Recall that the intrinsic resolution in the spectrum is limited by the total acquisition time; the longer the acquisition time, the better the resolution. The digital resolution is simply calculated as the sweepwidth divided by the number of data points. In general, it is best to set the resolution ($1/T_{acq}$) to be roughly equal to the linewidth of the majority of the peaks in the spectrum.
4. The transmitter frequency, or offset.

15.6.3.1 Setting the Initial Delay

Studies by Zhu and co-workers [177] have shown that significant baseline curvature can occur with digital Fourier transforms unless the initial point is collected at a time equal to either zero, one-half, or one full dwell time. Due to the finite length of the radio-frequency pulses it is not possible to sample the data at $t_o = 0$. Consequently only one-half or one full dwell time are viable options. Setting the initial delay to one-half of the dwell time is advantageous because aliased peaks will be inverted in sign relative to non-aliased peaks and are therefore easily recognized in the spectrum. In addition, both alias and non-aliased peaks can be phased to give pure absorption mode lineshapes.

The reason why aliased peaks are inverted is illustrated in Fig. 15.26. Immediately after the excitation pulse all magnetization is aligned along the real axis. The magnetization then precesses at a rate that is defined by the frequency difference between the chemical shift of a resonance and the transmitter frequency. Therefore the magnetization is displaced from the real axis when the first data point is sampled at $t_o = \tau_{dw}/2$. The zero- and first-order phase corrections required to rotate a non-aliased peak back to the *positive* real axis are 90° and -180°, respectively [177]. When the same phase correction is applied to aliased magnetization, it is rotated to the *negative* real axis. This occurs because the aliased magnetization has precessed more than 90° when the data sampling occurs. Therefore, the *apparent* frequency of the aliased peak in the spectrum places the peak a position such that the phase shift rotates the magnetization to the negative real axis. Consequently, both aliased and non-aliased peaks have the same pure absorption mode lineshape, but the aliased peaks are inverted.

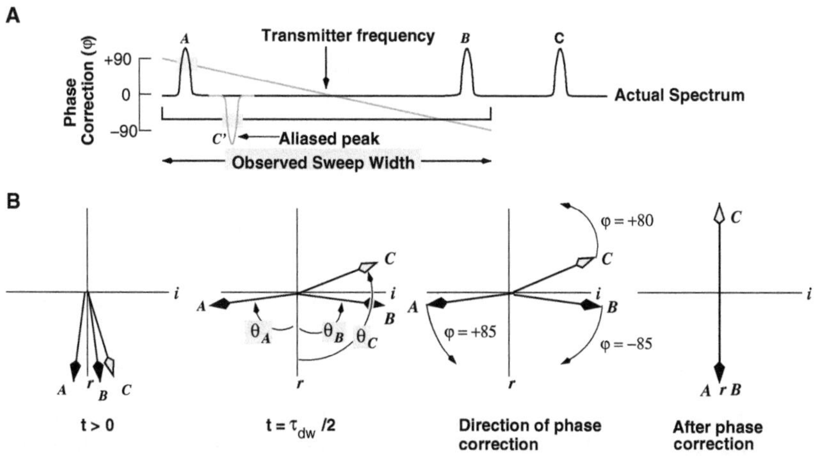

Figure 15.26. *Inversion of aliased peaks.* Panel A shows the entire spectrum, consisting of three resonances. The transmitter is placed midway between *A* and *B* and the sweepwidth was chosen to include resonances *A* and *B*, but not *C*. Resonance *C* is therefore aliased into the spectra, and its aliased position is indicated by *C'*. Panel A also shows the phase change (φ) that would be applied to each point of the spectrum during processing assuming a zero-order phase change of 90° and a first-order phase correction of -180°. Peak *A* would receive a phase correction of $\varphi = +85°$, the aliased peak C would receive a phase change of $\varphi = +80°$ while peak *B* would receive a phase correction of $\varphi = -85°$. Panel B shows the evolution of the magnetization shortly after the excitation pulse (t > 0) and at the time of the first sampling ($t = \tau_{dw}/2$). The real axis (*r*) points down and the imaginary axis (*i*) points to the right. The angular rotation of the magnetization of the ith peak, θ_i, during this time period depends on the frequency difference between the transmitter frequency and the resonance frequency. However, the phase correction applied during processing depends on the position of the peak in the spectrum. The effect of this phase correction on the magnetization is shown in the two rightmost sections of Panel B. Resonances *A* and *B* are rotated in opposite directions with respect to each other and both are returned to the positive real axis by their respective phase corrections. Resonance *C* receives almost the same phase correction as *A* and is rotated counter-clockwise from its position to the negative real axis, giving an inverted peak.

When setting the delay for the initial time point in the indirectly detected dimensions it is necessary to take into account the finite pulse length of the pulses that bracket the delay period. A 90° pulse of length τ_{90} can be approximated by a δ function, followed by a delay of $2\tau_{90}/\pi$, as described by [14]. Since the evolution period generally begins and ends with a 90° pulse, the evolution period time is increased due to finite widths of the pulse, by an amount:

$$\delta = 4\tau_{90}/\pi \tag{15.19}$$

Consequently, it is necessary to subtract this time from a programmed delay in the pulse sequence. In addition, if the evolution period contains a 180° pulse, which occurs in both indirect evolution periods in the example in Fig. 15.25, then the programmed delay should also account for the 180° pulse. For example, the initial delay for each of the $t_1/2$ evolution periods in the CN-NOESY experiment should be programmed to have an initial time period of:

$$\frac{t_1}{2} = \frac{\tau_{dw}}{4} - \frac{2\tau_{90C}}{\pi} - \frac{\tau_{180C'}}{2} \quad (15.20)$$

assuming that the carbonyl 180° pulse is longer than the proton or nitrogen 180° pulses. The actual pulse sequence would be written as:

(p3 ph2) : f3	carbon pulse, phase $=\phi_2$
delay τ	t1/2 delay
(40μsec p1 * 2 x) : f1 (15μsec p2 * 2 x) : f2 (p5 * 2 x) : f5	H, N, and Co 180° pulses
delay τ	t1/2 delay
(p3 x) : f3	carbon pulse, phase $=x$

where $\tau = (\tau_{dw}/4) - (2\tau_{90C}/\pi) - (\tau_{180C'}/2)$. All pulse lengths, e.g. $P1$, refer to 90° pulses. The ^{1}H, ^{15}N, $^{13}C_{ali}$, and $^{13}C_{CO}$, transmitters are assigned to channels 1, 2, 3, and 5, respectively. The syntax: (δ p1 * 2x) : f1 indicates that a delay δ will occur before a pulse of length p1*2, with a phase of x, will be applied to the $f1$ channel. The 180° decoupling pulses in the middle of t_1 period should be applied simultaneously, usually requiring an additional programmed delay prior to the shorter two of the three 180° pulses. These delays make the centers of all pulses coincident. In the above example, p1 = 10 μsec, p2 = 35 μsec, and p5 = 50 μsec. Therefore the start of the proton pulse is delayed 40 μsec while the start of the ^{15}N is delayed by 15 μsec, relative to the start of the pulse on the carbonyl carbon. Chemical shift evolution is achieved by the subsequent increment of each τ delay by $\tau_{dw}/2$.

15.6.3.2 Optimizing the Dwell Time and Number of Points

In the indirectly detected dimensions it is necessary to consider how the FID should be sampled, i.e. how to define the spacings between the points (the dwell time, τ_{dw}) and the total number of points (n) that should be collected. The total acquisition time is related to these two parameters as follows:

$$T_{acq} = n\tau_{dw} \quad (15.21)$$

15.6.3.3 Setting n and τ_{dw} for Non-Constant Time Evolution

For non-constant are evolution, such as the t_1 or t_2 evolution in the CN-NOESY, it is useful to first define T_{acq} and then calculate n based on the desired sweepwidth. Since the FID decays with a time constant of T_2 (the spin-lattice relaxation time) there is very little point in collecting data for acquisition times that are significantly longer than $2 \times T_2$. Data collected beyond this point will be largely noise since the original signal has decayed to $1/e^2$, or to 13.5% of its initial value.

The total acquisition time may also be limited by the desire to *not* resolve scalar couplings. For example, crosspeaks from aliphatic and aromatic carbons in the CN-NOESY experiment will show a splitting of 35 Hz in the carbon dimension due carbon-carbon coupling. If the resolution in this dimension is sufficiently high then the resulting crosspeaks will show a multiplet structure due to the carbon-carbon coupling(s). The signal-to-noise ratio of a spectrum acquired with additional time points can actually be slightly less than if the spectrum was acquired with fewer time points and the same total number of scans, as illustrated in Fig. 15.27. In addition, the increase in the number of peaks from the resolved scalar couplings may complicate analysis of the spectrum. Consequently, the highest resolution that should generally be used in the carbon dimension of most experiments is approximately equal to the carbon-carbon coupling (35 Hz), or a total acquisition time of approximately 28 msec.

Once the total acquisition time is defined, then it is necessary to define the number of data points. If the dwell time is selected such that the sweepwidth covers the entire range of chemical shifts, then n will be at its maximum value since τ_{dw} will be at a minimum. However, in the case of ^{15}N and ^{13}C evolution it is often possible to use a narrower sweepwidth. Although this results in aliasing of some resonance lines, the increase in τ_{dw} decreases n. Since the total acquisition time in a multi-dimensional NMR experiment is directly proportional to n, it is best to use the smallest value of n as possible. For example, a CN-NOESY experiment, acquired with a total of 128 complex points in t_1 (^{13}C), 32 complex points in t_2 (^{15}N) with 16 scans, and a total data acquisition time (t_3) and recovery delay of 1 sec, would require 72.82 hours of data collection:

$$128 \times 2 \times 32 \times 2 \times 16 \times 1 \text{ sec} = 262,144 \text{ sec} = 72.82 \text{ hours}$$

where the factors of two represent quadrature detection (i.e. complex points) in each indirectly detected domain. A reduction in the number of ^{13}C points from 128 to 64, by increasing the dwell time by two, would reduce the above experiment time by a factor of two, a significant savings in time.

If aliasing is used to reduce n, the actual sweepwidth that is used has to be carefully chosen such that the aliased peaks do not overlap with the non-aliased peaks. In the case of ^{15}N, it is generally feasible to alias approximately 5-10% of the peaks at the edge of the ^{15}N spectrum without causing problems of peak overlap. The situation with the ^{13}C spectrum is much more favorable because the carbon chemical shifts are correlated with the proton chemical shifts. Fig. 15.28 shows a proton-carbon HSQC spectrum that was acquired without and with aliasing.

Extensive aliasing in the carbon dimension permits acquiring spectra with enhanced resolution within the same experimental time. For example, if 128 points are used to acquire an unaliased carbon spectrum with a sweepwidth of 9000 Hz, then the digital

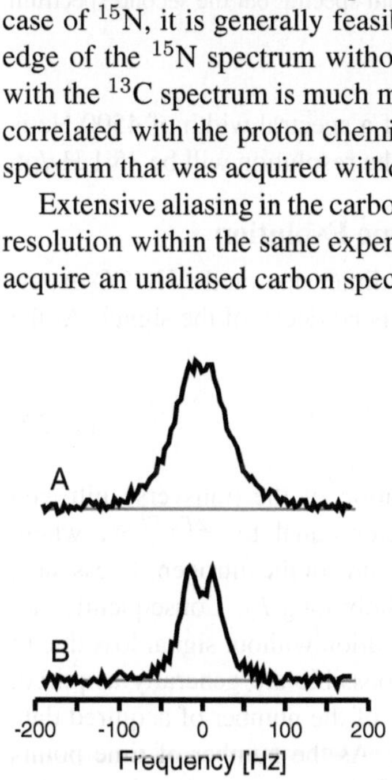

Figure 15.27 Optimization of indirect ^{13}C acquisition. A crosspeak corresponding to the α-carbon is shown in the carbon dimension (ω_1) of the CN-NOESY experiment. Two different acquisition times are shown, 28 msec (A) and 56 msec (B). The C_α-C_β coupling was set to 35 Hz. In both experiments, the *total* number of scans acquired are the same. Consequently, twice as many scans were acquired in Panel A, resulting in a more intense signal. The reduction of the intensity in panel B is due to spin-spin relaxation of the transverse carbon spin during the extended evolution period. The intensity difference between the two spectra will decrease as the T_2 of the carbon increases. The spectra shown here are typical for proteins in the 20-30 kDa range.

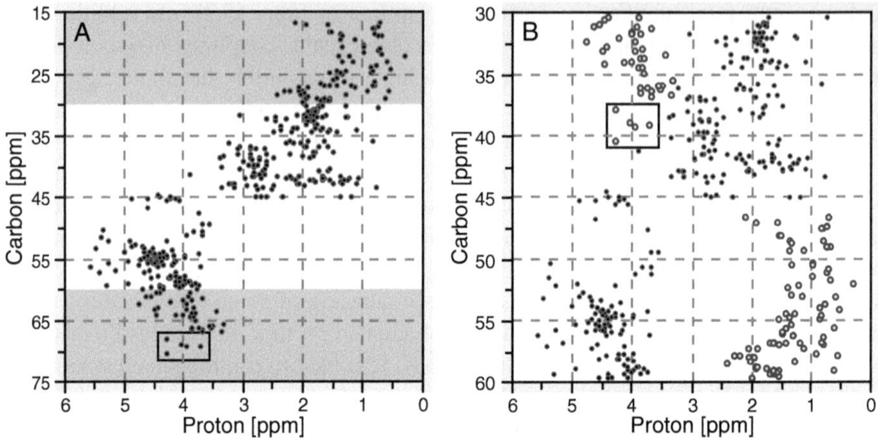

Figure 15.28. Aliasing of the carbon spectrum. Panel A shows a proton-carbon HSQC spectrum that was acquired without aliasing. Note that the carbon chemical shifts are correlated with the proton chemical shifts. This proportionality permits significant aliasing of the carbon dimension without peak overlap. The spectra were acquired at a carbon frequency of 150 MHz (ν_H = 600 MHz) with a total acquisition time of 14.22 msec. The spectrum in A was acquired with a dwell time of 111.1 μsec. The shaded regions in Panel A, from 15-30 and from 60-75 ppm, were aliased in Panel B. Panel B was acquired with a dwell time of 222.2 μsec, or with one-half the number of points that were used to collect the spectrum in A. The delay before acquiring the initial point was set to $\tau_{dw}/2$, therefore the aliased peaks are negative, and are represented by larger white circles. The boxed resonances are the same in both spectra. Note that the resolution, in terms of Hz/point, are identical in both spectra, but the second spectrum could be acquired in half the time.

resolution in the final spectrum will be 70.2 Hz/pt. If a spectral width of 4500 Hz is used with the same number of time increments, then the resolution will be 35.1 Hz/pt.

15.6.3.4 Setting n and τ_{dw} for Constant Time Evolution

In constant time evolution, such as in the nitrogen dimension on the HNCO experiment presented in Section 14.2.1 on page 282, there is no decay of the signal. At the end of the constant time period the density matrix is:

$$2N_zI_ycos(\omega_Ct_1)e^{i\omega_Nt_2}e^{i\gamma_NG5}e^{2T_N/T_{2N}} \tag{15.22}$$

note that the term representing decay due to relaxation of the transverse nitrogen spin does not depend on t_2 and is a constant factor equal to $e^{2T_N/T_{2N}}$, where $T_N = 1/(2J_{NCo})$ and T_{2N} is the spin-spin relaxation time of the nitrogen. In essence, the nitrogen resonance has zero linewidth or an infinitely long T_2. Consequently, acquiring additional time points will lead to higher resolution without signal loss due to relaxation. Therefore, the largest number of points possible are generally acquired, i.e. $n = 2T_N/\tau_{dw}$. Figure 15.29 illustrates the effect of the number of acquired data points on the intensity of the peaks in the spectrum. As the number of time points increases, the number of scans/point will decrease. However, increasing the number

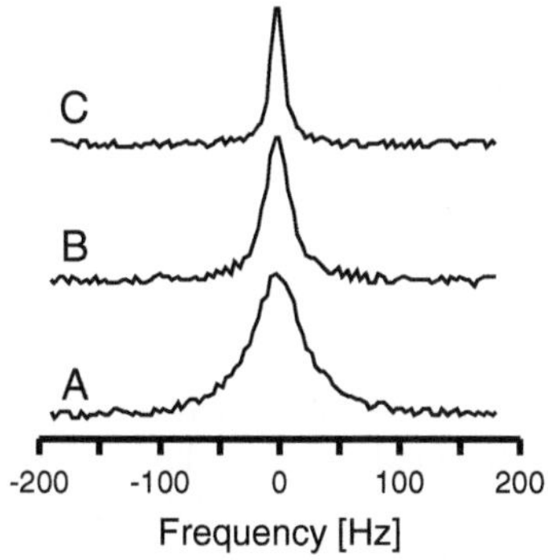

Figure 15.29 Effect of number of data points on signal intensity in constant time evolution. The effect of acquiring more data points in the case of constant time evolution, such as in the nitrogen evolution period of the HNCO experiment, is shown. All three spectra were acquired with the same total number of scans (i.e. same total acquisition time). The number of time points in spectrum B and C are 2 and 4 times the number of points in A, respectively. Note that the intensity of all three spectra are identical. Although increasing the number of points decreases the number of scans/point, the intensity of the peak increases due to a narrowing of the resonance line as the resolution improves.

of points narrows the resonance line by a factor that exactly compensates for the loss in intensity due to the decreased number of scans, therefore the line intensity remains constant.

15.6.3.5 Defining Transmitter Offsets

The observed frequency in the indirectly detected domain is the *difference* between the transmitter frequency and the resonance frequency of the spin. For example, if the chemical shift of an amide nitrogen was 120 ppm and the transmitter was set at 115 ppm, then the amide's precession frequency would be $5 \times 10^{-6} \times \nu_N$, where ν_N is the nitrogen frequency of the spectrometer.

For heteronuclear spins, such as carbon and nitrogen, the transmitter is usually set in the center of spectral region that is to be detected. In the case of the CN-NOESY experiment, the carbon transmitter ($^{13}C_{\rm Ali}$) would be set at approximately 45 ppm. The nitrogen transmitter would be set at approximately 115-120 ppm. Note that it may be necessary to change the location of the transmitter by a few ppm such that aliased peaks do not overlap with non-aliased peaks (see Section 15.6.3.2).

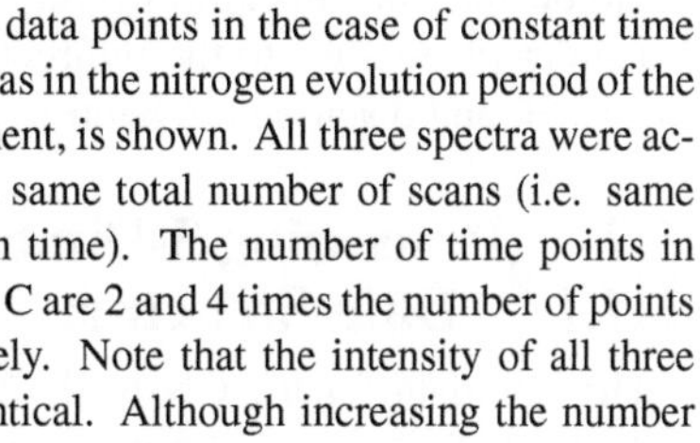

Figure 15.30. Shifting the apparent transmitter frequency. The apparent transmitter frequency in the indirectly proton dimension can be changed by applying a first-order phase correction to the FID. The left section of the panel shows the x-y plane at the time of sampling the first point. The transmitter was set at 4.7 ppm and the sweep width was 6 ppm. Resonances with chemical shifts smaller that 1.7 ppm will be aliased, as indicated by the gray sector. The right section shows the effect of shifting the resonances by 1.7 ppm. The apparent transmitter frequency becomes 3.0 ppm and resonances from 0.0 to 6.0 ppm will not be aliased.

In the case of indirect proton detection, the transmitter is generally placed on the water resonance to facilitate suppression of the water signal and minimize artifacts from the strong water resonance. Consequently, placement of the proton transmitter may not be ideal for a particular spectral range. For example, it may be desirable to sample only the aliphatic region of the proton spectrum, from 0 to 6 ppm. If the transmitter is on water, at 4.7 ppm, a spectral width of 6 ppm would cover a range of chemical shifts from 7.7 to 1.7

ppm. Consequently aliphatic resonances with a chemical shift of less than 1.7 ppm would be aliased. This aliasing can be removed during processing by applying a large first-order phase shift to the indirectly detected free induction decay, as illustrated in Fig. 15.30 and discussed in Section 15.7.6.2.

15.7 Processing 3-Dimensional Data

15.7.1 Data Structure

The raw data for the CN-NOESY experiment would be collected by first cycling through all of the t_1 increments. Then, t_2 would be incremented and all of the t_1 delays would be repeated at the new value of t_2, etc. The *cosine* and *sine* modulated FIDs associated with quadrature detection are collected as adjacent pairs. In the case of the CN-NOESY experiment, a total of 64 increments in carbon (t_1) and 32 increments in Nitrogen (t_2) would generally be acquired. The order in which the FIDs are acquired is listed in Table 15.2. Each FID represents a proton NMR spectrum, usually consisting of 1024 complex points.

15.7.2 Defining the Spectral Matrix

To begin processing the data it is necessary to define a three-dimensional matrix to hold the final spectrum. In the case of the CN-NOESY, 64 complex points in ^{13}C, 32 complex points in ^{15}N, and 1024 complex points in ^{1}H would provide adequate digital resolution for the spectrum. Since each time domain point consists of a real (cosine modulated) and an imaginary (sine modulated) point, it would be necessary to begin with a matrix that is 128 x 64 x 512 points in size, as illustrated in Fig. 15.31. In the

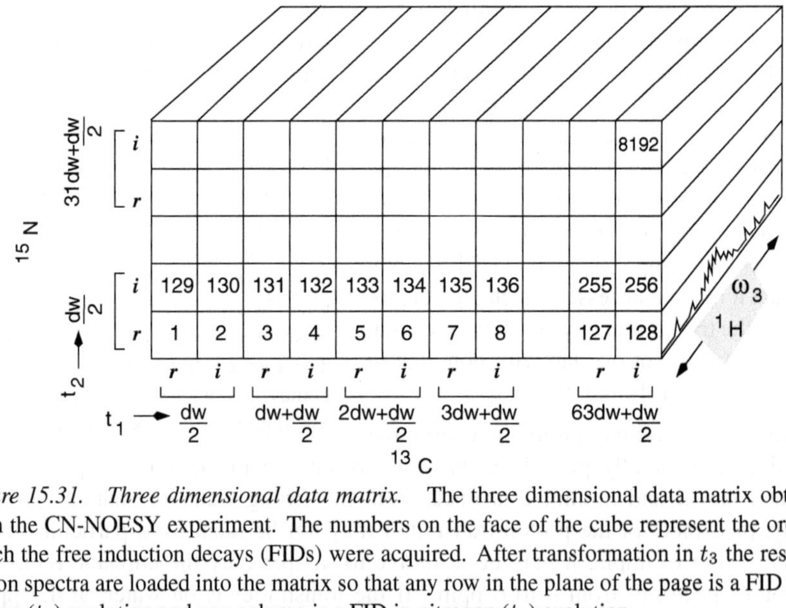

Figure 15.31. Three dimensional data matrix. The three dimensional data matrix obtained from the CN-NOESY experiment. The numbers on the face of the cube represent the order in which the free induction decays (FIDs) were acquired. After transformation in t_3 the resultant proton spectra are loaded into the matrix so that any row in the plane of the page is a FID in the carbon (t_1) evolution and any column is a FID in nitrogen (t_3) evolution.

case of the proton dimension, the aliphatic region of the spectrum would be discarded before storing the spectrum in the matrix, hence the reduction in size from 1024 to 512 points. Note that all dimensions are a power of two so that the fast Fourier transform algorithm can be used.

Table 15.2. Order of FIDs in a three-dimensional data set. The order of FIDs in the CN-NOESY is shown, assuming 64 carbon points and 32 nitrogen points. Quadrature detection was by States-TPPI hypercomplex method, thus separate *cosine* and *sine* modulated data is acquired for each time point.

FID #	Modulation	t_1 delay	Q. Phase	t_2 delay	Q. Phase
FID 1	$\cos(\omega_C t_1)\cos(\omega_N t_2)$	t_1=dw/2	$\phi_2 = x$	t_2=dw/2	$\phi_5 = x$
FID 2	$\sin(\omega_C t_1)\cos(\omega_N t_2)$	t_1=dw/2	$\phi_2 = y$	t_2=dw/2	$\phi_5 = x$
FID 3	$\cos(\omega_C t_1)\cos(\omega_N t_2)$	t_1=dw+dw/2	$\phi_2 = x$	t_2=dw/2	$\phi_5 = x$
FID 4	$\sin(\omega_C t_1)\cos(\omega_N t_2)$	t_1=dw+dw/2	$\phi_2 = y$	t_2=dw/2	$\phi_5 = x$
FID 127	$\cos(\omega_C t_1)\cos(\omega_N t_2)$	t_1=63dw+dw/2	$\phi_2 = x$	t_2=dw/2	$\phi_5 = x$
FID 128	$\sin(\omega_C t_1)\cos(\omega_N t_2)$	t_1=63dw+dw/2	$\phi_2 = y$	t_2=dw/2	$\phi_5 = x$
Now collect quadrature phase for ^{15}N					
FID 129	$\cos(\omega_C t_1)\sin(\omega_N t_2)$	t_1=dw/2	$\phi_2 = x$	t_2=dw/2	$\phi_5 = y$
FID 130	$\sin(\omega_C t_1)\sin(\omega_N t_2)$	t_1=dw/2	$\phi_2 = y$	t_2=dw/2	$\phi_5 = y$
FID 131	$\cos(\omega_C t_1)\sin(\omega_N t_2)$	t_1=dw+dw/2	$\phi_2 = x$	t_2=dw/2	$\phi_5 = y$
FID 132	$\sin(\omega_C t_1)\sin(\omega_N t_2)$	t_1=dw+dw/2	$\phi_2 = y$	t_2=dw/2	$\phi_5 = y$
FID 255	$\cos(\omega_C t_1)\sin(\omega_N t_2)$	t_1=63dw+dw/2	$\phi_2 = x$	t_2=dw/2	$\phi_5 = y$
FID 256	$\sin(\omega_C t_1)\sin(\omega_N t_2)$	t_1=63dw+dw/2	$\phi_2 = y$	t_2=dw/2	$\phi_5 = y$
Increment ^{15}N evolution delay, collect ^{13}C at new t_2					
FID 257	$\cos(\omega_C t_1)\cos(\omega_N t_2)$	t_1=dw/2	$\phi_2 = x$	t_2=dw+dw/2	$\phi_5 = x$
FID 258	$\sin(\omega_C t_1)\cos(\omega_N t_2)$	t_1=dw/2	$\phi_2 = y$	t_2=dw+dw/2	$\phi_5 = x$
FID 259	$\cos(\omega_C t_1)\cos(\omega_N t_2)$	t_1=dw+dw/2	$\phi_2 = x$	t_2=dw+dw/2	$\phi_5 = x$
FID 260	$\sin(\omega_C t_1)\cos(\omega_N t_2)$	t_1=dw+dw/2	$\phi_2 = y$	t_2=dw+dw/2	$\phi_5 = x$
FID 383	$\cos(\omega_C t_1)\cos(\omega_N t_2)$	t_1=63dw+dw/2	$\phi_2 = x$	t_2=dw+dw/2	$\phi_5 = x$
FID 384	$\sin(\omega_C t_1)\cos(\omega_N t_2)$	t_1=63dw+dw/2	$\phi_2 = y$	t_2=dw+dw/2	$\phi_5 = x$
Now collect quadrature phase for ^{15}N at new t_2					
FID 385	$\cos(\omega_C t_1)\sin(\omega_N t_2)$	t_1=dw/2	$\phi_2 = x$	t_2=dw+dw/2	$\phi_5 = y$
FID 386	$\sin(\omega_C t_1)\sin(\omega_N t_2)$	t_1=dw/2	$\phi_2 = y$	t_2=dw+dw/2	$\phi_5 = y$
FID 387	$\cos(\omega_C t_1)\sin(\omega_N t_2)$	t_1=dw+dw/2	$\phi_2 = x$	t_2=dw+dw/2	$\phi_5 = y$
FID 388	$\sin(\omega_C t_1)\sin(\omega_N t_2)$	t_1=dw+dw/2	$\phi_2 = y$	t_2=dw+dw/2	$\phi_5 = y$
FID 511	$\cos(\omega_C t_1)\sin(\omega_N t_2)$	t_1=63dw+dw/2	$\phi_2 = x$	t_2=dw+dw/2	$\phi_5 = y$
FID 512	$\sin(\omega_C t_1)\sin(\omega_N t_2)$	t_1=63dw+dw/2	$\phi_2 = y$	t_2=dw+dw/2	$\phi_5 = y$
Last two FIDS					
FID 8191	$\cos(\omega_C t_1)\sin(\omega_N t_2$	t_1=63dw+dw/2	$\phi_2 = x$	t_2=31dw+dw/2	$\phi_5 = y$
FID 8192	$\sin(\omega_C t_1)\sin(\omega_N t_2)$	t_1=63dw+dw/2	$\phi_2 = y$	t_2=31dw+dw/2	$\phi_5 = y$

15.7.3 Data Processing

15.7.4 Processing the Directly Detected Domain

The spectral matrix would be populated by transforming each of the directly detected free induction decays (FID) and then storing the resultant proton spectra in the appropriate cell of the matrix. Transformation of the FID to give the proton spectrum is identical to processing a one-dimensional spectrum (see Chapter 3) and involves the following steps:

1. Applying an apodization function.
2. Removal of water signal by linear prediction.
3. Fourier transformation.
4. Applying a phase correction to give pure absorption mode lineshapes. This phase correction is usually determined by transformation and phasing of the first FID. The same phase correction is applied to all spectra.
5. Discard the imaginary part of the spectrum, to conserve disk space.
6. Discard the aliphatic region of the proton spectrum since it does not contain any signal if only the amide protons were detected.

Fig. 15.31 shows the structure of the data matrix that is obtained after all of the directly detected FIDs are processed. The third dimension contains the amide proton spectrum for each value of t_1 and t_2. The amplitude of each spectrum in the third dimension will depend on the t_1 and t_2, as well as the quadrature phases. For example, the cell labeled 135 in Fig. 15.31 would have the following amplitude:

$$cos[\omega_C(3\tau_{dw} + \frac{\tau_{dw}}{2})]sin\omega_N[\frac{\tau_{dw}}{2}] \qquad (15.23)$$

The points in each row represent the free induction decay from evolution in t_1 (carbon), stored as alternating real (r) and imaginary (i) points. The points in each column are the free induction decay from evolution in t_2 (nitrogen), again with alternating real and imaginary points.

15.7.4.1 Processing the Indirectly Detected Domains

The steps involved in processing the indirectly detected domains are illustrated in Fig. 15.32 and summarized below:

1. Conversion of the 128 alternating real and imaginary pairs in carbon evolution to 64 complex points (Panel A→B, Fig. 15.32). Note that this operation, as well as the subsequent steps needs to be accomplished for all rows, i.e. a total of 64×512.
2. Linear prediction in the carbon dimension, typically an additional 25-50% points are predicted. (Panel B→C, Fig. 15.32).
3. Application of an apodization function. Typically, a phase shifted sin^2 function would be used (not shown in Fig. 15.32).

4. Fourier transformation of the points in t_1 (^{13}C). This will generate a complex spectrum of 128 points. The direction of data points for Fourier transformation are indicated on Panel C, Fig. 15.32.
5. Phase correction of the carbon spectra, since the initial delay was set to $\tau_{dw}/2$, a zero-order phase correction of 90° and a first-order phase correction of -180° would be applied.
6. The imaginary component is discarded to save disk space and the 128 point carbon spectrum is stored in the matrix. (Panel C→D, Fig.15.32).
7. Conversion of the 64 alternating real and imaginary pairs in the nitrogen evolution to 32 complex points, followed by linear prediction, as shown in Panel E of Fig.15.32. Note that this operation, as well as subsequent ones, have to be applied to all 128 $\times$ 512 columns in the matrix.
8. Application of an apodization function, followed by Fourier transformation of each column (t_2), generating the nitrogen spectrum (Not shown in Fig. 15.32).
9. Phase correction of the nitrogen spectra, since the initial delay was set to $\tau_{dw}/2$ a zero order-phase correction of 90° and a first-order phase correction of -180° is applied.
10. The imaginary component of the spectrum is discarded and the 64 point nitrogen spectrum is stored in the column of the matrix (Panel F, Fig. 15.32).
11. Additional phase adjustments in the carbon or nitrogen dimensions are performed. The required phase correction is obtained by loading a row (or column) that contains several peaks. It is often useful to sum several rows (or columns) such that peaks exist at near both edges of the spectrum so that the required first-order phase correction can be accurately defined. The spectrum is then converted from a real spectrum to a complex one using a Hilbert Transform and the phase is corrected. These phase values are then used to correct all of the rows (or columns).

15.7.5 Variation in Processing

Often it is advantageous to delay linear prediction of the ^{13}C dimension until after the ^{15}N dimension is transformed. The reason for this approach is to reduce the number of spectral lines that have to be predicted by linear prediction. Prior to the ^{15}N Fourier transform, each row of the data matrix will contain *all* of the ^{13}C resonances. Consequently, a large number of coefficient are required for linear prediction. After processing the ^{15}N dimension the peaks in the carbon dimension will be separated by their ^{15}N frequencies. Therefore, any given ^{15}N frequency will contain a small number of carbon resonance peaks, requiring fewer linear prediction coefficients.

If this approach is taken, the processing of the ^{13}C-time-domain data is initially performed without linear prediction or the use of an apodization function (steps 2 and 3 above). To generate the appropriate time domain data for linear prediction it is necessary to:

1. The real and imaginary components of the spectrum are regenerated using the Hilbert transform.

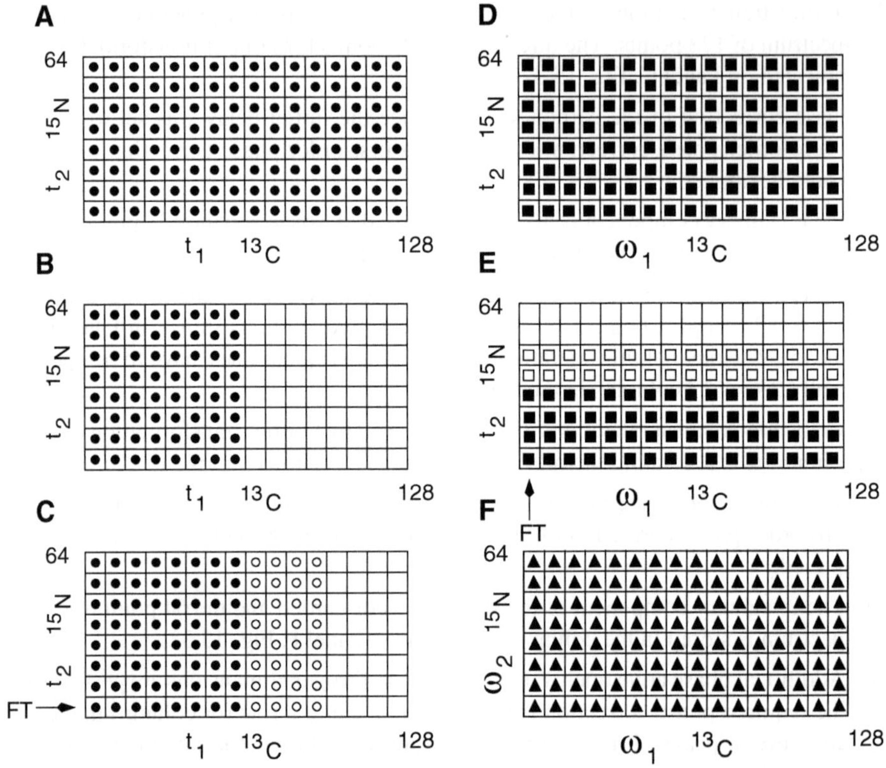

Figure 15.32. Processing indirectly detected domains. The grid in this figure represents the face of the data matrix, corresponding to the first point in the proton spectrum. The remaining 511 points of the proton spectrum are not shown, but would be behind the plane of the paper. When the axis is labeled with t, the data stored in that dimension represents time domain data. After Fourier transformation, the axis becomes labeled with ω. Panel A shows the matrix after immediately after loading the proton spectra, each dot represents the first point of the proton spectrum. Panel B shows the effect of converting the alternating real and imaginary carbon points to a complex number, halving the size of the data. Panel C shows linear prediction in the carbon dimension, the open circles represent predicted data points. Panel D shows the matrix after transformation of each row. The square symbols indicate that the carbon has been transformed. Panel E indicates that the nitrogen points have been converted to complex points and additional points (open squares) have been predicted. Panel F is the final matrix after transforming each column.

2. An inverse Fourier transformation is applied to regenerate the ^{13}C time domain data.

3. Linear prediction is applied to the FID, followed by apodization.

4. The FID is subject to Fourier transformation and after phasing (if required) the real portion of the spectrum is stored in the matrix.

15.7.6 Useful Manipulations of the Free Induction Decay

15.7.6.1 Reversal of the Frequency Axis

In a number of experiments the frequency axis of the indirectly detected domain is reversed as a consequence of how the chemical shift evolution was performed, i.e. resonances that should appear on the right of the spectrum are found on the left. The spectra can be reversed by simply taking the complex conjugate of the complex free induction decay:

$$e^{i\omega t} \rightarrow e^{-i\omega t} \tag{15.24}$$

This operation changes the sign of the frequency, converting ω to $-\omega$, which reverses the appearance of the spectrum.

15.7.6.2 Systematic Change in Resonance Frequencies

The resonance frequencies of all of the peaks in the spectrum can be shifted by applying a *first-order* phase correction, ϕ, to the free induction decay. This method was introduced by Bothner-by and Dadok and was originally used in one-dimensional difference spectroscopy [23]. It was later applied by Kay et al [83] to shift the indirectly detected proton frequency in in a three-dimensional NOESY-HMQC experiment.

The phase correction that is applied to each point of the FID is $n \times \phi/(N-1)$, where N is the total number of points in the FID and ϕ is the first-order phase correction that is specified to the processing software. Setting the evolution time t equal to $n\tau_{dw}$, the expression for the FID after the application of this phase shift is:

$$\begin{aligned} S(t) =& e^{i\omega(n\tau_{dw})} e^{in\phi/(N-1)} \\ =& e^{i\omega(n\tau_{dw})} e^{in\tau_{dw}\phi/(\tau_{dw}(N-1))} \\ =& e^{i(\omega+[\phi/(\tau_{dw}(N-1))])n\tau_{dw}} \end{aligned} \tag{15.25}$$

The frequency of the peaks has been shifted by $\phi/(\tau_{dw}(N-1))$.

Frequency shifting is commonly used in two applications. First, it can be used to un-alias peaks in indirectly detected domains, as discussed in Section 15.6.3.5. The second application of this technique is to place resonance peaks at the same *location* in a spectra. If two complementary triple-resonance experiments, such as the HNCA and HN(CO)CA experiments, were accidentally acquired with different carbon or nitrogen transmitter frequencies, then peaks with the same chemical shift will appear at different locations in the spectral matrix. Analysis of these two spectra is greatly facilitated if the peaks in one of the spectra can be shifted to same location as the corresponding peak in the other spectra.

Chapter 16

DIPOLAR COUPLING

16.1 Introduction

In Chapter 7, the coupling of nuclear spins through bonds was discussed. This scalar coupling is mediated by electrons and provides information on the conformation of the bonds joining the coupled spins. Scalar coupling is the basis of resonance assignment methods in isotopically labeled samples. Nuclear spins can also interact through space, via dipolar coupling. Dipolar coupling occurs when the magnetic field generated by one nuclear dipole affects the magnetic field at another nucleus. The magnitude of the dipolar coupling depends on the strength of the magnetic field generated by one spin and the size of the magnetic moment of the recipient spin. This form of coupling does not require the presence of connecting bonds; it is possible for spins on entirely different molecules to show dipolar coupling. Dipolar couping can be used to determine the distance between two spins, and in certain cases, the orientation of the inter-atomic vector relative to the applied B_o field. Consequently, dipolar coupling is a major source of information for structure determination by NMR spectroscopy.

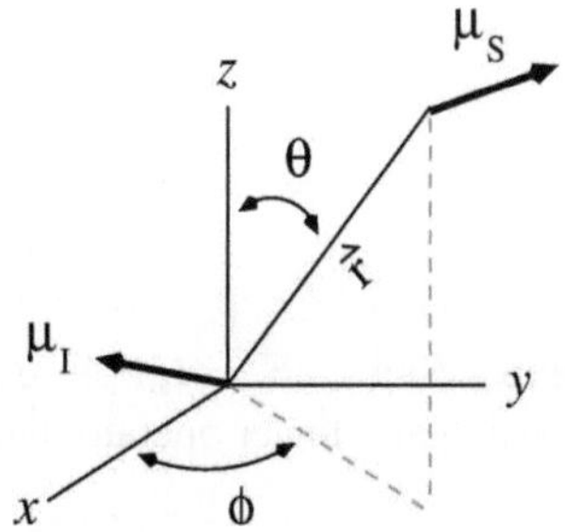

Figure 16.1. Dipolar coupling between spins. The orientation of two magnetic dipoles, μ_I and μ_S, separated by a distance $\vec{r}$ is shown. The angles θ and ϕ give the orientation of the inter-nuclear vector in polar coordinates.

16.1.1 Energy of Interaction

The overall energy of interaction between two magnetic dipoles, μ_I and μ_S, can be derived classically and is given by the following:

$$\mathcal{H} = \frac{\mu_I . \mu_S}{r_{IS}^3} - \frac{3(\mu_I . r_{IS})(\mu_S . r_{IS})}{r_{IS}^5} \tag{16.1}$$

where μ_I, and μ_S are the two coupled spins, separated by a distance r_{IS}. If r is written in polar coordinates $\vec{r} = (|r|sin\theta cos\phi, |r|sin\theta sin\phi, |r|cos\theta)$, and the magnetic dipoles are replaced by their quantum mechanical equivalent ($\mu = \gamma S$), then the energy of interaction becomes:

$$\begin{aligned} \mathcal{H} &= \frac{\gamma_I \gamma_S}{r^3}[I \cdot S - 3(I_x sin\theta cos\phi + I_y sin\theta sin\phi + I_z cos\theta) \\ &\quad \times (S_x sin\theta cos\phi + S_y sin\theta sin\phi + S_z cos\theta)] \end{aligned} \tag{16.2}$$

Converting the Cartesian transverse operators into raising and lowering operators provides considerable insight into the individual terms of this expression. Utilizing:

$$\begin{aligned} I_x &= \frac{I^+ + I^-}{2} \qquad I_y = -i\frac{I^+ - I^-}{2} \\ S_x &= \frac{S^+ + S^-}{2} \qquad S_y = -i\frac{S^+ - S^-}{2} \end{aligned} \tag{16.3}$$

and

$$cos\phi + isin\phi = e^{+i\phi} \qquad cos\phi - isin\phi = e^{-i\phi} \tag{16.4}$$

gives:

$$\begin{aligned} \mathcal{H} &= \frac{\gamma_1 \gamma_2}{r^3}[I \cdot S - 3(\frac{1}{2}sin\theta(I^+e^{-i\phi} + \frac{1}{2}I^-e^{+i\phi}) + I_z cos\theta)] \\ &\quad \times (\frac{1}{2}sin\theta(S^+e^{-i\phi} + S^-e^{+i\phi}) + S_z cos\theta)] \end{aligned} \tag{16.5}$$

expanding $I \cdot S = I_x S_x + I_y S_y + I_z S_z$ and converting these transverse operators to raising and lower operators gives the following six terms:

$$\begin{aligned} \mathcal{H} &= \frac{\gamma_I \gamma_S}{r^3}\left[F_Z + F_0 + F_1^+ + F_1^- + F_2^+ + F_2^-\right] \\ F_Z &= (3cos^2\theta - 1)I_z S_z \\ F_0 &= -\frac{1}{4}(3cos^2\theta - 1)(I^+S^- + I^-S_+) \\ F_{+1} &= \frac{3}{2}sin\theta cos\theta e^{-i\phi}(I_z S^+ + I^+ S_z) \\ F_{-1} &= \frac{3}{2}sin\theta cos\theta e^{i\phi}(I_z S^- + I^- S_z) \\ F_{+2} &= \frac{3}{4}sin^2\theta cos\theta e^{-2i\phi}(I^+S^+) \\ F_{-2} &= \frac{3}{4}sin^2\theta cos\theta e^{2i\phi}(I^-S^-) \end{aligned} \tag{16.6}$$

The effect coupled spins can be determined by applying the spin operators to the four basis states of two coupled spins:

$$\phi_1 = |\alpha\alpha> \quad \phi_2 = |\alpha\beta> \quad \phi_3 = |\beta\alpha> \quad \phi_4 = |\beta\beta> \tag{16.7}$$

Each spin operator can potentially stimulate one or more of the transitions that connect each of the four levels, as shown in Fig. 16.2.

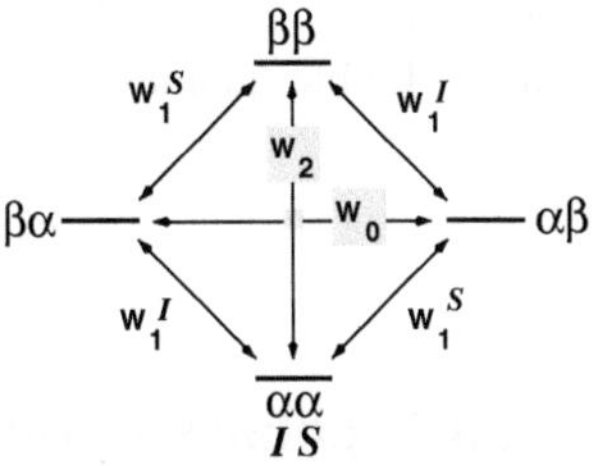

Figure 16.2 Energy level diagram for two coupled spins. The $\alpha\alpha$ state is the lowest energy. Single quantum transitions are indicated as W_1, the double and zero quantum transition are indicated as W_2, and W_0, respectively.

The first term, F_z, cannot cause transitions between energy levels because the individual states commute with the I_z and S_z operators. For example,

$$I_z S_z|\alpha\alpha> = \left(\frac{1}{2}\right)^2 |\alpha\alpha> \tag{16.8}$$

shows that the wavefunction is left unchanged by the application of F_z. However, this term will produce a splitting of the resonance lines in much the same manner as scalar coupling. In fact, the Hamilitonian that describes the evolution of the system, has the same form in both cases:

$$\mathcal{H}_{dipole} = DI_zS_z \qquad \mathcal{H}_{scalar} = JI_zS_z \tag{16.9}$$

therefore, the observed splitting will be the sum of both the dipolar and the scalar coupling if the spins are also scalar coupled.

The degree of splitting due to dipolar coupling can be calculated directly, using the above wavefunctions and F_z. For example, the two single-quantum transitions of the I spins in Fig. 16.2 have the following energy difference due to dipolar coupling:

$$\begin{aligned} E_{\alpha\alpha\to\beta\alpha} &= E_{\beta\alpha} - E_{\alpha\alpha} \\ &= \Delta\left[< \beta\alpha|I_zS_z|\beta\alpha > - < \alpha\alpha|I_zS_z|\alpha\alpha >\right] \\ &= \Delta[< \beta\alpha|\hbar^2(\frac{1}{2})(\frac{-1}{2})|\beta\alpha > - < \alpha\alpha|\hbar^2(\frac{-1}{2})(\frac{-1}{2})|\alpha\alpha >] \\ &= -\Delta\hbar^2\frac{1}{2} \end{aligned} \tag{16.10}$$

where $\Delta = \frac{\gamma_I\gamma_S}{r^3}\left[3cos^2\theta - 1\right]$. A similar calculation gives $E_{\alpha\beta\to\beta\beta} = +\Delta\hbar^2\frac{1}{2}$. Therefore, two resonance lines are observed at the following angular frequencies:

$$\omega = \omega_o \pm \frac{1}{2}\frac{\hbar\gamma_I\gamma_S}{r^3}\left[3cos^2\theta - 1\right] \tag{16.11}$$

where ω_o is the resonance frequency in the absence of dipolar coupling.

The second term, F_0, can cause zero-quantum transitions (W_o in Fig. 16.2), changing the state of *each* spin, thus there is no *net* change in the overall spin state of the system:

$$(I^+S^- + I^-S^+)|\alpha\beta >\to |\beta\alpha > \tag{16.12}$$

The terms F_{+1} and F_{-1} can cause single quantum transition from lower to higher energy states (F_{+1}) or from higher to lower energy states (F_{-1}), for example:

$$(I_zS^+ + I^+S_z)|\alpha\alpha >\to |\alpha\beta > + |\beta\alpha > \tag{16.13}$$

corresponding to the lower pair of single quantum transitions in Fig. 16.2. Finally, the terms F_{+2} and F_{-2} represent upward or downward double quantum transitions, respectively.

$$S_{1+}S_{2+}|\alpha\alpha> \rightarrow |\beta\beta> \quad (16.14)$$

16.1.2 Effect of Isotropic Tumbling on Dipolar Coupling

Although the contribution of the dipolar interaction to the energy of the spins can be quite large for spins with a *fixed* orientation, it is *completely* averaged to zero by isotropic rotation of a single molecule in solution, consequently there is no observed splitting of the resonance lines due to dipolar coupling. This can be easily shown by averaging $\Delta\omega$ over all angles [1]:

$$\begin{aligned}\overline{\Delta\omega} &= \frac{1}{4\pi}\int_0^{2\pi} d\phi \int_0^{\pi} \Delta\omega sin\theta d\theta \\ &= \frac{1}{2}\int_0^{\pi} \frac{\hbar\gamma_I\gamma_S}{r^3}(3cos^2\theta - 1)sin\theta d\theta \\ &= \frac{1}{2}\frac{\hbar\gamma_I\gamma_S}{r^3}\left[-3\frac{cos^3\theta}{3} + cos\theta\right]_0^{\pi} \\ &= \frac{1}{2}\frac{\hbar\gamma_I\gamma_S}{r^3}\left[-cos^3\theta + cos\theta\right]_0^{\pi} \\ &= 0 \end{aligned} \quad (16.16)$$

Note that the rate of rotational motion must be significantly faster than the frequency associated with the coupling, $\Delta\omega$, otherwise, the dipolar coupling will not be averaged and the couplings associated with the individual orientation of each molecule would be observed [2]. The dipolar splitting, $\Delta\omega$ can be easily calculated. The largest dipolar splitting is obtained when $\theta = 0$. In the case of the amide group in proteins:

$$\begin{aligned}\Delta\omega &= \frac{(1.054\times10^{-27})(2.67\times10^4)(-2.71\times10^3)}{(1.08\times10^{-8})^3}(3cos^2 0 - 1) \\ &= 120\times10^3 \text{ rad/sec} \\ \Delta\nu &= \Delta\omega/(2\pi) = 19.3 \text{ kHz}\end{aligned} \quad (16.17)$$

Given that rotational correlation times of typical proteins are generally shorter than 10^{-8} sec, corresponding to rotational frequencies of 10^8 Hz, the rotational motion is sufficiently rapid to average the dipolar coupling.

---

[1]The $sin\theta$ term in the integral is the probability of finding an angle θ on the surface of the unit sphere. The factor of $1/(4\pi)$ normalizes the integral to one, i.e.:

$$1 = \frac{1}{4\pi}\int_o^{\pi} sin\theta d\phi d\theta \quad (16.15)$$

[2]The spectrum obtained for a collection of immobilized spins is referred to as a 'powder pattern'.

Although there is no effect on the resonance frequency from dipolar coupling, the tumbling of the molecule generates a fluctuating magnetic field that can stimulate zero-quantum, single-quantum, and double-quantum transitions, providing a mechanism for nuclear spin relaxation, as discussed in Chapter 19. Since all possible orientations of the spins are sampled, the relaxation rates are *independent* of the relative orientation of the coupled spins. Consequently, it is only possible to obtain information on inter-proton distances from the measurements of these rates. Usually, the enhanced relaxation effects from dipolar coupling are measured by detecting changes in the population levels of the coupled spins. This effect was first observed by Overhauser [122], for electron-nuclear spin interactions and is usually referred to as the nuclear Overhauser effect, or NOE. Typically, distances of 5 Å or less can be readily measured. It is important to note that this distance information is entirely *local*, i.e. no information is obtained unless the spins are within 5 Å of each other.

16.1.3 Effect of Anisotropic Tumbling

When the orientation of molecules is anisotropic, then dipolar coupling can be exploited to obtain information on the *orientation* of the inter-nuclear vectors with respect to the external magnetic field. This orientation is obtained from the measured splitting of the resonance line due to the coupling:

$$\Delta\omega = \frac{\hbar\gamma_I\gamma_S}{r^3}[3cos^2\theta - 1] \qquad (16.18)$$

Usually, these measurements are performed on spins that are separated by a fixed distance (e.g. the amide NH pair, $r = 1.08$ Å), thus removing the dependence of the dipolar coupling on distance.

In samples that show complete ordering, the large splitting due to dipolar coupling (19.3 kHz for the amide group) would greatly complicate the analysis of the NMR spectra. Consequently, the coupling is attenuated by orienting only a small fraction of the molecules, usually one part in 10^4. Provided that the ensemble of molecules experiences the same averaged partial ordering a single averaged value for the coupling will be observed.

In general, the proteins under study are oriented by the inclusion of larger oriented macromolecules in the sample, such as filamentous bacteriophage. Consequently, only those proteins near the surface of the orientating media become ordered. To average this ordering over the entire ensemble of proteins it is necessary that the rate of translational diffusion is sufficiently rapid to distribute the anisotropy over all molecules. Given that the translational diffusion rate for most proteins studied by NMR are on the order of 10^5 cm^2/sec, this condition is easily met.

The small degree of ordering gives rise to observed dipolar couplings ranging from 1 to 50 Hz. Since the coupling arises from ordering a small fraction of molecules, it is generally referred to as the *residual dipolar coupling*, often abbreviated RDC. In contrast to the measurement of inter-proton distances, information from a RDC measurement is of a *global* nature; providing the orientation of the inter-nuclear vector with respect to a common axis, the applied magnetic field. Also note that this effect depends on $1/r^3$ and therefore can, in principal, obtain information on longer inter-atomic distances than with NOE effects. A more complete treatment of residual dipolar coupling is provided in Section 16.3 of this chapter.

16.2 Measurement of Inter-proton Distances

The reader is referred to the text by Neuhaus and Williamson for a comprehensive discussion on the use of dipolar coupling in structural analysis [118]. The text that follows is focused on the use of dipolar coupling to measure inter-proton distances in larger protein in the absence of internal motion.

Random isotropic tumbling of the entire molecule will generate fluctuations in the local magnetic field because the dipolar coupling depends on the relative orientation of the two coupled spins (see Eq. 16.6). These field fluctuations can provide a relaxation mechanism for the dipolar coupled spins if fluctuations exist at the frequency of zero-, single- and double-quantum transitions. The rate of the transition is given by [152]:

$$W_k = \zeta_k \frac{\gamma^4 \hbar^2}{r^6} J(k\omega) \tag{16.19}$$

where W_k is the rate of the zero- ($k = 0$), single- ($k = 1$), or double-quantum transitions ($k = 2$), $\zeta_0 = 0.1, \zeta_1 = 0.15, \zeta_2 = 0.6$, $J(k\omega)$ is the value of the spectral density function at $\omega = k \times \omega_s$, and ω_s is the resonance frequency.

The spectral density function, $J(\omega)$, is described in more detail in Chapter 19. For a molecule undergoing random isotropic motion the spectral density function is:

$$J(\omega) = \frac{\tau_c}{1 + \omega^2 \tau_c^2} \tag{16.20}$$

where τ_c is the rotational correlation time, or the average time it takes the molecule to rotate one radian. The rotational correlation time is proportional to the molecular weight of the protein and is approximately 1 nsec for each 2.6 kDa of mass.

The spectral density functions for small and large proteins are shown in Fig. 16.3. A large protein has a high spectral density at $\omega = 0$ and therefore its rotational motion will be efficient at stimulating zero-quantum transitions. In contrast a small protein will have a spectral density function that is nearly identical at $\omega = 0$, $\omega = \omega_s$, and at $\omega = 2\omega_s$. Consequently all three transition rates will be similar, and less than the zero-quantum transition rate for a large protein.

Since the transition rates, W_k, are proportional to $1/r^6$, they can be used to obtain the inter-proton distance (r), provided the contribution from the spectral density function to the transition rate can be estimated to some degree of accuracy. The transition

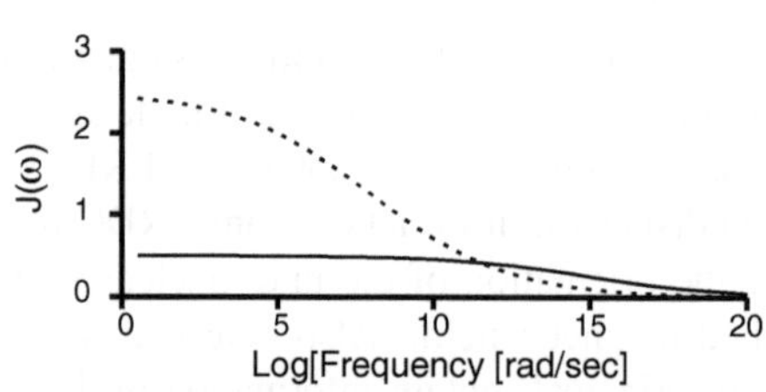

Figure 16.3 Effect of molecular weight on the spectral density function. The spectral density functions for a small protein (2.6 kDa, solid line) and a large protein (20 kDa, dashed line) are shown. Note that the spectral density function for the larger protein has a greater intensity at lower frequencies.

rates are obtained experimentally by measuring the rate by which the non-equilibrium population of one spin affects the population of energy levels of a dipolar coupled spin. There are several ways of generating a non-equilibrium population of energy levels. In the simple example of a one-dimensional NOE experiment, continuous RF-radiation is applied to a single resonance line (nuclear spin transition). The applied RF will equalize the ground and excited states of the irradiated spin by causing saturation of the single quantum transitions. The NOE is then obtained by measuring the effect of this saturation on the population levels of the coupled spin. Although this method works well for small molecules, it is generally not suitable for large proteins because the large number of overlapping resonance lines preclude the selective saturation of any single spin transition.

An illustration of the changes in population levels in the 1D-NOE experiment is shown in Fig. 16.4. Irradiation of the I spin equalizes the populations of the states that are connected by the single-quantum transitions of the I spin:

$$n_{\alpha\alpha} = n_{\beta\alpha} \qquad n_{\alpha\beta} = n_{\beta\beta} \tag{16.21}$$

In larger proteins the dipolar coupling between the spins results in efficient zero-quantum transitions, consequently the population levels of the states connected by the zero quantum transitions become equal, i.e. $n_{\beta\alpha} = n_{\alpha\beta}$. Therefore, the intensity of the transition associated with the S spin will be altered because of the coupling to the I spin, as illustrated in Fig. 16.4.

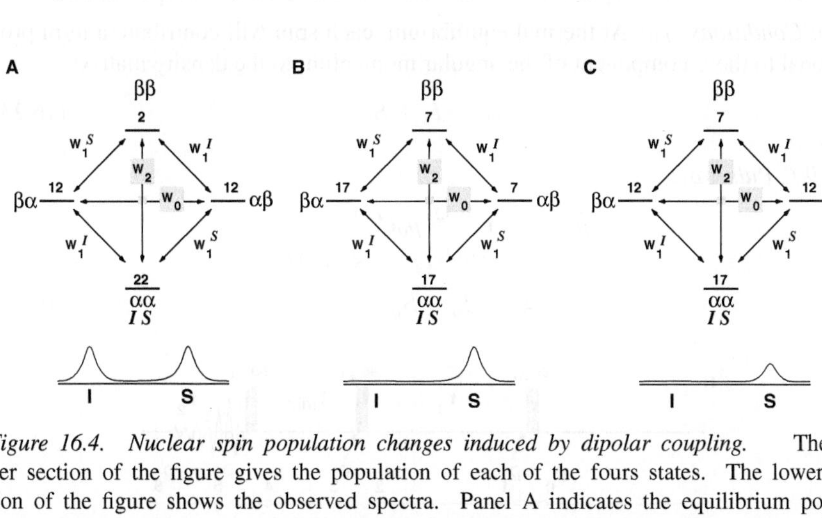

Figure 16.4. Nuclear spin population changes induced by dipolar coupling. The upper section of the figure gives the population of each of the fours states. The lower section of the figure shows the observed spectra. Panel A indicates the equilibrium populations. The intensity of a particular resonance line is given by the population difference between the connected levels. For example, the intensity of the resonance line for the I spin is: $(n_{\alpha\alpha} - n_{\beta\alpha}) + (n_{\alpha\beta} - n_{\beta\beta}) = (22 - 12) + (12 - 2) = 20$. Panel B illustrates the effect of saturating the I spins in the *absence* of dipolar coupling. Note that the population difference across both single quantum I transitions is zero while the population of the S spins is unaffected and is still equal to 20. Panel C illustrates the change in the population of the S spins due to efficient zero quantum transitions that equalize the populations of the $\beta\alpha$ and $\alpha\beta$ states. The net population of the S spins is reduced from 20 to 10.

16.2.1 NOESY Experiment

It is generally not feasible to saturate a single resonance line in a crowded spectrum. However, the equivalent effect is achieved in the multi-dimensional NOESY experiment by first generating a non-equilibrium population of spin I that is a function of its resonance frequency, e.g. $cos(\omega_I t)$. This non-equilibrium population is then transferred to the S spin via dipolar coupling. The degree of perturbation of the S spin population is directly proportional to the non-equilibrium state of the I spins. If the population levels of the I spins were equal to $cos(\omega_I t_1)$ during the first time period of a two-dimensional experiment, then the population of the S spins would become proportional to $cos(\omega_I t_1)$. Subsequent excitation of the S spins and detection of the precessing magnetization would generate a signal of the following form:

$$S(t_1, t_2) = \eta e^{i\omega_I t_1} e^{i\omega_S t_2} \tag{16.22}$$

where η is related to the quantum transition rates, which are proportional to $\frac{1}{r^6}$. Fourier transformation of this signal will produce a crosspeak at (ω_I, ω_S) with intensity η.

The process by which chemical shift information is transferred from one spin to another is most easily understood by the analysis of a simple two-dimensional dipolar coupled experiment, the 2D-NOESY experiment. The pulse sequence for this experiment is shown in Fig. 16.5 and the product operator description of this experiment is discussed below.

16.2.1.1 Product Operator Treatment of the NOESY Experiment

Initial Conditions. ρ_0*:* At thermal equilibrium, each spin will contribute a term proportional to the z-component of the angular momentum to the density matrix:

$$\rho_0 = I_z + S_z \tag{16.23}$$

First 90°_x *Pulse,* ρ_1*:*

$$\begin{aligned} \rho_1 &= e^{-i\beta J_x} \rho_0 e^{i\beta J_x} \\ &= {}^{-i\beta J_x}(I_z + S_z)e^{i\beta J_x} \\ &= -I_y - S_y \end{aligned} \tag{16.24}$$

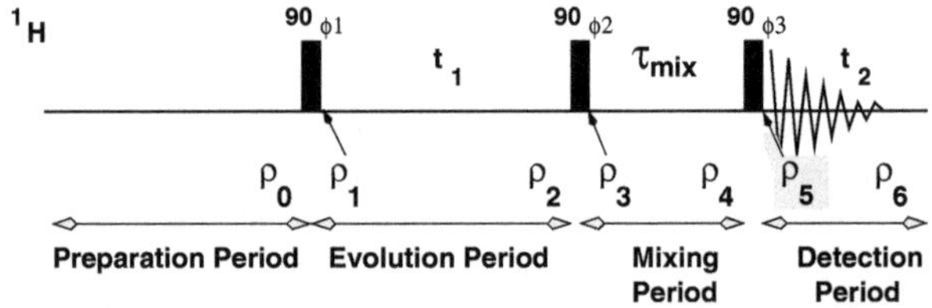

Figure 16.5. Pulse sequence for a simple two-dimensional NOESY. All pulses are applied as high power 90° proton pulses. See text for a description of pulse and receiver phases. The density matrix at various points in the sequence ($\rho_0 - \rho_6$), are discussed in the text. Quadrature detection is obtained by shifting the phase of the first pulse by 90° and acquiring an additional scan with the same t_1 values. A version of this sequence that is suitable for data acquisition in H_2O is shown in Fig. 15.9.

where $J = I + S$, i.e. the spin angular momentum of both spins. The pulse flip angle, β is equal to $\pi/2$.

Evolution during t_1, ρ_2*:* Evolution of the system during the t_1 period can be described using the evolution operator. Since I and S are not scalar coupled, each can be treated separately.

$$\begin{aligned} -I_y &\rightarrow -\left[I_y \cos(\omega_I t_1) - I_x \sin(\omega_I t_1)\right] e^{-t_1/T_2} \\ -S_y &\rightarrow -\left[S_y \cos(\omega_S t_1) - S_x \sin(\omega_S t_1)\right] e^{-t_1/T_2} \end{aligned} \tag{16.25}$$

The 2nd 90°_x *Pulse,* ρ_3*:*

$$\begin{aligned} \rho_3 &= -\left[I_z \cos(\omega_I t_1) - I_x \sin(\omega_I t_1)\right] e^{-t_1/T_2} \\ &\quad - \left[S_z \cos(\omega_S t_1) - S_x \sin(\omega_S t_1)\right] e^{-t_1/T_2} \end{aligned} \tag{16.26}$$

The Mixing Time, exchange of Magnetization (ρ_4)*:* At the beginning of the mixing time the z-magnetization of the I spins is $\cos(\omega_I t_1)$. Note that this represents a non-equilibrium population difference across the single quantum transitions and that this population difference is related to the chemical shift of the *I* spins.

In addition to the I_z and S_z terms, the density matrix also contains off-diagonal terms (remember that I_x represents a density matrix with non-zero off-diagonal terms). For the moment we are interested in describing the effects of dipolar coupling on the populations of the spins, thus we will ignore the off-diagonal terms in the density matrix. These terms can be removed with either phase cycling or pulse-field gradients, as discussed in the following section.

The exchange of magnetization between the I and S spins occurs because the non-equilibrium population of the *I* influences the population of the *S* spin, and vice-versa. At the end of the mixing time the change in the population levels can be represented as:

$$\begin{aligned} \rho_4(\tau) &= I_z \left[\alpha \cos(\omega_I t_1) + \eta \cos(\omega_S t_1)\right] e^{-t_1/T_2} \\ &\quad + S_z \left[\alpha \cos(\omega_S t_1) + \eta \cos(\omega_I t_1)\right] e^{-t_1/T_2} \end{aligned} \tag{16.27}$$

Where η is the fraction of magnetization that is transferred from one spin to another and α represents magnetization that has remained associated with each spin. Explicit expressions for η are provided in the following section.

After the third 90°_x *pulse,* ρ_5*:* This pulse returns the longitudinal magnetization to the transverse plane for detection during t_2.

$$\begin{aligned} \rho_5 &= -I_y \left[\alpha \cos(\omega_I t_1) + \eta \cos(\omega_S t_1)\right] e^{-t_1/T_2} \\ &\quad - S_y \left[\alpha \cos(\omega_S t_1) + \eta \cos(\omega_I t_1)\right] e^{-t_1/T_2} \end{aligned} \tag{16.28}$$

Detection in t_2, ρ_6*:* During detection, each term evolves with the chemical shift associated with the transverse spin operator, giving rise to the following detected signal (assuming quadrature detection in t_2):

$$
\begin{aligned}
S(t_1, t_2) \quad \propto \quad & [\alpha \cos(\omega_I t_1) + \eta \cos(\omega_S t_1)]\, e^{-t_1/T_2} e^{-i\omega_I t_2} \\
& + [\alpha \cos(\omega_S t_1) + \eta \cos(\omega_I t_1)]\, e^{-t_1/T_2} e^{-i\omega_S t_2} \quad (16.29)
\end{aligned}
$$

This time-domain signal contains two self-peaks, with intensity α, and two crosspeaks, with intensity η.

16.2.1.2 Phase Cycle in the NOESY Experiment

The required phase cycle in the NOESY experiment can be easily obtained by analysis of coherence changes that occur during the experiment. The coherence level changes in the NOESY experiment are shown in Fig. 16.6.

The desired coherence changes associated with paths A and B for each pulse in the sequence are as follows:

Pulse	1	2	3
Coherence change (A)	+1	-1	-1
Coherence change (B)	-1	+1	-1

The phase cycle associated with each pulse is obtained by considering all of the coherence changes that need to be retained in the experiment. Recall that a phase shift of:

$$\phi = k\frac{2\pi}{N} \quad (16.30)$$

will select only the following coherences: $\Delta p + nN; n \in I$.

Therefore, in the case of the first pulse, a 2-step phase cycle will retain both the +1 and the -1 change in coherence. The second pulse in the sequence has the same desired changes in coherence, therefore $N = 2$ as well. Note that phase cycling this pulse removes the transverse component of the magnetization that remains after the second pulse. The final pulse must bring the zero quantum coherence down to -1 for detection; all other coherence changes should be removed by phase cycling of this pulse. A four step phase cycle will cancel contributions from single, double, and triple

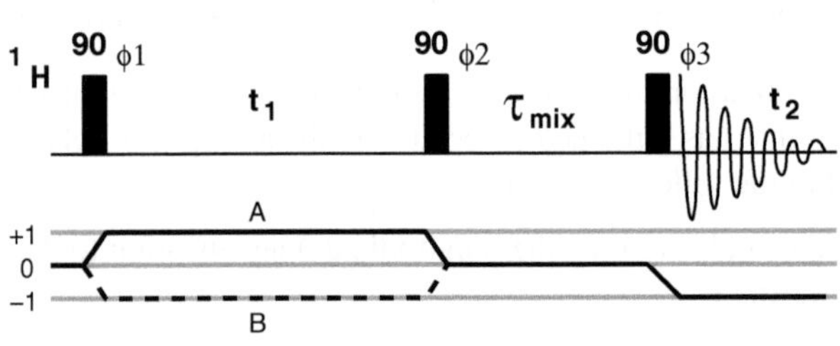

Figure 16.6 Coherence level changes in the NOESY experiment. The phase cycle for all three pulses is shown at the bottom of the figure. The receiver phase is obtained from the following expression: $\phi_{rec} = -\sum_{i=1}^{N} \Delta p_i \phi_i$.

Pulse	Δp	1	2	3	4	5	6	7	8	9	10	1	112	13	14	15	16
1	+1	0	2	0	2	0	2	0	2	0	2	0	2	0	2	0	2
2	-1	0	0	2	2	0	0	2	2	0	0	2	2	0	0	2	2
3	-1	0	0	0	0	1	1	1	1	2	2	2	2	3	3	3	3
ϕ_{rec}		0	2	2	0	1	3	3	1	2	0	0	2	3	1	1	3

quantum coherences that may have existed during the mixing time. The table in Fig. 16.6 summarizes the overall phase cycle for all three pulses.

16.2.2 Crosspeak Intensity in the NOESY Experiment

The crosspeak intensity in the NOESY spectrum can be easily obtained using standard differential equations to represent the changes in population that occur due to the zero, single, and double quantum transition rates.

The time dependence of the population of the ground state ($\alpha\alpha$) is given by:

$$\frac{dN_{\alpha\alpha}}{dt} = -(W_1^I + W_1^S + W_2)N_{\alpha\alpha} + W_1^I N_{\beta\alpha} + W_1^S N_{\alpha\beta} + W_2 N_{\beta\beta} \quad (16.31)$$

The time dependence of the three levels can be written in a similar fashion. These equations are just rate equations which describe the transfer of populations from one level to another due to transitions that are stimulated by the magnetic field fluctuations.

The longitudinal magnetization is found by calculating the differences between levels:

$$\begin{aligned} M_z^I &= N_{\alpha\alpha} + N_{\alpha\beta} - N_{\beta\alpha} - N_{\beta\beta} \\ M_z^S &= N_{\alpha\alpha} + N_{\beta\alpha} - N_{\alpha\beta} - N_{\beta\beta} \end{aligned} \quad (16.32)$$

This gives the following differential equations:

$$\begin{aligned} \frac{dM_z^I}{dt} &= -(W_0 + 2W_1 + W_2)M_z^I - (W_2 - W_0)M_z^S \\ \frac{dM_z^S}{dt} &= -(W_2 - W_0)M_z^I - (W_0 + 2W_1 + W_2)M_z^S \end{aligned} \quad (16.33)$$

Defining the following:

$$\begin{aligned} \rho_I &= W_0 + 2W_{1I} + W_2 \\ \rho_S &= W_0 + 2W_{1S} + W_2 \\ \sigma &= W_2 - W_0 \end{aligned} \quad (16.34)$$

gives the following differential equations that describe the time dependent changes in magnetization.

$$\begin{aligned} \frac{dI}{dt} &= -\rho_I(I - I_0) - \sigma(S - S_0) \\ \frac{dS}{dt} &= -\rho_S(S - S_0) - \sigma(I - I_0) \end{aligned} \quad (16.35)$$

These equations can be easily solved using Laplace transforms (see Appendix B), giving the final solution for one of the spins:

$$\begin{aligned} I(t) - I_o &= \frac{(S(0) - S_o)}{2}\left[e^{-(\rho+\sigma)t} - e^{-(\rho-\sigma)t}\right] \\ &+ \frac{(I(0) - I_o)}{2}\left[e^{-(\rho+\sigma)t} + e^{-(\rho-\sigma)t}\right] \end{aligned} \quad (16.36)$$

Under equilibrium conditions, $S(0) = S_o, I(0) = I_o$, there is no change in the magnetization, as expected.

16.2.3 Effect of Molecular Weight on the Intensity of NOESY Crosspeaks

The intensity of the NOESY crosspeak depends on the relative size of the cross- and self-relaxation rates and this dependence is shown in Fig. 16.7. For large proteins, the spectral density function is small at ω and 2ω, therefore $W_2 < W_1 < W_o$ and $\rho \approx W_o$ and $\sigma \approx -W_o$. Substituting these values into Eq. 16.36 shows that the crosspeaks are positive and increase as the molecular weight increases, as indicated in Fig. 16.7. For small molecules, the NOESY crosspeak is opposite in sign from the diagonal peak, and generally weaker. When $\omega\tau_c \approx 1$, the NOE crosspeaks have zero intensity, and it is not possible to obtain any information on inter-proton distances using this technique.

Equation 16.36 can be further simplified with the assumption that W_o is much greater than W_1 and W_2, which is certainly the case for most proteins. This assumption gives the following simplified equation for the transfer of magnetization between the two spins.

$$I(t) - I_o = \frac{(S(0) - S_o)}{2}\left[1 - e^{-2W_0 t}\right] + \frac{(I(0) - I_o)}{2}\left[1 + e^{-2W_0 t}\right] \qquad (16.37)$$

where $I(t)$ is the z-magnetization of the I spin at the end of the mixing time.

In the NOESY experiment, the magnetization at the very beginning of the mixing period is:

$$S(0) = cos(\omega_S t_1) \qquad I(0) = cos(\omega_I t_1) \qquad (16.38)$$

Therefore, the first term in Eq. 16.37 refers to the intensity of the *crosspeak* located at (ω_S, ω_I), while the second term will give the intensity of the *selfpeak*, located at (ω_I, ω_I).

In addition to peak intensity changes that are a result of dipolar coupling during τ_{mix}, the magnetization of both self- and crosspeaks will decay according to the spin-lattice relaxation rate, T_1 during the mixing time. Therefore the observed intensity of the crosspeak will be:

$$\eta(t_{mix}) = [1 - e^{-2W_o t_{mix}}]e^{-t_{mix}/T_1} \qquad (16.39)$$

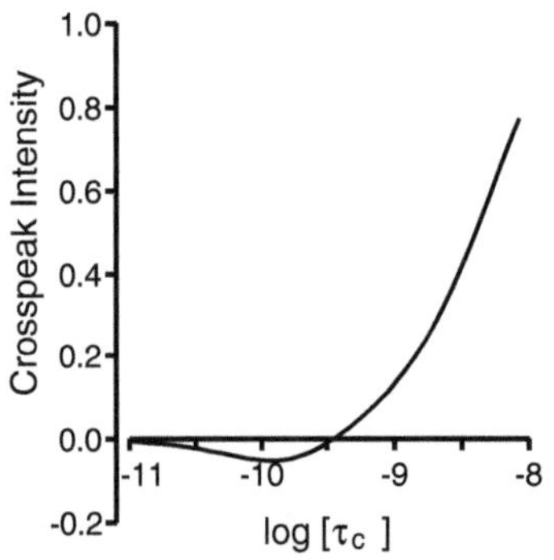

Figure 16.7 Effect of rotational correlation time on NOESY crosspeak intensity. The crosspeak intensity as a function of the rotational correlation time is shown. An inter-proton distance of 2 Å and a spectrometer frequency of 500 MHz was used in this calculation. For reference, a 20 residue peptide would have a rotational correlation time of $\approx$ 1 nsec ($\log \tau_c = -9$).

The intensity first increases due to the transfer of magnetization by dipolar coupling, and then decreases due to spin-lattice relaxation, as illustrated in Fig. 16.9. At short mixing times, this reduces to:

$$\eta(t_{mix}) = +2W_o t_{mix} \tag{16.40}$$

and the intensity of the crosspeak increases in a linear fashion at a rate of $2W_0$.

16.2.3.1 Dipolar Coupling Between Multiple Spins

The above treatment considered only two isolated spins. Given the high density of protons within folded proteins, this is clearly an oversimplification.

As a simple example, consider three coupled protons, I, S, J, then the set of differential equations that describe the effect of dipolar coupling on the magnetization of each spin is:

$$\begin{matrix} dI/dt \\ dJ/dt \\ dS/dt \end{matrix} = \begin{bmatrix} \rho_i & \sigma_{ij} & \sigma_{is} \\ \sigma_{ij} & \rho_j & \sigma_{sj} \\ \sigma_{is} & \sigma_{sj} & \rho_s \end{bmatrix} \begin{bmatrix} I \\ J \\ S \end{bmatrix} \tag{16.41}$$

For N-coupled spins, this generalizes to:

$$\frac{d\vec{M}}{dt} = -R\vec{M} \tag{16.42}$$

where M is now a vector that represents the magnetization of each spin. R is a matrix of exchange rates, or the complete relaxation matrix:

$$R_{ii} = \sum_{ij}(\rho_{ij}) + 1/T_{1i} \tag{16.43}$$

$$R_{ij} = \sigma_{ij} \tag{16.44}$$

The above equations can be used to calculate the complete relaxation behavior of the system for a given set of atomic positions. Usually, calculated intensities are compared to the measured intensities and the inter-proton distances are adjusted to minimize the difference between the measured and calculated intensities.

Calculation of the crosspeak intensities requires solving Eq. 16.42. This is equivalent to solving:

$$M(\tau_m) = e^{-R\tau_m}M(0) \tag{16.45}$$

A series expansion of $e^{-R\tau_m}$ gives:

$$e^{-R\tau_m} \approx 1 - R\tau_m + \frac{1}{2}R^2\tau_m^2 - ... \tag{16.46}$$

Unfortunately, this series does not converge rapidly, in fact, 10-13 terms are required for mixing times on the order of 200 msec, requiring considerable computation time. A computationally simpler solution is to write the rate equation as:

$$R = \chi\lambda\chi^T \tag{16.47}$$

where χ is the matrix of eigenvectors and λ are the eigenvalues of R. The series expansion produces:

$$e^{-R\tau_m} = 1 - \chi\lambda\chi^T\tau_m + \frac{1}{2}\chi\lambda\chi^T\chi\lambda\chi^T.... \quad (16.48)$$

$$= \chi e^{-\lambda\tau_m}\chi^T \quad (16.49)$$

It is much simpler now to expand the series since higher terms of the exponential function are calculated as a sum of a series instead of a product of matrices.

There are various computational approaches that have been developed to model the effect of multi-spin coupling on the intensity of NOE peaks and an excellent discussion of these approaches can be found in Ref. [22].

16.2.4 Experimental Determination of Inter-proton Distances

The goal is to convert the intensity of each crosspeak in the NOESY experiment to an inter-proton distance. The following provides a general summary of the approaches that have been employed to obtain these distances, listed in order of decreasing accuracy. Method 3 is probably the most common approach since method 2 requires a considerable amount of data fitting.

1. NOESY peak intensities are measured at several mixing times, from 100 to 300 msec, and a full relaxation matrix analysis is used to obtain the relaxation rates.

2. NOESY peak intensities are measured at several mixing times, and the data are fit to the complete equation (Eq. 16.39) for two isolated spins.

3. NOESY peak intensities are measured for a relatively short mixing time, e.g. 100 msec, and a linear relationship between the peak intensity and mixing time is assumed, with a slope of $2W_o$. This method suffers from poor signal-to-noise because of the small crosspeak intensity at short mixing times..

4. NOESY peak intensities are measured for a long mixing time, e.g. 400 msec in order to obtain intense peaks. A linear relationship between the peak intensity is again assumed, with a slope of $2W_o$.

In the case of the first method, the inter-proton distances are obtained directly from the R matrix. The other three methods provide an estimate of W_o. The inter-proton distance are then obtained from the following equation:

$$r = \sqrt[6]{\frac{1}{10}\frac{1}{W_o}\gamma^4\hbar^2 J(0)} \quad (16.50)$$

$J(0)$, the value of the spectral density at $\omega = 0$, is usually estimated from the crosspeak intensity for two protons that are separated by a known distance, such as the two H_α protons on glycine residues, provided they are immobilized in the protein.

16.2.4.1 Errors in the Measurement of Inter-Proton Distances

As one might expect, the first method discussed above yields the most accurate results, while the last produces the most inaccurate distances. The error in inter-proton distance that is associated with each method of data analysis is shown in Fig. 16.8. The relaxation matrix analysis produces the least error of all methods (Panel A).

Fitting the data to an exponential produces reasonably accurate inter-proton distances for distances that are less than 2.5 Å (Panel B). In contrast, measurements that rely on single time points produce inter-proton distances whose accuracy decreases substantially as the mixing time is increased. If the peak intensity at a mixing time of 200 msec is used to estimate W_o, then the error ranges from 15-30% (Panel C). If a single time point of 400 msec is used, then the error in distance can exceed 50%, i.e. protons that are separated by 4.5 Å appear as though they are only 3 Å apart (Panel D).

With the exception of analysis by the complete relaxation matrix, all methods produce measured experimental distances that are smaller than the true inter-proton distance. Or equivalently, the crosspeaks are more intense than they should be. This effect is due to a phenomenon called 'spin-diffusion' which is illustrated in Fig.16.9. In this example, spin A and C are separated by 4.5 Å, hence the direct dipolar coupling between A and C is weak. Consequently, the rate at which the crosspeak intensity increases with the mixing time will be slow. However, spins A and C are both close to a common spin, B, and magnetization from spin C can be effectively transferred to spin A by first transferring the magnetization from C to B, and then from B to A. This additional pathway for magnetization transfer will increase the intensity of the crosspeak between A and C, suggesting that A and C are closer than their actual separation.

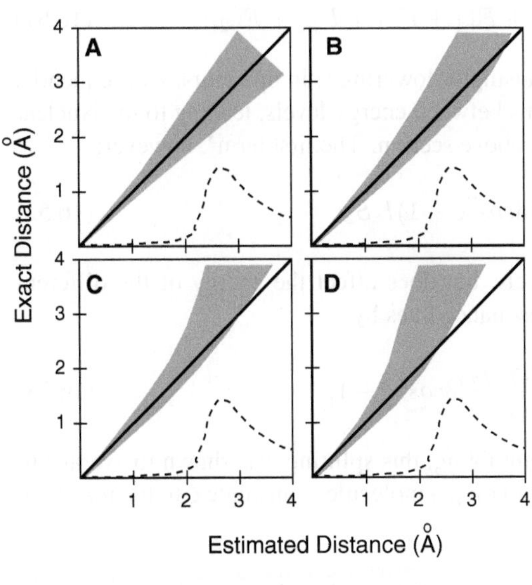

Figure 16.8 Distance errors in NOESY measurements. The distance estimated from NOESY crosspeaks is plotted versus the actual inter-proton distance. The dashed line shows the typical distribution of inter-proton distances in globular proteins. The gray area represent the range of errors in the measured distances in typical experiments. Note that shorter distances give more intense crosspeaks, generating a smaller error in the calculated distance. **A**: Full relaxation matrix used. The error in distance is due entirely to the intrinsic error in the crosspeak intensity. **B**: Measurement of initial rate using a single exponential fit. **C**:Single time point used, 200 msec. **D**: Single time point used, 400 msec mixing time. Adapted from Ref. [8].

Spin diffusion is much more efficient in larger proteins due to the large value of $J(\omega)$ at $\omega = 0$.

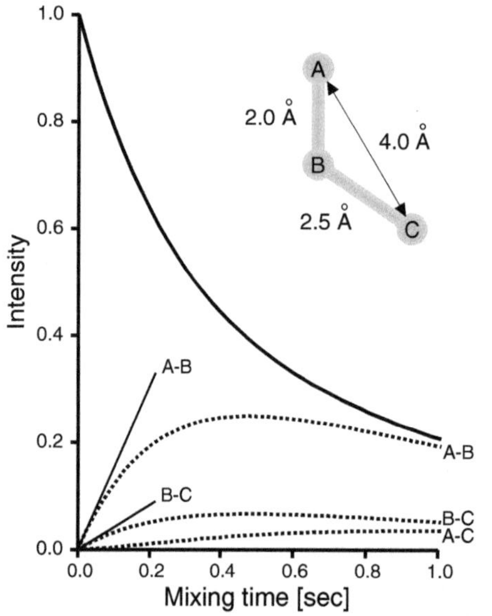

Figure 16.9 Peak intensities in NOESY spectra. The peak intensities as a function of the mixing time are shown for selfpeaks (solid curve) and crosspeaks (dotted). In this simulation, protons A and B were 2 Å apart, B and C 2.5 Å apart, and A and C were 4.0 Å apart and the spin-lattice relaxation time, T_1 was set to 0.5 sec. The time course of each of the associated crosspeaks are labeled. The initial slopes of the peak intensities for the crosspeaks between protons A and B, and between protons B and C are shown a solid lines. The ratio of these slopes is close to $(2.5/2.0)^6$. The intensity of the crosspeak between A and C rises slowly at early time points, but then increases in intensity due to relayed magnetization *via* proton B.

16.3 Residual Dipolar Coupling (RDC)

The dipolar Hamiltonian presented at the beginning of this chapter contained six terms:

$$\mathcal{H} = \frac{\gamma^2}{r^3}\left[F_Z + F_o + F_{+1} + F_{-1} + F_{+2} + F_{-2}\right] \tag{16.51}$$

The terms $F_o, F_{\pm 1}, F_{\pm 2}$ all contain raising or lowering spin operators. Consequently, they can enhance the rate of transitions between energy levels, leading to the Nuclear Overhauser Effect as discussed in the above section. The first term[3], however:

$$\frac{\gamma_I \gamma_S}{r^3}(3cos^2\zeta - 1)I_z S_z \tag{16.52}$$

cannot cause transitions between levels, but does affect the energy of the different states, cause a splitting, $\Delta\omega$, of the resonance lines by:

$$\Delta\omega = \frac{\hbar\gamma_I \gamma_S}{r^3}[3cos^2\zeta - 1] \tag{16.53}$$

If the system is undergoing isotropic tumbling, this splitting was shown to average to zero because the probability, $P(\zeta)$, of finding a molecule at an angle ζ to the magnetic

[3]In the following, ζ replaces θ as the angle between the magnetic field and the vector joining the two spins. This is to avoid confusion with the bond vector direction in the *molecular* coordinate system.

field is the same for all angles of ζ, i.e.:

$$\begin{aligned}\Delta\omega &= \frac{\hbar\gamma_I\gamma_S}{r^3}\int_0^\pi P(\zeta)(3cos^2\theta - 1)sin\theta d\theta \\ \Delta\omega &= \frac{\hbar\gamma_I\gamma_S}{r^3}P\int_0^\pi (3cos^2\theta - 1)sin\theta d\theta \\ &= 0 \end{aligned} \qquad (16.54)$$

However, if the isotropic distribution can be made anisotropic by partial orientation of the molecule, $P(\zeta)$ now depends on ζ, then the above integral will, in general, not be equal to zero. In this case, splitting of the spectral lines due to dipolar coupling will be observed.

16.3.1 Generating Partial Alignment of Macromolecules

In order to observe splitting of resonance lines from dipolar coupling it is necessary to induce partial ordering of the protein or nucleic acid molecules in the sample such that the ensemble average of the molecular orientation of the protein (or nucleic acid) is no longer zero. An anisotropic distribution can be obtained in one of two ways, by use of the intrinsic asymmetry of magnetic dipoles within the molecule, or by creating an anisotropic environment that induces alignment of the molecules.

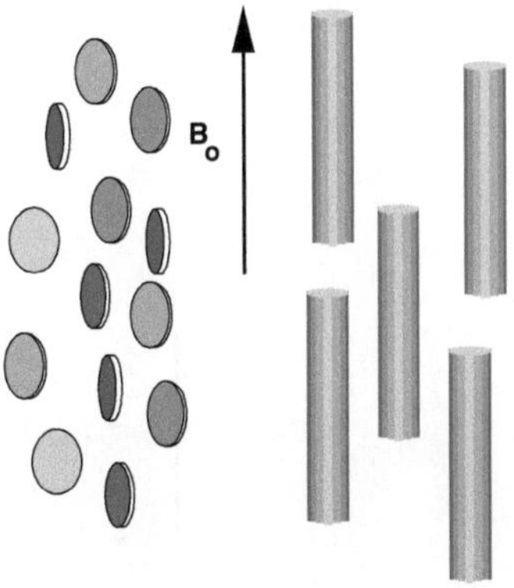

Figure 16.10. Alignment of bicells and filamentous bacteriophage in the magnetic field. Bicells (left) are coin shaped and usually align with their faces perpendicular to the magnetic field (B_o) while bacteriophage particles (right) are rod-like and align with their long axis in the direction of the field.

If the distribution of permanent and induced magnetic dipoles within the sample are strongly anisotropic then the molecules will assume a preferred orientation with respect to the magnetic field. In general, globular proteins are not sufficiently anisotropic to generate significant molecular alignment. The exception to this generalization are heme containing proteins, where the large number of delocalized electrons associated with the heme group provides a considerable anisotropic magnetic dipole. Nucleic acids, in particular double stranded DNA or RNA, align more readily in the magnetic field; the anisotropic magnetic dipoles of each nucleotide base add constructively due to extensive base stacking. However, the degree of alignment is quite small, leading to small couplings that may be difficult to measure.

Alignment can also be generated by placing the molecule of interest in an anisotropic environment. Two of the more popular methods of generating an anisotropic environment utilize lipid bicells [141, 156], which are flat lipid bilayers, or filamentous bacteriophage [72], both of which possess a large anisotropic magnetic dipole, and therefore become aligned in the magnetic field, as illustrated in Fig. 16.10. Although these media have large macroscopic viscosities they have little effect on the rotational correlation times of the proteins, therefore the observed linewidths are similar to those found in the absence of the orienting media.

In general, the desired degree of alignment is small, such that the observed dipolar couplings range from 5-15 Hz. Higher degrees of alignment can cause line broadening from unresolved proton-proton dipolar coupling. In addition, because the coupling between spins is given by the sum of the scalar and dipolar coupling, large dipolar couplings can increase or decrease the observed coupling, thus interfering with efficient polarization transfer during INEPT-like periods in pulse sequences.

The anisotropic distribution of the protein in the sample can can be generated passively or actively, depending on the type of interaction between the protein and the alignment media. In the case of bicells, the alignment is generally simply due to steric factors. The anisotropic shape of the protein causes certain orientations of the protein to become less populated, consequently, the probability distribution, $P(\zeta)$, is no longer uniform, and the dipolar coupling does not average to zero. In the case of filamentous bacteriophage, the surface of the phage particle has a net negative charge,

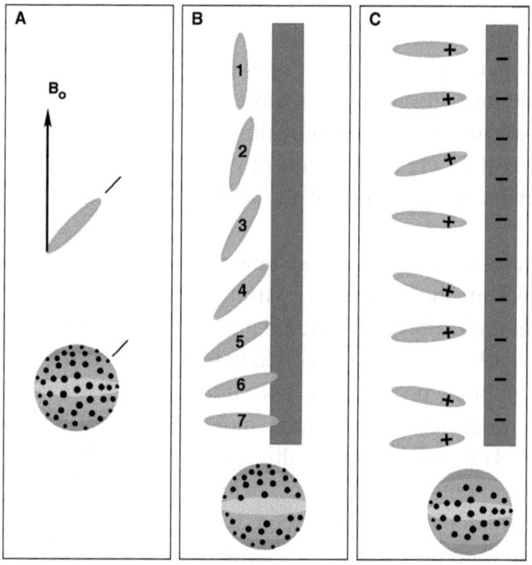

Figure 16.11. Generation of residual dipolar coupling by molecular alignment. Panel A shows a single protein molecule at an arbitrary orientation with respect to the magnetic field. The sphere in the lower part of the panel shows a dot for each orientation of the protein, illustrating that all orientations are possible in the absence of alignment media. Panel B shows the effect of alignment media on the orientations of the protein. The vertical solid rectangle represents the surface of a bicell. In the presence of the alignment media, orientations 6 and 7 cannot occur because of steric clashes with the surface of the bicell. Therefore the distribution becomes anisotropic, as illustrated on the lower sphere. Panel C illustrates how the negatively charged surface of the bacteriophage induces alignment of the protein with an asymmetric charge distribution. Note that the alignment of the proteins is opposite to that obtained by bicells, as illustrated on the sphere in the lower part of the panel.

consequently the alignment of molecules can be induced due to their electrostatic potential energy, as illustrated in Fig. 16.11.

Generally, it is advantageous to utilize more than one alignment media for the following reasons. First, the protein under study may interact strongly with the alignment media, thus preventing the averaging of the induced alignment over the ensemble. Second, certain orientation of internuclear vectors can give RDC values close to zero ($\zeta \approx 54°$). In this case it is unknown whether this value is due to rapid internal motion of the interacting atoms that averages the coupling or $\zeta = 54°$. Since the protein may orient differently in a different alignment media, it is possible to determine whether the zero RDC value in the first alignment media does indeed indicate $\zeta \approx 54°$.

16.3.2 Theory of Dipolar Coupling

This analysis follows closely that provided by Bax and Tjandra [13] which should be consulted for additional details. The splitting of resonance lines by dipolar coupling is given, Hz, as:

$$\Delta\nu(\zeta) = \frac{\hbar\gamma_I\gamma_S}{2\pi r^3}[3cos^2\zeta - 1] \qquad (16.55)$$

where ζ is the angle between the inter-nuclear vector and the magnetic field. It is convenient to write this as:

$$\Delta\nu(\zeta) = D_{max}\frac{1}{2}[3cos^2\zeta - 1] \qquad (16.56)$$

where $D_{max} = \hbar\gamma_1\gamma_2/\pi r^3$, the largest possible splitting, obtained when $\zeta = 0$.

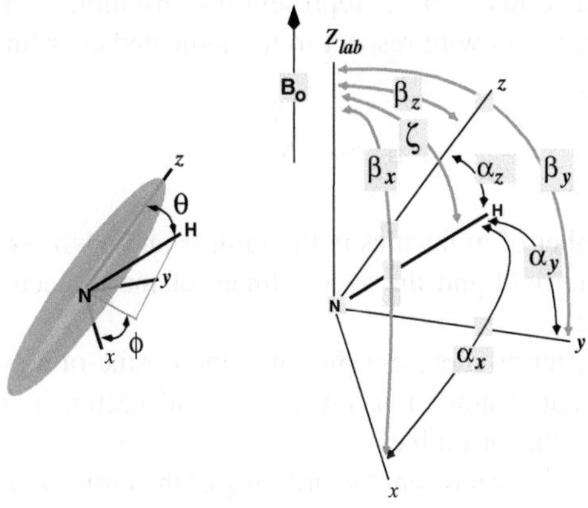

Figure 16.12. The coordinate system for the analysis of residual dipolar coupling. The left section of the diagram shows a single NH bond vector in a protein. The coordinate system that is shown refers to the molecular reference frame. The orientation of the NH bond vector in this frame is given by the polar angles θ and ϕ. The right section of the diagram shows the orientation of the molecular coordinate frame with respect to the external magnetic field, which defines the z-coordinate in the laboratory. The direction cosines between the NH bond vector and the molecular frame are given by $\alpha_x, \alpha_y, \alpha_z$. The direction cosines that relate the molecular frame to the direction of the magnetic field are given by β_x, β_y, and β_z.

The coordinates of the unit vector in the direction of the N-H bond, in the molecular coordinate frame, are given by the direction cosines [4] for each axis: $(cos\ \alpha_x,\ cos\ \alpha_y,\ cos\ \alpha_z)$. Similarly, the unit vector that defines the orientation of the magnetic field, also in the molecular coordinate system, is given by $(cos\beta_x, cos\beta_y, cos\beta_z)$. $cos\zeta$ can be obtained by taking the dot product of the NH bond vector and the vector that describes the magnetic field:

$$\begin{aligned} cos\zeta &= (cos\ \alpha_x, cos\ \alpha_y, cos\ \alpha_z) \cdot (cos\beta_x, cos\beta_y, cos\beta_z) \\ &= cos\ \alpha_x cos\beta_x + cos\ \alpha_y cos\beta_y + cos\ \alpha_z cos\beta_z \end{aligned} \qquad (16.57)$$

The coordinates of the inter-nuclear vector are fixed in the molecular frame. However, the orientation of the magnetic field with respect to the molecular frame varies from molecule to molecule over the entire sample. Therefore the average value of $cos\zeta$ is:

$$< cos\zeta >= cos\ \alpha_x < cos\beta_x > +cos\ \alpha_y < cos\beta_y > +cos\ \alpha_z < cos\beta_z > \qquad (16.58)$$

where the angled brackets (<>) represent the average over the entire ensemble of molecules in the sample.

The expression for dipolar coupling becomes (*cos* has been abbreviated using just its argument, i.e. $\alpha_x = cos\alpha_x$):

$$\Delta\nu(\zeta) = D_{max} \times$$

$$\left[\frac{3}{2} \left\{ \begin{array}{cccccc} \alpha_x^2 < \beta_x^2 > & + & \alpha_x\alpha_y < \beta_x\beta_y > & + & \alpha_x\alpha_z < \beta_x\beta_z > & + \\ \alpha_y\alpha_x < \beta_y\beta_x > & + & \alpha_y^2 < \beta_y^2 > & + & \alpha_y\alpha_z < \beta_y\beta_z > & + \\ \alpha_z\alpha_x < \beta_z\beta_x > & + & \alpha_z\alpha_y < \beta_z\beta_y > & + & \alpha_z^2 < \beta_z^2 > & \end{array} \right\} - \frac{1}{2} \right]$$

In the above sum, the terms $< \beta_i^2 >$ represent the ensemble average of the direction cosine of the magnetic field with respect to the indicated coordinate in the molecular coordinate frame, e.g.:

$$< \beta_x^2 >= \sum_i^N cos^2\beta_x^i \qquad (16.59)$$

where N are the number of molecules in the sample and $cos\beta_x^i$ is the direction cosine between the magnetic field and the x-coordinate of the molecular frame associated with the ith molecule.

In contrast, the α_i terms represent the direction cosine of a single bond vector in the molecular coordinate frame. For any given bond vector, α_i is *identical* for each and every molecule in the ensemble.

Since the terms $< \beta_i^2 >$ represent the ordering of the molecules in the sample, these averages are collected into what is generally referred to as the Saupe order matrix. The elements of this matrix are defined as:

$$S_{ij} = \frac{3}{2} < \beta_i\beta_j > -\frac{1}{2}\delta_{ij} \qquad (16.60)$$

[4] The direction cosine is the cosine of the angle between the vector and the indicated axis. In the case of a unit vector, the direction cosine gives the projection of that vector on to the axis, giving the coordinate of the vector in that direction.

or explicitly as

$$S = \frac{3}{2}\begin{bmatrix} <\beta_x^2> -\frac{1}{3} & <\beta_x\beta_y> & <\beta_x\beta_z> \\ <\beta_y\beta_x> & <\beta_y^2> -\frac{1}{3} & <\beta_y\beta_z> \\ <\beta_z\beta_x> & <\beta_z\beta_y> & <\beta_z^2> -\frac{1}{3} \end{bmatrix} = \begin{bmatrix} S_{xx} & S_{xy} & S_{xz} \\ S_{yx} & S_{yy} & S_{yz} \\ S_{zx} & S_{zy} & S_{zz} \end{bmatrix} \quad (16.61)$$

The expression for $\Delta\nu(\zeta)$ becomes:

$$\Delta\nu(\zeta) = D_{max}\sum_{i,j=1}^{3} S_{ij}\alpha_i\alpha_j \quad (16.62)$$

The Saupe order matrix is symmetric about the diagonal, $S_{xy} = S_{yx}$, and each element is clearly a real number. Consequently, it is possible to find a coordinate rotation, R, when applied to the molecular coordinate system, will diagonalize the Saupe order matrix. [5]

$$R \times \begin{bmatrix} S_{xx} & S_{xy} & S_{xz} \\ S_{yx} & S_{yy} & S_{yz} \\ S_{zx} & S_{zy} & S_{zz} \end{bmatrix} \times R^{-1} = \begin{bmatrix} S_{xx} & 0 & 0 \\ 0 & S_{yy} & 0 \\ 0 & 0 & S_{zz} \end{bmatrix} \quad (16.63)$$

The diagonal form of the order matrix is particular convenient, because the expression for dipolar coupling is greatly simplified, becoming:

$$\Delta\nu(\zeta) = D_{max}\left[S_{xx}\alpha_x^2 + S_{yy}\alpha_y^2 + S_{zz}\alpha_z^2\right] \quad (16.64)$$

Substituting the expression $S_{ii} = \left[\frac{3}{2} < \beta_{ii} >^2 - \frac{1}{2}\right]$ gives:

$$\Delta\nu(\zeta) = D_{max}\frac{3}{2}\left[< \beta_{xx} >^2 \alpha_x^2 + < \beta_{yy} >^2 \alpha_y^2 + < \beta_{zz} >^2 \alpha_z^2 - \frac{1}{3}\right] \quad (16.65)$$

For an isotropic system, the average value[6] of $cos^2\beta_{ii}$ is $\frac{1}{3}$. Therefore, if $< \beta_{ii} >^2 = \frac{1}{3}$ the dipolar coupling vanishes. Consequently, the *difference* in $< \beta_{ii} >^2$ from $\frac{1}{3}$ is of interest. This difference is generally referred to the alignment tensor, A, defined as follows:

$$A_{ii} = < \beta_{ii} >^2 - \frac{1}{3} \quad (16.66)$$

There are six possible forms of the diagonal alignment tensor since $\pm 90°$ rotations about any of the three orthogonal coordinate axis (x, y, z) will simply interchange

---

[5]This postulate can be easily shown for a 2×2 matrix. Consider the matrix $\tilde{A} = \begin{bmatrix} a & 0 \\ 0 & b \end{bmatrix}$. If the coordinate system is rotated by θ, then the matrix becomes:

$$\begin{aligned} \tilde{A}' &= R \times \tilde{A} \times R^{-1} \\ &= \begin{bmatrix} cos\theta & -sin\theta \\ sin\theta & cos\theta \end{bmatrix} \begin{bmatrix} a & 0 \\ 0 & b \end{bmatrix} \begin{bmatrix} cos\theta & sin\theta \\ -sin\theta & cos\theta \end{bmatrix} \\ &= \begin{bmatrix} acos^2\theta + bsin^2\theta & acos\theta sin\theta - bsin\theta cos\theta \\ acos\theta sin\theta - bsin\theta cos\theta & asin^2\theta + bcos^2\theta \end{bmatrix}. \end{aligned}$$

which is symmetric and real.

[6]The average is: $\frac{1}{2}\int_0^{\pi} sin\beta cos^2\beta d\beta = \frac{1}{3}$. The factor of 1/2 normalizes the probability distribution.

two diagonal elements of the tensor. For example, a 90° rotation about the x-axis simply interchanges A_{yy} with A_{zz}. To specify a single orientation, the diagonal order tensor is written such that $|A_{zz}| > |A_{yy}| > |A_{xx}|$.

Writing the orientation of the bond vector in polar coordinates:

$$\begin{aligned} cos^2\alpha_x &= x^2 = sin^2\theta cos^2\phi \\ cos^2\alpha_y &= y^2 = sin^2\theta sin^2\phi \\ cos^2\alpha_z &= z^2 = cos^2\theta \end{aligned} \qquad (16.67)$$

and substituting the alignment tensor for $< \beta_{ii} >^2$, gives:

$$\Delta\nu(\theta,\phi) = D_{max}\frac{3}{2}\left[A_{xx}sin^2\theta\cos^2\phi + A_{yy}sin^2\theta sin^2\phi + A_{zz}cos^2\theta\right] \qquad (16.68)$$

This equation can be further simplified using double angle formula [7] , as well as the relationship $A_{xx} + A_{yy} + A_{zz} = 0$, to give:

$$\begin{aligned} &\Delta\nu(\theta,\phi) = \\ &D_{max}\frac{3}{2}\left[\frac{1}{2}A_{xx}sin^2\theta[1+cos2\phi] + \frac{1}{2}A_{yy}sin^2\theta[1-cos2\phi] + A_{zz}cos^2\theta\right] \\ &D_{max}\frac{3}{2}\left[\frac{1}{2}(A_{yy}+A_{xx})sin^2\theta + A_{zz}cos^2\theta + \frac{1}{2}sin^2\theta cos2\phi(A_{xx}-A_{yy})\right] \\ &D_{max}\frac{3}{2}\left[\frac{1}{2}(-A_{zz})(1-cos^2\theta) + A_{zz}cos^2\theta + \frac{1}{2}sin^2\theta cos2\phi(A_{xx}-A_{yy})\right] \\ &D_{max}\frac{3}{2}\left[A_{zz}\frac{1}{2}(3cos^2\theta-1) + (A_{xx}-A_{yy})\frac{1}{2}sin^2\theta cos2\phi\right] \end{aligned} \qquad (16.69)$$

The axial component of the alignment tensor is defined as:

$$A_a = \frac{3}{2}A_{zz} \qquad (16.70)$$

and the rhombic component, which is a measure of asymmetry of the alignment of the molecule in the x-y plane of the molecular frame, is defined as:

$$A_r = A_{xx} - A_{yy} \qquad (16.71)$$

The observed dipolar splitting is thus:

$$D_{max}\left[A_a\frac{1}{2}(3cos^2\theta-1) + (A_r)\frac{3}{4}sin^2\theta cos2\phi\right] \qquad (16.72)$$

.

7

$$sin^2\phi = \frac{1-cos2\phi}{2} \qquad cos^2\phi = \frac{1+cos2\phi}{2}$$

Since the molecular alignment tensor cannot be directly measured, the axial component of the alignment tensor is often absorbed into D_{max} to give the following representation for the dipolar coupling:

$$\Delta\nu(\theta,\phi) = D_a \left[(3cos^2\theta - 1) + \frac{3}{2}Rsin^2\theta cos2\phi\right] \tag{16.73}$$

where D_a is the magnitude of the *residual dipolar coupling* tensor while R is defined as its rhombicity: $R = \frac{A_r}{A_a}$.

For proteins that are axially symmetric, e.g. ellipsoid in shape, then $R = 0$ because the x- and y-direction of the molecular coordinate system cannot be distinguished from each other. For non-axially symmetric molecules, the x- and y-values of the alignment tensor will differ from each other. Although there is no preferred orientation of the molecules in the x-y plane *per se*, there is a fixed relationship between the x- and y-axis in the molecular frame, thus the contribution of $cos^2\beta_x$ from one molecule in the ensemble to A_{xx} is associated with a defined contribution to A_{yy}, as illustrated in Fig. 16.13.

16.3.3 Measurement of Residual Dipolar Couplings

The maximum observed value of residual dipolar coupling depends on the gyromagnetic ratio of the coupling spins as well as the distance between them. The relative sizes of some couplings associated with peptide mainchain atoms are given in Table 16.1. For a given degree of alignment, the coupling between the α-proton and carbon is twice as large at the NH coupling. In contrast, the coupling between the carbonyl and the nitrogen is one-tenth that of the NH coupling, making it difficult to measure accurately.

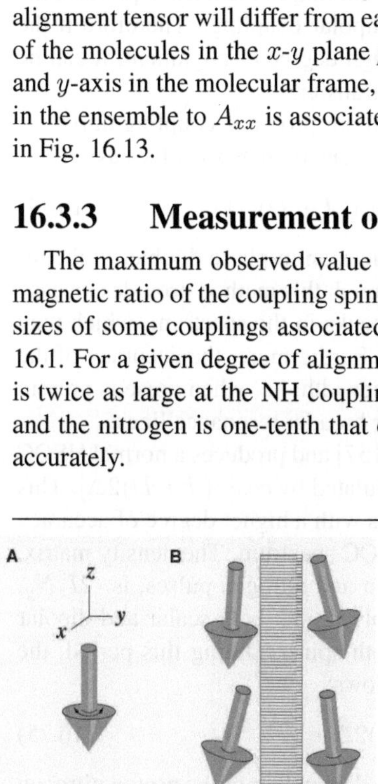

Figure 16.13 Alignment of axial and non-axial symmetric molecules. Panel A shows an axial symmetric molecule. The molecular coordinate frame is indicated above the molecule. Rotation about the z-axis does not change the alignment properties of the molecule. Panel B illustrates how this molecule would align in the presence of bicells or phage, note that rotation about the z-axis causes the angle between the x- or y-axis and the magnetic field to be identical, for *any* orientation of the molecule. Therefore, $A_{xx} = A_{yy}$ over the ensemble of molecules. Panel C shows three possible alignments for an non-axially symmetric molecule. In these examples, the y-axis of the molecular coordinate frame is perpendicular to the alignment media. In this particular orientation, rotations about the y-axis are not hindered by the alignment media and A_{yy}=0. However, the x-axis of the molecular frame can assume only a limited number of orientations with respect to B_o, giving $A_{xx} \neq 0$. In general, both A_{xx} and A_{yy} are non-zero and $A_{xx} \neq A_{yy}$.

The discussion here will focus on the measurement of couplings associated with N-H bonds as these are generally the easiest to obtain due to inexpensive ^{15}N labeling of proteins. In addition, this coupling is one of the largest because of the large gyromagnetic ratio of the proton ($D_{max} \propto \gamma_H \gamma_X$). Experiments directed at the measurement of additional couplings have been reviewed by Bax [13] and de Alba and Tjandra [48].

16.3.3.1 Measurement of D$_{NH}$

Pulse sequences for the measurement of D$_{NH}$ are shown in Fig. 16.14. These sequences will measure the sum of the scalar and dipolar coupling: $J_{NH} + D_{NH}$. Usually J is determined for *each* amide group by acquiring a separate data set in the absence of an orientating media. Note that during evolution in the INEPT periods, the system also evolves under the sum of scalar and dipolar coupling. Therefore if the dipolar coupling is particularly large, because of high degree of alignment, it can be difficult to find an optimal time, τ, for polarization transfer.

Sequence A is a standard HSQC experiment without proton decoupling during t_1 [157]. During this period the detected portion of the density matrix evolves as:

$$2I_zN_y \rightarrow 2I_zN_y cos(\omega_N t_1) cos(\pi(J + D)t_1) \qquad (16.74)$$

Fourier transformation of the resultant signal produces an in-phase doublet in the nitrogen dimension, with a separation of $J + D$. Although this method provides an easy method of obtaining D, it doubles the number of peaks in the spectrum, which may cause problems with analysis due to peak overlap. Furthermore, the accuracy of the measured splitting will depend somewhat on the linewidth, i.e. when noise is present it is more difficult to define the center frequency for broader resonance lines.

Sequence B was described by Tjandra and Bax [157] and produces a normal HSQC spectrum with the amplitude of the resonances modulated by $cos[\pi(J + D)2\Delta]$. This sequence provides a method of measuring couplings with a higher degree of accuracy and does not double the number of peaks in the HSQC spectrum. The density matrix, after the first INEPT period, followed by the proton and nitrogen pulses, is: $2I_zN_y$. During the next 2Δ period, the density matrix evolves with both scalar and dipolar coupling because of the 180° pulses applied to both spins. During this period, the observable part of the density matrix evolves as follows:

$$2I_zN_y \rightarrow 2I_zN_y cos(\pi(J + D)2\Delta)e^{-2\Delta/T_2} \qquad (16.75)$$

Chemical shift labeling by ω_N occurs during t_1 with decoupling of the proton nitrogen coupling by the 180° proton pulse.

Coupling	Relative Value
$N - H_N$	1.000
$C_\alpha - H_\alpha$	2.020
$C_\alpha - CO$	0.198
$CO - N$	0.121

Table 16.1 Mainchain residual dipolar couplings. Residual dipolar couplings, relative to that observed for the NH bond vector, are shown for the indicated atom pairs. These data were obtained by fitting observed couplings from the protein ubiquitin to its X-ray structure [13, 121].

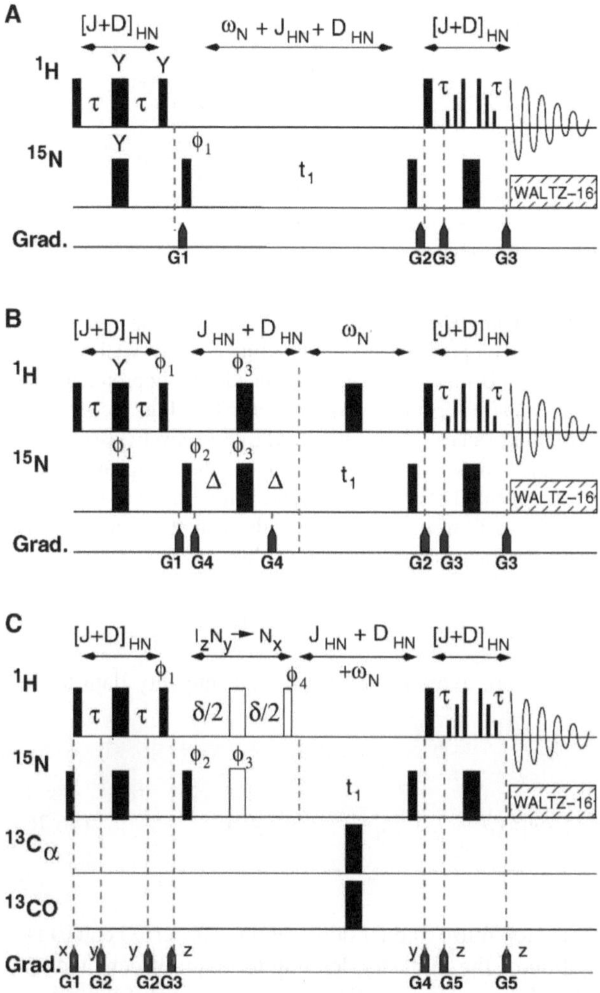

Figure 16.14 Pulse sequences for measuring D_{NH}. A is a ^{1}H-coupled HSQC, B is a J-modulated HSQC. Both A and B are to be used on ^{15}N labeled proteins and are from [157]. C is an inphase(IP)-antiphase(AP)-HSQC experiment from [121]. Narrow and wide bars represent 90° and 180° pulses, respectively. Pulse phases are along x, unless otherwise indicated. All three sequences utilize States-TPPI quadrature detection, water suppression using WATERGATE during the last INEPT period, and WALTZ-16 to decouple nitrogen during proton detection. The delay τ is set to 1/(4J). ***Sequence A:*** Phase cycling is $\phi_1 = x, -x, -x, x; \psi_{rec} = \phi_1$. Quadrature detection is obtained by varying ϕ_1. Gradients are sine-bell in shape, applied along the z-axis, with an amplitude of 25 G/cm. Durations are 2.5(G1), 1.0(G2), and 0.4 msec. (G3), respectively. G1 and G2 are zz-filters and G3 suppresses the water in the WATERGATE sequence.
Sequence B: The delay Δ is varied to modulate the intensity of the peaks by D_{NH}. The two 180° pulses in the center of the Δ period should be coincident, and the total time for evolution under scalar and dipolar coupling is $2\Delta + 2 \times \tau_{90N}/\pi$, where each Δ period includes one-half of the length of the 180° nitrogen pulse. Pulse phases are $\phi_1 = 8(y), 8(-y); \phi_2 = x, -x; \phi_3 = 2(x), 2(y), 2(-x), 2(-y); \psi_{rec} = x, 2(-x), 2(x), 2(-x), x, -x, 2(x), 2(-x), 2(x), -x$. Quadrature detection is obtained by shifting the phase of ϕ_2 and ϕ_3. Gradients are as in A, with the duration of G4 = 2.3 msec.
Sequence C: The carbon pulses are selective for the indicated nucleus, with length $\sqrt{3}/(2\Delta\nu)$, $\Delta\nu$ is the frequency difference between the carbon transmitter and the point of null excitation. The open bars, and the two $\delta/2$ delays generate the AP spectrum, see text. The delay δ is 5.3 ms. Phase cycle $\phi_1 = -y, y; \phi_2 = 2(x), 2(-x)$ for IP, $\phi_2 = 2(-y), 2(y)$, for AP. $\phi_3 = 4(x), 4(y), 4(-x), 4(-y); \phi_4 = 8(x), 8(-x); \psi_{rec} = x, 2(-x), x$ for IP, $\psi_{rec} = x, 2(-x), x, -x, 2(x), -x$ for the AP spectrum. Quadrature detection is obtained by varying either ϕ_2 (IP) or ϕ_2 and ϕ_3 (AP). Gradient are applied along the indicated axis at 25 G/cm with durations 2, 0.4, 2, 1, 0.4 msec, respectively.

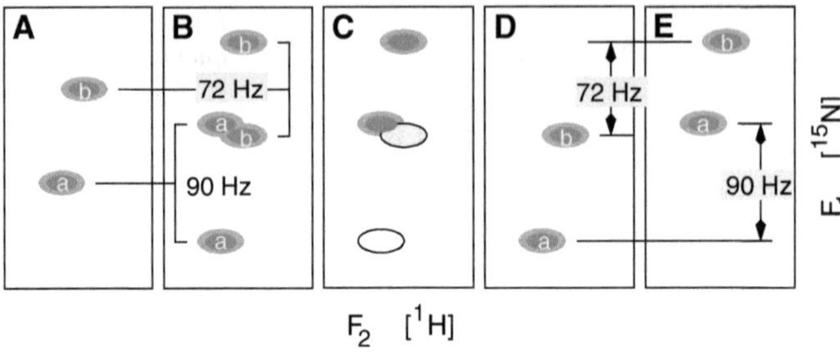

Figure 16.15. IPAP Spectra for Determining Dipolar Coupling. Panel A shows a portion of a standard HSQC of a protein, acquired with proton decoupling during t_1. Two well resolved peaks are shown, labeled 'a' and 'b'. Panel B shows the spectra obtained without proton decoupling, as would be obtained from the in-phase version of pulse sequence C in Fig. 16.14. Since 'a' and 'b' have different dipolar coupling values (-4 Hz and -22 Hz, respectively, assuming a scalar coupling of 94 Hz), their multiplets now overlap. Panel C is the the anti-phase HSQC spectrum obtained from the same experiment. The white ovals represent negative peaks. Panel D shows the spectrum obtained from $B - \kappa C$ while spectrum E is the sum of B and κC. The coupling is obtained by measuring the peak separation between multiplets in each spectra.

The scalar and dipolar couplings are extracted by fitting the intensity data to the following function:

$$I(2\Delta) = I_o cos[\pi(J + D)2\Delta]e^{-2\Delta/T_2}] + A \tag{16.76}$$

Pulse imperfections of the 180° proton pulse in the middle of the 2Δ period can cause systematic errors in the modulation of the intensity, leading to offset effects and apparent changes in the measured coupling. These artifacts are suppressed if the data is sampled over at least a single period of the $cosine$ function, see [157] for more details.

Sequence C was proposed by Ottiger *et al.* [121]. It restores the simplicity of measuring couplings by measuring splittings, but does not double the number of peaks in the spectrum, as in sequence A. This is accomplished by acquiring two independent data sets, in one case the scalar and dipolar coupling lead to in-phase splitting of the resonances in the nitrogen dimension, while in the other data set, this splitting is anti-phase. The addition of these two spectra gives a spectrum that contains one of the doublets while the subtraction of the two spectra gives a spectrum containing the other doublet, as illustrated in Fig. 16.15. This approach is usually referred to as IPAP (in-phase, anti-phase).

The anti-phase component is obtained by including an additional refocusing period in the pulse sequence, as indicated by the open bars in Panel C of Fig. 16.15. The evolution of the density matrix for the anti-phase version of the experiment proceeds as follows, beginning prior to the ϕ_2 nitrogen pulse and following only terms that lead to detectable magnetization after the last INEPT period:

$$\begin{aligned}
2I_zN_z &\overset{P90(-y)}{\longrightarrow} 2I_zN_x \\
2I_zN_x &\overset{\delta/2-180\delta/2}{\longrightarrow} N_y \\
N_y &\overset{t_1(J+D)}{\longrightarrow} -2I_zN_x sin(\pi[J+D]t_1) \\
-2I_zN_x sin(\pi[J+D]t_1) &\overset{t_1(\omega_N)}{\longrightarrow} \\
&- 2I_z sin(\pi[J+D]t_1)(N_x cos\omega_N t_1 + N_y sin\omega_N t_1)
\end{aligned} \tag{16.77}$$

Only the term containing $2I_zN_y$ will result in detectable magnetization, therefore the final signal will be:

$$sin(\pi[J+D]t_1) sin\omega_N t_1 \tag{16.78}$$

The signal obtained with the quadrature phase setting is:

$$sin(\pi[J+D]t_1) cos\omega_N t_1 \tag{16.79}$$

which gives the following complex signal:

$$S_{AP}(t_1) = sin(\pi[J+D]t_1)e^{i\omega_N t_1} \tag{16.80}$$

The Fourier transform of this signal gives a positive peak at $\omega_N + \pi(J+D)$ and a negative peak at $\omega_N - \pi(J+D)$.

The in-phase version of the experiment is just an proton-coupled HSQC experiment, and gives the following complex signal after quadrature detection:

$$S_{IP}(t_1) = cos(\pi[J+D]t_1)e^{i\omega_N t_1} \tag{16.81}$$

the Fourier transform of which are two positive peaks at $\omega_N \pm \pi[J+D]$.

Spectra containing only one component of the doublet are obtained by adding or subtracting the two spectra:

$$S_1 = S_{IP} + \kappa S_{AP} \qquad S_2 = S_{IP} - \kappa S_{AP} \tag{16.82}$$

where κ accounts for the difference in intensity between the two spectra. Intensity differences are due to several factors. First, the magnetization decays during the $\delta/2 - 180 - \delta/2$ period in the anti-phase version, causing a reduction in signal by approximately 5%, assuming a nitrogen T_2 of $\approx$ 100 msec. In addition, the fraction of magnetization that is converted from anti-phase to in-phase during the $\delta/2 - 180 - \delta/2$ period depends on the coupling between the nitrogen and proton, varying as $sin(\pi[J+D]\delta)$. For values of D that are within 10% of J_{NH}, the correction factor is small, within approximately 2-3%. Since both of these effects are small, a single κ value can be assumed for all resonances in the spectrum.

Note that this sequence can also be used to measure the coupling between the nitrogen and the carbonyl or C_α carbon by simply omitting the respective carbon refocusing pulse and decoupling the protons during t_1. The coupling can be measured from the splitting between doublets.

16.3.4 Estimation of the Alignment Tensor

Before the residual dipolar coupling can be used for structure determination it is necessary to determine the alignment tensor. In particular, five unknowns have to be determined: D_a, R, and the three rotation angles that specify the orientation of the molecular coordinate frame with respect to magnetic field direction. In the case of an unknown structure it is not possible to obtain the rotation angles from the measured residual dipolar couplings. Rather, these are obtained during refinement of the protein structure, as described in more detail below. The values for D_a and R are generally required prior to structure calculations, fortunately these can be determined directly from the distribution of observed dipolar couplings [38]. Alternatively, it is also possible to predict these parameters from initial structures, provided the alignment mechanism is due to steric factors (see [179]).

D_a and R are calculated from A_{xx}, A_{yy}, and A_{zz}, which can be determined directly from the distribution of couplings. The observed residual dipolar coupling is related to the alignment tensor, as follows:

$$\begin{aligned}\Delta\nu(\theta,\phi) &= D_{max}\frac{3}{2}\left[A_{xx}sin^2\theta\cos^2\phi + A_{yy}sin^2\theta sin^2\phi + A_{zz}cos^2\theta\right] \\ &= \left[D_{xx}sin^2\theta\cos^2\phi + D_{yy}sin^2\theta sin^2\phi + D_{zz}cos^2\theta\right] \qquad (16.83)\end{aligned}$$

where $D_{ii} = \frac{3}{2}D_{max}A_{ii}$.

The coupling with the largest absolute value corresponds to bond vectors parallel to the z-axis in the molecular frame ($\theta = 0°$), and is D_{zz}. The second largest coupling, which will be of opposite sign, must correspond to bond vectors parallel to the y-axis in the molecular frame ($\theta = 90°, \phi = 90°$) and is therefore D_{yy}. The value of D_{xx}

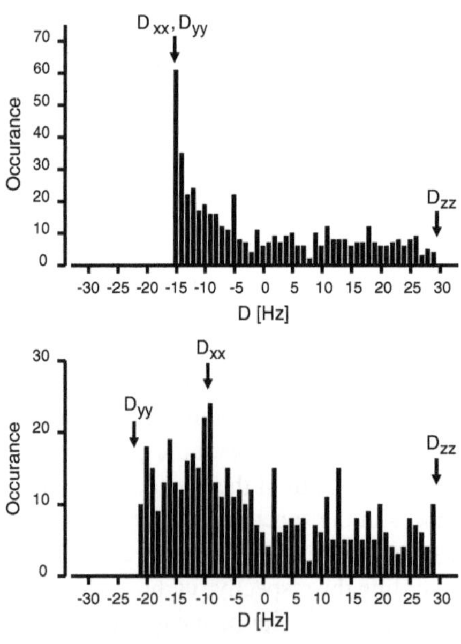

Figure 16.16 Obtaining the alignment tensor. Histograms of dipolar couplings are shown for an axially symmetric molecule (upper histogram) or an asymmetric molecule (lower histogram). A total of 250 random interatomic vectors are represented in each plot.

Axially Symmetric: $D_{zz} = 30$ Hz and $D_{yy} = D_{xx} = -15$ Hz. Assuming $D_{max} = 20$kHz, $A_a = 30/20 \times 10^3 = 1.5 \times 10^{-3}$.

Non-Axially Symmetric: $D_{zz} = -30$ Hz and $D_{yy} = -22$ Hz. The most frequently observed coupling is -8 Hz, giving D_{xx}. A_a is the same as above and the rhombicity is:

$$R = \frac{2}{3}\frac{(-8-(-22))}{+30} = \frac{2}{3}\frac{14}{30} = 0.31$$

can be found using two methods. First $D_{zz} + D_{yy} + D_{xx} = 0$, therefore if the first two terms are known, then the third can be calculated. Second, the most frequently observed dipolar coupling will correspond to D_{xx} because there are more orientations of the inter-nuclear bond vector that can give this coupling than any other orientation. The axial component of the alignment tensor is:

$$A_a = \frac{3}{2}A_{zz} = \frac{D_{zz}}{D_{max}} \tag{16.84}$$

and the rhombicity is:

$$R = \frac{A_r}{A_a} = \frac{A_{xx} - A_{yy}}{A_a} = \frac{2}{3}\frac{D_{xx} - D_{yy}}{D_{zz}} \tag{16.85}$$

The measured couplings are plotted as histograms, i.e. the number of occurrences for each measured coupling. Figure 16.16 illustrates the appearance of the histogram for an axially symmetric and a asymmetric molecule. Values of the alignment tensor that can be obtained from these distributions are given in the figure legend.

In order to be sure that the largest, smallest, and most frequent values are present in the data, it is advantageous to use couplings measured from as many different bond vectors as possible. If the measured couplings involve different atom types, ie H-N and N-C_α, then it is necessary to scale each type of coupling to that of the N-H dipolar interaction prior to analysis using the factors given in Table 16.1. For example, the measured C_α and H_α couplings would be reduced by a factor of 2.020 before combining them with the NH couplings.

Chapter 17

PROTEIN STRUCTURE DETERMINATION

Structure determination generates a molecular model of the protein or nucleic acid that is as consistent as possible with both the experimental data and known covalent and non-covalent features of the folded biopolymer. The most commonly used experimental constraints are:

- Inter-proton distances derived from NOESY experiments.
- Bond orientations determined from single bond residual dipolar couplings.
- Torsional angles from measurements of three bond J-couplings.
- Hydrogen bonds determined from amide exchange data.
- Peptide mainchain torsional angles from chemical shifts.

Non-experimental constraints consist of:

- Bond lengths.
- Bond angles.
- Torsional angles.
- Van der Waals Interactions.

The relative contribution of experimental and non-experimental constraints to

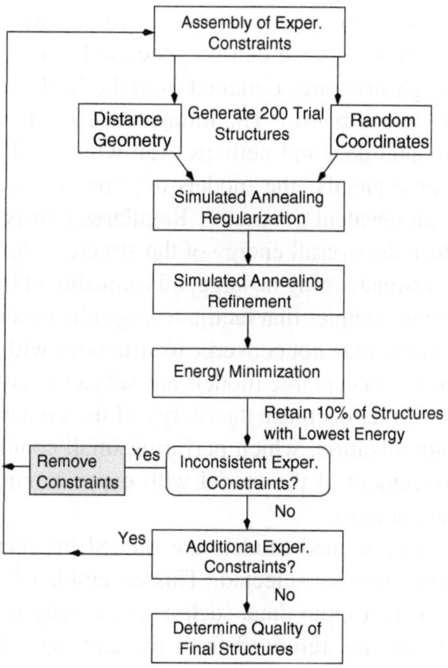

Figure 17.1. Overview of the structure determination process. Structure determination consists of a number of cycles that begin with the assembly of constraints, followed with building and refining models derived from the constraints, and then interpretation of the models to resolve errors or ambiguities in the experimental data.

the final structure are balanced by assigning an energy to both types of constraints and weighting the relative contribution of each type of constraint in an empirical manner:

$$E_{Total} = \kappa E_{Experimental} + E_{Non-experimental} \tag{17.1}$$

where κ is an empirical scaling factor.

Higher energies are associated with models whose structures gives the largest disagreement between the constraints and the structure. The refinement process seeks to create a structure that gives the lowest energy, and is therefore in best agreement with *both* experimental and non-experimental constraints.

The overall steps in structure determination are illustrated in Fig. 17.1 and outlined briefly in the following text. The first task is to assemble a collection of reliable structural constraints, these generally include inter-proton distances between protons that have unambiguous assignments. In addition, information on torsional angles from three-bond scalar (^{3}J) couplings can be utilized for the construction of the initial trial structures. Residual dipolar couplings, as well as hydrogen bonding information, are usually introduced during the latter stages of model building, when the structure is approaching its final form.

Following the assembly of constraints, 100 to 200 initial models are built for refinement. These can be generated from completely random atomic coordinates, or rough structures obtained from the NOE data via distance geometry, both methods are discussed below. The initial models generally show poor agreement with the experimental data and perhaps even with standard covalent and non-covalent interactions. Consequently, the models are "regularized" to produce structures that are constant with covalent geometry. Regularization is accomplished by moving the atoms to reduce the overall energy of the structure. Since some of the required changes in atomic coordinates may be large, this adjustment is usually performed using simulated annealing techniques that facilitate large changes in atomic coordinates. A number of the trial models may not converge to structures with acceptable energy and are discarded at this stage. Acceptable models are subject to additional refinement by simulated annealing to further decrease the energy of the system. The refinement is concluded with energy minimization, which performs small changes in atomic coordinates to maximize the agreement of the model with experimental data, as well as bonded and non-bonded interactions.

The refined models are ranked by energy and 5% to 10% of the lowest energy structures are selected. This ensemble of structures is inspected carefully to identify incorrect input data, such as incorrectly assigned NOEs, and such experimental constraints are removed from the data set. The ensemble of structures is also used to resolve ambiguities with the existing data, allowing the inclusion of more constraints in the next round of structure building. For example, in the case of inter-proton distances from NOE measurements, a proton (A) with a well resolved chemical shift may show an NOE to another proton (B) whose chemical shift is degenerate with a third proton (C). In the absence of a structure it is not possible to determine if the inter-proton distance corresponds to A-B or A-C. However, the ensemble of low-energy structures may make the choice clear by comparing the distances predicted from the model structures. Intermediate structures may also be useful in identifying hydrogen bond acceptors.

After the addition of more constraints, the entire process is repeated until all experimental constraints have been exhausted. The overall quality of the ensemble of lowest energy structures is then evaluated for compliance with covalent and non-covalent energy terms as well as for agreement to the experimental data.

17.1 Energy Functions

Several software packages are available for structure determination. The discussion of energy functions and refinement schemes is based on the X-PLOR package, which was originally developed by A. Brünger [26] for solving structures using both NMR and X-ray crystallography. X-PLOR has been recently updated by Tjandra *et al.* to include refinement using residual dipolar coupling as well as other experimental NMR constraints [145].

17.1.1 Experimental Data

17.1.1.1 NOE Constraints

The intensity of a crosspeak in NOESY spectra is related to the distance between the interacting protons. However, there is considerable uncertainty associated with converting the NOE peak intensities to distances. Consequently, it is common practice to specify both a lower bound (d_{lower}) and an upper bound (d_{upper}), with the assumption that the true distance lies between the two bounds. The upper and lower bounds are generally determined from the signal-to-noise ratio in the spectrum. For example, if the uncertainty in peak intensity is ΔI, and the measured intensity is I, then $d_{lower} \propto 1/(I + \Delta I)^6$ and $d_{upper} \propto 1/(I - \Delta I)^6$. Alternatively, the lower bound is often specified as the van der Waals radii of the atom.

X-PLOR provides several different energy functions for inter-proton distances. Two commonly used functions are described below and plotted in Fig. 17.2. One

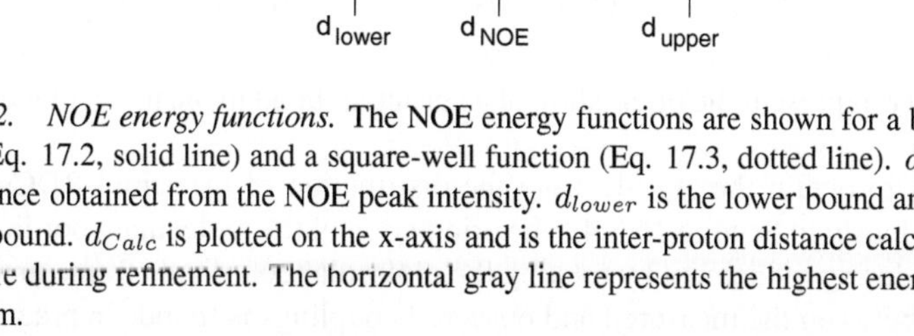

Figure 17.2. NOE energy functions. The NOE energy functions are shown for a bi-harmonic function (Eq. 17.2, solid line) and a square-well function (Eq. 17.3, dotted line). d_{NOE} (4 Å) is the distance obtained from the NOE peak intensity. d_{lower} is the lower bound and d_{upper} is the upper bound. d_{Calc} is plotted on the x-axis and is the inter-proton distance calculated from the structure during refinement. The horizontal gray line represents the highest energy possible for this term.

function is the biharmonic function:

$$E = K_{NOE}\frac{kT}{2c^2}(d_{\ Calc} - d_{NOE})^2 \tag{17.2}$$

where K_{NOE} is a user defined scaling factor, k is Boltzmann constant, T is the absolute temperature (°K) during the simulated annealing process, and $c = (d_{NOE} - d_{lower})$ if $d_{\ Calc} < d_{NOE}$, or $c = (d_{upper} - d_{NOE})$ if $d_{\ Calc} > d_{NOE}$. The factor, $kT/2c^2$, is not allowed to exceed a user specified level during the refinement process. This maximum in energy prevents incorrectly assigned NOE peaks from distorting the structure if the true distance between the protons that define the NOE is much larger than the distance assumed from the NOESY measurement.

The second function is a simple square-well with harmonic sides, defined as following:

$$E = K_{NOE}\Delta^{exp} \tag{17.3}$$

$$\begin{aligned} \Delta &= d_{calc} - (d_{upper} - d_{off}) && if\ d_{upper} - d_{off} < d_{calc} \\ &= 0 && if\ d_{lower} < d_{Calc} < d_{upper} - d_{off} \\ &= d_{lower} - d_{Calc} && if\ d_{Calc} < d_{lower} \end{aligned}$$

where the exponent, exp, is defined by the user, a typical value would be $exp = 2$. d_{off} allows global adjustment of the width of the region where E is zero, see Fig. 17.2. This energy function essentially assumes that all distances between d_{lower} and d_{upper} are acceptable in the final refined structure.

17.1.1.2 Residual Dipolar Coupling

The contribution of residual dipolar couplings (RDC) to the overall energy is:

$$E_{RDC} = \sum_{i=1}^{n_{RDC}} K_{RDC}(\Delta\nu_i^{Calc} - \Delta\nu_i^{Expt})^2 \tag{17.4}$$

where $\Delta\nu_i^{Calc}$ is the coupling calculated from the model, $\Delta\nu_i^{Expt}$ is the experimentally measured splitting, and the sum is over all observed dipolar couplings (n_{RDC}).

In order to compare calculated coupling to measured couplings it is necessary to know the orientation of the molecular coordinate system with respect to the magnetic field. In addition, the extent of alignment of the protein or nucleic acid (A_a, A_r) must also be known. If these are known then the expected dipolar coupling can be calculated from the molecular structure using the following equation:

$$\Delta\nu(\theta,\phi)^{Calc} = D_a\left[(3cos^2\theta - 1) + \frac{3}{2}Rsin^2\theta cos2\phi\right] \tag{17.5}$$

where θ and ϕ represent the orientation of a particular bond in the molecular coordinate system.

Methods to obtain the A_a, A_r from the distribution of measured RDCs was discussed previously in Section 16.3.4. The orientation of the molecular coordinate frame is obtained during refinement by rotating the molecular coordinate frame until the best agreement between the measured and observed couplings is found. In practice, this is accomplished in a clever manner by including a separate pseudo-molecule along with the molecule under refinement. This pseudo-molecule contains four atoms, one at the origin, and the other three at $x = 1$, $y = 1$, and $z = 1$, which represent the molecular coordinate system. During refinement the pseudo-molecule is free to rotate and will assume an orientation that minimizes the difference between the calculated and measured dipolar couplings, giving the orientation of the molecular axis system with respect to the magnetic field [48].

17.1.1.3 Torsional Angles

During the course of refinement, the expected three-bond J coupling, ${}^3J^{Calc}$ can be calculated from a torsional angle in the structure. Most 3-bond coupling constants, such as the coupling between the amide proton and the α-proton, are related to the torsional angle by the Karplus relationship [80]:

$$ {}^3J = A\cos^2\theta + B\cos\theta + C \tag{17.6} $$

The deviation of the calculated J-coupling constant from the measured constant contributes to the overall energy function as:

$$ E_J = \sum_{i=1}^{n_J} K_J (J_i^{Calc} - J_i^{Expt})^2 \tag{17.7} $$

where the sum is over all observed J-couplings, K_J is an empirical weighting factor, J_i^{Calc} is the coupling constant calculated from the torsional angle in the model (using Eq. 17.9), and J_i^{Expt} is the experimentally observed coupling constant.

In earlier versions of X-PLOR, the experimentally measured torsional angle was explicitly given, in which case the energy function is:

$$ E_{tor} = \sum_{i=1}^{n_{tor}} K_{tor} (\theta_i^{Calc} - \theta_I^{Expt})^2 \tag{17.8} $$

In this case, the torsional angle that most likely corresponds to the observed coupling constant would be specified by the user. For example, a $J_{H_N H\alpha}$ coupling of 4 Hz implies a torsional angle, ϕ of -60° (see Fig. 17.3).

Common Coupling Constants Used in Refinement: The J coupling between the H_N to the Hα proton is perhaps the most useful coupling constant because it provides information on the peptide backbone configuration (ϕ angle). This coupling is also relatively easy to measure because there is no passive coupling of the NH proton to other protons (except for Gly). Therefore COSY crosspeaks can be used to directly determine the coupling constant by measuring the splitting of the crosspeaks at the frequency of the amide proton. Accurate measurements of the coupling constant cannot be obtained from the splitting of COSY crosspeak unless the linewidth is less than the coupling constant ($\Delta\nu < J$), which is only true for proteins that contain 50 or fewer residues. For larger proteins, reliable couplings can be obtained by spectral simulation of the ω_2 slices, as described by Yang and Havel [175]. Also for larger proteins, TOCSY spectra can be used to obtain the J-couplings since the crosspeaks are

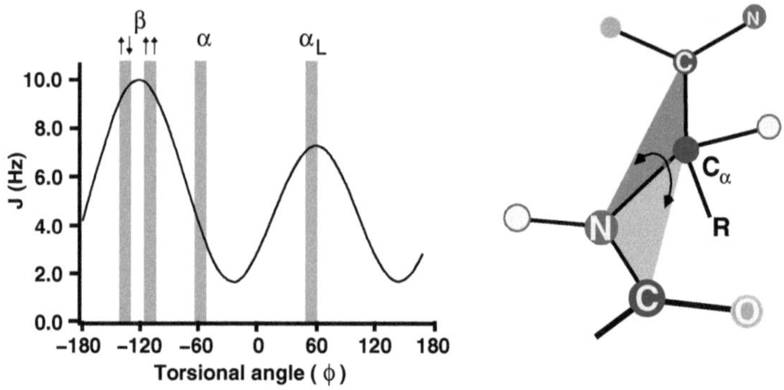

Figure 17.3. Relationship between secondary structure and $J_{N_H H\alpha}$. The observed ^{3}J-coupling versus the torsional angle, ϕ is shown. ϕ angles corresponding to common secondary structures are indicated by the vertical bars. The angle θ in Eq. 17.9 equals $\phi - 60°$. The ϕ torsional angle is the angle between the planes defined by C-N-C_α and N-C_α-C, as illustrated on the structure shown in the right-hand section of this figure.

in-phase doublets that can be fit to two Lorentzian lineshapes. Finally, if labeling with ^{15}N is feasible, the $J_{H_N H\alpha}$ coupling can be measured using the HNHA experiment, as discussed in Section 13.3.3.

The relationship between the observed coupling constant and the peptide ϕ angle is given by the following parametrized Karplus relationship:

$$^3J_{H_N H_\alpha} = 6.51\cos^2\theta - 1.76\cos\theta + 1.60 \tag{17.9}$$

where $\theta = \phi - 60°$, and the constants A, B, and C have been substituted with values determined by Vuister and Bax [162]. A number of different values for these constants have been obtained by other investigators [124] which give similar values for $^3J_{H_N H\alpha}$. The Karplus curve for $^3J_{H_N H_\alpha}$ is shown in Fig. 17.3.

Note that a single value of the ^{3}J-coupling constant can correspond to as many as four distinct values of ϕ, thus a measured coupling constant may not specify a unique value for ϕ. Furthermore, the observed coupling constant may be averaged by rotation about the $N - C_\alpha$ bond at a rate that is faster than $1/J$. For example, if an amino acid residue samples an α-helical configuration and a β-strand configuration with equal probability, then the observed coupling will be approximately 7 Hz, or the average of 10.0 Hz(β) and 4.0 Hz (α). Therefore, J-coupling constants that are approximately 7.0 Hz are not used in the initial model building because it is unclear whether it represents a single conformation with $\phi = -70°$ or is a result of conformational averaging. Once the conformation of the residue becomes established during refinement it may be possible to utilize these J-coupling constants as valid constraints for model building.

In addition to measuring $J_{H_N H_\alpha}$ it is often possible to measure the coupling between the H_α and H_β protons, defining the χ_1 torsional angle. This coupling can be obtained from either the fine structure of COSY crosspeaks or from the splitting in TOCSY crosspeaks. Alternatively, since the rate of magnetization transfer in a TOCSY experiment is proportional to $cos(\pi J \tau_{mix})$, the coupling constant can be estimated from the dependence of the crosspeak intensity on the mixing time. The

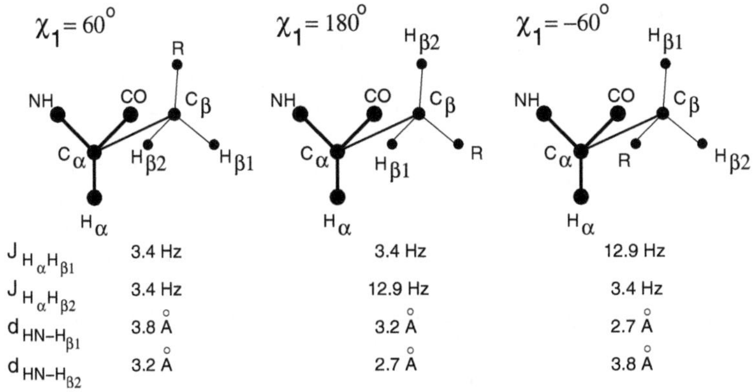

Figure 17.4. *Defining the χ_1 torsional angle.* The use of NOE information to define the $H_\alpha - H_\beta$ torsional angle. The three low energy conformations of the C_α-C_β bond are shown. The expected values for $J_{H_\alpha H_\beta}$ and the distance between the H_N and the H_β protons are given below each conformation in the diagram. A unique conformation of the C_α-C_β torsional angle can be defined by combining information from J-coupling and NOE measurements. The inter-proton distances are given for a residue a β-strand configuration. Slightly different distances would be obtained for an α-helical configuration. The naming of the β protons follows the nomenclature used in X-PLOR. The conversion from the XPLOR format to the IUPAC format [107] is as follows: is $H_{\beta 1} \rightarrow H_{\beta 2}$, and $H_{\beta 2} \rightarrow H_{\beta 3}$.

Karplus curves for H_α-H_β coupling are given by the following equations [103]:

$$\begin{aligned} J_{H_\alpha H_\beta 1} &= 9.5\cos^2(\chi_1 - 120°) - 1.6\cos(\chi_1 - 120°) + 1.8 \\ J_{H_\alpha H_\beta 2} &= 9.5\cos^2\chi_1 - 1.6\cos\chi_1 + 1.8 \end{aligned} \tag{17.10}$$

A minor difficulty with using this coupling constant to define this torsional angle arises from the fact that there are usually two β protons on a residue, and it is necessary to know stereo-specific assignments before the coupling constants can be used to define the correct torsional angle. However, measurement of the NOE between the amide proton and the H_β protons can often resolve this issue, as illustrated in Fig. 17.4.

17.1.1.4 Hydrogen Bonding

Amide protons that participate in hydrogen bonds are usually identified by virtue of slow amide hydrogen exchange rates as well as a small temperature dependence of the amide proton chemical shift (see Fig. 17.5).

The exchange rates are readily measured by replacing the solvent with D_2O and measuring the decrease in the intensity of amide proton resonances. The exchange rate follows first-order kinetics and the rate constant is obtained by fitting the peak intensity, $I(t)$, to the following equation:

$$I(t) = I_o e^{-k_{ex}t} \tag{17.11}$$

The amide exchange rates are both acid and base catalyzed, and pH dependence of the exchange rate, for a number of exchangeable protons, is presented in Fig. 15.2.

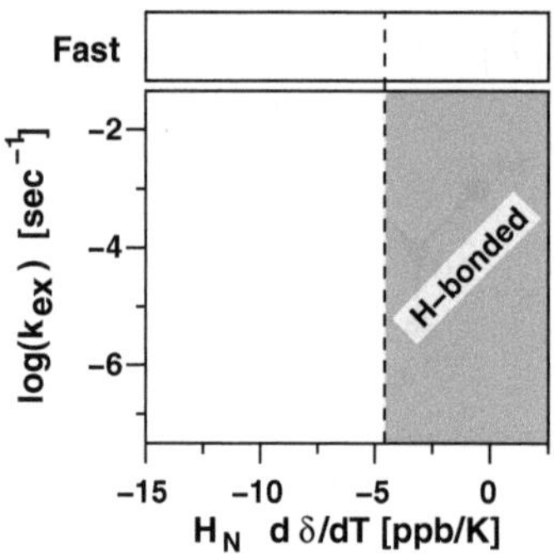

Figure 17.5 Identification of hydrogen bonds. The amide exchange rate versus the temperature dependence of the amide proton chemical shift. The range of each axis is typical for a folded globular protein at pH 5.0. An exchange rate of "fast" indicates that it was not possible to measure the rate due to rapid loss of the amide proton in D_2O. Amides that show an amide exchange rate less than 10^{-2}/sec and temperature dependence smaller than -4.5 ppb/K are likely to participate in the formation of a hydrogen bond. Amide protons that show a slow exchange rate but have a large temperature dependence of chemical shift may not be hydrogen bonded. Fig. adapted from [16].

Generally, if the observed exchange rate is 10 fold slower than the rate expected for a fully exposed amide it is reasonable to assume that the amide is *either* involved in a hydrogen bond or it is simply buried in a hydrophobic region of the protein. The participation of the amide proton in an hydrogen bond can be further substantiated by the temperature dependence of the amide proton chemical shift. Temperature coefficients that are less negative than $\approx$ 4.5 ppb/K are indicative of hydrogen bond formation [16], as illustrated in Fig. 17.5.

If a suitable hydrogen bond acceptor can be identified in preliminary models, for example a C=O group, then constraints involving the N-H and C=O atoms can be added to collection of constraints in the next round of structure generation.

In some cases it may be possible to unambiguously determine the presence of a hydrogen bond by detecting scalar coupling between the amide proton and the carbonyl carbon that are involved in the hydrogen bond. Since this coupling is quite weak, leading to long magnetization transfer times, experiments of this type are generally more successful with smaller proteins (e.g. < 20 kDa) because of their longer T_2 values [45, 164].

In terms of energy calculations, hydrogen bonding constraints are often represented as a pair of inter-proton distances, i.e. the distance between the amide proton and carbonyl oxygen (d_{H-O}) and the distance between the nitrogen and the carbonyl carbon (d_{NC_O}) to insure linearity of the hydrogen bond, giving the following energy function:

$$E_{h-bonds} = \sum_{i=1}^{n_{hbonds}} K_{H-bond} \left[\sum_{j=1}^{2} (d_{ij}^{Calc} - d_{ij}^{Expt})^2 \right] \tag{17.12}$$

where $j = 1$ represents the H-O distance and $j = 2$ represents the N-C distance and K_{H-bond} is an empirical scaling factor. K_{H-bond} is often set to the same value of K_{NOE} during refinement.

17.1.1.5 Chemical Shift Constraints

The chemical shift of backbone atoms, in particular the C_α, and carbonyl carbon, and to a lesser extent C_β, depend on the secondary structure of a residue, as indicated in Fig.17.6 [167]. Although many tertiary interactions also affect the chemical shift of these atoms, the change in chemical shift that is induced by the secondary structure

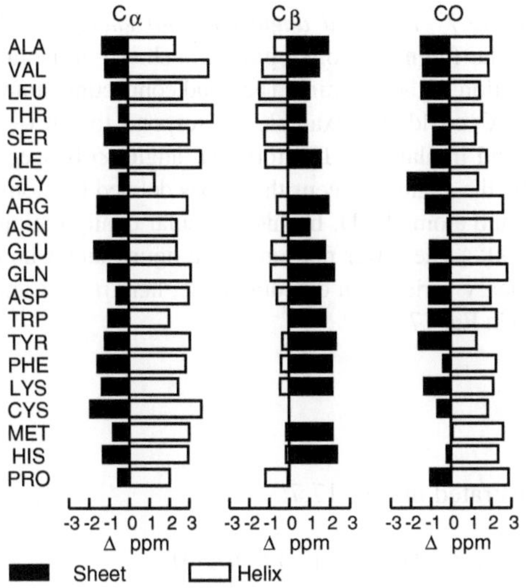

Figure 17.6 Effect of secondary structure on carbon chemical shifts. The deviation from the mean chemical shift of the C_α, C_β, and CO atoms for each residue type is illustrated by each horizontal bar. Filled bars represent residues in β-strand configuration while open bars represent residues in α-helical configuration. As an example, the chemical shift for the C_α of Ala is 52.42 ppm in a random coil, 51.15 ppm in a β-sheet, and 54.77 ppm in a helix. Therefore the change in chemical shift due to secondary structure is -1.27 ppm and 2.35 ppm from random coil for helix or sheet, respectively. Data from [52].

can provide a weak constraint during refinement (see [92]). For example, if an alanine residue in a protein showed C_α and CO shifts that were 3 ppm below the mean chemical shift for these atoms, then the ϕ and ψ torsional angles could be constrained to favor a β-strand conformation.

17.1.2 Covalent and Non-covalent Interactions

The energy associated with covalent and non-covalent interactions are defined by the following terms:

$$E_{covalent} = E_{bonds} + E_{angles} + E_{torsional} + E_{improper}$$
$$E_{non-bonded} = E_{van\ der\ Waals} + E_{Electrostatic} \qquad (17.13)$$

The covalent energy terms insure proper covalent bonding and molecular structure, including planarity of aromatic groups and the correct geometry of chiral centers.

Bond Lengths: Proper inter-atomic bond lengths are maintained during refinement with the following energy term:

$$E_{bonds} = \sum_{i}^{n_{bonds}} K_{bonds}(d_i^{Calc} - d_i^{Ideal})^2 \qquad (17.14)$$

where n_{bonds} are the number of covalent bonds in the structure, K_{bonds} is an empirical scale factor, d_i^{Calc} is the bond length calculated from the structure during refinement and d_i^{Ideal} is the ideal covalent bond length.

Torsional Angles: Torsional angles can also contribute to the energy function. They are used to maintain the planarity of aromatic rings, favor non-eclipsed configuration of atoms, and to define the geometry of chiral centers. In the case of X-PLOR, two energy functions can be used to specify torsional angles, the first of which is usually

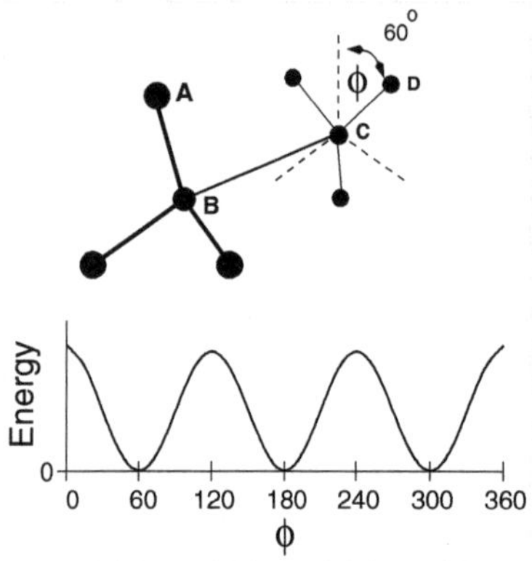

Figure 17.7 Torsional angle potential energy. A molecular fragment from a protein is shown in the upper section of the diagram. The bond connecting atoms B and C could, for example, correspond to the C_α-C_β bond in alanine. The torsional angle, ϕ is specified by the angle betweens the planes defined by atoms ABC and atoms BCD. In this particular configuration, $\phi = 60°$. The lower portion of the figure shows how the energy varies with dihedral angle, here $n = 3$ and $\delta = 0$ in Eq. 17.15.

used to insure non-eclipsed atoms, as illustrated in Fig. 17.7:

$$\begin{aligned} E_{torsional} &= \sum K_\phi \left[1 + cos(n\phi + \delta)\right] \;\; if \;\; n > 0 \\ &= \sum K_\phi (\phi - \delta)^2 \;\; if \;\; n = 0 \end{aligned} \quad (17.15)$$

where ϕ is the torsional angle calculated from the structure at some point during refinement, K_ϕ is the weighting factor, n is the multiplicity, and δ is the phase shift.

To specify that four atoms lie in a plane, such as in an aromatic ring, n would be set to zero and δ would be 180°. Any out of plane configurations would raise the energy due to a non-zero value of $(\phi - \delta)$. To specify three equally populated rotomers, for example a CH_3 group, a value of $n = 3$ and $\delta = 0°$ would be used, giving energy minima for torsional angles, $\phi = 60°, 180°$, and 300°, as illustrated in Fig. 17.7.

The second energy expression for torsionals angles is referred to as the *improper* energy term. It has exactly the same form as in Eq. 17.15 and is generally used to maintain chirality and planarity of groups within in the structure. The availability of two distinct potential functions permits a use of different scale factors for the two types of torsional angles.

van der Waals Interactions: The non-bonded energy term contains contributions from van der Waals interactions, usually encodes a pairwise standard 6-12 Lennard-Jones potential:

$$E_{vdw} = K_{vdw} \sum_{ij}^{n_{atom}} \frac{C_{12}}{d_{ij}^{12}} - \frac{C_6}{d_{ij}^6} \quad (17.16)$$

Electrostatic Energy: The simplest form of the electrostatic energy term is given by Coulomb's Law. However, given the uncertainty of the local dielectric constant as well as the absence of solvent and counter ions in most structure refinement protocols, this term is usually set to zero.

17.2 Energy Minimization and Simulated Annealing

Structures are refined by a combination of energy minimization and simulated annealing. The overall goal is to alter the atomic coordinates of the structure to attain a final set of atomic coordinates that give the lowest energy for both experimental and non-experimental energy functions.

17.2.1 Energy Minimization

The minimum energy of the structure can be found by moving the atoms in the direction defined by the gradient of the energy:

$$\xi_i = -\frac{\partial E}{\partial x_i} \tag{17.17}$$

During the energy minimization, multiple steps of adjustment of the atomic coordinates occur. The coordinate change at each step is calculated according to the following:

$$x_i' = x_i + \xi_i \tag{17.18}$$

ξ_i is recalculated after each step in the minimization process and becomes smaller and smaller as the system moves towards the minimum in energy. The minimization proceeds for either a set number of cycles or until ξ_i drops below a predetermined level.

If the energy function is smooth and has a single global minimum, then minimization will find the true global minimum and produce a structure that is as consistent as possible with the energy function. Unfortunately, the energy surface as a function of atomic coordinates is complex and multi-valued such that a simple minimization of the energy will inevitably reach a local minimum, not the true global minimum, as illustrated in Fig. 17.8.

17.2.2 Simulated Annealing

Simulated annealing is used to overcome the problem of the structure becoming trapped in a local energy minimum. This procedure receives its name because it simulates the annealing process in alloy formation in metals. Specifically, the metal is heated to high temperatures to facilitate atomic rearrangements and then cooled or annealed to more stable structures.

In the refinement of models the atoms in the protein are given a kinetic energy, as defined by the temperature of the system:

$$\frac{1}{2}\sum_{i=1}^{N} m_i v_i^2 = \frac{3}{2} N k_b T \tag{17.19}$$

Initially, the atoms are assigned a random velocity that depends on the temperature of the system [26]:

$$v = \left[\frac{m}{2\pi k_b T}\right]^{3/2} e^{-3m\delta^2/2k_bT} \tag{17.20}$$

where δ is a random number from 0 to 1, T is the temperature of the system, k_b is Boltzmann's constant, and m is the mass of the atom.

Since simulating annealing generally begins at high temperatures, the atoms will have high kinetic energy, and will be able to transverse the energy barrier between minima, as illustrated in Fig. 17.8. To insure the the system will converge on a minimum, the temperature of the system is slowly lowered at the end of the molecular mechanics calculation. Provided the temperature is lowered slowly, it is very likely that the system will anneal to a global energy minimum.

The motion, or trajectory, of the atoms during simulating annealing are determined by molecular mechanics calculations. Given a set of initial coordinates and velocities, x_o and v_o, as well as the energy of the system, the coordinates at a time Δt are calculated using Newtonian mechanics using the following expression:

$$x' = x_o + v_o \Delta t - \nabla E \frac{\Delta t^2}{2m} \quad (17.21)$$

where $-\nabla E$ is equal to the force applied to the atoms. E represents all, or a subset, of the experimental, covalent, and non-covalent energy terms discussed above. Generally, the time step, Δt is a fraction of a psec and 50-200 steps are performed at any given temperature. Multiple cycles of molecular mechanics are usually performed with any given refinement protocol. Each cycle will generally use different energy scaling factors as well as a number of other parameters, such as the van der Waals radii of atoms. Several examples of refinement protocols are presented below.

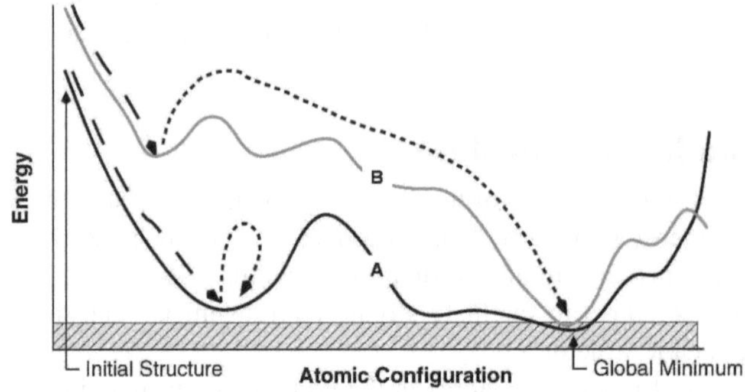

Figure 17.8. Energy changes during simulated annealing. The solid lines shown the energy of a structure as a function of its atomic coordinates. Two trial structures are shown, drawn in black (structure A) or gray (structure B). The energy of the structures immediately after generation by distance geometry or starting from random coordinates are shown on the far left of the plot. The dashed lines shows the change in energy due to regularization followed by energy minimization. Both structures reach a local minimum with reasonable covalent geometry after regularization. The dotted lines show the changes in energy that occur during additional refinement by simulated annealing. In the case of structure B (gray), the energy barriers between each local minimum can be transversed due to the high kinetic energy of the atoms during annealing, thus B eventually finds the global minimum in energy. The energy barriers surrounding the local minimum for structure A are too high, thus structure A is found at the local minimum after refinement. The gray stippled region indicates the range of energies that are considered to be acceptable after refinement; only a very small subset of all possible atomic configurations possess the indicated range of energies.

17.3 Generation of Starting Structures

17.3.1 Random Coordinates

In this case the trial structures are obtained by first generating a random distribution of atoms. The advantage of using random coordinates as starting trial structure is that a sizable portion of conformational space is sampled, increasing the likelihood that all structures that are consistent with the experimental data will be found. The disadvantage of this approach is that only a small number of the trial structures will be give acceptable agreement with the experimental data. Consequently, it is necessary to begin each cycle of refinement with a large number of trial structures.

A simulated annealing scheme to regularize structures after generating random coordinates is shown in Fig. 17.9. In this process the scaling factors for E_{NOE}, E_{Bond}, E_{angle}, and E_{vdw} are set to low values to allow the atoms in the structure to essentially move freely in three dimensional space, establishing positions that are defined by the inter-proton distances from NOE measurement. The scaling factor for bonds is raised from 0.00005 to 0.01 during repeated cycles of molecular dynamics, to establish the correct bonding pattern in the protein. After the scaling factor for bonds reaches 0.01, three molecular dynamics calculations are performed. During these calculations the scaling factor for bonds, bond angles, and van der Waals forces are gradually increased. The structures that are produced by this scheme will generally require additional regularization, using the procedure outlined in the following section on distance geometry.

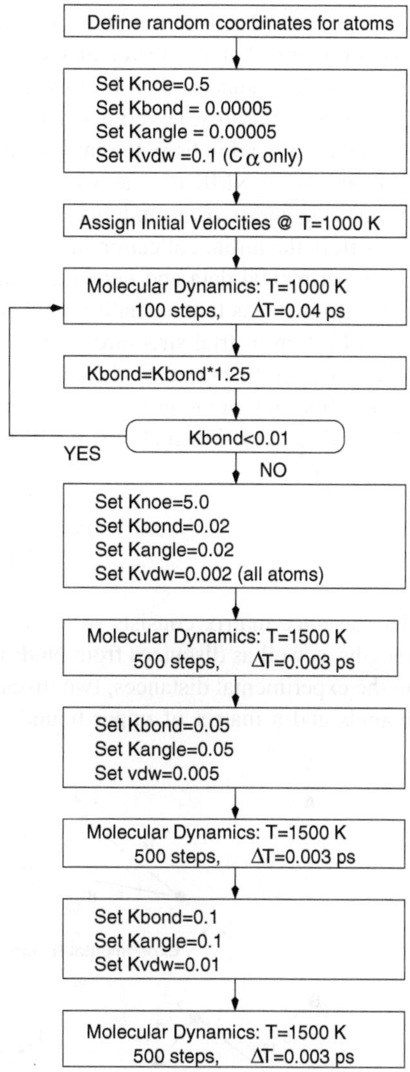

Figure 17.9. Simulated annealing from random coordinates. The simulated annealing scheme to proceed from a random collection of atoms to rough structures is shown. The flowchart was based on the *random.inp* script described by Brünger [26].

17.3.2 Distance Geometry

Distance geometry can also be used to generate trial structures for refinement purposes. This approach is based on the premise that any three dimensional structure can be defined as a set of inter-atomic distances. Therefore, a set of inter-atomic distances can be converted, or *embedded* into three dimensional space, to give the atomic coordinates of the protein. The process of converting distances to coordinates is associated with a branch of mathematics called distance geometry, hence its name. Note that structures that are mirror images of each other will have exactly the same set of inter-atomic distances, therefore structures generated by distance geometry can be of the opposite chirality, i.e. composed of D-amino acids.

In the application here, the inter-atomic distances that are used include inter-proton distances from NMR data as well as known bond lengths. The advantage of using distance geometry over random coordinates is that a larger number of candidate structures from the initial collection of trial structures will be in reasonable agreement with the experimental data and known covalent bond distances. The disadvantage of distance geometry is that a smaller number of different conformations will be present in the collection of trial structures. Therefore, there is a small possibility that the final structures will be biased by the constraints that are used to build the initial models.

The first step in the process is to generate a matrix, the distance matrix, with inter-atomic distances. For n atoms, the distance matrix (D) will be an $n \times n$ matrix, e.g. for 3 atoms:

$$D = \begin{bmatrix} d_{11} & d_{21} & d_{31} \\ d_{12} & d_{22} & d_{32} \\ d_{13} & d_{32} & d_{33} \end{bmatrix} \tag{17.22}$$

The distance matrix consists of the known distances associated with covalent bond lengths as well as distances from NOE measurements. To account for the uncertainty in the experimental distances, two distances matrices are generated, a matrix of lower bounds and a matrix of upper bounds. This matrix is often 'smoothed' by adjust-

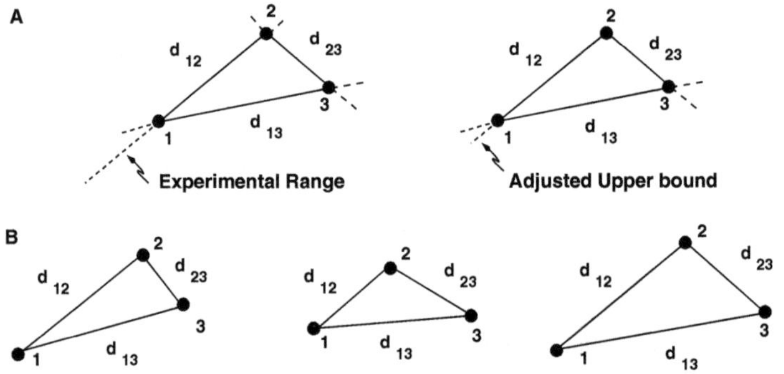

Figure 17.10. Smoothing of the distance matrix. Panel A shows how the distance matrix is smoothed by altering upper and lower bounds such that inter-proton distances are consistent. The dotted lines indicate the upper bound associated with each inter-proton distance. In this example the original upper bound associated with d_{12} (left triangle) is much longer than possible, given the values of d_{13} and d_{23}. Consequently d_{12} is decreased, or smoothed, as shown in the right triangle. Panel B shows three random distances that were generated from the smoothed distances. Each of these three distances will generate slightly different starting structures for refinement.

ing upper and lower bounds that are inconsistent with other inter-proton distances, as illustrated in Fig. 17.10.

After smoothing of the upper and lower bound matrices, the distance matrix that will give rise to one trial structure is generated by selecting a random distance that lies between the lower and upper experimental distances, as illustrated in panel B of Fig. 17.10. After the distance matrix is generated it is necessary to verify that it can be be embedded in a three dimensional space in order to determine the atomic coordinates. Not all distance matrices are embeddable in a three dimensional space. For example, the following representation of a tetrahedron is not embeddable in a *two*-dimensional space, but can be embedded in a three dimensional space:

$$D = \begin{bmatrix} 0 & 1 & 1 & 1 \\ 1 & 0 & 1 & 1 \\ 1 & 1 & 0 & 1 \\ 1 & 1 & 1 & 0 \end{bmatrix} \tag{17.23}$$

In general, if it can be proved that any four points satisfy the quadrangle inequality (i.e. the inter-atomic distances can generate a "real" tetrahedron) then the entire matrix can be embeddable. Usually, if all triplets of points satisfy the triangle equality:

$$d_{12} + d_{23} \geq d_{13}$$

then the matrix is likely to be embeddable.

Once the distance matrix is generated and shown to be embeddable, then there are various computationally stable methods of calculating the atomic coordinates and the translation from the distance matrix to coordinates is straight-forward. Note that the chirality of a structure cannot be specified by inter-atomic distances; structures that are mirror images of each other will have identical inter-proton distances.

17.3.3 Refinement

The embedded structures are usually subject to regularization to remove poor covalent geometry. A simulated annealing protocol for the initial regularization of structures from distance geometry is presented in Fig. 17.11. At this point in the refinement process it is usually possible to identify mirror image structures by virtue of their higher energy because of incorrect geometry at chiral centers.

Once the initial structures have been regularized, they are generally subject to simulated annealing refinement. A typical protocol for refinement is given in Fig. 17.12. The major difference between this protocol and that used for regularization of structures is that the simulated annealing is initiated at lower temperatures (T=1000 K). In addition, experimental constraints from residual dipolar couplings are introduced towards the end of this protocol. Constraints from RDCs should not be used in earlier steps because it is easy for these constraints to distort the local structure instead of providing information on global alignment.

The protocol begins at T=1000 K and slowly increases the van der Waals scale factor (K_{vdw}) from 0.003 to 4.0 over about 15 cycles of molecular dynamics. In addition, the temperature is lowered in steps of 50 K, until it reaches 300 K. During this phase of the refinement, constraints from residual dipolar couplings do not contribute to the energy; $K_{rdc} = 0$. The second phase of refinement is accomplished by 20 cycles of molecular mechanics calculations at constant temperature. The scale factor for RDCs is increased from 0.001 to 0.02 over these cycles. Finally, the structure is subject to 200 cycles of energy minimization, which would include all experimental constraints as well as bonded and non-bonded interactions.

The structures obtained from this step can be rank ordered in terms of energy and the lowest 5-10% of the structures serve as trial structures that can be used to detect possible errors in the experimental constraints or facilitate the inclusion of additional

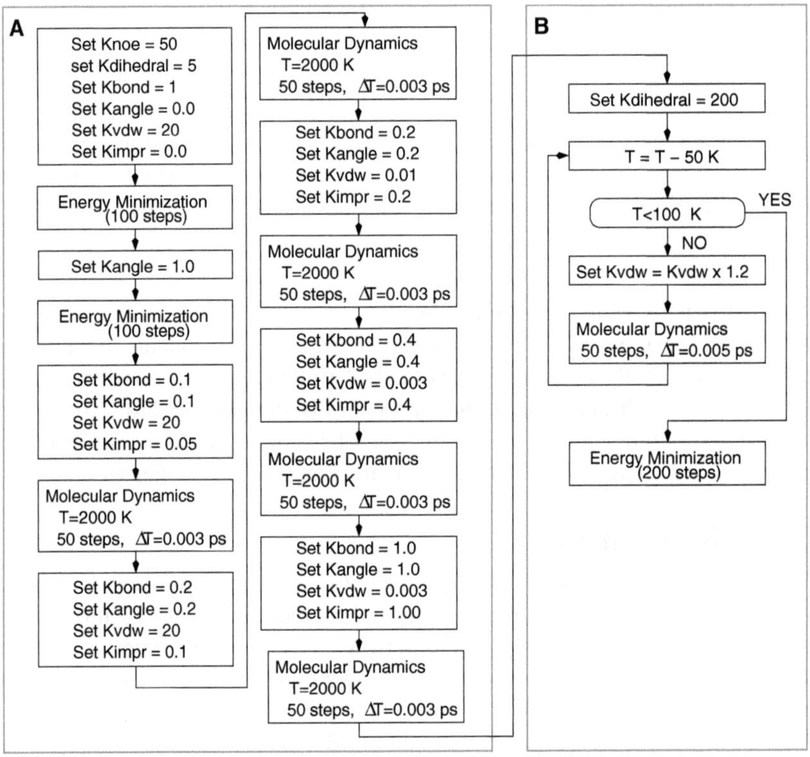

Figure 17.11. Protocol for regularization of distance geometry structures. This flowchart is based on the *dgsa.inp* protocol given by Brünger [26]. The initial part of the protocol, shown in part A of the figure, is applied to a structure and its mirror image, with the goal of determining the correct chirality based on the energy of the structure. First, bad van der Waals contacts are removed by setting $K_{vdw} = 20$. Subsequently the scaling factors for bonds, bond angles, and improper torsional angles ($impr$) are slowly increased. The latter term is generally used to specify chiral centers. The structure with the lowest energy is usually of the correct geometry at chiral centers. These models are then taken to part B of the protocol, where the temperature is gradually reduced from 2000 K to a final temperature of 100 K in approximately 20 steps. In each of these steps the scaling factor for van der Waals is increased from its initial value of 0.0003 to a final value of 4.0 when T=100 K.

experimental data. Depending on the number of changes in experimental constraints it may not be necessary to repeat the entire structure determination process, beginning with the generation of new trial structures. If the number of additional constraints is a few percent of the total number, and they would result in minor modifications of the structure, then it may be possible to simply repeat the refinement step with the additional constraints.

17.4 Illustrative Example of Protein Structure Determination

The following section outlines the process of structure determination for a 130 residue protein, rho130 [25]. This protein was chosen as an example because it consists of two sub-domains, an amino-terminal domain that is largely α-helical, and a carboxy-terminal domain that is largely β-sheet (see Fig. 17.13, Panel D). The number of contacts between the two domains is very limited, thus it is difficult to obtain a large number of inter-proton distances to constrain the structure of one domain with respect to the other. In this case the importance of including constraints from residual dipolar couplings is quite apparent, giving a precise definition of the relative orientation of the two domains with respect to each other.

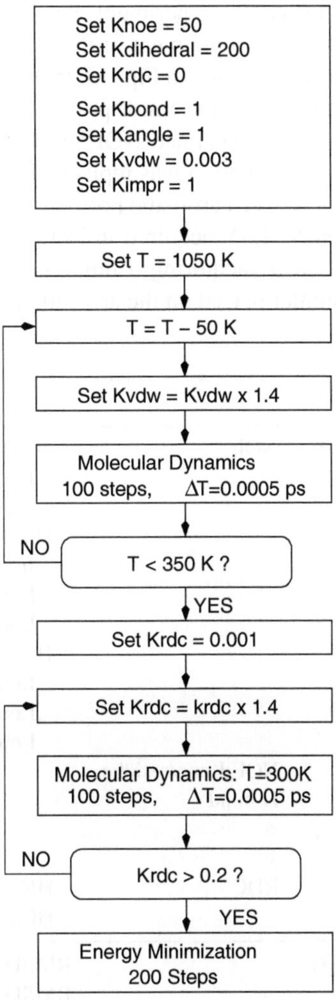

Figure 17.12. Final refinement protocol. This protocol was adapted from Brünger's *refine.imp* [26] with the inclusion of constraints from residual dipolar coupling as suggested by de Alba and Tjandra [48].

Four major steps (A-D) in the process of protein structure determination are depicted here. The constraints that were used at each step are listed in Table 17.1 and the resultant structures are shown in Fig. 17.13. At the beginning of each step, 200 trial structures were built using distance geometry. These were regularized using the procedure shown in Fig. 17.11 and further refined by the procedure shown in Fig. 17.12. Four of the lowest energy models that were obtained after the final refinement protocol are shown in Fig. 17.13.

The initial models (A) were constructed from approximately 900 inter-proton distances, which were obtained from NOESY crosspeaks involving unambiguously assigned resonance. The amide-amide distances were obtained from a ^{15}N separated three-dimensional NOESY spectrum. The amide-aliphatic distances were also obtained from this experiment. Since the aliphatic region of the proton spectrum is quite crowded the assignment of the aliphatic peaks was aided by the CN-NOESY experiment, as described in Fig. 15.25, which gives the carbon shifts of the aliphatic protons that are close to the amide protons. Distances between aliphatic protons were obtained using a ^{13}C separated NOESY experiment, similar to the ^{15}N NOESY, except that nitrogen excitation was replaced by excitation of aliphatic carbons. Inter-proton distances between aromatic protons were largely obtained from a two-dimension proton-proton NOESY acquired in D_2O, however a small number of such constraints were obtained by acquiring a three-dimensional ^{13}C separated NOESY with the carbon transmitter placed on the aromatic carbon region.

			A	B	C	D
I	NOE	H_N-H_N				
		Local	147	167	184	184
		Long	38	42	53	53
		H_N-H_C				
		Intra	238	238	308	308
		Local	245	247	379	379
		Long	53	55	137	137
		H_C-H_C				
		Intra	163	163	159	159
		Local	14	14	52	52
		Long	39	39	118	118
	NOE:Long/residue		1.0	1.0	2.4	2.4
	H-bond		0	36	43	43
	ϕ		30	80	91	91
	χ_1		0	0	23	23
	RDC	HN	0	0	0	63
		HC_α	0	0	0	49
II		RMSD (β)	1.3 Å	0.90 Å	0.70 Å	0.39 Å
		RMSD (α & β)	7.0 Å	6.75 Å	1.40 Å	0.43 Å

Table 17.1. Constraints used in determining structure of Rho130. The constraints used at each stage in structure determination (A through D, see Fig. 17.13) of structure determination are given in section I of this table. The NOE constraints are divided into amide-amide (H_N-H_N), amide-aliphatic (H_N-H_C), and aliphatic-aliphatic or aliphatic-aromatic or aromatic-aromatic (H_N-H_C). Each of these categories is further divided into intra-residue, local, and long-range distances. A local distance constraint involves residues that are within four residues of each other in the primary sequence. Long range distances involve residues that are more than four residues from each other. The average number of long range distance constraints/residue are also given. The number of H-bonds are also listed. The number of torsional constraints constraining the N-C_α bond (ϕ) and the C_α-C_β bond (χ_1) are also listed. Finally, the number of HN and HC_α residual dipolar couplings are listed. Section II of this table gives the root-mean-squared-difference (RMSD) between each of the four low-energy structures. The first entry is the RMSD for alignment of the β-domain while the second entry is for aligning both domains.

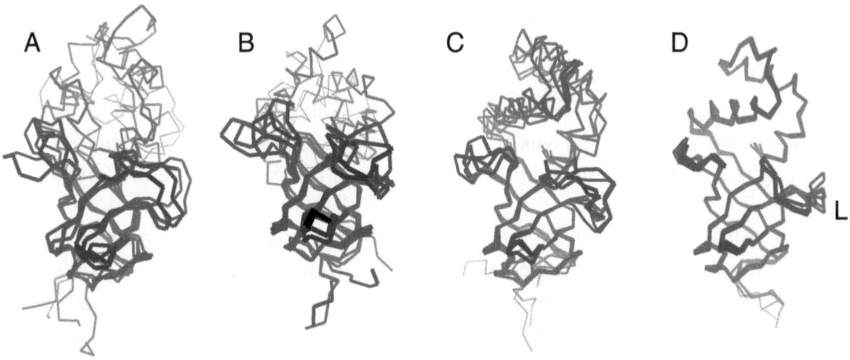

Figure 17.13. Stages in the structure determination of Rho130. Each panel shows the superposition of the four lowest energy structures after final energy minimization. The β-strand domain, which forms the lower part of the structure in this presentation, was used to define the alignment. Panel A shows the set of structural models generated from the initial set of constraints. Panel B and C show improvement in the structural models by the addition of hydrogen bonds as well as torsional angle and additional distance constraints. Panel D shows the effect of utilizing residual dipolar couplings as additional constraints. Note that the relative alignment of the two sub-domains is not well defined until Panel C and becomes more precise with the addition of the RDCs.

Hydrogen bond constraints could not be utilized during stage A of structure determination. Although potential hydrogen bond donors had been identified by virtue of slow amide exchange kinetics, the identification of the acceptors required initial models. A similar situation also existed for the use of ϕ torsional angle constraints. Although most of the coupling constants had been measured for the three bond H_N-H_α coupling, only approximately one-third (30/91) were above 9 Hz and could be used without concern of conformational averaging of the coupling constant.

Residual dipolar couplings for the H-N and H-C_α bond vectors were obtained from samples aligned using filamentous bacteriophage. Both couplings were obtained by measuring the oscillation of peak intensity as a function of time and then fitting the data to a damped cosine function to determine the value of the coupling constant. In the case of H-N coupling, pulse sequence B in Fig. 16.14 was used. The C-H couplings were measured using a modification of the HA(CACO)NH experiment as described by Hitchens et al [73]. These constraints were not used for refinement until the final stage, when reasonably accurate structures became available.

The lowest energy model structures that were obtained from step A showed good agreement within each sub-domain of the protein. The structural similarity can be characterized by the root-mean-squared deviation RMSD. The RMSD is a measure of the similarity of one structure to another and is defined as follows:

$$RMSD = \frac{1}{N} \sum_{i=1}^{N\ Atoms} \sqrt{[(x_{ij} - x_{ik})^2 + (y_{ij} - y_{ik})^2 + (z_{ij} - z_{ik})^2]} \qquad (17.24)$$

where x_{jk} represents the x-coordinate of the kth atom in molecule j and x_{ik} is the x-coordinate of the same atom in molecule i.

The smaller the RMSD, the more similar the structures are to each other. In the case of the sub-domains, the RMSD is about 1.3 Å (Table 17.1). However, the relative orientation of the two domains was poorly determined, as indicated by the high RMSD for aligning the entire protein and the obvious poor alignment of structures in Panel A of Fig. 17.13.

The set of lowest energy structures from step A were inspected and a small number of inter-proton distances were added after using these structures to resolve ambiguities in the assignment of NOE crosspeaks. In addition, a total of 36 hydrogen bond acceptors were identified on the basis that the residues were in regular secondary structure. Finally, most of the ϕ torsional constraints could be used for the next stage since it was clear that the residues were in regular secondary structure. The refined structures from step B are shown in Fig. 17.13. There is a significant improvement in the sub-domain structure; the RMSD for the β region dropped from 1.3 Å to 0.9 Å. In addition, there is a modest decrease in the RMSD for overall alignment.

The structures from step B were sufficiently well defined to allow the assignment of a large number long-range distance constraints: 11 amide-amide, 82 amide-aliphatic, and 79 aliphatic-aliphatic. In addition, it was possible to identify 7 more hydrogen bonds and utilize 11 more ϕ torsional constraints. A number of torsional angle constraints, involving the C_α-C_β (χ_1) bond, could also be incorporated at this time.

The resultant structures are shown in Panel C of Fig. 17.13. The lowest energy structures were very well defined within each sub-domain, with an RMSD of 0.7 Å. The alignment of the overall protein was poorer showing an RMSD of 1.4 Å. However, the overall structure was quite acceptable. The increase in the quality of the structures in step C is due almost entirely to the increase in the number of long-range inter-proton distances, from 1.0/residue in B to 2.4/residue in stage C.

The difficulty in determining the relative orientation of the two sub-domains is not surprising given the fact that there are few distance constraints between the domains and that the information from these NOE derived distances is entirely of a local nature. The inclusion of constraints from residual dipolar coupling, as shown in D, provided information on the independent alignment of each sub-domain with respect to the direction of the applied magnetic field. A small number of such constraints are sufficient to fix the relative orientation of each sub-domain, as shown in panel D of Fig. 17.13.

The inclusion of residual dipolar couplings also increase the precision of the local geometry. The conformation of an inter-strand loop region of the protein, marked with an "L" in Panel D of Fig, 17.13, is poorly defined in C because of the lack of experimentally measured inter-proton distances in this region of the protein. However, the inclusion of several RDCs from this region causes the structures to converge to a common configuration. The RDCs also increase the overall precision of the structure, reducing the RMSD to 0.42 Å.

In summary, with a sufficient number of experimental constraints, it is possible to obtain structures of proteins in solution that rival structures obtained from high-resolution X-ray crystallography.

Chapter 18

EXCHANGE PROCESSES

18.1 Introduction

The exchange between two or more environments can have a profound effect on the appearance of the resonance lines of the exchanging species. Under favorable conditions it may be possible to obtain information on both the forward and reverse rate-constants for the exchange reaction as well as the equilibrium population of each environment from the changes in the NMR spectrum.

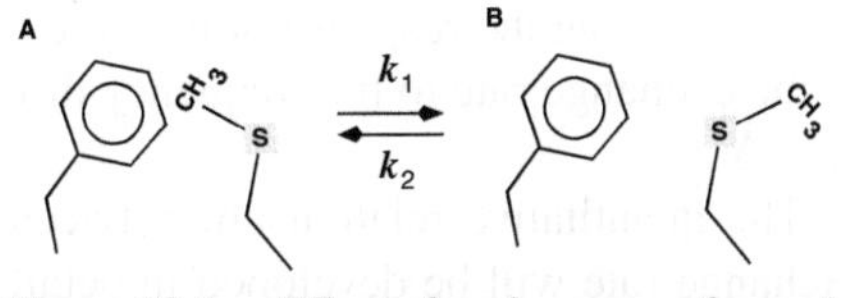

Figure 18.1. Effect of exchange on the environment of a Spin. In this illustration the methionine residue exists in two conformations. Since the environment of the methyl group is different in each environment, two different resonance frequency will be observed, ω_A and ω_B. In conformation A, the methyl group is found above an aromatic ring and will experience a ring current shift. In conformation B, the methyl group is removed from the aromatic ring, resulting in a change in its chemical shift. Note that the relaxation properties of the methy group may also differ between the two environments.

When a spin exchanges between environments its spectral properties may change if the chemical shift or relaxation properties of the spin are different in each environment. Figure 18.1 shows how a conformational change in a protein can lead to a change in the chemical shift of the methyl group of a methionine residue.

Since the exchange phenomenon involves a change in the chemical environment of a spin, leading to a chemical shift change, it is often referred to as *chemical exchange*. Usually, exchange processes are studied when there are *non*-covalent changes in the molecule. However, changes in the covalent structure of a molecule can also lead to chemical exchange.

In this chapter, two general aspects of exchange will be considered. First, we will discuss the effects of a spin sampling two distinct environments on its spectral properties. Here the goal is to measure the rate constants for exchange and/or the equilibrium constant. In the latter part of the chapter we will extend these studies to the investigation of the kinetics and binding affinity of ligands.

18.2 Chemical Exchange

The kinetics of the exchange reaction are defined by the following scheme:

$$A \underset{k_2}{\overset{k_1}{\rightleftharpoons}} B$$

The kinetic rate constant for the conversion of A to B is k_1 and the rate constant for the reverse reaction is k_2, giving an overall equilibrium constant, $K_{eq} = k_1/k_2$. The fraction of the system found in each conformation is:

$$p_A = \frac{k_2}{k_1 + k_2} = \frac{1}{1 + K_{eq}} \qquad p_B = \frac{k_1}{k_1 + k_2} = \frac{K_{eq}}{1 + K_{eq}} \tag{18.1}$$

To characterize the different time scales of exchange it is useful to define an apparent exchange rate, $k_{ex} = k_1 + k_2$, and a frequency difference between the two states, $\Delta\omega = \omega_A - \omega_B$. Table 18.1 illustrates how the relationship between the apparent exchange rate, k_{ex}, and the frequency separation, $\Delta\omega$, will affect the observed spectrum. Note that the response of the system to chemical exchange depends on the ratio of the exchange rate to the frequency difference of the spins in each environment, i.e. $k_{ex}/\Delta\omega$.

The quantitative relationship between the measured spectral parameters and the exchange rate will be developed in detail in the remainder of this chapter. However, before developing these expressions it is useful to consider in qualitative fashion two illustrative cases, fast exchange and slow exchange, to gain an understanding of the general features of chemical exchange.

Fast Exchange ($k_{ex} >> \Delta\nu$): Under conditions of fast exchange, a single resonance line is observed. The averaging of the chemical shift occurs because the spins do not exist in either environment long enough to establish an associated resonance frequency (See Fig. 18.2). Consequently, the spin processes at a population averaged resonance

Exchange Rate		α^1	*Observed Spectrum*	*Experimental Technique*
Very slow	$k_{ex} << \Delta\nu$	0	Two Resonances	Exchange Spectroscopy
Slow	$k_{ex} < \Delta\nu$	<1	Two Broadened Resonances	Linewidth Measurements
Intermediate	$k_{ex} \approx \Delta\nu$	1	Complex Lineshape	Lineshape Analysis
Fast	$k_{ex} > \Delta\nu$	>1	Single Broadened Resonance	Spin-Echo R(τ_{cp})
Very fast	$k_{ex} >> \Delta\nu$	2	Single Resonance	Spin-lock R(ω_e)

[1] α is defined in Section 18.5.

Table 18.1. Summary of the effects of exchange on the properties of the NMR spectrum. The left columns describe the exchange in terms of the relationship between the frequency separation ($\Delta\nu = \frac{1}{2\pi}(\omega_A - \omega_B)$) and the apparent exchange rate constant ($k_{ex} = k_1 + k_2$). The remaining columns describe the effect of exchange on the spectrum and the appropriate experimental technique to characterize the exchange rate. A more comprehensive description of the methodologies that can be used to measure chemical exchange can be found in [123].

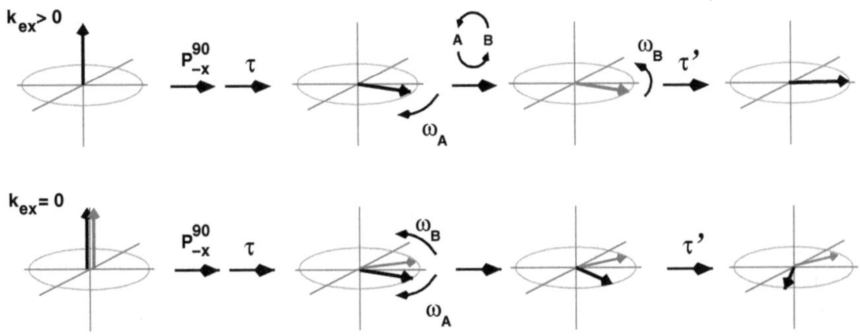

Figure 18.2. Effects of fast exchange on the environment of a spin. The upper part of the figure follows a single spin as it precesses during fast exchange. The lower part of the figure shows two spins, one in conformation A (black) and one in conformation B (gray), precessing in the absence of chemical exchange. In both cases the coordinate frame is rotating at a frequency that is midway between ω_A and ω_B and $p_A = p_B$. The left most part of the figure represents the system before the excitation pulse. Following from left to right, after the 90° pulse and a short period τ, the spins have precessed at a frequency that is representative of their environment. In the case of fast exchange, the spin that was in environment A is now found in environment B. Consequently its precessional frequency is now ω_B and the spin precesses counter-clockwise at ω_B for the next τ' period. At the end of this period the magnetization is found along the y-axis. Consequently no net precession has occurred and the *observed* resonance frequency is $\frac{1}{2}[\omega_A + \omega_B]$. In contrast, the non-exchanging spins ($k_{ex} = 0$) remain in their original environment and will continue to precess in the same direction for an additional period of τ'. Note that $\tau \neq \tau'$ because exchange is a random process, but the average time between exchange events, $\bar{\tau}$ and $\bar{\tau'}$ will be the same if $p_A = p_B$.

frequency , ω_{obs}, and exhibits a population averaged spin-spin relaxation rate, T_2^{avg}:

$$\omega_{obs} = p_A\omega_A + p_B\omega_B \qquad \frac{1}{T_2^{avg}} = \frac{p_A}{T_2^A} + \frac{p_B}{T_2^B} \tag{18.2}$$

where p_A and p_B are the populations of each environment.

Slow Exchange ($k_{ex} << \Delta\nu$): Under conditions of slow exchange, the rate of exchange is slower than the frequency difference, in Hz, of the spin in each environment. Thus the exchange process is incapable of averaging the chemical shifts while the spins are precessing. Consequently, two resonance lines are observed, one line from the fraction of population of the spins that are found in conformation A and one line from the fraction that are found in conformation B. Since the environment of a spin determines its absorption frequency two separate resonance lines are observed, one at ω_a and one at ω_b. The integrated intensity of each line is equal to the fraction of spins in conformation A and B, respectively.

Although slow exchange has no effect on the position of the resonance lines, the exchange process reduces the lifetime of the spin within a particular environment. Consequently, the resonance linewidth increases by an amount that is proportional to the rate of exchange. This line broadening can be understood by considering the effects of exchange on the magnetization as it precesses in the x-y plane after excitation

(See Fig. 18.3). After excitation, the spins in environment A will precesses in the x-y plane at a frequency of ω_A. However, as time passes, some of these spins will change their environment and will begin precessing at ω_B. This change in environment results in a loss of *coherent* magnetization of the spins in conformation A, resulting in a decay of the transverse magnetization at a rate faster than the intrinsic spin-spin relaxation rate.

Since the system is at equilibrium, an equal number of spins also convert from conformation B to conformation A. This magnetization does not contribute to the magnetization originally associated with ω_A since the conversion from B to A occurs randomly. Consequently, the phase of the magnetization coming from environment B is incoherent with respect to the phase of the magnetization originally associated with environment A.

The rate at which the magnetization decays during the exchange process, $1/T_{2A}$, is given by the sum of the intrinsic spin-spin decay rate ($1/T_2^{oA} = R_2^{oA}$) and the rate at which conformation A is converted to conformation B (k_1). A similar expression can be written for the spins in environment B:

$$\frac{1}{T_2^A} = \frac{1}{T_2^{oA}} + k_1 = R_2^{oA} + k_1 \qquad \frac{1}{T_2^B} = \frac{1}{T_2^{oB}} + k_2 = R_2^{oB} + k_2 \tag{18.3}$$

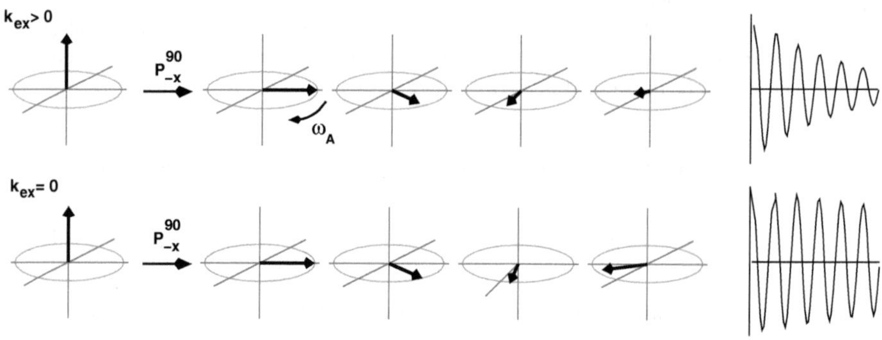

Figure 18.3. Effect of slow exchange on the relaxation rate of transverse magnetization. The top of the figure shows the bulk magnetization of spins in conformation 'A', undergoing slow exchange with conformation "B". The lower part of the figure shows the same spins in the absence of exchange. In both cases, the coordinate system is rotating at a frequency that is midway between ω_A and ω_B. The left-most section of the figure shows the bulk magnetization before the excitation pulse. The following panels show the spin precessing clockwise at a rate ω_A. The right-most section of the figure shows the observed free induction decay for the spins at ω_A. In the top section of the figure ($k_{ex} > 0$) the spins leave conformation 'A' and return from conformation 'B' a random time later. Since the precessional frequencies, ω_A and ω_B are different, the phase of the returning magnetization is no longer the same as those spins that remained in conformation A, causing randomization of the transverse magnetization. This reduces the signal, as illustrated by the shortening of the arrow that represents the x-y magnetization. This loss of magnetization increases the rate of decay of the transverse magnetization, shortening the observed T_2. In the absence of exchange, the transverse magnetization decays at the intrinsic T_2, as shown in the lower part of the figure.

The increase in T_2 leads to broadening of the resonance line by each rate constant:

$$\pi\Delta\nu_{1/2}^{A} = \frac{1}{T_2^{oA}} + k_1 \qquad \pi\Delta\nu_{1/2}^{B} = \frac{1}{T_2^{oB}} + k_2 \tag{18.4}$$

18.3 General Theory of Chemical Exchange

The effect of exchange on the NMR lineshape can be analyzed in a number of ways. The traditional method, attributed to H. M. McConnell [109], utilizes an analysis of the steady-state magnetization while the measurement frequency is changed (continuous wave spectroscopy, CW). More recent approaches include the use of Laplace transforms [63], or relaxation matrix analysis [30]. Both of these latter methods determine the time dependence of M_A and M_B, from which the NMR spectra can be obtained by the usual method of Fourier transformation. Although the latter approaches are elegant, the analysis via steady-state magnetization is more straightforward and provides an simple expression for the lineshape that is suitable for direct fitting of experimental lineshapes by least-squares methods.

The fundamental equation that describes the effect of chemical exchange on the magnetization is given by:

$$\begin{aligned} \frac{dM_A}{dt} &= -k_1 M_A + k_2 M_B \\ \frac{dM_B}{dt} &= +k_1 M_A - k_2 M_B \end{aligned} \tag{18.5}$$

The amount of magnetization in environment “A” is decreased when the system exchanges from A to B, but is increased when conformation B returns to conformation A.

The above time dependent changes in spin populations can be combined with chemical shift precession and relaxation using an analysis based on continuous-wave spectroscopy. In CW spectroscopy, the sample is irradiated at a constant frequency and the energy absorption as a function of the magnetic field strength is recorded. Under the conditions of irradiation along the x-axis, the following equation defines the time-evolution of the spins in the transverse (x-y) plane:

$$\frac{dM}{dt} = -M[\frac{1}{T_2} + i(\omega_s - \omega)] + iI_z\omega_1 \tag{18.6}$$

where ω is the independent variable (frequency of irradiation) and ω_s is the resonance frequency of the spin. This equation is identical to that presented for the free precession of $M_x + iM_y$, at a rate of ω_s, after a radio-frequency pulse (see Chapter 1) with the addition of the $iI_z\omega_1$ term. This term represents the steady-state conversion of longitudinal (I_z) magnetization by the excitation field (ω_1), applied along the x-axis. The factor of i indicates that this field produces magnetization along the y-axis, which is defined here to be the real, or absorption, component of the magnetization.

If Eq. 18.5 is combined with Eq. 18.6, then the following two modified Bloch equations are obtained:

$$
\begin{aligned}
\frac{dM_A}{dt} &= -M_A[\frac{1}{T_2^A} + i(\omega_A - \omega)] - k_1 M_A + k_2 M_B + ip_A\omega_1 \\
\frac{dM_B}{dt} &= -M_B[\frac{1}{T_2^B} + i(\omega_B - \omega)] + k_1 M_A - k_2 M_B + ip_B\omega_1
\end{aligned} \quad (18.7)
$$

Note that I_z has been replaced by p_A and p_B, the fraction of the molecules in state A and state B, respectively. Under steady-state conditions, $dM_A/dt = dM_B/dt = 0$, giving:

$$
\begin{aligned}
M_A[\frac{1}{T_2^A} + i(\omega_A - \omega) + k_1] - k_2 M_B &= ip_A\omega_1 \\
M_A[-k_1] + M_B[\frac{1}{T_2^B} + i(\omega_B - \omega) + k_2] &= ip_B\omega_1
\end{aligned} \quad (18.8)
$$

To simplify the manipulations, the following definitions will be used:

$$
\begin{aligned}
\gamma_A &= \frac{1}{T_2^A} + k_1 & \gamma_B &= \frac{1}{T_2^B} + k_2 \\
\Omega_A &= \omega_A - \omega & \Omega_B &= \omega_B - \omega
\end{aligned}
$$

giving,

$$
\begin{aligned}
M_A[i\Omega_A + \gamma_A] - M_B k_2 &= ip_A\omega_1 \\
M_A[-k_1] + M_B[i\Omega_B + \gamma_B] &= ip_B\omega_1
\end{aligned} \quad (18.9)
$$

Solving for M_A gives:

$$
M_A = i\omega_1 \frac{p_A(i\Omega_B + \gamma_B) + p_B k_2}{(i\Omega_A + \gamma_A)(i\Omega_B + \gamma_B) - k_1 k_2} \quad (18.10)
$$

M_B is obtained by interchanging the A and B labels and k_1 and k_2, giving the following expression for the total magnetization:

$$
M_A + M_B = i\omega_1 \frac{p_A(i\Omega_B + \gamma_B + k_1) + p_B(i\Omega_A + \gamma_A + k_2)}{(i\Omega_A + \gamma_A)(i\Omega_B + \gamma_B) - k_1 k_2} \quad (18.11)
$$

The observed, or absorption mode spectrum, is given by the imaginary part of this function.

Although Eq. 18.11 is suitable for use in data fitting, it is informative to simplify this equation to obtain a qualitative understanding of the lineshape. If we assume that the intrinsic T_2 is very long (i.e. $1/T_2 = 0$, therefore, $\gamma_A = k_1$, $\gamma_B = k_2$) then:

$$
M_A + M_B = i\omega_1 \frac{p_A(i\Omega_B + k_2 + k_1) + p_B(i\Omega_A + k_1 + k_2)}{(i\Omega_A + k_1)(i\Omega_B + k_2) - k_1 k_2} \quad (18.12)
$$

Using the relationships, $p_A = k_2/(k_1 + k_2)$ and $p_B = k_1/(k_1 + k_2)$, as well as the fact the $p_A = p_B = \frac{1}{2}$, and further assuming that the kinetic rate constants are the same for both the forward and reverse reactions (i.e. $k_1 = k_2 = k$) the following result is obtained:

$$M_A + M_B = i\omega_1 \frac{\frac{i}{2}\Omega_A + k + \frac{i}{2}\Omega_B + k}{(i\Omega_A + k)(i\Omega_B + k) - k^2} \tag{18.13}$$

Expanding and collecting real and imaginary terms:

$$M_A + M_B = i\omega_1 \frac{\frac{i}{2}(\Omega_A + \Omega_B) + 2k}{(-\Omega_A\Omega_B) + ik(\Omega_A + \Omega_B)} \tag{18.14}$$

Rationalizing the denominator by multiplying with $-(\Omega_A\Omega_B) - ik(\Omega_A + \Omega_B)$ and keeping the imaginary term gives:

$$M_A + M_B = \omega_1 \frac{\frac{1}{2}k(\Omega_A - \Omega_B)^2}{(\Omega_A\Omega_B)^2 + k^2(\Omega_A + \Omega_B)^2} \tag{18.15}$$

Note that:

$$\begin{aligned}(\Omega_A - \Omega_B) &= (\omega_A - \omega) - (\omega_B - \omega) = \omega_A - \omega_B \\ (\Omega_A + \Omega_B) &= (\omega_A - \omega) + (\omega_B - \omega) = 2(\overline{\omega} - \omega)\end{aligned} \tag{18.16}$$

where $\overline{\omega} = \frac{1}{2}(\omega_A + \omega_B)$, or the average chemical shift. The average shift can be defined to be zero, implying $\omega_A = -\omega_B$. If we define $\omega_A - \omega_B = \Delta\omega$, then $\omega_A = +\Delta\omega/2$ and $\omega_B = -\Delta\omega/2$.

These relationships further simplify the expression for the total lineshape, I(ω):

$$I(\omega) \propto \frac{1}{2}k(\Delta\omega)^2 \frac{1}{(\omega_A - \omega)^2(\omega_B - \omega)^2 + 4k^2\omega^2} \tag{18.17}$$

18.3.1 Fast Exchange Limit

Under conditions of fast exchange, $k >> \Delta\omega$, the second term in the denominator of Eq. 18.17 will dominate, giving a single resonance line at $\omega = 0$, half-way between ω_A and ω_B ($\omega = p_A\omega_A + p_B\omega_B$, $p_A = p_B$ in this example). The linewidth can be obtained by investigating the behavior of the lineshape near $\omega = 0$. Here, $\omega_A - \omega$ can be approximated as ω_A, a similar approximation can be taken for ω_B.

$$\begin{aligned}I(\omega) &\propto \frac{1}{2}k(\Delta\omega)^2 \frac{1}{(\omega_A)^2(\omega_B)^2 + 4k^2\omega^2} \\ &\propto \frac{1}{2}k(\Delta\omega)^2 \frac{1}{(\Delta\omega/2)^2(\Delta\omega/2)^2 + 4k^2\omega^2} \\ &\propto \frac{1}{8k}(\Delta\omega)^2 \frac{1}{\frac{1}{k^2}(\Delta\omega)^4(\frac{1}{8})^2 + \omega^2}\end{aligned} \tag{18.18}$$

This gives a rate of decay equal to:

$$\frac{1}{T_2} = \frac{(\Delta\omega)^2}{8k} \tag{18.19}$$

In the general case, when $k_1 \neq k_2$ the observed rate of decay of transverse magnetization due to exchange is [155]:

$$R_2 = \frac{1}{T_2} = p_A p_B (\Delta\omega)^2 / k_{ex} \tag{18.20}$$

This equation is equivalent to Eq. 18.19, when the two rate constants are equal.

18.3.2 Slow Exchange Limit

Under conditions of slow exchange, $k << \Delta\omega$, we anticipate the presence of two resonance lines, one at ω_A and the other at ω_B, as predicted from the previous qualitative analysis in Section 18.2. If the behavior of the general lineshape function in the vicinity of ω_A is considered, then the lineshape function reduces to:

$$\begin{aligned} I(\omega) &\propto \frac{1}{2}k(\Delta\omega)^2 \frac{1}{(\omega_A - \omega)^2(\omega_B - \omega_A)^2 + 4k^2\omega_A^2} \\ &\propto \frac{1}{2}k(\Delta\omega)^2 \frac{1}{(\omega_A - \omega)^2(\Delta\omega)^2 + 4k^2(\Delta\omega/2)^2} \\ &\propto \frac{1}{2}k \frac{1}{(\omega_A - \omega)^2 + k^2} \end{aligned} \tag{18.21}$$

The above transformations assume that when $\omega \approx \omega_A$ both $(\omega_B - \omega)$ and ω^2 are slowly varying functions of ω, i.e. we can approximate $(\omega_B - \omega)$ as $(\omega_B - \omega_A)$, and ω^2 as ω_A^2. On the other hand, $(\omega_A - \omega)$, is a rapidly varying function when $\omega \approx \omega_A$ and must be treated exactly.

Recall that the lineshape for a free induction decay that decays with a rate of $\frac{1}{T_2}$ is:

$$I(\omega) \propto \frac{1}{\omega^2 + [\frac{1}{T_2}]^2} \tag{18.22}$$

Therefore, the contribution of exchange to the decay of the transverse magnetization is k, i.e.

$$R_2 = \frac{1}{T_2} = k \tag{18.23}$$

as predicted from the earlier qualitative analysis.

18.3.3 Intermediate Time Scales

The lineshape for intermediate exchange rates is obtained by evaluation of Eq. 18.11. Numerical simulations are presented in Fig. 18.4. This simulation shows the predicted transition from two peaks in the case of slow exchange to a single peak under conditions of fast exchange.

Note that in the intermediate time scale, the resonance lines may be so extensively broadened such that no peak is observed in the actual spectrum. This situation will also hold true in a NOESY-type experiment if the dipolar coupled spins are exchanging between environments.

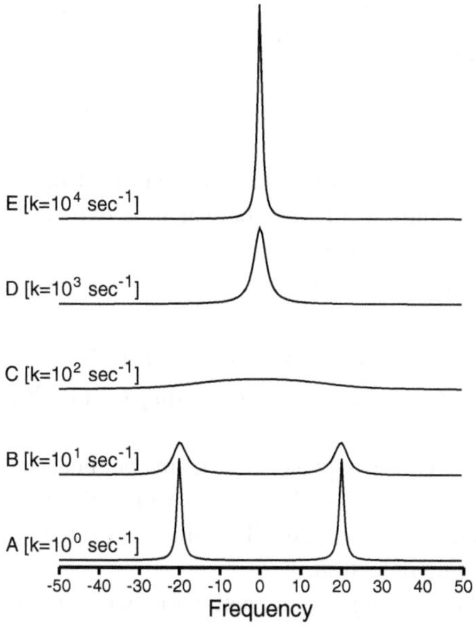

Figure 18.4 Effect of chemical exchange on lineshape. The lineshape is shown over a wide range of exchange rates. In this simulation, the population of both states were set to $\frac{1}{2}$. The frequency separation between ω_A and ω_B, $\Delta\omega$, was set to 40 Hz. The exchange rates vary from 1 $\sec^{-1}$ to 10^4 $\sec^{-1}$. Spectrum A corresponds to very slow, B to slow, C to intermediate, D to fast, and E to very fast exchange rates, as defined in Table 18.1.

18.4 Measurement of Chemical Exchange

18.4.1 Very Slow Exchange: $k_{ex} << \Delta\nu$

When the exchange rate is very slow, a resonance line is observed for each conformation and the intensity of each line is proportional to the population of spins occupying each state. Thus the equilibrium constant for the system can be directly obtained from the integrated line intensities:

$$K_{eq} = \frac{I_B}{I_A} \tag{18.24}$$

If the exchange rate is much less than R_2, then the exchange event has little or no effect on the observed linewidth. Consequently, the individual rate constants cannot be obtained from an analysis of the lineshape. However, if the exchange rate is faster than R_1, it may be possible to measure the rate constants by detecting the exchange of longitudinal magnetization between the two environments. Two-dimensional exchange spectroscopy is a common method of detecting chemical exchange of longitudinal magnetization [79, 112]. The spin is first frequency labeled with the resonance frequency of one environment (e.g. ω_A), allowed to exchange to the other environment, and then frequency labeled with the resonance frequency of the new environment (ω_B). The presence of a crosspeak at ω_A and ω_B shows that exchange has occurred between the two environments and the time dependence of the intensity of this exchange peak can be used to determine k_{ex}.

A simple homonuclear pulse sequence for the measurement of exchange is shown in Fig. 18.5. This experiment is identical to the homonuclear NOESY experiment that detects exchange of magnetization via dipolar coupling. A more detailed product operator description of this experiment is provided in Chapter 16. The density matrix immediately after the second pulse in the sequence is given by:

$$\rho = I_{zA}^{o} cos(\omega_A t_1) + I_{zB}^{o} cos(\omega_B t_1) \quad (18.25)$$

where $I_{zA}^{o} \propto p_A$ and $I_{zB}^{o} \propto p_B$, representing the equilibrium magnetization. Chemical shift evolution during t_1 generates a non-equilibrium state between the two environments, permitting the detection of chemical exchange.

The time domain signal will consist of four components,

$$\begin{aligned} S(t_1, t_2) &= I_{AA} cos(\omega_A t_1) cos(\omega_A t_2) + I_{BB} cos(\omega_B t_1) cos(\omega_B t_2) + \\ & I_{AB} cos(\omega_A t_1) cos(\omega_B t_2) + I_{BA} cos(\omega_B t_1) cos(\omega_A t_2) \end{aligned} \quad (18.26)$$

resulting in two selfpeaks and two crosspeaks. The crosspeaks arise from chemical exchange. For example, the peak at (ω_A, ω_B), with intensity I_{AB}, arises from spins that were in environment 'A' during t_1, exchanged to environment 'B' during the exchange period, and remained in environment 'B' during detection.

The differential equations that describe the exchange of magnetization between the two states are:

$$\frac{dI_{zA}(t)}{dt} = (-R_{1A} - k_1) I_{zA} + k_2 I_{zB} \quad (18.27)$$

$$\frac{dI_{zB}(t)}{dt} = k_1 I_{zA} + (-R_{1B} - k_2) I_{zB} \quad (18.28)$$

where R_1 is the spin-lattice relaxation rate of the spin in each environment.

These equations can be easily solved using Laplace transforms (see Appendix C). Assuming that the rate of spin-lattice (R_1) relaxation is the same for both environ-

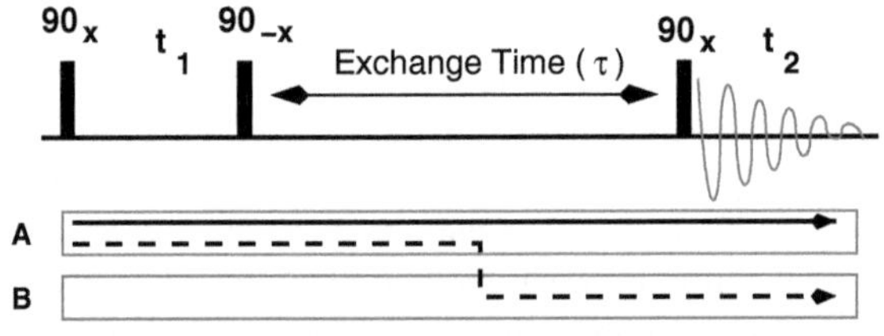

Figure 18.5. Pulse sequence to measure exchange of longitudinal magnetization. During the t_1 period the magnetization of the spins becomes labeled with the frequency appropriate for their environment. The second pulse returns a component of the magnetization to the z-axis. During the exchange period, τ, the spins may exchange environments and thus are detected at the other frequency after returning to the x-y plane by the action of the third pulse. The lower part of the figure illustrates the fate of two spins that originally existed in environment A (upper gray box). During t_1 both spins will precesses at ω_A. However, within the exchange period one spin (solid arrow) remains in environment A while the other (dotted arrow) exchanges to environment B. The former spin will contribute to the selfpeak at (ω_A, ω_A) while the latter will contribute to the crosspeak at (ω_A, ω_B).

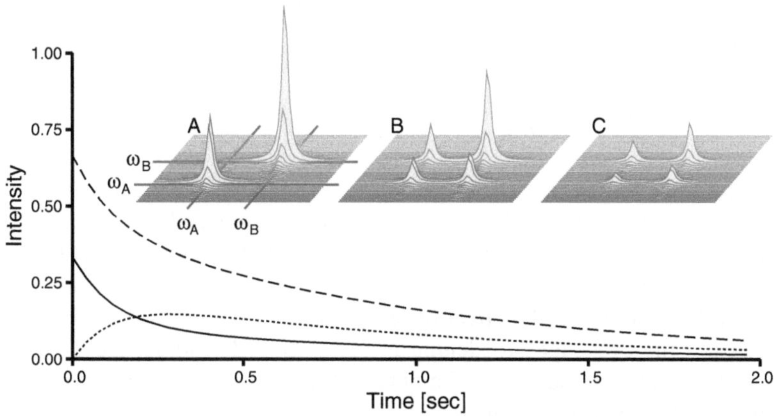

Figure 18.6. Line intensities in 2D-exchange spectroscopy. In this simulation the spin-lattice relaxation rate is 1 sec^{-1}, k_1 is 5 sec^{-1}, and k_2 is 2.5 sec^{-1} ($p_A = 0.33$, $p_B = 0.67$). The time dependence of the intensity of self- and crosspeaks are shown in the lower part of the spectrum. The 'A' and 'B' selfpeaks are illustrated by the solid and dashed lines, respectively, while the crosspeak intensities are drawn with a dotted line. The upper part of the figure shows the two-dimensional spectra for $t = 0$ sec (spectrum A), $t = 0.25$ sec (B), and $t = 1$ sec (C).

ments, the intensity of the self- and crosspeaks at time τ are given by:

$$I_{AA}(\tau) = p_A\left[p_A + p_B e^{-k_{ex}\tau}\right] e^{-R_1\tau} \qquad (18.29)$$

$$I_{BB}(\tau) = p_B\left[p_B + p_A e^{-k_{ex}\tau}\right] e^{-R_1\tau} \qquad (18.30)$$

$$I_{BA}(\tau) = I_{BA}(\tau) = p_A p_B\left[1 - e^{-k_{ex}\tau}\right] e^{-R_1\tau} \qquad (18.31)$$

The intensity of the selfpeaks decrease at a rate that is the sum of the exchange rate constants ($k_{ex} = k_1 + k_2$), while the crosspeaks grow at the same rate. The intensity of both the self- and crosspeaks decrease at the rate of the spin-lattice relaxation (R_1). Therefore, to detect significant crosspeak intensity it is necessary that $k_{ex} \geq R_1$. Note that the intensities of the crosspeaks are the same ($p_A p_B$), regardless of the relative populations of the two environments.

The time dependence of the self- and crosspeaks are illustrated in Fig. 18.6. This simulation illustrates the fact that the intensities of the crosspeaks can exceed that of selfpeaks if $k_{ex} > R_1$.

18.4.2 Slow Exchange: $k_{ex} < \Delta\nu$

In the case of slow exchange it is possible to obtain both the equilibrium constant as well as the rate constants for the exchange process. The equilibrium association constant for the exchange process can be obtained from the integrated intensities of each line, or the ratio of the forward and reverse rate constants:

$$K_{eq} = \frac{I_B}{I_A} = \frac{\frac{k_1}{k_1+k_2}}{\frac{k_2}{k_1+k_2}} = \frac{k_1}{k_2} \qquad (18.32)$$

and the individual rate constants can be obtained by measuring the increase in the linewidth due to the exchange process:

$$\pi \Delta \nu_{1/2}^{A} = \frac{1}{T_2^A} + k_1 \qquad \pi \Delta \nu_{1/2}^{B} = \frac{1}{T_2^B} + k_2 \tag{18.33}$$

Determining the rate constants from the linewidth requires knowledge of the intrinsic spin-lattice relaxation rate, T_2. This value can be calculated from relaxation theory (see Chapter 19), or estimated from spectra acquired at low temperature when k_1 and/or k_2 *may* become insignificant. If the latter approach is used it is necessary to take into the account for the effect of temperature on T_2, as discussed in Chapter 15. Finally, if the exchange broadening is due to ligand binding, the T_2 values in the absence of exchange can simply be obtained from spectra obtained with a ligand concentration of zero (see Section 18.6).

18.4.3 Slow to Intermediate Exchange: $k_{ex} \approx \Delta\nu$

When the exchange rate is of the same order as the frequency difference between the two environments it is necessary to calculate the lineshape explicitly using Eq. 18.11 and extract the exchange parameters from the spectra *via* data fitting.

18.4.4 Fast Exchange: $k_{ex} > \Delta\nu$

18.4.4.1 Lineshape Analysis

Under conditions of fast exchange a single resonance line is observed at a frequency that is given by the weighted average of the chemical shifts of the two individual environments:

$$\omega = p_A \omega_A + p_B \omega_B \tag{18.34}$$

The observed spin-spin relaxation rate, $R_2^{obs} (= 1/T_2^{obs})$ is given by:

$$R_2^{obs} = R_2^o + \frac{p_A p_B (\Delta\omega)^2}{k_{ex}} \tag{18.35}$$

where R_2^o is the spin-spin relaxation rate in the absence of exchange.

It is difficult to obtain any information regarding the kinetic rate processes by analysis of the lineshape since $\Delta\omega$, p_A, and p_B, are usually unknown. If ω_A and ω_B are known, which is often the case in ligand binding experiments, the population of each state can be deduced from the position of the line.

18.4.4.2 Measuring k_{ex}: Relaxation Dispersion Experiments

Under conditions of fast exchange the apparent exchange rate constant, k_{ex}, can be obtained from relaxation dispersion experiments, or the change in the relaxation rate due to a change in the magnetic field strength. In a relaxation dispersion experiment, the spins are subject to a transverse magnetic field, ω_T of varying strength while in the transverse (x-y) plane. The observed spin-spin relaxation rate depends on *both* the chemical exchange rate and the strength of the applied field. Because of this dependence it is possible to determine k_{ex} without relying on any other information.

In practice, the spin-spin relaxation rate, R_2, is measured at a number of different transverse field strengths and the exchange rate is obtained by fitting the measured dispersion curve, $R_2(\omega_T)$, to theoretical models.

Two methods are employed to generate magnetic fields of different strengths for dispersion experiments. The first employs a series of 180° pulses, spaced τ_{cp} apart, as illustrated in Fig. 18.7. The initial implementation of this sequence by Carr and Purcell utilized the same phase for the 90° excitation and 180° refocusing pulses [27]. A later modification by Meiboom and Gill reduced the effects of imperfect 180° pulses by phase shifting the excitation pulse with respect to the 180° pulses [111]. In recognition of the individuals who developed this method, the sequence is usually referred to as a CPMG sequence. This pulse train refocuses the effects of magnetic field inhomogeneities on the decay of transverse magnetization and is therefore useful for measuring T_2. The 180° pulses in the CPMG sequence generate a spin-echo at a point half-way between each pulse. The amplitude of this echo decays according to the spin-spin relaxation rate, as illustrated in Fig. 18.7.

In the absence of exchange the $\tau - 180^o - \tau$ sequence will refocus any precession due to chemical shift evolution or magnetic field inhomogeneities since the precessional frequency for any given spin is constant over the duration of the experiment.

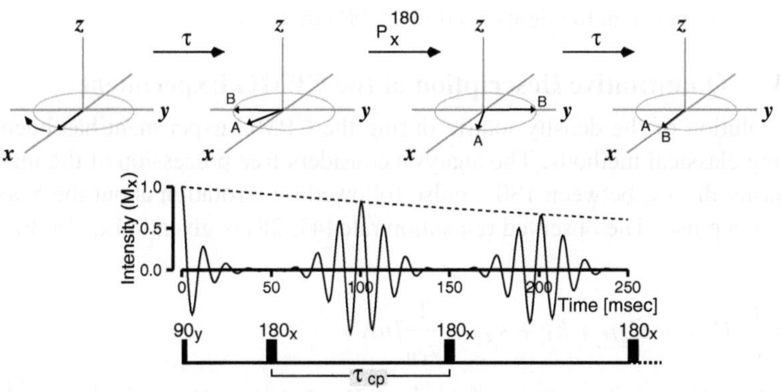

Figure 18.7. The generation of a spin-echo by a train of CPMG pulses. The top section shows the evolution of the transverse magnetization during the CPMG sequence. In this example the transverse magnetization dephases rapidly due to an inhomogeneous magnetic field. This dephasing is reversible and is refocused by a 180° pulse. The left-most section shows the magnetization immediately after the 90° pulse. During the first time period, τ, spins A and B precess at different rates due to differences in the the local magnetic field. In this example spin B precesses faster than spin A in the rotating frame. The 180° pulse rotates each spin about the x-axis. Spin A, which was lagging behind spin B, is placed ahead of spin B by the 180° pulse. In the next time period, the spins precess in the same direction as before, and at the same rate. After a period τ all of the spins will refocus to the x-axis because spins with slower precessional rates do not have to precess as far as those with higher rates. The effect of a train of such pulses, is shown in the lower part of the figure. A series of 180° pulses, spaced τ_{cp} apart, will generate a series of echos spaced τ_{cp} apart. The amplitude of which will decay according to the intrinsic spin-spin relaxation rate, R_2, as indicated by the dotted line. If chemical exchange occurs the decay rate will increase and may depend on τ_{cp}, depending on the size of k_{ex}.

However, in the case of exchange, the precessional frequency no longer is constant. Consequently, the height of the spin-echo will decrease with a rate equal to the sum of the spin-spin relaxation rate and the rate of exchange: $R_2 = R_2^o + R_{ex}$.

The second method of measuring relaxation dispersion is to apply a continuous B_1 field while the spins are transverse. Since this RF-field forces the magnetization to remain aligned with B_1, this method is referred to as *spin-locking*. This measurement of relaxation is referred to as T_1 relaxation in the rotating frame, or $T_{1\rho}$. In practice, $T_{1\rho}$ is measured at different field strengths by simply changing the strength of the B_1 field: $\omega_T = \omega_1 = \gamma B_1$. Larger spin-locking fields can be obtained using off-resonance B_1 fields.

Both CPMG methods and $T_{1\rho}$ techniques reduce the contribution of chemical exchange to the observed transverse relaxation rate. The degree of attenuation depends on the relationship between the rate of exchange and the applied field strength. A qualitative description of the effect of the transverse field strength on the contribution of chemical exchange to dephasing of the transverse magnetization is illustrated in Fig. 18.8. This figure shows the effect of increasing the transverse field strength at a fixed value of k_{ex}. In the case of CPMG methods the field strength is defined by the rate of 180° pulses, i.e. the more closely spaced pulses the higher the field. If the pulses are applied without any inter-pulse delay then a continuous B_1 is generated, which is equivalent to a spin-locking field. Thus field strengths during spin-locking are generally higher than the fields used in CPMG methods.

18.4.4.3 Quantitative Description of the CPMG Experiment

The evolution of the density matrix during the CPMG experiment has been analyzed using classical methods. The analysis considers free precession of the magnetization during the τ_{cp} between 180° pulse followed by a rotation about the x-axis of 180° by each pulse. The observed relaxation rate [47, 28] is given in Eq. 18.36.

$$\begin{aligned} R_2 &= \frac{1}{2}\left[R_{2A} + R_{2B} + k_1 + k_2\right] - \frac{1}{\tau_{cp}} ln\lambda^+ \\ \lambda^+ &= ln\left[(D_+ cosh^2\xi - D_- cos^2\eta)^{1/2} + (D_+ sinh^2\xi - D_- sin^2\eta)^{1/2}\right] \\ &= (1/2)cosh^{-1}\left[D_+ cosh2\xi - D_- cos2\eta\right] \end{aligned} \tag{18.36}$$

where,

$$D_\pm = \frac{1}{2}\left[\pm 1 + \frac{\psi + 2(\Delta\omega)^2}{\sqrt{\psi^2 + \zeta^2}}\right] \qquad \xi = \frac{\tau_{cp}}{\sqrt{8}}\left[+\psi + \sqrt{\psi^2 + \zeta^2}\right]^{1/2}$$

$$\eta = \frac{\tau_{cp}}{\sqrt{8}}\left[-\psi + \sqrt{\psi^2 + \zeta^2}\right]^{1/2} \qquad \zeta = 2\Delta\omega(R_{2A} - R_{2B} + k_1 - k_2)$$

$$\psi = (R_{2A} - R_{2B} + k_1 - k_2)^2 - (\Delta\omega)^2 + 4k_1k_2$$

Eq. 18.36 is valid for all exchange rates. Due to the complexity of the full expression it is helpful to gain an understanding of the response of the system by considering

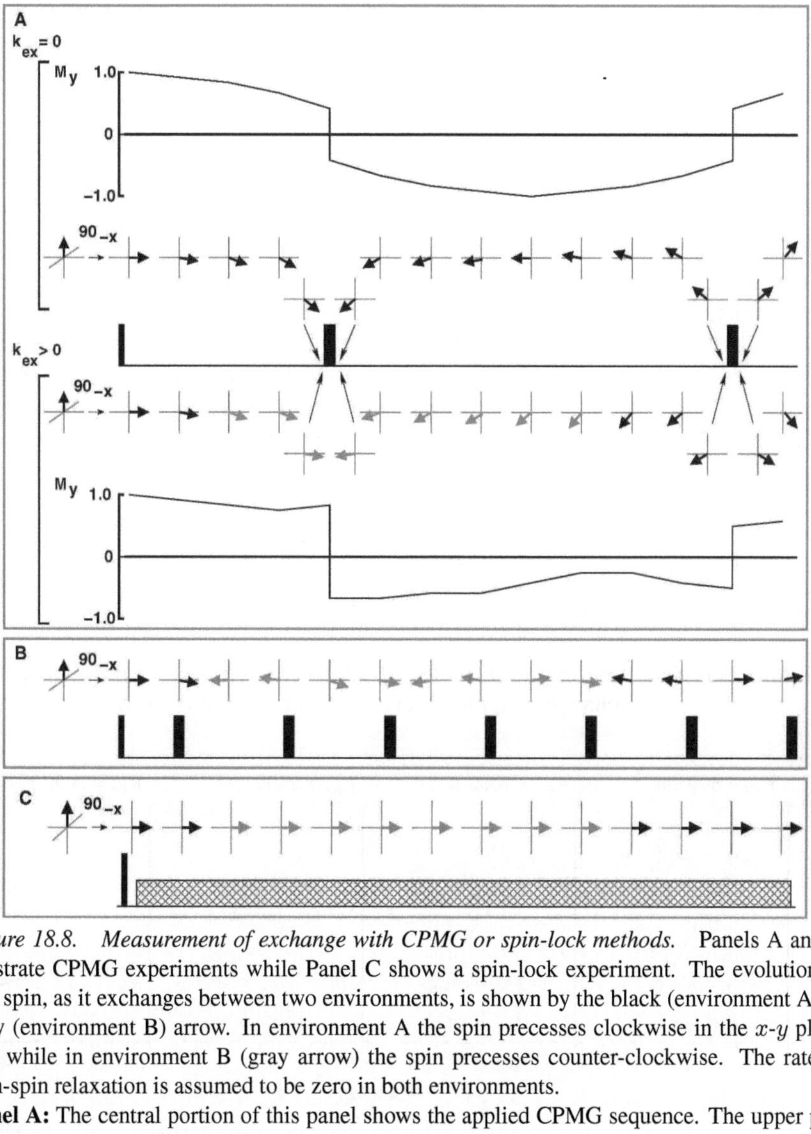

Figure 18.8. Measurement of exchange with CPMG or spin-lock methods. Panels A and B illustrate CPMG experiments while Panel C shows a spin-lock experiment. The evolution of one spin, as it exchanges between two environments, is shown by the black (environment A) or gray (environment B) arrow. In environment A the spin precesses clockwise in the x-y plane and while in environment B (gray arrow) the spin precesses counter-clockwise. The rate of spin-spin relaxation is assumed to be zero in both environments.

Panel A: The central portion of this panel shows the applied CPMG sequence. The upper part of the figure shows the evolution of a spin that does not undergo chemical exchange ($k_{ex} = 0$). The y-component of the magnetization is plotted along with the trajectory of the spin in the x-y plane. The magnetization that is precessing in the x-y plane is refocused by the first 180° pulse to give a maximum signal at τ_{cp}. In the presence of exchange ($k_{ex} > 0$), the spin begins in environment A and exchanges to B prior to the first 180° pulse. The change in precessional frequency interferes with the refocusing effect, causing an attenuation of the signal.

Panel B: The closer spacing of the 180° pulses makes it more likely that a spin will exist in a single environment during any τ_{cp} period, thus the refocusing is more effective at reducing attenuation of the signal due to exchange.

Panel C: A continuous on-resonance B_1 field along the y-axis is applied to the spins. In the rotating frame the field along the z-axis disappears, leaving only the B_1 field. The spins remain aligned, or *locked*, along B_1 in analogy to the alignment along B_o in the laboratory frame.

the simpler limiting case of fast exchange, which was derived by Luz and Meiboom [101] and by Allerhand and Thiele [4]:

$$R_2 = \frac{R_2^A + R_2^B}{2} + \frac{p_A p_B (\Delta\omega)^2}{k_{ex}} \left[1 - \frac{2}{k_{ex}\tau_{cp}} \tanh \frac{k_{ex}\tau_{cp}}{2}\right] \tag{18.37}$$

This simplification is remarkably robust in predicting the effect of exchange for experimentally accessible values of τ_{cp} (see Fig. 18.9).

The first part of the fast-exchange formula is just the averaged spin-spin relaxation rate that would be observed in the presence of very fast exchange. The second term gives the increase in the relaxation rate that is due to chemical exchange. When the exchange is very fast, $k_{ex} >> \tau_{cp}$, the observed increase in the relaxation rate is equal to that found for free precession in the presence of fast exchange (Eq. 18.20): $p_A p_B (\Delta\omega)^2 / k_{ex}$. In contrast, when the rate of the 180° pulses becomes much faster than the exchange rate, the effect of chemical exchange on the linewidth disappears completely because:

$$\begin{aligned} &lim_{\tau_{CP}\to 0} \left[1 - \frac{2}{k_{ex}\tau_{cp}} tanh(\frac{k_{ex}\tau_{cp}}{2})\right] = 0 \\ &(lim_{x\to 0}\frac{1}{x}tanh(x) = 1) \end{aligned} \tag{18.38}$$

A more complete description of the relationship between the CPMG pulse delay, the rate of chemical exchange, and the observed relaxation rate is shown in Fig. 18.9.

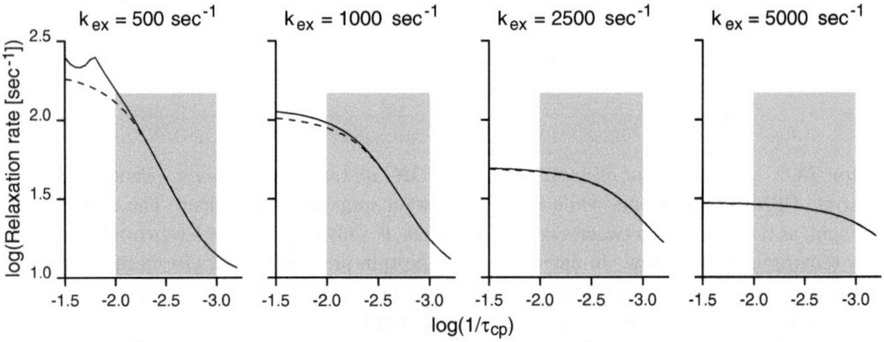

Figure 18.9. The dependence of the observed spin-spin relaxation rate on the frequency of CPMG pulses. The relaxation rate, R_2^o, in the absence of exchange, is 10 sec^{-1}. This corresponds to the intersection of the x-axis with the y-axis. The frequency separation between the two states, $\Delta\omega/2\pi$ was 100 Hz and equal populations of both states were assumed ($p_A = p_B$). The solid line shows the relaxation rate calculated with the complete formula (Eq. 18.36), while the dotted line shows the result from the fast-exchange approximation, Eq. 18.37. The sum of the exchange rates ($k_{ex} = k_1 + k_2$), from left to right, are 500 sec^{-1}, 1000 sec^{-1}, 2500 sec^{-1}, and 5000 sec^{-1}. The shaded area corresponds to τ_{cp} values (10 msec, left side; 1 msec, right side) that are experimentally accessible for proteins in the 20 – 40 kDa range. Note that the fast-exchange approximation offers a good estimation for the relaxation dispersion within the range of typical values of τ_{cp}.

In general, as the exchange rate increases, the dispersion, or change in R_2 versus τ_{cp} decreases. In the case of a relatively slow exchange rate, $k_{ex} = 500$ sec^{-1}, the R_2 due to exchange is large for long τ_{cp} values and is almost completely attenuated as τ_{cp} is decreased. In contrast, when the exchange rate is much faster than $\Delta\omega$ ($k_{ex} =$ 5000 sec^{-1}, right panel) there is very little effect of τ_{cp} on the observed R_2. For very fast exchange, e.g. $k_{ex} = 100 \times \Delta\omega$ there is no dispersion and it is not possible to obtain information on the exchange rate with CPMG techniques.

The range of experimentally feasible τ_{cp} times is reduced by limitations on the pulse rate that can be generated by the instrument and by the intrinsic relaxation properties of the sample. Although the τ_{cp} times shown in Fig. 18.9 range from 31 msec to 0.5 msec, intervals much longer than 10 msec cannot be used with moderately sized proteins because of the relatively rapid spin-spin relaxation rate causes the transverse magnetization to decay before a single cycle of the CPMG sequence can be applied. Longer τ_{cp} values, that would extend the shaded area in Fig. 18.9 to the left, can be obtained for smaller proteins because of their long spin-spin relaxation rates (see [114]).

Values of τ_{cp} much shorter than 1 msec can cause sample heating problems. Furthermore, the analytical expressions shown above do not take into account finite pulse widths, thus the effect of τ_{cp} intervals that are of the same order as the 180° pulse lengths are not well represented by existing theories.

18.4.5 Measurement of Exchange Using CPMG Methods

With the advent of widespread isotopic labeling, most exchange measurements are performed on ^{15}N or ^{13}C labeled samples using two-dimensional HMQC-like pulse sequences. The exchange properties of the heteronuclear spin are investigated and the attached proton is simply used to increase the polarization of the signal and the resolution of the spectrum.

Two widely used pulse sequences for the measurement relaxation dispersion are shown in Fig. 18.10. These sequences transfer polarization from the proton to the nitrogen (segment labeled "Polarization Transfer" in Fig. 18.10), and the nitrogen magnetization becomes transverse at the beginning of the application of the CPMG pulses. After the CMPG period, the magnetization is frequency labeled with the nitrogen frequency (segment labeled "ω_N" in Fig. 18.10) and then returned to the proton magnetization for detection using sensitivity enhancement (segments labeled "Sen. Enh." and "Coh. Sel.", in Fig. 18.10).

To measure the relaxation dispersion, a series of time points at different values of T are collected and the spin-spin relaxation rate, R_2 is obtained by fitting the data to a simple exponential decay,

$$I(T) = I_o e^{-R_2 T} \tag{18.39}$$

R_2 values are measured for a number of τ values ($\tau_{cp} = 2\tau$) to produce the dispersion curve, i.e. R_2 as a function of τ_{cp}.

A key problem with the use of heteronuclear sequences to monitor exchange is the presence of several different states of the magnetization while the spins are transverse. For example, the transverse magnetization associated with a nitrogen spin, N_y evolves

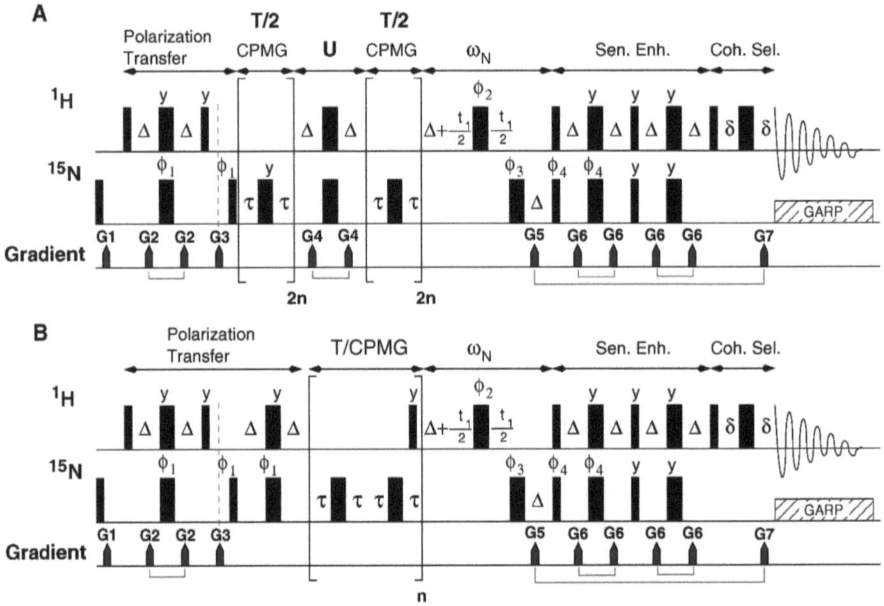

Figure 18.10. Two-dimensional nitrogen-proton correlated pulse sequences for the measurement of CPMG relaxation dispersion, $R_2(\tau_{cp})$ These sequences are described in [114]. Sequence A can be used with any value of τ while in sequence B τ is restricted to a multiple of 1/(2J) (e.g. 5.4 msec, 10.8 msec, etc.). In both cases $\tau_{cp} = 2\tau$. The total relaxation time, T, can be multiples of 8τ for sequence A and multiples of 4τ for sequence B. The initial intensity, $I(T = 0)$ is obtained by omitting the sequence in brackets. Both sequences begin by removing any nitrogen magnetization with a 90° nitrogen pulse, followed by a gradient pulse (G1). This is followed by a polarization transfer period which is a single INEPT in (A) and refocused INEPT in (B). The CPMG segment, $\tau - 180^o - \tau$, is repeated an even number of times. In the case of sequence A, the U-element is placed in the middle of the CPMG sequence to average the N_x and N_yI_z relaxation rates. Following the CPMG period, the magnetization is delivered to the ω_N segment for frequency labeling. This is followed by a sensitivity enhanced refocusing period that returns the magnetization to the amide proton. The last segment of the sequence, $\delta - 180^o_H - \delta$, containing gradient G7, selects the desired coherence level. Narrow bars represent 90° pulses and wide bars represent 180° pulses. The delay Δ is set to $1/(2J_{NH}) = 2.7$ msec, $\delta = 384$ μsec. Pulse phases are: $\phi_1 = x, -x$, $\phi_2 = 4(x)4(-x)$, $\phi_3 = x, x, y, y, -x, -x, -y, -y$, $\phi_4 = x, -x$, receiver= x, -x, -x, x. Gradients are typically applied as sine-shaped, with lengths of 1.0, 0.4, 2, 0.5, 1.8, 0.6, and 0.184 msec. Gradient strengths are $G1_{xyz} = 8$, $G2_{xyz} = 6$, $G3_z = 15$, $G4_{xyz} = 7$, $G5_{xyz} = 24$, $G6_z = 21$, $G7_{xyz} = 24$ G/cm. Selection of N- and P-coherences is accomplished with gradients G5 and G7. For each value of t_1, spectra are acquired with the above pulse phases and gradients. A second FID is then acquired with the amplitude of G5 and phase ϕ_4 inverted. The spectra are processed as described in Section 12.4. Note that the ratio of the length of G5 to G7 = γ_H/γ_N. Gradient pairs G2 and G6 serve to remove pulse imperfections of the 180° pulses. Gradient G3 is a zz-filter, retaining magnetization in the N_zI_z state.

due to proton-nitrogen J-coupling to give N_xI_z,

$$N_y \rightarrow N_y sin(\pi JT) - N_xI_z cos(\pi JT) \qquad (18.40)$$

The spin-spin relaxation rates for N_y and N_xI_z are not equal, thus the observed R_2 rates depend on the fraction of the time the spin is in the N_y state and the fraction of the time in the N_xI_z state. Therefore the observed relaxation rate will depend on the length of the CPMG pulse train, T. This leads to an oscillatory variation of the magnetization during the measurement of R_2, making it difficult to fit the data to a single exponential. Consequently, it is necessary to insure that these different relaxation rates are averaged during the measure of T_2.

In the case of sequence A in Fig. 18.10, the pulse element labeled 'U', interchanges the N_y and N_xI_z states, averaging the relaxation properties over the entire CPMG sequence. In the case of sequence B, the τ time is chosen to be a multiple of 1/(2J), e.g. $\tau = m/(2J_{NH})$, therefore over the interval $\tau - 180° - \tau$, the magnetization is completely refocused and the spin has spent an equal amount of time in both states. Note that the construction of sequence 'A' limits the values of T to multiples of 8τ while the sequence 'B' limits the values of T to multiples of 4τ. Therefore, sequence B should be used to measure R_2 for long τ_{cp} times since more values of T can be obtained for any given value of τ, allowing a more reliable determination of the relaxation rate.

18.4.5.1 Measurement of k_{ex} with $T_{1\rho}$ Spin-Lock Methods

The CPMG methods discussed above are effective for measuring exchange rates on the order of 5 to 50 times $\Delta\omega$. As shown in Fig. 18.9, $R_2(\tau_{cp})$ becomes independent of τ_{cp} for fast exchange rates, making it difficult to obtain k_{ex} from the dispersion of R_2. Higher effective pulse rates can be obtained by replacing the CPMG sequence with a continuous B_1 field oscillating at frequency ω:

$$\begin{aligned} M(t) &= B_1 cos(\omega t) \\ &= \frac{\omega_1}{\gamma} cos(\omega t) \end{aligned} \qquad (18.41)$$

If the strength of this field is larger than the frequency difference between the two environments, $\Delta\omega$, the spins will remain aligned, or locked, in the direction of the effective magnetic field in the rotating frame.

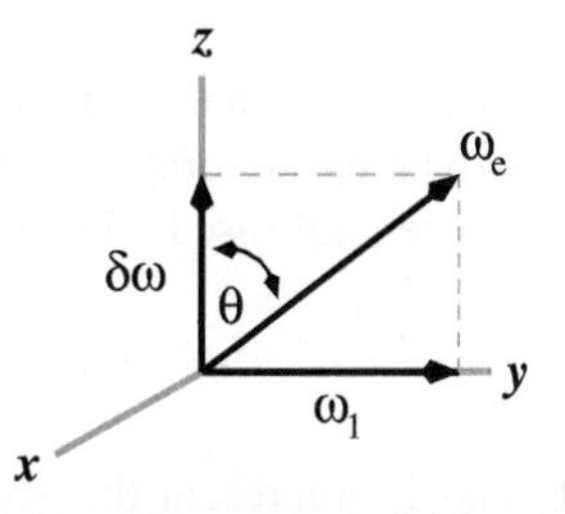

Figure 18.11. The magnetic fields present during a $T_{1\rho}$ measurement. ω_1 is the strength of the B_1 field, applied along the y-axis. $\Delta\omega$ is the remaining static field along the z-axis and is the difference between the resonance frequency of a spin (ω_o) and the frequency of the applied B_1 field (ω). ω_e is the effective field strength. The angle between the z-axis and the effective field is $\theta = tan^{-1}(\omega_1/\delta\omega)$. θ is 90° when the B_1 field is on-resonance, and approaches zero as ω becomes more distant from ω_o. $\delta\omega$ is the frequency separation between the resonance frequency and the transmitter frequency.

The fields present in the rotating frame are shown in Fig. 18.11. As discussed in Section 1.3.2.1, the change in coordinate frame causes the generation of a fictitious field that opposes the static B_o field. If the precessional frequency of a spin equals that of the B_1 field, then the effective field felt by the spin is just B_1, along the direction of the applied RF-field. If the precessional frequency of the spin differs from the frequency of the B_1 field then a z-component of the magnetic field remains because the fictitious field no longer cancels the static field at the nucleus. The effective magnetic field strength, in frequency units ($\omega = \gamma B$), is just the vector sum of the two fields.

Chemical exchange between the two environments causes relaxation of the transverse magnetization due to the rapid fluctuations in the local magnetic field at the nucleus. Since the relaxation occurs while the magnetization is aligned with the locking field, which is stationary in the rotating frame, the relaxation time is referred to as $T_{1\rho}$, in analogy to spin-lattice relaxation that occurs while the spins are aligned along the static B_o field.

The exchange rate, k_{ex}, is obtained from the dispersion of $R_{1\rho}$ versus B_1 in a manner analogous to varying τ_{cp} in CPMG experiments. The strength of the B_1 field is varied by either increasing the applied RF-power (on-resonance methods) or by moving the frequency away from the resonance frequency of the spin (off-resonance methods). In the off-resonance case the effective spin-locking field, ω_e, is given as the vector sum of both the spin-lock field (ω_1) and the field generated by the resonance off-set, ($\delta\omega = \omega - \omega_o$),

$$\omega_e = \sqrt{\omega_1^2 + \delta\omega^2} \tag{18.42}$$

consequently, much larger spin locking fields can be generated by off-resonance techniques.

The relaxation rate in the rotating frame due to chemical exchange, $R_{1\rho}$ ($= 1/T_{1\rho}$) has been derived by Deverell et al [51] for the specific case of $p_A = p_B$ and more recently for the general case by Davis et al [47]. Under the assumption of fast exchange:

$$R_{1\rho} = R_1 cos^2\theta + R_2 sin^2\theta + p_A p_B(\Delta\omega^2)\frac{k_{ex}}{k_{ex}^2 + \omega_e^2} sin^2\theta \tag{18.43}$$

where R_1 and R_2 represent the average spin-lattice and spin-spin relaxation rates of the two states. The angle between the effective magnetic field and the z-axis, θ, is given by:

$$\theta = tan^{-1}\frac{\omega_1}{\delta\omega} \tag{18.44}$$

where $\delta\omega$ refers to the frequency difference between the B_1 field and the resonance frequency of the spin.

If the spin-lock is on resonance ($\delta\omega = 0$) then the above simplifies to:

$$R_{1\rho} = R_2 + p_A p_B(\Delta\omega^2)\frac{k_{ex}}{k_{ex}^2 + \omega_1^2} \tag{18.45}$$

The largest change in the dispersion curve will occur when $k_{ex} \approx \omega_e$. The relationship between $R_{1\rho}$, k_{ex}, and the intensity of the field in the rotation frame is illustrated in Fig. 18.12 for typical values of ω_1 (on-resonance), and for ω_e (off-resonance). In the case of on-resonance methods, the experimentally available field strengths allow characterization of k_{ex} rates on the order of 10^3 sec^{-1} (see Fig. 18.12), which is slightly

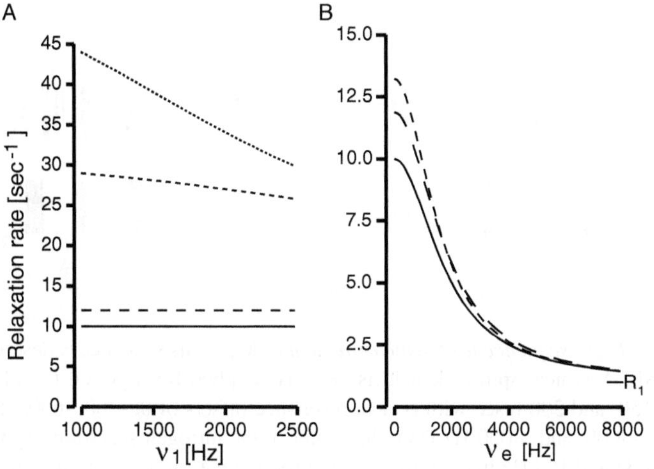

Figure 18.12. Measurement of exchange with $T_{1\rho}$. The observed relaxation rate ($T_{1\rho}$) is shown for on-resonance (A) and off-resonance (B) spin-locking fields. In the case of on-resonance spin-locking, the field strength corresponds to the field strength of the applied B_1 field, in Hz ($\nu_1 = \gamma B_1/2\pi$). In the case of the off-resonance experiment, the effective field strength is given: $\nu_e = \sqrt{\Delta\nu^2 + \nu_1^2}$, $\nu_1 = 2000$ Hz. The spin-lattice (R_1) and spin-spin (R_2) relaxation rates were 1 sec^{-1} and 10 sec^{-1}, respectively. In both panels, the solid line gives the expected response in the absence of exchange. Simulations with increased exchange rates are drawn with increased dash length. For example. in panel A, the dotted line, short dashed, and long dashed curves correspond to $k_{ex} = 2500$, 5000, and 50000 sec^{-1}, respectively. In panel B, the exchange rates correspond to 5000 and 50000 sec^{-1}. Note that the two slower exchange rates in panel A correspond to the two fastest exchange rates used to illustrate measurements with CPMG methods (Fig. 18.9), illustrating the fact that spin-lock methods are more sensitive to faster motions.

faster than CPMG methods. At very high exchange rates, $k_{ex} >> \omega_1$, $R_{1\rho}$ becomes independent of ω_1, thus making it impossible to obtain k_{ex} from the essentially flat dispersion curve. In contrast, much higher field strengths can be obtained with off-resonance methods, allowing the characterization of exchange rates from 10^3 sec^{-1} to 10^4 sec^{-1}, as illustrated in Fig. 18.12.

18.4.5.2 Measurement of $R_{1\rho}$

There are a number of published sequences that can be used to measure $R_{1\rho}$ [123, 178]. The sequence by Zinn-Justin et al [178] is shown in Fig. 18.13. The sequence begins with a refocused INEPT [1] sequence that transfers the proton polarization to the N_x state just prior to the 90° nitrogen pulse with phase ϕ_3. The 90° nitrogen pulse

[1]Formally, an INEPT sequence transfers in-phase (e.g. I_x) magnetization from an sensitive spin (e.g. protons) to anti-phase magnetization ($2I_yS_z$) of an insensitive spin, such as nitrogen. A refocused INEPT transfers in-phase magnetization from the sensitive spin to in-phase magnetization of the insensitive spin, i.e. $I_x \rightarrow S_x$, and vice-versa.

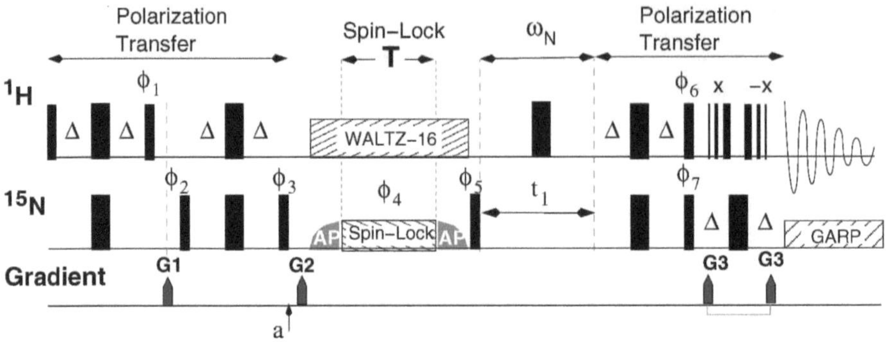

Figure 18.13. Pulse sequence for the measurement of $R_{1\rho}$. This sequence is described in more detail in [178]. A typical spin-lock field is 1800 Hz applied for a period (T) of 10, 20, 30, 50, 70, 100, 150, and 200 msec with typical resonance offsets of 0, -300, -600, -1000, -1400, -2000, -3000, -4000, and -8000 Hz from the center of the nitrogen spectrum. Typical spectral widths are 2 kHz and 10 kHz in the nitrogen and proton dimension, acquired with 128 and 512 complex points in t_1 and t_2, respectively [117]. The wide and narrow bars represent high-power 90° and 180° pulses, respectively. The two adiabatic pulses, labeled AP, are applied for 3 msec. Zinn-Justin et al [178] used amplitude modulated pulsed that are trapezoidal in shape. Mulder et al [117] describes adiabatic pulses that are both phase and amplitude modulated, which are more difficult to implement but give somewhat better alignment of the spins along the effective field. The ^{15}N frequency is shifted to the off-resonance position just before the application of the first adiabatic pulse and returned to the middle of the ^{15}N spectrum after the second. The delay Δ is set to $1/(2J_{NH})$. The phase cycle is ϕ_1 = 2(y), 2(-y); ϕ_2 = x, -x; ϕ_3 = 2(y,-y,-y,y), 2(-y,y,y,-y); ϕ_4 = 8(x), 8(-x), 8(y), 8(-y); ϕ_5 = x, -x; ϕ_6 = 4(x), 4(-x); ϕ_7 = 2(y), 2(-y). The receiver phase is (x,-x,-x,x), 2(-x,x,x,-x), (x,-x,-x,x). Quadrature detection if the ^{15}N dimension is obtained using the States method applied to ϕ_5. Gradient G1 is a zz-filter, retaining I_zN_z, G2 is a z-filter, retaining N_z, gradients G3 are for water suppression using WATERGATE (see Section 15.2.2.4). Gradient strengths are 3.5, 4.5, and 12.5 G/cm, respectively.

(ϕ_3) returns the pure nitrogen magnetization to the z-axis, aligned along the static magnetic field at point *a* in the sequence. The magnetization is then aligned along the effective field by the application of an adiabatic pulse[2]. This sequence uses a simple adiabatic pulse for alignment of the spins, a more complicated adiabatic pulse is discussed by Mulder et al [117]. In the sequence discussed here the adiabatic pulse is simply a ramp in the B_1 field over a period of 3 msec. Initially, the effective magnetization is along the z-axis (B_o). As the B_1 field increases the effective field tips towards the y-axis. If the rate of B_1 increase is slow enough, the magnetization will remain aligned along the effective field at all times. At the end of the adiabatic pulse, the magnetization is aligned along the direction of the applied off-resonance spin lock field. The spin-lock field is applied for a variable time T, and at the end of the spin lock period the magnetization is returned to the z-axis by another adiabatic pulse. In

[2]In theory, an adiabatic pulse changes the direction of the magnetization without loss or gain of energy by the spins. Here, we refer to adiabatic pulses loosely, i.e. any pulse that modifies the direction of the magnetization by causing the spins to follow the effective field in the rotating frame.

this case the B_1 field is slowly returned to zero, and the the magnetization follows the effective field back to the z-axis. The remainder of the pulse sequence consists of a frequency labeling period (ω_N), and a refocused INEPT period to return the magnetization back to the amide proton. The last 180° proton pulse has been replaced by a 3-9-19 WATERGATE sequence for water suppression (see Section 15.2.2.4).

A series of two dimensional spectra are acquired with different resonance offsets and a fixed value of ω_1, the strength of the B_1 field. For each offset value the decay of the magnetization during the spin-lock period is obtained by systematically varying the spin-lock time (T) and fitting the peak intensities to the following equation:

$$I(t) = I_o e^{-R_{1\rho}T} \tag{18.46}$$

Determination of the exchange rate, k_{ex}, requires fitting of the dispersion data to the following equation:

$$R_{1\rho} = R_1 cos^2\theta + R_2 sin^2\theta + p_A p_B(\Delta\omega^2)\frac{k_{ex}}{k_{ex}^2 + \omega_e^2} sin^2\theta \tag{18.47}$$

Note that the separate populations, p_A and p_B, or the frequency separation, $\Delta\omega$, cannot be obtained from this data fitting since the product of all three of these terms occurs in Eq. 18.47.

As indicated in Eq. 18.47, $R_{1\rho}$ depends on θ, R_1, R_2, $p_A p_B \Delta\omega^2$, k_{ex}, and ω_e. Of these, θ, and ω_e can be calculated directly since the frequency difference, $\delta\omega$, between the transmitter and the resonance line is known. The remaining parameters are unknown. Since R_1 is not affected by exchange, it can be measured independently using the techniques discussed in Chapter 19. Therefore, the remaining parameters that have to be obtained by data fitting are: $k_{ex}, p_A p_B \Delta\omega^2$, and R_2.

Instead of determining R_1 directly, the $T_{1\rho}$ experiment can be modified to measure R_1 concurrently with the measurement of $R_{1\rho}$, as described by Palmer and co-workers [3]. In this case, the fitted parameters become:(R_1-R_2), k_{ex}, and $p_A p_B \Delta\omega^2$.

18.5 Distinguishing Fast from Slow Exchange

The use of Eq. 18.37, to analyze CPMG data, or Eq. 18.43 to analyze $T_{1\rho}$ data, assumes that the exchange is fast, i.e. $k_{ex} > \Delta\omega$. Often, the time scale for exchange can be easily deduced from the NMR spectrum. If two separate resonance lines are observed for a single spin, then the system is clearly in slow exchange. However, the converse is not necessarily true, namely the presence of a single resonance line does not prove that the system is in fast exchange. It is possible that the system is in slow exchange, but one resonance line is of such low intensity (e.g. $p_A << p_B$) that only one is readily visible in the spectrum. Alternatively, the second resonance line is not resolved in the spectrum. Therefore, before CPMG and $T_{1\rho}$ measurements are used to characterize the exchange processes it is important to verify the presence of fast exchange.

18.5.1 Effect of Temperature

Observing the effect of temperature on the spectrum is the simplest method of determining the time scale of chemical exchange. Exchange rates will increase with temperature, with the rate of increase dependent on the activation enthalpy, $\Delta H^{\dagger}$, for exchange:

$$k_{ex} \propto e^{-\Delta H^{\dagger}/RT} \quad (18.48)$$

In the case of slow exchange, resonance lines will broaden as the temperature is increased, therefore line broadening at elevated temperatures implies that the system is transitioning from slow exchange at low temperature to intermediate exchange at higher temperatures.

Unfortunately, line narrowing at higher temperature does not necessarily imply that the time scale is transitioning from intermediate to fast exchange. Line-narrowing occurs at higher temperatures regardless of chemical exchange because the rotational correlation time of the protein decreases with temperature, which also causes line narrowing. Unless the activation enthalpy for exchange is much larger than the temperature dependence of the rotational correlation time, it may be difficult to verify the existence of fast exchange from the temperature dependence of the linewidth.

18.5.2 Magnetic Field Dependence

Fast exchange can be distinguished from slow exchange by varying the magnetic field strength. When the exchange is very slow the resonance line will be broadened by the exchange rate constant, k_1,

$$\Delta\nu = \frac{1}{\pi T_2} + k_1 \quad (18.49)$$

where T_2 is the spin-lattice relaxation time in the absence of exchange. The increase in linewidth due to exchange is clearly *independent* of the magnetic field strength.

Conversely, when the exchange is very fast, the observed linewidth is:

$$\Delta\nu = \frac{1}{\pi T_2} + \frac{1}{\pi}\frac{p_A p_B (\Delta\omega)^2}{k_{ex}} \quad (18.50)$$

The increase in linewidth due to exchange depends on the *square* of the magnetic field strength, due to the $\Delta\omega^2$ term.

Palmer and co-workers [114] have defined a dimensionless parameter, α, to characterize the time scale of chemical exchange.

$$\alpha = \frac{d ln R_{ex}}{d ln \Delta\omega} \quad (18.51)$$

where R_{ex} is the contribution of exchange to the overall spin-spin relaxation rate: R_2=R_2^o+R_{ex}.

An analytical expression for α can be easily obtained from an approximate expression for R_{ex} that is valid over all time scales, provided $p_A >> p_B$ [155]:

$$R_{ex} = \frac{p_A p_B k_{ex}}{1 + (k_{ex}/\Delta\omega)^2} \quad (18.52)$$

gives[3] the following for α:

$$\alpha = \frac{2/(k_{ex}/\Delta\omega)^2}{1 + (k_{ex}/\Delta\omega)^2} \tag{18.53}$$

Therefore α varies from 0 for very slow exchange to 2 for very fast exchange. Intermediate exchange, where $k_{ex} \approx \Delta\omega$ is characterized by an α value of 1.

Experimentally, α is obtained by measuring R_{ex} at two different static field strengths, B_{o1} and B_{o2} and approximating α as: [4]

$$\alpha \approx \left[\frac{B_{o2} + B_{o1}}{B_{o2} - B_{o1}}\right]\left[\frac{R_{ex2} - R_{ex1}}{R_{ex2} + R_{ex1}}\right] \tag{18.54}$$

An α value of zero ($R_{ex2} - R_{ex1} = 0$) indicates that the exchange is not dependent on the magnetic field and is therefore slow.

18.6 Ligand Binding Kinetics

The binding of a ligand, L, to a protein, P, generally causes a change in the environment of the nuclear spins on either the protein, the ligand, or both. Since a change in environment usually results in a change in chemical shift, ligand binding is equivalent to chemical exchange between two environments. In this case the two states correspond to the bound (PL) or free protein (P). The reaction for ligand binding to a single site is:

$$\mathrm{P} + \mathrm{L} \underset{k_{off}}{\overset{k_{on}}{\rightleftharpoons}} \mathrm{PL}$$

The equilibrium binding constant for this reaction is given by:

$$K_{eq} = \frac{k_{on}}{k_{off}} \tag{18.55}$$

where k_{on} is the second order rate constant for the binding of the ligand to the protein, and k_{off} represents the rate at which the ligand leaves the protein.

The kinetic rate equation in this situation is:

$$\begin{aligned} \frac{d[P]}{dt} &= -k_{on}[L][P] + k_{off}[PL] \\ \frac{d[PL]}{dt} &= +k_{on}[L][P] - k_{off}[PL] \end{aligned} \tag{18.56}$$

Note that the only difference between these equations and those given for general exchange in Eq. 18.11 is that the forward rate constant k_1, now becomes $k_{on}[L]$. Therefore, any of the previous described techniques can be applied to the study of ligand binding kinetics.

There are two key differences between general exchange and ligand binding:

---

[3] Recall $\frac{dlnu}{dx} = \frac{1}{u}\frac{du}{dx}$, therefore $\frac{dlnR}{dln\Delta\omega} = \frac{dR}{R}\frac{\Delta\omega}{d\Delta\omega}$.

[4] $\alpha = \frac{dR}{R}\frac{\Delta\omega}{d\Delta\omega} = \frac{dR}{R}\frac{B_o}{dB_o}$.

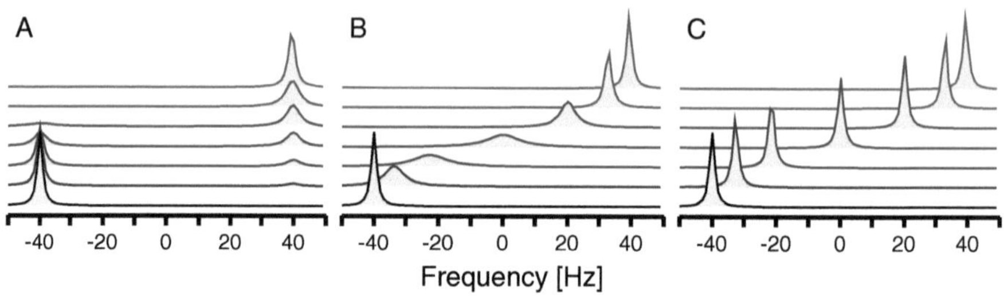

Figure 18.14. Effects of Ligand Binding on NMR Lineshapes. This figure illustrates the effect of slow (A), intermediate (B), or fast (C) exchange on the spectrum of a resonance whose frequency is changed as a result of ligand binding. All three panels correspond to an equilibrium binding constant, $K_{eq} = k_{on}/k_{off}$, of 10^4 M^{-1} but differ in the on- and off-rates. The on-rates were 10^5 $M^{-1}sec^{-1}$ (A),10^7 $M^{-1}sec^{-1}$ (B), and 10^9 $M^{-1}sec^{-1}$(C). The free and bound resonance positions are separated by 80 Hz ($\nu_A = -40$ Hz, $\nu_B = +40$ Hz). Spectra were simulated with ligand concentrations of 0, 10 μM, 30 μM, 100 μM, 300 μM, 1 mM, and 10 mM, increasing from front to back. These concentrations give the following fractions of liganded protein (p_B): 0.00, 0.09, 0.23, 0.50, 0.75, 0.91, and 0.99, respectively.

1 The forward rate, k_1, can be conveniently changed by the ligand concentration, since $k_1 = k_{on}$[L]. Thus the time scale of exchange can be controlled, to some extent, by the varying the ligand concentration.

2 The chemical shifts of the unliganded and fully liganded states can be measured by acquiring spectra in the absence and presence of ligand. Consequently it is possible to obtain $\Delta\omega$, and in some cases a direct measurement of the equilibrium constant by acquiring a series of spectra at different ligand concentrations.

Figure 18.14 shows the effect of chemical exchange of the observed spectra. Three different scenarios are shown, slow (A), intermediate (B), and fast (C) exchange. Approaches to obtain the kinetic rate constants for each of these time scales are briefly discussed below.

18.6.1 Slow Exchange

In the case of slow exchange the resonance for the unliganded state simply disappears and a resonance line at the position of the liganded state appears. At any given ligand concentration the fraction of protein with bound ligand can be calculated from the integrated intensities of the two lines:

$$p_B = \frac{I_B}{I_A + I_B} \qquad (18.57)$$

The dependence of p_B on the *free* ligand concentration follows the ligand binding equation:

$$p_B = \frac{K_{eq}[L]}{1 + K_{eq}} \qquad (18.58)$$

K_{eq} can be obtained by direct fitting of this function. Alternatively, it can be linearized with a Scatchard plot,

$$\frac{p_B}{[L]} = K_{eq} - p_B K_{eq} \tag{18.59}$$

If the binding is non-cooperative and to a single site, then the slope of this line is $-K_{eq}$.

The individual rate constants can be obtained by direct measurement of the linewidth at low ligand concentrations. At low ligand concentrations the lifetime of the unliganded state is shortened when the protein binds a ligand, therefore the observed spin-spin relaxation rate becomes:

$$R_2^A = R_2^{oA} + k_{on}[L] \tag{18.60}$$

The relaxation rate, R_2^{oA} in the absence of exchange is known, therefore a plot of R_2^A versus the *free* ligand concentration will have a slope of k_{on}.

18.6.2 Intermediate Exchange

In the intermediate exchange time regime the resonance line becomes very broad as ligand is added and slowly migrates from the unliganded position towards the position of the fully liganded protein. Under some conditions, the line can disappear completely during the titration, making it difficult to identify the resonance position of the fully bound species. Due to the complex nature of lineshape it is necessary to analyze these data using the complex expression for chemical exchange that was presented in Section 18.4.3. The only modification to this equation is to replace k_1 with $k_{on}[L]$. Generally, several spectra are acquired over a range of ligand concentrations and values of k_{on} and k_{off} are found that minimize the difference between the observed lineshape and that predicted from Eq. 18.11.

18.6.3 Fast Exchange

In the case of fast exchange, the exchange is sufficiently fast that the chemical shift of the observed line, δ, is essentially equal to the weighted average of the initial and final states:

$$\delta = p_A \delta_A + p_B \delta_B \tag{18.61}$$

At any given ligand concentration, the amount of bound protein, p_B, is easily found:

$$p_B = \frac{\delta - \delta_A}{\delta_B - \delta_A} \tag{18.62}$$

The equilibrium binding constant can be obtained using the methods described above for slow exchange.

Obtaining the individual rate constants is more difficult in the case of fast exchange. However, since k_{ex} is now dependent on the ligand concentration,

$$k_{ex} = k_{on}[L] + k_{off} \tag{18.63}$$

it is possible to use non-saturating amounts of ligand such that the exchange rate is in the range that is suitable for CPMG or $T_{1\rho}$ measurements. Once k_{ex} is obtained, then the rate constants can be obtained using the equilibrium binding constant.

18.7 Exercises

1. Show that the CPMG sequence will refocus evolution due to chemical shift evolution or magnetic field inhomogeneities.

2. The H_α proton of a histidine residue in a protein shows a chemical shift of 5.0 ppm when the sidechain is protonated and 4.0 ppm when it is deprotonated. If this histidine has a pK_a of 6.5, what is the chemical shift of the H_α proton at this pH? State any assumptions that you make.

3. The resonance line from a proton on a small organic ligand has a linewidth of 1 Hz. The affinity constant for the binding of this ligand to a protein is $10^6\ M^{-1}$. The NMR spectrum of a solution of 10 μM protein and 6 μM ligand is obtained and the *position* of the resonance line from the ligand is unchanged, but the linewidth is increased to 10 Hz. Determine k_{on} for the binding of this ligand.

18.8 Solutions

1. This can easily be seen using a product operator analysis. During the first τ period evolution due to chemical shift is: $I_y \rightarrow I_y cos(\omega\tau) - I_x sin(\omega\tau)$. The 180° pulse along x converts this to $-I_y cos(\omega\tau) - I_x sin(\omega\tau)$. This magnetization evolves for an additional τ period to give: $-cos(\omega\tau)[I_y cos(\omega\tau) - I_x sin(\omega\tau)] - sin(\omega\tau)[I_x cos(\omega\tau) + I_y sin(\omega\tau)]$. This can be simplified as: $-I_y[cos^2(\omega\tau) + sin^2(\omega\tau] + I_x[cos(\omega\tau)sin(\omega\tau) - cos(\omega\tau)sin(\omega\tau)] = -I_y$.

2. We assume the system is in fast exchange, i.e. the rate of protonation/deprotonation is much faster that the chemical shift difference. Therefore, the observed chemical shift will be the weighted average of the two states. Since the pH = pK_a, both states are equally populated and the chemical shift will be:

$$4.5 = \frac{1}{2}4.0 + \frac{1}{2}5.0 \tag{18.64}$$

3. Since the position of the line did not move, the system is under slow exchange. The increase in linewidth of 9 Hz is equal to k_1, which is equal to $k_{on}[L]$. The free ligand concentration is obtained from the ligand binding equation:

$$\frac{[PL]}{[PL]+[P]} = \frac{K_A[L]}{1+K_A[L]} \tag{18.65}$$

where $[L]$ is the concentration of free ligand. In general, this can be found by solving a standard quadratic equation found in may physical biochemistry texts. Here, the ligand concentration is such that 50% of the protein has ligand bound and therefore $[L] = 1\ \mu$M. Therefore the kinetic on-rate is $9/(*1 \times 10^{-6}) = 9 \times 10^6$ sec^{-1}M^{-1}.

Chapter 19

NUCLEAR SPIN RELAXATION AND MOLECULAR DYNAMICS

19.1 Introduction

The principal goal of NMR relaxation studies is to characterize molecular motion. Two of the more common applications of these measurements are:

1. Investigate relative motion of domains within a protein. In this case, the rotational diffusion constant is obtained for each domain. For example, the motion of each domain in the protein calmodulin has been characterized using relaxation [33].
2. Determining the extent and rate of internal motion at specific sites within a protein. A comprehensive analysis of the relaxation properties of the amide group in a number of proteins appeared more than a decade ago [36, 84]. More recent studies have focused on the relaxation properties of methyl-containing side chains to characterize the dynamics of hydrophobic regions in proteins [76].

In all of the above applications, the desired information is obtained by comparing measured relaxation parameters to those calculated from models of the motion. Therefore, it is important to be able to predict the relaxation of the spins with reasonable accuracy. Consequently, the relaxation rates of heteronuclear spins, such as ^{15}N and ^{13}C, are generally measured because their relaxation is dominated by the dipolar coupling to the attached proton and by their own chemical shift anisotropy. The effect of both of these interactions on relaxation can be calculated with a high degree of accuracy. In contrast, the relaxation properties of a ^{1}H spin will depend on interactions with multiple surrounding protons. Clearly this situation is very difficult to model without accurate coordinates of the protons within the proton.

Although the major thrust of relaxation studies is to characterize molecular motion, knowledge of relaxation rates is also important for defining experimental parameters. For example, the proton T_1 relaxation time usually limits the delay before acquiring the next FID while the T_2 relaxation time of both protons and heteronuclear spins determines the length of time that free induction decay should be sampled. These issues are discussed in more detail in Chapter 15.

19.1.1 Relaxation of Excited States

All excited systems eventually return to the ground state by relaxation. In the case of NMR we will be concerned with spin-lattice relaxation (T_1), which restores thermal equilibrium, and spin-spin relaxation (T_2), which is the loss of coherent transverse magnetization. Both of these processes require transitions between the excited and ground states, as illustrated in Fig. 19.1.

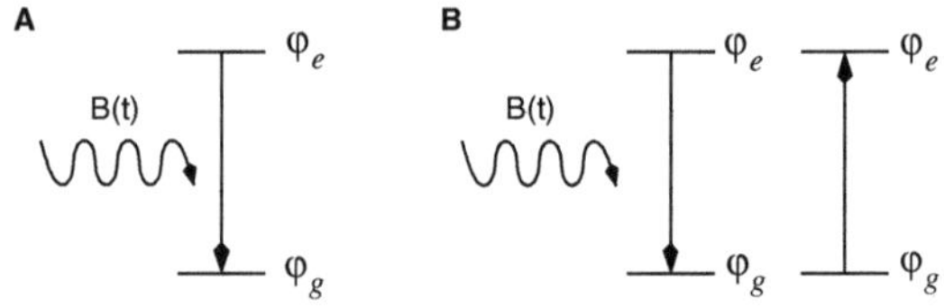

Figure 19.1. Relaxation of the excited state. Nuclear spin relaxation is a consequence of time-dependent fluctuations in the magnetic field. These fluctuations can cause the net loss of energy from the excited state (part A) or enhance the rate of spin-spin exchange (part B). The former contributes to both T_1 and T_2 relaxation while the latter only contributes to T_2 relaxation.

In addition to spin-lattice and spin-spin relaxation times, we will also investigate the relationship between dipolar coupling and the Nuclear Overhauser Effect. The Nuclear Overhauser Effect (NOE) measures how perturbation of the ground and excited state populations of one spin affects the populations of another coupled spin. If the two coupled spins are unlike, such as the ^{15}N-^{1}H group, then this effect is referred to as the heteronuclear NOE (hnNOE). The hnNOE is used to characterize the motional properties of ^{15}N-H and ^{13}C-H bond vectors and will be discussed in detail in this chapter. If the two coupled spins are protons, then the NOE measurement can be used to obtain inter-proton distances for structure determination (see Chapter 16).

Transition between the excited and ground states can occur by two independent mechanisms, spontaneous emission and stimulated emission. Spontaneous emission is due to coupling between the excited state and the blackbody radiation field. The rate of spontaneous emission is proportional to the third-power of the absorption frequency. In the case of optical transitions ($\nu = 4^{15}$ Hz), spontaneous emission occurs rapidly, with a rate on the order of 10^9 sec^{-1}. However, at typical proton NMR resonance frequencies, in the megahertz range, the rate of spontaneous emission is essentially zero. Therefore, the major source of relaxation for excited NMR spins is from *stimulated* emission.

Stimulated emission requires the presence of an oscillating electromagnetic field whose frequencies are matched to the absorption frequencies of the NMR transitions. In the case of T_1 relaxation, these field fluctuations have to be orthogonal to the B_o field such that they can return the magnetization to the ground state to restore thermal equilibrium. In the case of T_2 relaxation, there is no net loss of energy by the system, but simply a dephasing of the transverse magnetization. Consequently, T_2 relaxation is caused by fluctuations in the magnetic field in the z-direction (parallel to B_o). Since static inhomogeneities in the magnetic field can also cause dephasing of the transverse magnetization, T_2 relaxation can also be caused by motions with frequencies much lower than the transition energies.

The oscillating fields are created by random rotational motion of the molecule or by internal motions within the molecule. Such motions can generate a time dependent magnetic field by two mechanisms. First, if the electron shielding is anisotropic, then the effective field at the nucleus will vary depending on the orientation of the electrons with respect to the external field. Second, if two spins are coupled via dipolar coupling, then the relative orientation of the spins affects the strength of the dipolar field. If the temporal changes in the magnetic field strength contain frequencies that match the resonance frequencies then efficient relaxation can occur. The intensity of the magnetic field fluctuations as a function of the frequency is represented by the spectral density function, J(ω). These magnetic field fluctuations cause transitions between the energy levels at a rate W.

The first part of the chapter will develop the connection between molecular motion and the observed relaxation parameters, T_1, T_2, and hnNOE. The origin of the magnetic field fluctuations will be described using classical models. The frequency components associated with these motions will be determined by calculating $J(\omega)$. The transition rates for stimulated emission, W, will be obtained from $J(\omega)$ using a simple quantum mechanical analysis. Once the transition rates are obtained, it is possible to calculate the expected values for the relaxation parameters, T_1, T_2, and hnNOE. The relationship between molecular motion and the experimental data can be summarized as:

$$Molecular\ Motion \rightarrow J(\omega) \rightarrow W \rightarrow T_1, T_2, hnNOE$$

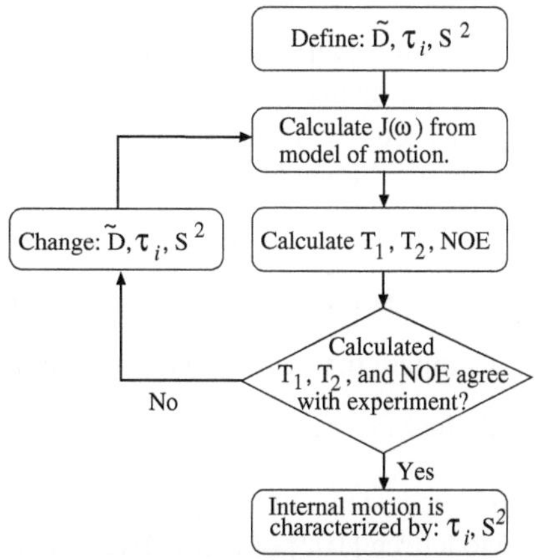

Figure 19.2. Flow chart for analysis of relaxation data. This flow chart indicates how parameters that describe internal motion are obtained from relaxation measurements.

The second part of the chapter will discuss how the motional parameters are obtained from the experimentally measured relaxation parameters. It is difficult to work 'backwards' from the measured relaxation parameters to a description of the molecular motion because the number of unknowns can exceed the number of experimental measurements. For example, to truly describe the internal motion of a C_α-H bond vector in a protein you would have to specify the rotational diffusion constant of the protein as well as the rate and extent of change in the ϕ and ψ angles of the residue. This gives a total of five parameters if the protein is spherical and 9 if the protein tumbles as an asymmetric particle, with three diffusion coefficients[1]. For a spherical N-residue protein, the total number of unknowns would be $4N + 1$. However, the experimental data consist of three observations, T_1, T_2, and the hnNOE for each residue, giving a total of $3N$ experimental measurements. In theory, the number of data points

[1]The complete description of anisotropic diffusion requires specification of the diffusion rates about all three axis (D_x, D_y, and D_z), and the three Euler angles to orient the coordinate system of the diffusion tensor in the molecular frame.

can be increased by performing relaxation measurements at different field strengths. However, the range of the available field strengths is small, therefore the data is highly correlated and does not provide a sufficient amount of independent information.

The current approach for the analysis of molecular motion begins with a general model of internal motion that contains only two or three parameters per residue. These parameters are obtained from the relaxation data by following the scheme that is diagrammed in Fig. 19.2 consisting of the following steps:

1. Collect T_1, T_2, and hnNOE data at one or more magnetic field strengths.
2. Define a parameterized model for the molecular motion. The parameters are generally the overall diffusion rate of the protein, $\tilde{D}$, a time constant that characterizes the rate of internal motion, τ_i, and a limited range over which the internal motion can occur, with this range defined by S^2 (order parameter). The spectral density function depends on these parameters, i.e. $J(\omega) = \mathcal{F}(\tilde{D},\ \tau_i,\ S^2)$.
3. Select a set of initial guesses for $\tilde{D}$, τ_i, and S^2 and calculate $J(\omega)$.
4. Calculate the expected T_1,T_2, and heteronuclear NOE from $J(\omega)$.
5. Compare the measured relaxation parameters to those calculated.
6. Adjust $\tilde{D}$, τ_i, and S^2 until the calculated and measured relaxation rates agree.

In practice, steps 2-6 are performed with the assistance of a computer program, such as Modelfree [102] or NORMAdyn [126].

19.2 Time Dependent Field Fluctuations

The time dependent magnetic field fluctuations are generated by two principal mechanisms, the anisotropic chemical shift (chemical shift anisotropy, CSA) associated with a single spin and the dipolar coupling between nearby spins. Both of these mechanism operate simultaneously for all types of spins, however dipolar coupling is the principal relaxation mechanism for the proton because of its relatively small chemical shift anisotropy.

19.2.1 Chemical Shift Anisotropy

The external B_o field is reduced at the nucleus by the surrounding electron density, by a shielding factor σ, giving an observed chemical shift of $\omega_s = \gamma(1-\sigma)B_o$. If the electron density is the same in all directions (isotropic) then a change in the orientation of the spin will have no effect on the shielding of the nucleus, and therefore no effect on the magnetic field at the nucleus.

If the shielding is anisotropic, then different orientations of the molecule will generate different magnetic fields at the nucleus. The actual field will depend on the orientation of the molecule with respect to the external magnetic field. To characterize the anisotropic nature of the shielding we will define a chemical shift tensor, $\tilde{\sigma}$, which will give the chemical shift for any given orientation of the molecule with respect to the B_o field. The chemical shift tensor is related to the shielding as follows: $\tilde{\delta} = (1-\tilde{\sigma})$, where $\tilde{\sigma}$ is the shielding tensor.

For arbitrary orientations of the molecule the chemical shift tensor takes the form:

$$\tilde{\delta} = \begin{bmatrix} \delta_{xx} & \delta_{xy} & \delta_{xz} \\ \delta_{yx} & \delta_{yy} & \delta_{yz} \\ \delta_{zx} & \delta_{zy} & \delta_{zz} \end{bmatrix} \quad (19.1)$$

In one particular orientation of the molecule with respect to the magnetic field, the principal axis system (PAS), the chemical shift tensor takes on a simple form:

$$\tilde{\delta}_{PAS} = \begin{bmatrix} \delta_{xx} & 0 & 0 \\ 0 & \delta_{yy} & 0 \\ 0 & 0 & \delta_{zz} \end{bmatrix} \quad (19.2)$$

The diagonal form of the chemical shift tensor gives directly the chemical shift that would be observed if the magnetic field were along the x-axis (δ_{xx}), y-axis (δ_{yy}), or the z-axis (δ_{zz}) of the principal axis system (see Fig. 19.3). Note that the a coordinate transformation can always be found that produces a diagonal chemical shift tensor.

If the molecule tumbles rapidly and isotropically in solution, then the observed *isotropic* shift is simply the average of all three components:

$$\delta_{iso} = \frac{1}{3}[\delta_{xx} + \delta_{yy} + \delta_{zz}] \quad (19.3)$$

The asymmetry of the tensor is defined as:

$$\eta = \frac{\delta_{yy} - \delta_{xx}}{\delta_{zz}} \quad (19.4)$$

and the chemical shift anisotropy (CSA) is usually defined as:

$$\Delta\delta = \delta_{zz} - (\delta_{xx} + \delta_{yy})/2 \quad (19.5)$$

In many cases the shielding, and therefore the chemical shift, shows axial symmetry, i.e. two of the three values in Eq. 19.2 are equal. The unique tensor element, which is parallel to the symmetry axis, is defined as $\delta_{\parallel}$. The two equivalent elements are defined as $\delta_{\perp}$. In this case, the CSA is $\Delta\delta = \delta_{\parallel} - \delta_{\perp}$ and the asymmetry is zero.

19.2.1.1 Transition Energies

The relationship between the chemical shift tensor and the observed frequency can be obtained by considering the energy of a particular state using the following Hamiltonian:

$$\mathcal{H} = \gamma \vec{B} \tilde{\delta} \vec{S} \quad (19.6)$$

If the molecule is oriented such that the z-axis of the PAS is aligned along the magnetic field, the energy of each state is obtained as follows:

$$\mathcal{H} = \gamma \begin{bmatrix} 0 & 0 & B_o \end{bmatrix} \begin{bmatrix} \delta_{xx} & 0 & 0 \\ 0 & \delta_{yy} & 0 \\ 0 & 0 & \delta_{zz} \end{bmatrix} \begin{bmatrix} S_x \\ S_y \\ S_z \end{bmatrix} = \gamma B_o \delta_{zz} S_z \quad (19.7)$$

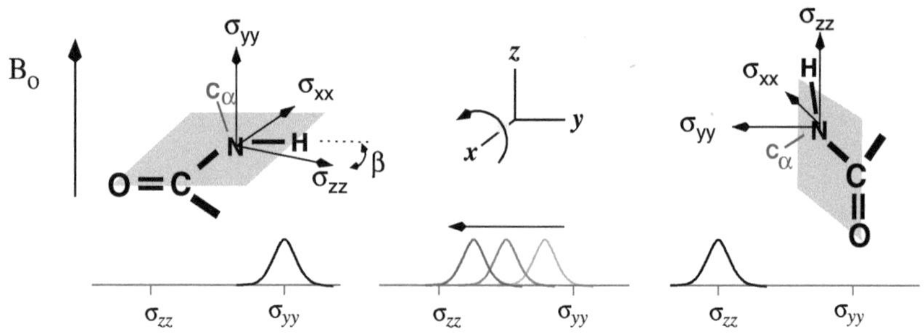

Figure 19.3. Effect of orientation on the amide chemical shift. The orientation of the peptide group affects the observed chemical shift of the amide nitrogen. The chemical shift can be obtained from the chemical shift tensor. This tensor, in the principal axis system is superimposed on the molecular structure. The z-axis of the PAS lies in the plane of the peptide bond (shaded gray) and is rotated away from the N-H bond vector by the angle β. The x-axis of the PAS is in the plane of the peptide bond and the y-axis is perpendicular to the peptide plane. If the tensor is axially symmetric, $\delta_{zz} = \delta_{\parallel}$ and $\delta_{\perp}$ lies in the plane defined by δ_{xx} and δ_{yy}. In the leftmost structure the orientation of the amide group is such that δ_{yy} is parallel to the static field (B_o). Therefore the measured chemical shift is δ_{yy}, as illustrated in the spectrum in the lower part of the figure. As the molecule rotates about the x-axis the peak position will move from δ_{yy} to δ_{zz}, as shown in the middle of the diagram. After a 90° rotation, the z-component of the chemical shift tensor becomes parallel to the static field, and the observed chemical shift is δ_{zz}.

The energy of the transition is, as usual, given by the energy difference between the states:

$$\Delta E = \gamma B_o \delta_{zz}(\frac{1}{2} - \frac{-1}{2}) = \gamma B_o \delta_{zz} = \omega_o \delta_{zz} \tag{19.8}$$

The units of δ are in ppm relative to a reference compound, giving a transition frequency in rad/sec. Since the contribution of the CSA to relaxation depends on the *range* of frequencies sampled as the molecule reorients, the actual chemical shift is not important.

19.2.1.2 Effect of Molecular Orientation

The rotation of atoms with anisotropic chemical shifts will result in a change in the observed chemical shift. The effect of molecular rotation on the chemical shift can be calculated by applying the rotation matrix, R, to the chemical shift tensor: $\tilde{\delta}' = R^T \tilde{\delta} R$. For example, rotation of the amide group shown in Fig. 19.3 about the x-axis by an angle θ would have the following effect on the chemical shift tensor:

$$\begin{aligned}
\tilde{\delta}' &= R_x^{\theta T} \tilde{\delta} R_x^{\theta} \qquad (19.9) \\
&= \begin{bmatrix} 1 & 0 & 0 \\ 0 & cos\theta & -sin\theta \\ 0 & sin\theta & cos\theta \end{bmatrix} \begin{bmatrix} 100 & 0 & 0 \\ 0 & 150 & 0 \\ 0 & 0 & 100 \end{bmatrix} \begin{bmatrix} 1 & 0 & 0 \\ 0 & cos\theta & sin\theta \\ 0 & -sin\theta & cos\theta \end{bmatrix}
\end{aligned}$$

Table 19.1. ^{15}N Amide Chemical Shift Tensors.

Compound	δ_{xx}	δ_{yy}	δ_{zz}	$\Delta\delta$	η
AlaAla	65.3	78.1	215.5	-144	0.06
AcGlyAla	44.6	85.1	229.4	-165	0.16

19.2.1.3 Observed Chemical Shift Tensors

Amide Nitrogen: The chemical shift tensor for an amide group in a number of dipeptide model compounds has been determined from solid-state NMR studies [120] and the principle axis tensor values are given in Table 19.1. For calculations of relaxation rates, the tensor is generally assumed to be axially symmetric ($\eta = 0$) with an average $\Delta\delta$ value of 165 ppm. Based on the values presented in Table 19.1 this is clearly an approximation and highly quantitative studies of relaxation will require the determination of the CSA tensor at individual sites within the protein. These values can be obtained by measuring the combined effect of chemical shift anisotropy and dipolar coupling on relaxation. These experiments are beyond the scope of this text, details can be found in [59].

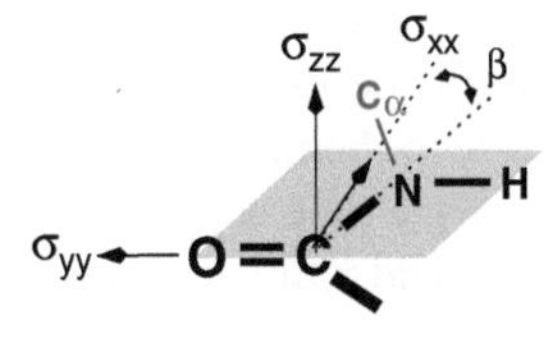

Figure 19.4. Carbonyl chemical shift tensor. The orientation of the carbonyl chemical shift tensor is shown. δ_{xx} and δ_{yy} both lie in the peptide plane while δ_{zz} is perpendicular to the peptide plane. β is the angle between the C-N bond and the δ_{xx} component of the chemical shift tensor.

Carbonyl Carbon: The carbonyl carbon also possesses a large CSA because of its highly asymmetric bonding environment. The orientation of the tensor is shown in Fig. 19.4. Tensor values have been determined for a number of compounds [165] and values for the tensor components in AlaGly are: $\delta_{xx} = 248$ ppm, $\delta_{yy} = 173$ ppm, $\delta_{zz} = 91$ ppm. Relaxation studies of the carbonyl carbon are generally not performed because of the lack of an attached proton. However, the large CSA tensor of the carbonyl carbon contributes to its relaxation, especially at higher magnetic fields. The enhanced transverse relaxation rate can reduce the sensitivity of multi-nuclear experiment that utilize the carbonyl carbon for magnetization transfer, such as the HNCO or HNCOCA experiment (see Chapter 14).

19.2.2 Dipolar Coupling

Dipolar coupling arises when the magnetic field of one nuclear spin affects the local magnetic field of another spin. The magnetic field generated by the I spin is given by the classic equation for a dipole field:

$$B_{dipole}(t) = \frac{\mu}{r(t)^3}(3cos^2\theta(t) - 1)$$
$$= \frac{\gamma\hbar}{r(t)^3}(3cos^2\theta(t) - 1)$$

where θ and r are defined in Fig. 19.5.

The energy associated with this additional magnetic field is obtained from $E = \hbar\omega = \hbar\gamma_S B_{dipole}$:

$$\mathcal{H} = \hbar^2 \frac{\gamma_S \gamma_I}{r(t)^3}(3cos^2\theta(t) - 1) \quad (19.10)$$

The intensity of the dipole field, B_{dipole}, depends on both the orientation of the two spins and the distance between them. If either of these properties are time dependent then the magnetic field will vary with time.

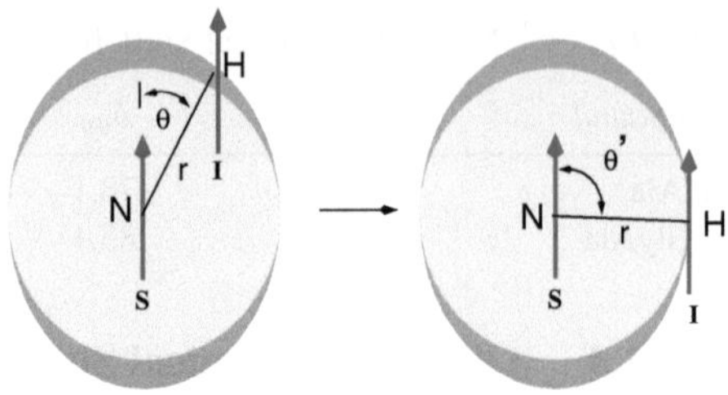

Figure 19.5. Dipole-dipole coupling. The magnetic field generated by the amide nitrogen generates an additional field, B_{dipole} at the amide proton. The strength of this field depends on the relative orientation of the two spins, as illustrated by the gray crescent shapes. When $\theta = 0$, B_{dipole} is at a maximum. The dipole field is zero when $\theta \approx 54°$, and is negative for $0 < |\theta| < 54°$. Molecular tumbling changes the relative orientation of the two spins, resulting in a change of B_{dipole}.

Most relaxation studies utilize the relaxation properties of heteronuclear spins, such as ^{15}N or ^{13}C, to characterize molecular motion. The advantages of using these spins for relaxation studies are two-fold. First, the relaxation of the heteronuclear spin is dominated by the attached proton due to the $1/r^3$ dependence. Consequently, dipolar fields that are generated by other protons are considered negligible. Secondly, the distance is fixed to the bond length, consequently only fluctuations in the angular term contribute to time dependent field fluctuations. The production of protein samples with isolated ^{15}N-H and ^{13}C-H spin pairs is briefly discussed in Section 19.6.

19.2.3 Frequency Components from Molecular Rotation

Both chemical shift anisotropy and dipolar coupling generate magnetic field fluctuations at the nucleus. In order for these motions to be effective at stimulating nuclear transitions the random molecular motion must contain field fluctuations at the appropriate frequencies to stimulate transitions. Consequently it is necessary to determine the frequency spectrum of the fluctuating magnetic fields.

The rate, W, at which a transition occurs is equal to the rate of change of the probability of finding a system in a particular state. In the following, the transition rate as a function of frequency, $W(\omega)$, will be obtained using a simple quantum mechanical treatment to illustrate the general features of the calculation. A more detailed treatment is beyond the scope of this text and the reader is referred to Solomon [152] or Abragam [1] for a more comprehensive treatment.

Assume a simple one spin system with a wave function:

$$\Phi = c_g\phi_g + c_e\phi_e \quad (19.11)$$

and the probability, p, of finding the system in the ground or excited state is $c_g^*c_g$ or $c_e^*c_e$, respectively.

Under the influence of a time varying magnetic field the rate of change of the population of the ground state, W(ω), would be:

$$W(\omega) = \frac{dp_g}{dt} = c_g^* \frac{dc_g}{dt} + \frac{dc_g^*}{dt} c_g \tag{19.12}$$

The time dependent Schrödinger Equation provides a method of calculating W.

$$i\hbar \frac{d\Phi}{dt} = \mathcal{H}\Phi \tag{19.13}$$

The Hamiltonian contains terms that relate to geometrical properties of the molecule, e.g. the relative orientation of dipoles, as well as terms that relate to the spin operators. To simplify the treatment the Hamiltonian will be divided into two terms:

$$\mathcal{H} = \mathcal{A} \cdot \mathcal{F} \tag{19.14}$$

where $\mathcal{A}$ represents time independent spin operators and $\mathcal{F}$ represents geometrical terms and other known constants. In general, $\mathcal{A}$ is defined by the type of interaction (e.g. dipolar or CSA) while $\mathcal{F}$ is defined by the type of motion.

Using the orthonormality relationship, $< \phi_g|\phi_g >= 1, < \phi_g|\phi_e >= 0$, and assuming that $c_g = 0$ and $c_e = 1$ (i.e. the system is completely in the excited state), gives the following for dc_g/dt:

$$i\hbar \frac{dc_g}{dt} =< \phi_g|\mathcal{A}\mathcal{F}|\phi_e > e^{i\omega t} \tag{19.15}$$

This equation can be integrated to give:

$$c_g = \frac{1}{i\hbar} \int_0^t < \phi_g|\mathcal{A}\mathcal{F}|\phi_e > e^{i\omega t} dt \tag{19.16}$$

Therefore the expression for the transition probability, neglecting the second term, is:

$$\begin{aligned} W =& c_g^* \frac{dc_g}{dt} \\ =& \frac{-1}{\hbar^2} \int_o^t < \phi_e|\mathcal{A}\mathcal{F}(t)|\phi_g > e^{-i\omega t} dt < \phi_g|\mathcal{A}\mathcal{F}(t')|\phi_e > e^{i\omega t'} \end{aligned} \tag{19.17}$$

Defining $\tau = t - t'$ gives:

$$\begin{aligned} W &= \frac{1}{\hbar^2} \int_o^t < \phi_e|\mathcal{A}\mathcal{F}(t)|\phi_g >< \phi_g|\mathcal{A}\mathcal{F}(t-\tau)|\phi_e > e^{-i\omega\tau} d\tau \\ &= \frac{1}{\hbar^2} < \phi_e|\mathcal{A}|\phi_g >< \phi_g|\mathcal{A}|\phi_e > \int_o^t \mathcal{F}(t)\mathcal{F}^*(t-\tau) e^{-i\omega\tau} d\tau \\ &= \frac{1}{\hbar^2} \kappa^2 \int_o^t \mathcal{F}(t)\mathcal{F}(t-\tau) e^{-i\omega\tau} d\tau \end{aligned} \tag{19.18}$$

In the above, the time-invariant spin operators have been taken outside the integrals and $\kappa = < \phi_e|\mathcal{A}|\phi_g >$.

Since $\mathcal{F}(t)$ is a random variable, $\mathcal{F}(t-\tau) = \mathcal{F}(t+\tau)$. It is customary to use the $\mathcal{F}(t+\tau)$ form. The function $\mathcal{F}(t)\mathcal{F}^*(t+\tau)$ is an autocorrelation function. The autocorrelation function measures how rapidly knowledge of a prior orientation is lost by the reorienting molecule. For random fluctuations the autocorrelation function does not depend on a specific origin of time, only on the time difference between instances in time. Consequently, t can be set to zero and the auto-correlation depends only on τ.

For a collection of spins, the average auto-correlation function over all of the spins is written as:

$$G(\tau) = <\mathcal{F}(0)\mathcal{F}^*(\tau)> \tag{19.19}$$

where the angled brackets represent the average over all of the spins. The auto-correlation function can also be written as the product of the normalized autocorrelation function, $g(\tau)$, and the average value of $\mathcal{F}^2$:

$$G(\tau) = <\mathcal{F}^2(0)> g(\tau) \tag{19.20}$$

To obtain the transition rates at specific frequencies it is necessary to extract the frequency components from the autocorrelation function by Fourier transformation. However, it is apparent from Eq. 19.18 that W is the Fourier transform of $G(\tau)$, therefore W already represents that frequency spectrum of the transition rates and no further manipulation is required.

The Fourier transform of the normalized autocorrelation function is given a special name, the spectral density function, $J(\omega)$, because it represents the *density* of fluctuations at different frequencies. It is defined as follows:

$$J(\omega) = \int_0^\infty g(\tau)e^{-i\omega\tau}d\tau \tag{19.21}$$

Therefore, the complete expression for the transition probabilities is:

$$W(\omega) = \frac{\kappa^2}{\hbar^2} < \mathcal{F}^2(0) > J(\omega) \tag{19.22}$$

The $\mathcal{F}^2$ term depends on the average value of the geometric terms and κ depends on the Hamiltonian. Using the amide nitrogen as an example, and assuming an axial symmetric chemical shift tensor:

$$\kappa^2/\hbar^2 \quad \propto \quad \frac{\hbar^2\gamma_N^2\gamma_H^2}{r^6} \quad for\ dipolar\ coupling \tag{19.23}$$

$$\kappa^2/\hbar^2 \quad \propto \quad (\gamma_N B_o(\delta_\parallel - \delta_\perp))^2 \quad for\ CSA \tag{19.24}$$

Equation 19.23 follows directly from Eq. 19.10. Equation 19.24 indicates, as expected, that the contribution of the CSA to the relaxation rate depends on the total range of magnetic fields that are sampled by the nucleus as it reorients. A concise derivation is provided by Luginbuhl and Wüthrich [100].

The relationship between the spectral density function and the type of molecular motion will be discussed in Section 19.4. For illustrative purposes, the autocorrelation

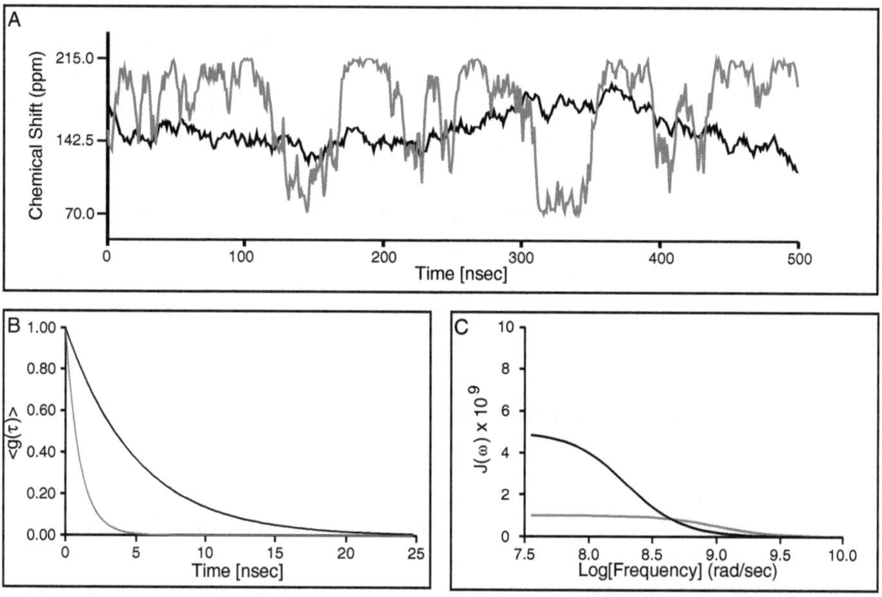

Figure 19.6. Magnetic field fluctuations and the spectral density function. The relationship between random magnetic field fluctuations, the auto-correlation function, and the spectral density function is shown. An axially symmetric chemical shift tensor with principal values of $\delta_\perp = 70$ ppm and $\delta_\parallel = 215$ ppm was used in the simulations. The fluctuations in chemical shift that occur due to random isotropic rotation of the amide group are shown in Panel A. The rotational correlation times are 1 nsec (gray line) and 5 nsec (black line). The resultant autocorrelation functions are shown in Panel B. Panel C shows the spectral density functions calculated from $< g(\tau) >$. High frequency motions (gray curve) result in smaller values of J(ω) at $\omega = 0$. However, J(ω) extends to higher frequencies.

and spectral density functions for random isotropic tumbling due to Brownian motion are given here:

$$g(\tau) = e^{-\tau/\tau_c} \qquad J(\omega) = \frac{\tau_c}{1 + \omega^2\tau_c^2} \tag{19.25}$$

where τ_c is the rotational correlation time. It is the time required for a molecule to rotate, on average, 1 radian. The rotational correlation time is related to the isotropic rotational diffusion constant:

$$\tau_c = 1/6D_{iso} \tag{19.26}$$

and proportional to the size of the protein and the viscosity of the solution: $\tau_c = 4\pi\eta a^3/3kT$, where a is the radius of the protein, and η is the viscosity. It follows that τ_c is *approximately* 1 nsec for each 2.6 kDa of protein mass at T = 300 K.

An illustration of the relationship between rotational motion of the particle, magnetic field fluctuations due CSA, and the spectral density is shown in Fig. 19.6. A small molecule, which has a short correlation time, will experience rapid fluctuations in the magnetic field (Panel A, gray curve). These fluctuations will rapidly become uncorrelated, leading to a rapid decrease in the autocorrelation function (Panel B).

Such rapid fluctuations contain both high and low frequency components, thus the resultant spectral density is broad and extends over a large frequency range (Panel C). In contrast, a large particle tumbles slowly, experiencing magnetic field fluctuations of lower frequencies (Panel A, black curve), hence the spectral density function will have a higher intensity at lower frequencies (Panel C).

19.3 Spin-lattice (T_1) and Spin-spin (T_2) Relaxation

The above discussion developed the connection between molecular motion and magnetic field fluctuations at the nucleus. In this section explicit formulas for the effect of these fluctuations on the measured relaxation parameters (T_1, T_2, and hnNOE) of the excited spins will be given.

19.3.1 Spin-lattice Relaxation

19.3.1.1 Isolated Spin

We first treat an isolated spin that is coupled to the lattice. This analysis is not very useful from a molecular point of view since the details of the interactions between the spin and the lattice are obscure. However, the discussion is included to show how to mathematically treat a relaxation process that relaxes to a *non-zero* equilibrium state.

The lattice acts as a thermal reservoir, accepting energy from the excited spins as they return to thermal equilibrium. Time dependent changes in the molecular structure of the lattice generate magnetic field fluctuations that stimulate transitions at a rate of W. At any given time the longitudinal magnetization is proportional to the population difference between the ground and excited states:

$$I_z(t) = N_g(t) - N_e(t) \tag{19.27}$$

The change in the population of the ground and excited states is described by the following standard rate equations:

$$\frac{dN_g}{dt} = -WN_g + WN_e \tag{19.28}$$

$$\frac{dN_e}{dt} = +WN_g - WN_e \tag{19.29}$$

These equations, as written, indicate that the population of the ground and the excited state are equal [2]. Since this is clearly not the case it is necessary to alter the above equation such that they represent relaxation toward the equilibrium values of the populations:

$$\frac{d(N_g - N_g^o)}{dt} = \frac{dN_g}{dt} = -W(N_g - N_g^o) + W(N_e - N_e^o) \tag{19.30}$$

$$\frac{d(N_e - N_e^o)}{dt} = \frac{dN_e}{dt} = +W(N_g - N_g^o) - W(N_e - N_e^o) \tag{19.31}$$

[2] At equilibrium, $dN_g/dt = 0$, therefore $WN_g = WN_e$, or $N_g = N_e$.

The above satisfies the boundary condition; when dN_g/dt=0, $(N_g - N_g^o) = (N_e - N_e^o)$, which is only true if $N_g = N_g^o$ and $N_e = N_e^o$.

Subtraction of Eq. 19.31 from Eq. 19.30 gives the following for the change in the longitudinal magnetization.

$$\frac{d(I_z - I_z^o)}{dt} = -2W(I_z - I_z^o) \tag{19.32}$$

This expression can be solved using standard methods to give:

$$I_z(t) = I_z^o + Ce^{-2Wt} \tag{19.33}$$

The constant C is obtained from the initial conditions. Assuming that the initial value of I_z is 0, i.e. immediately after a 90° pulse, then $I_z(0) = 0 = I_z^o + Ce^0$, giving $C = -I_z^o$, and therefore:

$$I_z(t) = I_z^o(1 - e^{-2Wt}) = I_z^o(1 - e^{-t/T_1}) \tag{19.34}$$

In conclusion, the spin-lattice relaxation of an isolated spin is characterized by a single decay time, T_1, whose inverse is twice the transition rate, W.

19.3.1.2 Dipolar Coupled Spins

To evaluate the effect of dipolar coupling on the relaxation properties of two coupled spins, I and S, it is necessary to expand the energy level diagram to include coupling between the spins. Since there are two possible spin states for each spin (α, β), the energy diagram will require four levels, as shown in Fig. 19.7. In this diagram, W_n refers to the *rate* of single quantum ($n = 1$), zero quantum ($n = 0$), and double quantum ($n = 2$) transitions. Single quantum transitions connect levels that differ in the spin state of one spin, zero quantum transitions simply interchange the spin state of the two coupled spins, and double quantum transitions require that both spins change their state at the same time. Therefore, single quantum transitions will be stimulated by fields that fluctuate at the energy difference between states connected by single quantum transition, e.g. ω_I and ω_S. Zero quantum transitions will be stimulated by field fluctuations that occur at a frequency that corresponds to the energy difference between the $\beta\alpha$ and $\alpha\beta$ states, $(\omega_I - \omega_S)$. Double quantum transitions will be stimulated by field fluctuations that generate frequencies of 2ω, or the sum of the resonance frequencies of both spins $(\omega_I + \omega_S)$.

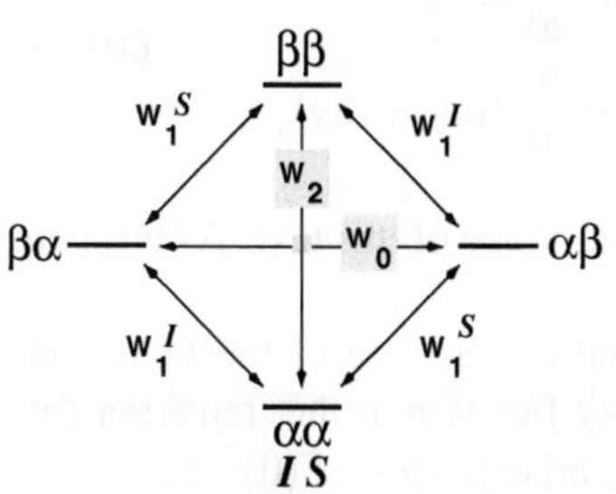

Figure 19.7 Energy level diagram for two coupled spins. The lowest energy state corresponds to both spins being in the $\alpha(m_z = +1/2)$ state. The four single quantum transitions, labeled with W_1^I or W_1^S, connect states that differ in the spin state of only one spin. There is one zero quantum transition (W_0) and one double quantum transition (W_2), corresponding to the net change in the magnetization of both spins by zero or two units, respectively.

The rate of change of the longitudinal magnetization for the I and S spins is obtained following the same procedure as discussed above for a single isolated spin. This approach was first described by Solomon [152] and the final results are often referred to as the Solomon Equations.

First, the longitudinal magnetization is expressed as differences in populations across both single quantum transitions:

$$I_z = (N_{\alpha\alpha} - N_{\beta\alpha}) + (N_{\alpha\beta} - N_{\beta\beta}) \qquad S_z = (N_{\alpha\alpha} - N_{\alpha\beta}) + (N_{\beta\alpha} - N_{\beta\beta}) \quad (19.35)$$

The rate of change of each population is given by standard rate equations, taking into account relaxation towards equilibrium values, for example:

$$\begin{aligned}\frac{d(N_{\alpha\alpha} - N^o_{\alpha\alpha})}{dt} &= -(W_2 + W_1^S + W_1^I)(N_{\alpha\alpha} - N^o_{\alpha\alpha}) + W_1^I(N_{\beta\alpha} - N^o_{\beta\alpha}) \\ &+ W_1^S(N_{\alpha\beta} - N^o_{\alpha\beta}) + W_2(N_{\beta\beta} - N^o_{\beta\beta}) \quad (19.36)\end{aligned}$$

Separate expressions are obtained for $N_{\alpha\beta}$, $N_{\beta\alpha}$, and $N_{\beta\beta}$. The time dependence of I_z and S_z are obtained by adding together the appropriate expressions for the changes in populations. A complete derivation is given by Neuhaus and Williamson [118]. The final result is:

$$\begin{aligned}\frac{dI_z}{dt} &= -(W_o + 2W_1^I + W_2)(I_z - I_z^o) - (W_2 - W_o)(S_z - S_z^o) \\ \frac{dS_z}{dt} &= -(W_2 - W_o)(I_z - I_z^o) - (W_o + 2W_1^S + W_2)(S_z - S_z^o) \quad (19.37)\end{aligned}$$

The term associated with the same spin, $W_o + 2W_1 + W_2$, is termed the longitudinal relaxation rate, ρ, and the term that connects the two different spins, $W_2 - W_0$, is called the cross-relaxation rate, σ. The Solomon Equation can be written in terms of ρ and σ:

$$\begin{aligned}\frac{dI_z}{dt} &= -\rho_I(I_z - I_z^o) - \sigma(S_z - S_z^o) \\ \frac{dS_z}{dt} &= -\sigma(I_z - I_z^o) - \rho_S(S_z - S_z^o) \quad (19.38)\end{aligned}$$

The individual transition rates are obtained by evaluation of Eq. 19.22:

$$\begin{aligned} W_0 &= d^2\frac{1}{10}J(\omega_I - \omega_S) \qquad & W_1^I &= d^2\frac{3}{20}J(\omega_I) \\ W_1^S &= d^2\frac{3}{20}J(\omega_S) \qquad & W_2 &= d^2\frac{6}{10}J(\omega_I + \omega_S) \end{aligned} \quad (19.39)$$

where $d^2 = (\hbar^2\gamma_I^2\gamma_S^2/r^6)$. The derivation is beyond the scope of this text. A complete description is provided by Solomon [152].

As anticipated from Eq. 19.23, these rates depend on the size of the field fluctuations (d^2) and the magnitude of the spectral density function at the transition frequency, $J(\omega_n)$. The numerical factors, such as (1/10), arise from $< \mathcal{F}(0)^2 >$.

19.3.2 Spin-lattice Relaxation of Like Spins

Proton-proton relaxation is a common example of relaxation of like spins. In this case $\rho_I = \rho_S = \rho$ and $\omega_I = \omega_S = \omega_H$. Since the CSA of the proton is quite small, the relaxation can be described by just considering dipolar coupling. The solution to the Solomon equations is given in Appendix C, and the result is:

$$I_z(t) - M_o = \frac{e^{-(\rho+\sigma)t}}{2}\left[I_z(0) + S_z(0) - 2M_o\right] + \frac{e^{-(\rho-\sigma)t}}{2}\left[I_z(0) - S_z(0)\right] \quad (19.40)$$

The above equation shows that the relaxation of a spin is, in general, multi-exponential. However, if *both* spins are inverted at the beginning of the measurement ($I_z(0) = S_z(0) = -M_o$), the time dependence of I_z and S_z simplifies to:

$$I_z(t) = M_o(1 - 2e^{-(\rho+\sigma)t}) \qquad S_z(t) = M_o(1 - 2e^{-(\rho+\sigma)t}) \quad (19.41)$$

and the longitudinal magnetization relaxes with a single time constant:

$$\begin{aligned}\frac{1}{T_1} &= \sigma + \rho = (W_2 - W_0) + (W_0 + 2W_1 + W_2) = 2(W_1 + W_2) \\ &= \frac{6}{20} d^2 \left[J(\omega_H) + 4J(2\omega_H)\right] \quad (19.42)\end{aligned}$$

19.3.3 Spin-lattice Relaxation of Unlike Spins

The relaxation of ^{15}N and $^{13}C_\alpha$ nuclei (S spins) by their attached protons (I spins) is a common example of relaxation of unlike spins. In general, the relaxation properties of the heteronuclear spins will also be multi-exponential. However, if the proton is saturated during the recovery of the heteronuclear magnetization, a single relaxation rate is found. The solution to the Solomon equations in this case is given in Appendix C, and the result is:

$$S_z(t) = \left[S_z^o + \frac{\sigma}{\rho_s} I_o\right](1 - e^{\rho_s t}) \quad (19.43)$$

Therefore, the spin-lattice relaxation (T_1) due to dipolar coupling is given by:

$$\begin{aligned}\frac{1}{T_1^{Dipole}} &= W_o + 2W_1^S + W_2 \\ &= d^2 \frac{1}{10}\left[J(\omega_I - \omega_S) + 3J(\omega_S) + 6J(\omega_I + \omega_S)\right] \quad (19.44)\end{aligned}$$

In the case of nuclei with significant CSA , it is necessary to add the contribution of the CSA to the spin-lattice relaxation [100]:

$$\frac{1}{T_1^{CSA}} = \frac{2}{15}\omega_S^2 \Delta\sigma^2 J(\omega_S) \quad (19.45)$$

The factor (2/15) arises from $< \mathcal{F}(0)^2 >$, $\Delta\sigma$ is the CSA, and ω_S is the resonance frequency of the S spin. $J(\omega_S)$ appears because field fluctuations that can cause single quantum transition of the heteronuclear spin provide a mechanism for the excited state to release energy and return to the ground state.

The overall spin-lattice relaxation rate (R_1) is the sum of the rate due to dipolar coupling and CSA. Using the amide group as an example:

$$R_1 = \frac{d^2}{10}\left[J(\omega_H - \omega_N) + 3J(\omega_N) + 6J(\omega_H + \omega_N)\right] + \frac{2}{15}\omega_S^2 \Delta\sigma^2 J(\omega_N) \quad (19.46)$$

19.3.4 Spin-spin Relaxation

Spin-spin relaxation involves the loss of coherent transverse magnetization. This can occur by the conversion of transverse magnetization to longitudinal magnetization and by the loss of coherence of the initially coherent population of spins. Therefore, the decay of transverse magnetization is caused by zero, single, and double quantum transitions. Single and double quantum transitions cause the conversion of transverse magnetization to longitudinal magnetization. Zero quantum transitions cause two spins to interchange their spin-states (e.g. $\alpha\beta \rightarrow \beta\alpha$), altering their precessional frequencies, which leads to dephasing of the transverse magnetization.

The relationship between the spin-spin relaxation rate and the spectral density functions is more difficult to derive than the equivalent formula for spin-lattice relaxation. However, the overall approach is the same. A four level diagram is used to describe the possible states of the system. In this case the levels correspond to transverse magnetization instead of populations. The transition rates that connect each level become more complicated because it is necessary to represent each of the states as a linear sum of the basis vectors associated with longitudinal magnetization. The final results will be presented here, the derivation for dipolar coupling can be found in Solomon [152] and for chemical shift anisotropy in [100].

19.3.4.1 Spin-spin Relaxation of Like Spins

In contrast to spin-lattice relaxation, the relaxation of transverse magnetization is always mono-exponential. In the case of the protons, the CSA can be safely ignored. However, as discussed in Chapter 18, the spin-spin relaxation rate can be affected by chemical exchange. Therefore the complete equation for the relaxation rate is:

$$R_2 = \frac{3}{20}d^2\left[3J(0) + 5J(\omega_H) + 2J(2\omega_H)\right] + R_{ex} \quad (19.47)$$

The spin-spin relaxation rate is generally larger than the spin-lattice relaxation rate because of the $J(0)$ term. However, under conditions of fast motion the spin-spin and spin-lattice relaxation rates due to dipolar coupling become equal. The equivalence of T_1 and T_2 under these conditions is often referred to as the fast motion, or extreme narrowing, limit.

The equivalence of the relaxation rates occurs because $J(0) = J(\omega) = J(2\omega)$ when $\omega\tau_c << 1$, causing the spectral density to become independent of ω. Setting all spectral densities to the same value, J, and calculating R_1 and R_2, shows the two relaxation rates to be equal:

$$\begin{aligned} R_1 &= \frac{6}{20}d^2\left[1J + 5J\right] = \frac{3}{2}d^2 J \\ R_2 &= \frac{3}{20}d^2\left[2J + 5J + 2J\right] = \frac{3}{2}d^2 J \end{aligned} \quad (19.48)$$

19.3.4.2 Spin-spin Relaxation of Unlike Spins

The spin-spin relaxation rate of a heteronuclear spin is affected by dipolar coupling to the attached proton [152], chemical shift anisotropy [100], as well as chemical exchange (see Chapter 18). Using the amide group as an example (S=^{15}N, I=^{1}H):

$$\begin{aligned} R_2 &= \frac{d^2}{20}\left[4J(0) + J(\omega_H - \omega_N) + 3J(\omega_N) + 6J(\omega_H) + 6J(\omega_H + \omega_N)\right] \\ &\quad + \frac{1}{45}\omega_N^2 \Delta\sigma^2 \left[4J(0) + 3J(\omega_N)\right] + R_{ex} \end{aligned} \tag{19.49}$$

where $d^2 = \gamma_H^2 \gamma_N^2 \hbar^2 / r^6$.

The $J(0)$ term, which is present for both the dipolar coupling and CSA terms, represents the dephasing of the transverse magnetization by an inhomogeneous local magnetic field. The $J(\omega_N)$ and $J(\omega_N + \omega_H)$ terms represent relaxation to the ground state which also destroys the transverse magnetization. The $J(\omega_H - \omega_N)$ term represents zero-quantum mutual proton-nitrogen spin flips, which cause dephasing of the nitrogen magnetization. Finally, the presence of the $J(\omega_H)$ term indicates that transitions of the coupled proton affects the coherence of the transverse nitrogen magnetization.

19.3.5 Heteronuclear NOE

The heteronuclear NOE is a measure of the change in the steady state populations of the heteronuclear, or S-spin, when the attached proton (I-spin) is saturated. The relationship between the transition rates, W_n, and the heteronuclear NOE (hnNOE) is easily derived from the time dependence of $S_z(t)$:

$$S_z(t) = \left[S_z^o + \frac{\sigma}{\rho_s} I_z^o\right](1 - e^{\rho_s t}) \tag{19.50}$$

under steady-state conditions ($t \rightarrow \infty$), this reduces to:

$$S_z = \left[S_z^o + \frac{\sigma}{\rho_s} I_z^o\right] \tag{19.51}$$

The equilibrium population of the I spins, I_z^o, can be converted to S_z^o using the ratio of the gyromagnetic ratios:

$$S_z = \left[S_z^o + \frac{\sigma}{\rho_s}\frac{\gamma_I}{\gamma_S} S_z^o\right] \tag{19.52}$$

Usually the ratio of the signal intensity, which is proportional to S_Z, in the presence and absence of saturating the I spin transitions is defined as the heteronuclear NOE enhancement. Using the amide group as an example:

$$\begin{aligned} \frac{S_z}{S_z^o} &= 1 + \frac{\sigma}{\rho_N}\frac{\gamma_H}{\gamma_N} \\ &= 1 + \frac{\gamma_H}{\gamma_N} d^2 \frac{1}{10}\left[6J(\omega_H + \omega_N) - J(\omega_H - \omega_N)\right] T_1^N \end{aligned} \tag{19.53}$$

Note that T_1^N includes the contribution of the CSA to the spin-lattice relaxation. Also of interest is that the hnNOE enhancement for ^{15}N is less than 1 because of the negative value for γ_N.

19.4 Motion and the Spectral Density Function

The actual form of the spectral density function depends on the nature of the molecular motion. A number of different models have been used to represent internal motion. These are described in some detail by Luginbuhl and Wüthrich [100] and the references contained within. As the complexity of the model increases, so does the number of parameters that have to be determined from the rather limited experimental data. Consequently, the discussion here will be restricted to simple models of internal motion that are coupled to two simple models of overall rotation.

The overall rotation of the protein is assumed to be uncorrelated with the internal motion. Thus, the autocorrelation function can be written as a product of the autocorrelation function for overall rotation, $g_{rot}(\tau)$ and that for internal motion, $g_i(\tau)$:

$$g(\tau) = g_{rot}(\tau) \times g_i(\tau) \tag{19.54}$$

The derivation of the auto-correlation functions is beyond the scope of this text, the reader is referred to Luginbuhl and Wüthrich [100] for more details.

19.4.1 Random Isotropic Motion

The autocorrelation and spectral density function for isotropic random rotation was discussed above in Section 19.2.3. In summary, the rate of random Brownian rotational motion is characterized with a single diffusion coefficient, D_{iso}, giving a rotational correlation time of: $\tau_c = \frac{1}{6D}$. The autocorrelation function for random isotropic motion is:

$$g_{rot}(\tau) = e^{-\tau/\tau_c} \tag{19.55}$$

The spectral density function is the Fourier transform of the above, giving[3]:

$$J(\omega) = \int_0^\infty g(\tau) e^{-i\omega\tau d\tau} = \frac{\tau_c}{1 + \omega^2 \tau_c^2} \tag{19.56}$$

19.4.2 Anisotropic Motion - Non-spherical Protein

If the protein is non-spherical, then the rate of rotational diffusion will differ in each direction. The relationship between the shape of a molecule and its rotational diffusion can be expressed as a tensor relationship. As with the chemical shift tensor, there will be one orientation of the molecule in which the diffusion tensor has only diagonal elements:

$$\begin{bmatrix} D_x & 0 & 0 \\ 0 & D_y & 0 \\ 0 & 0 & D_z \end{bmatrix} \tag{19.57}$$

[3] In many treatments a factor of 2/5 is included in the spectral density function. Here, this factor is absorbed into $< \mathcal{F}^2(0) >$.

If the protein is axially symmetric, then two of the above will be equal. Usually the z-axis is defined to be the symmetry axis, and the above diffusion tensor becomes:

$$\begin{bmatrix} D_{\perp} & 0 & 0 \\ 0 & D_{\perp} & 0 \\ 0 & 0 & D_{\parallel} \end{bmatrix} \tag{19.58}$$

Note that a total of four parameters are required to define this tensor, the values of each component, $D_{\perp}$ and $D_{\parallel}$, as well as the two angles that rotate the molecule into the coordinate frame in which the tensor is diagonal.

The autocorrelation function for axially symmetric diffusion is given by [169]:

$$g_{rot}(\tau) = \frac{1}{4}(3cos^2\beta - 1)^2 e^{-\tau/\tau_1} + 3sin^2\beta\ cos^2\beta\ e^{-\tau/\tau_2} + \frac{3}{4}sin^4\beta\ e^{-\tau/\tau_3} \tag{19.59}$$

where $\tau_1 = 1/(6D_{\perp})$, $\tau_2 = 1/(5D_{\perp} + D_{\parallel})$, $\tau_3 = 1/(2D_{\perp} + 4D_{\parallel})$ and β is the angle between the bond vector (e.g. N-H) and the axis of symmetry (see Fig. 19.8).

19.4.3 Constrained Internal Motion

Internal motion of a portion of the molecule can also cause magnetic field fluctuations and therefore contribute to the relaxation of the spins. The contribution of the frequency and extent of internal motion to the autocorrelation function have been characterized for a number of different types of motions (see Luginbuhl and Wüthrich [100]). We will discuss one widely used method, the model-free approach, initially defined by Lipari and Szabo [97] and extended by Clore, *et al.* [37]. In this treatment the details of the motion are not considered, rather it is assumed that rapid internal motion will lead to a decay of the autocorrelation function by an amount equal to the generalized order parameter, S^2 and a time constant for internal motion of τ_i.

19.4.3.1 Simple Model-Free Analysis (SMF)

In this model, internal motion is described by a single correlation time, τ_f and order parameter S_f^2. The motion is assumed to be much faster than the overall rotation of

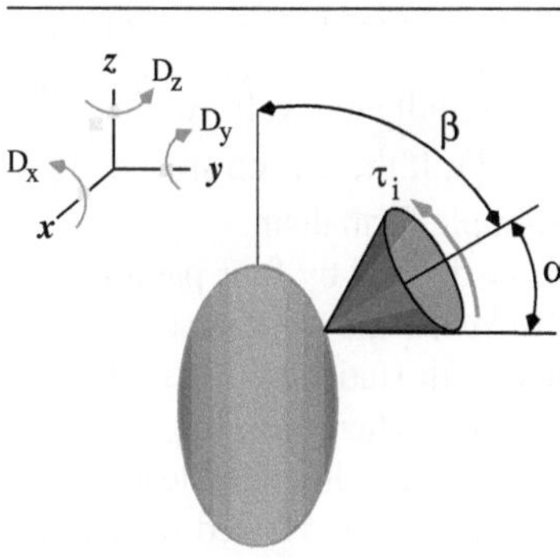

Figure 19.8 Framework for the analysis of internal motion. The rotational diffusion of the entire protein (large oval) is characterized by three rotational diffusion constants, D_x, D_y, and D_z. In the case of a spherical molecule these are all equal and $\tau_c = 1/6D_{iso}$. For a molecule with axial symmetry, as illustrated here, $D_{\parallel} = D_z$ and $D_{\perp} = D_x = D_y$. The cone protruding off of the protein represents internal motion of the bond. The extent of internal motion is characterized by the semi-angle of the cone, α, and the rate that the N-H group diffuses within this cone is characterized by τ_i, the correlation time for internal motion. The orientation of the bond with respect to the z-axis is given by β.

the protein. The autocorrelation function for internal motion is:

$$g_i(\tau) = S_f^2 + (1 - S_f^2)e^{-\tau/\tau_f} \quad (19.60)$$

If the motion is assumed to consist of random motion *on the surface* of a cone, then the order parameter is given by:

$$S_f^2 = \left[< \frac{1}{2}(3cos^2\alpha - 1) > \right]^2 \quad (19.61)$$

where the angled brackets represent the average over all molecules and α is the semi-angle of the cone. If the motion is assumed to consist of random motion *within a cone* of semi-angle α, then the order parameter, S^2, is approximated by [100]:

$$S_f^2 = \left[\frac{1}{2}(1 + cos\alpha)cos\alpha \right]^2 \quad (19.62)$$

Regardless of the particular model used, it is clear that as the bond-vector becomes more restricted, i.e. as α decreases, the order parameter approaches 1. In contrast, as the motion on the cone becomes less restricted, α increases and for completely random motion, the order parameter approaches zero. In summary:

$S^2 = 1$	No internal motion	$S^2 = 0$	Random internal motion.

19.4.3.2 Extended Model-Free Analysis (EMF)

It was noted in the earlier analysis of relaxation data that the simple model-free formalism could not account for the relaxation data for a subset of residues in a protein [37]. The systematic discrepancy between measured and calculated relaxation data was thought to be due to the presence of an additional slow internal motion, with its own correlation time, τ_s, and amplitude, S_s^2. The rate of slow motion, is assumed to be faster than the overall rotation of the molecule, i.e. $\tau_f < \tau_s << \tau_c$, such that the internal motion remains uncorrelated with the overall rotational motion.

Due to the presence of slower motion, the autocorrelation now decays with two time constants:

$$g_i(\tau) = S_f^2 S_S^2 + (1 - S_f^2)e^{-\tau/\tau_f} + S_f^2(1 - S_S^2)e^{-\tau/\tau_S} \quad (19.63)$$

This equation shows that the correlation function decays rapidly to a value of $(1 - S_f^2)$ and then undergoes a slower decay to a level of $S_f^2(1 - S_S^2)$. If there is no slow motion ($S_S^2 = 1$), the extended formalism reverts back to the simple formalism.

Note that the representation of internal motion is now defined by four parameters. Since it is likely that the number of parameters will exceed the number of experimental observations it will be impossible to determine with statistical tests whether the additional parameters have improved the fit of the data. Therefore, the extended model-free formalism should only be considered if relaxation data is available at multiple field strengths. Even so, the presence of slow motion should be viewed with some skepticism because it can be erroneously generated by a failure to account for the anisotropic diffusion properties of the molecule, as discussed by Schurr *et al.* [143].

19.4.4 Combining Internal and External Motion

Assuming that the internal motion is not correlated with the overall rotational motion, the correlation function is simply the product of the correlation function of internal and external motion. This assumption is valid if the rate of internal motion is at least 10 fold faster than the rotational correlation time.

In the case of isotropic tumbling and the simple model-free formalism:

$$\begin{aligned} g(\tau) &= e^{-\tau/\tau_c}\left[S^2 + (1-S^2)e^{-\tau/\tau_i}\right] \\ &= (1-S^2)e^{-\tau/\tau_{mix}} + S^2 e^{-\tau/\tau_c} \\ J(\omega) &= (1-S^2)\frac{\tau_{mix}}{1+\omega^2\tau_{mix}^2} + S^2\frac{\tau_c}{1+\omega^2\tau^2} \end{aligned} \tag{19.64}$$

$$\tau_{mix} = \tau_i\tau_c/(\tau_i+\tau_c) \tag{19.65}$$

In the case of tumbling of an axially symmetric molecule with the simple model-free formalism:

$$\begin{aligned} g(\tau) &= \left[\sum_{k=1}^{3} A_k e^{-\tau/\tau_k}\right]\left[S^2 + (1-S^2)e^{-\tau/\tau_i}\right] \\ &= S^2\left[\sum_{k=1}^{3} A_k e^{-\tau/\tau_k}\right] + (1-S^2)\left[\sum_{k=1}^{3} A_k e^{-\tau/\tau_k'}\right] \end{aligned} \tag{19.66}$$

$$J(\omega) = S^2\left[\sum_{k=1}^{3}\frac{A_k\tau_k}{1+\omega^2\tau_k^2}\right] + (1-S^2)\left[\frac{A_k\tau_k'}{1+\omega^2(\tau_k')^2}\right] \tag{19.67}$$

the A_k, τ_k, are defined as follows:

k	A_k	τ_k
1	$\frac{1}{4}(3cos^2\beta - 1)^2$	$1/6D_\perp$
2	$3sin^2\beta\, cos^2\beta$	$1/(5D_\perp + D_\parallel)$
3	$\frac{3}{4}sin^4\beta$	$1/(2D_\perp + 4D_\parallel)$

and $\tau_k' = \tau_k\tau_i/(\tau_k + \tau_i)$. Both of the above equations are easily modified to include the extended model-free formalism.

19.5 Effect of Internal Motion on Relaxation

The effect of internal motion on the T_1, T_2, and the heteronuclear NOE are shown in Fig. 19.9. Panels A and B illustrate the effect of changing τ_i (A) and S^2 (B) on T_1 and T_2. Panels C and D show the effect of these changes on the heteronuclear NOE. General trends are listed below along with explanations that are based on the relationship between $J(\omega)$, τ_i, and S^2, as illustrated in Fig. 19.10.

1. The T_1 and T_2 are not sensitive to changes in τ_i (Panel A, Fig. 19.9). This occurs because $J(\omega_N)$ dominates T_1 relaxation and $J(0)$ dominates T_2 relaxation. Neither of these are appreciably changed when τ_i is changed at a fixed value of S^2, as shown in Fig. 19.10.

2. Changes in S^2 cause an increase in *both* T_1 and T_2 by a similar amount; the ratio of T_1 to T_2 remains essentially constant (Panel B, Fig. 19.9). Decreasing S^2 lowers the value of the spectral density at J(0) and $J(\omega_N)$, as illustrated in Fig. 19.10. This decrease is proportional to S^2 and roughly the same at both $\omega = 0$ and ω_N. Consequently, the overall rotation of the molecule is less effective at enhancing T_1 and T_2 relaxation. Although $J(\omega)$ associated with internal motion increases as S^2 increases, it is less effective at relaxation because its maximum value, which occurs at $\omega = 0$, is approximately τ_i, which is smaller than τ_c.
3. The heteronuclear NOE is very sensitive to changes in τ_i (Panel C, Fig. 19.9). The heteronuclear NOE depends on $J(\omega_N + \omega_H) - J(\omega_H - \omega_N)$. This difference is affected by the spectral density associated with the internal motion, especially if the inflection point for $J(\omega)$ is close to these two frequencies.
4. The heteronuclear NOE is also sensitive to changes in the order parameter (Panel D, Fig. 19.9). In this case the change in NOE is largely due to changes in T_1 since the NOE $\propto [J(\omega_N + \omega_H) - J(\omega_H - \omega_N)] \times T_1$.

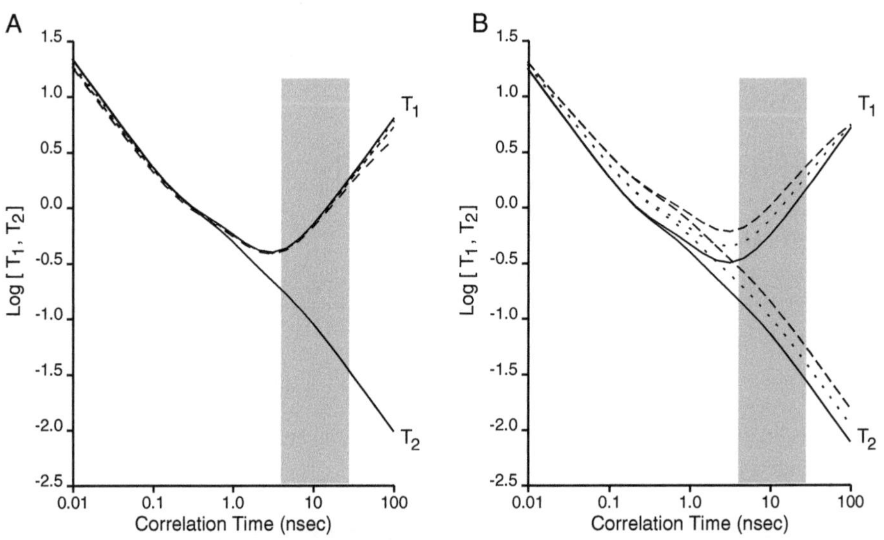

Figure 19.9. Effect of internal motion on ^{15}N relaxation. The effect of internal motion on T_1 and T_2 (panels A and B) and on the heteronuclear NOE (panels C and D). Panels C and D are given on the following page. In each plot the gray box indicates the range of overall correlation time (τ_c) that are typically studied, spanning protein sizes from 10 kDa to 60 kDa. In panels A and C, the order parameter, S^2, was set to 0.8 and the correlation time for internal motion, τ_i was 1 (solid), 10 (dotted), 30 (dashed), or 80 psec (long dash). In the case of T_2, the different curves are indistinguishable. In panels B and D, the correlation time for internal motion was fixed at 30 psec and the order parameter was 1.0 (solid), 0.7 (dotted), and 0.5 (dashed). The horizontal line marked 0.65 indicates a cut-off value that is used to determine whether a residue can be used for determining the rotational diffusion tensor, see Section 19.7.1.

Fig. 19.9, continued.

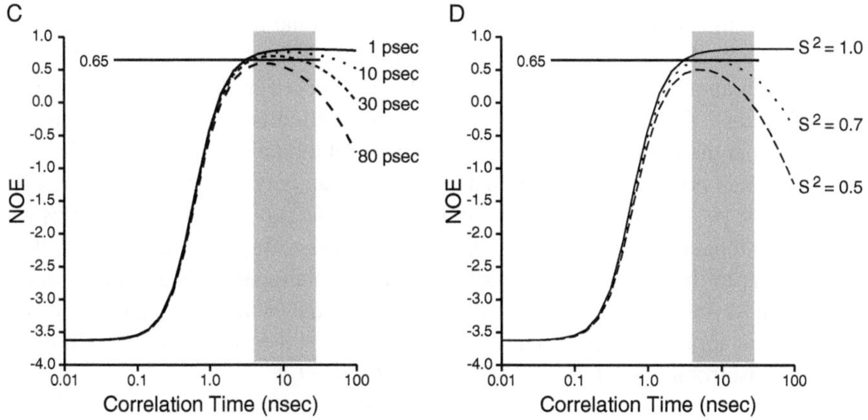

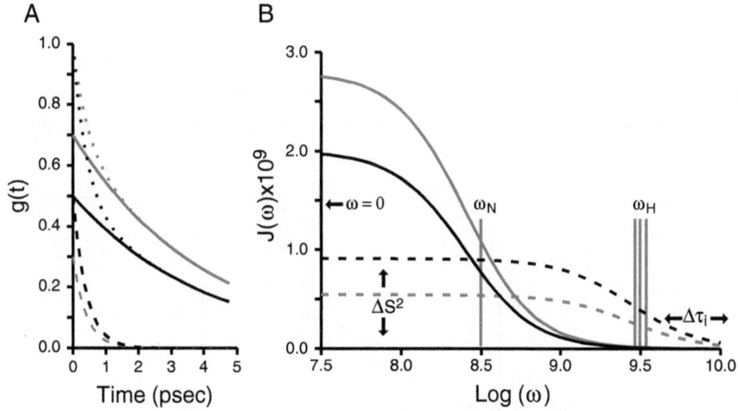

Figure 19.10. Effect of internal motion on the autocorrelation and spectral density functions. Panel A shows the autocorrelation function and Panel B shows the resultant spectral density functions. Two different order parameters were used, $S^2 = 0.7$ (gray curves) and $S^2 = 0.5$ (black curves). In both cases the overall rotational correlation time, τ_c, was 4 nsec and the correlation time for internal motion was 400 psec. In Panel A the solid lines show the correlation function attributed to overall motion and the dashed lines show the correlation function for internal motion. The dotted line shows the sum of both functions. In Panel B the same representation is used, but the sum of the two spectral densities is not shown. The spectral density function associated with internal motions (dashed lines) have been scaled up by a factor of 5. The frequencies ω_N, $\omega_H + \omega_N$, ω_H, and $\omega_H - \omega_N$ are indicated by the vertical bars, assuming $\omega_H = 2\pi \times 500$ MHz. In general, changing the order parameter changes the height of J(ω), while changing the internal correlation time changes the location of the inflection point in J(ω), as indicated by the double arrows ($\leftarrow \quad \rightarrow$).

19.5.1 Anisotropic Rotational Diffusion

If the protein is non-spherical, then its rate of rotational motion will depend on the axis of rotation. In the case of an axially symmetric ellipsoid, as illustrated in Fig. 19.8, rotation about the $D_{\|}$ axis (z-axis) will occur more rapidly than rotation about the x- or y-axis. Consequently, the rotational correlation time about the z-axis will be shorter than the rotational time associated with rotation about any other axis ($\tau_c = 1/6D$). An N-H bond that is aligned with the z-axis can only change its orientation by rotation of the protein around the x- or y-axis. Consequently, those amide groups will experience a longer effective correlation time. In contrast, those amide groups that are perpendicular to the z-axis will experience a shorter rotational correlation time due to the more rapid rotational diffusion rate about the z-axis. When the angle of the N-H bond vector is approximately equal to 54.7°, the rotational correlation time will be the same as the isotropic rate, $D_{iso} = (1/3)(D_x + D_y + D_x)$.

The effect of the bond orientation on the T_1 and T_2 for axially symmetric diffusion is illustrated in Fig. 19.11. Panel A shows that the T_1 is longer for amide bond vectors aligned along the principle axis ($D_{\|}$) and shorter for those perpendicular to principle axis. The opposite occurs for T_2, shorter values are found for N-H bonds aligned with $D_{\|}$. This is simply due to the different dependence of T_1 and T_2 on $J(\omega)$. For N-H bond vectors with $\beta < 54.7°$ the rotational correlation time is longer, therefore, the spectral density function is increased at $\omega = 0$ and decreased for higher frequencies. Since T_2 relaxation is dominated by $J(0)$, the spin-spin relaxation rate is enhanced

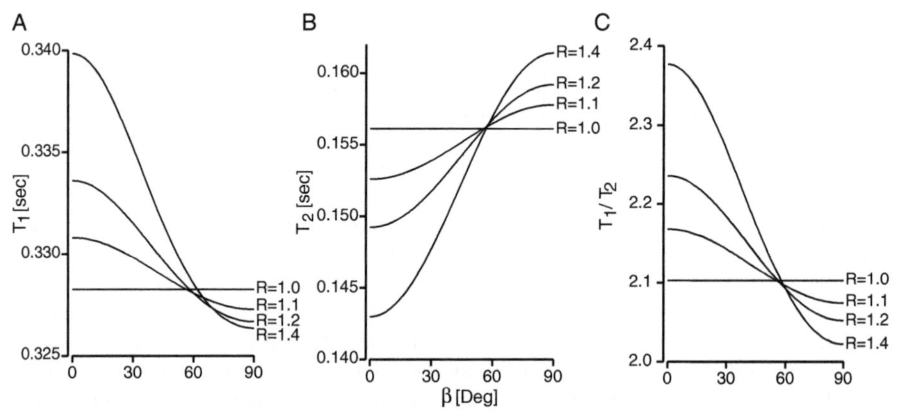

Figure 19.11. Effect of anisotropic diffusion on T_1 and T_2. In this simulation, the isotropic diffusion constant was set to 4.37×10^7 sec^{-1}, giving an isotropic rotational correlation time, τ_c^{iso}, of 3.81 nsec. This correlation time is appropriate for a $\approx$ 55 residue protein. The anisotropy of the diffusion tensor, $R = D_{\|}/D_{\perp}$, was set to 1.0, 1.1, 1.2, and 1.4 and is indicated next to each curve. An angle of 0° indicates that the N-H bond vector is parallel to the principal ($D_{\|}$) axis while an angle of 90° indicates that the N-H bond vector is perpendicular to the principal axis. For residues with $\beta < 54.7°$, the apparent τ_c is longer than τ_c^{iso}, while for residue with $\beta >$ 54.7°, the apparent τ_c is shorter than τ_c^{iso}. The T_1, T_2, and ratio of T_1/T_2, as a function of β are shown in panels A, B, and C, respectively.

and T_2 shortens. Conversely, the decrease in the spectral density function for $\omega > 0$ reduces the efficiency of spin-lattice relaxation and the T_1 lengthens.

The ratio of T_1/T_2 is shown in Panel C in Fig. 19.11. Note that this ratio is constant for a spherically symmetric protein (R = 1.0) and depends on β for a non-spherical protein. Consequently, this ratio can be used to indicate the presence of non-spherical symmetry and allow the determination of the diffusion constants, $D_{\parallel}$ and $D_{\perp}$.

19.6 Measurement and Analysis of Relaxation Data

In this section the process of data acquisition and analysis will be described using the model-free formalism that was outlined in Fig 19.2. The overall process consists of the following steps:

1. Acquire T_1, T_2, and heteronuclear NOE data, preferably at multiple field strengths.
2. Obtain an estimate of the diffusion constants for the protein from the ratio of T_1 to T_2. Utilize statistical tests to determine if non-isotropic diffusion models are necessary to account for the observed data.
3. For each residue in the protein, determine the parameters for internal motion, this would include the order parameter, internal correlation time, as well as a contribution from chemical exchange. Statistical tests are used to determine how many parameters are required to account for the observed relaxation data.

The following text focuses on the application of the above steps to the measurement of the relaxation properties of the amide nitrogen. In this case the protein is prepared in the absence of a ^{13}C enriched carbon source such that dipolar coupling between the amide nitrogen and the α- and carbonyl-carbon cannot occur.

The same approach to characterizing molecular motion could also be used to characterize the relaxation of *isolated* ^{13}C-H groups. However, if the protein is uniformly labeled with ^{13}C it would be necessary to incorporated the additional carbon-carbon dipole-dipole couplings into the analysis. Similarly, the relaxation properties of CH_2 and CH_3 groups are complicated by multiple carbon-proton couplings as well as carbon-carbon couplings. One approach to solve this problem is to prepare the sample using ^{13}C-methyl labeled pyruvate, and to select out the $^{13}CHD_2$ isotopomer for relaxation studies (see [77]).

19.6.1 Pulse Sequences

Pulse sequences that can be used to measure the amide nitrogen T_1, T_2, and heteronuclear NOE are shown in figs. 19.12, 19.13, and 19.14. These sequences utilize coherence selection to generate quadrature detection and they employ sensitivity enhancement to increase the signal to noise. In addition, they are designed to avoid saturation of the water magnetization in order to reduce the saturation of the amide protons that would arise from exchange with the water protons. Additional details regarding the measurement of ^{15}N relaxation data can be found in Skelton et al [149].

There are two concepts that are common to all of the experiments, the removal of cross correlation and the effects of water saturation on the intensities of the amide resonance peaks. Hence, both of these will discussed prior to the discussion of the individual pulse sequences.

19.6.1.1 Correlation between Dipolar and CSA Field Fluctuations

During the relaxation delay period, T, the nitrogen spins will relax due to dipolar coupling and CSA. The contribution of the dipolar field from the amide proton to field fluctuations at the nitrogen nucleus will depend on the spin state of the attached proton. In one case, the dipolar field will augment field fluctuations from the CSA while in the other case attenuation will occur. The relaxation effects of dipolar coupling and the CSA are said to be cross-correlated. A consequence of this cross-correlation is that one-half of the nitrogen spins will relax faster than the other half, depending on the spin state of their coupled proton. To insure that all nitrogen spins relax at the same rate, the proton spin state is inverted during the measurement of the nitrogen T_1 and T_2 by a series of 180° pulses applied to the amide protons. These are indicated in Fig. 19.12 by the Gaussian shaped pulses in the relaxation delay period, T.

19.6.1.2 Radiation Damping & Preventing Saturation of the Water

Saturation of the water protons will lead to a decrease in the intensity of the amide protons due to chemical exchange between the amide proton and saturated water protons. In most experiments, proton exchange simply leads to a decrease in signal intensity. However, in relaxation measurements, water saturation can lead to incorrect measurements of relaxation parameters [54, 69]. In the case of T_1 and T_2 measurements, the level of water saturation increases as the relaxation delay period, T, increases because the proton pulses present during this period can saturate the water. Consequently, the intensity of the amide peaks will decrease as T is lengthened, causing an apparent decrease in the amount of magnetization. Therefore, the magnetization will 'decay' faster than it should and the measured relaxation times will be shorter than their true values. In the case of heteronuclear NOE experiment, the intensity of the resonances in the control experiment, which is performed without saturation of the attached amides protons, will be effected by transfer of saturation from the water.

To prevent saturation of the water it is important to keep the water magnetization orientated along the plus z-axis as much as possible during the measurement. It can be difficult to control the direction of the water magnetization during the experiment because of radiation damping. Radiation damping is the *self*-relaxation of transverse magnetization by an oscillating magnetic field in the receiver coil. This oscillating field is induced by the precession of bulk magnetization in the transverse plane. Consequently, the field fluctuations are exactly the correct frequency to stimulate emission. Radiation damping is very efficient at relaxing water because of the high concentration of protons. Consequently, once the water resonance is placed in the x-y plane, it will rapidly relax back to the z-axis due to radiation damping.

Radiation damping can be reduced by actively dephasing the water magnetization with magnetic field gradients. More efficient suppression of radiation damping occurs if the following two rules are followed:

1. In the case of a spin echo sequence: $P_{90} - \tau - P_{180} - \tau$, the first gradient pulse should be applied immediately after the first 90° pulse. The second gradient pulse, which will refocus the water, should be placed immediately before the end of the second τ period. Consequently the water will remain dephased throughout the entire period.

2. When water is placed on the minus z-axis by a pulse, the pulse should be followed by a gradient to dephase any residual transverse magnetization generated by pulse imperfections.

As an example of the application of the above principles, we will follow the water magnetization through the T_1 experiment (see Fig. 19.12). The first 90° pulse places the water on the $-y$-axis. The immediate application of gradient G2 prevents radiation damping. The water is flipped to the $+y$-axis by the proton 180° pulse, and is refocused by the second G2 gradient pulse immediately before the proton 90° pulse with a phase of y. The subsequent *selective* pulse along $-x$ places the water along the $-z$-axis. The gradient G3 will dephase any traces of water magnetization that may remain in the x-y plane. The subsequent proton 180° pulse places the water on the $+z$-axis. During the relaxation delay period, T, the selective pulses do not excite the water resonance. The proton 180° pulse just prior to G6 inverts the water magnetization. Gradient G6 will dephase any residual transverse water magnetization. The first 90° pulse of the sensitivity enhancement period will place water along the $+y$-axis. The final proton 90° pulse places the water on the $-z$-axis and the final 180° pulse returns the water to the $+z$-axis, as desired.

19.6.2 Measuring Heteronuclear T_1

The sequence for measuring T_1 is shown in Fig 19.12. In this experiment the polarization of the pure nitrogen longitudinal magnetization, N_z, is increased by transfer of the more intense proton polarization to the nitrogen spins via two INEPT periods[4]. During the relaxation delay time, T, the nitrogen magnetization will return to the level appropriate for the equilibrium state of the nitrogen magnetization, proportional to γ_N. Values for the relaxation recovery period (T) range from 0 to $\approx$1 sec, depending on the expected T_1. Typically, data are acquired for 4 to 5 different values of T, with one or two duplicate points to assess the error in the fitted T_1 time.

For the purposes of data fitting, it is more convenient to have the magnetization relax to zero, rather than the equilibrium nitrogen polarization. In this case it is only necessary to fit two parameters, the initial amplitude and T_1. If the magnetization that enters the relaxation delay time is inverted on alternate scans, and the receiver phase is also inverted, then the resultant signal from adding both of these scans together will produce a curve that decays to zero as T$\rightarrow \infty$. To demonstrate that this is the case, we first show that inverting the phase of ϕ_1 inverts the signal. Following the density matrix from *a* to *b* in the sequence (see Fig. 19.12):

$$\begin{aligned}\gamma_H I_z N_z \quad &\stackrel{\phi_1=x}{\longrightarrow} \quad -\gamma_H I_z N_y \rightarrow \gamma_H N_x \rightarrow -\gamma_H N_z \\ &\stackrel{\phi_1=-x}{\longrightarrow} \quad +\gamma_H I_z N_y \rightarrow -\gamma_H N_x \rightarrow +\gamma_H N_z\end{aligned}$$

[4]Formally, an INEPT sequence refers to transfer of polarization from a more sensitive spin to an insensitive spin. Here, the term is used to describe a pulse sequence element that inter-converts in-phase and anti-phase magnetization, i.e. $I_y \rightarrow -2I_x S_z$ or $2I_z S_y \rightarrow -S_x$.

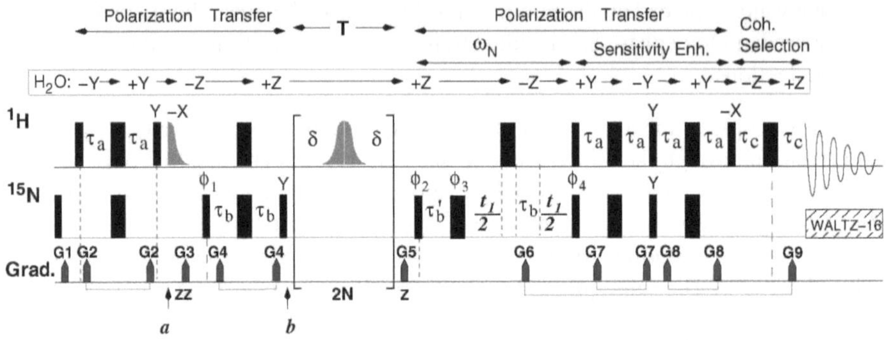

Figure 19.12. Pulse sequence for the measurement of $^{15}N\ T_1$. In this experiment the relaxation of the longitudinal nitrogen magnetization occurs during the period labeled 'T'.

Orientation of Water Magnetization: This is indicated in the box above the pulse sequence.

Pulses: Narrow solid bars are 90° pulses along the x-axis, unless indicated otherwise. Proton and nitrogen 90° pulses are of duration P_{90}^{H} and P_{90}^{N}, respectively. Wide solid bars are 180° pulses. The non-rectangular gray pulses are selective pulses. The first is a selective 90° pulse, such as a 1 msec Gaussian pulse, applied with the transmitter frequency on water. This pulse places the water on the $-z$-axis. The second selective pulse, applied an even number of times (2N) during the T period, is a 550 μsec square pulse that is applied 2 kHz away from the water resonance. Since the null excitation point for this pulse is 2 kHz away from the transmitter frequency the water line is not inverted. Decoupling of nitrogen during t_2 is accomplished using WALTZ-16 with a field strength, γB_1 of $\approx$ 1 kHz. This decouples over a bandwidth of 1.8 kHz, or 36 ppm, assuming a 500 MHz (ν_H) spectrometer.

Delays: The delay τ_a is set to less than $1/(4J_{NH})$ (e.g. 2.25 msec) to reduce signal loss due to relaxation. The τ_b delays are set to exactly $1/(4J_{NH})$, or 2.75 msec. The delay $\tau_b' = \tau_b + 2P_{90}^{H}$ and accounts for evolution of the nitrogen magnetization during the proton 180° pulse during the t_1 time evolution period. The delay δ is 2.5 msec and the delay τ_c is sufficiently long to apply G9 and allow for recovery of the gradient, $\approx$ 0.4 msec.

Phase Cycling and Coherence Selection: The phase cycle is: ϕ_1 = x, -x; ϕ_2 = y; ϕ_3 = 2(x), 2(y), 2(-x), 2(-y); ϕ_4 = x. The receiver phase is: ϕ_{rec} =x, -x, -x, x. Quadrature detection in ω_1 is obtained by coherence selection using gradients G6 and G9. Two FIDs are acquired for each t_1 period and the second FID is acquired with ϕ_4 = -x and inverting the sign of G6. Axial peaks are shifted to the edge of the spectrum by shifting the phase of ϕ_2 and ϕ_{rec} by 180° for each increment in t_1.

Gradients: Values used by Farrow *et al.* [54] are G1 (1 msec, 5 G/cm), G2 (0.5 msec, 4 G/cm), G3 (1 msec, 10 G/cm), G4 (0.5 msec, 8 G/cm), G5 (1 msec, 5 G/cm), G6 (1.25 msec, 30 G/cm), G7 (0.5 msec, 4 G/cm), G8 (0.5 msec, 4 G/cm), G9 (0.125 msec, 27.8 G/cm). Note that the ratio of G6/G9 must be $\propto \gamma_N/\gamma_H$. G1 purges any nitrogen magnetization, gradients G2, G4, G7, and G8 remove imperfections from the 180° pulses, G3 is a zz-filter, retaining only I_zN_z, Gradient G5 is a z-filter, and insures that only nitrogen magnetization along the z-axis is returned to the amide proton for detection.

For each of these cases the nitrogen magnetization relaxes as follows:

$$I_{\phi_1=x}(t) = (-\gamma_H - \gamma_N)e^{-t/T_1} + \gamma_N \quad (19.68)$$
$$I_{\phi_1=-x}(t) = (\gamma_H - \gamma_N)e^{-t/T_1} + \gamma_N \quad (19.69)$$

After two scans, with inversion of the receiver during the second scan, the sum of these two signals is:

$$\begin{aligned} I(t) &= I_{\phi_1=x}(t) - I_{\phi_1=-x}(t) \\ &= [(-\gamma_H - \gamma_N)e^{-t/T_1} + \gamma_N] - [(\gamma_H - \gamma_N)e^{-t/T_1} + \gamma_N] \\ &= -2\,\gamma_H e^{-t/T_1} \end{aligned} \quad (19.70)$$

19.6.3 Measuring Heteronuclear T_2

Spin-spin relaxation times can be measured using two different experimental approaches, $T_{1\rho}$ and CPMG methods. Methods that utilize CPMG sequences are more widely used because they yield T_2 directly while $T_{1\rho}$ requires knowledge of the T_1 in order to obtain T_2 (see below). In addition, CPMG methods generally place smaller demands on the hardware.

CPMG versus $T_{1\rho}$ - A word of warning: CPMG methods require the application of 180° nitrogen pulses at regular intervals while the transverse magnetization relaxes. As discussed in more detail below, these pulses remove artifacts associated with the measurement of T_2. Unfortunately, because the CPMG sequence utilizes *multiple* 180° pulse it is very sensitive to resonance offset effects. This is particularly true when the carbon T_2 is measured because of the larger frequency range, $\Delta\nu$, compared to the amide nitrogen. The error introduced by off-set effects depends on a number of factors, including the magnetic field strength, which affects $\Delta\nu$, the length of the 180° pulses, which affects the bandwidth of the pulse, and the spacing between the pulses. Due to the complexity of relationship between the error in T_2 and the experimental parameters the reader is referred to the articles by Ross *et al.* [138] and by Korzhnev *et al.* [88] for a detailed description. In some cases these errors can be *substantial*, as large as 10% in the case of measuring ^{15}N T_2 and as large as 20% in the case of ^{13}C T_2 measurements. The errors in T_2 are propagated to the estimation of the order parameter, S^2, and the correlation time for internal motion, τ_i, as discussed by Ross *et al.* [138].

If CPMG methods are to be used for measuring T_2 then the following points should be kept in mind:

1. Perform the measurement at low magnetic fields, i.e. 500 MHz spectrometer, to reduce the frequency offset.
2. Use the shortest 180° pulse length as possible, to increase the bandwidth.
3. Consider acquiring the data for two different transmitter frequencies, positioned at 1/3 and 2/3 the width of the spectrum.

Even if the above precautions are taken, it would be useful to calculate the error introduced by the offset, and if significant, correct the measured rates accordingly.

19.6.3.1 Measuring ^{15}N T_2 with CPMG Methods

A pulse sequence to measure T_2 is shown in Fig. 19.13. In this experiment the relaxation of the transverse nitrogen magnetization, N_x, is measured during the relaxation delay period T. Typical values of T range from 0 to 150 msec, depending on the average T_2 of the protein under study. The decay of the transverse magnetization depends on several factors, its intrinsic decay rate due to dipolar coupling and CSA, dephasing due to chemical exchange, and the presence of an inhomogeneous magnetic field:

$$R_2 = (1/T_2) = R_2^{dipole} + R_2^{CSA} + R_2^{ex} + R_2^{\Delta B_o} \tag{19.71}$$

The contribution of an inhomogeneous field, ΔB_o, to the T_2 is completely removed by the use of the CPMG sequence during the relaxation delay period, as indicated in Fig. 18.7. In addition, if the rate of chemical exchange is slower than rate of the 180° pulses (e.g. $k_{ex} << 1/\tau_{CPMG}$), then the CPMG sequence will also refocus any dephasing due to chemical exchange. Consequently, the observed relaxation rate is the sum of three terms:

$$R_2 = R_2^{dipole} + R_2^{CSA} + R_2^{ex} \tag{19.72}$$

where R_2^{ex} is the contribution of fast-exchange processes to R_2.

Suppression of Water Saturation: The T_2 experiment utilizes similar schemes to avoid saturation of the water magnetization. The water magnetization is maintained along the $+z$-axis during the relaxation delay time (T) by inverting the phases of the pair of 180° proton pulses within each CPMG segment. Consequently, there is no net rotation of the water magnetization. Note that it is not possible to apply a gradient pulse after the first proton 180° pulse because the nitrogen magnetization is transverse and would be dephased by the gradient.

19.6.3.2 Measuring ^{15}N T_2 with $T_{1\rho}$

In this case, the transverse magnetization relaxes in the rotating frame while aligned along the applied B_1, or 'spin-lock' field. Resonances whose frequency is not equal to the transmitter frequency experience an effective field of:

$$B_{eff} = \frac{\omega_e}{\gamma} = \frac{\sqrt{\omega_1^2 + \delta\omega^2}}{\gamma} \tag{19.73}$$

where $\delta\omega$ is the frequency difference between the position of a resonance line and the transmitter frequency.

Although $T_{1\rho}$ measurements suffer greatly from off resonance effects, the relaxation measurement is much easier to correct than with CPMG techniques. Peng *et al.* [127] give a clear and detailed analysis of relaxation in the rotation frame. The end result is that the observed relaxation is approximately equal to[5]:

$$\frac{1}{T_{1\rho}} = \frac{1}{T_1} cos^2\theta + \frac{1}{T_2} sin^2\theta \tag{19.74}$$

[5]In the case of $T_{1\rho}$ measurements, the frequency arguments to $J(\omega)$ are altered by ω_e, i.e. $J(\omega_N)$ is replaced by $J(\omega_N + \omega_e)$. Since ω_e is in the kHz range, $\omega_N + \omega_e \approx \omega_N$.

where θ is the angle between the effective magnetic field and the z-axis and is given by:

$$\theta = tan^{-1}\frac{\omega_1}{\delta\omega} \tag{19.75}$$

$T_{1\rho}$ can be measured using the same pulse sequence as described for measuring T_2 (Fig. 19.13) with the CPMG sequence replaced by a $\approx$ 1-2 kHz B_1 field. This approach will cause a reduction in the amplitude of the signals that are off-resonance

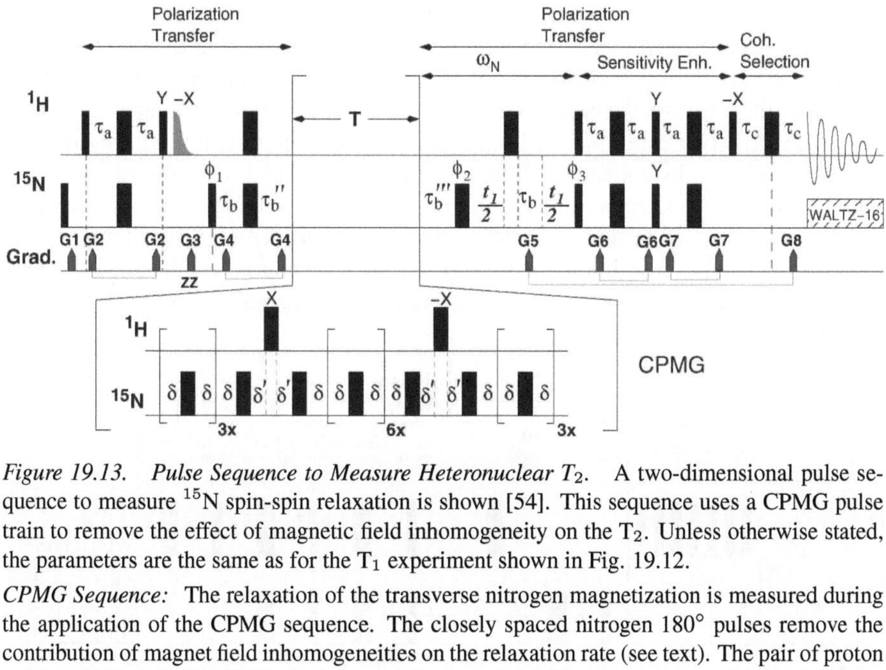

Figure 19.13. Pulse Sequence to Measure Heteronuclear T_2. A two-dimensional pulse sequence to measure ^{15}N spin-spin relaxation is shown [54]. This sequence uses a CPMG pulse train to remove the effect of magnetic field inhomogeneity on the T_2. Unless otherwise stated, the parameters are the same as for the T_1 experiment shown in Fig. 19.12.

CPMG Sequence: The relaxation of the transverse nitrogen magnetization is measured during the application of the CPMG sequence. The closely spaced nitrogen 180° pulses remove the contribution of magnet field inhomogeneities on the relaxation rate (see text). The pair of proton pulses within each CPMG block are required to remove dipole-CSA cross-correlation effects.

Delays: Delay τ_a is set to $< 1/(4J_{NH})$ and τ_b is set to be equal to $1/(4J_{NH})$. The delay τ_b'' is set to $\tau_b + 2 \times P_{90}^N/\pi$. This adjustment accounts for evolution of the nitrogen magnetization during the 90° nitrogen pulse (ϕ_1) ensuring that the $\tau_b - 180° - \tau_b''$ period completely refocuses the nitrogen magnetization such that it is aligned along the x-axis at the beginning of the CPMG pulse train. The delay τ_b'''$= \tau_b + 2P_{90}^H + 2 \times P_{90}^N/\pi$. This adjustment accounts for contribution of the proton 180° pulse length and the nitrogen 90° pulse (phase ϕ_3) to the evolution of the nitrogen magnetization. The delay δ, during the CPMG pulse train, is set to $\approx$ 500 μsec. The delay $\delta' = \delta - P_{90}^H$, accounts for the contribution of the proton 180° pulse to the 2δ delay.

Phase Cycle and Coherence Selection: $\phi_1 = x, -x$; $\phi_2 = 2(x), 2(y), 2(-x), 2(-y)$; $\phi_3 = x$. The receiver phase is: $\phi_{rec} = x, -x, -x, x$. For quadrature detection, two FIDs are acquired for each t_1 point and the phase of ϕ_3 and the amplitude of G5 were inverted for the second FID. Axial peaks are suppressed by incrementing ϕ_1 and the receiver phase by 180° for each t_1 pt.

Gradients: Values suggested by Farrow *et al.* [54] are: G1(1 msec, 5 G/cm), G2(0.5 msec, 4 G/cm), G3(1 msec, 10 G/cm), G4(0.5 msec, 8 G/cm), G5(1.25 msec, 30 G/cm), G6(0.5 msec, 4 G/cm), G7(0.5 msec, 4 G/cm), G8(0.125 msec, 27.8 G/cm). G5 and G8 are used for quadrature detection. G1 purges any nitrogen magnetization prior to the experiment, G2, G4, G6, and G7 gradient pairs remove artifacts from the 180° pulses. G3 is a zz-filter, retaining I_zN_z.

because they are not aligned along the effective field when the spin-lock is applied. Only the component of the magnetization along the effective field will remain coherent during the T delay. Any magnetization that is perpendicular to B_{eff} is dephased within 1-2 msec by the relatively inhomogeneous B_1 field of the pulse. Consequently, it is better to use methods, such as adiabatic pulses, that align the magnetization along the effective field, as discussed in Section 18.4.5.2.

19.6.3.3 Measuring the Heteronuclear NOE

The pulse sequence to measure the heteronuclear NOE is shown in Fig. 19.14. The experiment is performed twice, first with no proton saturation, and then with proton saturation, giving S_z^o and S_z, respectively. It is important that *all* parameters, including the number of scans and the receiver gain, are identical in both experiments. It is best to interleave the two experiments, i.e. collect a t_1 time point (FID) without proton saturation, followed by collection of the same time point with proton saturation.

It is also useful to keep in mind that in this experiment the magnetization does not originate on the amide proton, thus the sensitivity is considerably lower than the T_1 and T_2 experiments. In addition, the hnNOE for the N-H pair is less than one, further reducing the signal. Consequently, more scans should be acquired to reduce the error in the peak intensities.

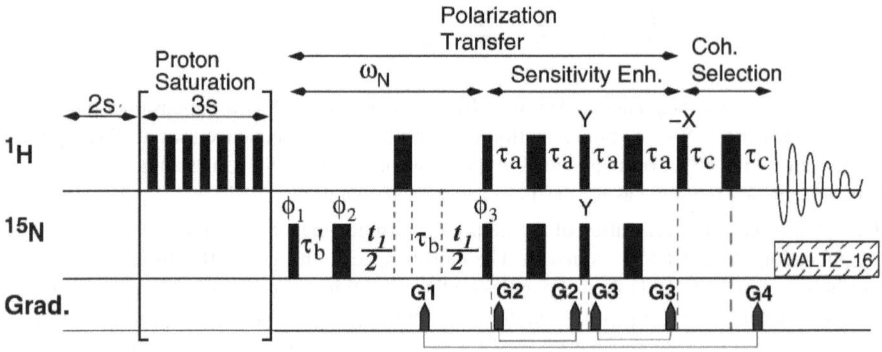

Figure 19.14. Pulse sequence for the measurement of the heteronuclear NOE. This sequence is described in more detail in [54]. Experimental parameters are the same as the T_1 experiment in Fig. 19.12. Saturation of the amide protons is accomplished by applying a series of 120° pulses spaced 5 msec apart for 3 sec prior to the first proton pulse [108]. The water saturation pulses are preceded by a 2 sec relaxation delay. For the control experiment with no proton saturation, the 120° proton pulses are not applied and the total relaxation delay is 5 sec.
Delays: The delay $\tau_b' = \tau_b + 2P_{90}^H$ and accounts for the evolution of the nitrogen spins during the 180° proton pulse (ϕ_3). τ_a and τ_b are defined in Fig. 19.12.
Phase Cycle and Coherence Selection: $\phi_1 = y$, $\phi_2 = x, y, -x, -y$, $\phi_3 = x$, $\phi_{rec} = x, -x$. Quadrature detection was accomplished by inverting the phase of ϕ_3 and the amplitude of G1 for the second FID acquired at each t_1 point. Axial peaks are shifted to the edge of the spectrum by shifting the phase of ϕ_1 and the receiver by 180° for each new t_1 value.
Gradients: G1(1.25 msec, 30 G/cm), G2(0.5 msec, 4 G/cm), G3(0.5 msec, 4 G/cm), G4(0.125 msec, 27.8 G/cm). Gradient pairs G2 and G3 remove any artifacts associated with imperfect 180° pulses.

19.7 Data Analysis and Model Fitting

The relaxation data will be used to determine the motional properties of the amide groups in the protein. As a reminder, this process involves the following steps:

1. Measuring the amplitude, or volume, of peaks in the spectra from T_1, T_2, and hnNOE spectra.
2. In the case of T_1 or T_2 measurements, the relaxation rate is obtained by fitting the data to the following equation:

$$I = I_o e^{-T/T_{1,2}}$$

   Duplicate measurements are used to determine the error in the relaxation rate by Monte Carlo methods, as illustrated in Fig. 19.17. In the case of the hnNOE measurement, the NOE is obtained by simply dividing the peak intensity obtained during saturation of the protons by the intensity obtained without saturation.
3. Defining a model that describes the global motion of the protein as well as internal motion of individual bond vectors. The parameters of this model include information on the rotation of the molecule (D_x, D_y, D_z), the extent (S^2) and rate (τ_i) of internal motion.
4. Calculate $J(\omega)$, and then T_1, T_2, and the hnNOE.
5. Adjust the parameters of the model until the difference between the experimental and calculated relaxation parameters is minimized.

There are a number of programs for the fitting of relaxation data [102, 126]. Consequently, the emphasis of this section is on the general approach to data analysis and the following section will discuss in detail the use of statistical tests in the data analysis.

19.7.1 Defining Rotational Diffusion

The first goal is to define the rotational diffusion properties of the molecule. If an incorrect model is used then systematic errors will occur in determining the order parameter(s) and internal correlation time(s). A comprehensive discussion of the errors that can occur when an axially symmetric protein is assumed to be spherical has been given by Schurr et al. [143]. In this study, the authors generated synthetic data using the simple model-free formalism with axially symmetric diffusion. These data were then used to obtain motional parameters from either the simple model-free formalism or the extended model-free formalism, *assuming a spherical protein.* To summarize their findings:

1. If the simple model-free formalism (τ_i, S^2) is used with noise-free data then the fitted parameters are in close agreement to the true parameters. The inclusion of typical noise in the T_1, T_2, and hnNOE measurement results in a modest error in S^2 ($\approx$5%) and more extensive errors in τ_i ($\approx$ 40%).
2. If the extended model-free formalism ($\tau_i, \tau_f, S_f^2, S_s^2$) is used to fit the relaxation data, then spurious slow motions are obtained, even though none existed in the initial model.

The rotational diffusion coefficient(s) are generally obtained from the ratio of the T_1 to T_2 because this ratio is relatively insensitive to internal motion [84]. Inspection of Fig. 19.9 shows that T_2 is completely insensitive to the internal correlation time while T_1 decreases slightly as the correlation time for internal motion increases (Panel A, Fig. 19.9). Both T_1 and T_2 are affected by the order parameter, however both relaxation times increase by similar amounts as S^2 decreases (Panel B), therefore their ratio remains essentially constant. The ratio of T_1 to T_2 is sensitive to the shape of the particle, as illustrated in Fig. 19.11, thus providing data that can be used to determine the diffusion properties of the molecule.

In the case of a non-symmetric molecule, it is necessary to determine both the orientation of the molecule and the size of the diagonal components of the diffusion tensor because this information is required in the calculations of $J(\omega)$. The process of determining the diffusion tensor is most easily understood using an axially symmetric (i.e. cylindrical) molecule as an example.

A sub-set of residues are selected for the process of determining the diffusion tensor. The selection criteria is given below, in Section 19.7.1.1. The structure of the protein is required in order to calculate β, the angle between the N-H bond vector and the principle rotation axis. However, the orientation of the principle axis with respect to the protein is unknown. The available coordinate file of the molecule, such as the PDB file, can have the molecule in any arbitrary orientation; the molecule need not have its long axis oriented in the direction of the z-axis. Therefore, prior to calculating T_1 and T_2 it is necessary to apply the appropriate rotation to the structure such that it becomes oriented with its long axis along the z-axis.

Because of axial symmetry, only two rotation angles are required to complete this rotation, as illustrated in Fig. 19.15. Therefore a total of four parameters are required to describe the rotational diffusion properties of an axially symmetric protein, the two rotation angles (θ, ϕ) that re-orient the molecule into its principle axis system, and the two diffusion constants, $D_{\|}$ and $D_{\perp}$.

In practice, a grid search is performed, varying θ, ϕ, $D_{\|}$, and $D_{\perp}$. For each of these values the T_1 and T_2 are calculated from the known structure and the calculated ratio is compared to the observed ratio. The values of θ, ϕ, $D_{\|}$, and $D_{\perp}$ that minimize this difference define the rotational diffusion properties of the molecule. Accurate initial guesses regarding the diffusion coefficient can accelerate this process. The hydrodynamic properties of a molecule can be calculated from the structure using the program HYDROPRO [60], and these values can be used for initial guesses to facilitate the search for the best parameters.

19.7.1.1 Selection of Residues for Determining Diffusion Tensor

Although the T_1/T_2 ratio is relatively insensitive to internal motion, the presence of significant amounts of internal motion will distort the ratio, leading to errors in the determination of the diffusion properties. In addition, the presence of chemical exchange will also lead to a decrease in T_2, again distorting the T_1/T_2 ratio. Therefore, the collection of residues that are used to determine the diffusion properties of the entire molecule should be essentially rigid; residues that are mobile or show exchange broadening should not be used in the determination of the diffusion tensor.

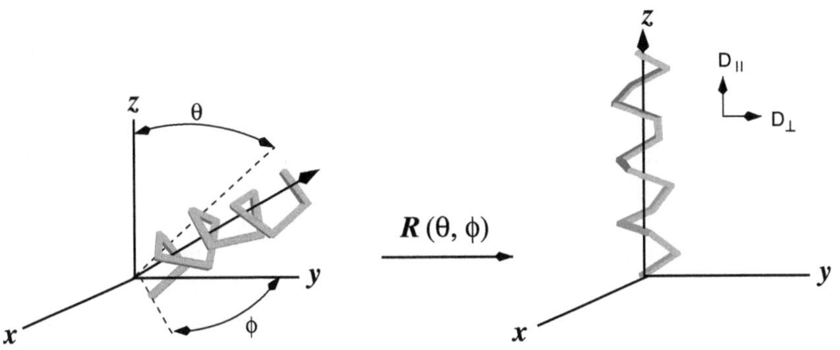

Figure 19.15. Determining the orientation of the diffusion tensor. The process of determining the orientation of the diffusion tensor is illustrated with a simple α-helical protein. The orientation of the molecule in the molecular frame is illustrated in the left part of the figure. The molecular frame is the internal x, y, and z-coordinates of the molecule, such as those contained in a protein database (pdb) file. The diffusion tensor is not diagonal for this particular orientation. This molecule is axially symmetric, thus only four parameters are required to define its hydrodynamic properties: two rotation angles (θ, ϕ) and two diffusion coefficients $(D_{\parallel}, D_{\perp})$. In this example the orientation shown in the left figure is transformed to the orientation shown in the right figure by a rotation of $\phi = 45°$ about the z-axis in the laboratory frame, followed by a $\theta = 40°$ rotation about the new x-axis. A total of six parameters are required to describe the hydrodynamic properties of an asymmetric molecule, three rotation angles (θ, ϕ, γ) and D_x, D_y, and D_z.

Several 'data filters' have been described [126, 158] for the exclusion of residues. The approach by Pawley [126] will be described in detail below.

1. Residues with ^{15}N hnNOE smaller than 0.65, are discarded. These residues obviously possess considerable internal motion. Note that this selection criteria should take into consideration the size of the protein. Fig. 19.9 shows the relationship between internal motion and the correlation time of the protein. In general, smaller proteins, with shorter correlation times, will show increased hnNOE values. In this case, higher values should be used as the cut-off.
2. Residues with T_2 times shorter than one standard deviation from the average T_2 (e.g. $T_2 < \bar{T_2} - \sigma$) are discarded. A short T_2 may indicate a contribution of chemical exchange to the T_2. However, since a short T_2 also occurs if the N-H bond vector is less than $\approx 54°$ from the z-axis (see Fig. 19.11), if the residue also has a longer than normal T_1 it should be retained.[6]
3. Once the diffusion properties are defined from the initial set of residues, the motional parameters of these residues are obtained from the model-free formalism and residues that do not fit the simple model, or show slow internal motion ($\tau_f > 600$ psec), are removed from the set and the diffusion constants are obtained from the remaining residues.

[6]This description is true for a prolate ellipsoid, the opposite trend will be observed for an oblate ellipsoid [126].

19.7.2 Determining Internal Rotation

The final step is to determine parameters that describe the internal motion. In this process, it is assumed that the diffusion coefficient(s) of the protein have been obtained as described above and are therefore fixed.

The number of parameters that are required to represent the relaxation data will depend on the model for internal motion. In addition to these parameters, T_2 relaxation may also contain a contribution from chemical exchange, R_{ex}. Therefore it is necessary to include this term when fitting the data. Consequently, in the simple model-free formalism a total of three parameters are required per residue (τ_f, S_f^2, R_{ex}) while a total of five are required for the extended model-free formalism, the two additional parameters are τ_S, and S_S^2.

Since the number of fitted parameters are often equal to, or even exceed, the number of experimental data points it may be possible to only fit a sub-set of the above parameters. Mandel, Akke, and Palmer [102] have suggested a hierarchical route, fitting increasingly more complicated models to the data (see Table 19.2).

The suitability of models 1-3 for describing the relaxation data can be tested using the χ^2 goodness-of-fit test. In addition, the partial F-test can be used to determine whether the inclusion of additional parameters, such as τ_f (model 2) or R_{ex} (model 3), improves the fit to the data. If data is acquired at only one magnetic field strength, then statistical tests cannot be applied to models 4 and 5. In this case, the model that gives the smallest χ^2 value is presumed to be correct. The hierarchical process of model selection has been semi-automated in the software package called MODELFREE, which is available from A. Palmer.

More recently, alternative methods of model selection have been applied to relaxation data [46, 34]. These newer methods are based on Akaike's information criteria [2] and Bayesian information criteria [144]. These methods provide a more straight-

Table 19.2. Model-free parameter fitting.

Model Number	Fitted Parameter	Degrees of Freedom[a] $\nu = N - m$	Comments
1	S^2	2	Simple model-free formalism, the correlation time for internal motion is assumed to be very fast, i.e. $\tau_f = 0$ and there is no chemical exchange ($R_{ex} = 0$).
2	S^2, τ_f	1	Simple model-free formalism with no chemical exchange ($R_{ex} = 0$).
3	S^2, R_{ex}	1	Simple model-free formalism, the correlation time for internal motion is assumed to be very fast, i.e. $\tau_f = 0$.
4	S^2, τ_f, R_{ex}	0	Simple model-free formalism, complete description.
5	S_f^2, S_S^2, τ_S	0	Extended model-free formalism, $\tau_f = 0$ and no chemical exchange.

[a] Assuming measurement of T_1, T_2 and hnNOE at a single field, $N = 3$, m = number of parameters.

forward method of model selection and generally provide a more accurate description of internal rotational correlation times [34].

19.7.3 Systematic Errors in Model Fitting

A number of assumptions and/or approximations are taken when calculating relaxation parameters. If incorrect assumptions are made, then systematic errors will be introduced into the determined order parameters and correlation times. The errors that are introduced by the inadequate description of the rotational diffusion properties of the protein have been discussed previously. These errors can be minimized by either direct determination or calculation of the diffusion constants.

A more serious source of error is the assumption of the value of the chemical shift anisotropy. Average values ranging from -165 to -170 ppm are commonly used in model fitting. However, it is apparent that the CSA varies from residue-to-residue due to residue specific variations in the electronic environment of the ^{15}N nucleus. In the case of the small protein, ubiquitin, the CSA values were shown to range from -120 ppm to over -200 ppm with 68.3% of the residues showing a CSA between -140 and -180 ppm [59]. The measured CSA for a number of residues in Ribonuclease H1 is shown in Fig. 19.16. Systematic errors in the order parameter (S^2) and the internal correlation time (τ_i) will occur if the incorrect CSA value are used in calculating the relaxation parameters. In general, the errors are largest for the order parameter, and

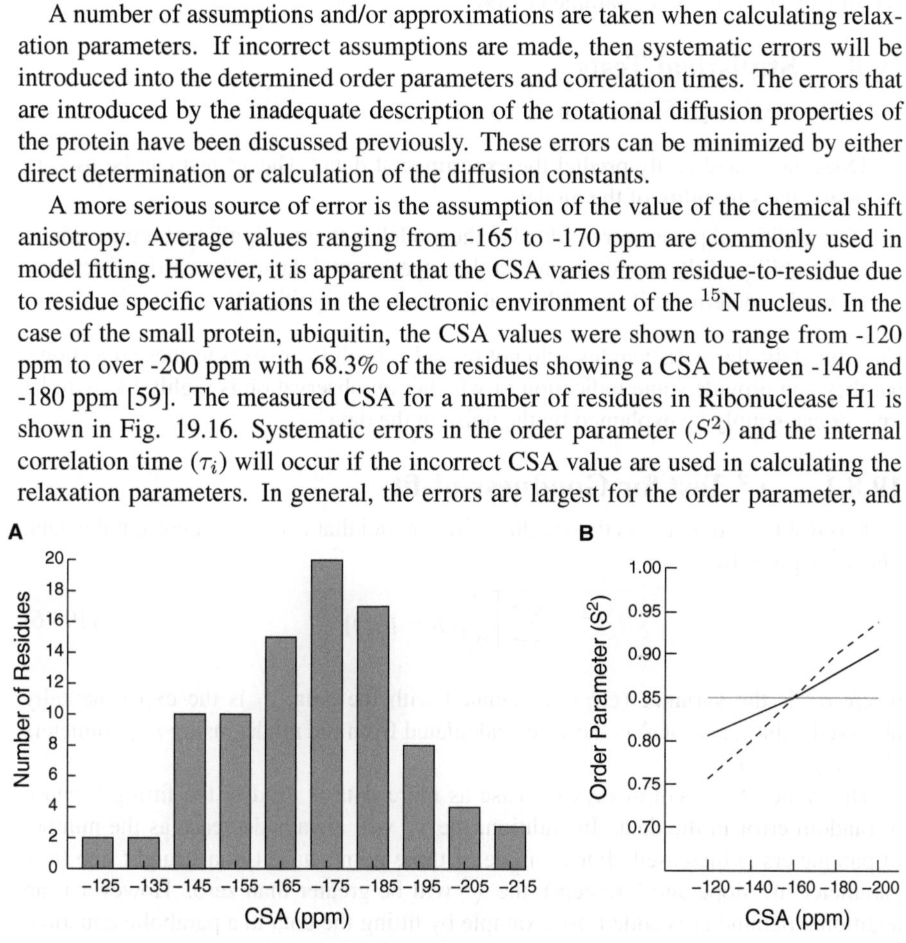

Figure 19.16. Variation of CSA for Ribonuclease H1. The distribution of CSA values in Ribonuclease H1 from *E. coli* are shown in Panel A [89]. The CSA values range from -129 ppm to -213 ppm. The mean CSA is -173 ppm and the distribution of CSA values can be approximated as a normal distribution with a standard deviation of 13 ppm. Panel B shows the effect of an incorrect CSA value on the order parameter. Two magnetic field strengths are shown, 400 MHz (solid black line) and 600 MHz (dashed line). Synthetic data was generated assuming $S^2 = 0.85$ (indicated by the horizontal gray line) and CSA values ranging from -120 to -200 ppm. The order parameter that was obtained by fitting the data to the simple model-free formalism is plotted, assuming that the CSA was equal to -160 ppm. If the true CSA is smaller then -160 ppm the order parameter will be underestimated, while a CSA higher than the -160 will give an over-estimate of the order parameter. This error depends on the magnetic field strength. Adapted from Fushman and Cowburn [58].

increase as the magnetic field strength increases, as illustrated in Panel B of Fig. 19.16. In general, the range of CSA values is sufficiently small as to not cause significant problems at magnetic field strengths $\leq$ 600 MHz. However, at higher field strengths, $\geq$ 800 MHz, it will be necessary to take this effect into account for a quantitative determination of the order parameter [89].

19.8 Statistical Tests

During the fitting of relaxation data, two key questions arise:

1. Does the model really predict the experimental data? The χ^2 test can be used to assess the suitability of the model.
2. If an additional parameter is added to the model, is there a significant improvement in the ability of the model to predict the experimental data? The partial F-test can be used to determine if the additional parameter is significant.

Unfortunately, the statistical tests do not provide absolute answers to these questions, but they can provide some indication of whether an observation is highly likely to be true, or can simply be explained by the noise in the data.

19.8.1 χ^2 Test for Goodness-of-fit

This test is used to assess the validity of the model that is used to represent the data. The χ^2 is given by:

$$\chi^2 = \sum_{i=1}^{N} \left[\frac{1}{\sigma_i^2} [y_i - y(\vec{p})]^2 \right] \tag{19.76}$$

where σ^2 is the variance (error) associated with the data, y_i is the experimentally observed values, and $y(\vec{p})$ is the value calculated from the model, using n, parameters $\vec{p} = p_1, p_2, ..., p_n$.

The value of χ^2 will always increase as more data is used in the fitting because of random error in the data. In addition, the χ^2 will always decrease as the number of parameters is increased. For example, if three points are fit to a straight line (two parameter fit: slope and intercept), the χ^2 will be greater than zero. However if an additional parameter is added, for example by fitting the data to a parabolic equation, the χ^2 will decrease to zero. To account for the influence of the number of data points and parameters on the χ^2, the χ^2 is normalized by dividing by ν, the number of degrees of freedom, to give the reduced χ^2:

$$\chi^2_\nu = \chi^2/\nu \tag{19.77}$$

Where the number of degrees of freedom, $\nu = N - n$, is the number of data points (N) minus the number of parameters (n) in the model.

19.8.1.1 Fitting Relaxation Data

T_1 and T_2 Data: In the case of fitting the raw T_1 or T_2 relaxation data to a single exponential two parameters are required, the amplitude ($p_1 = I_o$) and the time constant

($p_2 = T$). The reduced χ^2 is:

$$\chi^2_\nu = \frac{1}{N-2}\sum_{i=i}^{N}\left[\frac{1}{\sigma_i^2}[I_i - I_o e^{-t_i/T_1}]^2\right] \tag{19.78}$$

where, I_i is the measured intensity of the peak at time t_i, and N is the total number of T values used in the experiment.

Diffusion Constants: The ratio, T_1/T_2, is sensitive to anisotropic rotational diffusion of the molecule. For the case of an axially symmetric molecule there are a total of four parameters that have to be obtained from the relaxation data: $D_\|$, $D_\perp$, and the rotational angles, θ and ϕ, to specify the orientation of the the $\|$ and $\perp$ axis with respect to the protein structure. The χ^2_ν in this case is:

$$\chi^2 = \sum_{i=1}^{N\,res}\frac{1}{N_{res}-4}\left[\frac{((T_1/T_2)^{obs} - (T_1/T_2)^{calc})^2}{\sigma_i^2}\right] \tag{19.79}$$

where N_{res} are the number of residues that are used in the fitting.

Internal Motion In the case of defining the motion of a single residue, a typical parameter set would be the order parameter ($p_1 = S^2$), and the correlation time for internal motion ($p_2 = \tau_i$). The χ^2 is calculated as:

$$\chi^2_\nu = \frac{1}{3N-2}\sum_{i=1}^{N}\left[\frac{(T_1^{obs} - T_1^{calc})^2}{\sigma^2_{T1}} + \frac{(T_2^{obs} - T_2^{calc})^2}{\sigma^2_{T2}} + \frac{(NOE^{obs} - NOE^{calc})^2}{\sigma^2_{NOE}}\right] \tag{19.80}$$

where N is the number of magnetic field strengths. This assumes that that hydrodynamic properties of the molecule have already been defined.

19.8.1.2 Interpretation of χ^2_ν

The χ^2 is simply the ratio of the difference between the experimental measurement and the model ($y_i - y(\vec{p})$) divided by the error in the measurements, σ. If the model accounts for the data, then the expectation is that χ^2 will be close to 1.0 because the differences between the data and the model should be the same as the error in the data. If the model does not account for the experimental observations, the χ^2 will be larger than one. Alternatively, if the error in the data is significantly over-estimated, then χ^2 can be significantly less than one.

The χ^2 can be used to assess how well the model fits the experimental data by comparing the actual χ^2 to expected χ^2 values. The expected χ^2 values are obtained from the integral of the χ^2 distribution and selected values of this integral are shown in Table 19.3. The reader should consult Bevington [17] for more information and any statistics text for a more extensive probability table. The probabilities listed in Table 19.3 are the probabilities that the model *can* account for the data, given random variation of the experimental data. Consider the following example.

Table 19.3. The χ^2 distribution is shown for different probabilities, P, and degrees of freedom, ν.

P	0.95	0.80	0.50	0.30	0.10	0.01	0.001
ν=1	0.004	0.064	0.455	1.074	2.706	6.635	10.83
ν=2	0.052	0.223	0.693	1.204	2.303	4.605	6.91
ν=3	0.117	0.335	0.789	1.222	2.084	3.780	5.42
ν=4	0.178	0.412	0.839	1.220	1.945	3.319	4.61
ν=5	0.229	0.469	0.870	1.213	1.847	3.017	4.10
ν=6	0.273	0.512	0.891	1.205	1.774	2.802	3.74
ν=8	0.342	0.574	0.918	1.191	1.670	2.511	3.27
ν=10	0.395	0.618	0.934	1.178	1.599	2.321	2.96
ν=15	0.484	0.687	0.959	1.115	1.487	2.039	2.51
ν=20	0.543	0.729	0.967	1.139	1.421	1.878	2.27

Example Problem: Five data points are used to fit a T_1 decay curve. The resultant reduced χ^2_ν is found to be 4.0. Determine the probability that this χ^2_ν occurred randomly.

Answer. In this case the number of degrees of freedom, ν, is 3 and the observed χ^2_ν is found between the $P = 0.01$ ($\chi^2_\nu = 3.78$) and $P = 0.001$ ($\chi^2_\nu = 5.42$). Based on these probabilities one could state that the chances that this poor of a fit could occur randomly is between 1.0 and 0.1%. Since the event is very unlikely, the model is probably incorrect. Alternatively, the actual error in the data, σ, has been under-estimated.

19.8.2 Test for Inclusion of Additional Parameters

In many cases the model that is being used to fit the data can be extended by one or more additional parameters. For example, the model for isotropic diffusion can be extended to an axially symmetric molecule, replacing one parameter (D_{iso}) with four ($D_\parallel$, $D_\perp$, θ, ϕ). Extension to a fully asymmetric molecule will require 6 parameters. In the case of fitting models for internal motion, initially the data are fit using only the order parameter, S^2, and then extended by addition of either the internal correlation time τ_i, or a chemical exchange term, R_{ex} (see Table 19.2).

In all cases the inclusion of an additional parameter will always decrease the χ^2_ν. If the additional parameter(s) decrease the χ^2_ν by more than what would be expected from the random error in the data, then the inclusion of the additional parameter is justified.

The test for the inclusion of additional parameters is called the partial F-test. It is related to the F-test, which is used to determine if variances in two data sets are equal. In the case of the partial F-test, the F statistic is given by:

$$F_{(m-n),N-n} = \frac{(\chi^2_m - \chi^2_n)/(n-m)}{\chi^2_n/(N-n)} \tag{19.81}$$

Table 19.4. The F distribution is shown for a probability of 5% as a function of the two degrees of freedom, $\nu_1 = m - n$, and $\nu_2 = N - n$.

$\nu_1 = m - n$	1	2	3	4	5
$\nu_2 = N - n$	161.4	199.5	215.7	224.6	230.2
ν_2=2	18.5	19.0	19.2	19.3	19.3
ν_2=3	10.1	9.6	9.1	9.1	9.0
ν_2=4	7.7	6.9	6.4	6.3	6.2
ν_2=5	6.6	5.8	5.4	5.2	5.1
ν_2=6	6.0	5.1	4.8	4.5	4.4
ν_2=7	5.6	4.7	4.3	4.1	4.0
ν_2=8	5.3	4.5	4.1	3.8	3.7
ν_2=10	5.0	4.1	3.7	3.5	3.3
ν_2=15	4.5	3.7	3.3	3.1	2.9

for the case where N data points are initially fit to m parameters and then subsequently fit to n parameters. For example, $m = 1$ for isotropic diffusion and $n = 4$ for axially symmetric diffusion. This statistic is essentially the ratio of the difference in the reduced χ^2 values (corrected for the change in degrees of freedom) to the new reduced χ^2.

We expect the reduced F statistic to be large if additional parameters improved the fit because this indicates that the drop in χ^2 is much larger than the average deviation of the data from the new model. The F-distribution is given in Table 19.4 for a number of different degrees of freedom. This table is given for a probability of 5%, therefore, if the calculated F statistic is *larger* than the values, then the probably that the improvement in the model occurred by chance is less than 5%, i.e. there is a good chance that the improvement is legitimate. The following provides an example of how to use Table 19.4.

Example Problem: Ten residues have been selected to evaluated the rotational diffusion properties of a protein. The data are fit to the isotropic, axially symmetric, and anisotropic models and the following χ^2 (non-reduced) were obtained:

Model	# Para.	χ^2	$F_{(a,b)}$ Isotropic	Axial
Isotropic	1	9.0	-	
Axial	4	1.5	$F_{(3,6)}$=(7.5/3)/(1.5/6)= 10.0	-
Aniso.	6	0.8	$F_{(5,4)}$=(8.2/5)/(0.8/4)= 8.2	$F_{(2,4)}$=(0.7/2)/(0.8/4)=1.75

Which model of rotational diffusion is best described by the data?

Answer: The partial F-value for comparing each model to the others is given in the above table. For the comparison of axial to isotropic rotation, $F_{3,6} = 10.0$. Comparing the anisotropic model to the isotropic model gives $F_{5,4} = 8.2$ and to axially symmetric model gives $F_{2,4} = 1.75$. The F statistics for 5% probability are obtained from Table 19.4. In the case of axial versus isotropic, the F statistic is 4.8 for $\nu_1 = 3$ and $\nu_2 = 6$. Since the observed value of 10.0 is greater than 4.8, there is *less* than a 5% change that this occurred randomly. Similarily, the anisotropic model also provides a statistically significant improvement in the model. In comparing the anisotropic to the axially symmetric model, the calculated F statistic of 1.75, is less than the $F_{2,4}$ value of 6.9 in Table 19.4, therefore the chance that the improvement occurred by random is greater than 5%, consequently the more complicated asymmetric model should be adopted with caution.

19.8.2.1 Error Estimation of Raw Data

To calculate the χ^2 it is necessary to estimate the variance of the data σ_i. In the case of fitting to raw data, such as in the T_1, T_2 and heteronuclear NOE experiments the error should be estimated from N measurements at a single relaxation time (T) and calculated as:

$$\sigma^2 = \frac{1}{N-1}\sum_{i=1}^{N}(y_i - \bar{y})^2 \tag{19.82}$$

where the degrees of freedom, ν, is N-1, one degree of freedom was used for determining the mean value, $\bar{y}$. Alternatively, the average noise in the spectrum can be used as an estimate of σ.

Since the hnNOE is the ratio of two measurements, its error, σ_{noe}, is given the following [17]:

$$\frac{\sigma^2_{noe}}{[NOE]^2} = \frac{\sigma^2}{(S_o)^2} + \frac{\sigma^2}{(S_Z)^2} \tag{19.83}$$

where σ is the error in the raw data, S_o is the intensity of the ^{15}N magnetization in the absence of proton saturation, S is the intensity in the presences of proton saturation, and NOE is S/S_o.

19.8.3 Alternative Methods of Model Selection

As discussed by Chen et al [34] the χ^2 and F-test may fail to select the correct model. Alternative tests for model selection are Akaike's information criteria (AIC) [2] and Bayesian information criteria (BIC) [144]. To compare models using these techniques, one calculates either the AIC or BIC for all models and then selects the model with the lowest values. The AIC and BIC are defined as follows (see [34]):

$$\text{AIC} = \chi^2 + 2k \tag{19.84}$$

$$\text{BIC} = \chi^2 + klnM \tag{19.85}$$

where k is the number of independent parameters and M is the number of independent relaxation measurements. Studies by Chen et al [34] with simulated data suggest that the BIC method provides a more accurate method of model selection.

19.8.4 Error Propagation

It is necessary to propagate the uncertainties in the raw relaxation data to the fitted parameters. In the case of the T_1, T_2, and hnNOE measurements, these errors are required for calculating χ^2 values when fitting the diffusion constants and parameters associated with internal motion. Error propagation to the final fitted parameters, such

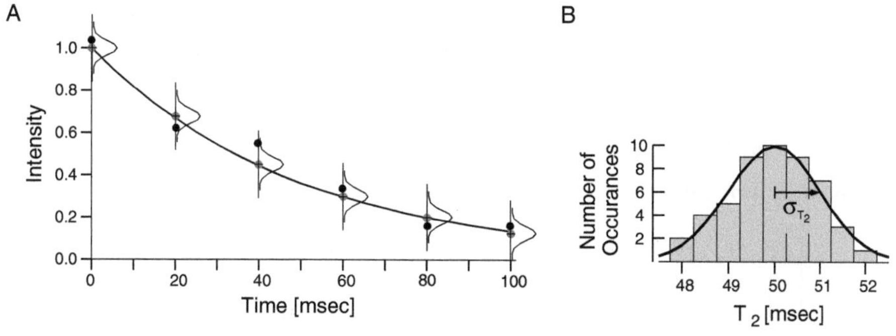

Figure 19.17. Error estimate of T_2 *by Monte Carlo Methods.*
Panel A: The peak intensity was measured as a function of the relaxation delay and the observed values are indicated by the gray dots. The values of the initial intensity, I_o, and the decay time, T_2, that give the smallest χ^2 for the observed data are obtained by standard methods. These values define the solid curve.
A Gaussian probability distribution is sketched at each data point. The width of this distribution is defined by the error in the data, σ, and defines the probability of selecting values for the construction of fictitious data sets. One such data set is indicated by the black filled circles.
Panel B: The distribution of T_2 values that were obtained after fitting 50 fictitious data sets is shown. The error in T_2 is obtained from the standard deviation of this distribution.

as the diffusion constants ($D_{||}$, $D_{\perp}$) and the parameters that characterize internal motion (S^2, τ_i) is also important in order to determine if the differences in parameters are statistically significant.

The best way to propagate errors in fitting non-linear models is by Monte Carlo methods. As the name suggests, this method generates a large number of random data sets to estimate the error in the final fitted parameters. The fictitious data are obtained by selecting data values at random, such that the selected values reflect a normal probability distribution with a mean equal to the observed value and the standard error equal to the estimated error in the data, σ. In the fictitious data sets the most probable values will be close to the observed value, and values that are more distant from the observed value are less likely to appear. Each one of these fictitious data sets are fit to the model and the distribution of the fitted parameters is used to obtain an estimate of the error of each parameter. This process is illustrated in Fig.19.17 for obtaining the error in T_2.

19.9 Exercises

1. The chemical shift tensor for an amide nitrogen in the principal axis system is given by:

$$\begin{bmatrix} 100 & 0 & 0 \\ 0 & 150 & 0 \\ 0 & 0 & 100 \end{bmatrix}$$

Calculate the chemical shift tensor for a rotation of 90° about the x-axis and give the chemical shift for this orientation of nitrogen, assuming the magnetic field is along the z-axis.

19.10 Solutions

1. The transformation is described by Eq. 19.9:

$$\tilde{\delta}' = \begin{bmatrix} 1 & 0 & 0 \\ 0 & 0 & -1 \\ 0 & 1 & 0 \end{bmatrix} \begin{bmatrix} 100 & 0 & 0 \\ 0 & 150 & 0 \\ 0 & 0 & 100 \end{bmatrix} \begin{bmatrix} 1 & 0 & 0 \\ 0 & 0 & 1 \\ 0 & -1 & 0 \end{bmatrix} = \begin{bmatrix} 100 & 0 & 0 \\ 0 & 100 & 0 \\ 0 & 0 & 150 \end{bmatrix}$$

Rotation about the x-axis by 90° has interchanged the z- and y-values of the chemical shift tensor. The observed chemical shift will be 150 ppm.

Appendix A
Fourier Transforms

A Fourier transformation is a mathematical technique that determines the amplitude and frequencies of oscillatory signals contained within a time dependent function. In the case of NMR, the Fourier transform is used to converts the time domain NMR signal, or FID, to a frequency domain signal, otherwise known as the NMR spectrum. For example, if the time dependent signal is given by the following equation,

$$f(t) = Acos(\omega t) \tag{A.1}$$

then the Fourier transform of this signal will provide both the frequency of the oscillation, (ω), as well as the amplitude of the component at that frequency (A).

The Fourier transformation algorithm that is used in most NMR processing software is the fast Fourier transform, or FFT [44]. From the perspective of this text, the most important feature of this technique is the requirement for 2^N data points in the time domain data.

A.1 Fourier Series

The Fourier series is a useful starting point to understand some concepts of Fourier transforms. The Fourier series describes any time dependent *periodic* function, $f(t)$, in terms of discrete frequency components: $\omega, 2\omega, 3\omega \ldots$. If a function has a period of 2π then it can be represented by the following sum, or Fourier series representation of $f(t)$:

$$f(t) = \frac{a_o}{2} + \sum_{n=1}^{\infty} a_n cos(nt) \; + \; b_n sin(nt) \tag{A.2}$$

The above indicates that we can represent an arbitrary function as a linear combination of a series of basis functions ($cosine$ and $sine$), just as we might write an arbitrary vector in terms of its x-, y-, and z-components. As with any other set of basis functions, these functions are orthogonal:

$$\int_{-\pi}^{\pi} sin(mt)sin(nt)dt = \pi\delta_{mn} \qquad \int_{-\pi}^{\pi} cos(mt)cos(nt)dt = \pi\delta_{mn}$$
$$\int_{-\pi}^{\pi} sin(mt)cos(nt)dt = 0 \tag{A.3}$$

$\delta_{mn} = 1$ if $m = n$, otherwise, $\delta_{mn} = 0$.

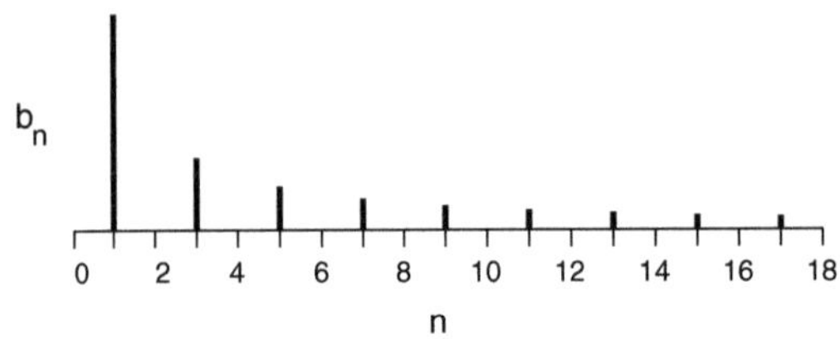

Figure A.1 Fourier components of a square wave. The coefficients, b_n are plotted as a function of frequency, ω_n. The strongest frequency component corresponds to ω, the second strongest is 3ω, etc. Note that only discrete values of ω are allowed since a square wave is a periodic function.

These orthogonal relationships can be used to calculate the coefficients of the Fourier series representation of $f(t)$:

$$a_n = \frac{1}{\pi}\int_{-\pi}^{\pi} f(t)cos(nt)dt \quad (A.4)$$

$$b_n = \frac{1}{\pi}\int_{-\pi}^{\pi} f(t)sin(nt)dt \quad (A.5)$$

a_n and b_n are the amplitudes of the various frequency components that sum to give $f(t)$. Note that b_o is always zero since $sin(0) = 0$.

For well behaved functions, as n gets larger both a_n and b_n approach zero. For example, coefficients for the Fourier series representation of a square wave are given in Fig. A.1 and the Fourier representation of the square wave is given in Fig. A.2. How well a Fourier series represents its function depends on the nature of the function. In general, functions with sharp edges (i.e. square wave) require a large number of terms to adequately represent the function, as illustrated in Fig. A.2.

A.2 Non-periodic Functions - The Fourier Transform

Once the function becomes non-periodic it is necessary to utilize a continuous transform, called the Fourier transform:

$$F(\omega) = \frac{1}{\sqrt{2\pi}}\int_{-\infty}^{\infty} f(t)e^{i\omega t}dt \quad (A.6)$$

In this case, the non-periodic time domain signal $f(t)$ is represented by a continuous function in the frequency domain, $F(\omega)$. $F(\omega)$ gives the amplitude of the frequency components which

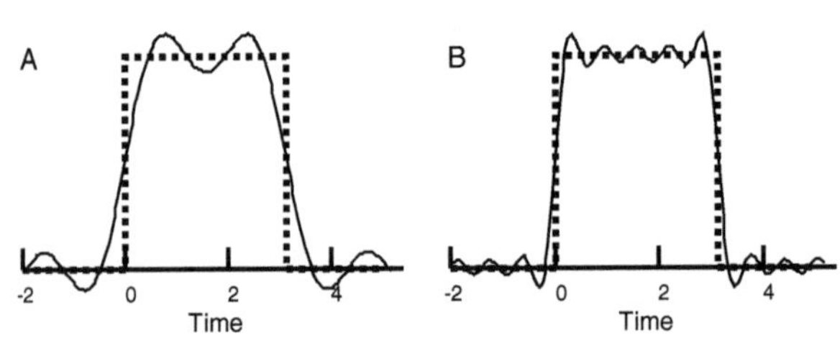

Figure A.2 Fourier series of a square wave. The Fourier series that represents a square wave is shown as the sum of the first 3 terms of the series (panel A) and as the sum of the first 7 terms of the series (panel B). Note that the representation of the square wave becomes more accurate as the number of terms increases. The square wave is drawn with a dotted line while the Fourier sums are shown as a continuous line.

are present in the time domain signal. If $F(\omega)$ exists, then it is possible to calculate the inverse Fourier transform of a function:

$$f(t) = \frac{1}{\sqrt{2\pi}} \int_{-\infty}^{\infty} F(\omega) e^{-i\omega t} dw \tag{A.7}$$

A.2.1 Examples of Fourier Transforms

A.2.1.1 Cosine and Sine

These two functions describe the time evolution of the detected magnetization from the magnetic dipoles in the x-y plane as they precess about B_o. The Fourier transforms of these two functions are shown in Fig. A.3

Cosine.

$$f(t) = cos(\omega_o t) \tag{A.8}$$

$$\begin{aligned} F(\omega) &= \int_{-\infty}^{\infty} cos(\omega_o t) e^{i\omega t} dt \\ &= \int_{-\infty}^{\infty} cos(\omega_o t)[cos(\omega t) + i sin(\omega t)] dt \end{aligned} \tag{A.9}$$

The complex term is zero since sin is an odd function, its integral from $-\infty$ to $+\infty$ is zero. Therefore:

$$F(\omega) = \int_{-\infty}^{\infty} cos(\omega_o t) cos(\omega t) dt \tag{A.10}$$

this integral can be evaluated by taking the limits to be from $-a$ to $+a$:

$$\begin{aligned} F(\omega) &= \int_{-a}^{+a} cos(\omega_o t) cos(\omega t) dt \\ &= \frac{1}{2} \int_{-a}^{+a} \left[cos([\omega_o - \omega]t) + cos([\omega_o + \omega]t) \right] dt \\ &= \frac{sin(\omega_o - \omega)a}{(\omega_o - \omega)} + \frac{sin(\omega_o + \omega)a}{(\omega_o + \omega)} \end{aligned} \tag{A.11}$$

This function becomes large when $\omega = \pm\omega_o$. In the limit, as $a \to \infty$, it consists of two δ functions, one at $+\omega_o$ and a second at $-\omega_o$ [1].

This result can also be obtained by expressing $cos(\omega t)$ as a sum of complex exponentials:

$$\begin{aligned} F(\omega) &= \int_{-\infty}^{\infty} cos(\omega_o t) e^{i\omega t} dt \\ &= \frac{1}{2} \int_{-\infty}^{\infty} \left[e^{i\omega_o t} + e^{-i\omega_o t} \right] e^{i\omega t} dt \\ &= \frac{1}{2} \left[\delta(\omega - \omega_o) + \delta(\omega + \omega_o) \right] \end{aligned} \tag{A.12}$$

[1] A delta function is a special function that is infinitely narrow and infinitely high at a single point. For example, such a peak at $x = 2$ would be written as $\delta(x - 2)$; the delta function is found at the value of x that makes the argument, $x - 2$, zero, i.e. $\delta(0) = \infty$.

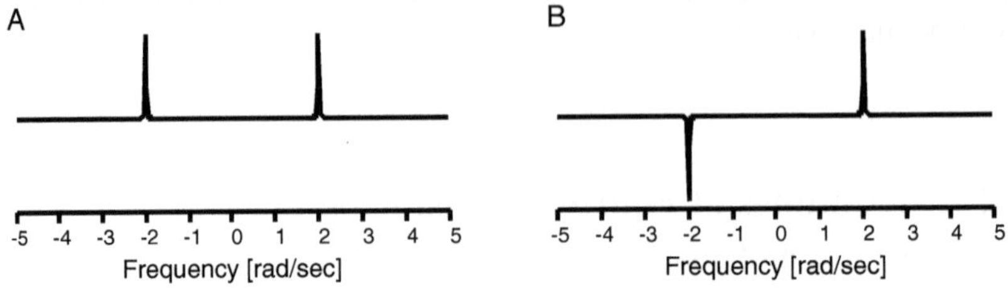

Figure A.3. Fourier transform of $cos(\omega t)$ and $sin(\omega t)$. Fourier transform of $cos(\omega t)$ (Panel A) and $sin(\omega t)$ (Panel B). All of these peaks are delta functions. Also note that the Fourier transform of $sin(\omega t)$ is imaginary.

Sine. The Fourier transform of *sine* is very similar to that of a cosine:

$$\begin{aligned} F(\omega) &= \int_{-\infty}^{\infty} sin(\omega_o t) e^{i\omega t} dt \\ &= \frac{1}{2i} \int_{-\infty}^{\infty} \left[e^{i\omega_o t} - e^{-i\omega_o t} \right] e^{i\omega t} dt \\ &= \frac{1}{2i} \left[\delta(\omega - \omega_o) - \delta(\omega + \omega_o) \right] \end{aligned} \qquad (A.13)$$

Note that this function is imaginary.

A.2.1.2 Square-Wave

A single pulse that is centered about zero with a width of $2a$ has the following Fourier transform:[2]

$$\begin{aligned} F(\omega) &= \frac{1}{\sqrt{2\pi}} \int_{-a}^{a} 1 e^{i\omega t} dt \\ &= \frac{e^{i\omega a}}{i\omega\sqrt{2\pi}} - \frac{e^{-i\omega a}}{i\omega\sqrt{2\pi}} \\ &= \frac{\sqrt{2} sin(\omega a)}{\omega\sqrt{\pi}} \end{aligned} \qquad (A.14)$$

This function is called a *sinc* function and its shape is shown in Fig. A.4. The width of this function, measured as the distance between the first zero-crossing points, is:

$$\Delta\omega = \frac{2\pi}{a} \qquad (A.15)$$

The position of the first zero-crossing point is $\frac{\pi}{a}$ in rad/sec or $\frac{1}{2a}$ in Hz[3]. Note that as

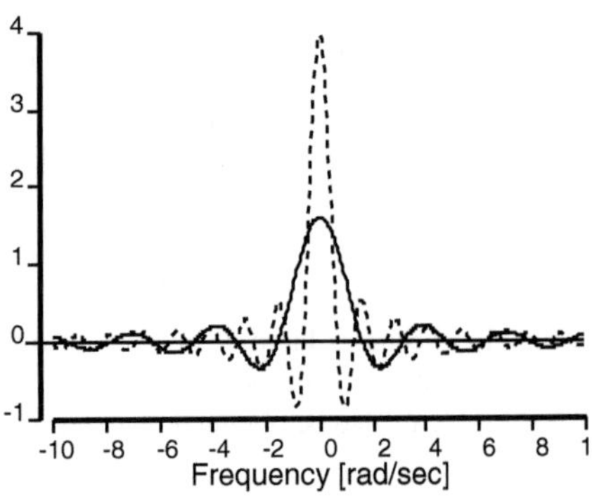

Figure A.4. Effect of pulse width on Sinc function. Sinc functions for two different values of a are shown. The solid line corresponds to $a = 2$ and the dotted line to $a = 5.0$.

[2] $\int e^{ax} dx = \frac{e^{ax}}{a}$.

[3] Frequencies are given in two common units, rad/sec and Hz. The former is usually associated with the symbol ω while the latter is associated with the symbol ν or f. The two are related as follows: $\omega = 2\pi\nu$.

the value of *a decreases* (the square pulse becomes narrower) the width of the sinc function *increases*. This inverse relationship between a function and its Fourier transform is quite general and of importance in both NMR spectroscopy and X-ray diffraction.

A.2.1.3 Exponential Decay

NMR signals decay exponentially: e^{-at}. The Fourier transform of this function is:

$$\begin{aligned} F(\omega) &= \int_0^\infty e^{-at} e^{i\omega t} dt \\ &= \int_0^\infty e^{-(a-i\omega)t} dt \\ &= \frac{1}{a - i\omega} \\ &= \frac{a + i\omega}{(a - i\omega)(a + i\omega)} \\ &= \frac{a + i\omega}{(a^2 + \omega^2)} \end{aligned} \tag{A.16}$$

This function is called a Lorentzian line shape, with a real and an imaginary, or dispersion component, as shown in Fig. A.5. Since modern NMR spectrometers record data in complex form, it is possible to observe both of these components. In fact, most raw spectra are a linear combination of the real and imaginary functions. Phase correction of the NMR spectrum can convert this mixed form to a pure lineshape. Because the real component has a narrower linewidth and has a Maximum at the Resonance frequency, it is usually the component that is plotted in spectra.

In the case of NMR, $a = 1/T_2$, where T_2 is the spin-spin relaxation time. This gives for the real part:

$$F(\omega) = \frac{T_2}{1 + T_2^2\omega^2} \tag{A.17}$$

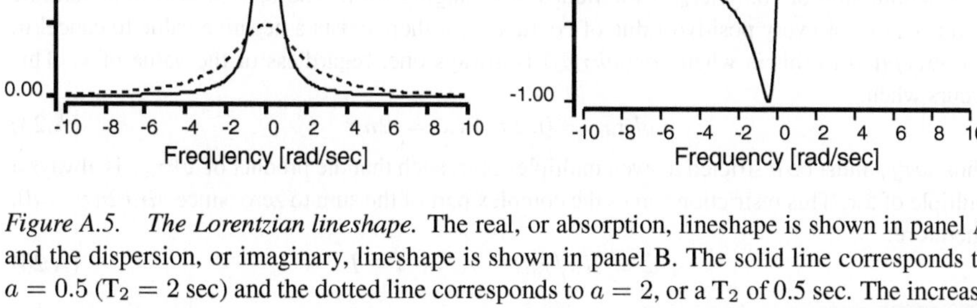

Figure A.5. The Lorentzian lineshape. The real, or absorption, lineshape is shown in panel A and the dispersion, or imaginary, lineshape is shown in panel B. The solid line corresponds to $a = 0.5$ ($T_2 = 2$ sec) and the dotted line corresponds to $a = 2$, or a T_2 of 0.5 sec. The increase in linewidth for the shorter T_2 results in a decrease in peak height, however the total area on the peak remains the same.

Note the dependence of the linewidth on the T_2. The smaller (shorter) T_2 is, the broader the line. In fact, the full width of this line at half-height is:

$$\Delta\nu = \frac{1}{\pi T_2} \qquad \text{(A.18)}$$

A.2.1.4 Fourier Transform of a Comb Function

A comb function is just an infinite series of equally spaced delta functions, similar to the teeth of the comb. Comb functions occur in both the time domain and the frequency domains in NMR spectroscopy. The digital NMR signal, or free induction decay (FID) is simply the product of a comb function, with teeth spaced τ_{dw} apart, and the continuous signal from the probe. Similarly, the spectrum is simply a convolution of the NMR spectrum and a comb function, with the teeth spaced $1/\tau_{dw}$ apart. Convolution functions are explained in more detail in Section A.2.3. The Fourier transform of a comb function in the time domain is:

$$\begin{aligned} F(\omega) &= \int_o^\infty f(t)e^{i\omega t}dt \\ &= \int \sum_{k=-\infty}^{k=+\infty} \delta(t - k\tau_{dw})\, e^{i\omega t}dt \\ &= \sum_{k=-\infty}^{k=+\infty} \int \delta(t - k\tau_{dw})\, e^{i\omega t}dt \end{aligned} \qquad \text{(A.19)}$$

The integral of a product of a delta function with another function is simply the value of that function at the position of the delta function, i.e.:

$$\int \delta(t - k\tau_{dw})\, e^{i\omega t}dt = e^{i\omega(k\tau_{dw})} \qquad \text{(A.20)}$$

Therefore the Fourier transform is:

$$F(\omega) = \sum_{k=-\infty}^{k=+\infty} e^{i\omega(k\tau_{dw})} \qquad \text{(A.21)}$$

The above is an infinite sum of complex functions $e^{i\omega k\tau_{dw}}$. This function can be represented as $cosine$ and $sine$ functions[4]:

$$F(\omega) = \sum_{k=-\infty}^{k=+\infty} [cos(\omega k\tau_{dw}) + i\; sin(\omega k\tau_{dw})] \qquad \text{(A.22)}$$

Now, summation of $cos(\omega k\tau_{dw})$ for values of k ranging from $-\infty$ to $+\infty$ will in general be zero because for every positive value of $cos(\omega k\tau_{dw})$, there exists a negative value to cancel it. The *exception* to this is when $cos(\omega k\tau_{dw})$ is always one, regardless of the value of k. This occurs when:

$$\omega k\tau_{dw} = 0, 2\pi, 4\pi ... = 2n\pi \qquad \text{(A.23)}$$

Thus, $\omega\tau_{dw}$ must be restricted to even multiples of π, such that the product of $\omega k\tau_{dw}$ is always a multiple of 2π. This restriction forces the complex part of the sum to zero, since $sin(2n\pi) = 0$. Therefore:

$$\omega = 2l\pi/\tau_{dw} \qquad l \in I \qquad \text{(A.24)}$$

[4] Recall $e^{i\theta} = cos\theta + isin\theta$.

or in Hz ($\omega = 2\pi\nu$):

$$\nu = l/\tau_{dw} \qquad\qquad l \in I \tag{A.25}$$

Equation A.25 defines another comb function, with teeth spaced $1/\tau_{dw}$ apart. Note the reciprocal relationship between the teeth spacing. The Fourier transform of a comb function in time, with teeth spaced τ apart, will generate a comb function in frequency, with the teeth spaced $1/\tau$ apart.

A.2.2 Linearity

If the Fourier transform of two functions, $g(t)$ and $h(t)$ are known, then the Fourier transform of any linear combination of these two functions is just the sum, or combination, of their individual transforms:

$$G(\omega) = \frac{1}{\sqrt{2\pi}} \int_{-\infty}^{\infty} g(t)e^{i\omega t}dt \tag{A.26}$$

$$H(\omega) = \frac{1}{\sqrt{2\pi}} \int_{-\infty}^{\infty} h(t)e^{i\omega t}dt \tag{A.27}$$

$$aG(\omega) + bH(\omega) = a\frac{1}{\sqrt{2\pi}} \int_{-\infty}^{\infty} g(t)e^{i\omega t}dt + b\frac{1}{\sqrt{2\pi}} \int_{-\infty}^{\infty} h(t)e^{i\omega t}dt \tag{A.28}$$

A.2.3 Convolutions: Fourier Transform of the Product of Two Functions

A common task is to calculate the Fourier transform of a product of two functions. If the Fourier transform of each of the individual functions is known then the Fourier transform of the product is the *convolution* of the two individual Fourier transforms.

$$F \otimes G(\omega) = \frac{1}{\sqrt{2\pi}} \int_{-\infty}^{\infty} G(\omega')F(\omega - \omega')d\omega' \tag{A.29}$$

The convolution of two functions can be difficult to visualized. Consider the example shown in Fig. A.6, where $g(x)$ is some complex shape (a cat) located at the origin and $h(x)$ is a delta function located at $x = 2$, i.e. $h(x) = \delta(x-2)$. The convolution of g and h is particularly straight-forward since $h(x)$ is a delta function. Evaluating $g \otimes h$ at x=2:

$$\begin{aligned} g \otimes h(x=2) &= \int_{-\infty}^{\infty} g(x')h(2-x')dx' \\ g \otimes h(x=2) &= \int_{-\infty}^{\infty} g(x')\delta([2-x'] - 2)dx' \\ g \otimes h(x=2) &= \int g(x')\delta(-x')dx' \\ &= g(0) \end{aligned} \tag{A.30}$$

The last step used the fact that $\delta(-x')$ is only non-zero when $x' = 0$. Therefore, the convolution of $g(x)$ with $h(x)$ has essentially moved the shape $(g(x))$ from the origin to the position of the delta function $(h(x))$.

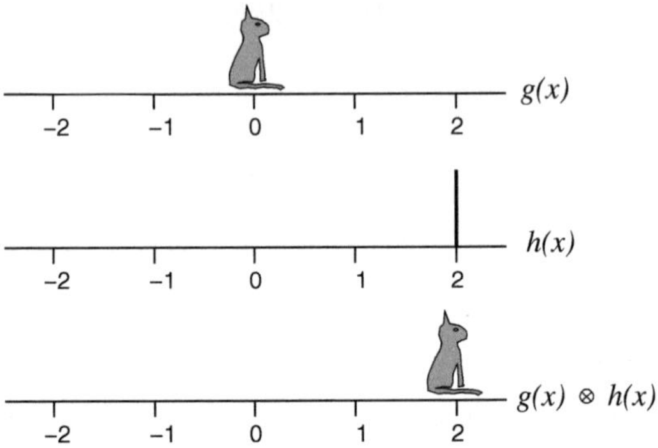

Figure A.6. Convolution of two functions. The convolution of one function, a cat at the origin ($g(x)$), with a delta function at $x = 2$ ($h(x)$), moves the first function such that it is centered at $x = 2$.

As another example, consider the tail of the cat, which is located at $x = 0.25$ in $g(x)$ and at $x = 2.25$ in $g \otimes h$:

$$\begin{aligned} g \otimes h(x = 2.25) &= \int_{-\infty}^{\infty} g(x')h(2.25 - x')dx' \\ g \otimes h(x = 2.25) &= \int_{-\infty}^{\infty} g(x')\delta(2.25 - x' - 2)dx' \\ g \otimes h(x = 2.25) &= \int_{-\infty}^{\infty} g(x')\delta(0.25 - x')dx' \end{aligned} \quad \text{(A.31)}$$

The last integral is only non-zero for $x' = 0.25$, therefore:

$$g \otimes h(x = 2.25) = g(0.25) \quad \text{(A.32)}$$

A.2.3.1 Convolution with a Lattice Function

In the above example, $h(x)$, was a single δ function. A particularly useful extension of this example occurs when $h(x)$ is a series of equal spaced δ functions (a comb function). In this case the convolution of g and h places an image of g at each position of the delta functions of h (see Fig. A.7).

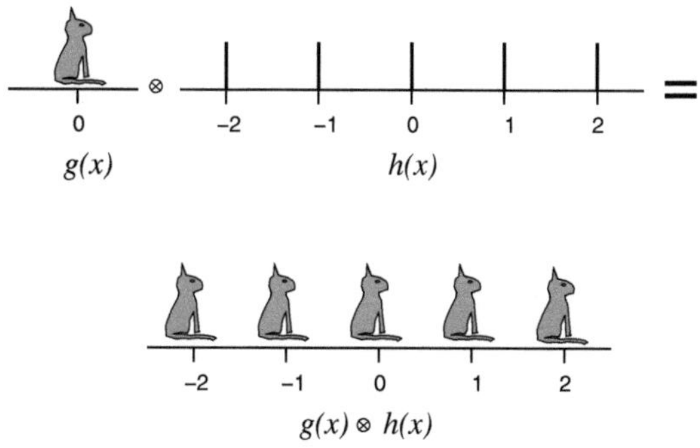

Figure A.7. Convolution with a comb function. The convolution of a function, $g(x)$, with a comb function, $h(x)$, produces an image of $g(x)$ at every tooth of the comb.

A.2.3.2 Application of Convolution in NMR Spectroscopy

The detected NMR signal is:

$$cos(\omega_o t)e^{-t/T_2} \tag{A.33}$$

where ω_o is the resonance frequency and T_2 is the spin-spin relaxation time,or the time constant for decay of the FID.

To calculate the Fourier transform of the FID the following steps are taken:

1 The Fourier transform of a cosine function, or pure harmonic wave, is:

$$\begin{aligned} f(t) &= cos(\omega_o t) \\ G(\omega) &= \delta(\omega - \omega_o) + \delta(\omega + \omega_o) \end{aligned} \tag{A.34}$$

$G(\omega)$ consists of two delta functions, one at $\omega = +\omega_o$ and the other at $\omega = -\omega_o$.

2 The Fourier transform of the second function is (real part only):

$$\begin{aligned} f(t) &= e^{\frac{-t}{T_2}} \\ H(\omega) &= \frac{T_2}{1 + {T_2}^2\omega^2} \end{aligned} \tag{A.35}$$

3 The convolution of these functions produces a Lorentzian line positioned at $\delta(\omega - \omega_o)$ and $\delta(\omega + \omega_o)$, as shown in Fig. A.8.

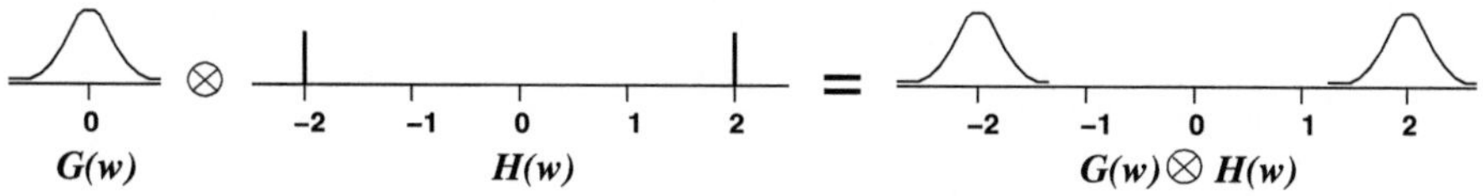

Figure A.8. Fourier transform of $cos(\omega_o t)e^{-t/T_2}$. The Fourier transform of the product of $cos(\omega_o t)$ and e^{-t/T_2} is given by the convolution of their respective transforms.

Appendix B
Complex Variables, Scalars, Vectors, and Tensors

B.1 Complex Numbers

Complex numbers are widely used in NMR theory as a convenient method of keeping track of the frequency and phases of pulses and signals. For example, suppose an excitation pulse consisted of one of the two signals shown in Fig. B.1. These are both oscillatory functions, but with different starting points, or phase shifts. Complex numbers can be used to represent both the frequency as well as any phase shift associated with the signal. For example, the solid line can be represented as $e^{i\omega t}$ while the dotted line can be represented as $e^{i(\omega t - 120°)}$.

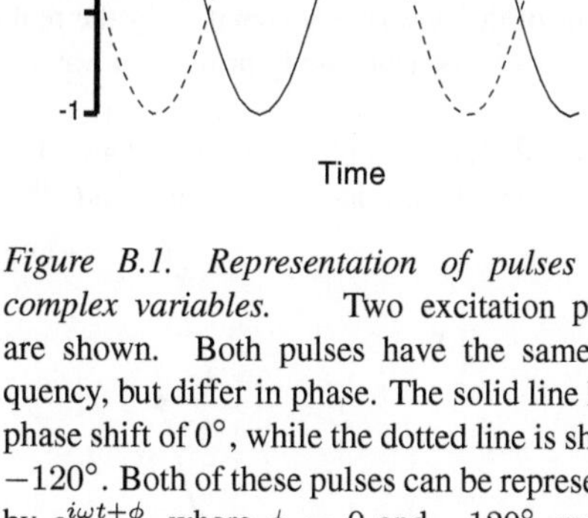

Figure B.1. Representation of pulses with complex variables. Two excitation pulses are shown. Both pulses have the same frequency, but differ in phase. The solid line has a phase shift of 0°, while the dotted line is shifted −120°. Both of these pulses can be represented by $e^{i\omega t + \phi}$, where $\phi = 0$ and $-120°$, respectively.

The symbol i is defined as $i = \sqrt{-1}$. A complex number is of the form $z = a + ib$, where a and b are real numbers. There are various ways to represent complex numbers. one of the most useful is a Argand diagram, shown in Fig. B.2. In the Argand diagram the real (a) and imaginary (b) components of the complex number define the position of the point in a Cartesian system with the real and imaginary axis as the two basis, or coordinate axis. It is also possible to represent complex numbers in polar coordinates. The two representations are, of course, equivalent. The latter representation is more useful in NMR because the phase of the NMR signal can be obtained directly from the size of θ.

The representation of the complex number in polar coordinates gives the following definition:

$$z = r(cos\theta + isin\theta) \tag{B.1}$$

which leads directly to Euler's identity:

$$e^{i\theta} = cos\theta + isin\theta \tag{B.2}$$

This can be seen from the series expansion of $e^{i\theta}$, $cos\theta$, and $sin\theta$.

$$cos\theta = 1 - \frac{\theta^2}{2!} + \frac{\theta^4}{4!} - \dots \quad sin\theta = \frac{\theta}{1!} - \frac{\theta^3}{3!} + \frac{\theta^5}{5!} \dots \tag{B.3}$$

$$\begin{aligned} e^{i\theta} &= 1 + \frac{i\theta}{1!} + \frac{(i\theta)^2}{2!} + \frac{(i\theta)^3}{3!} + \frac{(i\theta)^4}{4!} + \dots \\ &= 1 + \frac{i\theta}{1!} + \frac{i^2\theta^2}{2!} + \frac{i^3\theta^3}{3!} + \frac{i^4\theta^4}{4!} + \dots \\ &= 1 + \frac{i\theta}{1!} + \frac{-\theta^2}{2!} + \frac{-i\theta^3}{3!} + \frac{\theta^4}{4!} + \dots \\ &= \left[1 + \frac{-\theta^2}{2!} + \frac{\theta^4}{4!} + \dots\right] + \left[\frac{i\theta}{1!} + \frac{-i\theta^3}{3!} + \dots\right] \\ &= cos\theta + isin\theta \end{aligned}$$

B.2 Representation of Signals with Complex Numbers

An electrical signal, whether it is a pulse or the induced current from the probe, consists of a time dependent oscillation of voltage. This oscillation can be characterized by an amplitude (A), frequency(ω), and phase (ϕ). Complex numbers provide a way of representing all three of these parameters in a concise manner:

$$S(t) = Ae^{i(\omega t + \phi)} \tag{B.4}$$

This function can be imagined as a vector that rotates in the Argand plane with a frequency ω. In terms of the *observable* signal, only the real component of the complex function exists. Its imaginary component simply provides a way of incorporating the phase shift of the signal.

Phase Shifts. Application of a phase shift, ψ, to the above signal can be evaluated by simply taking the product of the signal and $e^{i\psi}$:

$$e^{i(\omega t + \phi + \psi)} = e^{i\psi} e^{i\omega t + \phi} \tag{B.5}$$

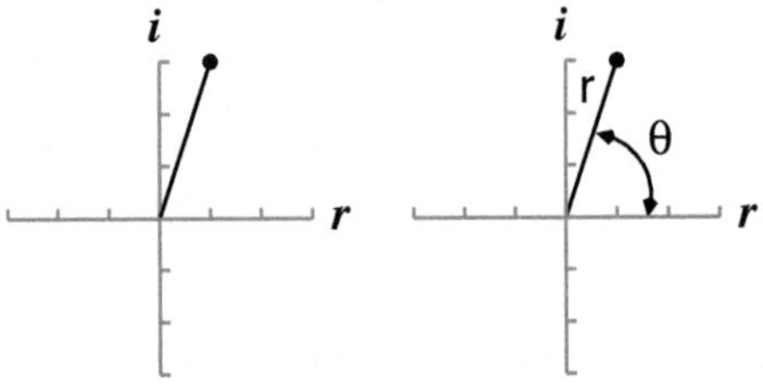

Figure B.2. Representation of complex numbers with an Argand diagram. An Argand diagram is shown on the left and the representation in polar coordinates is shown on the right. The real and imaginary axis are labeled with r and i, respectively. In polar coordinates, r $= \sqrt{a^2 + b^2}$ and $\theta = atan(b/a)$, where a is the real coordinate and b is the imaginary coordinate. Note that this 'r' refers to the length of the vector from the origin to the position of the point, not the real axis. In this particular example, $a = 1, b = 3$, r $= \sqrt{10}$, and $\theta = 71.56°$.

A sine function is simply a cosine function that has been shifted by 90°:

$$\begin{aligned} cos(\omega t) &= Re[e^{i\omega t}] \\ sin(\omega t) &= Re[e^{i(\omega t+\pi/2)}] \end{aligned} \tag{B.6}$$

where $Re[a + ib]$ indicates the real value of the complex number, or a.

Linear combinations of $e^{i\omega t}$ and $e^{-i\omega t}$ can also be used to define signals that are represented by $cos(\omega t)$ and $sin(\omega t)$:

$$cos\theta = \frac{e^{i\theta} + e^{-i\theta}}{2} \qquad sin\theta = \frac{e^{i\theta} - e^{-i\theta}}{2i} \tag{B.7}$$

The linear combination of the two complex numbers simply serves to cancel the complex part of the number, leaving only the real part. Nevertheless, it can be useful to think of $cos(\omega t)$ signal as the addition of two complex vectors, one that rotates counter clockwise ($e^{i\omega t}$) and the other that rotates clockwise ($e^{-i\omega t}$). Similarly, $sin(\omega t)$ can be thought of as the difference between these same two vectors, except that the difference is multiplied by $\frac{1}{i}$ which interchanges the real and imaginary axis.

B.3 Scalars, Vectors, and Tensors

B.3.1 Scalars

Scalars are numbers whose values are independent of a rotation of the coordinate systems. They are often a physical property of a system, such as the total energy. For example, the energy of a magnetic dipole in a magnet field is a scalar and is given by:

$$E = \vec{\mu} \cdot \vec{B} \tag{B.8}$$

This equation represents the projection of $\vec{\mu}$ on $\vec{B}$ and will always be invariant to rotation of the coordinate system.

The invariance of scalars to rotations can be easily shown. If the magnetic dipole, $\vec{\mu}$, and the magnetic field, $\vec{B}$ are subject to a rotation, using the rotation operator R, their representation after the rotation is as follows (see following section for transformation laws of vectors):

$$\begin{aligned} \vec{\mu'} &= R\vec{\mu} \\ \vec{B'} &= R\vec{B} \end{aligned}$$

The energy of the system is given as:

$$\begin{aligned} E &= \vec{\mu'} \cdot \vec{B'} \\ &= (R\vec{\mu}) \cdot (R\vec{B}) \\ &= RR^{-1}\vec{\mu} \cdot \vec{B} \\ &= \vec{\mu} \cdot \vec{B} \end{aligned} \tag{B.9}$$

Recall that $V^T \cdot R = R^{-1} \cdot V$, where V is a vector, and that the rotation operator is unitary, $RR^{-1} = 1$.

B.3.2 Vectors

Vectors, of course, describe the size and direction of a physical entity, for example, the applied magnetic field is defined by three components, B_x, B_y, and B_z, each of which give the projection of the magnetic field in the x-, y-, and z-axis, respectively. The appearance of a vector is changed by a rotation of the coordinate system. For example, if the magnetic field is defined to be along the z-axis in one coordinate frame: $\vec{B} = |B|(0, 0, 1)$, then a rotation of the *coordinate system* about the x-axis by 90° will cause the magnetic field to be aligned along the y-axis in the new coordinate frame, with component $\vec{B} = |B|(0, 1, 0)$. Note that vector itself has not changed, simply its representation by the coordinate system.

The transformation of a vector from one coordinate frame to another can be written as:

$$V' = RV \tag{B.10}$$

where V is the vector in the original frame, R is a rotation matrix, and V' is the representation of the vector in the new coordinate frame.

B.3.3 Tensors

Tensors are mathematical entities that are useful for describing the anisotropic properties of physical systems. Recall that J-coupling arises from an alteration in the local magnetic field of one spin due to the magnetic dipole of the coupled spin. This effect is transmitted through the intervening electrons that join the two atoms. Since the electron distribution is, in general, anisotropic, the coupling should depend on the relative orientation of the two spins with respect to the electron distribution. This orientational information is conveniently represented by a *tensor*. This tensor is written as a 3 × 3 array:

$$\tilde{J} = \begin{bmatrix} J_{xx} & J_{xy} & J_{xz} \\ J_{yx} & J_{yy} & J_{yz} \\ J_{zx} & J_{zy} & J_{zz} \end{bmatrix} \tag{B.11}$$

Tensors that describe physical interactions are symmetric, i.e. $J_{yx} = J_{xy}$, $J_{zx} = J_{xz}$, and $J_{yz} = J_{zy}$, therefore there are only six independent components that define a tensor.

The Hamiltonian, or energy, due to the J- or scalar-coupling between two spins, I and S, is given as:

$$\mathcal{H} = \vec{I} \cdot \tilde{J} \cdot \vec{S} \tag{B.12}$$

The two terms of the right, $\tilde{J} \cdot \vec{S}$ represent the local magnetic field at the I spin due to the S spin:

$$\begin{aligned} \vec{B}_{loc} &= \tilde{J} \cdot \vec{S} \\ &= \begin{bmatrix} J_{xx} & J_{xy} & J_{xz} \\ J_{yx} & J_{yy} & J_{yz} \\ J_{zx} & J_{zy} & J_{zz} \end{bmatrix} \begin{bmatrix} S_x \\ S_y \\ S_z \end{bmatrix} \\ &= \begin{bmatrix} J_{xx}S_x + J_{xy}S_y + J_{xz}S_z \\ J_{yx}S_x + J_{yy}S_y + J_{yz}S_z \\ J_{zx}S_x + J_{zy}S_y + J_{zz}S_z \end{bmatrix} \end{aligned} \tag{B.13}$$

The energy of interaction between the I spin and this local magnetic field is $\vec{I} \cdot B_{loc}$.

B.3.3.1 Transformation properties of Tensors

The rotation of the coordinate system transforms a tensor as follows, using the example of the coupling tensor:

$$J' = R \cdot J \cdot R^{-1} \tag{B.14}$$

where J' is the form of the tensor in the new coordinate systems.

The transformation law for tensors can be proven using the above expression for the coupling energy, $\mathcal{H} = \vec{I} \cdot \tilde{J} \cdot \vec{S}$. Since the energy is a scalar, it must be invariant to rotations. Beginning with the transformed spin angular momentum vectors and the tensor, it is possible to show that this is equal to the same expression written with the *untransformed* vectors and tensors:

$$\begin{aligned}\mathcal{H} &= \vec{I'}\,\tilde{J'}\,\vec{S'} \\ &= [R\,\vec{I}]\,[R\,\tilde{J}\,R^{-1}]\,[R\,\vec{S}] \\ &= R\,\vec{I}\,R\,\tilde{J}\,\vec{S} \\ &= R\,R^{-1}\,\vec{I}\,\tilde{J}\,\vec{S} \\ &= \vec{I}\,\tilde{J}\,\vec{S} \end{aligned} \tag{B.15}$$

If a tensor is symmetric, then there is a coordinate transformation that will produce a diagonal form of the tensor. The orientation of the coordinate system in which the tensor is diagonal is often called the principal axis system, or PAS. The J-coupling tensor in the coordinate system is:

$$J_{PAS} = \begin{bmatrix} J_{xx} & 0 & 0 \\ 0 & J_{yy} & 0 \\ 0 & 0 & J_{zz} \end{bmatrix} \tag{B.16}$$

Note that six components are still required to specify this form of the tensor, J_{xx}, J_{yy}, J_{zz}, and the three angles that describe the orientation of the principle axis system.

As an example, consider the coupling between the carbonyl carbon and the amide nitrogen. The coupling between the nuclear spins is due to the electrons in the σ bond as well as in the π orbital that is formed from the individual p_z orbitals (see Fig. B.3). The later bond gives the peptide bond its partial double bond character.

If the coordinate system was defined such that the y-axis was along the C-N bond and the z-axis was in the direction of the p_z orbitals, then the J-coupling tensor would have the following form:

$$J_{NC} = \begin{bmatrix} 0 & 0 & 0 \\ 0 & J_{yy} & 0 \\ 0 & 0 & J_{zz} \end{bmatrix} \tag{B.17}$$

where J_{yy} would be related to the electron density in the σ bond and J_{zz} would be related to the electron density in the π bond. Since the bonding σ and π orbitals are in the y- and z-direction, these components of the J-coupling tensor would be larger than the x-component, which has been set to zero in this simple example.

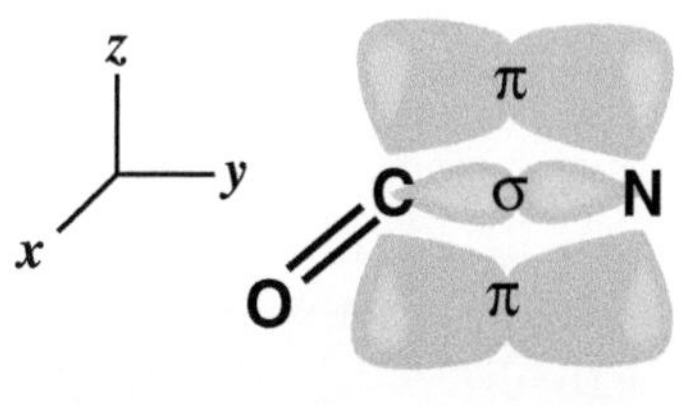

Figure B.3. Effect of molecular orbitals on J-coupling. The effect of the anisotropic distribution of molecular orbitals on the J-coupling tensor is illustrated using the coupling between the nitrogen and carbonyl carbon.

With the [illegible] angular momentum vectors and the tensor, it is possible to show that [illegible]

[illegible]

[illegible]

Appendix C
Solving Simultaneous Differential Equations: Laplace Transforms

Simultaneous linear differential equations often describe the time dependence of systems. For example, in the case of chemical exchange between two sites the following equations describe the time dependence of the concentrations of A and B:

$$\frac{dA}{dt} = -k_1[A] + k_2[B] \qquad \frac{dB}{dt} = +k_2[A] - k_2[B] \tag{C.1}$$

Examples of exchange reactions in NMR include chemical exchange (Chapter 18) and exchange of magnetization by dipolar coupling (Chapters 16 and 19).

C.1 Laplace Transforms

A straight forward method of solving simultaneous differential equations is by the use of Laplace transforms. This method automatically allows the inclusion of initial conditions and generally yields solutions in closed form that are suitable for simulations or data fitting. The Laplace transform, $\mathcal{L}$, of a function, $F(t)$, is:

$$\mathcal{L}[F(t)] = f(s) = \int_o^{\infty} e^{-st}F(t)dt \tag{C.2}$$

Table C.1. Inverse Laplace transforms.

$f(s)$	$F(t)$
$\frac{1}{s-a}$	e^{at}
$\frac{1}{(s-a)(s-b)}$	$\frac{1}{a-b}(e^{at} - e^{bt})$
$\frac{s}{(s-a)(s-b)}$	$\frac{1}{a-b}(ae^{at} - be^{bt})$

The Laplace transform of the derivative of a function is:

$$\mathcal{L}[F'(t)] = sf(s) - F(o) \tag{C.3}$$

where $F(0)$ is the value of F at $t = 0$. Differential equations are solved by first obtaining the Laplace transform of the equations, solve for $f(s)$ and then obtain the solution, $F(t)$ from the inverse transform. A few useful inverse transforms are shown in Table C.1.

C.1.1 Example Calculation

As a simple example, consider the following equation:

$$\begin{aligned} \frac{dF(t)}{dt} &= -a\,F(t) \\ sf - F_o &= -af \\ f(s+a) &= F_o \\ f &= \frac{F_o}{s+a} \end{aligned}$$

Using the inverse transform given in Table C.1 gives the final solution:

$$F(t) = F_o e^{-at}$$

Returning to the general case of two simultaneous equations:

$$\frac{dU}{dt} = a_{11}U + a_{12}V \tag{C.4}$$

$$\frac{dV}{dt} = a_{21}U + a_{22}V \tag{C.5}$$

Taking the Laplace transform of both gives:

$$s\,u - U_o = a_{11}u + a_{12}v \tag{C.6}$$

$$s\,v - V_o = a_{21}u + a_{22}v \tag{C.7}$$

Eliminating v gives:

$$-U_o(a_{22} - s) + V_o a_{12} = u\left[(a_{11} - s)(a_{22} - s) - a_{21}a_{12}\right]$$

Solving for u:

$$u = \frac{-U_o(a_{22} - s) + V_o a_{12}}{[a_{11}a_{22} - a_{11}s - a_{22}s + s^2 - a_{21}a_{12}]} \tag{C.8}$$

The denominator is a quadratic function with two roots, λ_1, and λ_2:

$$\begin{aligned} \lambda_1 &= \frac{(a_{11} + a_{22}) \pm \sqrt{(a_{11} + a_{22})^2 - 4(a_{11}a_{22} - a_{21}a_{12})}}{2} \\ \lambda_2 &= \frac{(a_{11} + a_{22}) \pm \sqrt{(a_{11} - a_{22})^2 + 4a_{21}a_{12}}}{2} \end{aligned} \tag{C.9}$$

Note: $(a_{11} + a_{22})^2 - 4a_{11}a_{22} = (a_{11} - a_{22})^2$.

Therefore,

$$u = -U_o a_{22}\frac{1}{(s-\lambda_1)(s-\lambda_2)} + U_o\frac{s}{(s-\lambda_1)(s-\lambda_2)} + V_o a_{12}\frac{1}{(s-\lambda_1)(s-\lambda_2)} \tag{C.10}$$

Taking the inverse transform gives:

$$U(t) = U_o\left[-a_{22}\frac{e^{\lambda_1 t} - e^{\lambda_2 t}}{\lambda_1 - \lambda_2} + \frac{\lambda_1 e^{\lambda_1 t} - \lambda_2 e^{\lambda_2 t}}{\lambda_1 - \lambda_2}\right] + V_o a_{12}\frac{e^{\lambda_1 t} - e^{\lambda_2 t}}{\lambda_1 - \lambda_2} \tag{C.11}$$

The equation for V(t) is obtained by simply interchanging U and V as well as the indices on a_{ij}:

$$V(t) = V_o\left[-a_{11}\frac{e^{\lambda_1 t} - e^{\lambda_2 t}}{\lambda_1 - \lambda_2} + \frac{\lambda_1 e^{\lambda_1 t} - \lambda_2 e^{\lambda_2 t}}{\lambda_1 - \lambda_2}\right] + U_o a_{21}\frac{e^{\lambda_1 t} - e^{\lambda_2 t}}{\lambda_1 - \lambda_2} \tag{C.12}$$

C.1.2 Application to Chemical Exchange

The correspondence between the general coefficients, a_{ij}, and the kinetic rate constants is as follows:

$$\begin{array}{cc} a_{11} = -R_A - k_1 & a_{12} = +k_2 \\ a_{21} = +k_1 & a_{22} = -R_B - k_2 \end{array} \tag{C.13}$$

The intensity of the two selfpeaks (I_{AA}, I_{BB}) and the two crosspeaks (I_{AB}, I_{BA}) can be obtained if we define the magnetization associated with the 'A' environment as U(t) and that associated with the 'B' environment as V(t).

If the spin-lattice relaxation rates of both environments is the same, then:

$$\lambda_1 = -k_{ex} - R \tag{C.14}$$

$$\lambda_2 = -R \tag{C.15}$$

$$\lambda_1 - \lambda_2 = -k_{ex} \tag{C.16}$$

The magnetization that gives rise to the selfpeak at (ω_A, ω_A) begins in environment 'A' and remains in the same environment after the mixing time:

$$\begin{aligned} I_{AA}(t) &= I_A^o(k_2 + R)\frac{e^{(-k_{ex}-R)t} - e^{-Rt}}{-k_{ex}} + \frac{(-k_{ex} - R)e^{(-k_{ex}-R)t} - (-R)e^{-Rt}}{-k_{ex}} \\ &= I_A^o\frac{\left[k_2 + (k_{ex} - k_2)e^{-k_{ex}t}\right]e^{-Rt}}{k_{ex}} \\ &= I_A^o(f_A + f_B e^{-k_{ex}t})\,e^{-Rt} \end{aligned} \tag{C.17}$$

The intensity of the crosspeak, I_{BA}, representing magnetization that began in environment B, but was transferred to environment A, is:

$$\begin{aligned} I_{BA}(\tau) &= I_B^o a_{12}\frac{e^{-\lambda_1 t} - e^{-\lambda_2 t}}{\lambda_1 - \lambda_2} \\ &= I_B^o\frac{k_2}{-k_{ex}}\left[e^{-k_{ex}t}e^{-Rt} - e^{-Rt}\right] \\ &= I_B^o\frac{k_2}{-k_{ex}}\left[e^{-k_{ex}t} - 1\right]e^{-Rt} \\ &= I_B^o f_A\left[1 - e^{-k_{ex}t}\right]e^{-Rt} \end{aligned} \tag{C.18}$$

The equation for the other self- and crosspeak are obtained by exchanging A and B. Substituting f_A for I_B^o and f_B for I_B^o, gives the complete solution for all four peaks:

$$I_{AA}(t) = f_A\left[f_A + f_B e^{-k_{ex}t}\right] e^{-Rt} \quad \text{(C.19)}$$

$$I_{BB}(t) = f_B\left[f_B + f_A e^{-k_{ex}t}\right] e^{-Rt} \quad \text{(C.20)}$$

$$I_{BA}(t) = f_B f_A\left[1 - e^{-k_{ex}t}\right] e^{-Rt} \quad \text{(C.21)}$$

$$I_{BA}(t) = f_A f_B\left[1 - e^{-k_{ex}t}\right] e^{-Rt} \quad \text{(C.22)}$$

C.1.3 Application to Spin-lattice Relaxation

C.1.3.1 Two Identical Spins

The Solomon equations for two coupled identical spins ($\rho_I = \rho_S$) are:

$$\frac{dI_z}{dt} = -\rho(I_z - I_z^o) - \sigma(S_z - S_z^o) \quad \text{(C.23)}$$

$$\frac{dS_z}{dt} = -\sigma(I_z - I_z^o) - \rho(S_z - S_z^o) \quad \text{(C.24)}$$

Making the substitution, $U(t) = I_z(t) - I_z^o$ and $V(t) = S_z(t) - S_z^o$ gives:

$$\frac{dU}{dt} = -\rho U - \sigma V \qquad \frac{dV}{dt} = -\sigma U - \rho V \quad \text{(C.25)}$$

Therefore,

$$a_{11} = -\rho \qquad a_{12} = -\sigma \quad \text{(C.26)}$$

$$a_{21} = -\sigma \qquad a_{22} = -\rho \quad \text{(C.27)}$$

The two roots are:

$$\lambda_1 = -(\rho + \sigma) \qquad \lambda_2 = -(\rho - \sigma) \quad \text{(C.28)}$$

and $\lambda_1 - \lambda_2 = -2\sigma$.

The time dependence of $U(t)$ is:

$$\begin{aligned} U(t) &= U_o\left[-a_{22}\frac{e^{\lambda_1 t} - e^{\lambda_2 t}}{\lambda_1 - \lambda_2} + \frac{\lambda_1 e^{\lambda_1 t} - \lambda_2 e^{\lambda_2 t}}{\lambda_1 - \lambda_2}\right] + V_o a_{12}\frac{e^{\lambda_1 t} - e^{\lambda_2 t}}{\lambda_1 - \lambda_2} \\ &= \frac{e^{\lambda_1 t}}{\lambda_1 - \lambda_2}\left[U_o(\lambda_1 - a_{22}) + V_o a_{12}\right] + \frac{e^{\lambda_2 t}}{\lambda_1 - \lambda_2}\left[U_o(a_{22} - \lambda_2) - V_o a_{12}\right] \\ &= \frac{e^{\lambda_1 t}}{-2\sigma}\left[U_o(-\sigma) + V_o(-\sigma)\right] + \frac{e^{\lambda_2 t}}{-2\sigma}\left[U_o(-\sigma) - V_o(-\sigma)\right] \\ &= \frac{e^{\lambda_1 t}}{2}\left[U_o + V_o\right] + \frac{e^{\lambda_2 t}}{2}\left[U_o - V_o\right] \end{aligned} \quad \text{(C.29)}$$

Substituting $U(t) = I_z(t) - I_z^o$, $V(t) = S_z(t) - S_z^o$, and setting the equilibrium values (I_z^o, S_z^o) to M_o gives:

$$I_z(t) - M_o = \frac{e^{-(\rho+\sigma)t}}{2}\left[I_z(0) + S_z(0) - 2M_o\right] + \frac{e^{-(\rho-\sigma)t}}{2}\left[I_z(0) - S_z(0)\right] \quad \text{(C.30)}$$

The above equation shows that the relaxation of the spin is, in general, multi-exponential. However, if *both* spins inverted at the begining of the measurement ($I_z(0) = S_z(o) = -M_o$), the time dependence of I_z and S_z are:

$$I_z(t) = M_o(1 - 2e^{-(\rho+\sigma)t}) \qquad S_z(t) = M_o(1 - 2e^{-(\rho+\sigma)t}) \tag{C.31}$$

and the longitudinal magnetization relaxes with a single time constant:

$$\frac{1}{T_1} = \sigma + \rho = (W_2 - W_0) + (W_0 + 2W_1 + W_2) = 2(W_1 + W_2) \tag{C.32}$$

C.1.4 Spin-lattice Relaxation of Two Different Spins

In general the spin-lattice relaxation of a heteronuclear spin that is coupled to a proton is also multi-exponential. However, if the attached proton is saturated while the longitudinal magnetization is changing, the relaxation becomes a single exponential, with a rate constant of ρ_S. This result is conveniently obtained using Laplace transforms.

If the protons are saturated, $I_z = 0$ and dI_z/dt=0. Therefore the Solomon equation becomes:

$$\frac{dS_z}{dt} = -\sigma I_z^o - \rho_s(S_z - S_z^o) \tag{C.33}$$

Substituting $V(t) = S_z(t) - S_z^o$ gives:

$$\frac{dV}{dt} = -\sigma I_z^o - \rho_s V \tag{C.34}$$

The Laplace transform of this is:

$$sv - V_o = \frac{-\sigma I_z^o}{s} - \rho_s\, v \tag{C.35}$$

Solving for v:

$$v = \frac{-V_o}{(s + \rho_s)} + \sigma I_o \frac{1}{s(s + \rho)} \tag{C.36}$$

Taking the inverse transform, and substituting $S_z(t) - S_Z^o$, gives

$$S_z(t) - S_z^o = e^{-\rho_s t}\left[S_z(0) - S_Z^o - \frac{\sigma}{\rho_s}I_o\right] + \frac{\sigma}{\rho_s}I_o \tag{C.37}$$

The usual initial condition for the relaxation measurement is to begin with $S_z(0) = 0$, giving:

$$\begin{aligned} S_z(t) &= e^{-\rho_s t}\left[-S_z^o - \frac{\sigma}{\rho_s}I_o\right] + S_z^o + \frac{\sigma}{\rho_s}I_o \\ &= \left[S_z^o + \frac{\sigma}{\rho_s}I_o\right](1 - e^{\rho_s t}) \end{aligned} \tag{C.38}$$

Appendix D
Building Blocks of Pulse Sequences

Chapters 13 and 14 contain a number of complex pulse sequences. These sequences are generally constructed from small segments, with each segment performing a specific function. Consequently, the easiest way to understand many of these experiments is to first recognize each segment within the sequence and then consider the effect of each these segments on the flow of magnetization through the entire sequence. This appendix offers a brief review of the product operator treatment of the density matrix and provides a summary of the most common segments that are found in pulse sequences.

D.1 Product operators

D.1.1 Pulses

Pulses transform magnetization by a rotation of the magnetization by an angle of β about the axis of the applied magnetization. Usually the right-handed rule is adopted, i.e. the direction of the change of magnetization follows the curvature of the fingers when the pulse is applied along the direction of the thumb. For example:

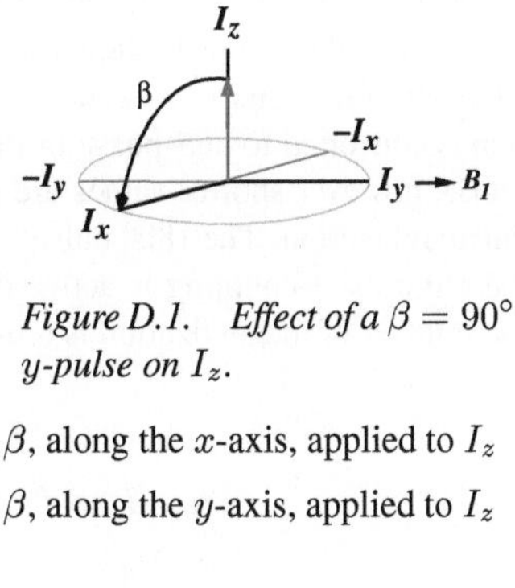

Figure D.1. Effect of a $\beta = 90°$ y-pulse on I_z.

$I_z \overset{P_{\beta x}}{\rightarrow} I_z cos(\beta) - I_y sin(\beta)$ is a pulse of flip angle β, along the x-axis, applied to I_z

$I_z \overset{P_{\beta y}}{\rightarrow} I_z cos(\beta) + I_x sin(\beta)$ is a pulse of flip angle β, along the y-axis, applied to I_z

D.1.2 Evolution by J-coupling

A spin, I, that is transverse and coupled to another spin, S, will oscillate between in-phase (e.g. I_x) and anti-phase magnetization ($2I_yS_z$) as follows (see Fig. D.2):

$$I_x \overset{J}{\rightarrow} I_x cos(\pi Jt) + 2I_yS_z sin(\pi Jt)$$
$$I_y \overset{J}{\rightarrow} I_y cos(\pi Jt) - 2I_xS_z sin(\pi Jt) \quad \text{(D.1)}$$

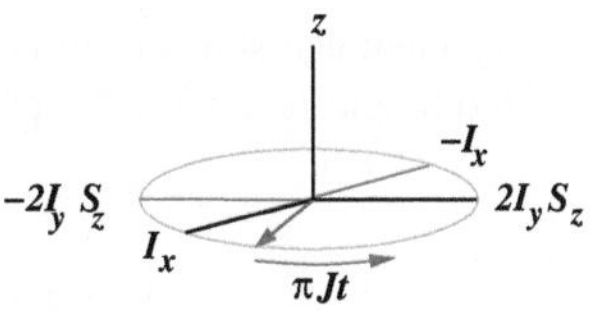

Figure D.2. Effect of scalar coupling on the evolution of the density matrix I_x.

Note that a density matrix corresponding to the product of two-transverse operators does not evolve under J-coupling, for example:

$$2I_xS_y \overset{J}{\rightarrow} 2I_xS_y \quad \text{(D.2)}$$

D.1.3 Evolution by Chemical Shift

Only transverse magnetization will evolve due to chemical shift. The evolution is equivalent to a rotation about the z-axis by an angle ωt, as illustrated in Fig. D.3. For example:

$$I_x \rightarrow I_x cos(\omega_I t) + I_y sin(\omega_I t) \quad \text{(D.3)}$$

Anti-phase terms evolve according to the chemical shift of the transverse spin:

$$2I_x S_z \rightarrow 2I_x S_z cos(\omega_I t) + 2I_y S_z sin(\omega_I t) \quad \text{(D.4)}$$

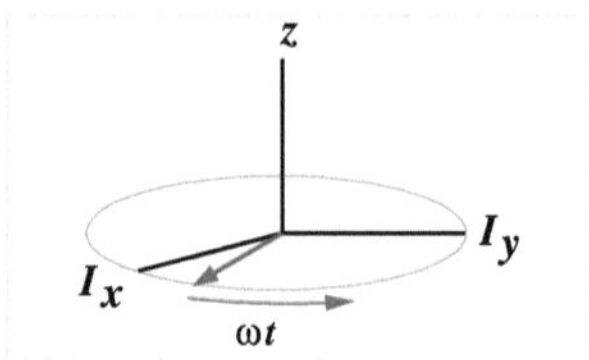

Figure D.3. Effect of chemical shift on the evolution of the density matrix.

D.2 Common Elements of Pulse Sequences

The following discusses a number of common elements that are found in a variety of heteronuclear and ^{15}N separated homonuclear sequences. In the following figures the narrow bars represent 90° pulses and the wider bars 180° pulses. All pulses are along the x-axis unless otherwise noted.

D.2.1 INEPT Polarization Transfer

The INEPT element, shown in Fig. D.4, transfers polarization from spin I to S and *vice-versa.* In the process in-phase magnetization is converted to anti-phase or the reverse. The delay Δ is set to 1/(4J), however shorter values are often used to reduce signal loss due to relaxation. The 180° pulses are applied to both types of spins therefore the J-coupling is active during the entire 2Δ period and the transfer of magnetization is represented by:

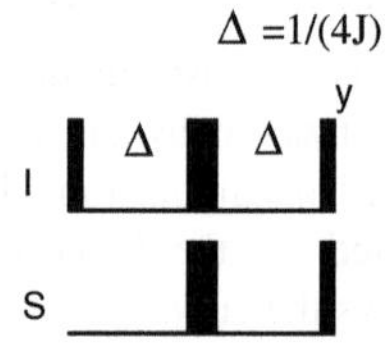

Figure D.4. INEPT polarization transfer.

$$\begin{aligned}
-I_y \quad &\xrightarrow{2\Delta} \quad -I_y cos(\pi J 2 \frac{1}{4J}) + 2I_x S_z sin(\pi J 2 \frac{1}{4J}) \\
&= -I_y cos(\pi/2) + 2I_x S_z sin(\pi/2) \\
&= 2I_x S_z \\
&\xrightarrow{P90_{I,S}} \quad 2I_z S_y
\end{aligned} \quad \text{(D.5)}$$

Note that chemical shift evolution of the transverse spin (e.g. I) is refocused by the centered 180° pulse:

$$\begin{aligned}
-I_y &\xrightarrow{\Delta} -I_y cos(\omega\Delta) + I_x sin(\omega\Delta) \\
&\xrightarrow{180_x} +I_y cos(\omega\Delta) + I_x sin(\omega\Delta) \\
&\xrightarrow{\Delta} cos(\omega\Delta)[I_y cos(\omega\Delta) - I_x sin(\omega\Delta)] \\
&+ sin(\omega\Delta)[I_x cos(\omega\Delta) + I_y sin(\omega\Delta)] \\
&= I_x[-cos(\omega\Delta)sin(\omega\Delta) + cos(\omega\Delta)sin(\omega\Delta)] \\
&+ I_y[cos^2(\omega\Delta) + sin^2(\omega\Delta)] \\
&= I_y
\end{aligned} \quad \text{(D.6)}$$

D.2.2 HMQC Polarization Transfer

The HMQC (heteronuclear multiple quantum) segment also transfers polarization, however the resultant magnetization is represented by a multiple quantum term because of the absence of the 90° pulse on the I-spins:

$$-I_y \xrightarrow{\Delta} 2I_xS_z \xrightarrow{P90_S} 2I_xS_y. \quad \text{(D.7)}$$

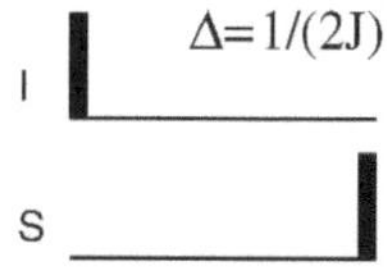

Figure D.5. HMQC polarization transfer.

the delay Δ is set to 1/(2J). Note that the chemical shift evolution of the I spins is *not* refocused by this segment. However, a 180° pulse is usually placed at a later time in the sequence, such as in the middle of the indirect time evolution domain, to refocus this chemical shift evolution.

D.2.3 Constant Time Evolution

A constant time segment permits the density matrix to evolve according to chemical shift, but during a fixed time period, 2T. The 180° pulse, which is initially centered in the 2T period is systematically moved in one direction by an amount $t/2$, generating a net time increment of t for chemical shift evolution. Evolution due to J-coupling would be prevented by the application of decoupling to the S-spins. The chemical shift evolution can be calculated in two ways, by a tedious product operator analysis or by simply keeping track of the phase change of the magnetization. The product operator analysis proceeds as follows, beginning from transverse magnetization[1]:

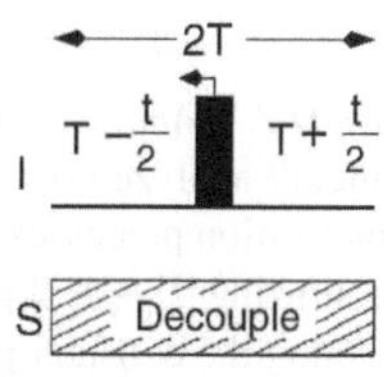

Figure D.6. Constant time evolution.

$$\begin{aligned} I_y &\stackrel{T-t/2}{\rightarrow} I_y cos(\omega[T-\frac{t}{2}]) - I_x sin(\omega[T-\frac{t}{2}]) \\ &\stackrel{P180_x}{\rightarrow} -I_y cos(\omega[T-\frac{t}{2}]) - I_x sin(\omega[T-\frac{t}{2}]) \quad \text{(D.8)} \end{aligned}$$

$$\begin{aligned} \stackrel{T+t/2}{\rightarrow} \quad & -cos(\omega[T-\frac{t}{2}])\left[I_y cos(\omega[T+\frac{t}{2}]) - I_x sin(\omega[T+\frac{t}{2}])\right] \\ & -sin(\omega[T-\frac{t}{2}])\left[I_x cos(\omega[T+\frac{t}{2}]) + I_y sin(\omega[T+\frac{t}{2}])\right] \\ = \quad & I_y\left[-cos(\omega[T-\frac{t}{2}])cos(\omega[T+\frac{t}{2}]) - sin(\omega[T-\frac{t}{2}])sin(\omega[T+\frac{t}{2}])\right] \\ & +I_x\left[cos(\omega[T-\frac{t}{2}])sin(\omega[T+\frac{t}{2}]) - sin(\omega[T-\frac{t}{2}])cos(\omega[T+\frac{t}{2}])\right] \\ = \quad & -I_y cos(\omega t) + I_x sin(\omega t) \end{aligned} \quad \text{(D.9)}$$

this is the same evolution that would be obtained by precession during a delay of $2\frac{t}{2}$: $I_y \rightarrow I_y cos(\omega t) - I_x sin(\omega t)$, with a simple inversion of sign.

[1]The final step uses the following trigonometric identities: $cos(\alpha-\beta) = cos\alpha cos\beta + sin\alpha sin\beta$, $sin(\alpha-\beta) = sin\alpha cos\beta - cos\beta sin\beta$.

The above evolution of the magnetization can be obtained in a more straight-forward manner by considering the phase of the magnetization. This approach is illustrated in Fig. D.7 and the relevant equations are given below.

$$\begin{aligned} I_y &\overset{T-t/2}{\longrightarrow} I_y e^{i\theta_1} : \theta_1 = \omega\left[T - t/2\right] \\ I_y e^{i\theta_1} &\overset{P180_x}{\longrightarrow} I_y e^{i(\pi-\theta_1)} \\ I_y e^{i\theta_1} &\overset{T+t/2}{\longrightarrow} I_y e^{i(\pi-\theta_1+\theta_2)} : \theta_2 = \omega\left[T + t/2\right] \end{aligned} \tag{D.10}$$

The π term in the final phase angle can be ignored since it simply reflects an inversion of the signal, therefore the total phase angle is: $-\theta_1 + \theta_2 = \omega\left[-(T - t/2) + (T + t/2)\right] = \omega t$.

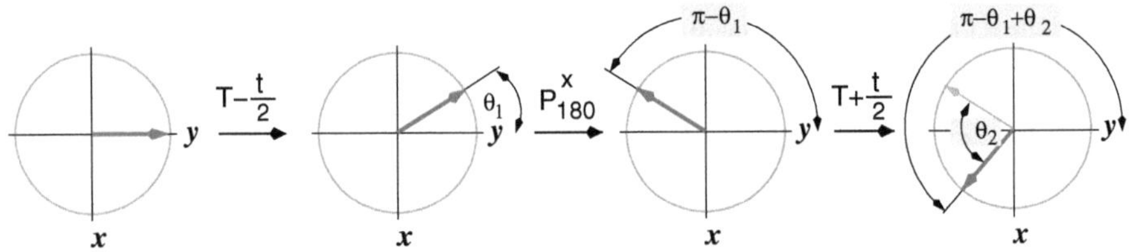

Figure D.7. Analysis of constant time evolution using phase angles. The use of phase angles to quickly analyze constant time evolution is illustrated. During the first delay, $T - t/2$, the magnetization precesses by an angle θ_1, the 180° pulse essentially negates this phase angle, and the following $T + t/2$ period adds θ_2 to the phase, giving a net rotation of $-\theta_1 + \theta_2 = \omega t$, neglecting the constant phase shift of π.

D.2.4 Constant Time Evolution with J-coupling

This pulse sequence element simultaneously permits chemical shift evolution to occur for a period equal to t and J-coupling to evolve for a period of $2T$. A typical application is to record the nitrogen (N-spin) chemical shift while generating anti-phase magnetization between the nitrogen and the carbonyl-carbon (C-spin), such as in the HNCO experiment.

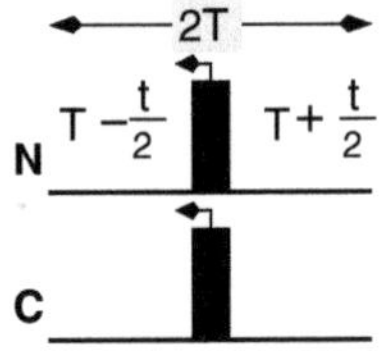

Figure D.8. Constant time with J-coupling evolution.

The analysis of the evolution of the chemical shift of the nitrogen is identical to that discussed above since the 180° pulse on the carbons cannot affect the nitrogen-spin. The evolution of the system due to J-coupling can be analyzed using the following two rules:

1. A single 180° pulse will reverse the direction of evolution due to J-coupling since it flips the spin-state of one of the coupled partners.
2. 180° pulses, when applied to both spins, do not change the evolution of the J-coupling. Thus, evolution due to J-coupling is active for the entire period of $2T$.

The second rule applies here, so the total evolution period for J-coupling is $2T$. Therefore the total evolution of the density matrix is:

$$\begin{aligned} N_y &\overset{\omega_N}{\rightarrow} N_y cos(\omega_N t) - N_x sin(\omega_N t) \\ &\overset{J_{NC}}{\rightarrow} -2N_x C_z cos(\omega_N t) - 2N_y C_z sin(\omega_N t) \end{aligned}$$

assuming $2T = 1/(2J_{NC})$. Note that the same result is obtained if evolution according to J-coupling is considered first.

D.2.5 Sequential Chemical Shift & J-coupling Evolution

Constant time evolution with J-coupling is an efficient way of simultaneously allowing evolution due to J-coupling and chemical shift. However, the period $2T$ is often too short for the desired time for chemical shift evolution. This is particularly true for CH groups, due to the strong J-coupling of 140 Hz, which gives an upper limit on the evolution time of 3.5 msec. Consequently, the chemical shift evolution of the protons is done separately from the J-coupling in a sequential fashion. The net result of the sequence shown to the right is:

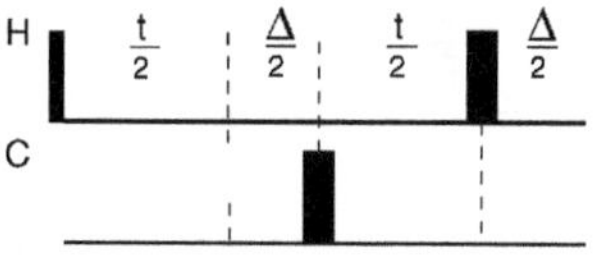

Figure D.9. *Sequential chemical shift and J-coupling evolution.*

$$I_z \xrightarrow{P90_H} -I_y \xrightarrow{\omega_H, J_{CH}} 2I_xC_z cos(\omega_H t) + 2I_yC_z sin(\omega_H t) \quad \text{(D.11)}$$

assuming $\Delta = 1/(2J)$.

This result can be readily obtained by following the phase changes due to evolution by chemical shift and J-coupling. The total evolution of proton magnetization by chemical shift is (neglecting the phase change of π, see Section D.2.3):

$$\begin{aligned} \Theta_{CS} &= -\theta_1 + \theta_2 \\ &= -\omega_H \left[\frac{t}{2} + \frac{\Delta}{2} + \frac{t}{2}\right] + \omega_H \left[\frac{\Delta}{2}\right] \\ &= -\omega_H t \end{aligned} \quad \text{(D.12)}$$

Evolution due to J-coupling is obtained in the same fashion, using the two rules described above in Section D.2.4. The phase will evolve in one direction prior to the 180° carbon-pulse. The 180° carbon-pulse, because it is applied to only one of the coupled spins, will reverse the sense of evolution for the period $t/2$. The 180° proton pulse reverses this change in direction during the last time period, $\Delta/2$, giving a total phase evolution of:

$$\begin{aligned} \Theta_J &= \theta_1 - \theta_2 + \theta_3 \\ &= \pi J \left[\frac{t}{2} + \frac{\Delta}{2} - \frac{t}{2} + \frac{\Delta}{2}\right] \\ &= \pi J \Delta \end{aligned} \quad \text{(D.13)}$$

If $\Delta = 1/(2J)$, then $-I_y \xrightarrow{J} I_y cos(\pi/2) + 2I_xC_z sin(\pi/2) = 2I_xC_z$. Of course, both chemical shift evolution and J-coupling occur at the same time, giving the final result shown in Eq. D.11.

The sequential evolution of chemical shift and J-coupling can reduce the overall sensitivity of the experiment because the proton magnetization is transverse for a total time of $t + \Delta$. Since the proton spin-spin relaxation rate (R_2) is usually fast, the extended time period can lead to signal loss due to relaxation. The amount of time the proton spends in the x-y plane can be reduced by the use of semi-constant time methods, as discussed below.

D.2.6 Semi-constant Time Evolution of Chemical Shift & J-coupling

The semi-constant time evolution [70] has the same outcome as the sequential evolution described above, evolution of chemical shift for a time t and evolution of J-coupling for a time Δ. The key difference is that the longest time the proton is transverse is equal to t, instead of $t + 1/(2J)$.

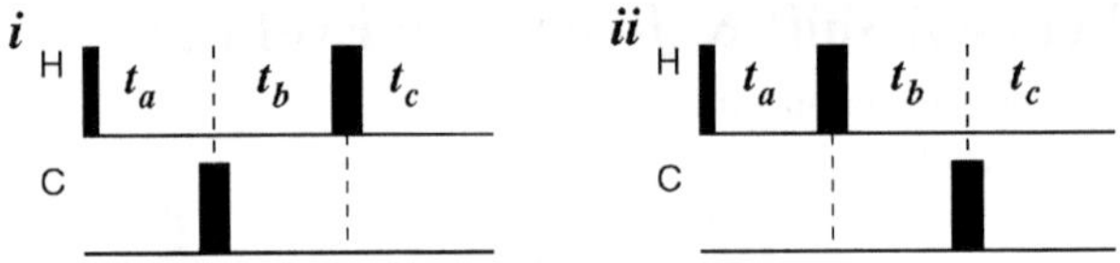

Figure D.10 *Semi-constant time evolution.* Two different ways of implementing this building block are shown (i and ii).

This is accomplished by incorporating the time required for evolution of J-coupling into the time allotted for chemical shift evolution.

Two versions of this pulse segment are used, and both are shown in Fig. D.10. Version i differs from ii in the order of the 180° pulses. The initial values (t^0) and the increments (δt) for each delay are given in Table D.1.

Considering the first version (i), the value of the delays for the nth data point are:

$$\begin{array}{lll} t_a = t_a^0 + n\delta t_a & t_b = t_b^0 + n\delta t_b & t_c = t_c^0 + n\delta t_c \\ \quad = \frac{\Delta}{2} + n\frac{\Delta t}{2} & \quad = 0 + n\left[\frac{\Delta t}{2} - \frac{\Delta}{2}\frac{1}{N}\right] & \quad = \frac{\Delta}{2} - n\frac{\Delta}{2}\frac{1}{N} \end{array} \qquad \text{(D.14)}$$

where Δt is the dwell time. Note that t_a and t_b increase with each data point while t_c decreases to zero length at the last data point of the FID (N).

The total evolution of the density matrix from chemical shift is:

$$\begin{aligned} \Theta_{CS} &= [\theta_a + \theta_b - \theta_c]\,\omega_H \\ &= \left[\frac{\Delta}{2} + n\frac{\Delta t}{2} + n\frac{\Delta t}{2} - n\frac{\Delta}{2}\frac{1}{N} - \frac{\Delta}{2} + n\frac{\Delta}{2}\frac{1}{N}\right]\omega_H \\ &= n\Delta t\omega_H \end{aligned} \qquad \text{(D.15)}$$

Similarly, the total phase due to evolution via J-coupling is:

$$\begin{aligned} \Theta_J &= \theta_a - \theta_b + \theta_c \\ &= \frac{\Delta}{2} + n\frac{\Delta t}{2} - n\frac{\Delta t}{2} + n\frac{\Delta}{2}\frac{1}{N} + \frac{\Delta}{2} - n\frac{\Delta}{2}\frac{1}{N} \\ &= \Delta \end{aligned} \qquad \text{(D.16)}$$

Therefore the total evolution of the density matrix is:

$$I_x \xrightarrow{J_{CH}} 2I_yC_z \xrightarrow{\omega_H} 2C_z\left[I_y cos\omega_H t - I_x sin\omega_H t\right] \qquad \text{(D.17)}$$

The derivation of the delays and increments is straight-forward. For example, with version i, the following constraints hold for the initial time point:

$$\begin{aligned} \Theta_{CS}(t=0) &= t_a^0 + t_b^0 - t_c^0 = 0 \\ \Theta_J(t=0) &= t_a^0 - t_b^0 + t_c^0 = \Delta \end{aligned} \qquad \text{(D.18)}$$

Table D.1. *Delays for semi-constant time evolution.*

Version	t_a	t_b	t_c
i	$t_a^0 = \frac{\Delta}{2}$	$t_b^0 = 0$	$t_c^0 = \frac{\Delta}{2}$
	$\delta t_a = \frac{\Delta t}{2}$	$\delta t_b = \frac{\Delta t}{2} - \frac{\Delta}{2}\frac{1}{N}$	$\delta t_c = -\frac{\Delta}{2}\frac{1}{N}$
ii	$t_a^0 = \frac{\Delta}{2}$	$t_b^0 = 0$	$t_c^0 = \frac{\Delta}{2}$
	$\delta t_a = -\frac{\Delta}{2}\frac{1}{N}$	$\delta t_b = \frac{\Delta t}{2} - \frac{\Delta}{2}\frac{1}{N}$	$\delta t_c = \frac{\Delta t}{2}$

$\Delta = 1/2J$, Δt = dwell time, N=index of last data point.

Adding these two equations together gives $t_a^0 = \Delta/2$ and defining $t_b^0 = 0$ gives $t_c^0 = \Delta/2$.

The increment in Θ_{CS} must be equal to the dwell time (Δt) and the increment in Θ_J must be zero.

$$\begin{aligned} \Delta\Theta_{CS} &= \delta t_a + \delta t_b - \delta t_c = \Delta t \\ \Delta\Theta_J &= \delta t_a - \delta t_b + \delta t_c = 0 \end{aligned} \tag{D.19}$$

Adding these two equations together gives $\delta t_a = \Delta t/2$. To fix the value of δt_b, the increment in t_c is set to -$\Delta/(2N)$, such that the delay t_c goes to zero at the last point ($n = N$) to minimize the total time of the segment. This gives $\delta t_b = \Delta t/2 - \Delta/(2N)$.

[illegible]

[illegible]

References

[1] A. Abragam. *The Principles of Nuclear Magnetism.* Oxford, 1961.

[2] H. Akaike. A new look at the statistical model identification. *IEEE transactions on automatic control*, AC-19:716–723, 1974.

[3] M. Akke and A.G. Palmer. Monitoring macromolecular motions on microsecond to millisecond time scales by $R_{1\rho}-R_1$ constant relaxation time NMR spectroscopy. *Journal of the American Chemical Society*, 118:911–912, 1996.

[4] A. Allerhand and E. Thiele. Analysis of Carr-Purcell spin-echo NMR experiments on multiple-spin systems. II. The effect of chemical exchange. *Journal of Chemical Physics*, 45:902–916, 1966.

[5] P. Andersson, B. Gsell, B. Wipf, H. Senn, and G. Otting. HMQC and HSQC experiments with water flip-back optimized for large proteins. *Journal of Biomolecular NMR*, 11:279–288, 1998.

[6] S. J. Archer, M. Ikura, D. A. Torchia, and A. Bax. An alternative 3D NMR technique for correlating backbone ^{15}N with side chain H_β resonances in larger proteins. *Journal of Magnetic Resonance*, 95:636–641, 1991.

[7] H. Barkhuijsen, R. de Beer, W. M. M. J. Bovee, and D. van Ormondt. Retrieval of frequencies, amplitudes, damping factors, and phases from time-domain signals using a linear least-squares procedure. *Journal of Magnetic Resonance*, 61:465–481, 1985.

[8] I.L. Barsukov and L.-Y. Lian. *Structure determination from NMR data I, NMR of Macromolecules:A Practical Approach, Editor G.C.K. Roberts*, pages 315–357. IRL Press, 1993.

[9] A. Bax, G. M. Clore, and A. M. Gronenborn. ^1H-^1H correlation via isotropic mixing of ^{13}C magnetization, a new three-dimensional approach for assigning ^1H and ^{13}C spectra of ^{13}C-enriched proteins. *Journal of Magnetic Resonance*, 88:425–431, 1990.

[10] A. Bax, R.H. Griffey, and B.L. Hawkins. Correlation of proton and ^{15}N chemical-shifts by multiple quantum NMR. *Journal of Magnetic Resonance*, 55:301–315, 1983.

[11] A. Bax and S. Grzesiek. Methodological advances in protein NMR. *Accounts of Chemical Research*, 26:131–138, 1993.

[12] A. Bax, M. Ikura, L.E. Kay, D.A. Torchia, and R. Tschudin. Comparison of different modes of 2-dimensional reverse-correlation NMR for the study of proteins. *Journal of Magnetic Resonance*, 86:304–318, 1990.

[13] A. Bax, G. Kontaxis, and N. Tjandra. Dipolar couplings in macromolecular structure determination. *Methods in Enzymology*, 339:127–174, 2001.

[14] A. Bax and D. Marion. Improved resolution and sensitivity in ^{1}H-detected heteronuclear multiple-bond correlation spectroscopy. *Journal of Magnetic Resonance*, 78:186–191, 1988.

[15] A. Bax and S.S. Pochapsky. Optimized recording of heteronuclear multidimensional NMR spectra using pulsed field gradients. *Journal of Magnetic Resonance*, 99:638–643, 1992.

[16] N.J. Baxter and M.P. Williamson. Temperature dependence of ^{1}H chemical shifts in proteins. *Journal of Biomolecular NMR*, 9:359–369, 1997.

[17] P. R. Bevington. *Data Reduction and Error Analysis for the Physical Sciences*. McGraw-Hill, 1992.

[18] F. Bloch. Nuclear induction. *Physical Review*, 70:460–474, 1946.

[19] F. Bloch and A. Siegert. Magnetic resonance for non-rotating fields. *Physical Review*, 57:522–527, 1940.

[20] G. Bodenhausen, H. Kogler, and R.R. Ernst. Selection of coherence-transfer pathways in NMR pulse experiments. *Journal of Magnetic Resonance*, 58:370–388, 1984.

[21] G. Bodenhausen and D. Ruben. Natural abundance ^{15}N NMR by enhanced heteronuclear spectroscopy. *Chemical Physics Letters*, 69:185–189, 1980.

[22] B. A. Borgias, M. Gochin, D.J. Kerwood, and T.L. James. Relaxation matrix analysis of 2D NMR data. *Progress in NMR spectroscopy*, 22:83–100, 1990.

[23] A.A. Bothner-By and J. Dadok. Useful manipulations of the free induction decay. *Journal of Magnetic Resonance*, 72:540–543, 1987.

[24] L. Braunschweiler and R.R. Ernst. Coherence transfer by isotropic mixing - application to proton correlation spectroscopy. *Journal of Magnetic Resonance*, 53:521–528, 1983.

[25] D.M. Briercheck, T.C. Wood, T.J. Allison, J.P. Richardson, and G.S. Rule. The NMR structure of the RNA binding domain of *E. coli* rho factor suggests possible RNA-protein interactions . *Nature Structural Biology*, 5:393–399, 1998.

[26] A. T. Brünger. *X-PLOR: Version 3.1, A System for X-ray Crystallography and NMR*. Yale University Press, 1992.

[27] H. Y. Carr and E. M. Purcell. Effects of diffusion on free precession in nuclear magnetic resonance experiments. *Physical Review*, 94:630–638, 1954.

[28] J.P. Carver and R.E. Richards. A general two-site solution for the chemical exchange produced dependence of T_2 upon the Carr-Purcell pulse sequence. *Journal of Magnetic Resonance*, 6:89–105, 1972.

[29] J. Cavanagh, W. J. Chazin, and M. Rance. The time dependence of coherence transfer in homonuclear isotropic mixing experiments. *Journal of Magnetic Resonance*, 87:110–131, 1990.

[30] J. Cavanagh, W. J. Fairbrother, A. G. Palmer III, and N. J. Skelton. *Protein NMR Spectroscopy*. Academic Press, 1996.

[31] J. Cavanagh, A.G. Palmer, P.E. Wright, and M. Rance. Sensitivity improvement in proton-detected 2-dimensional heteronuclear relay spectroscopy. *Journal of Magnetic Resonance*, 91:429–436, 1991.

[32] J. Cavanagh and M. Rance. Suppression of cross-relaxation effects in TOCSY spectra via a modified DIPSI-2 mixing sequence. *Journal of Magnetic Resonance*, A105 : 328, 1993.

[33] S.-L. Chang, A. Szabo, and N. Tjandra. Temperature dependence of domain motions of calmodulin probed by NMR relaxation at multiple fields. *Journal of the American Chemical Society*, 125:11379–11384, 2003.

[34] J. Chen, C.L. Brooks III, and P.E. Wright. Model-free analysis of protein dynamics: assessment of accuracy and model selection protocols based on molecular dynamics simulation. *Journal of Biomolecular NMR*, 29:243–257, 2004.

[35] G. M. Clore, A. Bax, P. C. Driscoll, P. T. Wingfield, and A. M. Gronenborn. Assignment of the side-chain ^{1}H and ^{13}C resonances of interleukin-1β using double- and triple-resonance heteronuclear three-dimensional NMR spectroscopy. *Biochemistry*, 29:8172–8184, 1990.

[36] G. M. Clore, P. C. Driscoll, P. T. Wingfield, and A. M. Gronenborn. Analysis of the backbone dynamics of interleukin-1β using two-dimensional inverse detected heteronuclear ^{1}H-^{15}N NMR spectroscopy. *Biochemistry*, 29:7387–7401, 1990.

[37] G. M. Clore, A. Szabo, A. Bax, L. E. Kay, P. C. Driscoll, and A. M. Gronenborn. Deviations from the simple two-parameter model-free approach to the interpretations of nitrogen-15 nuclear magnetic relaxation of proteins. *Journal of the American Chemical Society*, 112:4989–4991, 1990.

[38] G.M. Clore, A.M. Gronenborn, and A. Bax. A robust method for determining the magnitude of the fully asymmetric alignment tensor of oriented macromolecules in the absence of structural information. *Journal of Magnetic Resonance*, 133:216–221, 1998.

[39] R. T. Clubb, V. Thanabal, and G. Wagner. A constant-time three-dimensional triple-resonance pulse scheme to correlate intra-residue $^{1}H_{N}$, ^{15}N, and ^{13}C' chemical shifts in ^{15}N-^{13}C-labeled proteins. *Journal of Magnetic Resonance*, 97:213–217, 1992.

[40] R. T. Clubb, V. Thanabal, and G. Wagner. A new 3D HN(CA)HA experiment for obtaining fingerprint H_{N}-H_{α} crosspeaks in ^{15}N- and ^{13}C-labeled proteins. *J. Biomolecular NMR*, 2:203–210, 1992.

[41] R. T. Clubb and G. Wagner. A triple-resonance pulse scheme for selectively correlating amide $^{1}H_{N}$ and ^{15}N nuclei with the $^{1}H_{\alpha}$ proton of the preceding residue. *Journal of Biomolecular NMR*, 2:389–394, 1992.

[42] C. Cohen-Tannoudji, B. Diu, and F. Laloë. *Quantum Mechanics*. John Wiley & Sons, 1977.

[43] W. W. Conover. Magnetic shimming. *Acorn NMR: http://www.acornnmr.com*, 2004.

[44] J.W. Cooley and J.W. Tukey. An algorithm for the machine calculation of complex Fourier series. *Mathematics of Computation*, 19:297–301, 1965.

[45] F. Cordier and S. Grzesiek. Direct observation of hydrogen bonds in proteins by inter-residue $^{3H}J_{NC}$ scalar couplings. *Journal of the American Chemical Society*, 121:1601–1602, 1999.

[46] E.J. d'Auvergne and P.R. Gooley. The use of model selection in the model-free analysis of protein dynamics. *Journal of Biomolecular NMR*, 25:25–39, 2003.

[47] D.G. Davis, M.E. Perlman, and R.E. London. Direct measurements of the dissociation-rate constant for inhibitor-enzyme complexes via the $T_{1\rho}$ and T_2(CPMG) methods. *Journal of Magnetic Resonance*, 104B:266–275, 1994.

[48] E. de Alba and N. Tjandra. NMR dipolar couplings for the structure determination of biopolymers in solution. *Progress in Nuclear Magnetic Resonance Spectroscopy*, 40:175–197, 2002.

[49] D.E. Demco, P. Van Hecke, and J.S. Waugh. Phase-shifted pulse sequence for measurement of spin-lattice relaxation in complex systems. *Journal of Magnetic Resonance*, 16:467–470, 1974.

[50] C.E. Dempsey. Hydrogen exchange in peptides and proteins using NMR spectroscopy. *Progress in Nuclear Magnetic Resonance Spectroscopy*, 39:135–170, 2001.

[51] C. Deverell, R.E. Morgan, and J.H. Strange. Studies of chemical exchange by nuclear magnetic relaxation in the rotating frame. *Molecular Physics*, 18:553–559, 1970.

[52] J.F. Doreleijers, S. Mading, D. Maziuk, K. Sojourner, L. Yin, J. Zhu, J.L. Markley JL, and E.L. Ulrich. BioMagResBank (BMRB) database with sets of experimental NMR constraints corresponding to the structures of over 1400 biomolecules deposited in the protein data bank. *Journal of Biomolecular NMR*, 26:139–146, 2003.

[53] R. R. Ernst, G. Bodenhausen, and A. Wokaun. *Principles of Nuclear Magnetic Resonance in One and Two Dimensions*. Oxford, 1987.

[54] N. A. Farrow, R. Muhandiram, A. U. Singer, S. M. Pascal, C. M. Kay, G. Gish, S. E. Shoelson, T. Pawson, J. D. Forman-Kay, and L. E. Kay. Backbone dynamics of a free and a phosphopeptide-complexed Src homology 2 domain studied by ^{15}N relaxation. *Biochemistry*, 33:5984–6003, 1994.

[55] S. W. Fesik, H. L. Eaton, E. T. Olejniczak, E. R. P. Zuiderweg, L. P. McIntosh, and F. W. Dahlquist. 2D and 3D NMR spectroscopy employing ^{13}C-^{13}C magnetization transfer by isotropic mixing. Spin system identification in large proteins. *Journal of the American Chemical Society*, 112:886–888, 1990.

[56] R. Freeman. *Spin Choreography: Basic Steps in High Resolution NMR*. Spektrum Academic Publishers, 1997.

[57] D. B. Fulton, R. Hrabal, and F. Ni. Gradient-enhanced TOCSY experiments with improved sensitivity and solvent suppression. *Journal of Biomolecular NMR*, 8:213–218, 1996.

[58] D. Fushman and D. Cowburn. Nuclear magnetic resonance relaxation in determination of residue-specific ^{15}N chemical shift tensors in proteins in solution: protein dynamics, structure, and applications of transverse relaxation optimized spectroscopy. *Methods in Enzymology*, 339:109–126, 2001.

[59] D. Fushman, N. Tjandra, and D. Cowburn. Direct measurement of ^{15}N chemical shift anisotropy in solution. *Journal of the American Chemical Society*, 120:10947–10952, 1998.

[60] J. Garcia de la Torre, M.L. Huertas, and B. Carrasco. HYDRONMR: Prediction of NMR relaxation of globular proteins from atomic-level structures and hydrodynamic calculations. *Journal of Magnetic Resonance*, 147:138–146, 2000.

[61] H. Geen and R. Freeman. Band-selective radiofrequency pulses. *Journal of Magnetic Resonance*, 93:93–141, 1991.

[62] S. J. Glaser and G. P. Drobny. Assessment and optimization of pulse sequences for homonuclear isotropic mixing. *Advances in Magnetic Resonance*, 14:35–58, 1990.

[63] M. Goldman. *Quantum Description of High-Resolution NMR in Liquids*. Oxford Science Publications, 1988.

[64] N.K. Goto, K.H. Gardner, G.A. Mueller, R.C. Willis, and L.E. Kay. A robust and cost-effective method for the production of Val, Leu, Ile(δ1) methyl-protonated ^{15}N-, ^{13}C-, ^{2}H-labeled proteins. *Journal of Biomolecular NMR*, 13:369–374, 1999.

[65] S. Grzesiek, J. Anglister, and A. Bax. Correlation of backbone amide and aliphatic side-chain resonances in ^{13}C/^{15}N-enriched proteins by isotropic mixing of ^{13}C magnetization. *Journal of Magnetic Resonance*, B101:114–119, 1993.

[66] S. Grzesiek and A. Bax. Correlating backbone amide and side chain resonances in larger proteins by multiple relayed triple resonance NMR. *Journal of the American Chemical Society*, 114:6291–6293, 1992.

[67] S. Grzesiek and A. Bax. An efficient experiment for sequential backbone assignment of medium-sized isotopically enriched proteins. *Journal of Magnetic Resonance*, 99:201–207, 1992.

[68] S. Grzesiek and A. Bax. Improved 3D triple-resonance NMR techniques applied to a 31-kDa protein. *Journal of Magnetic Resonance*, 96:432–440, 1992.

[69] S. Grzesiek and A. Bax. The origin and removal of artifacts in 3D HCACO spectra of proteins uniformly enriched with ^{13}C. *Journal of Magnetic Resonance*, B102:103–106, 1992.

[70] S. Grzesiek and A. Bax. Amino-acid type determination in the sequential assignment procedure of uniformly ^{13}C/^{15}N-enriched proteins. *Journal of Biomolecular NMR*, 3:185–204, 1993.

[71] S. Grzesiek, P. Wingfield, S. Stahl, J. D. Kaufman, and A. Bax. Four-dimensional ^{15}N-separated NOESY of slowly tumbling perdeuterated ^{15}N-enriched proteins. Applications to HIV-1 Nef. *Journal of the American Chemical Society*, 117:9594–9595, 1995.

[72] M.R. Hansen, L. Mueller, and A. Pardi. Tunable alignment of macromolecules by filamentous phage yield dipolar coupling interactions. *Nature Structural Biology*, 5:1065–1074, 1998.

[73] T.K. Hitchens, S.A. McCallum, and G.S. Rule. A J(CH)-modulated 2D (HACACO)NH pulse scheme for quantitative measurement of $^{13}C_\alpha -^1 H_\alpha$ couplings in ^{15}N, ^{13}C-labeled proteins. *Journal of Magnetic Resonance*, 109:281–284, 1999.

[74] T.-L. Hwang, M. Kadkhodaei, A. Mohebbi, and A. J. Shaka. Coherent and incoherent magnetization transfer in the rotating frame. *Magnetic Resonance in Chemistry*, 30:S24–S34, 1992.

[75] M. Ikura, L. E. Kay, and A. Bax. A novel-approach for sequential assignment of ^{1}H, ^{13}C, and ^{15}N spectra of larger proteins. Heteronuclear triple-resonance 3-dimensional NMR-spectroscopy: Application to calmodulin. *Biochemistry*, 29:4659–4667, 1990.

[76] R. Ishima, J. M. Lois, and D. A. Torchia. Transverse ^{13}C relaxation of CHD_2 methyl isotopmers to detect slow conformational changes of protein side chains. *Journal of the American Chemical Society*, 121:11589–11590, 1999.

[77] R. Ishima, J.M. Louis, and D. A. Torchia. Optimized labeling of $^{13}CHD_2$ methyl isotopomers in perdeuterated proteins: Potential advantages for ^{13}C relaxation studies of methyl dynamics of larger proteins. *Journal of Biomolecular NMR*, 121:167–71, 2001.

[78] J. Jeener. Pulse pair techniques in high resolution NMR. *Ampere International Summer School, Basko Poljie, Yugoslavia*, 1971.

[79] J. Jeener, B. H. Meier, P. Bachmann, and R. R. Ernst. Investigation of exchange processes by 2-dimensional NMR-spectroscopy. *Journal of Chemical Physics*, 71:4546–4553, 1979.

[80] M. Karplus. Contact electron-spin coupling of nuclear magnetic moments. *Journal of Chemical Physics*, 30:11–16, 1959.

[81] L. E. Kay, M. Ikura, R. Tschudin, and A. Bax. 3-Dimensional triple-resonance NMR spectroscopy of isotopically enriched proteins. *Journal of Magnetic Resonance*, 89:496–514, 1990.

[82] L. E. Kay, P. Keifer, and T. Saarinen. Pure absorption gradient enhanced heteronuclear single quantum correlation spectroscopy with improved sensitivity. *Journal of American Chemical Society*, 114:10663–10665, 1992.

[83] L. E. Kay, D. Marion, and A. Bax. Practical aspects of 3D heteronuclear NMR of proteins. *Journal of Magnetic Resonance*, 84:72–84, 1989.

[84] L. E. Kay, D. A. Torchia, and A. Bax. Backbone dynamics of proteins as studied by ^{15}N inverse detected heteronuclear NMR spectroscopy: Application to staphylococcal nuclease. *Biochemistry*, 28:8972–8979, 1989.

[85] L. E. Kay, G. Y. Xu, A. U. Singer, D. R. Muhandiram, and J. D. Forman-Kay. A gradient-enhanced HCCH-TOCSY experiment for recording side-chain ^{1}H and ^{13}C correlations in H_2O samples of proteins. *Journal of Magnetic Resonance*, B 101:333–337, 1993.

[86] L. E. Kay, G. Y. Xu, and T. Yamazaki. Enhanced-sensitivity triple-resonance spectroscopy with minimal H_2O saturation. *Journal of Magnetic Resonance*, A 109:129–133, 1994.

[87] A. E. Kelly, H. D. Ou, R. Withers, and V. J. Dotsch. Low-conductivity buffers for high-sensitivity NMR measurements. *Journal of American Chemical Society*, 124:12013–12019, 2002.

[88] D. M. Korzhnev, E. V. Tischenko, and A. S. Arseniev. Off-resonance effects in ^{15}N T_2 CPMG measurements. *Journal of Biomolecular NMR*, 17:231–237, 2000.

[89] C.D. Kroenke, M. Rance, and A.G. Palmer III. Variability of the ^{15}N chemical shift anisotropy in *Escherichia coli* Ribonuclease H in solution. *Journal of the American Chemical Society*, 121:10119–10125, 1999.

[90] H. Kuboniwa, S. Grzesiek, F. Delaglio, and A. Bax. Measurement of H_N-H_α J couplings in calcium-free calmodulin using new 2D and 3D water-flip-back methods. *Journal of Biomolecular NMR*, 4:871–878, 1994.

[91] A. Kumar, G. Wagner, R. R. Ernst, and K. Wüthrich. Studies of J-connectivities and selective ^{1}H-^{1}H Overhauser effects in H_2O solutions of biological macromolecules by two-dimensional NMR experiments. *Biochemical Biophysical Research Communications*, 96:1156–1163, 2000.

[92] J. Kuszewski, J. Qin, A.M. Gronenborn, and G. M. Clore. The impact of direct refinement against $^{13}C_\alpha$ and $^{13}C_\beta$ chemical shifts on protein structure determination by NMR. *Journal of Magnetic Resonance*, 106B:92–96, 1995.

[93] W. E. Lamb. The theory of chemical shielding in atoms. *Physical Review*, 60:817, 1941.

[94] M.H. Levitt, R. Freeman, and T. Frenkiel. Broad-band heteronuclear decoupling. *Journal of Magnetic Resonance*, 47:328–330, 1982.

[95] Y.-C. Li and G.T. Montelione. Solvent saturation-transfer effects in pulsed-field-gradient heteronuclear single-quantum-coherence (PFG-HSQC) spectra of polypeptides and proteins. *Journal of Magnetic Resonance*, B101:315–319, 1993.

[96] J C Lindon and A G Ferrige. Digitization and data processing in Fourier transform NMR. *Progress in NMR Spectroscopy*, 14:27–66, 1980.

[97] G. Lipari and A. Szabo. Model-free approach to the interpretation of nuclear magnetic resonance relaxation in macromolecules. 1. Theory and range of validity. *Journal of the American Chemical Society*, 104:4546–4559, 1982.

[98] G. Lippens, C. Dhalluin, and J.-M. Wieruszeski. Use of water flip-back pulse in the homonuclear NOESY experiment. *Journal of Biomolecular NMR*, 5:327–331, 1995.

[99] M. Liu, X. Mao, C. Ye, H. Huang, J. K. Nicholson, and J. C. Lindon. Improved WATERGATE pulse sequences for solvent suppression in NMR spectroscopy. *Journal of Magnetic Resonance*, 132:125–129, 1998.

[100] P. Luginbuhl and K. Wüthrich. Semi-classical nuclear spin relaxation theory revisited for use with biological macromolecules. *Progress in Nuclear Magnetic Resonance Spectroscopy*, 40:199–247, 2002.

[101] Z. Luz and S. Meiboom. Nuclear magnetic resonance study of the photolysis of the trimethylammonium ion in aqueous solution - Order of the reaction with respect to solvent. *Journal of Chemical Physics*, 39:366–370, 1963.

[102] A. M. Mandel, M. Akke, and A. G. Palmer III. Backbone dynamics of *Escherichia coli* ribonuclease H1:Correlations with structure and function in an active enzyme. *Journal of Molecular Biology*, 246:144–163, 1995.

[103] A. De Marco, M. Llinas, and K. Wüthrich. Analysis of H-1 NMR spectra of ferrichrome peptides. 1. Non-amide protons. *Biopolymers*, 17:617–636, 1978.

[104] D. Marion, P.C. Driscoll, L.E. Kay, P.T. Wingfield, A. Bax, A.M. Gronenborn, and G.M. Clore. Overcoming the overlap problem in the assignment of ^{1}H NMR spectra of larger proteins by use of three-dimensional heteronuclear ^{1}H-^{15}N Hartmann-Hahn multiple quantum coherence and nuclear Overhauser-multiple quantum coherence spectroscopy: Application to interleukin 1β. *Biochemistry*, 28:6150–6156, 1989.

[105] D. Marion, M. Ikura, R. Tschudin, and A. Bax. Rapid recording of 2D NMR spectra without phase cycling. application to the study of hydrogen exchange in proteins. *Journal of Magnetic Resonance*, 85:393–399, 1989.

[106] D. Marion and K. Wüthrich. Application of phase sensitive two-dimensional correlated spectroscopy (COSY) for measurements of ^{1}H-^{1}H spin-spin coupling constants in proteins. *Biochemical Biophysical Research Communications*, 113:967, 1983.

[107] J.L. Markley, A. Bax, Y. Arata, C.W. Hilbers, R. Kaptein, B.S. Sykes, P.E. Wright, and K. Wüthrich. Recommendations for the presentation of NMR structures of proteins and nucleic acids. *Pure and Applied Chemistry*, 70:117–142, 1998.

[108] J.L. Markley, W.J. Horsley, and M.P. Klein. Spin-lattice relaxation measurements in slowly relaxing complex spectra. *Journal of Chemical Physics*, 55:3604, 1971.

[109] H.M. McConnell. Reaction rates by nuclear magnetic resonance. *Journal of Chemical Physics*, 28:430–431, 1958.

[110] M. A. McCoy and L. Mueller. Selective shaped pulse decoupling in NMR: Homonuclear [^{13}C] carbonyl decoupling. *Journal of the American Chemical Society*, 1992:2108–2112, 1992.

[111] S. Meiboom and D. Gill. Modified spin-echo method for measuring nuclear relaxation times. *Review of Scientific Instruments*, 29:688–691, 1958.

[112] B. H. Meier and R. R. Ernst. Elucidation of chemical-exchange networks by 2-dimensional NMR-spectroscopy - Heptamethylbenzenonium ion. *Journal of the American Chemical Society*, 101:6441–6442, 1979.

[113] B. A. Messerle, G. Wider, G. Otting, C. Weber, and K. Wüthrich. Solvent suppression using a spin lock in 2D and 3D NMR spectroscopy with H_2O solutions. *Journal of Magnetic Resonance*, 85:608–614, 1989.

[114] O. Millet, J.P. Loria, C.D. Kroenke, M. Pons, and A.G. Palmer III. The static magnetic field dependence of chemical exchange linebroadening defines the NMR chemical shift time scale. *Journal of American Chemical Society*, 122:2867–2877, 2000.

[115] N. Müller, R. R. Ernst, and K. Wuthrich. Multiple-quantum-filtered two-dimensional correlated NMR spectroscopy of proteins. *Journal of the American Chemical Society*, 108:6482–6492, 1986.

[116] D. R. Muhandiram and L. E. Kay. Gradient-enhanced triple-resonance three-dimensional NMR experiments with improved sensitivity. *Journal of Magnetic Resonance*, 1994:203–216, 1994.

[117] F.A.A. Mulder, P.J.A.van Tilborg, R. Kaptein, and R. Boelens. Microsecond time scale dynamics in the RXR DNA-binding domain from a combination of spin-echo and off-resonance rotating frame relaxation measurements. *Journal of Biomolecular NMR*, 13:275–288, 1999.

[118] D. Neuhaus and M. Williamson. *The Nuclear Overhauser Effect in Structural and Conformational Analysis*. VCH Publishers, Inc., New York, 1989.

[119] J. S. Nowick, O. Khakshoor, M. Hashemzadeh, and J. O. Brower. DSA: A new internal standard for NMR studies in aqueous solution. *Organic Letters*, 2003:3511–3513, 1992.

[120] T. G. Oas, C. J. Hartzell, F. W. Dahlquist, and G. P. Drobny. The amide ^{15}N chemical shift tensors of four peptides determined from ^{13}C dipole-coupled chemical shift powder patterns. *Journal of the American Chemical Society*, 109:5962–5966, 1987.

[121] M. Ottiger, F. Delaglio, and A. Bax. Measurement of J and dipolar couplings from simplified two-dimensional NMR spectra. *Journal of Magnetic Resonance*, 131:373–378, 1998.

[122] A. W. Overhauser. Paramagnetic relaxation in metals. *Physical Review*, 89:689–700, 1953.

[123] A.G. Palmer, C.D. Kronenke, and J.P. Loria. Nuclear magnetic resonance methods for quantifying microsecond-to-millisecond motions in biological macromolecules. *Methods in Enzymology*, 339:204–238, 2001.

[124] A. Pardi, M. Billeter, and K. Wüthrich. Calibration of the angular dependence of the amide proton-C alpha proton coupling constants, $^3J_{HN-\alpha}$, in globular protein. use of $^3J_{HN-\alpha}$ for identification of helical secondary structure. *Journal of Molecular Biology*, 180:741–751, 1984.

[125] S. L. Patt. Single- and multiple-frequency-shifted laminar pulses. *Journal of Magnetic Resonance*, 96:94–102, 1992.

[126] N. H. Pawley, C. Wang, S. Koide, and L. K. Nicholson. An improved method for distinguishing between anisotropic tumbling and chemical exchange in analysis of ^{15}N relaxation parameters. *Journal of Biomolecular NMR*, 20:149–165, 1988.

[127] J. W. Peng, V. Thanabal, and G. Wagner. 2D heteronuclear NMR measurements of spin-lattice relaxation times in the rotating frame of X-nuclei in heteronuclear HX spin systems. *Journal of Magnetic Resonance*, 94:82–100, 1991.

[128] K. Pervushin. Impact of Transverse Relaxation Optimized Spectroscopy TROSY on NMR as a technique in structural biology. *Quarterly Reviews of Biophysics*, 33:161–197, 2000.

[129] K. Pervushin, R. Riek, G. Wider, and K. Wüthrich. Attenuated T_2 relaxation by mutual cancellation of dipople-dipole coupling and chemical shift anisotropy indicates an avenue to NMR structures of very large biological macromolecules in solution. *Proceedings of the National Academy of Sciences of the United States of America*, 94:12366–12371, 1997.

[130] K. Pervushin, R. Riek, G. Wider, and K. Wüthrich. Transverse relaxation-optimized spectroscopy (TROSY) for NMR studies of aromatic spin systems in ^{13}C-labeled proteins. *Journal of the American Chemical Society*, 120:6394–6400, 1998.

[131] U. Piantini, O. W. Sorensen, and R. R. Ernst. Multiple quantum filters for elucidating NMR coupling networks. *Journal of the American Chemical Society*, 104:6800–6801, 1982.

[132] M. Piotto, V. Saudek, and V. Sklenar. Gradient-tailored excitation for single-quantum NMR-spectroscopy of aqueous solutions. *Journal of Biomolecular NMR*, 2:661–665, 1992.

[133] P. Plateau and M. Gueron. Exchangeable proton NMR without base-line distortion, using new strong-pulse sequences. *Journal of the American Chemical Society*, 104:7310–7311, 1982.

[134] J. A. Pople. Proton magnetic resonance of hydrocarbons. *Journal of Chemical Physics*, 24:1111, 1956.

[135] D. S. Raiford, C. L. Fisk, and E. D. Becker. Calibration of methanol and ethylene-glycol nuclear magnetic resonance thermometers. *Analytical Chemistry*, 51:2050–2051, 1979.

[136] M. Rance and J. Cavanagh. RF phase coherence in rotating-frame NMR experiments in isotropic solutions. *Journal of Magnetic Resonance*, 87:363–371, 1990.

[137] M.K. Rosen, K.H. Gardner, R.C. Willis, W.E. Parris, T. Pawson, and L.E. Kay. Selective methyl group protonation of perdeuterated proteins. *Journal of Molecular Biology*, 263:627–636, 1996.

[138] A. Ross, M. Czisch, and G. C. King. Systematic errors associated with the CPMG pulse sequence and their effect on motional analysis of biomolecules. *Journal of Magnetic Resonance*, 124:355–365, 1997.

[139] S. P. Rucker and A. J. Shaka. Broadband homonuclear cross polarization in 2D NMR using DIPSI-2. *Molecular Physics*, 68:509–517, 1989.

[140] M. Salzmann, K. Pervushin, G. Wider, H. Seen, and K. Wüthrich. TROSY in triple-resonance experiments: New perspectives for sequential NMR assignment of large proteins. *Proceedings of the National Academy of Sciences of the United States of America*, 95:13585–12590, 1998.

[141] C.R. Sanders and J.P. Schwonek. Characterization of magnetically orientable bilayers in mixtures of dihexanoylphosphatidylcholine and dimyristoylphosphatidylcholine by solid-state NMR. *Biochemistry*, 31:8898–8905, 1992.

[142] M. Sattler, J. Schleucher, and C. Griesinger. Heteronulear multidimensional NMR experiments for the structure determination of proteins in solution employing pulsed field gradients. *Progress in Nuclear Magnetic Resonance Spectroscopy*, 34:93–158, 1999.

[143] J.M. Schurr, H.P. Babcock, and B.S. Fujimoto. A test of the model-free formulas. Effects of anisotropic rotational diffusion and dimerization. *Journal of Magnetic Resonance*, 105:211–224, 1994.

[144] G. Schwarz. Estimating the dimension of a model. *The annals of statistics*, 6:461–464, 1978.

[145] C.D. Schwieters, J.J. Kuszewski, N. Tjandra, and G.M. Clore. The Xplor-NIH NMR molecular structure determination package. *Journal of Magnetic Resonance*, 160:65–73, 2003.

[146] A. J. Shaka, P. B. Barker, and R. Freeman. Computer-optimized decoupling scheme for wideband applications and low-level operation. *Journal of Magnetic Resonance*, 64:547–552, 1985.

[147] A. J. Shaka, J. Keeler, and R. Freeman. Evaluation of a new broadband decoupling sequence: WALTZ-16. *Journal of Magnetic Resonance*, 53:313–349, 1983.

[148] A. J. Shaka, C. J. Lee, and A. Pines. Iterative schemes for bilinear operators; applications to spin decoupling. *Journal of Magnetic Resonance*, 77:274–293, 1988.

[149] N.J. Skelton, A.G. Palmer III, M. Akke, J. Kordel, M. Rance, and W.J. Chazin. Practical aspects of two-dimensional proton-detected ^{15}N spin relaxation measurements. *Journal of Magnetic Resonance*, B 102:253–264, 1993.

[150] V. Sklenar, M. Piotto, R. Leppik, and V. Saudek. Gradient-tailored water suppression for ^{1}H-^{15}N HSQC experiments optimized to retain full sensitivity. *Journal of Magnetic Resonance*, A102:241–245, 1993.

[151] C. P. Slichter. *Principles of Magnetic Resonance*. Springer-Verlag, Berlin, 1990.

[152] I. Solomon. Relaxation processes in a system of two spins. *Physical Review*, 99:559–565, 1955.

[153] O.W. Sorensen, G.W. Eich, M.H. Levitt, G. Bodenhausen, and R.R. Ernst. Product operator-formalism for the description of NMR pulse experiments. *Progress in Nuclear Magnetic Resonance Spectroscopy*, 16:163–192, 1983.

[154] D. J. States, R. A. Haberkorn, and D. J. Ruben. A two-dimensional nuclear overhauser experiment with pure absorption phase in four quadrants. *Journal of Magnetic Resonance*, 48:286–292, 1982.

[155] T. J. Swift and R. E. Connick. NMR-relaxation mechanism of ^{17}O in aqueous solutions of paramagnetic cations and the lifetime of water molecules in the first coordination sphere. *The Journal of Chemical Physics*, 37:307–320, 1962.

[156] N. Tjandra and A. Bax. Direct measurement of distances and angles in biomolecules by NMR in a dilute liquid crystalline medium. *Science*, 278:1111–1114, 1997.

[157] N. Tjandra, S. Grzesiek, and A. Bax. Magnetic field dependence of nitrogen-proton J splittings in ^{15}N-enriched human ubiquitin resulting from relaxation interference and

residual dipolar coupling. *Journal of the American Chemical Society*, 118:6264–6272, 1996.

[158] N. Tjandra, P. T. Wingfield, S. Stahl, and A. Bax. Anisotropic rotational diffusion of perdeuterated HIV protease from ^{15}N NMR relaxation measurements at two magnetic field strengths. *Journal of Biomolecular NMR*, 8:273–284, 1996.

[159] J.R. Tolman, H.M. Al-Hashimi, L.E. Kay, and J.H. Prestegard. Structural and dynamic analysis of residual dipolar coupling data for proteins. *Journal of the American Chemical Society*, 123:1416–1424, 2001.

[160] L. A. Trimble and M. A. Bernstein. Application of gradients for water suppression in 2D multiple-quantum-filtered COSY spectra of peptides. *Journal of Magnetic Resonance*, B105:67–72, 1994.

[161] V. Tugarinov and L.E. Kay. Ile, Leu, and Val methyl assignments of the 723-residue malate synthase G using a new labeling strategy and novel NMR methods. *Journal of the American Chemical Society*, 125:13868–13878, 2003.

[162] G. W. Vuister and A. Bax. Quantitative J correlations: A new approach for measuring homonuclear three-bond J(H_nH_α) coupling constants in ^{15}N-enriched proteins. *Journal of the American Chemical Society*, 115:7772–7777, 1993.

[163] A.C. Wang, P. J. Lodi, J. Qin, G. W. Vuister, A. M. Gronenborn, and G. M. Clore. An efficient triple-resonance experiment for proton-directed sequential backbone assignments of medium-sized proteins. *Journal of Magnetic Resonance*, B105:196–198, 1994.

[164] Y.-X. Wang, J. Jacob, F. Cordier, P. Wingfield, S. Stahl, S. Lee-Huang, D. Torchia, S. Gresiek, and A. Bax. Measurement of $^{3H}J_{NC}$ connectivities across hydrogen bonds in a 30 kDa protein. *Journal of Biomolecular NMR*, 14:181–184, 1999.

[165] Y. Wei, D.-K. Lee, and A. Ramamoorthy. Solid-state ^{13}C NMR chemical shift anisotropy tensors of polypeptides. *Journal of the American Chemical Society*, 123:6118–6126, 2001.

[166] D. S. Wishart, C. G. Bigam, J. Yao, F. Abildgaard, H. J. Dyson, E. Oldfield, J. L. Markley, and B. D. Sykes. ^{1}H, ^{13}C, ^{15}N chemical shift referencing in biomolecular NMR. *Journal of Biomolecular NMR*, 6:135–140, 1993.

[167] D. S. Wishart and B. D. Sykes. The ^{13}C chemical-shift index: A simple method for the identification of protein secondary structure using ^{13}C chemical-shift data. *Journal of Biomolecular NMR*, 4:171–180, 1994.

[168] M. Wittekind and L. Mueller. HNCACB, a high-sensitivity 3D NMR experiment to correlate amide-proton and nitrogen resonances with the alpha- and beta-carbon resonances in proteins. *Journal of Magnetic Resonance*, B 101:201–205, 1993.

[169] D.E. Woessner. Nuclear spin relaxation in ellipsoids undergoing rotational brownian motion. *Journal of Chemical Physics*, 37:647, 1962.

[170] K. Wüthrich, M. Billeter, and W. Braun. Polypeptide secondary structure determination by nuclear magnetic resonance observation of short proton-proton distances. *Journal of Molecular Biology*, 180:715–740, 1984.

[171] K. Wüthrich and G. Wagner. Nuclear magnetic resonance of labile protons in basic pancreatic trypsin inhibitor. *Journal of Molecular Biology*, 130:1–18, 1979.

[172] T. Yamazaki, W. Lee, , M. Revington, D. L. Mattiello, F. W. Dahlquist, C. H. Arrowsmith, and L. E. Kay. An HNCA pulse scheme for the backbone assignment of ^{15}N, ^{13}C, ^{2}H-labeled proteins: Application to a 27-kDa *Trp* repressor-DNA complex. *Journal of the American Chemical Society*, 116:6464–6465, 1994.

[173] T. Yamazaki, W. Lee, C. H. Arrowsmith, D. R. Muhandiram, and L. E. Kay. A suite of triple resonance NMR experiments for the backbone assignment of ^{15}N, ^{13}C, ^{2}H labeled proteins with high sensitivity. *Journal of the American Chemical Society*, 116:11655–11666, 1994.

[174] D. Yang and L. E. Kay. Improved lineshape and sensitivity in the HNCO-family of triple resonance experiments. *Journal of Biomolecular NMR*, 14:273–276, 1999.

[175] J. X. Yang and T. F. Havel. An evaluation of least-squares fits to COSY spectra as a means of estimating proton-proton coupling constants. I. Simulated test problems. *Journal of Biomolecular NMR*, 4:807–826, 1994.

[176] O. W. Zhang, L. E. Kay, J. P. Olivier, and J. D. Forman-Kay. Backbone ^{1}H and ^{15}N resonance assignments of the N-terminal SH3 domain of DRK in folded and unfolded states using enhanced-sensitivity pulsed field gradient NMR techniques. *Journal of Biomolecular NMR*, 4:845–858, 1994.

[177] G. Zhu, D.A. Torchia, and A. Bax. Discrete Fourier transformation of NMR signals. The relationship between sampling delay time and spectral baseline. *Journal of Magnetic Resonance*, 105A:219–222, 1993.

[178] S. Zinn-Justin, P. Berthault, M. Guenneugues, and H. Desvaux. Off-resonance RF-field in heteronuclear NMR: Application to the study of slow motions. *Journal of Biomolecular NMR*, 10:363–372, 1997.

[179] M. Zweckstetter and A. Bax. Prediction of sterically induced alignment in a dilute liquid crystalline phase: Aid to protein structure determination by NMR. *Journal of the American Chemical Society*, 122:3791–3792, 2000.

Index

FOCUS ON STRUCTURAL BIOLOGY

1. S.R. Kiihne and H.J.M. de Groot (eds.): *Perspectives on Solid State NMR in Biology*. 2001 ISBN 0-7923-7102-X
2. J.N. Housby (ed.): *Mass Spectrometry and Genomic Analysis*. 2001 ISBN 0-7923-7173-9
3. C. Dennison: *A Guide to Protein Isolation*. 2003 ISBN 1-4020-1224-1
4. O.M. Becker and M. Karplus (eds.): *Guide to Biomolecular Simulations*. 2005 ISBN 1-4020-3586-1
5. G.S. Rule and T.K. Hitchens: *Fundamentals of Protein NMR Spectroscopy*. 2006 ISBN 1-4020-3499-7

MIX
Papier aus verantwortungsvollen Quellen
Paper from responsible sources
FSC® C105338

If you have any concerns about our products,
you can contact us on
ProductSafety@springernature.com

In case Publisher is established outside the EU,
the EU authorized representative is:
Springer Nature Customer Service Center GmbH
Europaplatz 3, 69115 Heidelberg, Germany

Printed by Libri Plureos GmbH
in Hamburg, Germany